CONTAMINANTS AND THE SOIL ENVIRONMENT IN THE AUSTRALASIA-PACIFIC REGION

Dr Kevin George Tiller

It is our privilege to dedicate this book on *Contaminants and the Soil Environment* Kevin Tiller who was our leader and great inspiration for the present contamina research team at CSIRO Division of Soils and the Cooperative Research Centre f Soil and Land Management. Dr Tiller, a leading international soil scientist on hea metal contaminants in the soils, died after a short illness in Adelaide. He was born Pinnaroo, South Australia. Following BSc and MSc degrees in Chemistry from t University of Adelaide, he travelled to the USA to study for his doctorate at Corn University. After the completion of his PhD degree he returned to CSIRO and initial studied the reactions of trace metals in agricultural soils.

During the 1970s Dr Tiller turned his attention to heavy metal pollution, pla micronutrients and the problem of cadmium and lead affecting the quality agricultural products. His pioneering research into lead levels in the Port Pirie ar and the effect of heavy metals on crop quality has had significant impact on Australi health and regulatory controls on food and soil quality. He was the recognised nation leader in environmental pollution by heavy metals and an international authori whose opinion was frequently sought and highly regarded. His contributions environmental research are recognised by the many invitations that he received lecture at major international conferences and meetings throughout the world.

Dr Tiller recognised the need for contaminants research in Australia long before was publicly acknowledged as a serious problem. His foresight led to the developme of much of the current research in Australia on soil contaminants. Kevin's untime death was a great loss to CSIRO, Australia and the international scientific communi

Contaminants and the Soil Environment in the Australasia-Pacific Region

Proceedings of the First Australasia-Pacific Conference on Contaminants and Soil Environment in the Australasia-Pacific Region, held in Adelaide, Australia, 18–23 February 1996

Edited by

R. NAIDU
CSIRO Division of Soils and CRC for Soil and Land Management, Private Mail Bag No. 2, Glen Osmond, Adelaide, SA 5064.

R. S. KOOKANA
CRC for Soil and Land Management and CSIRO Division of Soils, Private Mail Bag No. 2, Glen Osmond, Adelaide, SA 5064.

D. P. OLIVER
CSIRO Division of Soils and CRC for Soil and Land Management, Private Mail Bag No. 2, Glen Osmond, Adelaide, SA 5064.

S. ROGERS
CRC for Soil and Land Management and Dept. of Soil Sciences, Waite Agricultural Research Institute, University of Adelaide, Waite Road, Urrbrae, Adelaide, SA 5064.

M. J. McLAUGHLIN
CSIRO Division of Soils and CRC for Soil and Land Management, Private Mail Bag No. 2, Glen Osmond, Adelaide, SA 5064.

KLUWER ACADEMIC PUBLISHERS
DORDRECHT / BOSTON / LONDON

A C.I.P. Catalogue record for this book is available from the Library of Congress

ISBN 0-7923-3900-2

Published by Kluwer Academic Publishers,
P.O. Box 17, 3300 AA Dordrecht, The Netherlands.

Kluwer Academic Publishers incorporates
the publishing programmes of
D. Reidel, Martinus Nijhoff, Dr W. Junk and MTP Press.

Sold and distributed in the U.S.A. and Canada
by Kluwer Academic Publishers,
101 Philip Drive, Norwell, MA 02061, U.S.A.

In all other countries, sold and distributed
by Kluwer Academic Publishers Group,
P.O. Box 322, 3300 AH Dordrecht, The Netherlands.

Printed on acid-free paper

Printed and bound in Great Britain by
Hartnolls Limited, Bodmin, Cornwall

Contents

Foreword

Australasia and the Pacific region includes some countries that have some of the fastest rates of industrialisation in the world with annual average economic growth rates exceeding 8%. The surge in wealth in these countries has enabled some populations to achieve very much better standards of living, and life expectancy has risen with GDP per capita in nearly every instance.

One of the prices to be paid for such rapid growth however, is environmental deterioration. Air, water and soil pollution have been serious concerns for decades in the advanced economies of Japan, Taiwan and Korea. In the last decade the rate of growth in Thailand, Indonesia, parts of eastern China, and the major urban centres of India has given rise to an increasing catalogue of pollution and waste disposal problems. Much of this stems from the very rapid urbanisation that has occurred at the same time. Cities such as Bangkok, Beijing, Calcutta, Jakarta, Kuala Lumpur, Manilla, Seoul, and Shanghai, with population densities of several thousand per square kilometre, and often doubling in size every decade, cannot keep pace with the infrastructural demands imposed upon them.

At the other end of the scale are those countries or regions with very poor economies and GDPs per capita of less than US$1000 per annum. They are predominantly rural, but there is often an acute shortage of good agricultural land, and the pressure to produce increased food has led to widespread use of pesticides that are now being found in water supplies and ground waters. Processing of rural products also brings pollution to tributaries draining from rural districts with effluents from local mills and tanneries. Competition for land is most intense in the urban fringes of coastal and alluvial plains that support the highest population densities. The seeds of long-term pollution problems in soil and water resources are being sown that will threaten health and environmental integrity for decades if these issues are not vigorously attacked now.

These issues require solutions that are appropriate to the environmental, economic and social conditions of these countries. Tropical and subtropical climates, and the soils developed from them, produce reactions with pollutants that are outside the normal range of response found in the industrialised north, where most of the science and technology for remediation has developed. Economic and social solutions cannot depend upon the full range of regulatory and financial support that may have developed in more mature economies. This conference therefore meets a special need in bringing together for the first time those expert in the science and management of the specific reactions of pollutants in tropical, sub-tropical and semi-arid regions.

As a part of the Asia-Pacific, Australia is proud to host this first conference devoted to the management of soil pollutants in the region. Australia, also emerging from its development phase, shares many of the problems, lack of knowledge and urgent needs of the other countries in the region. This conference provides the opportunity to share and exchange experience of specific regional contamination problems and rectify gaps in current knowledge. We are confident that this initiative will be the first of a series of conferences that will document increasing knowledge and effective solutions to the pollution problems we all face.

Ann Hamblin
Director, CRC for Soil and Land Management

Roger Swift
Chief, CSIRO Division of Soils

Preface

The Australasia-Pacific Region supports approximately 50% of the world's population. The last 50 years has witnessed a rapid increase in the regional population, agricultural productivity, industrial activities and trade within the region. Both the demand for increased food production and the desire to improve the economic conditions have affected regional environmental quality. This volume presents an overview of the fate of contaminants in the soil environment; current soil management factors used to control contaminant impacts, issues related to sludge and effluent disposals in the soil environment; legal, health and social impacts of contaminated land, remediation approaches and strategies to manage contaminated land, some of the problems associated with environmental degradation in the Australasia-Pacific Region and steps that we need to take to safeguard our environment.

With increased use of chemicals in agriculture and industry, soil and water resources are becoming increasingly contaminated by elements and compounds either added intentionally or inadvertently to terrestrial ecosystems. For example, while use of fertilisers is essential to maintain and improve soil fertility, this practice may also introduce unwanted elements into soils due to the presence of impurities in some sources of fertiliser. Research studies in developed countries have identified Cd as a potential threat to the food chain, but little work has been carried out in developing countries to assess the impact on soil and crop quality of impurities in fertilisers. Water quality issues are closely related to land use and are therefore closely linked to the use of fertilisers, both organic and inorganic, in agriculture. While much research has been undertaken in developed countries on the impact and management of agricultural activities in relation to water quality, much work still needs to be undertaken in less developed economies.

The move to enhance crop production for export markets in the Australasia-Pacific region has also increased the use of pesticides. In some developing countries, certain highly toxic and carcinogenic pesticides are still being used e.g. DDT and benzene hexachloride (BHC), both of which have been banned from USA and Europe. These chemicals accounted for about 75% of pesticide use in Asian countries in 1976, according to the World Health Organisation. Pesticides which were once thought to degrade in soils are now being detected in surface and groundwaters in some regions, indicating that either applications are not according to manufacturers' instructions, or that the behaviour of these chemicals in soil is not fully understood.

Exploitation of mineral ores, processing and marketing is an ongoing activity which generates as much as 40% of the export income in many of the developing countries in the region. Mining operations may lead to significant challenges for the management of the natural environment during, and subsequent to, mining operations. Impacts of earlier mining activities in some countries now pose significant problems for management of reclaimed areas e.g. acid mine drainage. The revegetation and rehabilitation of mined lands requires a multi-disciplinary approach, with management required to produce structurally and ecologically stable ecosystems that are independently sustainable long after mining operations are completed.

Contaminated urban and industrial sites, a feature common in European and North American countries, are now also becoming a major issue in many south-east Asian countries. Industrialisation has often outpaced the development of appropriate guidelines for control of discharge of wastes to the environment. Indeed, regulations for protection of water and soil quality are seldom based on local information, where particular soil and climatic conditions may render international guidelines inappropriate. Hazards to health for any contaminant depend primarily on its fate and behaviour in the soil environment, which controls both mobility and bio-availability. While some information exists on the toxicological properties of heavy metal contaminants, knowledge of the hazardous effects of synthetic organic compounds and their behaviour in soil lags the rate at which they are synthesised and introduced into the environment. Development of management options to minimise or reverse the impacts of contaminants on human health requires a multi-disciplinary approach.

The impact of contaminants on human health is often a contentious issue with regulations often set in the absence of firm scientific data on behaviour, persistence and human toxicity. The extent of contaminated land in the Australasian-Pacific region and its management and remediation requires action by regional governments, scientists and national environmental protection agencies. This publication is the first step in the creation of a database of environmental degradation problems in the region with a view to prioritising appropriate research activities. We hope that this book will stimulate more work in the region and lead to more detailed and comprehensive assessments of soil contamination issues in many countries, and also lead to development of new remediation technologies specifically suited to the region's soils and environments.

The organising Committee is grateful for financial support provided by Australian Centre for International Agricultural Research, Commonwealth Environmental Protection Agency, South Australian Environmental Protection Agency, British Overseas Development Administration, AusAid, Horticultural Research and Development Corporation, Grains Research and Development Corporation, QANTAS, Cooperative Research Centre for Soil and Land Management and CSIRO Division of Soils.

We are most appreciative to Ms Chris Harford who provided excellent secretarial assistance, Mr Andrew Kerekes, Ms Ina Dumitrescu, Ms Maria Janowska and Ms Julie Smith for technical assistance and Greg Rinder and Bob Schuster for excellent graphic illustrations. We are also grateful to Dr Ken Lee, CSIRO Division of Soils for his contribution towards the editing and review of the country reports.

Finally, it is with regret that Dr Kevin G. Tiller, the inspiration for much of the work on soil contamination in Australia, cannot be with us to share the publication of this book. Kevin's untimely death was a great loss to CSIRO, Australia and the international scientific community. It is perhaps fortunate that he was able to complete his contribution to this book and share his personal perspectives on over 40 years of research into soil contamination issues. It is with pleasure that we dedicate this book to his memory.

Ravi Naidu
Rai Kookana
Danni Oliver
Steve Rogers
Mike McLaughlin

September 1995

List of Contributors

Barzi F, Environment Protection Agency, 40 Blackall Street, Barton ACT 2600, Australia

Bolan, N S, Soil Science Department, Massey University, Private Bag 11222, Palmerston North, New Zealand

Brusseau, M L, Soil, Water, and Environmental Science Department, and Hydrology and Water Resources Department, The University of Arizona, 429 Shantz Building No. 38, Tucson, Arizona 85721, USA

Burns, R G, Research School of Biosciences, University of Kent, Canterbury, Kent CT2 7NJ, UK

Cameron, K C, Soil Science Department, Post Office Box 84, Lincoln University, Canterbury, New Zealand

Chaney, R L, USDA-Agricultural Research Service, Beltsville, MD, USA

Chaudhry, M R, IWASRI, 13- West Wood Colony, Thokar Niaz Beg, Lahore, Pakistan

Chen, T B, Department of Biology, and Centre for Waste Recycling & Environmental Biotechnology, Hong Kong Baptist University, Kowloon Tong, Hong Kong

Chen, Z S, Department of Agricultural Chemistry, National Taiwan University, Taipei, Taiwan 106, Republic of China

Davis, G B, Centre for Groundwater Studies, CSIRO Division of Water Resources, Perth Laboratory, Private Bag PO, Wembley, WA 6014, Australia

Ellis, H K, Pollution and Risk Management Directorate, Ministry for the Environment, Post Office Box 10362, Wellington, New Zealand

Fowler, R, Australian Centre for Environmental Law, Faculty of Law, The University of Adelaide, Adelaide, SA 5005, Australia

Gangaiya, P, Chemistry Department, University of the South Pacific, Suva, Fiji

Helmke, P A, Department of Soil Science, University of Wisconsin-Madison, 1525 Observatory Drive, Madison, Wisconsin 53706, USA

Hunt, S, Tonkin & Taylor Limited, Post Office Box 5271, Wellesley Street, Auckland, New Zealand

Hussain, Q, Pakistan Agricultural Research Council, G-5/1, Islamabad, Pakistan

Ji, G L, Institute of Soil Science, Academia Sinica, Nanjing, China

Johnston, C D, Centre for Groundwater Studies, CSIRO Division of Water Resources, Perth Laboratory, Private Bag PO, Wembley, WA 6014, Australia

Khanif, Y M, Department of Soil Science, Faculty of Agriculture, Universiti Pertanian Malaysia, 43400 U.P.M. Serdang, Selangor Darul Ehsan, Malaysia

Kookana, R S, Cooperative Research Centre for Soil and Land Management and CSIRO Division of Soils, Private Bag No 2, Glen Osmond SA 5064, Australia

Koshy, K, Chemistry Department, University of the South Pacific, Suva, Fiji

Kumar, R, Department of Agriculture and Livestock, Post Office 417, Konedobu, NCD, Papua New Guinea

Kye-Hoon, K, Korea Institute of Nuclear Safety, Post Office Box 114 Yusong, Taejon, 305-600, Korea

Langley, A, Hazardous Substances Section, Public Health and Environmental Health Service, Post Office Box 6, Rundle Mall SA 5000, Australia

Lee, D Y, Department of Agricultural Chemistry, National Taiwan University, Taipei, Taiwan 106, Republic of China

Levett, M P, Department of Geography, University of Papua New Guinea, Post Office Box 320, University, NCD, Papua New Guinea

Lin, C F, Graduate Institute of Environmental Engineering, National Taiwan University, Taipei, Taiwan 106, Republic of China

Lo, S L, Graduate Institute of Environmental Engineering, National Taiwan University, Taipei, Taiwan 106, Republic of China

Manuweera, G K, Office of the Registrar of Pesticides, Department of Agriculture, Peradeniya, Sri Lanka

McGhee, I, Research School of Biosciences, University of Kent, Canterbury, Kent CT2 7NJ, UK

McLaren, R G, Department of Soil Science, Lincoln University, New Zealand

McLaughlin, M J, Cooperative Research Centre for Soil and Land Management, Private Bag No. 2, Glen Osmond SA 5064, Australia

Meek, T, Environmental Protection Authority, GPO Box 4395 QQ Melbourne, Victoria 3001, Australia

Morrison, R J, Environment Research Institute, University of Wollongong, New South Wales, Australia

Mubarak, A M, Chemical and Environmental Technology Division, Ceylon Institute of Scientific and Industrial Research, 363 Bauddhaloka Mawatha, Colombo 7, Sri Lanka

Naidu, R, CSIRO Division of Soils and CRC for Soil and Land Management, Private Bag No. 2, Glen Osmond, SA 5064, Australia

Oliver, D P, CSIRO Division of Soils, Private Bag No 2, Glen Osmond, SA 5064, Australia

Prasad, R, Council of Scientific & Industrial Research, Anusandhan Bhawan, Rafi Marg, New Delhi-110001, India

Rao, P S C, Inter-Disciplinary Program in Hydrologic Sciences, Post Office Box 110290, University of Florida, Gainesville, Florida 32611-0290, USA

Roberts, A H C, Soils and Fertiliser Group, AgResearch, Private Bag 3123, Hamilton, New Zealand

Rogers, S, Cooperative Research Centre for Soil & Land Management, Private Bag No 2, Glen Osmond, SA 5064, Australia

Salmijah, S, Department of Soil Science, Faculty of Agriculture, Universiti Pertanian Malaysia, 43400 U.P.M. Serdang, Selangor Darul Ehsan, Malaysia

Senviratne, R, Horticulture Research and Development Institute, Department of Agriculture, Peradeniya, Sri Lanka

Sharif, M, Natural Resources Division, Pakistan Agricultural Research Council, G-5/1, Islamabad, Pakistan

Sharpley, A N, USDA-ARS, Pasture Systems & Watershed Management Research Laboratory, Curtain Road, University Park, PA 16802-3702, USA

Singh, K, Department of Chemistry, University of Papua New Guinea, Post Office Box 320, University, NCD, Papua New Guinea

Smith, C J, CSIRO Division of Soils, GPO Box 639, Canberra, ACT 2601, Australia

Sumner M, Department of Crop and Soil Sciences, University of Georgia, Athens GA 30602, USA

Sun-Hoo, Y, College of Agriculture and Life Science, Seoul National University, 103 Sodundong, Suwon, 441-744, Korea

Taylor, G F, CSIRO Division of Exploration and Mining, Private Bag No 2, Glen Osmond, SA 5064, Australia

Tiller, K G, CSIRO Division of Soils, Private Bag No 2, Glen Osmond, SA 5064, Australia

Wang, T-P, Department of Soil Science, National Chung Hsing University, Taichung, Taiwan, Republic of China

White, R E, Department of Agriculture, The University of Melbourne, Parkville, Victoria 3052, Australia

Wang, Y P, Department of Soil Science, National Chung Hsing University, Taichung, Taiwan, Republic of China

Wong, J W C, Department of Biology, and Centre for Waste Recycling & Environmental Biotechnology, Hong Kong Baptist University, Kowloon Tong, Hong Kong

Wong, M H, Department of Biology, and Centre for Waste Recycling & Environmental Biotechnology, Hong Kong Baptist University, Kowloon Tong, Hong Kong

Yu, T R, Institute of Soil Science, Academia Sinica, Nanjing, China

Zuberi, F A, IWASRI, 13- West Wood Colony, Thokar Niaz Beg, Lahore, Pakistan

1

Soil contamination issues: past, present and future, a personal perspective

K.G. TILLER

1. Introduction

I am honoured to be asked to speak on this topic but am daunted by the huge amount of work that has been done over a very long period in a range of contexts and for many potentially toxic contaminants. Having a personal perspective as a term of reference simplifies the task to some extent because of the extensive literature on this topic. This has inevitably resulted in additional emphasis on my own work which in the earlier stages of my scientific career was related to problems connected to the soil chemistry of trace metals, their availability to plants and soil fertility. I will only briefly refer to organic contaminants. I believe that many of the principles and mechanisms being investigated then have continued relevance although modern-day research has focussed on problems of soil contamination in which different heavy metals may be of concern. Equally, some of the weaknesses of early research approaches are still evident in currently published papers. The only difference is that with the advance in knowledge and experience, deficiencies in experimentation cannot now so easily be justified. These will be discussed later.

Another obstacle to reviewing this topic is that the 'past' of some research laboratories is unfortunately the 'present' of others and equally the 'present' of some organisations will be the 'future' or may never be achieved by others. This observation is not necessarily a reflection of scientific skills, although sometimes this is deserved, but more often results from a lack of imagination or commitment to expand the boundaries of knowledge on the behalf of supervisors or students. On the other hand, some laboratories are frustrated by economic, political and management restraints which result in inadequate facilities or restriction of goals, which ensure that scientific approaches are stuck in a time warp and that local soil issues are not adequately pursued.

I will discuss a series of topics which, many others would agree, have been of importance to the development of our knowledge in soil chemistry and are still of current relevance to soil contamination and its environmental consequences. Those discussed in greater detail will reflect the enthusiasms of myself and co-workers. Some important research areas such as the application of sewage sludge to land to which we have contributed, will not be discussed in any detail (see Chapter 6). I am well aware that many of the views expressed below are far from unique and are shared by many others.

R. Naidu et al. *(eds): Contaminants and the Soil Environment in the Australasia-Pacific Region, 1–27.*

2. Past to present

2.1 Early Days

A re-examination of papers published during the 1950's and 1960's and earlier shows that many important concepts were established or at least raised as hypotheses at that time. With the benefit of hindsight, some of the ideas proposed were not well supported by the evidence provided; sometimes the results were misinterpreted but still encouraged others to take the research forward.

Quite early on it was recognised that the micronutrient cations like Zn and Cu behaved differently from the major cations such as Ca and Na. Jones et al. (1936) found that some of the Zn added to soils could not be extracted later with ammonium acetate. Peech (1941) obtained similar results. Hibbard (1940) showed that the ease of extraction of Zn from Zn-treated soil was pH dependent. Banerjee et al. (1953) found that a large proportion of the sorbed metal could only be extracted with acids. Scharrer and Hofner (1958) and others confirmed that that the 'fixation' of Zn and other trace metals was dependent on soil pH. Brown (1950) observed that some of the Zn reacted with the soil by a process much slower than could be attributed to weak electrostatic bonding. Several authors (Zende, 1955; Leydon and Toth, 1960; Delas et al., 1960) noted that heavy metal cations were largely bound in a difficultly exchangeable form after long contact with the soil. Apart from the dependence on soil pH, Bambergs and Balode (1957) made the important observation that the adsorption of Mn, Zn, and Cu was largely independent of the other major cations present. During this period it had been commonly observed that the availability of trace metals to plants decreased with increasing pH (e.g. Truog, 1946).

In summary, heavy metal reactions with soils were recognised about 40 years ago as being different from major soil cations with respect to strength of bonding, pH dependence, rates of reaction and that reactions of trace metal cations differed from each other.

2.2 The role of hydrolysed metal species

Several early research papers proposed that so called basic cation species could be involved in the electrostatic adsorption of metals by soils. The work of Bower and Truog (1940) had the greatest impact on subsequent thinking on the role of hydrolysed ionic species in adsorption processes. Their ideas were based on the observation that the estimations of cation exchange capacities (CEC) of soils were greatest for those metals which hydrolysed most readily. Almost certainly their observations were due to precipitation rather than adsorption because of the general relation to the respective solubility constants. This was confirmed by de Mumbrum and Jackson (1956b). Although for the wrong reasons, the idea of control of adsorption by hydrolysed species gained credence. Elgabaly and Jenny (1943) supported this view and concluded that the ions Zn^{2+}, $ZnCl^{+}$, and $ZnAc^{+}$ were preferentially adsorbed by clays. More recent work has, however, shown that the formation of ion pairs with Cl^- reduces adsorption of Zn. Higher CECs were also observed by Elgagaby and Jenny (1943) using Zn as the index cation but not to the extent described by Bower and Truog (1940), probably reflecting conditions which were less favourable for precipitation. Other studies around this period supporting Bower and Truog's hypothesis were affected by the combined problems of both high pH and metal concentrations and the lack of adequate concentrations of supporting electrolyte. These included works of Menzel and Jackson (1950a; 1950b), Brown (1950), Spencer and Gieseking (1954), de Mumbrum and Jackson (1956a) and Huser (1958). Others got on the bandwagon and concluded that hydrolysed metal ions controlled adsorption but without any supporting evidence.

The work of this period can be summarised as having established the possibility that adsorption of heavy metal cations by soils may be controlled by their hydrolysed ionic species, notwithstanding inappropriate experimental conditions, overzealous interpretation of poor, or sometimes no, appropriate data. Some early studies utilised sufficient trace metal to saturate the total exchange capacity of the adsorbent. The development of the ideas on the role of ion hydrolysis was a classic case of an error when repeated often enough becoming fact. Nevertheless it had to be admitted that metals do hydrolyse and there have been many observations of a strong pH dependence of metal reactions with a variety of soil surfaces and often with a concomitant observed release of protons. In addition, recognition of an ordered difference in adsorption affinity between metals related to their hydrolysis constants (e.g. Tiller, 1961, 1968; Grimme, 1969; Forbes et al., 1976; Kinniburgh et al., 1976; McKenzie, 1980) encouraged the view that hydrolysed ions in soil solutions controlled the adsorption process at low metal concentrations. This hypothesis was unacceptable to many, and still has its detractors because of the extremely low concentrations of the hydrolysed ions, but of course this low concentration could well be compensated for by a correspondingly much higher binding constant than would apply to unhydrolysed species (Naidu et al., 1994).

This research topic was pursued by Hodgson and Tiller over several years (Hodgson, 1960; Hodgson and Tiller, 1962; Tiller and Hodgson, 1962; Tiller et al., 1963; Hodgson et al., 1964; Hodgson, 1969). They studied the adsorption of mainly Co and Zn under standard conditions of excess Ca ions of constant ionic strength in order to virtually eliminate the electrostatic interactions of the trace metal ions with the pure mineral and separated soil clay fractions used. The initial molar ratios of Ca to trace metal in solution were about 10^5 and higher still at the completion of the 7 d reaction period. The intention was to use in most experiments initial metal concentrations that were relevant to soil solution, namely about 10^{-6} M. With the benefit of later knowledge, these levels were also too high. The need to work at realistically low concentrations could only be met at that time by use of radioisotopic procedures. The amount of Co or Zn adsorbed under these conditions was thus defined operationally as specifically adsorbed and was further fractionated by washing with non-radioactive metal, EDTA or more routinely, dilute acetic acid to identify the very small component that resisted repeated extraction. This very strongly retained fraction of bound Co could only be liberated by partial dissolution of the Fe oxide and silicate mineral substrates, from which it was postulated that following an intermediate adsorption reaction some of the Co migrated to within the crystal structure (Tiller and Hodgson, 1962). For experiments using higher metal concentrations, clay suspensions were washed with $CaCl_2$ before measurement, to eliminate any electrostatically bound trace metal. This approach was later developed further to measure operationally, and reproducibly, the separate components of specifically and nonspecifically bound metal (Tiller et al., 1979; 1984a; 1984b).

In summarising my and other papers available at that time, I (Tiller, 1961) suggested the relative bonding energy of some trace metals for several minerals was Cu>Zn>Co>Ni, Mn, and noted that this order approximated the hydrolysis constants of these cations and the second ionisation potentials of the ions (Pauling, 1960). In my early work the observed relative affinities of Co and Ni were sometimes anomalous; I assumed that this could be due in some soils to a highly specific interaction of Co with Mn (McKenzie, 1967) or favoured sorption sites for Ni in some Mg minerals when present (Tiller et al., 1984b). Tiller (1961) proposed that the high preference of these trace metals for mineral surfaces was related to their polarising power, which could either induce increased dissociation of surface hydroxyl groups on the mineral surface as part of the bonding process or promote hydroxo-metal species by polarisation of coordinated water molecules. There was no way of deciding which of these alternative processes operated or whether the alternative of enhanced hydrolysis occurred on the surface or in the equilibrium solution. The problem is well recognised (e.g. Quirk and Posner, 1975; Kinniburgh et al., 1976) and the evidence is still inconclusive but has been recently well reviewed by McBride (1991).

Because of their concerns and those of others too, Hodgson and Tiller investigated further the role of ion hydrolysis in the reactions of metals with clays (Hodgson et al., 1964) by using the observation (Tiller, 1961) that the pH dependence of adsorption of Co by several minerals was largely eliminated in the presence of Mg, a competing ion which also hydrolysed, in comparison with Ca which was much less hydrolysed. They hypothesised that one would not expect a net change in protons if one hydroxo-cation simply exchanged with another. Hodgson and Tiller's theoretical treatment of adsorption data was interpreted as favouring an adsorption mechanism based on the concept of adsorption of a hydrolysed ion rather than exchange of the divalent metal ion for a surface proton. This work did not completely eliminate the possibility of surface induced hydrolysis of the adsorbed divalent metal ion by dissociation of a water molecule within its aqueous sheath. In any case, adsorption of a hydroxo-cation would result in further cation hydrolysis in solution.

Hingston et al. (1970) later suggested that the specific adsorption of metals could arise from the bonding of a hydroxo-cation which could provide a hydroxyl ion which would combine with a surface proton and thus generate a negative site. We continued on the premise that high energy bonding, possibly chemisorption, of trace and heavy metals by soil minerals resulted from the reaction (specific adsorption) of metal hydroxo-species at specific sites on the main mineral surfaces or with mineral or organic impurities within the matrix. The specific sites on the main surfaces may also be due to defects arising from weathering processes (Hodgson, 1963). Hodgson and Tiller (1962) had showed radiographically that outer basal surfaces of naturally occurring vermiculite preferentially retained Co when applied in very low concentrations but when this outer layer was separated, retention of Co was restricted to edges and cracks and other surface imperfections. Enhanced sorption of Co arising from reactions in the presence of Li, compared with Na as supporting electrolyte (Tiller, 1961) and increased retention of Zn by silicate clay following various pretreatments with Li salts (K.G. Tiller unpublished data) could arise from defects resulting from the replacement of octahedral Mg ions exposed at edge surfaces by Li (Greene-Kelly, 1953).

2.3 Complexation reactions in soil solution

Complexation reactions of heavy metal cations by a great variety of ligands had been well recognised for decades in the realms of pure chemistry, and the chelating ability of materials of soil origin had been identified and studied (e.g. Schnitzer, 1969). Bloomfield (1948) had identified the key role of complexation of Fe and Al by components of leaf extracts which mobilise and assist leaching of these metals during the podzolisation process. Bremner et al. (1946) had suggested that some transition metals were complexed with solid phase organic matter. Mattson et al. (1948), quoted by Beckwith (1955), suggested that some complexing agents that formed in the plant root rhizosphere may react with metal cations and form soluble diffusible anions. Beckwith (1955) worked primarily with Cu and Mn and progressed these ideas to help explain specific micronutrient metal interactions and uptake by plants.

However, considerations of soil chemistry in relation to plant nutrition remained mainly in terms of chemical equilibria involving undissociated cations. When I shared a laboratory with Dr J.F. ('Skefe') Hodgson, among our many vigorous discussion topics, he became concerned in the late 1960s with the complexation reactions especially of the trace metal cations in soil solution and the ways in which they might be characterised. These considerations were triggered by observations such as the occurrence of Zn deficiency under adverse soil conditions, e.g. of both high clay content and pH, under which Cu was often nutritionally adequate. He set up a 'factory type' apparatus, occupying a large fume hood, in which various leaching schemes were tested involving hydrogen peroxide and other reagents, trying to limit any back reactions of metals released. These early experiments were not completely successful, but Dr Hodgson's study leave with Professor W.L. Lindsay's group at Fort Collins, Colorado, led to publications (Hodgson et al., 1965; 1966) which had an enormous impact on the future direction and thinking in soil chemistry and its applications in plant nutrition and environmental studies. The death of such a creative and innovative soil scientist at a young age was a great loss. The recognition and measurement by Hodgson and his collaborators of the important role of complexing by organic ligands in solution must be rated as one of the most significant advances in our understanding of heavy metal reactions in recent decades, with many applications in the area of soil contamination and its consequences (Harter and Naidu, 1995). They showed that an appreciable proportion of several trace metal cations in soil solution was not in simple ionic form.

In the following years, greater emphasis was given to ionic speciation in soil solution and the use of relevant thermodynamic data helped achieve these goals. Development of computer based models such as GEOCHEM (Sposito and Mattigod, 1979) and MINTEQ (Allison et al., 1991) provided another great leap forward and they are now in common use in many laboratories. Early limitations relating to the availability and reliability of stability constants appropriate for soil relevant conditions have been largely overcome or at least a consistent set of values are in use. For many years I corresponded with Professor L.G. Sillen, Sweden, who helped edit (Sillen and Martell, 1964) one of the first major compilations of stability data, in relation to a problem I had about some stability constants for metal ion hydrolysis. He wrote 'This is a field where more than half of the literature is a waste of paper (though it may be hard to decide which part one can believe in) and where most of the important experiments still remain to be done.' Clearly we have advanced since then but comments in the literature from time to time suggest that some constants may still be incorrect or inappropriate. The deficiency in data on both the composition in solution and the stability constants of naturally occurring organic ligands of both low and especially higher molecular weight is well recognised. This lack is still a real limitation although useful approximations based on sewage sludge fulvic acid or an assumed mixture of well characterised organic acids (the mixture model) calculated from soluble organic carbon, have been developed (Sposito, 1981) and incorporated into the speciation models. These approximations are most useful in applications involving similar soils or comparisons of the same soil under different environmental conditions or time scales but are likely to introduce errors in comparisons of soils varying in key properties or which have received appreciably different organic amendments.

Paradoxically, the increasing sophistication and ease of use of the computer-based models for ionic speciation provides access to scientists who are computer-literate yet inexperienced or even incompetent in soil chemistry. This could well lead to false conclusions with unreliable solutions to environmental problems.

2.4 Adsorption /desorption or precipitation/dissolution?

This somewhat controversial topic developed initially during the 1960's and the 1970's. The proposition that trace metal solubility was controlled by dissolution of simple compounds was attractive and, if proven, could have facilitated interpretation and prediction of trace metal behaviour from soil solution composition. This approach was stimulated largely by the research of Professor W.L. Lindsay and students in Colorado (e.g. Lindsay and Norvell, 1969; Norvell and Lindsay, 1969; Ma and Lindsay, 1990; Sachdev et al., 1992) and did much to promote the potential of soil chemistry to provide a better basis for understanding the reactions and the nutritional and environmental impacts of trace and toxic metals. There never was much doubt that soils with very high loadings of heavy metals could well have metal compounds which could control metal solubility, but in the majority of contaminated soils, trace metal compounds could not be identified. It is of some concern that the recent expansion of work into X-ray Absorption Fine Structure (EXAFS), the development of which needs to focus on very contaminated soils most likely to contain metal compounds in order to establish the utility of the technique, will add credence to the idea of a control of heavy metal solubility by compounds which may well not apply to the more normal range of soil contamination .

Results of solubility studies expressed as equations of pM^{2+}-pH with slopes of 2 were interpreted as evidence that some divalent compounds controlled solubility, although the compounds were never identified on the basis of published stability constants. Although it must be admitted that low concentrations of mineral compounds which are undetected by techniques such as X-ray diffraction could well control metal concentrations, some conceptual problems remained. Firstly, minerals containing trace metals formed in a heterogeneous soil weathering environment would likely contain various metals in unknown proportions as the result of isomorphous substitution for which solubility criteria would not be available. A solubility slope of 2 could be maintained but the intercept, that is the solubility at any fixed pH, would likely vary with parent materials and pedogenic environment, making prediction impossible until the intercept of the solubility line was established for the soil in question. Secondly, the adsorption of secondary ions on the surface of any fine grained trace metal compound, analogous to the adsorption of P and Si species on Fe oxides or Al species on silica or trace metals on some smectitic clays (Tiller, 1969) would also affect solubility. Thirdly, the solubility of any possible trace metal compound formed preferentially on the template of major soil minerals would be decreased. Tiller (1968) showed that Zn and Co hydroxides were 2-3 orders of magnitude less soluble when precipitated in small amounts in the presence of layer silicate clays. This can be contrasted with Bingham et al. (1964) who reported that the apparent solubility product of Zn hydroxide, when precipitated in bentonite suspensions in larger amounts (about 1 mmol g^{-1}) agreed with published data for pure Zn hydroxide systems. Furthermore, Tiller and Pickering (1974) showed that a layer lattice stevensite-like Zn silicate could be synthesised under normal temperature conditions which could well template on 2:1 layer lattice silicates in soils. The synthetic mineral alone, after ageing about seven years at 20°C in suspension, had the same solubility as low amounts of Zn hydroxide precipitated on montmorillonite (Tiller, 1983) but still much too soluble to explain Zn concentrations in soil solution. The apparent irrelevance of trace metal compounds in soils to their solution chemistry may be simply a matter of not having the appropriate constants for as yet unidentified mineral phases. However, the fact that most soil conditions are at best in quasi-equilibrium due to changes in temperature, leaching, biological activity and various additions, is a severe limitation to the application of constants relying on chemical (reversible) equilibrium.

A final difficulty is the more philosophical question discussed by Tiller (1983), namely, when does a limited development of a trace metal multilayer structure on the template of the basal surface of a layered alumino-silicate clay or a partial extension of a silicate clay's octahedral layer constitute a separate solid phase rather than an adsorbed state. Extension of the octahedral could take place by the specific adsorption of trace metal alone or in conjunction with silicic acid (Tiller, 1968a; 1968b).

Finally, the available data do not provide a consistent picture. Professor W.L. Lindsay and associates (references above) have published solubility data which they interpreted as having a slope of 2 for the relationship of pZn^{2+} versus pH. However, some studies had too few data points to confidently predict solubility slope and the data were equally consistent with a much lower slope. The work of Ma and Lindsay (1990), however, clearly conforms to a slope of 2 in confirmation of their hypothesis that solubility is controlled by a mineral phase such as franklinite. The lack of agreement between their results and others cannot be adequately explained, but may relate either to the approaches used by different workers and their assumptions, or possibly to something unique about the semi-arid Colorado soils used by Lindsay and associates.

Many research groups have proposed slopes of the same relationship for Zn in soil solution, namely, $pZn = a + b.pH$, in the range of 0.5 – 1.0 (Saeed and Fox, 1977; Sinhar et al., 1978; Curtin and Smillie, 1983; Geoffrey and Uren, 1983; Tiller, 1983; Naidu et al., 1995 and Dang et al., 1995). Slopes (b) were commonly about 0.7. These data were found using a great variety of soils (including Oxisols and Vertisols) and were achieved with soil pH values within the range 4–8. Quite recently, Helmke (1995), using a newly developed technique based on Donnan equilibrium applied to a sludge-treated soil in Wisconsin, showed still lower solubility slopes for some trace metals, e.g. 0.25 for Zn^{2+} which he ascribed to exchange and complexation phenomena. When Helmke's approach was applied to a large range of Australian soil types ranging in pH from about 5 to 8.5, an average Zn^{2+} vs pH slope of 0.74 was obtained (R. Naidu, unpublished data).

It is also recognised that chelation reactions of metals with organic matter (OM) and specific adsorption reactions of metal ions with Fe oxides (e.g. Forbes et al., 1976), can also release between one and two protons per mole of metal adsorbed. In MacBride's (1991) review of adsorption processes he concluded that solubility slopes of less than 2 in soils may reflect chemisorption processes. Brummer et al. (1983) concluded that adsorption processes in soil provided a major constraint to precipitation. Soils with high areas of adsorbing surface reduced metal concentrations to such an extent that the necessary saturation or super-saturation required for precipitation rarely occurs.

In summary, the problem as described by Hodgson et al. (1964) and developed further by MacNaughton and James (1974) may be that one set of experiments can be interpreted in various ways, for example, as simple ion exchange, or adsorption of the hydroxo-ion at the surface, or adsorption of the divalent ion with concomitant hydrolysis in solution or surface-encouraged hydrolysis or precipitation. There is a risk that the preferred interpretation could be largely intuitive and controlled to some extent by an arbitrary choice of model parameters which may maximise agreement between the predicted model and experimental results. I prefer the view that MOH^+ ions at normal soil solution concentrations are adsorbed with concomitant hydrolysis in solution in the weakly to alkaline range of soil pH.

2.5 *Use of model systems*

Model systems have provided the basis for increasing our knowledge of the chemistry of trace metals in soils for at least half a century. Different research groups at different time periods have emphasised different soil materials to simulate trace metal reactions in soils of importance to their region or their particular enthusiasm.

2.5.1 Organic matter

Much excellent work related to soil OM was based on peat (e.g. de Mumbrum and Jackson, 1956; Bloom and McBride, 1979) and separated fulvic and humic fractions as the model substances. Schnitzer (1969) published many papers on the reactions between fulvic acid and a large range of divalent cations which provided a sound understanding of the potential of OM to bind metals and a basis for explaining differences in affinity of different trace metals for soil surfaces. Many other authors also published on this subject, especially Stevenson who reviewed this field on several occasions (e.g. Stevenson, 1991). This work contributed significantly to early development of computer modelling of ionic composition in soil solution by providing some relevant although approximate stability constants likely to control metal solubility.

2.5.2 Silicate clay minerals

As far as the mineral component of the soil is concerned, studies of soil chemical reactions of trace metals were largely restricted in the 1960's and the 1970's to the layer silicate minerals, the minerals commonly of greatest abundance in the soil. Many studies focussed on smectite but others compared the reactivity of different layer silicate minerals (e.g. Menzel and Jackson, 1950a; Banerjee et al., 1953; Spencer and Gieseking, 1954; Tiller and Hodgson, 1962; Tiller, 1967; 1968a; 1968b; Farrah and Pickering, 1977; Garcia-Miragaya and Page, 1977; 1978). These studies provided insights into the nature of adsorption mechanisms, the importance of pH, the much stronger bonding of the heavy metal cations compared to the alkali and alkali earth cations, the reactivities of different adsorption surfaces, ionic strength, accompanying anion and reaction time. Some studies were limited in possible interpretation and relevance to real soil situations because of unrealistically high concentrations of reacting metal as well as the absence of competing major cations. Under the latter circumstances one would expect the behaviour of the metals to approach those of the normal exchangeable cations, especially the more weakly bound trace metal cations e.g. Cd, and particularly as solution concentrations increased for specified conditions of pH and ionic strength. Tiller et al. (1979) showed that except for one very reactive soil, the preference of Cd over Ca decreased from >10× for an ionic fraction of 10^{-3} of the CEC, to equality when about 10% of the cation exchange was saturated with Cd. Different silicate minerals exhibited a large range in the amounts adsorbed with increasing concentrations in solution, as reflected in standard adsorption isotherm plots.

2.5.3 Oxide minerals

In the meantime, following geochemical observations of the association of Co with Mn oxides by a research group in the CSIRO Division of Soils, Adelaide (Tiller, 1959; Taylor et al., 1964; Taylor et al., 1966), McKenzie (1967) showed that a range of trace metals were strongly adsorbed by Mn oxide minerals separated from soils. Stanton (1964), quoted in Stanton and Burger (1967), had earlier shown that goethite and lepidocrocite had an important role in the strong binding of Zn in soils, and Hodgson (1963) had reported the extraction of rather large amounts of Cu and Co from a subsoil by a hydrogen peroxide reagent. Since the soil was virtually free of OM the metals must have come from Mn oxides. Grimme (1968) found that the concentration of Cu in Fe oxides in some subsoils could be at least 30 times that of the bulk soil and that synthetic Fe oxides adsorbed a range of heavy metal cations. Although the ground work had already been laid, the review of Jenne (1968), in which he proposed that the hydrous oxides of Mn and Fe provided the main control of the reactions of heavy metals in soils, stimulated much subsequent research in many other laboratories. The numerous publications over the next years (e.g. Loganathan and Burau, 1973; Gadde and Laitinen, 1974; Murray, 1975; Forbes et al., 1976; Kinniburgh et al., 1976; Shuman, 1977; Kalbasi et al., 1978; Swallow et al., 1980; Tiller et al., 1984b) confirmed that soil oxides reacted strongly with a range of heavy metals, not just Co, by pH dependent reactions with protons being released on the basis of one or two moles per mole of metal adsorbed. Evidence was provided that the reactions were not just with the main external surfaces but also with surfaces of the microstructure of the oxides. Lead and Cu were found to react much more strongly than other metals and in larger amounts with both Fe and Mn oxides. Lead was much more strongly bound by Mn than Fe oxides and by haematite in preference to goethite (McKenzie, 1980). K. Norrish (unpublished data) found that microaggregations of Mn but not the Fe oxides in a sandy loam subsoil affected by long range atmospheric fallout from a Pb-Zn smelter, contained several per cent of Pb compared to about 300 mg Pb kg^{-1} in the bulk soil. The strong affinity of Co for Mn oxides in soils explained the very low phytoavailability to pastures and associated animal deficiency problems (Adams et al., 1969).

Since Fe and Mn oxides are ubiquitous in Australian soils, it is not surprising that early progress in developing models for the adsorption of microelements was made in Australia. The major contribution to our understanding of the adsorption by Fe oxides was based on the work of Dr A Posner and Professor J P Quirk and associates in the University of Western Australia. Early emphasis was on models for anion adsorption (reviewed by Hingston, 1970) but this was soon developed to include cations (Bowden et al., 1973). They proposed a generalised model for the adsorption on oxide surfaces with variable charge which took into account the effects of charge, ionic strength, ionic competition and pH. This model was developed further (Bowden et al., 1977; Barrow et al., 1981) with many other contributions by Barrow (e.g. Barrow, 1985, 1986). Models of the mechanisms of the sorption reactions of Zn and other metals with soils have been reviewed by Sposito (1984), McBride (1991) and Barrow (1994).

2.5.4 Soil clay fractions

Some of us chose the less theoretical and more pragmatic alternative of using clay fractions separated from soils. Tiller et al. (1963) attempted to confirm the importance of different mineral soil surfaces in relation to Co adsorption by using soil clay suspensions separated from 17 subsoils selected on the basis of their being dominated by different mineral species, including Fe oxides, kaolinite, montmorillonite, illite, vermiculite, and amorphous material. The suspensions were untreated except for saturation with Na^+ during dispersion and later by Ca^{2+}. Surprisingly, with the exception of one vermiculitic soil, the pH-adsorption curves of all soils were remarkably similar. Initial concentrations of Co in solution were about 10^{-6} M. Various treatments showed that this observation could not be ascribed to common factors such as OM, amorphous material, prior saturation of high energy sites by heavy metals, or later (Hodgson et al., 1969), removal of both the Fe oxides (1 – 6%) and the small amount of Mn oxides (0.2%) present. This result certainly did not encourage us to pursue these oxides as important models of reactions of metals with soils, an anomalous result for which no explanation was provided.

Many years later we continued to use clay fractions from real soils, to provide greater realism to our adsorption studies. The first of these (Tiller et al., 1979) showed that only Zn and not Ca competed with the strongly (specifically adsorbed) bound form of Cd and that the component of Cd (nonspecifically adsorbed) that could be extracted by a dilute salt of Ca was still strongly preferred by soil surfaces over Ca, with the preference being closely and inversely related to the degree of occupancy of sites by Ca. This accords with the accepted views on the heterogeneity of adsorption surfaces relating surface saturation with bonding energy. Hendrickson and Corey (1981) confirmed this result in a thorough evaluation of data from the literature and stressed the importance of measuring selectivity coefficients appropriate to natural conditions. Later work (Tiller et al., 1984b) using clay fractions separated from a great range of soils did confirm the results of many oxide model system studies, namely, that Fe oxides can strongly affect adsorption of several heavy metal cations. Four out of the eight soils studied exhibited similar sigmoidal pH versus adsorbed metal curves typical of adsorption by Fe oxides, with strong inflections in the pH $Ca(NO_3)_2$ range 5.5 – 6.5. Iron oxides were only detected by X-ray diffraction in two of these soils. Other smectitic and organic-rich soil fractions had quite different adsorption patterns, exhibiting appreciable adsorption at lower pH values where adsorption by the Fe oxide-dominated samples was negligible. Whether Fe oxides dominate soil adsorption processes or not could not be readily predicted by simple measures of amorphous or crystalline Fe oxide contents or their specific surface area (Tiller et al., 1995).

2.5.5 Limitations of model systems

Although these model system studies have clearly provided a good understanding of those soil surfaces thought to be important for reactions for trace metals, I have reservations as to how useful they are for the understanding of the behaviour of heavy metals in real field soils. I'm sure many share this view. Pristine reactive surfaces are most unlikely to be preserved in real soils. Monomeric silicic acid, phosphate ions, other anions, a range of organic compounds and Al and other trace metals will occupy reactive bonding sites in competition with the metal under consideration. Ionic strength will vary from soil to soil and between and within seasons. Often unspecified isomorphous substitutions in the oxide structures will also affect surface properties. Convenience of laboratory procedures has also dictated that realistic soil to solution ratios are rarely even approached. The assumption that a particular oxide surface dominates the overall reaction of metals within the complex and often interacting mixture of soil surfaces is very hard to justify especially as the kinetics of reactions may vary with different soil components.

Finally we need to recognise that the soil environment is at best in a quasi-equilibrium state. Each soil component will have different adsorption characteristics and the least reactive will tend to dilute the impact of the most reactive surfaces, e.g. as indicated by pH_{50} values (Tiller et al., 1984). In addition there seems to be no means of easily evaluating the influence of a Fe oxide on soil reactivity within a complex soil matrix (Tiller et al., 1995). Of course there are extreme situations where some simple models may provide meaningful predictions e.g. in some highly weathered Oxisols or sandy soils dominated by OM and even then only the most general of interpretations are possible and usually in terms of the dominant pH factor and the dominant complexing ions present. Further development of increasingly sophisticated models of metal reactions based on simple monominerallic surfaces seems very hard to justify in the modern pragmatic world which requires results of relevance, providing value for money, which can help solve real problems. Such research is of most use in the academic world where it provides valuable training in basic soil chemistry.

Most of the criticisms applied to simple oxide or clay mineral models also apply to soil clay fractions but some soil factors can at least be tested more realistically. Predictions of metal contaminant behaviour in soils with a variety of soil surfaces are complicated because various soil surfaces (silicate, oxidic, organic) react to varying soil environmental conditions differently and in different pH ranges.

In summary, some reaction models and many published studies to date provide a basis for interpreting and understanding the reactions of heavy metals with soils, but suffer many limitations, especially a lack of relevance to authentic soil conditions by reason of unrealistic surface areas, solution concentrations, time scales and often the absence of competing surfaces and sometimes the absence of other competing ions or molecules. In addition, adsorption was normally stressed whereas the important reaction for most environmental applications, such as uptake by crops, absorption in a child's stomach or leaching to water supplies, depends on desorption which for many contaminants under most relevant conditions involves hysteresis. Schindler (1975) commented, in discussing a model for the removal of metals from sea water, that the reader could be easily manipulated because of the difficulty of choosing relevant parameters and because different combinations of parameters may seem to be equally suitable. Future work aimed at solving actual environmental problems should be based on relevant soil conditions and preferably using real soils.

2.6 The use of radioisotopic techniques

The use of radioisotopic methods as applied to the major nutrients in agricultural chemistry (e.g. for P. Larsen, 1952) were transferred to potentially toxic heavy metals (e.g. Tiller et al., 1968; Tiller, 1972a; 1972b; Tiller and Wassermann, 1974; Beckwith et al., 1975). The L-value technique of Larsen was shown in our CSIRO laboratories to provide a useful reference procedure for estimating the *total* pool of heavy metal accessible to crops. The effectiveness of different extractants to access this same pool was validated by comparing specific activities of metals, e.g. Zn (Tiller et al., 1972b) in plants with the specific activities in chemical extractants. This pool was presumed to derive mainly from metals bound to the surfaces of soil minerals and OM. The L-values provided an estimate of the *total* pool of trace element (Q), as discussed later, but not of the metal's affinity for the soil (K_d) and thus, were of no use in predicting uptake of metals (or P) to crops except in groups of soils of nearly constant K_d. This was not always appreciated by many researchers at that time and some still nowadays. Radioisotopic work using ^{65}Zn-labelled soils in which the specific activities of Zn in plants and various soil extracts were compared, reassured our use of the EDTA type extractions rather than those involving weak acids for assessing phytoavailable forms of trace metals. This was confirmed for Cd (K.G. Tiller and E. Suwadji, unpublished data). Dilute acids can dissolve insoluble mineral forms of Zn that are unavailable to plants. The main limitation to the use of isotopic exchange techniques is the risk of false results if the labelled element under study is partially irreversibly adsorbed. This is a special problem when determining E-values, that is, by extracting radioactively labelled soils by solutions without a plant reference. The application of radioisotopes to micronutrient phytoavailability has been critically reviewed by Tiller (1979; 1981).

Isotopic techniques still have a useful role in soil/plant research, (e.g. Smolders and McLaughlin, 1996a; 1996b) but are unsuited for routine applications. Our laboratory has carried out soil and plant research using radioisotopes of Zn, Co, Mn, Ni, Cd and Pb. The isotopes of the latter metal proved unmanageable because of its strong interaction with all surfaces, including the container in which it was supplied!

2.7 The transition from agricultural to environmental chemistry

With increasing public and political concern in environmental issues, scientists transferred research interest from considerations of plant deficiencies of trace elements to the toxic effects of heavy metals, some of which were of course the same elements, e.g. Cu, Zn. This was a natural transition for many of us because of the accumulated knowledge of trace metal chemistry, existing procedures (extraction and analytical) and availability of suitable equipment. The transition was not always satisfactory. Some existing equipment often required experimentation at unrealistic loading levels in order to achieve satisfactory analytical sensitivity and extraction procedures were not always suitable (discussed later). Glasshouse experiments formerly found useful to differentiate soils either on the basis of nutrient deficiencies or whether they may respond to fertilisation or not, were already recognised as having limitations. In the environmental context, they were found to give completely false indications of a soil's ability to supply potentially toxic metals to a crop (e.g. Tiller and de Vries, 1978; Page and Chang, 1979). These studies showed that concentrations of heavy metals in vegetables derived from soils treated with sewage sludge were much higher when grown in pots, even pots containing 16 kg of soil (de Vries and Tiller, 1978) than occurred under real farm conditions on identical soils. In addition, the high, linear responses in metal concentrations in plants to an increasing dose rate of metal in the soil of the pot experiment did not occur under farm conditions. These results arise from the more efficient root exploration of soil under pot conditions compared to the field, as well as the special environmental conditions of the glasshouse. Clearly glasshouse experiments cannot be relied on to provide environmentally relevant conclusions on soil/crop pollution problems.

Some proliferation of 'formula' research carried over from agricultural applications by researchers and their graduate students has continued; namely, 'grow a crop in the glasshouse using a group of soils, hopefully differing in properties controlling K_d, extract the soil, analyse the crop, measure correlation coefficients and there you have it'. Sometimes the soils chosen may have different loadings of a heavy metal but have similar properties controlling K_d and, as would be expected, the correlations are excellent confirming the preference of the particular extraction used, when a range of chemical solutions would have given a similar result. This type of experiment has a role in studying processes, in determining general impact of major soil and crop factors (e.g. varietal and species screening), and in research and method training for students but we need a much better balance between glasshouse and more pragmatic field research. Ultimately, the only useful and usable results are those tested under real field conditions.

3. Challenges for the present and the future

3.1 Introduction

Increased public awareness of the environmental threats to animal and human health, to food and water quality and to the sustainability of land use will strengthen the trend of recent years for research on trace metal and organic contaminants to be transferred from agricultural to environmental applications. However, the funding agencies supporting environmental research will insist on research of direct relevance to real problems, and with identified outcomes.

In this section I will discuss the topics of relevant experimentation for environmental applications, the need for greater attention to bioavailability, including the problems of its establishment in a legal framework, and the urgent need for greater interdisciplinary and interorganisational cooperation.

3.2 Relevant experimentation for soil environmental investigations

This will mainly involve avoiding the errors of the past which are generally well recognised although often ignored. Recognition will need to be made of the complex interactions between different competing sorption surfaces, different contaminants competing for the available bonding sites, the role of soluble organic carbon and other constituents in soil solution and the impacts of biological activity. Dosing of soils with pure metal and other contaminant compounds to simulate soil contamination must be avoided unless long equilibration times are planned.

Work will often need to be carried out on moist soils as collected in the field, although difficult and inconvenient. This approach will ensure relevance but can also reveal significant scientific insights. A good example is the recent field work of McLaughlin et al. (1995) in which field soil samples were used to study the transfer of fertiliser-derived Cd from soil to farm grown potato tubers. This work not only solved the local problem of potato quality and provided new management solutions but also identified the soil chemical processes controlling uptake of Cd, namely, the mobilisation of Cd by Cl introduced in the irrigation water. The importance of $CdCl^+$ shown to control Cd availability to plants in this field study corroborates the conclusions of the glasshouse study of Bingham et al. (1984), but for different reasons. Bingham et al. (1984) used much higher soil Cd concentrations by adding Cd salts, so that increasing salinity raised both Cd^{2+} and $CdCl^+$ activities. Bingham et al. (1984) concluded increases in Cd^{2+} were responsible for the salinity effect on plant Cd. However, in most agricultural soils, Cd^{2+} activity will be strongly buffered by the soil solid phase, as shown by McLaughlin and Tiller (1994), so that the effect of salinity on plant Cd is due either to uptake of $CdCl^+$ ions, or greater diffusion of Cd^{2+} to the root. The field approach as exemplified by McLaughlin et al. (1995) can also provide relevant estimates of the K_d of desorption, in this case, of Cd. Of course, use of such soil-derived K_d values to model generally the leaching or uptake of Cd at other loadings of Cd requires assumption that the K_d remains constant over the range studied. This may not be the case.

Achievement of useful outcomes depends more on measurement of relevant soil parameters such as the desorption K_d within a relevant kinetic framework and less on concern as to whether the solubility is controlled by adsorption or solubility product of some complex or ill-defined solid phase. The K_d can be estimated from measurement of the contaminant of concern associated with the solid phase component. We use an exhaustive EDTA extraction for metals (Clayton and Tiller, 1979) and calculation of the undissocciated metal from the solution composition. If facilities do not permit this, then the total concentration of metal in solution will be a useful approxi-mation for those metal ions that are not too strongly complexed e.g. Cd in contrast to Cu, for which this approximation could not apply.

Use of model system studies in sponsored research will be less tolerated except in the context of student training. Enthusiastic interpretation of data in terms of adsorption equations should be restrained. Many earlier papers applied the Langmuir equation, for example, to reactions for which the basic assumptions did not apply, especially relating to the need for surface coverage to be unrelated to bonding energy, no interactions between adsorbate species, for adsorption in a monolayer and for reversible reactions at equilibrium. Many early papers showed an apparent agreement of data with the Langmuir equation for experiments where this would seem impossible. The applications and limitations of the use of the Langmuir and Freundlich equations were discussed in depth by Adamson (1960). The form of the Langmuir and Freundlich equations will still, however, continue to provide useful means of presenting research data.

Desorption rather than adsorption is essential to more adequately describe the actual soil reactions which control transport processes (leaching), uptake by crops or absorption of toxic substances associated with ingested soil in a child's digestive system. Since these latter processes are rarely reversible or at equilibrium, the kinetics of reactions may be essential for sound predictions of likely outcomes of contamination or remediation of contaminated soils. To this, an understanding of the kinetics of desorption is critically important yet a very rare commodity. This topic provides a major challenge for investigation in the future.

Soil chemists should be cautious in assuming that soil tests will always predict key processes controlling the transfer of toxic metals from soils to crops. Plant factors relating to the balance of nutrients, root integrity, and the exudates of acids and complexing substances can affect the uptake of toxic metals in ways unrelated to simple soil chemical characteristics. Good examples are the enhanced uptake of Cd by field wheat grown on Zn-deficient soils (Oliver et al., 1994) and the role of crop rotations where wheat following legumes, especially lupins, had increased Cd concentrations compared to wheat following other cereals or fallow (Oliver et al., 1993). Also, it is well recognised that uptake of heavy metals by crops is strongly dependent on crop species and variety e.g. Cd in varieties of potatoes (McLaughlin et al., 1994) and Cd in wheat varieties (Oliver et al., 1995).

3.3 Bioavailability of soil contaminants

3.3.1 Introduction

There are several difficulties relating to bioavailability, the question of definition of the concept as a basis for discussion and communication, the methodologies of measurement and their lack of universality, and the great reluctance of the regulatory authorities in many countries to translate the concept into a legislative framework. These aspects will be discussed in detail below. In this discussion I will mainly use the term 'bioavailable' although the terms 'available' and 'phytoavailable' are also in common use. The concept of bioavailability applies to all organisms whether microflora and microfauna, plants, animals or humans.

3.3.2 Concepts

Some of the difficulties arise because of confusing an understanding of the concept with approaches to its measurement. The bioavailable nutrient or contaminant is understood by some, for example, in terms of the concentration in soil solution in contact with the plant root at any point in time. This relies on the kinetics of desorption in response to crop uptake. Others more correctly prefer to view bioavailability in terms of the soil's ability to replenish the soil solution over some defined period, say one growing season or indefinitely. Many consider bioavailability in terms of some particular extraction procedure which has achieved acceptable correlation coefficients with the nutritional or environmental problem under investigation in their region or country. This latter approach has been quite satisfactory and well accepted on a local or regional basis for separating soils on the basis of possible nutritional disorders or their likelihood to respond to fertilisers (e.g. the use of DTPA (Lindsay and Norvell, 1978). Procedures developed in the agricultural sphere do not have universal acceptance in all countries, not necessarily because of any lack of usefulness but because of a certain inertia we all share in terms of our mental commitment as well as availability of equipment and the need to support existing data bases. Similarly we should also not expect that some arbitrarily chosen procedure found to be satisfactory for one purpose will be necessarily useful for another. I say 'arbitrary' because as stated earlier, several chemical extractants are often almost equally efficient in prediction, the correlation coefficient differing slightly (Tiller et al., 1972a). The predictive measure of a bioavailable contaminant relating to say uptake by crops or leaching in transport processes, will not be relevant to estimation of the potential absorption of a contaminant in a child's stomach from ingested soil or evaluating the need for remediating a contaminated soil which has exceeded the legal investigatory guidelines. Choices of test procedures may sometimes depend on analytical limitations or the need to assess a large range of elements but ultimately methods must be reliable and easy to use in routine operations.

3.3.3 Methodology

Many prefer to consider the quantification of bioavailability in terms of a quantity variable (Q), an intensity variable (I) and the ratio Q/I, analogous to a distribution coefficient (K_d), an approach originally proposed by Beckett (1964) for K. By this approach, the prediction of the impact of pollution on human health, environmental well being or the planning of remediation procedures may require estimates of either or all of the Q, I and Q/I parameters.

The variable Q should equate to the total potentially bioavailable or mobile concentration of the contaminant of interest and not the concentration calculated from complete digestion of the sample in strong acids. In some situations as contamination increases, this Q variable will approach in amount the true total concentration, as determined by complete digestion in strong acids or XRF because the proportion of contaminant held in unavailable forms within mineral structures becomes less significant. This situation occurred in the most contaminated area near the Pb-Zn smelter at Port Pirie, SA (Cartwright et al., 1977), at least for Cd and Zn. The measurement of the Q variable needs to recognise that the potentially labile component of the contaminant will be largely surface bound and will not reside within the crystal structures of soil minerals. Acids are unsuited for general soil application because of their at least partial neutralisation of calcareous soils and their ability to dissolve some soil minerals. We have carried out our assessment of the total loading of heavy metal contaminants in soils using an exhaustive EDTA procedure based on 0.05 or 0.1 M EDTA buffered at pH 6 (Clayton and Tiller, 1979). Extractions are carried out for 7 d, the period needed for complete extraction when metals have high affinity for soils (e.g. soil Pb bound by some soil types). A much shorter period may suffice for the evaluation of contamination by Cd, for example. Very contaminated soils require the higher concentration of EDTA because of reagent saturation.

The quantity parameter is inherent in many current EPA regulations, and is essential for calculations of input-output balances and for assessing total environmental impact, e.g. from atmospheric fallout, and recognising the maximum threat (the total potential amount likely to become mobile or bioavailable with time or under the most adverse conditions) but only provides part of the story.

The actual concentration of the active or soluble contaminant in the associated soil solution (I) will depend on the chemical environment as determined by electrolytes, complexing ligands, temperature etc. and the extent to which equilibrium has been achieved. Sometimes the 'equilibrium' solution concentration of the contaminant (I) should be defined in terms of a specific ionic species if it is known to control the surface reaction. The estimate of the I parameter, expressed either as a concentration in soil solution or as a proportion of Q, and its rate of replenishment, supports models predicting transport of contaminants, e.g. leaching to groundwater, uptake by crops and absorption by humans and animals from ingested soil.

Prediction of the environmental consequences of soil contamination is best achieved by simulating the appropriate soil condition, e.g. by using solutions that proxy for soil solution, such as dilute neutral salts, to predict uptake by crops or acidic solutions of appropriate pH values to simulate the pH changes that take place when contaminated soil passes through a child's digestive system. This is well recognised in principle but lack of research data or consensus as to the best approach is a barrier to implementation of widely accepted suitable tests for regulatory limits. The ideal approach for food crop quality, although unsuited to routine practice, is to separate the actual soil solution for analysis and assess ionic speciation.

As stated above, many published models are flawed because they rely upon Q/I measured as adsorption slopes rather than desorption without considering the kinetics involved and also assume that the value of Q/I (K_d) is linear over the concentration range of interest. Fortunately this provides an unknown safety factor in calculating the extent and rate of leaching of contaminants, possibly of several orders of magnitude in the case of some heavy metals because of adsorption hysteresis. In addition, modelling of contaminant transport now recognises the need for estimation of Q/I (K_d) as a function of the changing concentration of contaminant with depth and the different adsorbing materials in the sediment or soil layers.

The disadvantage of using some of the correlative procedures for pollution potential transferred from earlier agricultural applications e.g. dilute acids and short term extractions by dilute complexing agents, is that they are neither estimates of Q nor I and their usefulness only applies to the restricted range of soils on which the relationship was developed. As such they are of limited value in an environmental context except where a correlative role may suffice, for example, in grading and demarking environmental hazard in areas with similar soil properties. Numerous compromise leaching tests have been developed to identify, classify, and compare hazardous wastes and to provide solutions to problems of waste management (WTC, 1990). The commonly used USEPA Toxicity Characteristic Leaching Procedure (TCLP) is a good example and now has some value primarily because of its widespread acceptance and accumulated data base. Its drawbacks are that it estimates neither the total potential amount of bioavailable contaminant nor the relevant concentration of mobile contaminant in solution in equilibrium with the surface bound contaminant. The conditions imposed by the procedure, namely, a buffered acetate medium at pH 5, would rarely simulate conditions for any soil or sediment.

3.3.4 Solid state speciation

Characterisation of the total loading of soil contaminants by solid state speciation has remained a popular approach over recent decades despite recognition (Beckett, 1989) that the procedures commonly used lack specificity and the results are difficult or sometimes impossible to interpret. If the procedures are valid then one should be able to apportion the total loading of contaminant between different binding and mineral forms. As the extent of contamination increases, the procedures will tend to identify the different adsorbents rather than different mineral forms. Progress is being made (Zeien and Brummer, 1989) and more reliable approaches are being developed but every major scientific meeting on soil pollution has several, sometimes many, papers or posters which contribute little to this topic. Besides doubts about the validity of the procedures, rarely is the efficiency of the overall recovery of the initial amount of contaminant given. Also, rarely is any indication given of the way in which the data can be interpreted and utilised in practice. Of course, there are situations where a knowledge of the role of different bonding forms can be useful. Examples are, for instance, the identification of possible sinks in terms of soil and soil layers in landscapes subject to regional contamination or prediction of the impact of a changing soil environment on contaminant solubility such as the affect of hydrology on metals or As released from destabilised Fe or Mn oxides. In most situations, however, the bonding energy of the contaminant on a particular surface (as measured as K_d) rather than the amount in a particular form, determines the environmental impact. Accordingly I believe that there is rarely any benefit in solid state speciation in relation to soil contamination apart from any information that it can provide about the distribution of soil contaminant in terms of bonding energy. This can be achieved more easily and is less subject to misinterpretation. Such a fractionation could be based on three kinds of assessment, namely, weakly bound forms e.g. (a) by dilute salts of Ca or ammonium, (b) strongly adsorbed forms e.g. by some kind of exhaustive EDTA procedure, and (c) the mineral residue e.g. by difference or strong acids or XRF. The value of some existing solid state speciation methods resides mainly in the ability to deduce the above kind of information related to bonding energy. This fractionation based on bonding energy may also possibly be achieved by the kinetics of desorption. Admitting the usefulness of these procedures for some specific applications, there can clearly be no universal procedure suited to all soil types. A method developed for temperate light-textured soils high in OM, for example, would be unsuited to an Oxisol with very high concentrations of Fe and Mn oxides in addition to some OM.

3.3.5 A note on organic contaminants

I have little experience with the environmental chemistry of organic contaminants, e.g. pesticides, polyaromatic hydrocarbons (PAHs) etc, but the same approach will apply in relation to their impact.

The need to define the total loading (Q) is important, as defined by an efficient organic extraction procedure, to understand the maximum potential hazard. The organic contaminant bound to the soil colloids will equilibrate with the soil water and soil atmosphere. The resultant concentrations of the organic compound in soil solution and soil air (I variables) will determine important environmental processes, namely, losses to the atmosphere by gaseous diffusion through unfilled soil pores, leaching to groundwater, uptake by crops and the extent of microbial degradation. Organic compounds bound irreversibly and strongly to soil surfaces and the range of expected soil conditions will have negligible environmental impact. They should be viewed in the same way as metals which are within a mineral crystal lattice or adsorbed very strongly at or within the soil surfaces. The more labile component of the soil organic contaminant in soils must be able to be measured just as metals are, but the approach will be different. The separation in terms of environmental risk could be based on solvent extractions of different efficacies perhaps combined with some modifications of the soil properties controlling the binding of the organic matter.

Remediation of organic contaminated soil by microbial degradation will be directly related to the affinity of the organic compound for the soil; the compounds need to be in the free form for microbial or fungal access. This kind of research is under way internationally e.g. Welp and Brummer (1992) and recent work in our laboratory (S. Rogers, R. Kookana and D. Oliver – unpublished) showed that strychnine bound strongly to a slightly acidic soil resisted microbial degradation whereas in a more alkaline soil, in which the strychnine was more labile, degradation proceeded relatively rapidly. The strong sorption of an organic contaminant by soil colloids limits microbial interaction. This means that investigations of the degra-dation of organic compounds in pure media without the presence of soil may give a completely false and overly optimistic evaluation of the potential of microbial degradation as a means of soil remediation.

3.3.6 A legal framework for bioavailability

Most environmental scientists recognise that regulatory control of soil contamination should be in terms of the related concepts of bioavailability, solubility or mobility. They understand that a certain fraction of the contaminant, perhaps situated within a stable crystal structure or rendered insoluble by soil conditions, has no chance of environmental impact within the appropriate kinetic framework. Accordingly, it does not seem logical to prescribe regulatory limits in terms of the total concentration of contaminant. For example, the health limit for Pb in Australia (ANZECC/NHMRC, 1992) is 300 mg kg^{-1}, taking into account the most critical pathway of ingestion of soil by a child (Ryan and Chaney, 1993). If we could demonstrate with certainty that only a maximum of 20%, for example, of this Pb in a particular soil would be mobilised in the child's digestive system, then a total concentration of 1500 kg^{-1} should be viewed as safe. This approach can be extended to the rehabilitation of large areas of expensive existing contaminated land in the inner city areas, where attaining a goal of a concentration near natural background is financially unachievable. If no sensitive pollution pathways are envisaged in the long term, that is, no schools, child-care centres etc. are ever planned, then estimates of bioavailable contaminants in terms of the next most critical pathway, perhaps production of home-grown vegetables (Chaney and Ryan, 1993), may permit a safe limit of the order of thousands of mg kg^{-1} for Pb rather than hundreds. An equivalent difference would apply to other heavy contaminants but the benefit of assessing the bioavailable form would be less for the more labile elements such as Cd, although still important. The situation is more straightforward when considering limits for relatively uncontaminated land, e.g. cultivated agricultural land and pastures, where some countries will seek to limit levels of potential toxic metals to natural background levels. Assessment of rehabilitation of many sites in terms of bioavailability and for specific land use may determine that only minimal remediation is needed.

Some countries are investigating soils, crops, and groundwater on a regional basis to provide a better soil chemical basis for the legal evaluation of pollution risk, for example, in Germany (Pruess, 1995) where an extraction with NH_4NO_3, German standard DIN 19730, may simulate the concentrations of mobile heavy metals both in relation to uptake by crops and the potential for leaching (Thiele and Brummer, 1993). Precautionary and action values have been incorporated in the soil protection laws of Baden-Wurttemburg (UWT, 1993). This extracting solution has the benefit of removing a greater range of toxic elements in a matrix convenient for analysis than the commonly used dilute Ca salts. This method will still need to be used in conjunction with an estimation of total pollutant loading where an EDTA type of extraction could be used, as discussed above.

I believe that the scientific community themselves are largely to blame for the non-acceptance of bioavailability criteria in the guidelines and regulations of the environmental agencies because of the plethora of methods, procedures and attitudes that have been developed for environmental applications. This situation is most surprising considering the large literature available on this topic. What is needed is a fresh approach to this problem with some compromising on the details of the procedures but not the principles involved. Some old favourites would need to be discarded. Scientific focus would need to be specifically directed to problems identified by the agencies and carried out in close collaboration with them. It is also most surprising that despite the enormous financial benefits accruing from reduced rehabilitation costs carried out in relation to regulatory limits based on bioavailability, it is very difficult to secure funding for this important subject. At the very least, environmental protection agencies should retain some flexibility and provide some encouragement for studies on bioavailability pending the achievement of appropriate results. Scientists will otherwise be wasting their efforts if the basis of environmental investigation and cleanup is entrenched in legal concrete. Promising research such as the immobilisation of Pb in heavily contaminated sites by treatment with phosphate (Logan et al., 1995) will be a waste of time if the environmental authorities persist with guidelines based on total metal concentrations. Lower limits based on bioavailability when viewed in terms of total chemical analysis, may initially cause some public concern, especially to some environmentally sensitive groups. An educational program may be needed to explain the sounder basis for the assessment of soil contamination.

4. Concluding statement

Great opportunities exist for those environmental scientists trained in soil science. However continued or increased support from funding agencies and government authorities for their work will only be assured if we tackle environmental problems using the best science available and strive for relevant outcomes. Research approaches and methodology must avoid the errors of the past by using real world, relevant conditions in terms of contaminant concentration, realistic time scales, and keeping in mind the kinetics of the desorption processes which control the transfer of contaminants to crops, water and humans.

It is well recognised that teams of researchers with different disciplinary backgrounds are essential for real progress and relevant outcomes. Too often though, the interdisciplinary cooperation is notional, between agencies well separated in mental commitment and sometimes by distance; a marriage of convenience for the funding application. What is needed are arrangements which ensure a genuine 'working together'. A truly symbiotic team structure between, say, biologists, chemists, and physicists will provide a synergism, promoting a greater depth of scientific understanding as well as more pragmatic outcomes.

As well as this important interdisciplinary collaboration, it is also essential to work in cooperation (that is, towards the same goals), with State agencies who have final responsibility for guidelines and limits for soil remediation, soil, food and water quality. This will ensure a better focus of research leading to more relevant goals, a better basis for regulation and environmental law, and greater prospects for scientific results being put into practice.

Finally, although I cannot commit colleagues to the opinions expressed as my personal perspectives in this paper, I am deeply appreciative of the encouragement, support and active collaboration of numerous people, from my thesis supervisors, to many technicians and scientific colleagues over the last 40 years. They certainly helped achieve the scientific results, usually, but not always, appearing in publications, but more importantly, provided stimulating ideas and argument which helped crystallise my own views and attitudes. Of course, as always, most of the outcomes were truly team efforts.

References

Adams S N, Honeysett J L, Tiller K G and Norrish K 1969 Factors controlling the increase of cobalt in plants following the addition of a cobalt. Aust. J. Soil Res. 7, 29-42.

Adamson A W 1960 Physical Chemistry of Surfaces. Interscience Publishers, New York.

Allison J, Brown D S, and Novo-Gradac K J 1991 MINTEQ2/PRODEFA2, a geochemical assessment model for environmental systems. EPA/600/3-91/021 Rep., U.S. NTIS.

ANZECC/NHMRC 1992 Australian and New Zealand guidelines for the assessment and management of contaminated sites. Australian and New Zealand Environment and Conservation Council and National Health and Medical Research Council. January 1992.

Bambergs K and Balode A 1957 Adsorption of manganese, Zn and Cu in soil. Latvijas PSR Zinatnu Akad. Vestis No. 8, 45-56.

Banerjee D K, Bray R H and Melstead S W 1953 Some aspects of the chemistry of cobalt in soils. Soil Sci. 74, 421-431.

Barrow N J 1985 Reactions of anions and cations with variable-charge soils. Adv. Agron. 38, 183-230.

Barrow N J 1995 Mechanisms of reaction of Zn with soil and soil components. In Zn in Soils and plants. Ed. A D Robson. pp 15-32, Kluwer Academic Press, Dordrecht.

Beckett P H T 1989 The use of extractants in studies on trace metals in soils, sewage sludges, and sludge-treated soils. In Advances in Soil Science, Vol. 9. Ed. A B Stewart. pp 143-176.

Beckett P H T 1964 Studies on soil potassium. II. The immediate Q/I relations of labile potassium soil. J. Soil Sci. 15, 9-23.

Beckwith R S 1955 Metal complexes in soils. Aust. J. Agric. Res. 6, 685-698.

Bingham F T, Page A L and Sims J R 1964 Retention of Cu and Zn by H-montmorillonite. Proc. Soil Sci. Am. 28, 351-354.

Bloom P R and McBride M B 1979 Metal ion bonding and exchange with hydrogen ions in acid-washed peat. Soil Sci. Soc. Am. J. 43, 687-692.

Bloomfield C 1948 Mobilization of iron in podzol soil by aqueous leaf extracts. Chem. Indust. 259-260.

Bowden J W, Bolland M D A, Posner A M and Quirk J P 1973 Generalized model for anion and cation adsorption at oxide surfaces. Nature Phys. Sci. 245, 81-83.

Bowden J W, Posner A M and Quirk J P 1977 Ionic adsorption on variable charge mineral surfaces. Theoretical-charge development and titration curves. Aust. J. Soil Res. 15, 121-136.

Bower C A and Truog E 1940 Base exchange capacity determination as influenced by nature of cation employed and formation of basic exchange salts. Soil Sci. Soc. Am. Proc. 5, 96-89.

Bremner J M, Heintze S G, Lees H and Mann P J G 1946 Metallo-organic complexes in soil. Nature 158, 790-791.

Brown A L 1950 Zn relationships in aiken clay loam. Soil Sci. 69, 349-358.
Brummer G, Tiller K G, Herms U and Clayton P M 1983 Adsorption-desorption and/or precipitation/dissolution processes of Zn in soils. Geoda. 31, 337-354.
Cartwright B, Merry R H and Tiller K G 1975 Heavy metal contamination of soils around a lead smelter at Port Pirie, South Australia. Aust. J. Soil Res. 15, 69-81.
Chaney R L and Ryan J A 1993 Heavy metals and toxic organic pollutants in MSW-composts: Research results on phytoavailability, bioavailability, fate, etc. In Science and Engineering of Composting: Design, Environmental, Microbiological and Utilization Aspects. Eds. H A J Hoitink and H M Keener. pp 451-505, Renaissance Publications, Worthington, Ohio.
Clayton P M and Tiller K G 1979 A chemical method for the determination of the heavy metal content of soils in environmental studies. Div. Soils, CSIRO, Tech. Paper No. 41 pp. 17.
Curtin D and Smillie G W 1983 Soil solution composition as affected by liming and incubation. Soil Sci. Soc. Am. J. 47, 701-707.
Dang Y P, Tiller K G Dalal R C and Edwards D 1995 Zn speciation in soil solutions of Vertisols. Aust J. Soil Res. (in press).
de Mumbrum L E and Jackson M L 1956 Cu and Zn exchange from dilute neutral solutions by soil colloidal electrolytes. Soil Sci. 81, 353-357.
de Vries M P C and Tiller K G 1978 Sewage sludge as a soil amendment, with special reference to Cd, Cu, Mn, Ni, Pb and Zn – comparison of results from experiments conducted inside and outside a glasshouse. Env. Pollut. 16, 231-240.
Delas J, Delmas J and Demias C 1960 A study by means of isotopic dilution of Cu incorporated in soil during a period of 70 years. C.R. Acad. Sci. Paris 250, 3867-3869.
Elgabaly M M and Jenny H 1943 Cation and anion interexchange with Zn montmorillonite clays. J. Phys. Chem. 47, 399-408.
Farrah H and Pickering W F 1977 The adsorption of lead and Cd species by clay minerals. Aust. J Chem. 30, 1417-1422.
Forbes E A, Posner A M and Quirk J P 1976 The specific adsorption of Cd, Co, Cu, Pb, and Zn on goethite. J. Soil Sci. 27, 154-166.
Gadde R R and Laitinen H A 1974 Studies of heavy metal adsorption by hydrous iron and manganese oxides. Analyt. Chem. 46, 2022-2026.
Garcia-Miragaya J and Page A L 1977 Influence of exchangeable cation on the sorption of trace amounts of Cd by montmorillonite. Soil Sci. Soc. Am. J. 41, 718-721.
Garcia-Miragaya J and Page A L 1978 Sorption of trace quantities of Cd by soils with different chemical and mineralogical composition. Water, Air and Soil Pollution. 9, 289-299.
Greene-Kelly R 1953 The identification of montmorillonoids in clays. J. Soil Sci. 4, 233-237.
Grimme H 1968 Die Adsorption von Mn, Co, Cu, Zn, durch Goethit aus verdunnten Losungen. Z. Pflanzenernahr. Bodenk. 121, 58-65.
Harter R D and Naidu R 1995 Role of metal-organic complexation in metal sorption by soils. Adv. Agron. 55, 219-263.
Helmke P A, Salal A K and Yan L 1995 Measurement and behaviour of indigenous levels of the free, hydrated cations of Cu, Zn and Cd in the soil-water system. Third International Conference on the Biochemistry of Trace Elements, Paris, 1995 (in press).
Hendrickson L L and Corey R B 1981 Effect of equilibrium metal concentrations on apparent selectivity coefficients of soil complexes. Soil Sci. 131, 163-171.
Hibbard P L 1940 The chemical status of Zn in the soil with methods of analysis. Hilgardia 13, 1-29.
Hingston F J, Posner A M and Quirk J P 1970 Anion adsorption at oxide surfaces – the adsorption envelope. Search 1, 324-327.
Hodgson J F, Tiller K G and Fellows M 1969 Effect of iron removal on cobalt sorption by clays. Soil Sci. 108, 391-396.
Hodgson J F and Tiller K G 1962 The location of bound cobalt on 2 : 1 layer silicates. Clays Clay Miner. 9, 404-411.
Hodgson J F 1960 Cobalt reactions with montmorillonite. Soil Sci. Soc. Am. Proc. 24, 165-168.
Hodgson J F 1963 Chemistry of the micronutrient elements in soils. Adv. Agron. 15, 119-159.
Hodgson J F Geering H R and Norvell W A 1965 Micronutrient cation complexes in soil solution: partition between complexed and uncomplexed forms by solvent extraction. Soil Sci. Soc. Am. Proc. 29, 665-669.
Hodgson J F, Lindsay W L and Trierweiler J F 1966 Micronutrient cation complexing in soil solution. II. Complexing of Zn and Cu in displaced solution from calcareous soils. Soil Sci. Soc. Am. Proc. 30, 723-726.

Hodgson J F, Tiller K G and Fellows, M 1964 The role of hydrolysis in the reaction of heavy metals with soil-forming materials. Soil Sci. Soc. Am. Proc. 28: 42-46.
Huser R 1958 Adsorption of cobalt by kaolinite. Z. Planzenernahr. Dung. Bodenk. 80, 56-66.
Jeffery J J and Uren N C 1983 Cu and Zn species in the soil solution and the effects of soil pH. Aust. J. Soil Res. 21, 479-488.
Jenne E A 1968 Controls on Mn, Fe, Ni, Cu and Zn concentrations in soils and water: the significant role of hydrous Mn and Fe oxides. In Trace Inorganics in Water. Advances in Chemistry Series No. 73. American Chemical Society, Washington, D.C. pp 337-387.
Jones H W, Gall O E and Barnette R M 1936 The reaction of Zn sulfate with the soil. Fla. Agr. Expt. Sta. Bull. 298, 5-43.
Kalbasi M, Racz G J and Loewen-Rudgers L A 1978 Mechanism of Zn adsorption by iron and aluminium oxides. Soil Sci. 125, 146-150.
Kinniburgh D G, Jackson M L and Syers J K 1976 Adsorption of alkaline earth, transition, and heavy metal cations by hydrous oxide gels of iron and aluminium. Soil Sci. Soc. Am. Proc. 40, 796-799.
Larsen S 1952 The use of P^{32} in studies on the uptake of phosphorus by plants. Plant. and Soil 4, 1-10.
Leydon R F and Toth S J 1960 Behaviour of Zn sulfate as foliar applications and as soil applications in some New Jersey soils. Soil Sci. 89, 223-228.
Lindsay W L and Norvell W A 1969 Equilibrium relationships of Zn^{2+}, Fe^{3+}, Ca^{2+}, and H^{+} with EDTA and DTPA in soils. Soil Sci. Soc. Am. Proc. 33, 62-65.
Lindsay W L and W A Norvell 1978 Development of a DTPA soil test for Zn, iron, manganese and Cu. Soil Sci. Soc. Am. J. 42, 421-428.
Logan T J, Traina S J, Heneghan J and Ryan J A 1995 Effects of phosphate addition on bioavailability of lead in contaminated soil fed to rats and pigs. Third International Conference on the Biochemistry of Trace Elements (Theme B4), Paris, May 1995.
Loganathan P and Burau R G 1973 Sorption of heavy metal ions by a hydrous manganese oxide. Geochem. Cosmochem. Acta 37, 1277-1293.
Ma Q and Lindsay W L 1990 Divalent Zn activity in arid zone soils obtained by chelation. Soil Sci. Soc. Am. J. 54, 719-722.
Mattson S, Erikson E and Vahtras K Lantbr Hogsk. 1948 Ann. 15, 291 (quoted by Beckwith 1955)
McBride M B 1991 Processes of heavy and transition metal sorption by soil minerals. In Interactions at the soil colloid–soil solution interface. Eds. G H Bolt et al., pp 147-176. Kluwer Academic Publishers, Dordrecht, The Netherlands.
McKenzie R M 1967 The sorption of cobalt by manganese minerals on soils. Aust. J. Soil Res. 5, 235-246.
McKenzie R M 1980 The adsorption of lead and other heavy metals on oxides of manganese and iron. Aust. J. Soil Sci. 18, 61-73.
McLaughlin M J and Tiller K G 1994 Choro-complexation of Cd in soil solutions of saline/sodic soils increases phyto-availability of Cd. Transactions of 15th World Congress of Soil Science, Mexico. Vol. 3b, 195-196.
McLaughlin M J, Williams C M J, McKay A, Kirkham R, Gunton J, Jackson K J, Thompson R, Dowling B, Partington D, Smart M K and Tiller K G 1994 Effect of cultivar on uptake of Cd by potato tubers. Aust. J. Agric. Res. 45, 1483-1495.
McNaughton M G and James R O 1974 Adsorption of aqueous mercury (II) complexes at the oxide/water interface. J. Coll. Interface Sci. 47, 431-440.
Menzel R G and Jackson M L 1950a Mechanism of adsorption of hydroxy cupric ions by clays. Soil Sci. Soc. Am. Proc. 15, 122-124.
Menzel R G and Jackson M L 1950b Sorption of Cu from acid systems by kaolinite and montmorillonite. Trans. Intern. Congr. Soil Sci. Fourth Congr. Amsterdam 1950 I, 125-128.
Murray J W 1975 The interaction of metal ions at the manganese dioxide–solution interface. Geochem. Cosmochim. Acta 39, 505-551.
Naidu R, Bolan N S, Kookana R S and Tiller K G 1994 Ionic strength and pH effects on the sorption of cadmium and the surface charge of soils. J. Soil Sci. 45, 419-429.
Naidu R, Rengasamy P, De Lacy N J and Zarcinas B A 1995 Soil solution composition of some sodic soils. In Australian sodic soils: Distribution, properties and management. (Eds. R Naidu, M E Sumner and P Rengasamy) pp 155-161, CSIRO Australia.
Norvell W A and Lindsay W L 1969 Reactions of EDTA complexes of Fe, Zn, Mn, and Cu with soils. Soil Sci. Soc. Am. 33, 86-91.
Oliver D P, Hannam R, Tiller K G, Wilhelm N and Merry RH 1993 The effects of Zn fertilization on Cd concentrations in wheat-grain. J. Environ. Qual. 23, 705-711.

Oliver D P, Schulz J E, Tiller K G and Merry R H 1993 The effects of crop rotations and tillage practices on Cd concentration in wheat grain. Aust. J. Agric. Res. 44, 1221-1234.

Page A L and Chang A C 1979 Trace elements impact on plants during cropland disposal of sewage sludges. In Proceedings – Fifth National Conference on Acceptable Sludge Disposal Techniques. Information Transfer Inc., Rockville, Maryland. pp 91-96.

Pauling L 1960 The Nature of the Chemical Bond. Cornell University Press, Ithaca, New York.

Peech M 1941 Availability of ions in light soil as affected by soil reaction. Soil Sci. 51, 473-486.

Pruess A 1995 Action values for mobile (NH_4NO_3-extractable) trace elements in soils according to the German national standard (DIN 19730). Third International Conference on the Biochemistry of Trace Elements (Symposium C2), Paris, May 1995.

Quirk J P and Posner A M 1975 Trace element adsorption on soil minerals. In Trace Elements in Soil-Plant Systems. Eds. D J D Nicholas and A R Egan. pp 95-107. Academic Press Inc., New York.

Ryan J A and Chaney R L 1993 Regulation of municipal sewage sludge under the Clean Water Act Section 503: A model for exposure and risk assessment for MSW-compost. In Science and Engineering of Composting: Design, Environmental, Microbiological and Utilization Aspects. Eds. H A J Hoitink and H M Keener pp 422-450. Renaissance Publications, Worthington, Ohio.

Sachdev P, Lindsay W L and Deb D L 1992 Activity measurements of Zn soils of different pH using EDTA. Geoderma 55, 247-257.

Saeed M and Fox R L 1977 Relations between suspension pH and Zn solubility in acid and calcareous soils. Soil Sci. 124, 199-204.

Scharrer K and Hofner W 1958 the sorption of Zn in and its leaching from soils. Z. Pflanzenernahr. Dung. u. Bodenk. 81, 202-212.

Schindler P W 1975 Removal of trace elements from the oceans: a zero order model. Thalassia Jugosl. 11, 101-111.

Schnitzer M 1969 Reactions between fulvic acid, a humic compound and inorganic soil constituents. Soil Sci. Soc. Am. Proc. 33, 75-81.

Shuman L M 1977 Adsorption of Zn by Fe and Al hydrous oxides as influenced by aging and pH. Soil Sci. Soc. Am. J. 41, 703-706.

Sillen L G and Martell A E 1964 Stability Constants of Metal-Ion Complexes. Chem. Soc. (London) Spec. Publ. No. 17.

Sinha M K, Dhillon S K, Dhillon K S and Dyanand S 1978 Solubility relationships of iron, manganese, Cu, and Zn in alkaline and calcareous soils. Aust. J. Soil Res. 16, 19-26.

Smolders E and McLaughlin M J 1996 Effect of Cl on Cd uptake by Swiss chard in unbuffered and chelator buffered nutrient solutions. Plant and Soil (in press).

Smolders E and McLaughlin M J 1996 The effect of chloro-complexation of Cd on uptake of Cd by Swiss chard determined by a resin-buffered solution culture technique. Soil Sci. Soc. Am. J. (in press).

Spencer W F and Gieseking J E 1954 Cobalt adsorption and release in cation-exchange systems. Soil Sci. 78, 267-276.

Sposito G 1981 Trace metals in contaminated waters. Environ. Sci. Technol. 15, 396-403.

Sposito G 1984 The Surface Chemistry of Soils. Oxford Univ. Press., New York. 234p.

Sposito G and Mattigod S V 1979 GEOCHEM: A computer program for the calculation of chemical equilibria in soil solutions and other natural water systems. The Kearney Foundation of Soil Science, University of California, Riverside.

Stanton D A 1964 Studies on Zn in selected Orange Free State soils. D. Sc. Agr. Thesis, Univ. Orange Free State, Bloemfontein.

Stevenson F J 1991 Organic matter-micronutrient reactions in soil. In Micronutrients in Agriculture. 2nd. Edit. Ed. J J Mortvedt Soil Sci. Soc. of Amer, USA.

Swallow K C, Hume D N and Morel F M M 1980 Sorption of Cu and lead by hydrous ferric oxide. Environ. Sci. Technol. 14, 1326-1331.

Taylor R M and McKenzie R M 1966 The association of trace elements with manganese minerals in Australian soils. Aust. J. Soil Res. 4, 29-39.

Taylor R M, McKenzie R M and Norrish K 1964 The mineralogy and chemistry of manganese in some Australian soils. Aust. J. Soil Res. 235-248.

Thiele S and Brummer G W 1993 Bestimmung der mobilen Fractionen ausgewahlter Elemente (Cd, Ni, Co, Cr, As) in Oberboden durch $CaCl_2$ – and NH_4NO_3 Extractionen. Mitteilgn. Dsch. Bodenkundl. 72, 1313-1316.

Tiller K G 1967 Silicic and the reaction of Zn with clays. Nature (London) 214, 852.

Tiller K G, Gerth J and Brummer G 1984b The relative affinities of Cd, Ni and Zn for different soil clay fractions and goethite. Geoderma 34, 17-35.

Tiller K G 1968b The interaction of some heavy metals cations and silicic acid at low concentrations in the presence of clays. Trans. 9th Cong. I.S.S.S. (Adelaide) 2, 567-575.

Tiller K G and Pickering J G 1974 The synthesis of Zn silicates at 20°C and atmospheric pressure. Clays Clay Miner. 22, 409-416.

Tiller K G, Gerth J and Brummer G 1984a The relative Cd, Zn and Ni by soil clay fractions: procedures for partition of bound forms and their interpretation. Geoderma 34, 1-16.

Tiller K G, Gerth J and Brummer G 1984b The relative affinities of Cd, Ni and Zn for interactions at the soil colloid–soil solution surface different soil clay fractions and goethite. Geoderma 34, 17-35.

Tiller K G, Gerth J and Brummer G 1984c The relative Cd, Zn and Ni by soil clay fractions: procedures for partition of bound forms and their interpretation. Geoderma 34, 1-16.

Tiller K G, Hodgson J F and Peech M 1963 Specific sorption of cobalt by soil clays. Soil Sci. 95, 392-399.

Tiller K G, Honeysett J L and Hallsworth E G. 1969 The isotopically exchangeable form of native and applied cobalt in soils. Aust. J. Soil Res. 7, 43-56.

Tiller K G, Nayyar V K and Clayton P M 1979 Specific and non-specific sorption of Cd by soil clays as influenced by Zn and Ca. Aust. J. Soil Res. 17, 17-28.

Tiller K G, Smith L H and Self P G 1995 Surface area changes arising from deferration treatment of soil clays and their interpretation. Proceedings 10th International Clay conference. Adelaide, 1993 (in press).

Tiller K G 1961 Specific sorption of some heavy metal cations by pure and soil clays. PhD Thesis. Cornell University, USA.

Tiller K G 1968a Stability of hectorite in weakly acidic solutions. III. Adsorption of heavy metal cations and hectorite stability. Clay Miner. 7, 409-419.

Tiller K G and Hodgson J F 1962 The specific sorption of cobalt and Zn by layersilicates. Clays Clay Miner. 9, 393-403.

Tiller K G 1979 Application of isotopes to micronutrient studies. In International Symposium on the Use of Isotopes and Radiation in Research on Soil-Plant Relationships Proc. Symp. Colombo 1978 pp 359-372. I.A.E.A., Vienna.

Tiller K G 1981 The availability of micronutrients in paddy soils and its assessment by soil analysis including radioisotopic techniques. In Proc., Symp. Paddy Soils, Oct. 1980, Ed. Institute of Soil Science, Academia Sinica pp 273-289. Nanjing Springer-Verlag, Berlin.

Tiller, K G 1983 Micronutrients. In 'Soils: An Australian Viewpoint'. Ed. Division of Soils, CSIRO. pp. 365-387. CSIRO, Melbourne/Academic Press, London.

Tiller K G and Wassermann P 1973 The effect of flooding on the availability of Zn and Mn to rice. Z. Pflanzenernahr. Bodenkd. 136, 57-67.

Truog E 1946 Soil reaction influence on availability of plant nutrients. Soil Sci. Soc. Am. Proc. 11, 305-308.

UMT 1993 Umweltschutz in Baden-Wurttemberg Boden. Umweltministerium Baden-Wurttemberg, Dec. 1993, Druck+ Verlagshaus H. Daniel, Balingen.

WPC 1990 Compendium of Waste Leaching Tests. Waste Technology Centre, Environment Canada, Report EPS 3/HA/7, May 1990.

Welp G and Brummer G W 1992 Toxicity of organic pollutants to soil micro-organisms. In Effects of Organic Contaminants in Sewage Sludge on Soil Fertility, Plants and Animals. Eds Hall J E, Sauerbeck D R and L' Hermite P. pp 161-168.

Zeien H and Brummer G W 1989 Chemische Extractionen Zur Bestimmung von Schwermetall-bindungsformen in Boden. Mitteilgn. Dsch. Bodenkundl. Ges. 59, 505-510.

2
The fate of non-metal contaminants in the soil environment

R.E. WHITE AND A.N. SHARPLEY

1. Introduction

A contaminant is defined as a substance that is not natural to the soil, or a naturally occurring substance which has undesirable effects if its concentration exceeds a certain threshold. A number of non-metals have the potential to be contaminants in soil and water. Because of their high ionic potential, non-metals of Groups IIIA to VIA of the Periodic Table in their normal oxidation states form oxyanions (Sposito, 1984). Examples of non-metal oxyanions commonly found in the soil solution are borate ($B(OH)_4^-$), biocarbonate (HCO_3^-), nitrate (NO_3^-), silicate ($H_3SiO_4^-$), sulphate (SO_4^{2-}), phosphate ($H_2PO_4^-$), arsenate (AsO_4^{3-}) and selenate (SeO_4^{2-}). Other non-metals such as Cl and F occur as the monovalent anions Cl^- and F^-. The ubiquitous ion Cl forms soluble salts with the common cations in soil and is only a potential problem when the salt concentration of the soil solution or groundwater is high and specific Cl toxicity effects in plants may occur. On the other hand, F^- forms insoluble complexes with cations such as Al^{3+} and Ca^{2+} and is rarely an active contaminant except in some underground waters of the Great Artesian Basin. Of the other anions, HCO_3^- derives naturally from the dissolution of CO_2 in soil water and dissociation of the weak carbonic acid formed. Silicate is formed from mineral weathering and dissociation of the very weak acid $Si(OH)_4$. Both are benign. Sulphate as SO_4^{2-} and less commonly HSO_4^- forms complexes with the hydroxyl surfaces of Fe and Al oxides, the edge faces of crystalline clay minerals, and Al^{3+} and hydroxy Al species adsorbed on charged surfaces. Its appearance in high concentrations may result from processes such as the oxidation of metal sulphides exposed to air following the disturbance of marine sediments or mineral deposits, or from atmospheric deposition of ammonium sulphate over many years. Any potential contamination results from the H^+ ions produced during sulphide or NH_4^+ oxidation, rather than from the accompanying SO_4^{2-} anion.

The non-metals that commonly occur as contaminants in soil are NO_3^-, $H_2PO_4^-$, SeO_4^{2-}, AsO_4^{3-} and $B(OH)_4^-$. All except NO_3^- can occur in protonated forms, the proportions depending on the ambient pH. For example, the pK value for the hydrolysis reaction

$$B(OH)_3 + 2H_2O \Leftrightarrow B(OH)_4^- + H_3O^+ \tag{1}$$

is 9.23 at 20°C, whereas the pK for the dissociation reaction

$$H_3PO_4 \Leftrightarrow H_2PO_4^- + H^+ \tag{2}$$

is 2.2 also at 20°C (White, 1980). Selenate and AsO_4^{3-} protonate at intermediate pHs.

R. Naidu et al. *(eds): Contaminants and the Soil Environment in the Australasia-Pacific Region, 29–67.*

Boron occurs in all natural waters at concentrations ranging from a trace to several parts per million. Like Se, the range between optimal and toxic concentrations is very narrow. Nitrate, AsO_4^{3-} and SeO_4^{2-} occur in several oxidation states in soil and transformations between these states is by redox reactions primarily mediated by microorganisms. The most common forms of As found in aerobic systems are AsO_3^{3-} (arsenite) and AsO_4^{3-}, while reduction and biomethylation to form $(CH_3)_2HAsO_2$ (cacodylic acid) and the gases $(CH_3)_2HAs$ (dimethylarsine) and $(CH_3)_3As$ (trimethylarsine) occurs under reducing conditions in soils, aquatic sediments and activated sewage sludge (McGeehan and Naylor, 1994). The chemistry of Se is very similar to S with primary oxidation states of +6, +4, 0 and −2. The common forms in aerobic systems are SeO_4^{2-} and SeO_3^{2-} (selenite). Under reducing conditions Se°, metal selenides and methylated selenides (e.g. $(CH_3)_2Se$, dimethylselenide) are produced (Thompson-Eagle and Frankenberger, 1992). All these non-metal anions except NO_3^- are adsorbed on to mineral surfaces by ligand exchange, a specific adsorption reaction in which P shows the highest affinity for the surface and $B(OH)_4^-$ the least. All can be adsorbed non-specifically at positively charged sites where the strength of adsorption depends on the anionic charge, and the relative proportions of the ions adsorbed are also governed by mass action effects. These processes are described in more detail below.

Of these non-metals, NO_3^- and P are the most widespread contaminants mainly because of the inputs of N and P to agriculture, and the losses that occur in various forms from agricultural systems (see Chapter 5). The recycling of P through urban communities is also an important source of contamination of soil and water. This chapter therefore concentrates on these two ions and their fate in the soil-plant-animal-water system.

2. Problems associated with N and P contamination

Nitrogen is vulnerable to loss from the soil-plant-animal system when it exists in simple organic forms such as urea, or in the mineral forms NH_4^+ and NO_3^-. Nitrate is highly mobile and its transport to water is a concern because high NO_3^- concentrations have been linked to methaemoglobinaemia in infants, toxicities in livestock, and increased eutrophication of both fresh and saline waters (Amdur et al., 1991; Sandstedt, 1990). In the United States, the EPA has established a Maximum Contaminant Level (MCL) for NO_3^- in drinking water of 10 mg N L^{-1} (45 mg NO_3^- L^{-1}). In the European Community, the maximum acceptable concentration for potable water is 11.3 mg NO_3–N L^{-1} (50 mg NO_3^- L^{-1}) and there is a guide level of 5.7 mg NO_3–N L^{-1} at which infants must be given bottled water (Anon., 1983). Infants younger than 3–6 months are most sensitive to high NO_3^- because bacteria that live in their digestive tract can reduce NO_3^- to nitrite (NO_2^-). High NO_2^- concentrations in the blood cause the ferrous iron in haemoglobin to be oxidised to ferric, giving rise to methaemoglobin, which is inefficient in binding oxygen and transferring it around the body (Amdur et al., 1991). Nitrate can also be toxic to livestock if reduced to NO_2^- which gives rise to methaemoglobinaemia and abortions in cattle. The tolerance level for livestock is about 40 mg NO_3–N L^{-1}, which is higher than for humans. Concentrations of 40–100 mg NO_3–N L^{-1} in drinking water are considered risky unless the animal feed is low in NO_3^- and fortified with vitamin A (Sandstedt, 1990). Nitrate is also widely recognised as contributing to eutrophication in estuaries where it can stimulate excess algal growth. This can lead to low dissolved oxygen concentrations as the algae decompose and subsequently to adverse impacts on fish and shellfish populations.

Inorganic forms of P are generally immobile and P moves into water mainly attached to soil particles or in soluble organic forms. Wastewater in urban areas contains substantial amounts of inorganic P from detergents. Phosphorus is not considered to be directly toxic to humans or animals (Amdur et al., 1991) so that drinking water standards have not been established (USEPA, 1990). However, P accelerates the eutrophication of surface waters causing increased growth of undesirable algae and aquatic weeds which create oxygen shortages during their senescence and decomposition and restrict water use for fisheries, recreation and industry. Storages of drinking water in Australia (McComb and Davis, 1993), North America (Kotak et al., 1993) and many other parts of the world have experienced periodic massive surface blooms of cyanobacteria. These blooms cause a wide range of problems including fish kills, unpalatability of drinking water, and the formation of trihalomethane during water chlorination in treatment plants (Palmstrom et al., 1988; Kotak et al., 1994). Consumption of cyanobacterial blooms or water-soluble neuro- and hepa-toxins, released when these blooms die, can kill livestock and pose a serious health hazard to humans (Lawton and Codd, 1991; Martin and Cooke, 1994). Although N, C and P are required for eutrophication, most attention has focused on controlling P inputs to fresh waters. Fixation of atmospheric N_2 by blue-green algae means that N is usually not a limiting factor for freshwater algal or bacterial growth.

There is a great deal of information documenting and quantifying the environmental impact of P loss in North America (Omernik, 1977; Rast and Lee, 1978; Levine and Schindler, 1989), Europe (Szilagyi et al., 1987; Rekolainen, 1989; Ryding et al., 1990) and Oceania (Sharpley and Syers, 1979; McColl and Syers, 1981; Ward et al., 1990; Hodgkin and Hamilton, 1993). In south-western Australia, Hodgkin and Hamilton (1993) attributed the accelerated eutrophication of fresh-waters mainly to P leaching from nutrient-deficient sandy soils which have received relatively high rates of P to maintain optimal production. Problems associated with increased P inputs to aquatic ecosystems can be exacerbated by weather conditions. In England, for example, a combination of a mild winter, high mid-summer temperatures, sunshine, and calm weather in July during 1989/90 led to lower lake levels and a concentration of P inputs from runoff, such that dramatic increases in algal bloom frequency and intensity resulted (National River Authority, 1990). However, in other regions of Europe such as the Po Valley in Italy, lower soluble P concentrations in surface waters and reduced eutrophication levels have been attributed to sediments produced by the erosion of high P-fixing soils (Sequi et al., 1991). In the U.S., several small lakes have experienced algal blooms as soon as erosion and sediment input were reduced by conservation practices, although the internal cycling of P and light limitations imposed by suspended sediment in aquatic ecosystems complicate the process. These general examples illustrate the complex and dynamic nature of aquatic response to P inputs from soil.

3. Processes in the soil-plant-animal system

3.1 N inputs, soil forms and transformations

Nitrogen enters soil from the atmosphere, fertilisers, plant residues, animal manures and wastewater. There are a number of books and articles which describe the nitrogen cycle in detail (e.g. Anon., 1983; Jenkinson, 1990; Garrett et al., 1992). Soil organic N is the dominant pool in terrestrial ecosystems, being an order of magnitude greater than plant biomass N. Soil N contents range from 1 to 10 t N ha^{-1} and the N is concentrated in the surface horizons where the residues accumulate and biological activity is greatest. Inputs of N as fertiliser range from zero to >500 kg N ha^{-1} yr^{-1} for high yielding and high value crops (sugar cane, potatoes, orchard crops) and intensively managed pastures (for dairying), or from zero to *c.* 200 kg N ha^{-1} yr^{-1} for legume-based pastures (Syers, 1982). In underpopulated countries with relatively clean air such as Australia and New Zealand, atmospheric N inputs in rainfall and as dry deposition of ammonia and oxides of N are usually <10 kg ha^{-1} yr^{-1}. But in more densely populated countries and those with intensive animal industries such as in northwest Europe (excluding Ireland), atmospheric inputs other than via biological fixation can be as high as 40 kg N ha^{-1} yr^{-1} (Goulding, 1990). An example of N inputs and transformations of N between the main pools in a typical arable soil in southern England is shown in Figure 1a. Figure 1b shows similar information for a recently established white clover-ryegrass pasture in New Zealand.

Since nearly all the soil N is in organic form, the balance between mineralisation and immobilisation is crucial in determining the amount of mineral N. Nitrification then determines how much of the NH_4^+ is oxidised to NO_3^- which is the form susceptible to loss either through leaching or denitrification. In temperate climates, spring applications of fertiliser N can be vulnerable to loss through the leaching of NO_3^- and there is a strong correlation between the amount leached and rainfall in the weeks after application in both crop and pasture systems (Goss et al., 1993; Powlson, 1993). Studies in the UK also show that NO_3^- leaching losses can be large when organic manures and slurries are applied to freely draining grassland or arable soils in the autumn or early winter: from January onwards the losses are much lower (Smith and Chambers, 1993). So provided that the amount and timing of fertiliser or manure application are adjusted to the growth of the crop or pasture, direct losses of fertiliser or manure N by leaching or in runoff should be small. Soil NO_3^- in cropped soils, and in pasture soils where the main input of N is through N_2 fixation by legumes, is mainly derived from the net mineralisation of organic N. Continued high N inputs through fertilisers, symbiotic N fixation, or animal manures lead to an accumulation of potentially mineralisable organic N in soil (Anon., 1983; Addiscott et al., 1991).

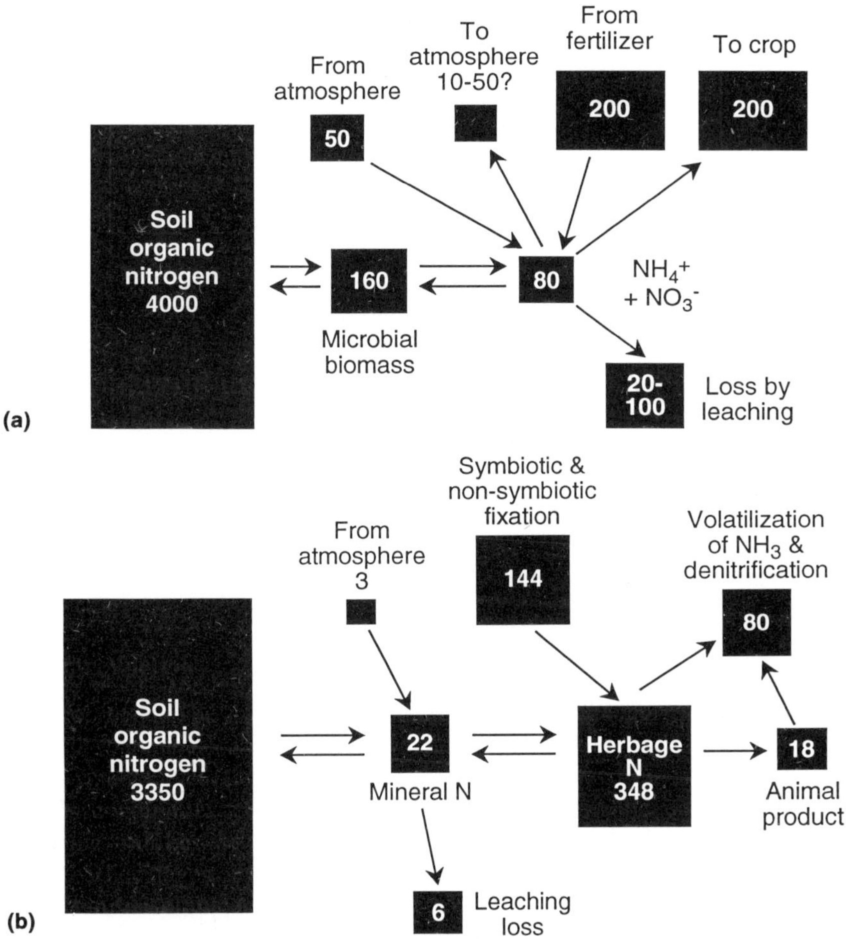

Figure 1. (a) N inputs, outputs and the main soil N components for an arable soil in southern England. The quantity of N is given in kg ha^{-1} for soil components or as kg ha^{-1} yr^{-1} for fluxes (after Powlson, 1993). (b) N inputs, outputs and the main soil N components for a soil under a clover-grass pasture in New Zealand. The quantity of N is given in kg ha^{-1} for soil components or as kg ha^{-1} yr^{-1} for fluxes (after Ruz-Jerez, 1991).

When plants are growing actively, much of the mineralised N is captured by the plant and temporarily immobilised. But there are two important situations where mineral N is at risk.

(a) When mineralisation continues after the uptake of N by the crop or pasture has slowed or ceased in late summer and autumn, especially in temperate regions (Garwood and Tyson, 1977; White et al., 1983). Residual values of 30–100 kg NO_3–N ha^{-1} to 1 m depth are common in arable fields in the U.K. (Powlson, 1993).

(b) When pastures receiving high N inputs are grazed at high stocking rates and 75–95% of the herbage N ingested is returned to the pasture in dung and urine patches. The proportion of the excretal N in urine increases as the N concentration in the animal's diet increases, but the range is 60–65% for dairy cattle to 70–75% for sheep (Haynes and Williams, 1993). Of this urinary N, 70–90% is present as urea which is rapidly hydrolysed in the soil to the unstable compound $(NH_4)_2CO_3$, which then hydrolyses according to the reaction:

$$(NH_4)_2CO_3 + H_2O \Leftrightarrow 2NH_4^+ + 2OH^- + CO_2 \quad (3)$$

Depending on temperature and moisture conditions, between 5 and 40% of the NH_4–N may be lost by volatilisation as NH_3; but most is oxidised to NO_3^-, resulting in high concentrations in very localised areas. The amount of mineral N in these patches is equivalent to 300–600 kg N ha^{-1} and far exceeds the uptake capacity of the pasture species (Ball and Ryden, 1984). Nitrogen in dung is almost entirely in organic forms which mineralise more slowly than urea. Nevertheless, high concentrations of NO_3^- (equivalent to 180–260 kg N ha^{-1}) can also accumulate under dung patches and be vulnerable to loss (Ryden, 1986). The problem created by localised high concentrations of mineral N derived from excreta (especially urine) is exacerbated by the non-random distribution of the excreta, especially its aggregation in stock camps and around sources of drinking water, where disproportionately large losses can occur through NH_3 volatilisation or NO_3^- leaching. However, the magnitude of these losses is not well known because many scientists sampling pasture soils have often avoided stock camps and watering points as being 'unrepresentative' of the paddock as a whole. Ruz-Jerez et al. (1995) estimated that for a rotationally grazed, clover-ryegrass pasture near Palmerston North, New Zealand that was fixing 144 kg N ha^{-1} yr^{-1}, only about 10% of the area was affected at any one time by urine patches but this area contributed 55% of the NO_3^- leached from the pasture. Working with grass swards in southern England fertilised with NH_4NO_3 at the rate of 420 kg N ha^{-1} yr^{-1}, Ryden et al. (1984) concluded that 5–6 times as much NO_3^- was leached below the root zone of the sward rotationally grazed by steers than from the sward cut for hay.

Wet soils predispose to NO_3^- loss by denitrification or leaching. Denitrification occurs in anaerobic zones (O_2 partial pressure <1%) through a series of reducing steps such as

$$NO_3^- \rightarrow NO_2^- \rightarrow NO \rightarrow N_2O \rightarrow N_2 \quad (4)$$

The predominant products of denitrification are the gases N_2O and N_2, with increasing reducing power (adequate fermentable C substrate and very low O_2 concentrations) favouring a lower ratio of N_2O/N_2. The ratio of N_2O/N_2 is also lowered as the NO_3^- concentration decreases. Some authors have reported that almost all the product of denitrification in acid soils ($pH \leq 5$) is N_2O (Tiedje et al., 1981), but most have reported the ratio of N_2O/N_2 to lie between 0.12 and 0.38 (Eichner, 1990; Ruz-Jerez et al., 1994). This variable ratio of N_2O/N_2 makes field measurements of denitrification very difficult and led to the development of the 'acetylene block' technique whereby the last reduction step in equation (4) is inhibited in an atmosphere of 10% C_2H_2, and all of the product of denitrification is measured as N_2O (Yoshinari et al., 1977).

Table 1. Nitrogen inputs and losses by leaching and denitrification for three contrasting pastures grazed by sheep in New Zealand.

Pasture	N input	N loss by denitrification (kg N ha^{-1} yr^{-1})	N loss by leaching (kg N ha^{-1} yr^{-1})	N loss/N input (%)
White clover-ryegrass	144[1]	3.4	5.8	6.4
Herbal ley (mixture of legumes, grasses and broadleaf species)	152[1]	4.4	7.3	7.7
Ryegrass	400[2]	19.3	41.1	15.1

[1]Symbiotic N_2 fixation.
[2]Urea applied in several applications.
(From Ruz-Jerez et al., 1994; 1995).

Addiscott and Powlson (1992) tried to partition the loss of N from ^{15}N-labelled fertiliser, applied in spring to winter wheat crops in southern England, between leaching and denitrification. Total N lost ranged between 1 and 35% of the fertiliser N, but leaching losses were generally low and on average only half the apparent losses by denitrification. However, data in Table 1 from Ruz-Jerez et al. (1994; 1995) show that on a well drained, fine sandy loam soil under three pasture types in New Zealand, denitrification was less than the leaching loss, irrespective of whether the main input of N was as urea fertiliser or through symbiotic fixation. There is evidence from the work of Garrett et al. (1992) in Northern Ireland of a reciprocal relationship between N leached and N denitrified over a large range of N inputs for pasture soils (Figure 2). This suggests that while the total N loss will depend on climatic and management factors, the balance between the two processes can switch according to the soil conditions (primarily whether the soil is well drained or not, and soil temperature). Consistent with this concept, the combined loss of N by denitrification and leaching from a drained, heavy clay soil under autumn-sown crops in southern England over eight years was comparable for ploughed and direct-drilled treatments (Goss et al., 1993). Barraclough et al. (1992) have questioned this concept of compensatory losses, but their leaching losses may have been underestimated because they used the geometric means of NO_3^- concentration derived from small, skewed sample sets (White et al., 1987). However, for recently established pasture, both denitrification and leaching as a proportion of the total N loss increased as N input increased (Ruz-Jerez et al., 1994; 1995).

3.2 *Phosphorus inputs, soil forms and their transformations*

Phosphate is applied to soil mainly as soluble fertilisers – single superphosphate (SSP), triple superphosphate (TSP), mono- and diammonium phosphate (MAP and DAP), slow release fertilisers such as reactive rock phosphates (RPR), their partially acidulated forms (PAPR) and mixtures of RPRs and SSP, and as animal manures, sewage water, sludges and slurries.

Soil P contents range from 500 to 2 500 kg P ha^{-1}. The P exists in inorganic and

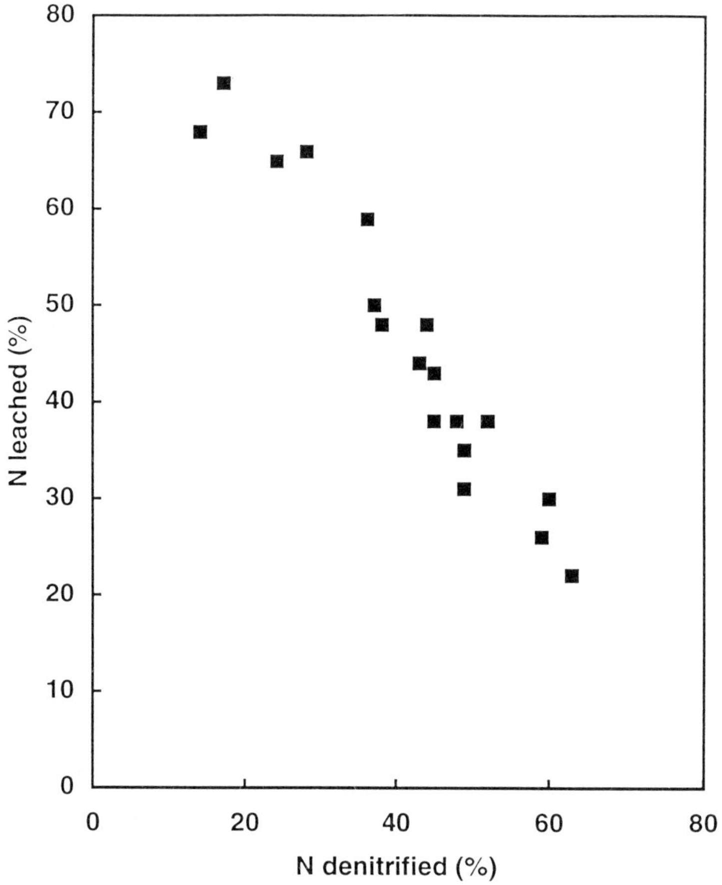

Figure 2. Relationship between leaching and denitrification for grazed grass and white clover-based pastures in Northern Ireland (after Garrett et al., 1992).

organic forms (Figure 3). In most agricultural soils 50 to 75% of the P is inorganic, although this fraction can vary from 10 to 90%. Inorganic P forms are dominated by hydrous sesquioxides, amorphous and crystalline Al and Fe compounds in acidic, non-calcareous soils and by Ca compounds in alkaline and calcareous soils. Organic P forms include relatively labile phospholipids, nucleic acids, inositols and fulvic acids, while more resistant forms are comprised of humic acids. The lability of these P fractions is based on the extent to which extractants of increasing acidity or alkalinity, applied sequentially, can dissolve the soil P.

When soluble P is applied to soil, the dominant reaction is adsorption by ligand exchange involving either OH or OH_2^+ groups at mineral surfaces (White, 1980), i.e.

$$MOH_2^+ + L^{-n}\ (aq) \Leftrightarrow ML^{1-n} + H_2O \tag{5}$$

$$MOH + L^{-n}\ (aq) \Leftrightarrow ML^{1-n} + OH^- \tag{6}$$

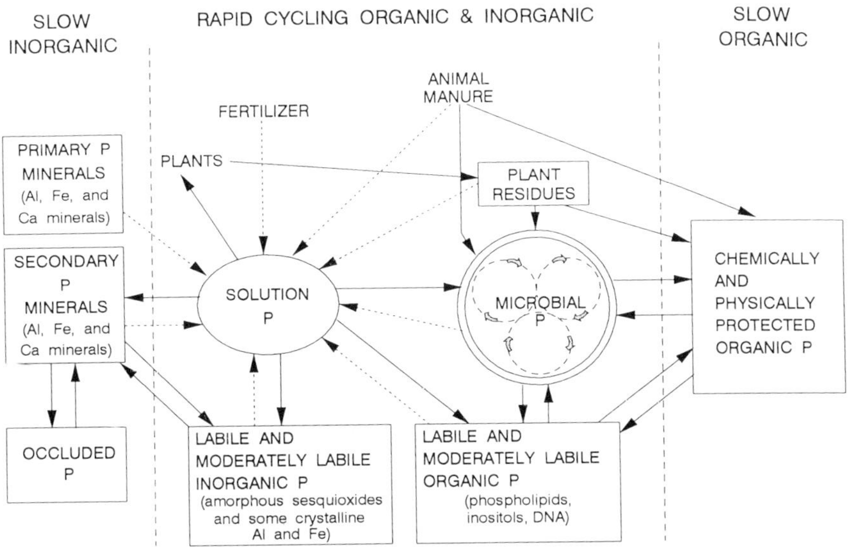

Figure 3. The soil P cycle: its components and measurable fractions (after Stewart and Sharpley, 1987).

where M stands for Al or Fe and L^{-n} (aq) stands for phosphate ($H_2PO_4^-$ or HPO_4^{2-}) in solution. Reaction (6) is more probable at high P concentrations in solution. After the initial rapid reaction, there is a gradual 'fixation' of the adsorbed P through secondary reactions which render a proportion of adsorbed P unavailable to plants (Barrow, 1978). In most soils, the P content of surface horizons is greater than the subsoil due to the sorption of added P and greater biological activity and accumulation of organic material at the surface. Soil P content varies with parent material, texture, and management factors such as rate and type of P applied and cultivation, and these factors also influence the relative amounts of inorganic and organic P.

With the application of P, the content of 'available' soil P, as measured by an empirical extraction technique or soil test, increases. This increase is a function of soil properties such as clay, organic C, Fe, Al and $CaCO_3$ content. Repeated application of fertiliser P can increase soil test P above levels required for optimum crop yields, as illustrated by the results of traditional fertiliser practice on light textured soils under pasture in Western Australia (Yeates, 1993; George and Bradby, 1993; Hodgkin and Hamilton, 1993). In 1989, in spite of 56% of the soils sampled in the Albany Harbour catchment in Western Australia having high soil tests (12 to >40 mg kg^{-1} Olsen P; Weaver and Prout, 1993), 90% of the fields still received P amendments. In areas of intensive livestock industries, concern focuses on manure application, where amounts of P added often exceed crop removal rates on an annual basis (Table 2). As a result, many areas with intensive confined animal operations in the Netherlands, Belgium, northeast US and Florida now have soil P contents that are of environmental rather than agronomic concern (Sharpley et al., 1994a).

Fixed P can be slowly released back into the soil solution when reserves of less

Table 2. Phosphorus balance and recovery of P in plant and animal products for the US and several European countries.

	Input $kg\,ha^{-1}\,yr^{-1}$	Output $kg\,ha^{-1}\,yr^{-1}$	Surplus $kg\,ha^{-1}\,yr^{-1}$	Recovery of P in: Plant uptake %	Animal uptake %	Total agriculture %
USA	39	13	26	56	15	33
Netherlands	143	55	88	69	24	38
E. Germany	79	8	71	59	10	11
W. Germany	84	29	55	76	34	35

Data for US adapted from National Research Council (1993) and for European countries from Iserman (1990).

strongly held P are exhausted (Johnston, 1989; Tiessen et al., 1993). As much as 70 to 85% of P exported from sandy soils under pasture in Western Australia originated from residual P from previous fertiliser applications (Ritchie and Weaver, 1993). Thus, the estimation of how long high P soils will remain above crop sufficiency levels is of economic and environmental importance to many farmers who are integrating manurial P into sustainable management systems. Analysis of soil test data from several cropping systems in North America showed the rate of decline in available P in high P soils when no further P was added varied from 0.1 to 30 mg $kg^{-1}\,yr^{-1}$, depending on soil type and management. With regular P applications, the importance of fixation processes is diminished as the soil's P sorption capacity slowly becomes saturated and a higher concentration of P is maintained in the soil solution (White, 1980). Similarly, several authors have found that the rate of decrease in available soil P with depletion by cropping when no P is added is inversely related to the soil's P buffering capacity (Holford, 1982; Aquino and Hanson, 1984) or P sorption saturation (available soil P / P sorption maximum; Sharpley, 1995a).

Even though inorganic P has generally been considered the major source of plant available P in soils, the incorporation of fertiliser P into soil organic P (McLaughlin et al., 1988) and lack of crop response to fertiliser P due to organic P mineralisation (Doerge and Gardner, 1978) emphasise the importance of organic P in soil P cycling. Sharpley (1985a) found organic P mineralisation (15 to 33 kg P $ha^{-1}\,yr^{-1}$) in several Oklahoma soils was not completely inhibited by fertiliser P application (20 to 28 kg P $ha^{-1}\,yr^{-1}$), with similar amounts of P contributed by both sources. Tate et al. (1991) also found that mineralisation of labile organic P was an important source of P to pasture in both low- and high-P fertility soils in New Zealand. Amounts of P mineralised range from 5 to 20 kg P $ha^{-1}\,yr^{-1}$ in temperate soils and from 67 to 157 kg P $ha^{-1}\,yr^{-1}$ in the tropics, where distinct wet and dry seasons and higher soil temperatures enhance microbial activity (Stewart and Sharpley, 1987). The importance of soil microbial biomass P (Figure 3) in P cycling has now been quantified (Stewart and Tiessen, 1987; McLaughlin et al., 1988). In a study of P cycling through soil microbial biomass in England, Brookes et al. (1982) measured annual P fluxes of 5 and 23 kg P $ha^{-1}\,yr^{-1}$ in soils under continuous wheat and permanent grass, respectively. Although the biomass P flux under continuous wheat was less than P uptake by the crop (20 kg P ha^{-1} yr^{-1}), the flux in the grassland soils was much greater than P uptake by the grass (12 kg P ha^{-1} yr^{-1}). Clearly, microbial P plays an important intermediary role in the short-term dynamics of organic P transformations, suggesting that management practices which maximise the build-up of organic matter during autumn and winter may reduce P requirements for plant growth during the following spring and early summer.

3.3 Plant and animal uptake of phosphorus

The recovery of P from soil amendments by plants in the year of application is usually less than 20% (Widdowson et al., 1982; Paynter and Dampney, 1991). The efficiency of P uptake depends on a number of environmental and management factors, such as soil temperature, moisture, aeration, pH, type and amount of clay content, nutrient status and microbial activity (Munson and Murphy, 1986; Kucey et al., 1989). Phosphorus uptake is reduced when soil temperatures are low during early spring and when soil moisture is limiting during drought periods. Soil compaction inhibits P uptake by reducing pore space and the availability of water and oxygen. The availability of P to plants is also reduced by the 'fixation' processes described above. Liming can increase P availability by stimulating mineralisation of organic P (Dalal, 1977) or may decrease P availability by the formation of insoluble calcium phosphates at pH > 6.5 (White and Taylor, 1977). A fall in pH or increased biological activity in the rhizosphere, including vesicular-arbuscular mycorrhizal associations with plant roots, can considerably enhance P uptake, especially on low P soils (Grinsted et al., 1982; Bolan, 1991).

A generalised P balance and efficiency of plant and animal recovery of P for the US and some European countries indicates the potential for P accumulation in agricultural systems (Table 2). Although the P inputs and outputs varied between countries, the relative proportions of P uptake in plant and animal products were similar. In spite of the relatively efficient recovery of P in crop production of 56–76%, the recovery in animal production was only 10–34%, so that total P recovery by agriculture was only 11–38%. Efficiency of P recovery in agriculture is dominated by animal performance because 76 to 94% of the total crop production is fed to animals (in addition to P additives). Animal-specific studies of P excretion rates substantiate this poor retention of P with values of 70–80% measured for dairy cows (Aarts et al., 1993), sheep (Haynes and Williams, 1993), feeder pigs (Archer, 1985), and 87% for poultry (Iserman, 1990). Clearly, agricultural systems which include confined animal operations can determine the overall efficiency of P recovery in agriculture, and thereby the magnitude of P surpluses and potential soil accumulations.

4. Losses from the soil-plant-animal system

The most benign pathway for N loss is by reduction of NO_3^- to N_2. No such pathway exists for P in soil. Concerns about N loss focus on NO_3^- movement through the soil profile to surface waters and groundwater, while for P the focus is on soluble and particulate P transport in surface runoff.

4.1 Nitrogen transport

Loss of N in water flow from soil is predominantly by NO_3^- leaching, even when large amounts of animal manure, slurry or sewage sludge are added to the soil (Anon., 1983). The rate of leaching is determined by a complex interaction of a number of factors including climate, irrigation, soil type, texture and structure, N inputs through N_2 fixation, fertiliser (amount and form) and animal wastes, the type of land use and management. Basically, however, NO_3^- leaching depends on the volume of water draining through the soil (out of the root zone) and the NO_3^- concentration in that water (White, 1988). Drainage (expressed as volume of water per unit area) is determined by the excess of precipitation over evaporation, corrected for any surface runoff and changes in soil storage. It also depends on the amount of supplemental irrigation. Apart from climatic controls, evaporation depends primarily on whether or not a crop is present. The source strength for NO_3^- leaching is determined by the balance between processes that input or produce NO_3^- in the soil and those that consume it. The reciprocal relationship between NO_3^- leaching loss and denitrification loss has been discussed (Figure 2). There are other complex interactions involving the site of NO_3^- formation in the soil, soil structure and water movement, all of which affect the temporal pattern of leaching and the amount of N leached.

When soluble N forms are applied to the soil surface and there is 'excess rainfall', NO_3^- can be leached by preferential flow to considerable depths and rapidly appear in drainage water (Addiscott and Cox, 1976; Barraclough et al., 1983). Preferential flow depends on the intensity of rainfall and the soil moisture content, and especially on the soil's structure. Soils with well developed macropores (worm holes and old root channels), or with cracks between aggregates, show marked preferential flow (Thomas and Phillips, 1979; Beven and Germann, 1982; White, 1985b). Soils high in sesquioxides, particularly in the subsoil, may have net positively charged surfaces at which NO_3^- ions can be adsorbed (Black and Waring, 1976; Wong et al., 1990); but most soils have net negatively charged surfaces from which NO_3^- ions are repelled. As a result, NO_3^- ions are excluded from very small pores and this reinforces the effect of preferential flow in accelerating NO_3^- leaching (Cameron and Wild, 1982). Even when soluble N is not applied to the soil surface, NO_3^- formed by mineralisation in the soil can be leached by preferential flow. This occurs predominantly in the surface horizons at the end of summer in temperate regions when NO_3^- production exceeds consumption, and NO_3^- ions are drawn to the outside of soil aggregates as they dry (White et al., 1983; White, 1985a). The effect is particularly marked at the end of unusually hot, dry summers (Garwood and Tyson, 1977; Foster and Walling, 1978; Scholefield et al., 1993). The stimulation of mineralisation which follows the ploughing of old pastures also markedly increases the supply of NO_3^- in the surface soil which is then vulnerable to leaching (Ryden et al., 1984).

Depending on the duration of a leaching event, NO_3^- that is readily accessible in the 'mobile water' is depleted and the concentration of NO_3^- in the drainage decreases (Magesan et al., 1995). Nitrate that continues to be generated inside soil aggregates is largely protected from leaching (Cunningham and Cooke, 1958; Wild, 1972). If water flow is intermittent, during the dry periods NO_3^- will diffuse to the exterior of soil aggregates in response to the concentration gradient, so that when the next leaching event occurs the NO_3^- concentration is initially high and then declines with time (Magesan et al., 1995). In soils of temperate regions, the rate of NO_3^- generation will generally subside to a low value as temperatures fall (<5°C) so that the supply of NO_3^- available for leaching decreases. The net effect of these processes is that as the winter season progresses, the NO_3^- concentration in drainage declines approximately exponentially with cumulative drainage (Harris et al., 1984; Haigh and White, 1986; Heng et al., 1991). Superimposed on this general trend are marked short-term fluctuations in concentration associated with individual leaching events, especially early in the season when the supply of NO_3^- is relatively high (Magesan et al., 1995). Similarly, if heavy rain falls shortly after fertiliser application in spring there can be a 'spike' in the concentration of NO_3^- in drainage (Macduff and White, 1984), resulting in an increase in leaching loss (Goss et al., 1993), but such events can also cause a significant loss of NO_3^- in surface runoff (Scholefield et al., 1993).

4.2 Phosphorus transport

4.2.1 Forms

The transport of P in runoff can occur in dissolved (DP) and particulate (PP) forms (Figure 4). Particulate P encompasses all phase forms, including P sorbed by soil particles and organic matter eroded during runoff, and constitutes the major proportion of P transported from cultivated land (75–90%). Runoff from grass or forest land carries little sediment and is therefore generally dominated by the dissolved form. While DP is, for the most part, immediately available for biological uptake (Walton and Lee, 1972; Nurnberg and Peters, 1984), PP can provide a long-term source of P for aquatic biota (Wildung et al., 1974; Carignan and Kalff, 1980).

4.2.2 Surface runoff

The transport of DP in surface runoff is initiated by desorption, dissolution, and extraction of P from soil and plant material (Figure 4). These processes occur as rainfall interacts with a thin layer of surface soil (1 to 5 cm deep) before a proportion leaves the field as runoff (Sharpley, 1985b). Although this depth is difficult to quantify, it is expected to be highly dynamic due to variations in rainfall intensity, soil tilth and vegetative cover. Several studies have shown the DP concentration of runoff is dependent upon Bray-I P content in the surface (0 to 5 cm) soil (Romkens and Nelson, 1974; Olness et al., 1975; Sharpley et al., 1986; Schreiber, 1988). For

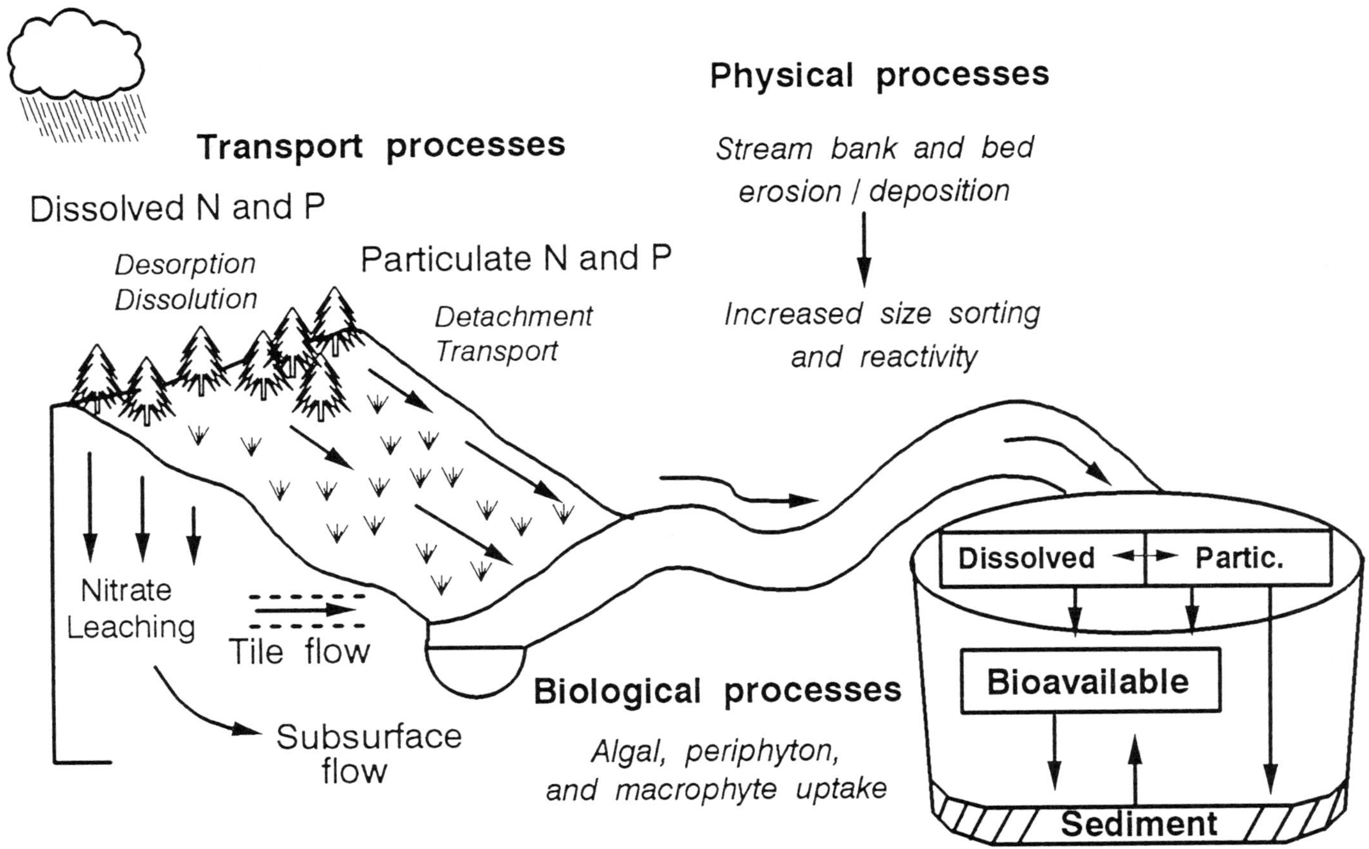

Figure 4. Processes involved in the transport of N and P from agricultural land (after Sharpley et al., 1993).

these studies, the DP–soil P relationship varied with management. On average, regression slopes (kg soil L^{-1}) for grass (6.0) were lower than for cultivated land (10.5). Even so, regression slopes varied from 4.1 to 7.0 for grass and from 8.3 to 12.5 for cultivated land, showing the risk of using a single or average relationship for recommending P inputs to satisfy water quality criteria.

Clearly, several soil and land management factors influence the relationship between dissolved P in runoff and soil P. These factors affect the volume of runoff as well as the release of soil P to runoff. As the sources of particulate P in streams include eroding surface soil, streambanks and channel beds, processes determining soil erosion also control PP transport. In general, the P content and reactivity of eroded particulate material is greater than that of the source soil due to preferential transport of clay-sized material (<2 mm). Sharpley (1985c) observed that the plant available P content of runoff sediment was on average 3-fold greater, and the total, inorganic and organic P forms 1.5-fold greater, than in the source soil. Readers are directed to Chapter 5 for detailed discussions on the impact of current soil and land management issues on soil and water quality.

4.2.3 Leaching

The P concentration of water percolating through soil is generally low due to sorption of P by P-deficient subsoils. Exceptions can occur in acid organic or peaty soils, where the adsorption affinity and capacity for P is low due to the predominantly negatively charged surfaces and the complexing of Al and Fe by the organic matter (Fox and Kamprath, 1971; Duxbury and Peverly, 1978; Miller, 1979; White and Thomas, 1981). Similarly, P is more susceptible to movement through sandy soils with low P sorption capacities and in soils which have become waterlogged, where conversion of Fe (III) to Fe (II) and organic P mineralisation occur (Ozanne et al., 1961; Gotoh and Patrick, 1974). Because of the variable path and time for water flow through a soil with subsurface drainage, factors controlling DP in subsurface waters are more complex than for surface runoff. Subsurface runoff includes tile drainage and natural subsurface runoff, where tile drainage is percolating water intercepted by artificial systems, such as mole and tile drains. In general, DP concentrations and losses in natural subsurface flow are lower than in tile drainage because the contact time with subsoil is longer for natural subsurface flow than for tile drainage. Increased sorption of P from percolating water accounted for lower total P losses of 0.5 kg ha^{-1} yr^{-1} from 1.0 m deep tiles draining a Brookston clay soil under alfalfa, compared with 1.1 kg ha^{-1} yr^{-1} from 0.6 m deep tiles (Culley et al., 1983). For the shallower drains, total P losses were about 1% of fertiliser P applied, whereas 1 m deep tiles exported about 0.6% of the 60 kg P ha^{-1} yr^{-1} that was applied.

Soil P content influences the loss of P in drainage water as it does surface runoff. For example, Sharpley et al. (1977) found that the DP loss in tile drainage during storm events from a dairy pasture in New Zealand was related to the amount of P extracted by 0.1 M NaCl from soil at tile drain depth (40 to 50 cm). A similar dependence of DP concentration in tile drainage on the P sorption-desorption properties of subsoil material was found for Histosols in Florida (Hortenstine and Forbes, 1972), New York (Cogger and Duxbury, 1984), Ontario (Nicholls and MacCrimmon, 1974), and for Haploquolls in Ontario and Michigan (Culley et al., 1983).

4.2.4 Changes during transport

Transformations between DP and PP occurring during transport in stream flow can alter both the amount and bioavailability of P entering a lake, compared to edge-of-field losses. These transformations are accentuated by the selective transport of fine materials which have a greater capacity to sorb or desorb P and will thus be important in determining the bioavailability of P transported. In addition, DP may be taken up by aquatic biota and PP deposited or eroded from the stream bed with a change in stream flow rate (Meyer, 1979; Vincent and Downes, 1980). The direction and extent of P exchange between DP and PP will depend on their relative concentration in stream flow, the sediments contacted, and the rate of stream flow. This exchange must be considered in assessing the impact of agricultural management on the potential biological productivity of a lake receiving P transported in runoff. Once sediment settles to the bottom of a lake, however, sediment P bioavailability will be increased if reducing conditions develop at the sediment-water interface (Syers et al., 1973; Nurnberg et al., 1986). For example, in a study of the P dynamics of two shallow hypereutrophic lakes in Indiana, Theis and McCabe (1978) found that the DP concentration of lake water was reduced by sorption during aerobic periods and increased by release of sediment P during anaerobic periods.

5. Amounts transported

5.1 Sources of nitrogen and phosphorus

Nitrogen and P transported from land provide a potential nutrient source for receiving water bodies. From the perspective of the water body, there are three types of source:

- diffuse or non-point sources – where there is no clearly identified origin of the nutrients, as is the case for soil erosion and general runoff and drainage from agricultural and forested land;
- point sources – where the nutrients originate from a discrete site such as an intensive animal industry (feedlot or processing plant), agricultural tile or pipe drains, or discharges in urban areas (a sewage treatment plant, stormwater drains or leaching from landfills);
- internal loadings – nutrients stored in the sediments of a lake or stream that can be released into the water.

Obviously, it is much easier to measure nutrient losses from point than diffuse sources. The proportion of dissolved N and P in point sources is usually much higher than in diffuse sources so the availability for growth of aquatic species is high. Typical N and P concentrations in various point source discharges are given in

Table 3. Typical nutrient concentrations for various point source discharges.

Discharge type	Nutrient concentration Total P ($mg L^{-1}$)	Total N (mg/L^{-1})
Domestic wastewater:		
– secondary treatment	8–10	20–30
– tertiary treatment	0.3–2	5–10
Pulp and paper mill	9	13
Cannery	2	76
Feedlot effluent	150	900
Dairy shed effluent	340	680
Fish farms	0.07	0.50
Irrigation drainage:		
– surface	0.4	1.6
– subsurface	0.05	5

From: Anon

Table 3. Generally, diffuse sources are more important as inputs of N to water bodies, and point sources are more important inputs of P. Point sources of P are particularly important during summer when diffuse source flows are relatively low. For example, the contribution of domestic wastewater to the total P input into the Murray-Darling river system in Australia ranges from 11% in a 'wet' year to 32% in a 'dry' year (Anon., 1995).

5.2 *Loadings*

Amounts of N and P transported from uncultivated or pristine land constitute the background loading which cannot be reduced. This input determines the natural status of a lake, which can be sufficient to cause eutrophication. There is little quantitative information on background losses from a given location prior to cultivation and P amendments, against which the impact of agricultural management on N and P loss in runoff can be assessed. This is a consequence of the expensive and labour-intensive nature of water quality studies which are site-specific and impossible to replicate because of spatial and temporal variations in climatic and edaphic conditions. Despite these problems, some generalisations can be made using published data on the effect of different systems of land use and associated N and P inputs on subsequent losses in runoff and drainage.

5.3 *Runoff*

Increases in N and P loss in surface runoff have been measured after the application of fertiliser or manure (Table 4). Losses are influenced by the rate and method of

Table 4. Effect of fertilizer and manure application on N and P loss ($kg\,ha^{-1}\,yr^{-1}$) in surface runoff.

Land use	Added		Loss				Reference and location
	N	P	NO_3–N	Total N	Soluble P ($kg\,ha^{-1}\,yr^{-1}$)	Total P	
Fertilizer							
Grass	0	0	1.4	–	0.02	0.22	McColl et al., 1977
	0	75	1.4	–	0.04	0.33	New Zealand
No till corn	189	0	3.4	10.7	0.70	2.0	McDowell and
Silage	170	30	5.5	10.1	0.80	1.8	McGregor, 1984;
No till corn	189	0	4.7	17.1	1.10	3.3	Mississippi
Grain	170	30	1.1	8.0	1.80	3.4	
Wheat	0	0	0.3	2.8	0.20	1.6	Nicolaichuk and
	50	54	0.3	4.4	1.20	4.1	Read, 1978; Canada
Grass	0	0	0.4	4.7	0.50	1.17	Sharpley and Syers,
	60	50	0.5	7.9	2.80	5.54	1979; New Zealand
Dairy manure							
Alfalfa	0	0	3.2	5.7	0.1	0.1	Young and Mutchler,
– spring	205	21	0.1	22.2	1.9	3.7	1976; Minnesota
– fall	285	55	0.2	37.6	4.8	7.4	
Corn	0	0	3.2	5.7	0.1	0.2	
– spring	205	21	0.1	5.4	0.2	0.6	
– fall	285	55	0.3	5.2	1.0	1.6	
Poultry manure							
Fescue	0	0	0.0	0.2	0.0	0.1	Edwards and Daniel,
	220	76	0.0	6.9	1.1	2.1	1992; Arkansas
	879	304	0.0	28.7	4.3	9.7	
Fescue	0	0	0.1	1.1	0.1	0.4	Westerman et al.,
	428	95	0.0	18.9	1.4	12.4	1983; North Carolina
Pig manure							
Fescue	0	0	0.0	0.2	0.1	0.1	Edwards and Daniel,
	214	19	0.0	5.8	1.5	1.5	1993; Arkansas
	435	38	0.0	12.6	4.8	4.8	

fertiliser or manure application, the form applied, the amount and timing of rainfall after application and vegetative cover. Nitrogen application generally has little effect on NO_3^- loss in runoff because the primary movement of NO_3^- from the site of N application is by infiltration and leaching. For the studies presented in Table 4, 3 to 11% of the applied N was lost in runoff, primarily as particulate organic N. This organic material is appreciably lighter than soil material and can be easily transported (Sharpley, 1985c). Although it is difficult to distinguish between losses of fertiliser, manure, and native soil nutrients without the use of expensive isotope tracers, losses of fertiliser N and P in runoff are generally estimated to be <5% of that applied. Losses of manurial N and P are slightly higher, ranging from 2 to 20% of that applied, particularly if rainfall immediately follows application. The greater proportion of N and P loss from manure than fertiliser may result from higher N and

P application rates for manure and often less flexibility in the timing of application than for fertiliser. This lack of flexibility often results from the continuous production of manure throughout the year and inadequate manure storage facilities.

5.4 Leaching

For the reasons given earlier, leaching losses of P are small (<1 kg ha^{-1} yr^{-1}), except in the special circumstances of sandy soils or very acid peaty soils receiving high inputs of soluble P, when losses may rise to 10 kg P ha^{-1} yr^{-1} (Larsen et al., 1958; McConaghy et al., 1971; Hodgkin and Hamilton, 1993).

Conversely, leaching losses of NO_3^- can be very large and also very variable. This variation is partly due to the wide range of combinations of drainage volume and source NO_3^- concentrations which are possible in the soil, and partly to the great difficulty in accurately measuring the losses because of spatial and temporal variability (Keeney, 1986; Powlson, 1993). The most reliable data come from experiments on underdrained fields where the spatial variability is integrated through the drain system and temporal variability can be monitored by frequent sampling of the drain flow at an intensity proportional to flow rate (Haigh and White, 1986; Heng et al., 1991; Goss et al., 1993). However, many soils are not suitable for mole and tile drains, and even where they are the value of production may not justify the investment in drainage. Thus, alternative methods of measuring NO_3^- leaching have been used, one of the best being to estimate the drainage flux from the water balance equation and sample the soil solution using suction cup samplers. Suction cup samplers work best in light textured soils (Webster et al., 1993) and may have limitations in clay loams and clays (Addiscott et al., 1991). Magesan et al. (1994) found that suction cup samplers provided reliable values of the flux concentration of non-reactive solutes such as Br^- and Cl^- in a silt loam soil under pasture, but were less reliable for NO_3^- concentration. Ever since the installation of the famous Rothamsted 'drain gauges' in 1870, which used undisturbed soil cores (Addiscott et al., 1991), lysimeters have been used extensively for measuring NO_3^- leaching (Chichester, 1977; Webster and Dowdell, 1984; Cameron et al., 1990). They have limitations of scale, especially for grazed pasture, and of maintaining a hydraulic head gradient through the soil similar to that in the field. Nor is there scope for lateral flow of water and solutes which can be important in many Australian duplex soils.

Given these problems of measurement, the amounts of N leached for different land uses and inputs are indicative only. Examples of N losses from contrasting land uses on mole and tile drained soils over several years are given in Tables 5 and 6. The N losses fluctuated markedly from year to year, but the range was comparable for the two systems despite the difference in management and forms of N input. In addition to the urea inputs for the pasture (see Table 5), symbiotic fixation of N_2 was comparable to that measured by Ruz-Jerez et al. (1995), i.e. 140–150 kg N ha^{-1} yr^{-1}. Generally on cropped soils, the amount of N leached during winter is not well correlated with the amount of 'starter' fertiliser applied in autumn (Goss et al., 1993), but on grazed pastures there is a clear dependence of N leached on fertiliser N

Table 5. Nitrate leaching loss during winter from paired paddocks (0.125 ha each) of white clover-rye grass pasture on a mole and tile drained Tokomaru silt loam (New Zealand).

	Paddock A		Paddock B	
Year	Drainage (mm)	NO_3-N leached (kg ha^{-1})	Drainage (mm)	NO_3-N leached (kg ha^{-1})
1988[1]	304	13	257	9
1989[2]	118	19	100	15
1990[3]	260	50	235	44
1991[4]	336	23	296	17

[1]No grazing.
[2]50 kg N ha^{-1} as urea in spring; no grazing.
[3]120 kg N ha^{-1} as urea to paddock B in May; intermittent grazing.
[4]No fertilizer and no grazing.
From Magesan et al. (1995).

Table 6. Nitrate leaching loss during winter from replicated plots (0.24 ha each) of a mole and tile drained Denchworth clay (southern England) under N fertilized winter crops, ploughed or direct drilled.

			NO_3-N leached	
Year	Fertilizer applied (kg N ha^{-1})	Drainage (mm)	Ploughed	Direct drilled (kg ha^{-1})
1985	285*	144	41	27
1986	130#	227	50	46
1987	100#	190	32	26
1988	194#	207	18	11

*46 kg N ha^{-1} applied in autumn; the remainder in spring.
#All fertilizer applied in spring.
From Goss et al. (1993).

applied (Barraclough et al., 1992; Scholefield et al., 1993). Further, there is a linear relationship between N leached and rainfall in the weeks following fertiliser application in spring for cropped soils, with approximately 3 kg N ha^{-1} being leached for every 10 mm rain up to 120 mm (Powlson, 1993; Goss et al., 1993).

6. Management strategies to minimise the adverse effects of N and P in the environment

Management strategies that minimise the environmental impact of N and P, particularly with respect to water quality, should encompass a number of factors including:

(a) reliable, yet realistic, threshold concentrations of N and P in surface and groundwater that should not be exceeded on environmental grounds;
(b) the identification of potential contaminant source areas so that more intensive management may be targeted to these areas;

Table 7. Nitrogen and phosphorus ratings for Victorian rivers and streams.

Rating	Total P concentration ($mg\,L^{-1}$)	Total N concentration ($mg\,L^{-1}$)
Degraded – poor	>0.05	>0.5
Moderate	0.025 – 0.05	0.35 – 0.5
Good – excellent	<0.025	<0.25

From: Anon

(c) the use of remedial measures to moderate the source effects and control the transport of N and P from a site;
(d) coordination of nutrient management strategies between sites and across catchment boundaries to maximise the benefit to the environment of site-specific actions; and
(e) the creation of favourable socio-economic conditions under which individual landholders, local authorities, state and federal agencies can work constructively towards a common goal of minimising the impact of contaminants in the physical environment.

7. Realistic water quality criteria

Water criteria for N and P have been established by the USEPA (1976). Acceptable limits for NO_3–N concentration for human and livestock consumption are 10 and 100 mg L^{-1}, respectively; NH_4-N concentrations above 0.5 and 2.5 mg L^{-1} may be harmful to humans and fish, respectively. For P, Sawyer (1947) and Vollenweider (1968) proposed critical DP and total P concentrations of 0.01 and 0.02 mg L^{-1}, respectively, which if exceeded, may accelerate the eutrophication of surface waters. These concentrations have been modified by state agencies in the USA, ranging as high as 0.05 to 0.1 mg L^{-1}, for DP and total P, respectively (USEPA, 1976). For example, Florida recently identified 0.05 mg L^{-1} as the concentration of DP allowable in drainage water entering the Everglades (USA vs. South Florida Water Management District, 1994). That state intends to reduce this concentration to 0.01 mg L^{-1} by the year 2000. In Australia, the states are in various stages of implementing nutrient management strategies for inland and coastal waters. An example of the nutrient rating of rivers and streams in Victoria is given in Table 7. In addition, a combined Federal-State body, the Murray-Darling Basin Commission, is addressing water quality issues in the whole of the basin (*c.* 115×10^6 ha) through its Algal Management Strategy.

Application of such water quality criteria should not be used as the sole determinant to guide nutrient amendments in agriculture where N and P loss in runoff is of concern. In some cases background concentrations of N and P in runoff from

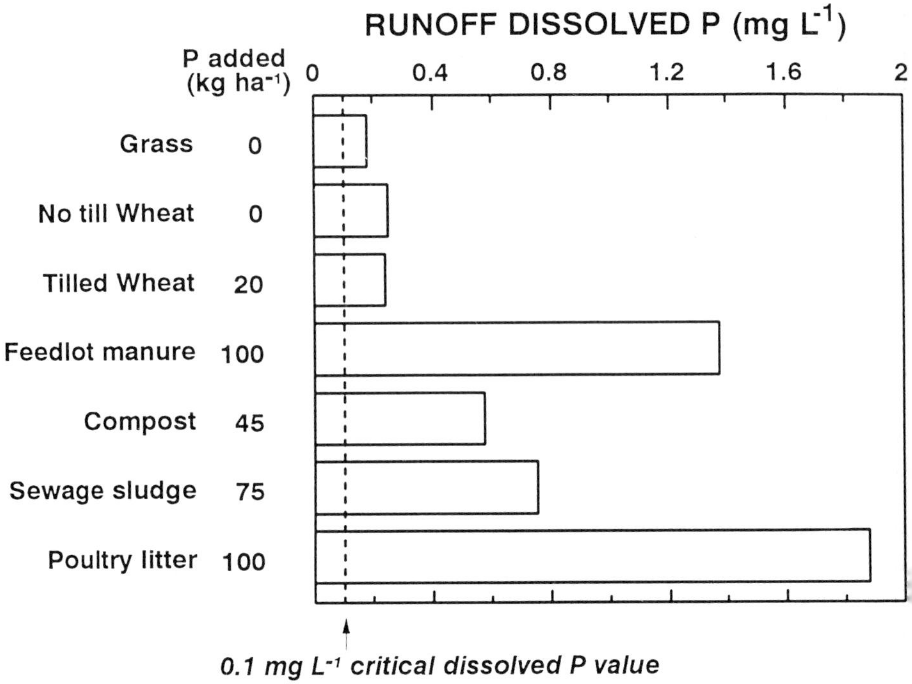

Figure 5. Concentration of dissolved P in runoff as a function of catchment and manure management in the Southern Plains relative to critical values associated with accelerated eutrophication (after Heathman et al., 1995; Jones et al., 1995, and Sharpley et al., 1986).

undisturbed areas may exceed the quality thresholds. For example, the mean annual DP concentration in runoff from several wheat catchments in Oklahoma and Texas receiving 20 kg P ha^{-1} yr^{-1} of fertiliser P (Smith et al., 1991a) and from grassed catchments receiving various types of manure (Heathman et al., 1995; Jones et al., 1995) all exceeded a 'generous' critical DP value of 0.1 mg L^{-1} (Figure 5). This was also the case for unfertilised native grass catchments (Sharpley et al., 1986). Thus, it is unlikely that any form of nutrient management would reduce DP in runoff below critical concentrations, particularly where large applications of manure P are regularly made. A more flexible approach considers the complex relationships between N and P loadings and the physical characteristics of affected catchments (leaching, runoff and erosion potential) and water bodies (mean depth and hydraulic residence time) on a site-specific and recognised water-use basis. Also, the use intended for the water will influence 'desired or tolerable' nutrient loadings. For example, lakes used principally for water supply, swimming, and multi-purpose recreation will benefit from low N and P loadings. However, lakes mainly used for fish production benefit from a moderate degree of biological productivity and thus tolerate higher N and P inputs.

Clearly, realistic water quality criteria to guide nutrient management within catchments should encompass more factors than just N and P concentrations in agricultural runoff. Unrealistic or unattainable criteria will not be adopted unless regulated. Thus, it is essential for long-term sustainable management of nutrients that workable water quality criteria are proposed initially. The phasing-in of environmental controls should then receive wider acceptance and compliance by farmers, without creating severe economic hardships within rural communities. In Western Australia, voluntary changes in farm practices based on improved land management are beginning to reduce the loss of P from agriculture (Weaver and Prout, 1993). These changes have included on-farm nutrient management such as soil testing, fertiliser management, and water and erosion control measures.

In order to meet water quality criteria, a system of buying and selling pollution credits within a given catchment, similar to that recently adopted for air quality control, has been suggested in several eastern states in the USA. Farmers who are able to limit N and P loss below recommended levels could sell credits to those unable to meet these levels. The number of credits allocated to a farmer could be linked to farm area, crop production, and where appropriate, the number of animals. As a result, N and P export from a catchment could be kept within predetermined limits by sharing the responsibility among farmers. It should be noted, however, that 'pollution trading' has been criticised by some environmental groups because it is perceived as allowing wealthier operations to buy the 'right to pollute.' Heated debate is likely to precede the adoption of pollution credits for agriculture, so careful planning to justify their value and need will be required.

8. Targeting remedial strategies

Strategies to minimise N and P loss in runoff will be most effective if sensitive or vulnerable source areas within a catchment are identified and targeted, rather than the strategy implemented over a broad area. In the absence of reliable and user-friendly models to predict nutrient behaviour on a catchment scale (see later), indexing approaches have been developed as a field tool to readily assess the vulnerability of sites to the loss of N by leaching and P by surface runoff. Initial site assessment involves determining whether leaching or runoff dominates water loss. If runoff is negligible and the potential for percolation is high, the N leaching index should be used to guide manure or fertiliser applications; if surface runoff potential is high, the P index system should be used.

Site vulnerability to N leaching can be ranked by considering textural and permeability characteristics affecting water movement through a soil profile, as in the index devised by Kissel et al. (1982) (Table 8). The site's textural class is determined by the texture of the soil horizon with the finest texture. Similarly, the permeability of the most limiting layer (lowest permeability) determines the permeability class. This simple index exemplifies one method of highlighting the potential for groundwater contamination by agricultural chemicals. However, such indices need to be evaluated and calibrated for the major soil types in each country to determine their applicability in new areas.

Table 8. An example of a nitrogen leaching index.

Textural class	Permeability class ($cm\,h^{-1}$)			
	50–15	14–5	4–1.5	<1.5
Sand to coarse sandy loam	High	Medium	Low	Very low
Loamy fine sand to sandy loam	Medium	Medium	Low	Very low
Loam to fine sandy loam	Low	Low	Low	Very low
Clay loam to loam	Very low	Very low	Very low	Very low

Adapted from Kissel et al. (1982).

Table 9. A P indexing system to rate the potential for P loss in runoff from site characteristics.

Site characteristic and weighting factor	Phosphorus loss potential (value)				
	None (0)	Low (1)	Medium (2)	High (4)	Very high (8)
Transport factors					
Soil erosion[1] (1.5)	Negligible	<10	10–20	20–30	>30
Runoff Class (0.5)	Negligible	Very low or low	Medium	High	Very high
Source factors					
Soil test P (1.0)	Negligible	Low	Medium	High	Excessive
Fertilizer P application rate[2] (0.75)	None applied	1–15	16–45	46–75	>76
Fertilizer P application method (0.5)	None applied	Placed with planter deeper than 5 cm	Incorporated immediately before crop	Incorporated >3 months before crop or surface applied <3 months before crop	Surface applied >3 months before crop
Organic P source application rate[2] (0.5)	None applied	1–15	16–30	30–45	>45
Organic P source application method (1.0)	None	Injected deeper than 5 cm	Incorporated immediately before crop	Incorporated >3 months before crop or surface applied <3 months before crop	Surface applied >3 months before crop

[1]Units for soil erosion are $Mg\,ha^{-1}$.
[2]Units for P application are $kg\,P\,ha^{-1}$.
Adapted from Lemunyon and Gilbert (1993).

Lemunyon and Gilbert (1993) developed a P indexing system to rank sites according to their vulnerability to P loss in runoff (Table 9). The index rates the source (measured by soil test P and fertiliser and manure inputs) and transport factors (runoff and erosion potential) of a site to provide an overall numerical ranking for site vulnerability to P loss in runoff. Each site characteristic affecting P loss is assigned a weighting, on the assumption that certain characteristics have a greater effect on P loss than others, and a value scale ranging from none to very high (see Table 9). To assess site vulnerability to P loss, the rating value for each site characteristic is selected from the table, multiplied by the appropriate weighting factor, and the weighted values of all the site characteristics summed to give the site's P loss potential. For different geographic areas, a user must establish a range of values for the site characteristics that make up the index. Such indices can account for actual P loss in runoff more accurately than single characteristics, such as the soil P test value (Sharpley, 1995b).

If these indices are soundly based scientifically and reliably reflect the potential for nutrient loss from a site, they can be used to identify a range of management options in any remedial strategy. While most freshwaters are P limited, there are notable exceptions where P controls will have marginal to no benefit. In some lakes and in many streams, plant productivity is limited by high turbidity either from anthropogenic or natural sources. In others, such as the high elevation lakes in the western USA, N is limiting (Eilers, 1991). To optimise control measures it is therefore often necessary to preferentially direct management actions to those catchments where the control of N or P will provide the greatest benefit. Management agencies are also often required to spend their limited resources on those P-sensitive lakes which have the highest public or ecosystem value. An integrated approach to water quality management is now being adopted, which considers the reality of achieving a significant reduction in P inputs, the use of the water, the uniqueness of the environmental resource, and the degree of threat, to identify priority lakes and catchments for the control of non-point source pollutants.

9. Source management

Source management of N mainly involves applying N at rates based on the soil's N status (measured by the quantity of mineral N in the soil profile at sowing and an estimate of the net N mineralisation potential), the crop's requirements, and the method and timing of application. Management of P sources involves the appropriate choice of fertiliser form, its placement, and soil test recommendations based on water quality rather than agronomic considerations to determine P application rates.

9.1 Forms of soil amendment

The choice of N fertiliser used by farmers is determined mainly by the cost per unit of N. For this reason, forms such as anhydrous ammonia, urea and the ammonium phosphates (MAP and DAP) are most widely used. For the same reason, slow release N fertilisers such sulphur-coated urea, isobutylidiene diurea and the ureaformaldehyde polymers are not used, except in specialist high value enterprises such as ornamental horticulture and sports turf. Nitrification inhibitors can be used to slow temporarily the rate of nitrification of NH_4^+ from fertiliser in soil, but this is costly to do on a large scale.

Slow release P fertilisers such as RPR and the partially acidulated forms (PAPR) may be suitable alternatives to the soluble P fertilisers SSP, TSP, DAP and MAP under certain conditions (Barnes and Kamprath, 1975; Khasawneh and Doll, 1978; Hammond et al., 1986; White et al., 1989). The agronomic value of RPR, and to a less extent PAPR, is influenced by particle size, soil, climate, and plant management factors. In Australia and New Zealand, direct application of RPR may be limited to pastures growing on soils of $pH_{H20} < 6$, under an average rainfall $> 800\,mm\ yr^{-1}$ which is reasonably evenly distributed throughout the year (Bolan et al., 1990). Even so, it may be possible to design agricultural management systems to take advantage of the lower cost of P in RPR compared to soluble fertilisers. For example, in soils of neutral pH, the efficiency of RPR utilisation may be improved by applying an initial heavy dressing of finely ground material, and including a fine-rooted legume in a crop rotation to generate low pH and Ca concentration in the rhizosphere, thus accelerating the rate of RPR dissolution. Further research on phosphate rock as an alternative P source is needed to evaluate cost/benefit ratios as a function of edaphic and climatic conditions, material pretreatment, and application methods for different cropping systems (Chien and Friesen, 1991). The main use of PAPR has been on acidic soils in tropical and New Zealand agricultural systems (Smyth and Sanchez, 1982; Hammond et al., 1986; Hedley et al., 1990), where reduced fertiliser costs are of prime concern. In Western Australia, Yeates and Clarke (1993) showed that less P was leached from sandy soils treated with S-fortified PAPR than from soluble P fertilisers used at rates of equal agronomic effectiveness. Thus, slow release PAPRs may be of environmental as well as agronomic benefit. An increasing adoption of 'low input' management and inclusion of forage legumes in U.S. crop rotations, along with the development of RPR sources, may increase the agronomic and economic benefit of phosphate rock.

9.2 Placement

Potential immobilisation of N by residues under no till may increase crop requirements for broadcast fertiliser N (Doran and Smith, 1987). Greater uptake of fertiliser N by crops can therefore be achieved with a surface banding application or subsurface knife applications. For example, a combined N-P-K application increased grain sorghum yields relative to broadcast application to a greater extent in reduced compared to conventional tillage systems in Kansas (Lammond, 1987).

Because of the normal immobility of P in soil, fertiliser placement is more critical for P than N. Several studies have shown a greater yield response to banded application of fertiliser P at low rates, compared to broadcast or mixing (Welch et al., 1966). For six soils having a hundred-fold range in P sorptivity, Holford (1989) found that fertiliser P effectiveness was consistently greater for shallow banding at 5 cm depth than for banding at 15 cm or broadcast applications. The effect of mixing P throughout the soil on P supply was almost independent of P sorptivity. This suggests that the importance of banding P fertiliser in maximising its effectiveness relates to its positional availability in the root zone, rather than to any reduction in chemical immobilisation in the soil (Holford, 1989). Wherever possible, subsurface placement of P away from the zone of removal in runoff will reduce the potential for P loss. It may be necessary periodically to plough no-till soils to redistribute surface P accumulations throughout the root zone. The practices of placement and no-till may indirectly reduce the loss of P by increasing crop uptake and yield, which affords greater vegetative protection of the surface soil from erosion.

9.3 Timing

Because soluble P is rapidly adsorbed by soil minerals, the timing of fertiliser application is more important for the efficient use of N than of P. Application of fertiliser N before or during periods when precipitation exceeds evaporation must be evaluated in terms of crop N requirements and the risk of leaching. However, in some systems such as rain-fed winter wheat in Nebraska, early application of N may be used to encourage partial leaching so that a major proportion of the stored water and N is available at similar depths in the root zone (Smika and Grabouski, 1976). Crop utilisation of N is usually greatest with split applications, when the time of N availability and crop uptake can be more closely synchronised. Although starter fertilisers may meet early season growth needs, final yields may not be affected as the amount of N in starter fertilisers placed near the seed is only a small part of the total crop requirement. Under irrigation, best N utilisation by corn is usually obtained with side-dressings at the time of greatest N uptake. Fertigation may also facilitate easy application of N at a rate matched to the demand of the crop. For pasture soils in temperate climates, the summer weather has a marked effect on the amount of leachable NO_3^- in the soil in autumn (Scholefield et al., 1993), so the efficiency of N utilisation can be improved by withholding fertiliser in dry years.

9.4 Animal inputs

Manipulation of dietary P intake by livestock is another aspect of source management receiving increasing attention. In the Netherlands, the concentration of P in manure decreased temporarily during the Second World War when concentrates and fertilisers were less available, and reductions in concentrate P contents are now being similarly implemented to help reduce the amounts of P excreted to land (Wadman et al., 1992). Morse et al. (1992) recorded a 23% reduction in excretion of P in faeces and a 17% reduction in total P excretion when the daily P intake of dairy cows was reduced from 82 to 60 g d^{-1}. When the dietary P intake increased from 82 to 112 g d^{-1}, excretion of P in faeces rose by 49% and total P excretion by 37%. Thus, there is a clear indication that amounts of excreted P can be reduced by carefully matching dietary P inputs to the animal's requirements, especially as P intakes above minimum dietary requirements do not seem to confer any milk yield advantage (Brodison et al., 1989; Morse et al., 1992).

10. Transport management

Loss of N and P by soil erosion and runoff may be reduced by changing land use (e.g. from cropping to permanent pasture), by strip cropping to maintain vegetation on a proportion of the land surface most of the year, by conservation tillage which retains greater amounts of plant residues on the soil surface, and by engineering structures such as contour banks and terraces which are designed to increase the infiltration of water and slow down overland flow. However, DP losses can be greater from no-till than from conventional till crops. Accumulation of crop residues and added P at the soil surface provides a source of P to runoff that would be more dispersed if the soil were tilled. Some have found that increased infiltration under reduced tillage systems increases NO_3^- leaching and groundwater concentrations (Sharpley and Smith, 1993), but others have found that NO_3^- concentrations and the amount of N leached are less under direct-drilled crops than crops established after ploughing (Goss et al., 1993). In the latter case, there was less mineralisation of soil organic N in the no-till soil and the soil-generated NO_3^- was mainly within aggregates where it was protected from leaching. However, this 'protected' NO_3^- was also more susceptible to denitrification than that in the tilled soil (Colbourn and Harper, 1987). With respect to pasture soils and NO_3^- losses, again there is conflicting evidence on the relative merit of encouraging a greater proportion of the surplus water to drain through the soil as opposed to running off the surface. Armstrong et al. (1983) reported no difference in N losses for a drained and undrained stagnogley soil under crop or grass, but Scholefield et al. (1993) found that mole and tile drainage increased NO_3^- leaching threefold compared to no underdrainage in a similar soil. The primary cause was greater mineralisation of organic N in the drained soil.

Additional measures to minimise N and P loss to streams by erosion and runoff include buffer strips, riparian zones, cover crops, and impoundments or small reservoirs. Although these practices are generally more efficient at reducing particulate than dissolved nutrient inputs, plant uptake of N and P can occur during subsurface drainage through riparian zones or forests. Several studies have indicated little decrease in lake productivity with reduced P inputs following implementation of conservation measures (Young and DePinto, 1982; Gray and Kirkland, 1986), with the lack of biological response being attributed to an increased bioavailability of P entering the lakes as well as internal recycling.

11. The role of simulation modelling

Longterm field studies that measure N and P movement reliably are costly, lengthy and labour intensive. Simulation models offer the possibility of extrapolating from site-specific measurements to predict more generally the effects of agricultural management on nutrient loss in runoff. Many models of nutrient cycling and plant root extraction have been developed for both practical management and the identification of basic research needs. A user can choose one suited to a particular problem or need, but herein lies one of the main difficulties associated with models:

the selection of the most appropriate model to obtain the level of detail required. Once an appropriate model is chosen, the major limitation is obtaining the necessary data for the variables and parameters to run the model. This most frequently limits model use. Not only do many models require detailed information on soil physical, chemical and biological properties, as well as crop and tillage operations, but the output will only be as reliable as the input data. In addition, lack of field validation often limits assessment of model accuracy. Consequently, the use of models to provide quantitative estimates of nutrient loss under specific environmental conditions is limited. Many models are only recommended for comparisons of management effects on nutrient behaviour.

11.1 Nutrient cycling

Much effort has been spent on simulating the complex processes involved in nutrient cycling and the influence of soil, climate and plant variables. For example, models have been developed to simulate specific processes such as denitrification (Rolston et al., 1980; Smith, 1981), ammonia volatilisation (Parton et al., 1983), N cycling through soil microbial biomass (Jenkinson and Parry, 1989), inorganic P (Mansell et al., 1977) and organic P transformations (Mishra et al., 1979), P availability (Barrow and Carter, 1978; Cox et al., 1981), and liming and soil nutrient availability (Kirk and Nye, 1986; Nye and Ameloko, 1987). More comprehensive models have been developed for N cycling (Frissel and Van Veen, 1981; Van Keulen and Seligman, 1987; Wolf et al., 1989), P cycling (Cole et al., 1977; Harrison, 1978), and the dynamics of soil C, N, P and S cycling (Parton et al., 1983; Sharpley and Williams, 1990).

11.2 Nutrient inputs

The Claassen and Barber (1976) model simulating soil nutrient flux to plant roots has been extended to assess the effect of fertiliser rate and placement on nutrient uptake by several crops (Anghioni and Barber, 1980; Kovar and Barber, 1987). A user-oriented model 'Decide' was developed by Bennett and Ozanne (1972) and Helyar and Godden (1976) to simulate the efficiency of residual fertiliser P, and give objective advice on fertiliser use to farmers. The model constructs a fertiliser response curve for each farming situation by combining research information with the farmer's knowledge of his own farming conditions, in terms of past soil fertility and future yield goals. An empirical model (PHOSUL-K) was developed by Yeates et al. (1984) to integrate research information on the P, S and K response of pastures in southwestern Australia, and to estimate optimum fertiliser applications. Weaver and Prout (1993) found that adoption of fertiliser recommendations from the PHOSUL-K model after soil testing would have reduced P applications by 50 to 60% in 1991 within the Albany Harbour catchment, Western Australia, without reducing pasture productivity. This reduction would have saved in excess of A$ 2 million in fertiliser costs.

11.3 Nutrient losses

Many agricultural non-point source models have been developed to simulate the loss of nutrients in surface runoff. More detailed information on these models and the approaches used is available in reviews by Leavesley et al. (1990) and Rose et al. (1990). In general, these models describe the loss of N and P in soluble and particulate or sediment-bound forms. Soluble P is predicted from equilibrium extraction coefficients or kinetic approaches (Sharpley and Smith, 1989). Because NO_3^- is easily leached through the soil profile, the amounts in surface soil and in runoff are not closely related (Sharpley et al., 1985), precluding accurate prediction of its loss in runoff. Particulate N and P loss can be described by enrichment ratio approaches (Sharpley, 1985c) enabling accurate predictions of N and P loss in runoff as a function of agricultural management to be made (Sharpley et al., 1991a; Smith et al., 1991b).

Modelling the leaching of NO_3^- through soil to surface water or groundwater has not been as successful because of the complexity of the processes involved and the heterogeneity of soil and geological layers through which the water and NO_3^- pass (White and Magesan, 1991). Detailed information on specific models and approaches used to simulate solute leaching is available in papers by Duffy et al. (1990), Vauchaud et al. (1990), and Wagenet et al. (1990). The accurate simulation of N loss from agriculture is a formidable task because N ranges in valence from +5 to −3, is present in both inorganic and organic forms, and exists in the solid, liquid and gaseous phases. Furthermore, N loss is strongly influenced by soil temperature, moisture, microbial activity and plant processes. Many leaching models have received limited field testing because of these complexities and a lack of adequate field data.

12. Looking to the future

Inputs of N and P to agricultural soils are essential to maintain crop yields and meet both local and world demand for food. For example, given a current world population of some 5.5 billion and grain production equivalent to some 350 kg per person per year, the amount of grain N required globally is 33×10^6 t annually (Jenkinson, 1990). If the world population expands to a predicted 8.5 billion by 2025, and current standards of nutrition are maintained, 51×10^6 t of grain N will be required. The massive inputs of N needed to sustain these levels of grain production (to say nothing of animal protein produced from grazing systems) will require very high standards of management if adverse effects on air and water quality are to be contained. Extensive research at Rothamsted and other centres has shown that little of the N fertiliser used for cropping accumulates in soil organic matter: only if N fertilisers increase the return of organic C to the soil do they increase the soil organic N content (Jenkinson, 1990). However, N fertilisers and N_2 fixing legumes generally do increase soil organic C and N contents under pasture, up to a new steady state level at which outputs balance inputs. With the intensification of both cropping and pastoral systems, therefore, losses of N in gas or liquid phases to the environment will increase. Phosphorus is much more conservative that N in the soil-plant-animal system so that if inputs exceed removals, soil accumulation can occur over many years. Such accumulations in the surface and subsoil can enhance the potential for loss in runoff or drainage.

On a global scale, losses of N to the atmosphere as NH_3 and N_2O can have an environmental impact as serious as the leaching of NO_3^- and the transport of N in runoff. Ammonia in the NH_4^+ form (mainly combined with SO_4^{2-}) contributes about one-third of the acidity in rainfall in regions of dense population and intensive animal industries, such as northwest Europe (White, 1987). Nitrous oxide released from the Earth's surface accounts for 5–10% of the 'greenhouse effect' through the absorption of infra-red radiation in the troposphere, and to the formation of ozone-depleting NO gas in the stratosphere (Warneck, 1988). At the terrestrial level, the greatest concern about N and P losses from rural and urban environments is for water quality. The problem in urban areas is primarily one of controlling point sources, for which the technologies already exist even though their implementation can be costly. Less success has been achieved in minimising non-point inputs in agricultural systems where the problems commonly arise because nutrient amendments are based on one component only. For example, manure applications have been based on crop N requirements and groundwater quality criteria, often ignoring potential accumulations of P. Thus, future research and its application should emphasise integrated nutrient management involving C, N and P. Research programs should also focus on holistic catchment management in which nutrient inputs and outputs are balanced on a large scale rather than at a field level.

To accomplish the goal of restricting undesirable N and P losses more effectively, we must look beyond pure soil science and agronomic research and involve scientists from other disciplines, such as meteorologists, limnologists, economists and rural sociologists. Better techniques are required for measuring gaseous N fluxes that can integrate small scale variability, in the same way that better areal averages of NO_3^- leaching fluxes are being sought. As most of the negative impacts of chemical amendments to soils are manifest in aquatic systems, limnologists will be best equipped to predict further impacts. Most importantly, however, the economic impact of any changes in the management of nutrient inputs on agricultural or rural communities must be considered. Future agricultural production systems will be successful only if the economic viability of farming communities is maintained.

Because many years are required to bring about a significant reduction in soil nutrient levels, time is not on our side. Also, once water quality is impaired it is usually not cost-effective to treat the water, and internal recycling of nutrients can support the growth of aquatic biota even when external inputs have ceased. Thus, to minimise non-point source inputs from agriculture, we must identify critical source areas of nutrients in a catchment in order to target cost-effective remedial strategies. Perhaps most crucial to any water quality improvement strategy is efficient transfer of research technology to the farmer. Effective implementation will involve education programs to overcome the perception by end-users of water that it is often cheaper to treat the symptoms of degradation rather than to control the causes. Unfortunately, the benefits of these programs are often not immediately visible to a concerned public. Future research and education programs should therefore emphasize the long-term economic and environmental benefits of preventative measures.

References

Aarts H F J, Biewinga E E and Van Keulen 1993 Dairy farming systems based on efficient nutrient management. Nether. J. of Ag. Sci. 40, 285-299.

Addiscott T M and Cox D 1976 Winter leaching of nitrate from autumn-applied calcium nitrate, ammonium sulphate, urea and sulphur-coated urea in bare soil. J. Agric. Sci. Camb. 87, 381-389.

Addiscott T M and Powlson D S 1992 Partitioning losses of nitrogen fertilizer between leaching and denitrification. J. Agric. Sci. Camb. 118, 101-107.

Addiscott T M, Whitmore A P and Powlson D S 1991 Farming, Fertilizers and the Nitrate Problem. CAB International,Wallingford, UK.

Amdur M O, Dull J and Klassen E D (eds.) 1991 Casarett and Doull's Toxicology. Fourth Ed. Pergamon Press, New York, USA.

Anghioni I and Barber S A 1980 Predicting the most efficient phosphorus placement for corn. Soil Sci. Soc. Am. J. 44, 1016-1020.

Anon 1983 The Nitrogen Cycle of the United Kingdom. A Study Group Report. The Royal Society, London, UK, 264p.

Anon 1995 Nutrient Management Strategy for Victorian Inland Waters. State Government of Victoria, Melbourne, p73.

Aquino B F and Hanson R G 1984 Soil phosphorus supplying capacity evaluated by plant removal and available phosphorus extraction. Soil Sci. Soc. Am. J. 48, 1091-1096.

Armstrong A C, Shaw K and Wilcockson S J 1983 Field drainage and nitrogen leaching: some experimental results. J. Agric. Sci. Camb. 101, 253-255.

Ball P R and Ryden J C 1984 Nitrogen relationships in intensively managed temperate grasslands. Plant Soil 76, 23-33.

Barnes J S and Kamprath E J 1975 Availability of North Carolina rock phosphate applied to soils. North Carolina Agric. Expt. Sta. Tech. Bull. 229.

Barraclough D, Hyden M J and Davies G P 1983 Fate of fertilizer nitrogen applied to grassland I. Field leaching results. J. Soil Sci. 34, 483-497.

Barraclough D, Jarvis S C, Davies G P and Williams J 1992 The relation between fertilizer nitrogen applications and nitrate leaching from grazed grassland. Soil Use and Management 8, 51-56.

Barrow N J 1978. The description of phosphate adsorption curves. J. Soil Sci. 29, 447-462.

Barrow N J and Carter E D 1978 A modified model for evaluating residual phosphate in soil. Aust. J. Agric. Res. 29, 1011-1021.

Bennett D and Ozanne P G 1972 CSIRO Division of Plant Industry. Annual Report, Canberra, Australia. 45 p.

Beven K and Germann P 1982 Macropores and water flow in soils. Water Resour. Res. 18, 1311-1225.

Black A S and Waring S A 1976 Nitrate leaching and adsorption in a Krasnozem from Redland Bay, Qld. III. Effect of nitrate concentration on adsorption and movement in soil columns. Aust. J Soil Res. 14, 189-195.

Bolan N S 1991 A critical review of the effect of mycorrhizal fungi on the uptake of phosphorus by plants. Plant Soil 132, 189-207.

Bolan N S, White R E and Hedley M J 1990 A review of the use of phosphate rocks as fertilisers for direct application in Australia and New Zealand. Aust. J. Exp. Agric. 30, 297-313.

Brodison J A, Goodall E A, Armstrong J D, Givens D I, Gordon F J, McCaughey W J and Todd J R 1989 Influence of dietary phosphorus on the performance of lactating dairy cattle. J. Agric. Sci. Camb. 112, 303-306.

Brookes P C, Powlson D S and Jenkinson D S 1982 Measurement of microbial biomass phosphorus in soil. Soil Biol. Biochem. 14, 319-329.

Cameron K C and Wild A 1982 Comparative rates of leaching of chloride, nitrate and tritiated water under field conditions. J. Soil Sci. 3, 649-657.

Cameron K C, Harrison D F, Smith N P and McLay C D A 1990 A method to prevent edge-flow in undisturbed soil cores and lysimeters. Aust. J Soil Res. 28, 879-886.
Carignan R and Kalff J 1980 Phosphorus sources for aquatic weeds: water or sediments. Science 207, 987-989.
Chichester F W 1979 Effects of increased fertilizer rates on nitrogen content of runoff and percolate from monolith lysimeters. J. Environ. Qual. 6, 211-217.
Chien S H and Friesen D K 1991 Phosphate rock for direct application. In Future directions for agricultural phosphorus research. Ed. F J Sikora, pp 1-18. National Fert. and Environ. Res. Cent., TVA Muscle Shoals, Alabama, USA.
Claassen N and Barber S A 1976 Simulation model for nutrient uptake from soil by a growing plant root system. Agron. J. 68, 961-964.
Cogger G and Duxbury J M 1984 Factors affecting phosphorus losses from cultivated organic soils. J. Environ. Qual. 13, 111-114.
Colbourn P and Harper I W 1987 Denitrification in drained and undrained clay soil. J. Soil Sci. 38, 531-539.
Cole C V, Innis G S and Stewart J W B 1977 Simulation of phosphorus cycling in semiarid grasslands. Ecology 58, 1-15.
Cox F R, Kamprath E J and McCollum R E 1981 A descriptive model of soil test nutrient levels following fertilization. Soil Sci. Soc. Am. J. 45, 529-532.
Culley J L B, Bolton E F and Bernyk V 1983 Suspended solids and phosphorus loads from a clay soil: I. Plot studies. J. Environ. Qual. 12, 493-498.
Cunningham R K and Cooke G W 1958 Soil nitrogen. II. Changes in levels of inorganic nitrogen in a clay loam soil caused by fertiliser application, by leaching and uptake by grass. J. Sci. Food Agric. 9, 317-324.
Dalal R C 1977 Soil organic phosphorus. Adv. Agron. 29, 83-117.
Doerge T and Gardner E H 1978 Soil testing for available P in southwest Oregon. In Proc. 29th Ann. Northwest Fertilizer Conf. pp 143-152, Beaverton, Oregon, USA.
Doran J W and Smith M S 1987 Organic matter management and utilization of soil and fertilizer nutrients. In Soil fertility and organic matter as critical components of production systems. Eds. R F Follett, J W B Stewart and C V Cole. pp 5372 Spec. Publ. No. 19, Soil Sci. Soc. Am., Madison, Wisconsin, USA.
Duffy C J, Kincaid C and Huyakorn P 1990 A review of groundwater models for assessment and prediction of nonpoint-source pollution. In Proc. Int. Symp. Water Quality Modeling of Agricultural Non-point Sources, part 1. Ed. D G DeCoursey. pp 253-278. USDA-ARS 81. US Government Printing Office, Washington DC, USA.
Duxbury J M and Peverly J H 1978 Nitrogen and phosphorus losses from organic soils. J. Environ. Qual. 7, 566-570.
Edwards D R and Daniel T C 1992 Potential runoff quality effects of poultry manure slurry applied to fescue plots. Trans. Am. Soc. Agric. Eng. 35, 1827-1832.
Edwards D R and Daniel T C 1993 Runoff quality impacts of swine manure applied to fescue plots. Trans. Am. Soc. Agric. Eng. 36, 81-80.
Eichner J M 1990 Nitrous oxide emissions from fertilized soils: summary of data available. J. Environ. Qual. 19, 272-280.
Eilers J M 1991 Are lakes in the Cascade Mountains receiving high ammonium deposition? Northwest Sci. 65, 238-247.
Foster I D L and Walling D E 1978 The effects of the 1976 drought and autumn rainfall on stream solute levels. Earth Surf. Process. 3, 393-406.
Fox R L and Kamprath E J 1971 Adsorption and leaching of P in acid organic soils and high organic matter sand. Soil Sci. Soc. Am. Proc. 35, 154-156.
Frissel M J and Van Veen J A 1981 Simulation of nitrogen behavior of soil-plant systems. Centre for Agricultural Publication and Documentation. Wageningen, The Netherlands.
Garwood E A and Tyson K C 1973 Losses of nitrogen and other plant nutrients to drainage from soil under grass. J. Agric. Sci., Camb. 80, 303-312.
Garrett M K, Watson C J, Crawford J, Steen R W J and Smith R V 1992 The nitrogen economy of grazed grassland. Proceedings of the Fertiliser Soc., Peterborough, UK.
Goss M J, Howse K R, Lane P W, Christian D G and Harris G L 1993 Losses of nitrate-nitrogen in water draining from under autumn-sown crops established by direct drilling or mouldboard ploughing. J. Soil Sci. 44, 35-48.

Gotoh S and Patrick Jr. W H 1974 Transformations of iron in a waterlogged soil as influenced by redox potential and pH. Soil Sci. Soc. Am. Proc. 38, 66-71.

Goulding K W T 1990 Nitrogen deposition to land from the atmosphere. Soil Use and Management 6, 61-63.

Gray C B J and Kirkland R A 1986 Suspended sediment phosphorus composition in tributaries of the Okanagan Lakes, BC. Water Res. 20, 1193-1196.

Grinsted M J, Hedley M J, White R E and Nye P H 1982 Plant-induced changes in the rhizosphere of rape (*Brassica napus* var Emerald) seedlings. New Phytol. 91, 19-29.

Haigh Roslyn A and White R E 1986 Nitrate leaching from a small underdrained grassland clay catchment. Soil Use and Manage. 2, 65-70.

Hammond L L, Chien S H and Mokwunye A V 1986 Agronomic value of unacidulated and partially acidulated phosphate rocks indigenous to the tropics. Adv. Agron. 40, 89-140.

Harris G L, Goss M J, Dowdell R J, Howse K R and Morgan A 1984 A study of mole drainage with simplified cultivation for autumn-sown crops on a clay soil. 2. Soil water regimes, water balances and nutrient loss in drainage water 1978-80. J. Agric. Sci. Camb. 102, 561-581.

Harrison A F 1978 Phosphorus cycles of forest and upland grassland systems and some effects of land management practices. In Phosphorus in the Environment: Its Chemistry and Biochemistry. pp 175-195. CIBA Foundation Symp. 57, Elsevier/North-Holland, Amsterdam, The Netherlands.

Haynes R J and Williams P H 1992 Long term effect of superphosphate on accumulation of soil phosphorus and exchangeable actions on a grazed, irrigated pasture site. Plant Soil 42, 123-133.

Haynes R J and Williams P H 1993 Nutrient cycling and soil fertility in the grazed pasture ecosystem. Adv. Agron. 49, 119-199.

Heathman G C, Sharpley A N, Smith S J and Robinson J S 1995 Poultry litter application and water quality in Oklahoma. Fert. Res. 37 (in press).

Hedley M J, Hussin J A and Bolan N S 1990 New approaches to phosphate fertilization. In Phosphorus requirements for sustainable agriculture in Asia and Oceania. Ed. J W B Stewart. 28 p. Int. Rice Res. Institute, Los Banos, Philippines.

Helyar K R and Godden D P 1976 The phosphorus cycle – What are the sensitive areas? In Prospects for Improving Efficiency of Phosphorus Utilization. Ed. G J Blair. pp 23-30. Australia Reviews in Rural Sci. III, Armidale, Australia.

Heng L K, White R E, Bolan N S and Scotter D R 1991 Leaching losses of major nutrients from a mole-drained soil under pasture. NZ J Agric. Res. 34, 325-334.

Hodgkin E P and Hamilton B H 1993 Fertilizers and eutrophication in southwestern Australia: setting the scene. Fert. Res. 36, 95-103.

Holford I C R 1982 Effects of phosphate sorptivity on long-term plant recovery and effectiveness of fertilizer phosphate in soils. Plant Soil 64, 225-236.

Holford I C R 1989 Efficacy of different phosphate application methods in relation to phosphate sorptivity in soils. Aust. J. Soil Res. 27, 123-133.

Hortenstine C C and Forbes R B 1972 Concentration of nitrogen, phosphorus, potassium and total soluble salts in soil solution samples from fertilized and unfertilized histosols. J. Environ. Qual. 1, 446-449.

Iserman K 1990 Share of agriculture in nitrogen and phosphorus emissions into the surface waters of Western Europe against the background of their eutrophication. Fert. Res. 26, 253-269.

Jenkinson D S 1990 An introduction to the global nitrogen cycle. Soil Use and Manage. 6, 56-61.

Jenkinson D S and Parry L C 1989 The nitrogen cycle in the Broadbalk wheat experiment: a model for the turnover of nitrogen through the soil microbial biomass. Soil Biol. Biochem. 21, 535-541.

Johnston A E 1989 Phosphorus cycling in intensive arable agriculture. In Phosphorus Cycles in Terrestrial and Aquatic Ecosystems, Regional Workshop 1 Europe, Czerniejewo Poland. Proc. SCOPE/UNEP Workshop. Ed. H Tiessen. pp 123-136, Saskatchewan Institute of Pedology, Canada.

Jones O R, Willis W M, Smith S J and Stewart B A 1995 Nutrient cycling of cattle feedlot manure and composted manure applied to Southern High Plains drylands. In Impact of animal manure and the land-water interface. Ed. K Steele. Lewis Pubications, pp 265-272. CRC Press, Boca Raton, Florida, USA.

Keeney D 1986 Sources of nitrate to groundwater. CRC Critical Reviews in Environmental Control 16, 257-304.

Khasawneh F E and Doll E C 1978 The use of phosphate rock for direct application to soils. Adv. Agron. 30, 159-206.

Kirk G J D and Nye P H 1986 A simple model for predicting the rates of dissolution of sparingly soluble calcium phosphates in soil. I. The basic model. J. Soil Sci. 37, 529-540.

Kissel D E, Bidwell O W and Kientz J F 1982 Leaching classes of Kansas soils. Kansas Agric. Expt. Stat. Bull. 641, 3-11.

Kotak B G, Prepas E E and Hrudey S E 1994 Blue green algal toxins in drinking water supplies: research in Alberta. Lake Line 14, 37-40.

Kotak B G, Kenefick S L, Fritz D L, Rousseaux C G, Prepas E E and Hrudey S E 1993 Occurrence and toxicological evaluation of cyanobacterial toxins in Alberta lakes and farm dugouts. Water Res. 27, 495-506.

Kovar J L and Barber S A 1987 Placing phosphorus and potassium for greatest recovery. J. Fert. Issues 4, 1-6.

Kucey R M N, Janzen H H and Leggett M E 1989 Microbially mediated increases in plant-available phosphorus. Adv. Agron. 42, 199-228.

Lammond R E 1987 Comparison of fertilizer solution placement methods for grain sorghum under two tillage systems. J. Fert. Issues 4, 43-47.

Larsen J E, Langston R and Warren G F 1958 Studies on the leaching of applied labelled phosphorus in organic soils. Soil Sci. Soc. Am. Proc. 22, 558-560.

Lawton L A and Codd G A 1991 Cyanobacterial (blue-green algae) toxins and their significance in UK and European waters. J. Inst. Wat. Environ. Managt. 5, 460-465.

Leavesley G H, Beasely D B, Pionke H B and Leonard R A 1990 Modeling of agricultural nonpoint-source surface runoff and sediment yield – a review from the modeler's persepctive. In Proc. Int. Symp. Water Quality Modeling of Agricultural Non-point Sources, Part 1. Ed. D G DeCoursey. pp 171-196, USDA-ARS 81, US Government Printing Office, Washington DC, USA.

Lemunyon J L and Gilbert R G 1993 Concept and need for a phosphorus assessment tool. J. Prod. Agric. 6, 483-486.

Levine S L and Schindler D W 1989 Phosphorus nitrogen and carbon dynamics of experimental lake 303 during recovery from eutrophication. Can. J. Fish. Aquat. Sci. 46, 2-10.

Macduff J H and White R E 1984 Components of the nitrogen cycle measured for cropped and grassland soil-plant systems. Plant Soil 76, 35-47.

Magesan G N, White R E, Scotter D R and Bolan N S 1994 Estimating leaching losses from sub-surface drained soils. Soil Use and Manage. 10, 87-93.

Magesan G N, White R E and Scotter D R 1995 Nitrate leaching from a drained sheep-grazed pasture 1. Experimental results and environmental implications. Aust. J. Soil Res. (in press).

Martin A and Cooke G D 1994 Health risks in eutrophic water supplies. Lake Line 14, 24-26.

McColl R H S and Syers J K 1981 The effects of agriculture. In The effects of land use on water quality, a review. Eds R H S McColl and H R Hughes. p59 Water and soil miscellaneous publication, 23. Ministry of Works and Development, Wellington, New Zealand.

McColl R H S, White E and Gibson A R 1977 Phosphorus and nitrate runoff in hill pasture and forest catchments, Taita, New Zealand. N.Z. J. Mar. Freshwater Res. 11, 729-744.

McComb A J and Davis J A 1993 Eutrophic waters of southwestern Australia. Fert. Res. 36, 104-114.

McConaghy S, McAllister J S V and Dickson D A 1971 Residual value of phosphates in relation to soil conditions. In Residual Value of Applied Nutrients. MAFF Tech. Bull. No. 20. pp 80-88. HMSO, London, UK.

McDowell L L and McGregor K C 1984 Plant nutrient losses in runoff from conservation tillage corn. Soil Tillage Res. 4, 79-91.

McGeehan S L and Naylor D V 1994 Sorption and redox transformation of arsenite and arsenate in two flooded soils. Soil Sci. Soc. Am. J. 58, 337-342.

McLaughlin M J, Alston A M and Martin J 1988 Phosphorus cycling in wheat-pasture rotations. III. Organic phosphorus turnover and phosphorus cycling. Aust. J. Soil Res. 26, 343-353.

Mansell R S, Selim H M and Fiskell J G A 1977 Simulated transformations and transport of phosphorus in soil. Soil Sci. 124, 102-109.

Meyer J L 1979 The role of sediments and bryophytes in phosphorus dynamics in a headwater stream ecosystem. Limnol. Oceanog. 24, 365-375.

Miller M 1979 Contribution of nitrogen and phosphorus to subsurface drainage water from intensively cropped mineral and organic soils in Ontario. J. Environ. Qual. 8, 42-48.

Mishra B, Khanna P K and Ulrich B 1979 A simulation model for organic phosphorus transformation in a forest soil ecosystem. Ecol. Model. 6, 31-46.

Morse D, Head H H, Wilcox C J, van Hern H H, Hissem C D and Harris B Jr 1992 Effects of concentration of dietary phosphorus on amount and route of excretion. J. Dairy Sci. 75, 3039-3045.

Munson R D and Murphy L S 1986 Factors affecting crop response to phosphorus. In Phosphorus for Agriculture, A Situation Analysis. Ed. R E Wagner pp 9-24. Potash and Phosphate Institute, Atlanta, Georgia, USA.

National Research Council 1993 Soil and water quality: an agenda for agriculture. National Academy Press, Washington DC, USA.
National River Authority 1990 Toxic blue-green algae. Water Quality Services. No. 2 Pub. Nat. River Auth., UK.
Nicholaichuk W and Read D W L 1978 Nutrient runoff from fertilized and unfertilized fields in western Canada. J. Environ. Qual. 7, 542-544.
Nicholls K H and MacCrimmon H 1974 Nutrients in subsurface and runoff waters of the Holland Marsh, Ontario. J. Environ. Qual. 3, 31-35.
Nurnberg G K and Peters R H 1984 Biological availability of soluble reactive phosphorus in anoxic and oxic freshwaters. Can. J. Fish. Aquat. Sci. 41, 757-765.
Nurnberg G K, Dillon P J and McQueen D J 1986 Internal phosphorus load in an oligotrophic precambrian shield lake with an anoxic hypolimnion. Can. J. Fish. Aquat. Sci. 43, 574-580.
Nye P H and Ameloko A T 1987 Predicting the rate of dissolution of lime in soil. J. Soil Sci. 38, 641-649.
Olness A E, Smith S J, Rhoades E D and Menzel R G 1975 Nutrient and sediment discharge from agricultural watersheds in Oklahoma. J. Environ. Qual. 4, 331-336.
Omernik J 1977 Nonpoint source-stream nutrient level relationships: a nationwide study. EPA-600/3-77-105. Corvallis, Oregon, USA.
Ozanne P G, Kirton D J and Shaw T C 1961 The loss of phosphorus from sandy soils. Aust. J. Agric. Res. 12, 409-423.
Palmstrom N, Scarlson R E and Cooke G D 1988 Potential links between eutrophication and formation of carcinogens in drinking water. Lake Reserv. Manage. 4, 1-15.
Parton W J, Anderson D W, Cole C V and Stewart J W B 1983. Simulation of soil organic matter formations and mineralization in semiarid agroecosystems. In Nutrient Cycling in Agricultural Ecosystems. Eds. R R Lowrance, R L Todd, L E Asmussen and R A Leonard. pp 533-550. The University of Georgia, College of Agric. Expt. Sta., Special Publ. 23. Athens, Georgia, USA.
Paynter R M and Dampney P M R 1991 The effect of rate and timing of phosphorus fertilizer on the yield and phosphate offtake of grass grown for silage at moderate to high levels of soil phosphorus. Grass For. Sci. 46, 131-135.
Powlson D S 1993 Understanding the soil nitrogen cycle. Soil Use and Management 9, 86-93.
Rast W and Lee G F 1978 Summary analysis of the North American (U.S. Portion) ECD eutrophication project: nutrient loading – lake response relationships and trophic state indices. EPA 600/3-78-008, U.S. EPA, Corvallis, Oregon, USA.
Rekolainen S 1989 Phosphorus and nitrogen load from forest and agricultural areas in Finland. Aqua Fennica 19, 95-107.
Ritchie G S P and Weaver DM 1993 Phosphorus retention and release from sandy soils of the Peel-Harvey catchment. Fert. Res. 36, 115-122.
Rolston D E, Sharpley A N, Toy D W, Hoffman D L and Broadbent F E 1980 Denitrification as affected by irrigation frequency of a field soil. EPA-600/2-80-066, US Environment Protection Agency, Ada, Oklahoma, USA.
Romkens M J M and Nelson D W 1974 Phosphorus relationships in runoff from fertilized soil. J. Environ. Qual. 3, 10-13.
Rose C W, Dickenson W T, Ghadiri H and Jorgensen S E 1990 Agricultural nonpoint-source runoff and sediment yield water quality models: modeler's perspective. In Proc. Int. Symp. Water Quality Modeling of Agricultural Non-point Sources, Part 1. Ed. D G DeCoursey. pp.145-170, USDA-ARS 81. US Government Printing Office, Washington DC, USA.
Ruz-Jerez B E 1991 Dynamics of nitrogen in three contrasting pastures grazed by sheep. PhD thesis, Massey University, New Zealand. 208 p.
Ruz-Jerez B E, White R E and Ball P R 1994 Long-term measurement of denitrification in three contrasting pastures grazed by sheep. Soil Biol. Biochem. 26, 29-39.
Ruz-Jerez B E, White R E and Ball P R 1995 A comparison of nitrate leaching under clover-based pastures and nitrogen-fertilized grass grazed by sheep. J. Agric. Sci., Camb. (in press).
Ryden J C, Ball P R and Garwood E A 1984 Nitrate leaching from grassland. Nature 311, 50-53.
Ryden J C 1986 Gaseous losses of nitrogen from grassland. In Nitrogen Fluxes in Intensive Grassland Systems. Eds. H G van Meer, J C Ryden and G C Ennik. pp 59-73. Martinus Nijhoff, Dordrecht, The Netherlands.
Ryding S O, Enell M and Wennberg L 1990 Swedish agricultural nonpoint source pollution: A summary of research and findings. Lake and Reserv. Mgt. 6, 207-217.
Sandstedt C A 1990 Nitrates: sources and their effects upon humans and livestock. The American University, Washington DC, USA.

Sawyer C N 1947 Fertilization of lakes by agricultural and urban drainage. J. New England Water Works Assoc. 61, 109-127.

Scholefield D, Tyson K C, Garwood E A, Armstrong A C, Haskins J and Stone A C 1993 Nitrate leaching from grazed grassland lysimeters: effects of fertilizer input, field drainage, age of sward and patterns of weather. J. Soil Sci. 44, 601-613.

Schreiber J D 1988 Estimating soluble phosphorus (PO_4-P) in agricultural runoff. J. Miss. Acad. Sci. 33, 1-15.

Sequi P, Ciavatta C and Antisari L V 1990 Phosphate fertilizers and phosphorus loadings to rivers and seawater. Agrochim. 35, 200-211.

Sharpley A N 1985a Phosphorus cycling in unfertilized and fertilized agricultural soils. Soil Sci. Soc. Am. J. 49, 905-911.

Sharpley A N 1985b Depth of surface soil-runoff interaction as affected by rainfall, soil slope and management. Soil Sci. Soc. Am. J. 49, 1010-1015.

Sharpley A N 1985c The selective erosion of plant nutrients in runoff. Soil Sci. Soc. Am. J. 49, 1527-1534.

Sharpley A N 1995a Residual availability of phosphorus in manured soils. Soil Sci. Soc. Am. J. 59 (in press).

Sharpley A N 1995b Identifying sites vulnerable to phosphorus loss in agricultural runoff. J. Environ. Qual. 25 (in press).

Sharpley A N and Smith S J 1989 Prediction of soluble phosphorus transport in agricultural runoff. J. Environ. Qual. 18, 313-316.

Sharpley A N and Smith S J 1993 Wheat tillage and water quality in the Southern Plains. Soil Tillage Res. 30, 33-48.

Sharpley A N and Syers J K 1979 Phosphorus inputs into a stream draining an agricultural watershed. II. Amounts and relative significance of runoff types. Water, Air and Soil Pollut. 11, 417-428.

Sharpley A N and Williams J R 1990 EPIC-Erosion/Productivity Impact Calculator. 1. Model documentation. USDA Technical Bull. 1768. US Govt Printing Office, Washington DC, USA. 235 p.

Sharpley A N, Daniel T C and Edwards D R 1993a Phosphorus movement in the landscape. J. Prod. Agric. 6, 492-500.

Sharpley A N, Smith S J and Menzel R G 1986 Phosphorus criteria and water quality management for agricultural watersheds. Lake Reserv. Managt. 2, 177-182.

Sharpley A N, Sims J T and Pierzynski G M 1994a Innovative soil phosphorus indices: assessing inorganic phosphorus. In Soil testing: prospects for improving nutrient recommendations. Eds. J Havlin, J Jacobsen, P Fixen and G Hergert. pp 115-142. Soil Sci. Soc. Am. Spec. Pub. 40, Am. Soc. Agron., Madison, Wisconsin, USA.

Sharpley A N, Smith S J, Berg W A and Williams J R 1985 Nutrient runoff losses as predicted by annual and monthly soil sampling. J. Environ. Qual. 14, 354-360.

Sharpley A N, Smith S J, Williams J R, Jones O R and Coleman G A 1991 Water quality impacts associated with sorghum culture in the Southern Plains. J. Environ. Qual. 20, 239-244.

Sharpley A N, Tillman R W and Syers J K 1977 Use of laboratory extraction data to predict losses of dissolved inorganic phosphate in surface runoff and tile drainage. J. Environ. Qual. 6, 33-35.

Smika D E and Grabouski P H 1976 Anhydrous ammonia applications during fallow for winter wheat production. Agron. J. 68, 919-922.

Smith K A 1981 A model of denitrification in aggregated soils. In Simulation of nitrogen behavior of soil-plant systems. Eds M J Frissel and J A van Veen. pp 259-266, Centre for Agric. Pub. and Documentation, Wageningen, The Netherlands.

Smith K A and Chambers B J 1993 Utilizing the nitrogen content of organic manures on farms – problems and practical solutions. Soil Use and Manage. 9, 105-112.

Smith S J, Sharpley A N, Naney J W, Berg W A and Jones O R 1991a Water quality impacts associated with wheat culture in the Southern Plains. J. Environ. Qual. 20, 244-249.

Smith S J, Sharpley A N, Williams J R, Berg W A and Coleman G A 1991b Sediment-nutrient transport during severe storms. In Fifth Interagency Sedimentation Conference. Eds. S S Fan and Y H Kuo. pp 48-55. Federal Energy Regulatory Commission, Washington DC, USA.

Smyth T J and Sanchez P A 1982 Phosphate rock and superphosphate combinations for soybeans in a Cerrado Oxisol. Agron. J. 74, 730-735.

Sposito G 1984 The Surface Chemistry of Soils. Oxford University Press, New York, USA, 234p.

Stewart J W B and Sharpley A N 1987 Controls on dynamics of soil and fertilizer phosphorus and sulfur. In Soil fertility and organic matter as critical components of production. Eds. R F Follett, J W B Stewart and C V Cole. pp 101-121. Soil Sci Soc Am Spec. Pub. 19, Am. Soc. Agron., Madison, Wisconsin, USA.

Stewart J W B and Tiessen H 1987 Dynamics of soil organic phosphorus. Biogeochem. 4, 41-60.

Syers J K 1982 Introduction. In Nitrogen Fertilizers in New Zealand Agriculture. Ed. P B Lynch. pp 11-17. NZ Inst. of Agric. Sci., Wellington, New Zealand.

Syers J K, Harris R F and Armstrong D E 1973 Phosphate chemistry in lake sediments. J Environ. Qual. 2, 1-15.

Szilagyi F, Somloydy L and Koncsos L 1987 Chemical, biological and mass balance studies of the Kis-Balaton Control System. VITUKI Report No. 7612/3/8. Budapest, Hungary. 137p.

Tate K R, Spier T W, Ross D, J Parfitt R L, Whale K N and Cowling J C 1991 Temporal variations in some plant and soil P pools in two pasture soils of different P fertility status. Plant Soil 132, 219-232.

Theis T L and McCabe P J 1978 Phosphorus dynamics in hypereutrophic lake sediments. Water Res. 12, 677-685.

Thomas G W and Phillips R E 1979 Consequences of water movement in macropores. J. Environ. Qual. 8, 149-152.

Thompson-Eagle E T and Frankenberger W T Jr 1992 Bioremediation of soils contaminated with selenium. Advances in Soil Science 17. Ed. B A Stewart. pp 261-310. Springer Verlag, New York, ISA.

Tiedje J M, Firestone M K, Firestone M R, Betlach M R, Kaspar H F and Sorenson J 1981 Use of nitrogen 13 in studies of denitrification. In Recent Developments of Biological and Chemical Research with Short-lived Isotopes. Eds. R A Krohn and J W Root. pp 295-317. Advances in Chemistry, Am. Chem. Soc., Washington DC, USA.

Tiessen H, Abetcoe M K, Salcedo I H and Owusu-Bennoah E 1993 Reversibility of phosphorus soption by ferruginous nodules. Plant Soil 153, 113-121.

USA vs South Florida Water Management District 1994 U.S. District Court / Southern District, Case number 88-1880-CIV.

US Environmental Protection Agency 1976 Quality criteria for water. US Govt Printing Office, Washington DC, USA.

US Environmental Protection Agency 1990 National water quality inventory. 1988 Report to Congress. Office of Water. US Govt Printing Office, Washington DC, USA.

Van Keulen H and Seligman N G 1987 Simulation of water use, nitrogen nutrition and growth of a spring wheat crop. Simulation Monographs, PUDOC, Wageningen, The Netherlands. 310p.

Vauchaud G, Vauclin M and Addiscott T M 1990 Solute transport in the vadose zone: a review of models. In Proc. Int. Symp. Water Quality Modeling of Agricultural Non-point Sources, Part 1. Ed D G DeCoursey. pp 81-104. USDA-ARS 81. US Govt Printing Office, Washington DC, USA.

Vincent W F and Downes M T 1980 Variation in nutrient removal from a stream by water cress (Nasturtium officinale R. BR.). Aquatic Bot. 9, 221-235.

Vollenweider R A 1968 Scientific fundamentals of the eutrophication of lakes and flowing waters with particular reference to nitrogen and phosphorus. OECD Rep. DAS/CSI/68.27. Paris, France. 182 p.

Wadman W P, Sluijsmans C M J and De La Lande Cremer L C N 1987 Value of animal manures: changes in perception. In Animal Manure on Grassland and Fodder Crops. Ed. H G Van der Meer. pp 1-16. Martinus Nijhoff, Dordrecht, The Netherlands.

Wagenet R J, Shaffer M J and Green R E 1990 Predictive approaches for leaching in the unsaturated zone. In Proc. Int. Symp. Water Quality Modeling of Agricultural Non-point Sources, Part 1. Ed. D G DeCoursey. pp 63-80. USDA-ARS 81. US Govt Printing Office, Washington DC, USA.

Walton C P and Lee G F 1972 A biological evaluation of the molybdenum blue method for orthophosphate analysis. Tech. Int. Ver. Limnol. 18, 676-684.

Ward J C, O'Connor K F and Wei-Bin G 1990 Phosphorus losses through transfer, runoff and soil erosion. In Phosphorus requirements for sustainable agriculture in Asia and Oceania. Ed J W Stewart. pp 167-183. Int. Rice Res. Institute, Los Banos, Philippines.

Warneck P 1988 Chemistry of the Natural Atmosphere. Academic Press, San Diego, USA.

Weaver D M and Prout A L 1993 Changing farm practice to meet environmental objectives of nutrient loss to Oyster Harbour. Fert. Res. 36, 177-184.

Webster C P and Dowdell R J 1984 Effect of drought and irrigation on the fate of nitrogen applied to cut permanent grass swards in lysimeters: leaching losses. J. Sci. Food Agric. 35, 1105-1111.

Webster C P, Shepherd M A, Goulding K W T and Lord E 1993 Comparisons of methods for measuring the leaching of mineral nitrogen from arable land. J. Soil Sci. 44, 49-62.

Welch L F, Mulvaney D L, Boone L V, McKibben C E and Pendleton J W 1966 Relative efficiency of broadcast versus banded phosphorus for corn. Agron. J. 58, 283-287.

Westerman P W, Donnely T L and Overcash M R 1983 Erosion of soil and poultry manure – a laboratory study. Trans. Am. Soc. Agric. Eng. 26, 1070-1078, 1084.

White R E, Hedley M J, Bolan N S and Gregg P E H 1989 Recent developments in the use of phosphate fertilizers on New Zealand pastures. Agric. Sci. 2, 26-32.

White R E 1985a A model for nitrate leaching in undisturbed structured clay soil during unsteady flow. J. Hydrol. 79, 37-51.

White R E and Magesen G N 1991 A stochastic-empirical approach to modelling nitrate leaching. Soil Use and Management 7, 85-94.

White R E and Taylor A W 1977 Effect of pH on phosphate adsorption and isotopic exchange in acid soils at low and high levels of soluble phosphate. J Soil Sci. 28, 48-61

White R E and Thomas G W 1981 Hydrolysis of aluminium on weakly acidic organic exchangers: implications for phosphorus adsorption. Fert. Res. 2, 159-167.

White R E, Haigh R A and Macduff J H 1987 Frequence distributions and spatially dependent variability of ammonium and nitrate concentrations in soil under grazed and ungrazed grassland. Fert. Res. 11, 193-208.

White R E 1987 Introduction to the Principles and Practice of Soil Science. 2nd Ed. Blackwell Scientific, Oxford, UK. 244 p.

White R E 1988 Leaching In Advances in Nitrogen Cycling in Agricultural Ecosystems. Ed J R Wilson. pp 193-211, CAB Int., Wallingford, UK.

White R E 1980 Retention and release of phosphate by soil and soil constituents. In Soils and Agriculture. Ed. P B Tinker. Critical Reports on Appl. Chem., Vol 2, pp 71-114. Blackwell Scientific, Oxford, UK.

White R E 1985b The influence of macropores on the transport of dissolved and suspended matter through soil. In Advances in Soil Science 3, Ed. B A Stewart. pp 95-120. Springer Verlag, New York, USA.

White R E, Wellings S R and Bell J P 1983 Seasonal variations in nitrate leaching in structured clay soils under mixed land use. Agric. Water Manage. 7, 391-410.

Widdowson F V, Penny A and Hewitt MV 1982 Results from the Woburn Reference experiment. III. Yields of the crops and the recoveries of N, P, K and Mg from manures and soil, 1975-1979. Rothamsted Experimental Station Report for 1981, Part 2. Harpenden, UK.

Wild A 1972 Nitrate leaching under bare fallow at a site in northern Nigeria. J. Soil Sci. 23, 315-324.

Wildung R E, Schmidt RL and Gahler AR 1974 The phosphorus status of eutrophic lake sediments as related to changes in limnological conditions – total, inorganic, and organic phosphorus. J. Environ. Qual. 3, 133-138.

Wolf J, De Wit C T and Van Keulen H 1989 Modeling long-term crop response to fertilizer and soil nitrogen. I. Model description and application. Plant Soil 120, 11-22.

Wong M T F, Hughes R and Rowell D L 1990 The retention of nitrate in acid soils from the tropics. Soil Use and Management 6, 72-74.

Yeates J S 1993 Changing fertilizer practices. Fert. Res. 36, 135-140.

Yeates J S and Clarke M F 1993 Developing alternatives to phosphate fertilizers of high water solubility. Fert. Res. 36, 141-150.

Yeates J S, Deeley D M, Clarke M F and Allen D 1984 Modifying fertilizer practices. J. Agric. West. Aust. 25, 87-91.

Yoshinari T, Haynes R and Knowles R 1977 Acetylene inhibition of nitrous oxide reduction and measurement of denitrification and nitrogen fixation in soil. Soil Biol. Biochem. 9, 177-183.

Young R A and Mutchler C K 1976 Pollution potential of manure spread on frozen ground. J. Environ. Qual. 5, 174-179.

Young T C and J V DePinto 1982 Algal availability of particulate phosphorus from diffuse and point sources in the lower Great Lakes basin. Hydrobiologia 91, 111-119.

3
Fate of contaminants in the soil environment: metal contaminants

PHILIP A. HELMKE AND RAVENDRA NAIDU

1. Introduction

The 90 chemical elements that occur naturally in soils have a very wide range of abundance. Ninety-nine percent of the mass of most soils is made up of only eight elements: H, C, O, Mg, Al, Si, Ca, Fe. The remaining elements exist in such small proportions in typical soils that in most situations they would be regarded as insignificant impurities. Yet nearly all of the elements are important to us in some way. Thirty elements are essential to life.

Metals added to soils as contaminants are subject to the same types of reactions as are the indigenous elements. Additional specific reactions may occur because of the addition of exotic components or extraordinary concentrations of indigenous components as part of the contaminant. Most of the metal reactions that affect the bioavailability of elements or their transformations in the soil-water system occur via the solution phase. Detailed consideration of the factors that affect the concentrations of elements in the solution phase is therefore central for an understanding of the behaviour of elements in the soil system.

The potential reactions that affect the solution concentrations of the hydrated free metal ions (Figure 1) offer a paradigm useful to study the behaviour of metals in the soil system. The free ion activity is selected as a central species because for most metals it is the form that is bioavailable to plants and microorganisms. The concentrations of metals in soil solution are affected by solubility phenomena, ion exchange and adsorption-desorption processes, complexation and redox reactions. In the absence of biotic effects, these processes can often be quantitatively modelled using the familiar chemical concepts of solubility products and other equilbrium constants. The uptake of elements and their return to soils by plants and microorganisms can be an important factor in element behaviour in soils but this reaction pathway is difficult to model because of the nonequilibrium nature of biotic reactions. The application of equilibrium models is limited by the scarcity of information about the kinetics of metal reactions in soils. For most metals, a quantitative understanding of the basic processes that control metal behaviour in the soil-water system still eludes researchers (Evans, 1989) or is not necessarily relevant to field conditions (see Chapter 1).

Detailed information on the behaviour of elements in the soil system requires sophisticated analytical techniques to study the solid and solution phases in the soil system. The task of understanding the behaviour of elements in the soil-water system is made extremely difficult by several factors.

R. Naidu et al. *(eds): Contaminants and the Soil Environment in the Australasia-Pacific Region, 69–93.*

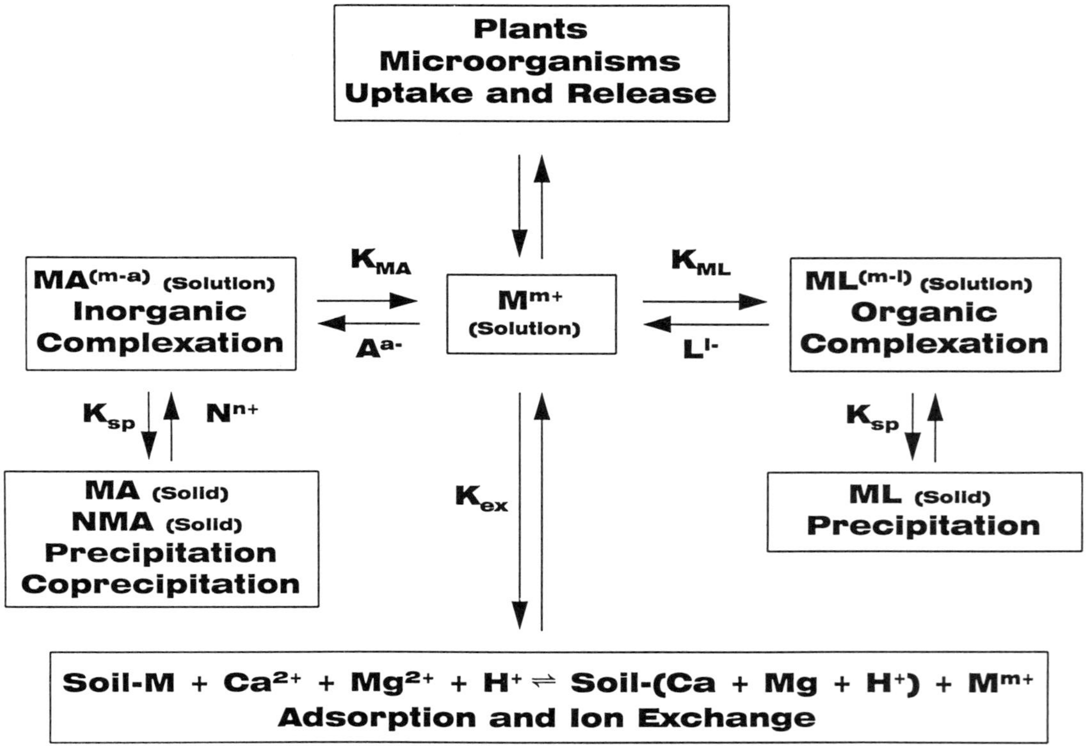

Figure 1. Potential reactions that affect the free ion concentration of metals in the soil-water system.

These factors include the:

- low concentrations of most elements in soils;
- small size of the most reactive soil particles;
- noncrystalline nature of many soil solids;
- intimate association of inorganic and organic soil compounds;
- impact of soil microorganisms and plants on chemical reactions; and
- wide variety of soils found in the various environments of our planet.

These factors confound many modern instrumental approaches that have been used very successfully to characterise solid phases in material science and other disciplines.

Water participates in most soil reactions and this limits the application of techniques that require a high vacuum, such as mass and electron spectrometric techniques and those that use lasers or ion beams as probes to characterise solid phases. The abundance of iron (Fe) in soils requires that it be removed before soils are studied by nuclear magnetic and electron paramagnetic resonance techniques. The required sample pretreatments often alter the samples in unknown or unpredictable ways.

A singularly successful technique in the study of the solid phase of soils is X-ray diffraction. It has been largely responsible for our current understanding of the soil solid phase, especially the nature and reactions of clay minerals, because it is applicable to small particles even in the presence of water. Normal X-ray diffraction, unfortunately, is not a trace technique and is limited to the major minerals in soil. The recent availability of high intensity X-ray sources at synchrotron research centres and the development of extended X-ray absorption fine structure (EXAFS) and related techniques is extending the use of X-rays to study the chemical bonds of trace elements in the soil-solid phase. These advancements are starting to provide information on the types of complexes formed between elements and the soil-solid phase, including such detailed information such as bond angle, bond distance, and the identity of next nearest atomic neighbours.

The limitations of physical techniques to characterise the soil-solid phase has resulted in that much of our knowledge of how elements behave in the soil-water system was inferred using data gathered from analysis of the water phase of soils. The general approach is to combine data from the water phase of soils with information from model systems that attempt to mimic the behaviour of particular soil components. This approach is limited because soil components are not pure compounds as used in most model systems and because the speciation of the dissolved elements in the solution phase of soils is generally unknown. There is also no method for obtaining a 'true' soil solution (Wolt, 1994).

Many techniques can be used to determine the total concentration of elements in soils. Total concentrations of elements however have very limited usefulness in studies determining the chemical reactions of elements in soils, although they are the primary basis for environmental regulations (see Chapter 1).

The consequence of these experimental limitations is that much of our knowledge of how elements behave in the soil-water system is still inexact and subject to controversy. The evidence is overwhelming that reaction phenomena such as precipitation-dissolution, charged surfaces, adsorption-desorption, ion exchange, speciation, etc. occur in the soil system. There are excellent data on model systems that confirm quantitative models of these phenomena. An important goal of modern research on element behaviour in soil-water systems is to develop and confirm a quantitative understanding of these phenomena in whole soils. Computer implemented models provide convenient ways to use the extensive quantitative data to make predictions of element behaviour.

2. Speciation of elements in the water phase of soils

Knowledge of the factors affecting the concentrations of elements dissolved in the solution phase of soils is central to the understanding of the behaviour of elements in the soil system. Elements in the solution phase may exist as free, hydrated ions and as dissolved species complexed with organic or inorganic ligands. The dissolved complexes may have positive, negative, or zero ionic charge. The free ions and potentially the complexed species interact with the soil-solid phases by precipitation-dissolution (solubility) phenomena or by adsorption-desorption (ion exchange) phenomena (Figure 1). Plant roots and microorganisms interact with the dissolved species, and root and microbial exudates can affect the solubility of soil compounds. The competing equilibria represent a dynamic system.

Only a small fraction of the total amount of each element present in the soil system is exchangeable and an even smaller fraction is water soluble. Water saturation extracts or displacement methods are preferred to sample the soil solution in fundamental studies of the exchangeable or soluble fractions of elements in soils. Chemical extractants greatly modify the natural soil water equilibrium. Chemical extractants are not well suited to investigate the fundamental effects of various soil properties (such as pH, ionic strength, competing ions, nature of organic matter, soil-solid phases, etc.) on the solubility behaviour of elements in soils at realistic metal concentrations. Procedures based on chemical extracts are used to estimate soil fertility factors but their success depends on extensive empirical correlations between the results from soil extracts and nutrient data from plants. The correlations often depend on soil type and plant species.

Two types of soluble complexes are formed between metals and complexant ligands in soil solution. Outer sphere complexes, or ion pairs, are relatively weak electrostatic associations formed between a hydrated cation and ligand. One or both of the charged species retains a hydration shell. Often such interactions are described in the literature as non-specific adsorption. Inner sphere complexes or specific interactions are strong associations between metals and ligands. They are characterised by a covalent bond between the metal and ligand.

A vast array of ligands which vary widely with soil type is found in soil solutions. Most ligands are the conjugate bases of weak acids. The extent of metal complexation therefore depends on the solution pH and on the concentrations of the competing metals and ligands. The reaction of a metal ion, M^{m+}, with an anionic ligand, L^{a-}, to form a 1 : 1 complex can be represented by:

$$M^{m+} + L^{a-} = ML^{(m-a)} \tag{1}$$

$$K_{ML} = \frac{(ML^{(m-a)})}{(M^{m+})(L^{a-})} \tag{2}$$

where K_{ML} is the stepwise formation constant for the complex.

The expression for the formation of polynuclear multiligand complexes can be written:

$$qM^{m+} + rL^{a-} = M_qL_r^{(qm-ra)} \tag{3}$$

$$\beta_{ML} = \frac{M_qL_r^{(qm-ra)}}{(M^{m+})^q(L^{a-})^r} \tag{4}$$

where β_{ML} is the overall formation constant for the complex. The concentration of dissolved metal is then the sum of the concentration of the hydrated free metal cation plus that of all of the individual complexed species.

The speciation of individual elements in the solution phase can be estimated by computer implemented models such as MINTEQA2 (Allison et al., 1991; Allison and Brown, 1995), SOILCHEM (Sposito and Coves, 1995), and GEOCHEM-PC (Parker et al., 1995) if the total concentrations of all elements and ligands are known. A serious limitation of the computer implemented models is that data for many possible naturally occurring complexes are not included in the programs because their formation constants are not known. This is especially the case for metal complexes with dissolved soil organic matter (Holm et al., 1995a; 1995b; Mattigod, 1995).

The total content of metals and inorganic ligands in the solution phase can be determined by several techniques (atomic absorption spectrophotometry, AAS; inductively coupled plasma optical emission spectroscopy, ICP-OES; inductively coupled plasma mass spectroscopy, ICP-MS; and high performance liquid chromatography, HPLC). The situation for organic compounds is not as simple. The total concentrations of some of the simpler aliphatic and aromatic organic ligands can be determined by HPLC in soil solutions. The more structurally complex compounds contained in the fulvic and humic acid fractions cannot be individually determined. The concentrations of these compounds are usually expressed collectively as total dissolved organic carbon.

The potential utility of computer implemented speciation models is shown by the following example. The total indigenous concentrations of selected elements in a water saturation extract of a Plano silt loam (Typic Arguidoll) at pH 6.3 are given in Table 1 (Helmke and Salam, 1996). The predicted speciation of this solution as a function of pH is shown in Figure 2. The values of the formation constants used are those tabulated in MINTEQA2. The optional composite ligand model in MINTEQA2 was used to estimate the metal speciation with dissolved organic matter. All of the dissolved organic matter was assumed to be fulvic acid with a molecular weight of 750 g $mole^{-1}$.

Table 1. Total concentrations of selected elements and compounds in a saturation water extract of a silt loam soil (Typic Arguidoll) (Helmke and Salam, 1996). The concentration of fulvic acid is estimated from the content of total organic carbon and assuming the molecular weight of fulvic acid is 750 g $mole^{-1}$.

Element	Total Concentration -log [Molar]
H	6.3
Ca	3.12
Mg	3.20
K	4.21
Na	4.11
Cu	6.42
Cd	7.92
Zn	5.26
SO_4^{2-}	3.49
Fulvic acid	3.55

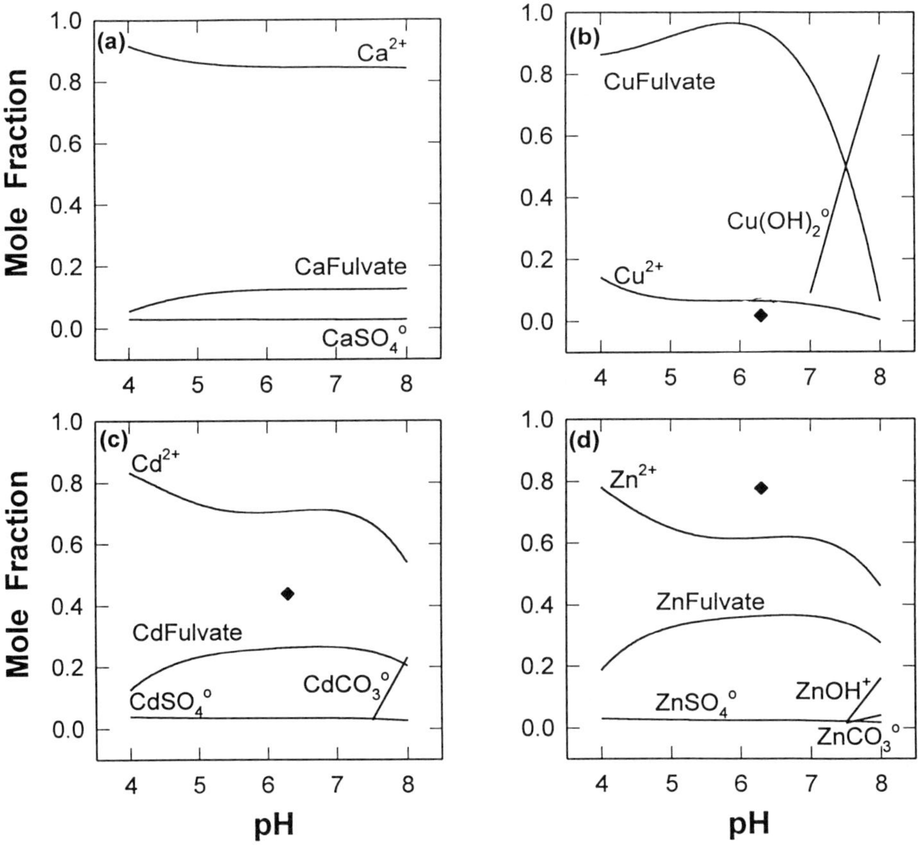

Figure 2. The effect of pH on the speciation of the soil solution given in Table 1 as calculated from MINTEQA2. The diamond symbol represents the free ion activity of the metal at pH 6.3 that was experimentally measured by Donnan membrane equilibrium analysis (Helmke and Kadir, 1996).

The calculations indicate that almost all of the dissolved Ca and most of the dissolved Cd and Zn in the water extract should exist as the free cation. Most of the dissolved Cu is in the form of Cu fulvate complexes except at higher values of pH where the hydroxide ion is an effective competitor to fulvate for the complexation of Cu. Sulphate ion pairs are minor components for all of the metals. The free ion concentrations of all of the metals decrease with increasing pH.

The measured proportions of the total Cu, Cd, and Zn concentrations that are free hydrated cations at pH 6.3 are shown by the solid diamond symbols in Figure 2 (Helmke and Salam, 1996; Helmke et al., 1996). The agreement between experimental and computed values is poor except for Cu, an element that occurs mostly in complexed forms in soil solutions. The constants for metal-dissolved organic matter interactions used in MINTEQA2 were determined by lanthanide ion probe spectroscopy on dissolved organic matter from the Suwannee River (Susetyo et al., 1991). Much better agreement between the estimated and calculated free ion activities would likely result if the constants for metal-dissolved organic matter interactions were determined for soil derived organic matter and on a soil-by-soil basis.

This calculation only shows the estimated equilibrium speciation of the extract if the pH of the extract was varied after it was separated from the soil. It does not show the effects of varying the pH of the soil-water extraction process. Such calculations are possible in principle as most of the computer implemented models contain the necessary ion exchange and adsorption subroutines along with the solubility and complexation subroutines (Goldberg, 1995). Unfortunately they do not in general contain the constants for the exchange and adsorption reactions because they are not yet available. These could be supplied from research data, especially if the free ion concentrations of the metals could be determined in the water extract or soil solution.

Analytical procedures that determine speciation of elements can be divided into two groups (Florence and Batley, 1980). One group defines the speciation of elements into terms of operationally defined procedures such as labile, inert, etc. by using selective extractant or chemical procedures. The second group attempts to measure the concentration or activity of thermodynamically defined species by species specific instrumental techniques, which are sometimes connected with well-defined chemical separation procedures.

Most species specific analytical techniques of the second group have insufficient sensitivity or selectivity to measure the indigenous concentrations of single metal species in the soil solution. Ion selective electrodes and anodic stripping voltammetry can determine the activity of the free cations of selected elements under limited conditions (Hirsch and Banin, 1990). Dialysis, ion exchange (Holm et al., 1995a; 1995b) and Donnan equilibrium across ion exchange membranes (Fitch and Helmke, 1989; Helmke et al., 1996) are new techniques that have sufficient sensitivity and selectivity to measure indigenous levels of the free ion activities of trace elements in uncontaminated soils.

3. Ion exchange and adsorption reactions of metals in soils

Soil particles commonly have an electric charge, especially those found within the clay size fraction (<2mm). Ions in the soil solution are attracted to the charged surfaces. Both humic and mineral substances contribute to the charged surfaces. Structural, or permanent, charges are associated with the phyllosilicate clay minerals where the charge results from charge imbalances due to isomorphous substitution of Al for Si and Mg for Al. Unlike soils with permanent charge, variable charge soils exhibit the property of being able to alter the sign and amount of charge in response to changes in the composition of the ambient soil solution. The major factors affecting variable charge are pH, ionic strength, nature of the index ions, and ligands in the soil solution capable of specific adsorption. Because H^+ and OH^- are the dominant potential determining ions, surface charge is governed by pH and hence the frequently used but technically incorrect term 'pH dependent charge'. The variable charge is associated with reactions of protons at the edges of phyllosilicates, the surfaces of oxide and (oxy)hydroxide minerals, and with certain functional groups within humic substances, such as carboxylic acid groups.

The charged sites on soil compounds are important for the retention of elements. Cations and anions in the soil solution can neutralize the intrinsic charges on the surfaces of soil particles by forming either inner sphere or outer sphere complexes.

As discussed above the electrical charge on the surfaces of oxides and hydroxides results from proton association and dissociation reactions with the surface [S]. The proton dissociation reactions can be represented by the two protolysis reactions:

$$\overline{S}-OH_2^+=\overline{S}-OH^0+H^+ \tag{5}$$

$$\overline{S}-OH^0=\overline{S}-O^-+H^+ \tag{6}$$

These protolysis reactions are analogous to the ionisation reactions of weak acids in solution. The relative proportions of each of the surface sites depend on the pH of the solution. The charge balance σ_s for the surface functional groups is

$$\sigma_s=(\overline{S}-OH_2^+)+(\overline{S}-O^-) \tag{7}$$

Extensive research on the effect of pH on the behaviour of the surface charge in synthetic oxide systems reveals that most oxide systems have a net positive charge at pH values less than 7 (Parks, 1965). However, most soils consist of a mixed system with both permanent and variable charge surfaces (van Raij and Peech, 1972; Naidu et al., 1990). In such systems, the whole soil generally has a net negative charge at field pH values. The effect of soil pH on the surface charge characteristics of variable charge soils is illustrated in Figure. 3 which shows that increasing soil pH results in an increase in the net negative charge in the system. Decreasing the soil pH results in an increase in the net surface positive charge. The pH at which the net charge is zero is defined as the point of zero net charge (PZNC).

Certain ligands, such as phosphate, sulphate, silicate, fluoride and organic ions, are specifically adsorbed onto the surfaces of the sesquioxide components in soils irrespective of the net charge of the soil. This results in an increase in the net

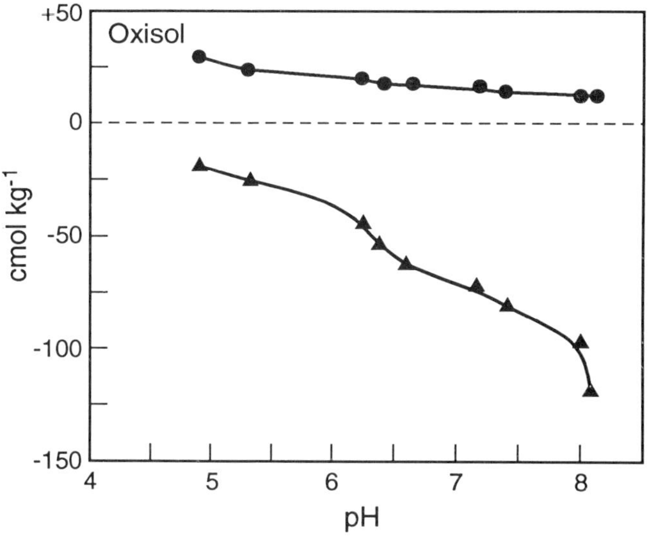

Figure 3. Effect of soil pH on the surface negative (▲) and surface positive (●) charge characteristics of an Oxisol (Naidu et al., 1990).

negative charge on the surfaces due to a lowering of the PZNC. Most soils from surface horizons that contain organic matter also have such low values of the PZNC that it is seldom reached under field pH conditions. The PZNC of soils generally increases with increasing depth due to the decrease in organic matter content. These changes in the surface charge density of soil particles affect the extent of metal adsorption-desorption reactions in the soil system (Naidu et al., 1994b).

The majority of the organic material in soils and waters are humic substances which are polyfunctional, acidic, macromolecular compounds. The charge sites in soil organic matter (SOM) arise primarily from the ionisation of carboxylic acid groups as shown in Equation (8) and secondarily by phenolic and enolic –OH functional groups.

$$\overline{SOM}-R-COOH = R-COO^{-} + H^{+} \tag{8}$$

Metal ion binding to soil organic matter is strongly pH dependent due in part to the competition between metal ions and protons for the available binding sites and in part to the effect of pH on the electric charge of the humic acid molecules. In a recent review on 'Role of metal-organic complexation in metal adsorption by soils', Harter and Naidu (1995) conclude that soil organic matter is the most important soil component for the retention of heavy metal ions in soils.

3.1 *Cation exchange reactions*

Most soils have a net negative charge. Cations in the soil solution are attracted by electrostatic forces to the negatively charged edges and surfaces of soil particles.

Attracted cations that retain their water of hydration form only outer sphere complexes. The absence of covalent bonds between the cations and the solids results in such weak associations that the adsorbed cations can be easily exchanged with other cations in cation exchange reactions. Cation exchange reactions are generally rapid, stoichiometric, and readily reversible. Cation exchange reactions are important mechanisms for the retention of alkali and alkaline earth elements in soils. The charge of the ion and the radius of the hydrated ion affect the selectivity of adsorption by the charged surface. In general, for ions having the same charge, ions with the larger hydrated radius are selectively adsorbed as shown in the series below:

$Cs^+ > Rb^+ > K^+ > Na^+ > Li^+$ and

$Ba^{2+} > Sr^{2+} > Ca^{2+} > Mg^{2+}$

The many ion exchange selectivity coefficients reported in the literature have been determined in the presence of complexant ligands but the effects of ion-pair formation and complex ions have been generally ignored. Comparisons between reported selectivity coefficients for various systems and exchange reactions can therefore not be made with confidence (Sposito et al., 1983). Very little information is available on the values of the coefficients for ion exchange between dominant and trace cations in soil-water systems.

3.2 *Specific adsorption at mineral surfaces*

Although the most abundant cations in soil solution (Ca^{2+}, Mg^{2+}, Na^+, and K^+) are weakly retained by soils as exchangeable cations, most trace cations are strongly retained by soils. Cations that are strongly retained form inner sphere complexes with variable charged surfaces by the formation of covalent bonds. Important exceptions are K^+, Rb^+, and Cs^+ cations which are selectively adsorbed onto the constant charged surfaces of some clay minerals, especially mica, vermiculite and smectites, because these cations easily lose their hydration shell of water molecules. The specific adsorption of metal ions by the formation of inner sphere complexes with variable charged mineral surfaces occurs most readily for metals that hydrolyse in water. These include most of the transition elements and some other divalent trace metal cations such as Hg^{2+} and Pb^{2+} and the rare-earth elements. Unlike the alkali and alkaline earth metals, ionic radius is not adequate as a single predictor of adsorption affinity for transition metals since their electron configuration plays a very important role in the complexes of these cations. The adsorption reactions generally involve the formation of an inner sphere complex between the hydroxometal ion and the deprotonated negatively charged surfaces of Al, Mn, and Fe. The relative affinities of the metals tend to follow the *Irving-Williams* order:

$Hg^{2+} > Pb^{2+} > Cu^{2+} > Zn^{2+} > Ni^{2+} \sim Co^{2+} > Cd^{2+}$

The elements Mo, V, W, and Cr may occur as the fully hydrolysed oxyanions MoO_4^{2-}, HVO_4^{2-}, WO_4^{2-}, and CrO_4^{2-} in soils. The mechanism of adsorption of these and other oxyanions is thought to involve the replacement of $-OH$ groups on variable charge mineral surfaces by the oxyanion in a process called ligand exchange. The extent of oxyanion adsorption onto variable charge surfaces is pH dependent because speciation of the oxyanion varies with pH as does the charge on the variable charge surface.

3.3 Factors influencing heavy metal adsorption

The properties of the soil and composition of the soil solution determine the dynamic equilibrium between trace metals in solution and the soil-solid phase (Figure 1). Trace metals interact with particulates by: metal ion adsorption at surface sites; ion exchange with clay minerals; binding with organic coated particulate matter or organic colloidal material; and by adsorption of metal-ligand complexes. All of these interactions are controlled by solution pH and ionic strength, nature of metal species, dominant cation and inorganic and organic ligands present in the soil solution. The effects of these parameters have been discussed in detail by numerous reviewers (e.g. Adriano, 1986) and therefore only a brief overview is presented in the following sections:

3.3.1 pH

Soil pH has a marked effect on the adsorption of metals (Tiller and Hodgson, 1962; Forbes et al., 1976; Kinniburgh et al., 1977; Kinniburgh and Jackson, 1981; Christensen, 1984; Brümmer et al., 1988; Naidu et al., 1994a). Generally, adsorption increases from <20 to 100% within the narrow pH range of 3 to 5 for all metals (Naidu et al., 1994a) except those that form oxyanions. This pH range, called the adsorption edge, generally occurs two to three pH units below the value of the pK for the hydroxide solubility constant of the hydrated metal ion. Barrow (1986) reports that the adsorption edge is partly a function of the experimental conditions because the position of the adsorption edge varies with the relative concentrations of adsorbate and adsorbent. Adsorption is virtually complete at pH values less than the value of pK_1 for the first hydrolysis constant of M^{2+}. However, the effect of pH on adsorption of heavy metals is dependent on the initial solution concentration, the nature of soil constituents and the metal retention capacity of soils. For example, Naidu et al. (1994a) reported that in the highly Cd retentive Australian Xeralfs and Xererts soils there was little effect of pH on Cd adsorption at low initial Cd concentrations ($<6\,\mu mol\ L^{-1}$) since most of the added Cd was adsorbed (Figure 4). In these soils, the adsorption edge was only apparent when the initial solution concentration was increased to more than $10\,\mu mol\ L^{-1}$. In contrast to the Xererts and Xeralfs, the adsorption edge was prominent even at initial solution Cd concentrations of $3\,\mu mol\ L^{-1}$ in Oxisols.

Increased adsorption of metal ions with increasing values of pH is due to both increased negative surface charge density (Naidu et al., 1994a) and the increased concentration of the MOH^+ species in the soil system (Hodgson et al., 1964). As

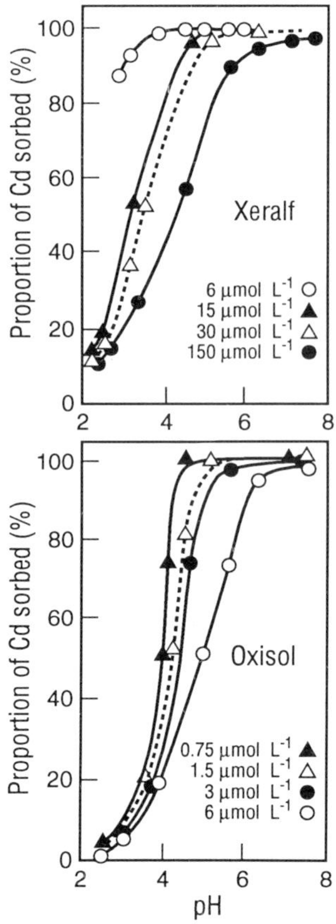

Figure 4. Effect of pH on the proportion of Cd sorbed by soils at various solution Cd concentrations (Naidu et al., 1994a) by a Xeralf and an Oxisol.

discussed above, at pH values less than the value of the PZNC, an increase in soil pH leads to a marked increase in the negative electrostatic potential of a variable charge system. The net effect of such a change in the potential is that cation adsorption will be favoured. However, the precise nature of interaction between metal ions and the soil particle surface is unclear as different mechanisms have been proposed to explain the increasing metal adsorption with pH. Using ΔpH_{50} values, Tiller et al. (1979) concluded that adsorption is controlled by metal-ion hydrolysis (see Chapter 1). The rapid increase in adsorption of Zn within the adsorption edge was attributed to the abrupt change in the concentration of $ZnOH^+$ species as the pH increases towards the pK_1 of the hydrolysis of Zn^{2+} (Barrow, 1986). However, recent studies on Cd adsorption by a wide range of soils and Cd speciation data suggest that the concentration of MOH^+ is negligible compared to M^{2+} at the pH value of the adsorption edge, indicating that other mechanisms may be involved in metal adsorption reactions (Naidu et al., 1994a). More research is needed to identify the precise mechanism for the adsorption of heavy metals by soils.

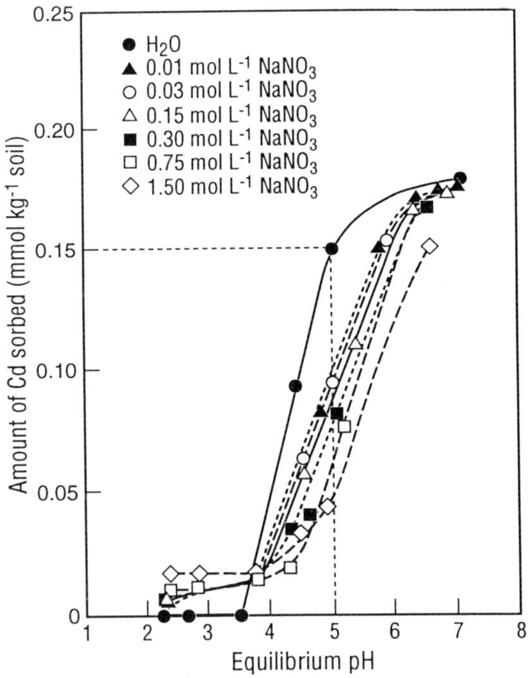

Figure 5. Effect of pH on the amount of Cd sorbed by an Oxisol in background electrolytes ranging in ionic strength from 0 (water) to 1.5 mol L^{-1} $NaNO_3$ (Naidu et al., 1994a).

For metals that form oxyanions, increasing the pH decreases their adsorption by soils. For example, Griffin et al. (1977) showed that adsorption of Cr(VI) by kaolinite and montmorillonite increased with decreasing pH. Similar results were obtained by James and Bartlett (1983) for adsorption of Cr(VI) by various soils and $Fe(OH)_3$. They attributed this phenomenon to the variations in anion exchange sites on clay colloids resulting from changes in surface charge density with pH.

3.3.2 Ionic strength

The effects of ionic strength on the adsorption of M^{2+} by soils are not as well understood as the effects of pH (Petruzzelli et al., 1985; Naidu et al., 1994a). The amount of M^{2+} adsorbed decreases with increasing ionic strength for minerals with permanent surface charge density (Garcia-Miragaya and Page, 1976; Schindler et al., 1987; Boekhold et al., 1993). The effects of ionic strength on the adsorption of metals by soils depend on the surface properties of the soil particles. For example, the adsorption of Cd increases with increasing ionic strength in an Oxisol at pH values below the PZNC (Figure 5) but it decreases with increasing ionic strength in soils with a net negative charge (Naidu et al., 1994a). The effect of ionic strength is also more pronounced in variable charge soils compared to its effect in soils where the permanent charge dominates. This is consistent with the results of Barrow (1986) who showed that the effects of ionic strength on cation adsorption vary with pH. This occurs because of the effect ionic strength has on the electrostatic potential in the plane of adsorption, which is most pronounced in soils with variable charge density.

3.3.3 Index cation and solution composition

The adsorption of cations by soils is a competitive process. The degree to which any element is adsorbed depends on the concentrations and identity of the other cations in the system. For example, the presence of other divalent cations (e.g. Ca^{2+}, Co^{2+}, Cr^{2+}, Cu^{2+}, Ni^{2+} and Pb^{2+}) retards the adsorption of Cd by soils due to competition between Cd and the other metal ions (Garcia-Miragaya and Page, 1976; Homann and Zasoski, 1987). A greater proportion of dissolved Cd is adsorbed when the competing ion is Na^+ than when it is Ca^{2+} because of the competition for adsorption sites between Ca^{2+} and Cd^{2+} ions (Tiller et al., 1979; Zhu and Alva, 1993; Naidu et al., 1994a). Even when the pH and ionic strength are maintained constant, Ca^{2+} had a marked retardation effect on Cd^{2+} adsorption for a range of Australian soils (Naidu et al., 1994a). Increasing the concentration of Ca^{2+} in the background electrolyte from 10^{-3} to 10^{-2} M in the Australian soils reduced the adsorption of Cd^{2+} by a sandy loam soil by 67%. Decreased adsorption of Cd^{2+} in the presence of Ca^{2+} appears to be a general phenomenon (Cavallaro and McBride, 1978; Tiller et al., 1979). Christensen (1987) reports that Zn^{2+} had the greatest inhibitory effect on Cd^{2+} adsorption. This phenomenon has practical implications on the behaviour of trace metals in soils because the concentration of several metal cations (e.g. Na^+, Ca^{2+}, Mg^{2+} etc.) in soil solution ranges from <0.01 M in Oxisols to >0.1 M in saline sodic soils. Such changes in ionic strength and concentration of metal ions can enhance the mobility of trace metals in soils. Soil amelioration procedures such as liming and application of gypsum also introduce Ca^{2+} ions into the soil system which compete with trace metals for adsorption sites.

3.3.4 Ligands

3.3.4.1 INORGANIC

Ligands play important roles in numerous soil chemical processes. Numerous studies document the role of dissolved organic ions on processes such as mineral weathering, control of dissolved trace metal concentrations, and dissolution and plant availability of trace metal nutrients (see Harter and Naidu, 1995). The soil solution also contains inorganic ligand ions such as chloride (Cl^-), sulphate (SO_4^{2-}), phosphate (PO_4^{3-}) and nitrate (NO_3^-). The concentrations of these ligand ions vary considerably with soil type. For example, the concentration of Cl^- ranges from >0.1 M in saline sodic soils to <0.001 M in strongly weathered soils. Such variations in ligand ion concentrations modify both the nature of the metal species in soil solution and their interactions with soil particles. This leads to marked variability in the bioavailability of metals in soils. McLaughlin et al. (1994) observed increased uptake of Cd by potatoes in soils irrigated with saline water, which was attributed to elevated Cl^- concentrations in the soil solution. High concentrations of Cl^- in soils also increase the mobility of Cd through soils because of the formation of Cd chloride complexes (Doner, 1978).

The nature of ligand ions also controls the competitive interaction between metal ions for adsorption sites. Puls and Bohn (1987) found that changing the background electrolyte from 0.01 M $Ca(ClO_4)_2$ to 0.01 M $CaCl_2$ solution changes the selectivity order for kaolinite from $Cd^{2+} > Zn^{2+} > Ni^{2+}$ to $Zn^{2+} > Ni^{2+} = Cd^{2+}$ in the pH range 5.5 to 7.5. Based on computer speciation (MINTEQA2) they attributed the reduced adsorption of Cd^{2+} relative to Zn^{2+} in 0.01 M $CaCl_2$ to the presence of $CdCl^+$ complexes which have a lower affinity for adsorption sites. These results suggest that solution composition can influence the affinity of Cd^{2+} and Zn^{2+} for adsorption sites. This has implication for applications of sewage sludges and farming systems that use irrigation waters. Since Cl^-, SO_4^{2-} and Ca^{2+} are major inorganic ions in irrigation waters and sewage sludge leachates, their application might affect Cd dynamics in the soil-plant system.

The influence of ligand ions on the reactions of Cd^{2+} and Zn^{2+} can be quite complex and dependent on the soil type. For example, specific interactions of negatively charged ions like SO_4^{2-} and PO_4^{3-} with strongly weathered soils may increase the negative charge on soil surfaces and enhance adsorption of metal cations such as Cd^{2+} (Naidu et al., 1994b). In contrast, the formation of metal-SO_4 complexes in soil solution may decrease the adsorption of metal ions, especially in highly negatively charged soils. Such interactions may either enhance or retard metal mobility in soils.

3.3.4.2 DISSOLVED ORGANIC MATERIAL

Dissolved organic material consists of a range of low molecular weight organic compounds and water soluble humic and fulvic acids. The humic and fulvic acids are dominant except for some cases in the rhizosphere where low molecular weight organic compounds may dominate. Interactions of metal ions with organic matter have been extensively studied. The classic case is the reaction of Cu with humic and fulvic acids. Stable complexes of Cu with organic matter are common in all soils that contain organic matter. Many computer implemented models base metal speciation on the presence of fulvic acids in soil solution but the values of their reaction constants with metals need to be improved. In contrast to the case for humic and fulvic acids, limited effort has been directed towards understanding the role of low molecular weight organic acids in the dynamics of metals in soil solution. This appears to be partly related to the commonly held but incorrect belief that such organic material is rapidly degraded in soils, and therefore short lived in soil solution, and partly to the analytical difficulties in determining the low concentrations of organic acids found in the bulk soil solution.

The concentrations of dissolved organic material in the vicinity of plant roots may be quite high due to root exudates and microbial activity (Bolan et al., 1994;

Harter and Naidu, 1995). The concentration of citric acid in the rhizosphere of white lupin (*Lupinus albus* L.) proteoid roots reaches millimolar levels and greatly enhances the bioavailability and uptake of soil P, Cu, Fe, Mn, and Zn (Braum and Helmke, 1995). The role of plant and microbial exudates on metal reactions in soils is an area which lacks information and is an active area of research.

3.4 Adsorption by soil organic matter

Metals can react with soil organic matter by ion exchange, complexation, and precipitation reactions. Complexation reactions are thought to dominate but the difficulties involved in studying soil organic matter prevent accurate separation of the processes in most experiments. The extent of retention of metals by soil organic matter at pH 5.8 decreases in the order (Schnitzer and Khan, 1978):

$$Fe = Al = Cr = Hg = Pb = Cu > Cd > Zn > Ni > Co > Mn$$

Complexation reactions of metals by soil organic matter involve formation of inner sphere complexes between the cation and the coordinating functional groups in soil organic matter. Multi-ligand complexes can form with the simpler organic acids but 1 : 1 complexes are principally formed with the ligands contained in the structurally complex macromolecules found in soil organic matter, although two or more donor atoms in the macromolecule may be involved in the complexation reaction. The electron donors in soil organic matter are the electronegative nonmetallic elements such as O, N and S. The important binding groups in the soil organic matter include: $-COOH$ (carboxyl), $-OH$ (enolic and phenolic), $-SH$ (thiol), $-NH_2$ (amino), $=O$ (carbonyl) and $-S$ (thioether).

Models have been developed that interpret the acidity and binding of metals to humic materials in terms of discrete binding sites with unique constants (Westall et al., 1995) or by assuming a continuous distribution of binding constants (Dzombak et al., 1986). Models that include electrostatic effects allow data to be extrapolated to different pH and ionic strength conditions (Milne et al., 1995). Most studies of metal complexation by organic matter have been done on organic matter from natural waters rather than soils. The formation constants for metal fulvates in the computer implemented model MINTEQA2 are based on values measured for organic matter dissolved in river water (Susetyo et al., 1991). More research is needed to develop suitable models that quantify metal retention by soil organic matter.

3.5 Precipitation and dissolution of metals

The equilibrium solubility and precipitation of a mineral MA in water is described by its solubility product, K_{sp}.

$$M_xA_y = xM^{m+} + yA^{a-} \tag{9}$$

The equilibrium constant for the reaction in Equation (9) is the product of the activity of its constituent ions in solution.

$$K = (M^{m+})^x (A^{a-})^y \tag{10}$$

The activities of the constituent ions depend on the speciation of the dissolved ions and on the ionic strength of the solution. The activities of the constituent ions can be estimated by computer implemented models from their total concentrations if all of the relevant elements and their species with their associated equilibrium constants are known.

The importance of precipitation/dissolution phenomena in controlling the solubility of trace elements in soil solution is highly controversial. The indigenous concentrations of most trace elements are much lower than those predicted by the solubility of known minerals (Lindsay, 1979). Many trace elements occur at such low concentrations in soils that they do not form separate mineral phases. Trace elements may occur as impurities in major minerals (Mattigod et al., 1981; Wersin, 1989).

Franklinite ($ZnOFe_2O_3$) has been proposed to control the solubility of Zn in soil solutions (Lindsay, 1979; Ma and Lindsay, 1990; 1993). Similar reactions have been proposed for Cd (Workman and Lindsay, 1990). Measurements of the free ion activities of Zn and Cd in soil-water systems as a function of pH were used to support these proposals. The measurements were made by a competitive chelation technique. The validity of the key assumptions in the competitive chelation analytical techniques used to measure the values of the free ion activities has not been tested (Oglesby, 1995). These proposals should therefore be considered as hypotheses until additional evidence is gathered.

Simple solubility phenomena cannot explain the dependence of the activities of Cu, Zn, and Cd on the activities of Ca plus Mg in natural and sewage sludge amended soils (Typic Arguidoll) reported by Helmke et al. (1996). The activities of the indigenous elements in water and Ca plus Mg saturation extracts were determined by Donnan membrane equilibrium analysis coupled with graphite furnace AAS. The activities of all three trace elements increased with increasing activities of Ca plus Mg. The effect of Ca plus Mg on the activities of the trace elements was greatest for Cd (Figure 6) and less for Zn and Cu. These results clearly show that the free ion activities of Cu, Zn, and Cd in water extracts of soils are strongly affected by competitive ion effects characteristic of ion exchange, adsorption, and complex ion phenomena.

The concentrations of trace metals in contaminated soils may be several orders of magnitude higher than the indigenous concentrations. Precipitation of the metals as secondary minerals may occur in these situations. The most important of these precipitates are the oxides, (oxy)hydroxides, carbonates and phosphates. Sulphides may be important under reducing conditions. Exotic metal compounds may form in soils if unusual anions or compounds are in the contamination.

Common metals that may form hydroxides at sufficiently high values of pH and metal concentrations include: Fe^{3+}, Al^{3+}, Cu^{2+}, Zn^{2+}, Cd^{2+} and Pb^{2+}. The most common carbonate in soils is calcite ($CaCO_3$). Some trace elements, such as Cd, can

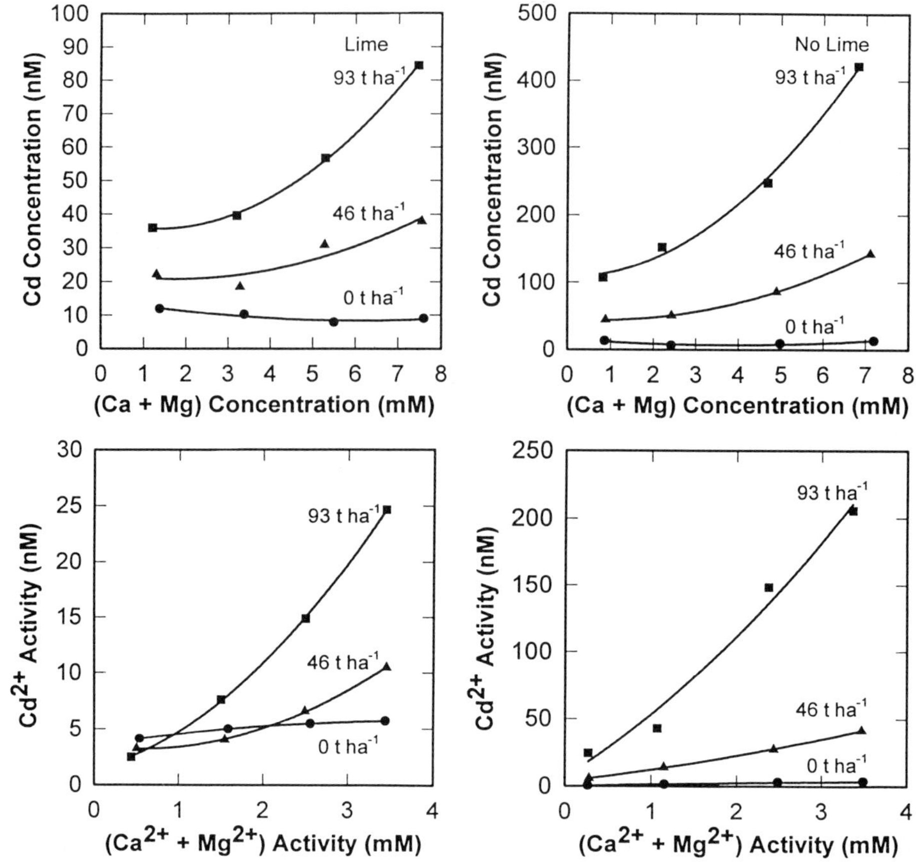

Figure 6. Relationship between the total dissolved concentrations of Cd and (Ca + Mg) and between the free ion activities of Zn and (Ca + Mg) in saturation extracts of soils at three levels of sewage sludge application to natural and limed soils.

be incorporated into the calcite structure to form a mixed solid. The solubility of carbonates depends on the partial pressure of CO_2. The partial pressure of CO_2 in air is about 0.0003 atmospheres but the value in soil air ranges from 10 to 50 times this value. Sulphide can form in soils under reducing conditions through the biological reduction of SO_4^{2-} ions to HS^- ions. Sulphate ions, SO_4^{2-}, can act as terminal electron acceptors during the metabolism of some anaerobic microorganisms. Metals expected to occur as sulphides under reducing conditions include Cd^{2+}, Zn^{2+}, Hg^{2+}, Fe^{2+} and Ag^+.

Soil organic matter may also form precipitates with trace elements as illustrated in Figure 1. The solubility of organic matter is greatly affected by metal ions. The solubility of organic matter decreases when it forms complexes with Fe, Al, Ca, and Mg. This phenomenon is the basis for the common mode of action for all of the methods used to extract soil organic matter (strong base, EDTA, Chelex 100 cation exchange resin). Precipitation of the metals at high pH or their complexation with EDTA or Chelex 100 resin greatly reduces their free ion activity in solution and strips the metals from the soil organic matter. When the functional groups on the soil organic matter are dissociated, the structural units assume a stretched configuration due to mutual repulsion of the negatively charged functional groups. Divalent and especially trivalent cations can form linkages between functional groups. This greatly decreases the solubility of humic substances. The effects of trace elements on the solubility of organic matter are largely unexplored but are probably similar to those of the major elements.

The solubility of the compounds discussed above depends on the pH and the total composition of the soil solution. The total concentration of a metal in solution may be much higher that the free ion activity of the element needed to form a precipitate. Computer implemented models such as MINTEQA2 and SOILCHEM are especially useful to estimate the solution concentrations needed to exceed the solubility product of minerals because they can account for the speciation of elements in solution. The major limitation of this approach is the tacit assumption that the system is in equilibrium. Kinetic constraints and not thermodynamics may control the extent of mineral reactions even when the system is far from equilibrium.

4. Contaminant transport in soils

The previous discussions show that the reactions of heavy metals in the soil environment are affected by many soil chemical and biological variables. The transport of metals and their fluxes in soils are also affected by the soil physical properties including particle size, pore size distribution, bulk density, temperature, moisture regime, and aeration. The transport and bioavailability of metals depends on the degree of partitioning of the metal compounds between the solution and solid phases. The solid phases consist of both stationary and mobile states. Suspended particles, especially the highly-charged high-surface area clay and colloid particles, play a significant role in the transport of contaminants that are highly insoluble in soil water but have a high affinity for soil colloids. These small particles have the potential to adsorb high concentrations of contaminant metals. It is wrong to assume that many toxic metals do not move in the soil environment solely because the colloidal fraction of the soil solid phase has a high affinity for metals.

Computer models describing the transport of metal contaminants in unsaturated soils as affected by the potential reactions of metals in soils are not yet available. This is due to the complexity of the soil system, uncertainty regarding the appropriate constants for many of the reaction pathways, and the difficulty in determining the fluxes and transport properties of metals in the unsaturated zone. Researchers are attempting to produce geochemically coupled transport models for the saturated zone. These may serve as a future base for similar models in the unsaturated zone.

The concept of distribution coefficients (K_d) is used to estimate the mobility of metals in soil systems. Distribution coefficients are defined as the ratio of the con-

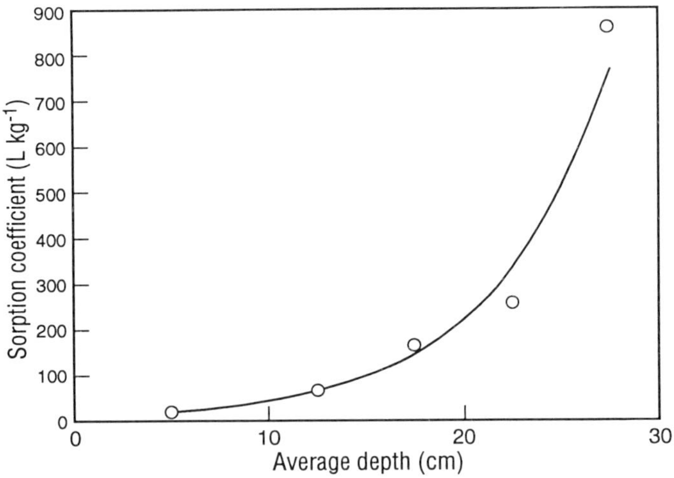

Figure 7. Relationship between Cd sorption coefficient and soil depth in a Xerert from South Australia (modified from Kookana and Naidu, 1995).

centration of the metal of interest in the soil solid phase to the concentration of metal in the water phase. The units on the ratio are volume per mass, such as L kg^{-1}. Elements that have a very low affinity for the soil solid phase and thus are highly mobile in soils have values of K_d near zero, e.g. Cl^-. Values of K_d are determined by either batch or column tests but are highly empirical and they are only applicable to the conditions used during their measurement.

Karickhoff et al. (1979) proposed how batch experiments could give values of distribution coefficients that would allow prediction of contaminant migration in unsaturated soils. However, Reynolds et al. (1982) reported that batch adsorption studies for Cs and Sr underestimated the migration of these elements in columns. Coles and Ramspott (1982) also found much more rapid migration of ruthenium (Ru) in the field than expected from batch studies. Such discrepancies may be due to both physical and chemical heterogeneity of the soil environment (Kookana and Naidu, 1995) and to colloid assisted transport of contaminants. Kookana and Naidu (1995) highlight that vertical soil chemical, biological and physical heterogeneity is seldom considered in modelling contaminant transport behaviour. Vertical heterogeneity in soil physical and chemical properties can result in significant variations in adsorption of metals within a soil profile. The data in Figure 7 show that the Cd adsorption coefficient increases by a factor of 50 from values near 18 in the surface layer to 860 L kg^{-1} at a depth of 30 cm. Predictions on contaminant mobility on the basis of K_d values and other properties of the surface soils would grossly underestimate the retardation effects of the highly Cd retentive subsurface soils. These results emphasise the limitations of existing models on contaminant transport and point to the need for a multidisciplinary approach to developing transport models.

Trace metals which are typically immobile in soils due to their strong bonding to soil particles may also move through the surface and subsurface soil environment in association with mobile colloids. Using laboratory column studies, Kaplan et al. (1993) showed that the rate of colloid-associated arsenate transport was 21 times that of dissolved arsenate. The role of colloidal particles as vehicles of contaminant transport has been well recognised in studies of the transport of radionuclides in soils (McCarthy and Zachara, 1989). Based on field data, McDowell-Boyer et al. (1986) report that viruses, bacteria and clay minerals can migrate considerable distances and that small particles and macromolecules assist the transport of radionuclides.

The role of colloids in surface runoff as a vehicle for trace metal transport was recently demonstrated in field measurements by Cooper et al. (1994) in Central and Northwest Regions Water Quality Program in New South Wales, Australia. These investigators studied the environmental impacts of trace metals from rural and agricultural practices on inland waters. They collected samples at 25 cm below the water surface on a monthly basis during September 1991 to June 1994. Time series plots for a selected site are given in Figure 8. Gaps in the data between April and October 1992 occur because no samples were collected during this period. The data in Figure 8 suggest that peak metal concentrations correspond to high flow conditions. The effect was most pronounced for As, Cu and Cr where concentrations increased during corresponding periods of high flow. The high concentrations in the water samples were related to suspended colloidal material.

Colloids as a vehicle of contaminant transport are often not considered because of the scarcity of information on the soil chemical and mineralogical conditions conducive to the generation of stable colloid suspensions. Puls and Powell (1992) report that colloid transport is highly dependent upon colloid stability. They found that iron oxide colloids were not only mobile, but that under some hydrogeochemical conditions they were transported faster than tritiated water, a conservative tracer. Matijevic (1973) reported that colloid stability is governed by particle mineralogy and surface charge density and by the thickness of the electrical double layer. Among the factors that influence colloid stability, adsorption of dissolved organics and humic substances can significantly alter particle surface charge density and increase the stability and mobility of colloid particles. Rapid mobility of both layer silicate clays and organic matter through the vadose zone has been observed (Jardine et al., 1989). Naidu et al. (1993) reported the presence of high concentrations of dissolved organics and suspended mineral particles in catchments in the Mount Lofty Ranges, South Australia, which facilitated the transport of metal ions, Fe and Al, in association with minerals, colloids and dissolved organics in subsurface waters.

5. Summary

The major mechanisms of metal retention in soils are: adsorption reactions where metal ions or species form inner or outer sphere complexes with the surfaces of organic and mineral components of soils; and the formation of inorganic and organic

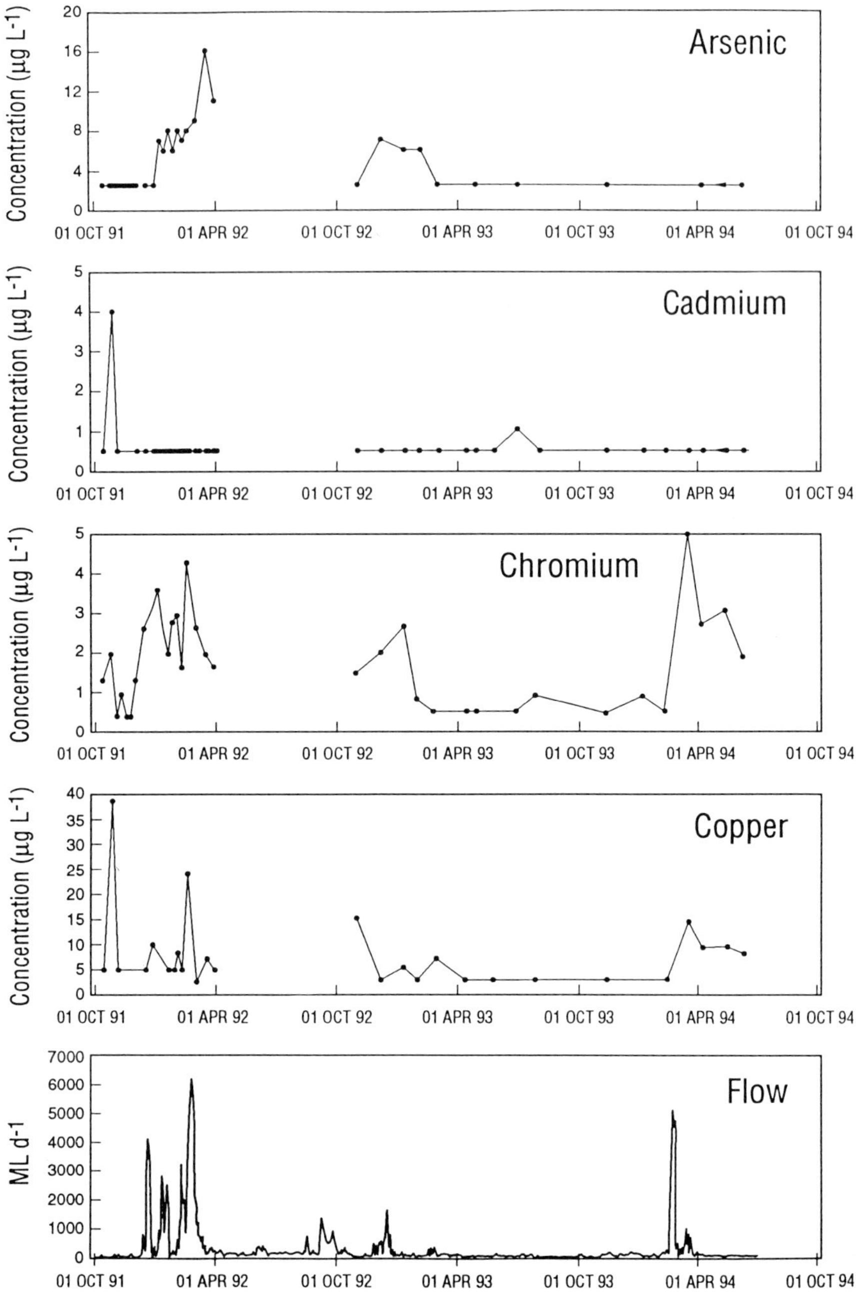

Figure 8. Time series plots illustrating the concentrations of As, Cd, Cr, Cu and volume of water flow in a catchment in Central North West region in New South Wales, Australia (Cooper et al., 1994).

precipitates. Any number of precipitates or adsorbed species may form in the soil-water system. However, the activity of the free metal ion at equilibrium is controlled by the reaction that results in its lowest value. Computer implemented models are available to estimate the equilibrium solution species and mineral assemblages in the soil-water system. The interpretation of such calculations is limited by the current lack of knowledge of reaction kinetics in the soil-water system.

References

Adriano D C 1986 Trace elements in the Terrestial Environment. Springer-Verlag, New York. p 533.

Allison J D and Brown D S 1995 MINTEQA2/PRODEFA2 – A geochemical speciation model and interactive preprocessor. In Chemical Equilibrium and Reaction Models. Eds. R H. Loeppert, A P Schwab and S Goldberg. pp 241-252. Soil Science Society of America-American Society of Agronomy, Madison, Wisconsin, USA.

Allison J D, Brown D S and Novo-Gradac K J 1991 MINTEQA2/PRODEFA2, a geochemical assessment model for environmental systems. EPA/600/3-91/021. US NTIS.

Barrow N J 1986 Testing a mechanistic model. IV. Describing the effect of pH on Zn retention by soils. J. Soil Sci. 37, 295-302.

Boekhold A E, Temminghoff E J M, and van der Zee S E A T M 1993 Influence of electrolyte composition and pH on cadmium sorption by an acid sandy soil. J. Soil Sci. 44, 85-96.

Bolan N S, Naidu R, Mahimairaja S and Baskaran S 1994 Influence of low molecular-weight organic acids on the solubilization of phosphorus. Biol. Fert. Soils 18, 311-319.

Braum S M and Helmke P A 1995 White lupin utilizes soil phosphorus that is unavailable to soybean. Plant and Soil 176, 95-100.

Brümmer G, Gerth J, and Tiller K G 1988 Reaction kinetics of the adsorption and desorption of Ni, Zn and Cd by goethite I. Adsorption and diffusion of metals. J. Soil Sci. 39, 35-52.

Cavallaro N and McBride M B 1978. Copper and cadmium adsorption characteristics of selected acid and calcareous soils. Soil Sci. Soc. Am. J. 42, 550-556.

Christensen T H 1984 Cadmium soil sorption at low concentrations: Effect of time, cadmium loading, pH and calcium. Water, Air and Soil Pollut. 21, 105-114.

Christensen T H 1987 Cadmium sorption at low concentrations: V. Evidence of competetion by other heavy metals. Water, Air and Soil Pollut. 34, 293-303.

Coles D G and Ramspott L D 1982 Migration of ruthenium-106 in a Nevada test site aquifer: Discrepancy between field and laboratory results. Science. 215, 1235-1237.

Cooper B, Graice J and Kimber J 1994 Central and North West Regions Water Quality Program. 1993/94 report on trace metals monitoring. NSW Department of Water Resources TS 94.090. p46.

Doner H E 1978 Chloride as a factor in mobilities of Ni(II), Cu(II) and Cd(II) in Soil. Soil Sci. Soc. Am. J. 42, 882-885.

Dzombak D A, Fish W and Morel F M M 1986 Metal-humate interactions 1. Discrete ligand and continous distribution models. Environ. Sci. Technol. 20, 669-675.

Evans L J 1989 Chemistry of metal retention by soils. Environ. Sci. Technol. 23, 1046-1056.

Fitch A and Helmke P A 1989 Donnan equilibrium/graphite furnace atomic absorption estimates of soil extract complexation capacities. Anal. Chem. 61, 1295-1298.

Forbes E A, Posner A M and Quirk J P 1976 The specific adsorption of divalent Cd, Co, Cu, Pb, and Zn on goethite. J Soil Sci. 27, 154-166

Florence T M and Bately G E 1980 Chemical speciation in natural waters. CRC Crit. Rev. Anal. Chem. August. pp 219-229.

Garcia-Miragaya J and Page A L 1976 Influence of ionic strength and inorganic complex formation on sorption of trace amounts of cadmium by montmorillonite. Soil Sci. Soc. Am. J 40, 658-663.

Goldberg S 1995 Adsorption models incorporated into chemical equilibrium models. In Chemical Equilibrium and Reaction Models. Eds. R H. Loeppert, A P Schwab and S Goldberg. pp 75-95. Soil Science Society of America-American Society of Agronomy, Madison, Wisconsin, USA.

Griffin R A, Au A K and Frost R R 1977 Effect of pH on adsorption of chromium from landfill leachate by clay minerals. J Environ. Sci. Hlth. 12, 431-449.

Harter R D and Naidu R 1995 Role of metal-organic complexation in metal sorption by soils. Adv. Agron. 55, 219-263

Helmke P A and Salam A K 1996 Free ion activities of Cu^{2+}, Cd^{2+}, Zn^{2+} as a function of Ca^{2+} plus Mg^{2+} free ion activities in water extracts of soil. Soil Sci. Soc. Am. J. (Submitted).

Helmke P A, Salam A K and Li Y. 1996. Measurement and behavior of indigenous levels of the free, hydrated cations of Cu, Zn and Cd in the soil-water system. In Proceedings of the Third International Conference on the Biogeochemistry of Trace Elements in the Environment. Ed. R. Prost. Paris, France. (In Press).

Hirsch D and Banin A 1990 Cadmium speciation in soil solutions. J. Environ. Qual. 19, 366-372.

Hodgson J F, Tiller K G and Fellows M 1964 The role of hydrolysis in the reaction of heavy metals with soil-forming materials. Soil Sci. Soc. Am. Proc. 28, 42-46.

Holm P E, Christensen T H, Tjell J C and McGrath S P 1995a Heavy metals in the environment: Speciation of cadmium and zinc with application to soil solutions. J. Environ. Qual. 24, 183-190.

Holm P E, Anderson S and Christensen T H 1995b Speciation of dissolved cadmium: Interpretation of dialysis, ion exchange, and computer (GEOCHEM) methods. Geochem. Cosmochem. Acta. 29, 803-809.

Homann P S and Zasoski R J 1987 Solution composition effects on cadmium sorption by forest soil profiles. J Environ. Qual. 16, 429-433.

James B R and Bartlett R J 1983 Behaviour of chromium in soils: VI. Adsorption and reduction of hexavalent forms. J Environ. Qual. 12, 177-181.

Jardine P M, Wilson G V, Luxmoore R J and McCarthy J F 1989 Transport of inorganic and natural organic traces through an isolated pedon in a forest watershed. Soil Sci. Soc. Am. J. 53, 317-323.

Kaplan D I P, Bertsch, P M, Adriano D C and Miller W P 1993 Soil-borne mobile colloids as influenced by water flow and organic carbon. Environ. Sci. Technol. 27, 1193-1200.

Karickhoff S W, Brown D S and Scott T A 1979 Sorption of hydrophobic pollutants on natural sediments. Water Res. 13, 241-248.

Kinniburgh D G, Sridhar K and Jackson M L 1977 Specific adsorption of zinc and cadmium by iron and aluminium hydrous oxides, p. 231-239. Hanford Life Science Symposium 15th on biological implications of metals in the environment, Hanford.

Kinniburgh D G and Jackson M L 1981 Cation adsorption by hydrous metal oxides and clay. In Adsorption of Inorganics of Solid-Liquid Interfaces. Eds. M A Anderson and A J Rubin. pp 91-160. Ann Arbor Science, Ann Arbor, Michigan.

Kookana R S and Naidu R 1995 Vertical heterogeneity in soil properties and contaminant transport through soil profiles. In IAH-ICH Volume on Shallow Groundwater Systems. Ed. I Simmers. Heise Verlag. In Press.

Lindsay WL 1979 Chemical equilibria in soils. Wiley Interscience, New York. 448 p.

Ma Q and Lindsay W L 1990 Divalent zinc activity in arid zone soils obtained by chelation. Soil Sci. Soc. Am J. 54, 719-722.

Ma Q Y and Lindsay W L 1993 Measurement of free $zinc^{2+}$ activity in uncontaminated and contaminated soils using chelation. Soil Sci. Soc. Am. J. 57, 963-967.

Matijevic E 1973 Colloid stability and complex chemistry. J Coll. Interface Sci. 43, 217-244.

Mattigod S V 1995 Validation of geochemical equilibrium models. In Chemical Equilibrium and Reaction Models. Eds. R H Loeppert, A P Schwab and S Goldberg. pp 201-218. Soil Science Society of America-American Society of Agronomy, Madison, Wisconsin, USA.

Mattigod S V, Sposito G and Page A L 1981 Factors affecting the solubilities of trace metals in soils. In Chemistry in the soil environment. Eds. R H Dowdy, J A Ryan, V V Volk and D E Baker. pp. 203-221. American Society of Agronomy, Madison, WI, USA.

McCarthy J F and Zachara J M 1989 Subsurface transport of contaminants: Binding to mobile and immobile phases in groundwater aquifers. Environ. Sci. Technol. 23, 496-504.

McDowell-Boyer L M, Hunt J R and Sitar N 1986 Particle transport through porous media. Water Resour. Res. 22, 1901-1921.

McLaughlin M J, Palmer L T, Tiller K G, Beech T A and Smart M K 1994 Increased soil salinity causes elevated cadmium concentrations in field-grown potato tubers. J Environ. Qual. 23, 1013-1018.

Milne C J, Kinniburgh D G, DeWit J C M, Van Riemsdijk W H and Koopal L K 1995 Analysis of proton binding by a peat humic acid using a simple electrostatic model. Geochim. Cosmochim. Acta 59, 1101-1112.

Naidu R, Syers J K, Tillman R W and Kirkman J H 1990 Effect of liming and added phosphate on charge characteristics of acid soils. J Soil Sci. 41, 157-164.

Naidu R, Williamson D R, Fitzpatrick R W, and Hollingsworth I 1993 Effect of land use on the composition of throughflow water immediately above the clayey B horizons in the Warren Catchment. Aust. J. Exper. Agric. 33, 239-244.

Naidu R, Bolan N S, Kookana R S and Tiller K G 1994a Ionic strength and pH effects on the surface charge and sorption of cadmium by soils. J. Soil Sci. 45, 419-429
Naidu R, deLacy N J, Kookana R S, Bolan N S and Tiller K G 1994b Effect of inorganic ligands on adsorption of cadmium by soils. In Proceedings of the 15th International Soil Science Congress, Acapulco, Mexico. International Society of Soil Science.
Oglesby J 1995 Competitive-chelation methods for estimating free metal ion activities. Soil Sci. Soc. Am. J. 59, 959.
Parks G A 1965 The isoelectric points of solid oxides, solid hydroxides and aqueous hydroxo complex systems. Chem. Rev. (Lond.) 65, 177-198.
Parker D R, Norvell W A and Chaney R L 1995 GEOCHEM-PC – A chemical speciation program for IBM and compatible personal computers. In Chemical Equilibrium and Reaction Models. Eds. R H. Loeppert, A P Schwab and S Goldberg. pp 253-269. Soil Science Society of America-American Society of Agronomy, Madison, Wisconsin, USA.
Petruzelli G, Guidi G and Lubrano L 1985 Ionic strength effects on heavy metal adsorption by soils. Commun. in Soil Sci. and Plant Anal. 9, 971-986.
Puls R W and Bohn H L 1987 Sorption of cadmium, nickel and zinc by kaolinite and montmorillonite suspensions. Soil Sci. Soc. Am. J. 52, 1289-1292.
Puls R W and Powell R M 1992 Transport of inorganic colloids through natural aquifer material: Implications for contaminant transport. Environ. Sci. Technol. 26, 614-621.
Reynolds W D, Gillham R W and Cherry J A 1982 Evaluation of distribution coefficients for the prediction of strontium and cesium migration in a uniform sand. Can. Geotech. J. 19, 92-103.
Schildler P W, Liechti P and Westall J C 1987 Adsorption of copper, cadmium and lead from aqueous solution to kaolinite/water interface. Nether. J Agric. Sci. 35, 219-230.
Schnitzer M and Kahn S U 1978 Soil Organic Matter. Elsevier, Amsterdam. 319 p.
Sposito G, Holtzclaw K M, Charlet L, Jouany C and Page A L 1983 Sodium-calcium and sodium-magnesium exchange on Wyoming bentonite in perchlorate and chloride background ionic media. Soil Sci. Soc. Am. J. 47, 51-56.
Sposito G and Coves J 1995 SOILCHEM on the Macintosh. In Chemical Equilibrium and Reaction Models. Eds. R H. Loeppert, A P Schwab and S Goldberg. pp 271-287. Soil Science Society of America-American Society of Agronomy, Madison, Wisconsin, USA.
Susetyo W, Carreira L A, Azarraga L V and Grimm D M 1991 Fluorescence techniques for metal-humic interactions. Fresenius J. Anal. Chem. 339, 624-635.
Tiller K G and Hodgson J F 1962. The specific sorption of cobalt and zinc by layer silicates. Clays Clay Miner. 9, 393-403.
Tiller K G, Nayyar V K and Clayton P M 1979 Specific and non-specific sorption of Cd by soil clays as influenced by zinc and calcium. Aust. J Soil Res. 17, 17-28.
van Raij B and Peech M 1972 Electrochemical properties of some oxisols and alfisols of the tropics. Soil Sci. Soc. Am. Proc. 36, 587-593.
Wersin P, Charlet L, Karthein R and Stumm W 1989 From adsorption to precipitation: Sorption of Mn^{2+} on $FeCO_3$. Geochim. Cosmochim. Acta. 53, 2787-2796.
Wolt J D 1994 Soil Solution Chemistry. John Wiley & Sons, New York, USA. 345 p.
Westall J C, Jones J D, Turner G D and Zachara J M 1995 Models for association of metals ions with heterogeneous environmental sorbents 1. Complexation of Co(II) by leonardite humic acid as a function of pH and $NaClO_4$ concentration. Environ. Sci. Technol. 29, 951-959.
Workman S M and Lindsay W L 1990 Estimating divalent cadmium activities measured in arid-zone soils using competitive chelation. Soil Sci. Soc. Am. J. 54, 987-993.
Zhu B and Alva A K 1993 Differential adsorption of trace metals by soils as influenced by exchangeable cations and ionic strength. Soil Sci. 155, 61-66.

4

Transport and fate of organic contaminants in the subsurface

MARK L. BRUSSEAU AND RAI S. KOOKANA

1. Introduction

The transport and fate of contaminants in subsurface systems has become one of the major research areas in the environmental/hydrological/earth sciences. This interest has been instigated by concerns associated with the effects of human activities on the environment. Numerous examples may be cited that illustrate the adverse impact of human activities on the subsurface environment, such as contamination of soil and groundwater by chemicals associated with industrial and commercial operations, service stations, waste disposal facilities, and agricultural production. An overview of the impact of such activities on the quality of the soil environment in the Australasia-Pacific Region is presented in Chapters 14 to 25.

An understanding of how contaminants move in the subsurface soil environment is required to address environmental problems. For example, such knowledge is needed to evaluate the probability that contaminants associated with a chemical spill will reach an aquifer and contaminate groundwater. Such knowledge is also required to develop and evaluate methods for cleaning up contaminated soils and aquifers. Of equal importance, knowledge of contaminant transport and fate is necessary to design 'pollution-prevention' strategies.

A tremendous amount of research on the transport and fate of contaminants in porous media has been generated in response to these needs. This research includes the results of theoretical studies designed to pose and evaluate hypotheses, the results of experiments designed to test hypotheses and investigate processes, and the development and application of mathematical models useful for integrating theoretical and experimental results and for evaluating complex systems. The purpose of this paper is to briefly review, in a selective manner, some of the major aspects of transport and fate of organic contaminants in soil and groundwater. Volatilisation and runoff aspects have not been covered here and readers are referred to reviews by Taylor and Spencer (1990) on volatilisation and by Leonard (1990) on surface transport. Some sections of this chapter are based on a recent review (Brusseau, 1994).

2. Factors influencing transport

The results of laboratory, field, and mathematical experiments indicate that the transport of organic contaminants at the field scale is characteristically nonideal. In other words, often-used assumptions of homogeneity and linear, instantaneous mass

R. Naidu et al. *(eds): Contaminants and the Soil Environment in the Australasia-Pacific Region, 95–124.*

transfer and reactions are often invalid. Furthermore, it is possible, even probable, that more than one factor may contribute to nonideal transport of organic solutes at the field scale. Some of the major factors that influence contaminant transport will be briefly discussed in the following sections.

2.1 Sorption mechanisms

Sorption of organic compounds in soils, including both adsorption on and absorption in the soil matrix, is a factor that strongly influences chemical transport in the environment. A variety of physical/chemical mechanisms, such as van der Waals forces, H-bonding, dipole-dipole interactions, ion exchange, covalent bonding, ligand exchange, cation bridging, and water bridging are responsible for sorption of organic compounds in soils. Several of these mechanisms or forces may operate in tandem, causing the sorption of a compound to be complex. The importance of these mechanisms for various types of organic compounds has recently been discussed by Koskinen and Harper (1990). Characteristics such as polarity, ionic nature, functional group content, and solubility of the organic compounds determine the nature of bonding mechanisms, and the extent as well as the strength of their sorption by the soil solid phase.

The retention of non-ionic, low-polarity organic compounds in soils is generally considered to be governed by entropy driven 'hydrophobic' interactions, wherein the solute 'partitions' between the aqueous phase and the soil organic matter (OM) (Chiou et al., 1979; Karickhoff et al., 1979). The role of OM as the controlling sorptive phase for many systems has led to the use of sorption (K) coefficients based on organic carbon: $K_{oc} = f_{oc} K$, where f_{oc} is the fraction of organic carbon. Several such equations have been compiled and discussed by Gerstl (1989) and Green and Karickhoff (1990). These equations are generally applicable for non-ionic organic compounds of limited aqueous solubility ($< 10^{-3}$ M) and for systems where OM controls sorption. There are a number of simplifying assumptions inherent to these equations and they should be viewed as nothing more than gross approximations of solute behaviour.

The sorption of non-ionic, low-polarity organic compounds has been represented by the use of a number of analogues or model systems. These phenomenological models have been used to investigate sorption mechanisms and to provide a means of estimating sorption parameters. Generally, use of the phenomenological models is based on the assumption that the OM components of a soil are the predominant sources of retention of non-ionic, low polarity organic compounds. The first significant analogue was that of an 'oily phase', where sorption of organic compounds by soil OM was likened to a partitioning of an organic solute between water and an immiscible liquid. The immiscible liquid analogue (e.g. octanol/water partitioning) has been used extensively to represent sorption by soil. Another analogue, the use of which has increased recently, is that of the stationary phases used in chromatography, where the mass transfer of solute between mobile and stationary phases is considered to be representative of sorption by soil. Whereas different types of phases have been suggested, the most widely used are those employed in reversed phase high performance liquid chromatography (RPLC). Polymer solution theories and polymer permeation theories have recently been used to explain the process of sorption of organic compounds by soils.

Recently, quantitative structure-activity relationship (QSAR) analysis was used by Brusseau (1993a) to evaluate the validity of the phenomenological models discussed above. The properties exhibited by the model systems (octanol/water, RPLC sorbent, polymer) were compared to those determined for the sorption of 29 organic compounds by two soils. The correlation patterns exhibited by the soil data were similar to the patterns exhibited by the polymer systems and dissimilar to those exhibited by the octanol and RPLC systems. In addition, the correlation pattern between the sorption rate coefficient and molecular connectivity was similar to that between polymer diffusion coefficients and molecular connectivity. Hence, it appears that the polymer analogue may be the most appropriate of the three models for representing both equilibrium and nonequilibrium sorption by soil. This is consistent with the results of recent research, which has demonstrated that rate limited sorption/desorption of low polarity organic compounds is caused by constrained diffusion within soil OM for systems wherein organic carbon controls sorption (Brusseau and Rao, 1991; Brusseau et al., 1991a).

2.2 *Soil and solution properties affecting sorption of organic compounds*

Organic matter and clay contents of a soil are undoubtedly the main substrates for sorption of organic chemicals (Green and Karickhoff, 1990; Koskinen and Harper, 1990). As discussed above, the organic fraction of the solid phase in soil has often been reported as the controlling phase for the sorption of organic compounds, especially for non-ionic organic compounds. Clay fractions in soils can also be important for sorption of organic compounds, especially since their contents are usually an order of magnitude higher than OM in soils. The physical/chemical characteristics of OM and clay are important factors for sorption of organics. The OM in soils differs widely in terms of the degree of humification as well as in their composition (e.g. the proportions of fats, waxes, resins, proteins and carbohydrates present in OM) (Stevenson, 1982). For example, humic materials of grassland soils are rich in humic acids whereas fulvic acids dominate in forest soils. The sorption affinities/capacities of humic and fulvic acids have been found to be different. The type of organic material and its decomposition stage both have been recognised as important in determining the sorption of some organic compounds. It is apparent that OM in different soils can behave quite differently in terms of the retention of organic compounds and consequently the transferability of data between soils may not always be possible.

The extent as well as reversibility of sorption processes in soils can also be greatly influenced by other factors such as pH, solute concentration, or the presence of organic cosolvents and competitive ions. The speciation of ionisable organic compounds is dependent on the pH of the system. For example, organobasic com-

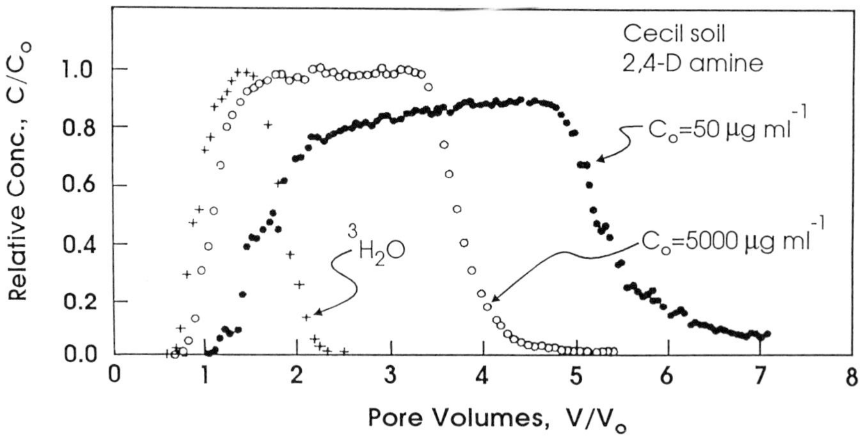

Figure 1. Illustration of the effect of non-linear sorption on contaminant transport. Breakthrough curves for 2,4-dichlorophenoxyacetic acid and tritiated water transport though a loamy soil; figure reproduced from Rao and Davidson (1979) with permission from Elsevier Science.

pounds (e.g. amines, s-triazines) are cations at low pH and this can result in increased sorption. Acidic compounds (e.g. phenols, aliphatic acids, phenoxy acid herbicides, sulphonylurea herbicides), which are proton donors and are predominantly anionic at neutral pHs and above, usually show a decrease in sorption with increasing pH. In addition, ion exchange capacities of both OM and clays change with pH. The influence of cosolutes and cosolvents on sorption will be discussed in a forthcoming section.

2.3 *Non-linear, rate-limited sorption/desorption processes*

Most solute transport models, especially for field scale applications, include the assumption that the distribution of solute between liquid and solid phases can be described as a linear process. Such a condition would be reflected, for example, in linear sorption isotherms (a regression of sorbed phase concentrations against solution phase concentrations). Numerous laboratory experiments have shown that sorption is often linear for low polarity organic compounds. However, this assumption should be evaluated for each case because non-linear isotherms have been reported (Ball and Roberts, 1991a; Weber et al., 1992). Non-linear isotherms are the norm rather than the exception for many polar organic chemicals. Several equations exist for describing sorption isotherms (see Travis and Etnier, 1981 and Kinniburgh, 1986 for reviews).

It is well known that non-linear sorption can cause asymmetrical breakthrough curves and concentration-dependent retardation factors (Crittenden et al., 1986; Brusseau and Rao, 1989). This was clearly demonstrated by Rao and Davidson (1979), who reported breakthrough curves for transport of 2,4-dichlorophenoxyacetic acid through two soils (Figure 1). Retardation increased as the influent solute

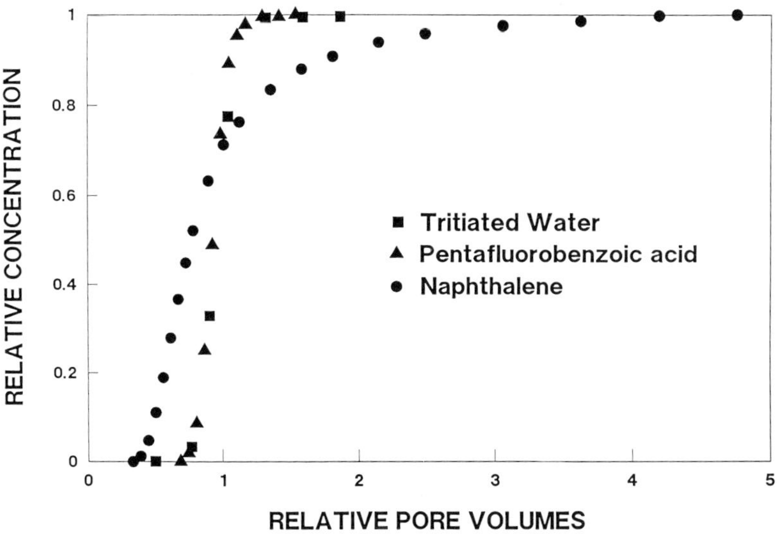

Figure 2. Illustration of the effect of rate-limited sorption/desorption on contaminant transport. Relative pore volume is defined as pore volume divided by the retardation factor; the retardation factors are: tritiated water = 1, pentafluorobenzoic acid = 1, naphthalene = 4.8. The use of relative pore volumes facilitates direct comparison of breakthrough curves by normalizing for differences in retardation. Data from Brusseau et al. (1991a).

concentration decreased. In addition, the elution waves were much more dispersed than were the injection waves. Recent research has examined the relative importance of non-linear sorption compared to other factors such as soil heterogeneity, rate limited sorption, and transformation (Nkedi-Kizza et al., 1989; Rabideau and Miller, 1994; Brusseau, 1995).

Most field-scale solute transport models include the 'local equilibrium assumption', in which sorption is considered instantaneous. Based on laboratory experiments, it is known that sorption/desorption of many organic compounds by soils and aquifer materials can be significantly rate limited (Curtis et al., 1986; Lee et al., 1988; Ball and Roberts, 1991a; Brusseau et al., 1991a; Brusseau et al., 1991b; Kookana et al., 1992; Kookana et al., 1993).

The mechanisms responsible for rate-limited sorption/desorption have been a focus of recent research. For low polarity organic compounds, the current view is that nonequilibrium is caused by constrained intrasorbent diffusion. While rate-limited specific chemical reactions (chemisorption) are not likely to be important for most low polarity organic compounds, such reactions can be important for ionic and polar organic compounds.

Numerous experimental and theoretical studies have shown that rate-limited sorption/desorption can cause nonideal transport. This nonideality can take the form of asymmetrical breakthrough curves exhibiting early breakthrough and tailing, as well as decelerating plumes (temporally increasing R values). An example of the effect of rate-limited sorption/desorption on transport is presented in Figure 2,

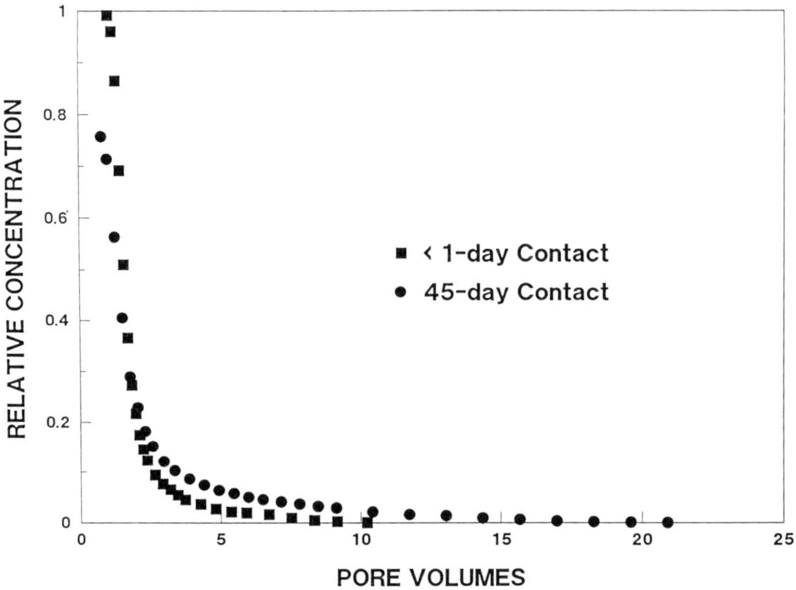

Figure 3. Illustration of the effect of contact time on contaminant desorption/elution. For one experiment naphthalene was in contact with the sandy soil for a few hours prior to elution. In the other experiment, naphthalene was in contact with the soil for 45 days. Data from Brusseau et al. (1991a).

which shows breakthrough curves for transport of several solutes through a sandy soil. Note that the breakthrough curves for the nonreactive tracers (^3H_2O and pentafluoro-benzoate) are sharp and symmetrical, signifying ideal behaviour. In contrast, the breakthrough curve for the sorbing organic solute (naphthalene) is asymmetrical. This suggests that the nonideality factor involves the sorption process.

A phenomenon that, in many cases, is related to rate-limited sorption/desorption is so-called contaminant 'aging'. Recent research has shown that contaminants that have been in contact with porous media for long times are much more resistant to desorption, extraction, and degradation. For example, contaminated samples taken from field sites exhibit solid:aqueous distribution ratios that are much larger than those measured or estimated based on spiking the porous media with the same contaminant (e.g. adding contaminant to uncontaminated sample) (Steinberg et al., 1987; Pignatello et al., 1990; Smith et al., 1990; Scribner et al., 1992). In addition, the magnitudes of the desorption rate coefficients determined for previously contaminated media collected from the field have been shown to be much smaller than the values obtained for spiked samples (Steinberg et al., 1987; Connaughton et al., 1993). These field-based observations are supported by laboratory experiments wherein measured values of desorption rate coefficients have been observed to decrease with increasing time of contact prior to desorption (Karickhoff, 1980; McCall and Agin, 1985; Coates and Elzerman, 1986; Brusseau et al., 1991a).

The effect of 'aging' on contaminant desorption is illustrated in Figure 3. Here, the elution curve obtained for a case where the contaminant was in contact with the porous medium for several hours (usual experimental conditions) is compared to a case with a 45-day contact period. It is evident that the elution curve for the long contact case exhibits greater tailing, reflecting a greater resistance to removal.

2.4 *Multiphase behaviour*

Immiscible organic liquids may exist in portions of the contaminated subsurface at many hazardous waste sites. Examples include gasoline, diesel fuel, and aviation fuel associated with various types of fuel stations, chlorinated solvents associated with cleaning facilities, and creosote associated with wood treatment operations. The movement of immiscible liquids in porous media, which has become a major component of contaminant transport research, has received considerable attention in the petroleum engineering field. It is well known that when an immiscible liquid moves through a porous medium, a fraction of the liquid can become entrapped thereby forming a residual saturation. Very large pore-water velocities (i.e. hydraulic gradients) are usually required to displace residual quantities of immiscible liquids (Wilson and Conrad, 1984; Willhite, 1986; Hunt et al., 1988). Hence, the primary means of removal will be dissolution into water and volatilisation into the soil atmosphere. The residual liquid, therefore, serves as a long-term source of contaminant. Readers are directed to Chapter 12 for a detailed description on transport behaviour of non-aqueous phase liquids.

Most models developed to simulate the dissolution of immiscible liquids are based on an assumption that the phase transfer is instantaneous. The results of several laboratory experiments have suggested that mass transfer between immiscible liquid and water is relatively rapid (van der Waarden et al., 1971; Fried et al., 1979; Schwille, 1988; Miller et al., 1990). Other investigations, however, have shown that liquid-liquid mass transfer can be significantly rate limited, especially under conditions that may be found in the field (Hunt et al., 1988; Powers et al., 1991; Powers et al., 1992; Brusseau, 1992a). If the dissolution of immiscible liquid into groundwater is rate limited, the time required to remove the contaminant mass could be greatly increased.

While some of the most widely studied immiscible liquids are composed of a single component (e.g. trichloroethene), many others (e.g. gasoline, diesel fuel, coal tar) are multicomponent liquids. Knowledge of the partitioning behaviour of multicomponent liquids is essential to the prediction of their impact on groundwater quality. The partitioning of components into water is controlled by the aqueous solubility of the component and the composition of the immiscible liquid. The actual maximum concentration of a contaminant in a water phase in contact with a multicomponent immiscible liquid will be less than the aqueous solubility. Hence, ignoring the multicomponent effect will lead to erroneous estimates of aqueous phase concentrations.

Many organic contaminants of interest are volatile (e.g. gasoline components, chlorinated solvents). For systems that are fully saturated with water (e.g. aquifers), this is of no special concern. However, it is of great significance for the vadose zone which, because it is usually not completely saturated with water, contains a gas phase. Consequently, a description of the transport of organic contaminants in the vadose zone will be incomplete without consideration of transport in the gas phase. Research in this area has increased greatly with the advent of 'soil venting' or 'vacuum extraction' as a popular means of remediating vadose zone contamination. (See Chapter 12).

One aspect of gas phase transport that has been most often considered is gas phase diffusion. For transport in flowing water systems, it is usually assumed that longitudinal diffusion is of minor importance compared to advective transport. Conversely, gas phase diffusive flux can be of great significance because gas phase diffusion coefficients are much larger than aqueous diffusion coefficients. The relatively rapid transport by gas phase diffusion can lead to larger than expected zones of contamination if such transport is ignored (Baehr, 1987; Sleep and Sykes, 1989). Advective transport in the gas phase, caused either by external pressure gradients (e.g. soil venting) (Massmann, 1989) or by gas phase density differences (Falta et al., 1989; Mendoza and Frind, 1990), can also be important.

The first generation of mathematical models designed to simulate gas phase contaminant transport were based on assumptions of ideal transport (Johnson et al., 1987; Baehr et al., 1989). Similarly to the case of water phase transport, it is to be expected that gas phase transport models based on ideal transport will be invalid for most real conditions. A generation of models that incorporate nonideal transport phenomena are beginning to appear in the literature. For example, a gas-phase advection model that included rate-limited sorption was presented by Brown and Rolston (1980). A model that included a dual-porosity description for aggregated soils was presented by Gierke et al. (1992). A model presented by Brusseau (1991a) was developed to explicitly simulate gas phase transport influenced by multiple non-ideality factors. The model described gas phase advection-dispersion in a structured or heterogeneous porous medium, rate-limited transfer of contaminant between water-saturated regions of low permeability and water in direct contact with the gas phase, mass transfer between the water and gas phases, and rate-limited sorption/desorption.

2.5 Multi-component solutions and facilitated transport

Many contaminated sites have multiple contaminants in the aqueous phase. The existence of multi-component solutions leads to the possibility of nonideal behaviour wherein one component may influence the behaviour of another. Naturally occurring subsurface components associated with the fluid phase may also influence the behaviour of contaminants. When multiple components are present in solution, a primary question to be addressed is the occurrence of antagonistic or synergistic interactions among the components, and between the components and the solid and aqueous phases. The presence of co-components can affect the behaviour of contaminants in several ways, resulting in the following three phenomena (Brusseau, 1993): (a) competitive solid-phase interactions; (b) cooperative solid-phase interactions; and (c) 'solubilisation'. The first and third phenomena reduce sorption and thus enhance the transport of solutes, whereas cooperative interactions have the opposite effect.

Competitive sorption, where sorption of a solute is reduced by the presence of a cosolute, has been investigated by several researchers and their results have generally shown no competition for non-ionic, low polarity organic solutes such as naphthalene and chlorinated benzenes (Karickhoff et al., 1979; Chiou et al., 1983). Indeed, non-competition is considered a defining characteristic of the sorption of nonionic, low-polarity organic solutes by a 'partitioning' mechanism (Chiou et al., 1983). However, some researchers have reported relatively small decreases in sorption resulting from competition (MacIntyre and deFur, 1985; Abdul and Gibson, 1986; McGinley et al., 1993). The sorption of ionic organic compounds can be greatly influenced by competitive effects in multi-component solutions.

The enhanced sorption of low-polarity organic solutes in the presence of cationic surfactants is one example of cooperative sorption in multi-component solutions. The cationic surfactant is readily sorbed by the soil, which effectively increases the sorptive phase for the organic solute, thereby increasing sorption (Smith and Bayer, 1967; Jaynes and Boyd, 1990). Cooperative sorption of low-polarity organic solutes in the presence of other low-polarity organic solutes has been studied by few researchers. Brusseau (1991b) investigated the effect of a low-polarity cosolute (tetrachloroethene) on the sorption of three low-polarity organic chemicals (naphthalene, p-xylene, 1,4-dichlorobenzene) by two aquifer materials with small organic carbon contents ($<0.03\%$). In all cases, the sorption of the primary solute was enhanced by 1.5 to 3 times in the presence of high concentrations of tetrachloroethene.

Competitive and cooperative sorption result primarily from solute-sorbent interactions. In contrast, solubilisation results from interactions between the solute and the co-component. For our purposes, solubilisation is defined as an increase in the amount of contaminant that can reside in the aqueous phase of a multi-component system beyond the solubility limit. Examples of solubilisation include the effect of miscible solvents (i.e. the 'co-solvent' effect), surfactants, dissolved organic matter (DOM), and colloids (the 'colloid' effect) on the solubility, sorption, and transport of low-polarity organic solutes. The presence of a co-solvent (e.g. ethanol), anionic or nonionic surfactant, DOM, or colloid has been shown to increase the apparent solubility of many low-polarity organic compounds. This, in turn, causes reduced sorption and retardation, thereby 'facilitating' the transport of the low-polarity organic compound. Complexation by colloids, DOM, or surfactants can lead to facilitated transport of inorganic compounds. Further discussion of these phenomena may be found in recent reviews (McCarthy and Zachara, 1989; Sabatini and Knox, 1992; Brusseau, 1993b).

2.6 *Transformation reactions*

Many contaminants may undergo various types of transformation reactions during transport. Examples of such reactions include radioactive decay, hydrolysis, and

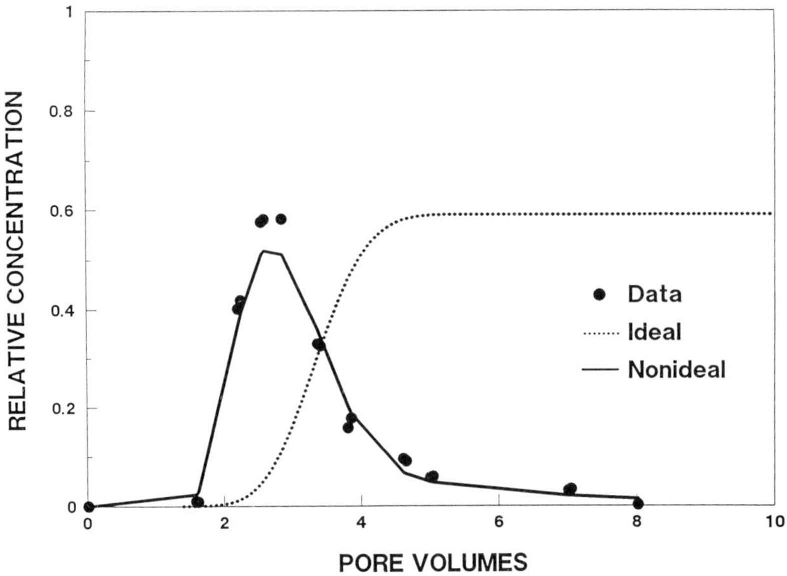

Figure 4. Illustration of the effect of microbial growth on biodegradation and transport of 2,4-dichlorophenoxyacetic acid through a soil column partially saturated with water. The nonideal simulation accounts for rate-limited sorption and microbial growth, whereas the ideal curve is based on instantaneous sorption and a non-growing microbial population. Figure adapted from Estrella et al. (1993).

biodegradation. A recent review of mathematical models designed to simulate transport of solutes influenced by transformation reactions revealed that the vast majority were based on the use of a first order transformation reaction (Brusseau et al., 1992a). First order kinetics, wherein the rate is a function only of solute concentration, accurately describes some transformation reactions (e.g. radioactive decay, pH-independent hydrolysis) but not others. Biodegradation is a good example of a transformation process that can be described as first order only for highly constrained conditions. Often, the rate of biodegradation of a contaminant will depend on many other factors (e.g. biomass growth, electron acceptor availability, sorption, environmental conditions) in addition to contaminant concentration. Contaminant transport under these conditions will not be describable with a model based on ideal transport (i.e. first order reaction).

The effect of nonsteady state microbial populations on contaminant transport is an example of transformation related nonideality. This effect is illustrated in Figure 4, which shows the transport of a pesticide (2,4-dichlorophenoxyacetic acid, 2,4-D) through soil packed in a column (Estrella et al., 1993). The soil contained a bacterial population capable of degrading the 2,4-D as demonstrated by separate microcosm experiments. A simulated breakthrough curve obtained for ideal conditions, including first order biodegradation, is also shown in Figure 4. Clearly, the biodegradation and transport of 2,4-D could not be simulated with the ideal model. One condition associated with the assumption of first order biodegradation kinetics is that there is no microbial growth. This condition is inconsistent with the data, which shows a steadily decreasing concentration of 2,4-D in the effluent to a point of about 8 pore volumes, whereupon 2,4-D was no longer detectable. It is important to note that the decline and disappearance of 2,4-D in the effluent occurred while the solution containing 2,4-D was still being injected into the column. This behaviour indicates the occurrence of temporally non-uniform biological activity. A model that accounted for, among other factors, microbial growth could successfully simulate the data (see Figure 4). Results similar to these have been reported by Chen et al. (1992).

The effects of interactions between sorption and biodegradation, two rate-controlled processes, on transport is another example of nonideal transport. Recent research performed with batch (i.e. non-flowing) systems has suggested that the rate of biodegradation can be controlled by the rate of desorption or mass transfer (Rijnaart et al., 1990; Robinson et al., 1990; Scow and Alexander, 1992), depending on the relative time scales of the two processes. When the rate of desorption or mass transfer is slower than that of biodegradation, the transport of the contaminant will not follow the behaviour expected based on ideal conditions. This has been shown by Brusseau and colleagues, who have investigated the influence of rate-limited sorption/desorption (Angley et al., 1992; Estrella et al., 1993) and heterogeneous porous media combined with rate-limited sorption/desorption (Brusseau et al., 1992b) on the transport of contaminants undergoing biodegradation.

2.7 Spatial variability of soil properties

The subsurface is inherently heterogeneous and, consequently, physical, chemical, and biological properties are spatially variable. A major parameter of interest is hydraulic conductivity. The influence of spatially variable hydraulic conductivity fields on water flow and contaminant transport in porous media has been a major research topic for many years. The impact of hydraulic-conductivity variability on transport is often discussed in terms of the 'scale effect', wherein apparent dispersivity values measured for field-scale transport are usually much larger than those measured for transport in packed columns (Fried, 1975; Anderson, 1979). A clear description of how spatially variable hydraulic conductivities could cause enhanced solute dispersion (compared to that measured in the laboratory) was given in the early 1960's (Theis, 1963; Warren and Skiba, 1964). Since then, a tremendous amount of research has demonstrated that the additional dispersion, often called macro- or full-aquifer dispersion, observed for field-scale transport of nonreactive solute is often primarily a result of large-scale hydraulic conductivity variability (Schwartz, 1977; Gelhar et al., 1979; Molz et al., 1983).

Nonideal transport, often referred to as physical nonequilibrium, can result from a structured or heterogeneous flow domain at the macroscopic scale (10^{-3} to 1 m). The existence of regions of smaller hydraulic conductivity within the flow domain creates a spatially variable velocity field, with minimal flow and advection occurring through the small conductivity domains. Due to the small advective flux, these domains act as sink/source components, with rate-limited diffusional mass transfer between the advective and nonadvective domains causing dispersion of the solute front. These sink/source regions can take various forms, including the internal porosity of aggregates, dead-end pores, the bulk matrix of fractured media, and the small hydraulic conductivity microlayers or laminae typically found in aquifers of sedimentary origin.

During the last few decades, the effect of aggregated or macroporous media on solute transport has been well documented (see Brusseau and Rao, 1990 for a recent review). Miscible displacement studies performed with well characterised aggregated or macroporous media packed in columns clearly demonstrate that these structures cause asymmetrical and tailed breakthrough curves and enhanced dispersion (Biggar and Nielsen, 1962; van Genuchten and Wierenga, 1977; Rao et al., 1980; Nkedi-Kizza et al., 1983). Experiments performed with undisturbed cores or lysimeters demonstrate that naturally occurring structures can cause nonideal solute transport (McMahon and Thomas, 1974; Richter and Jury, 1986; Seyfried and Rao, 1987). It has also been shown that nonideal behaviour can occur during transport of solutes through fractured media (Grisak et al., 1980; Bibby, 1981; Neretniks, 1983).

Due to the heterogeneity of subsurface systems, it is logical to expect sorption to be spatially variable. Several field-scale investigations have shown that this is indeed the case (Pickens et al., 1981; Elabd et al., 1986; Mackay et al., 1986a; Rao et al., 1986; Wood et al., 1987; Bunzl and Schimmack, 1988; Robin et al., 1991). Depending on the form of the spatial variability, non-uniform sorption may cause nonideal transport. Smith and Schwartz (1981), with a series of modelling exercises, evaluated the influence of spatially variable cation exchange capacity (i.e. variable retardation) on solute transport. They assumed that sorption was negatively correlated with hydraulic conductivity. Dispersion was observed to increase as the variability in retardation increased. However, they concluded that this effect would be of secondary importance in comparison to the effect of spatially variable hydraulic conductivity. Spatially variable sorption was proposed as one cause of the enhanced dispersion exhibited by Li, in comparison to Br, in a recent natural gradient experiment performed at Cape Cod, MA. In a theoretical analysis of the effect of coupled physical and chemical heterogeneity, Garabedian et al. (1988) observed enhanced dispersion compared to the uniform sorption case when a negative correlation between sorption and hydraulic conductivity was assumed. Similar results, based on theoretical and modeling exercises, have been reported by several other researchers (Bahr, 1986; Valocchi, 1989; Cvetković and Shapiro, 1990; Andricevic and Foufoula-Georgiou, 1991; Kabala and Sposito, 1991).

There have been few experimental investigations of the impact of spatially variable sorption on solute transport. One such study was reported by Brusseau and Zachara (1993), who investigated the transport of Co^{2+} in a column packed with layers of two media of differing hydraulic conductivities and sorption capacities. The sorption capacities of the two media differed by about a factor of three. The asymmetrical breakthrough curve obtained for transport of a nonreactive tracer through the column demonstrated the effect of the physical heterogeneity on transport (see Figure 5A). The breakthrough curve obtained for transport of Co^{2+} was

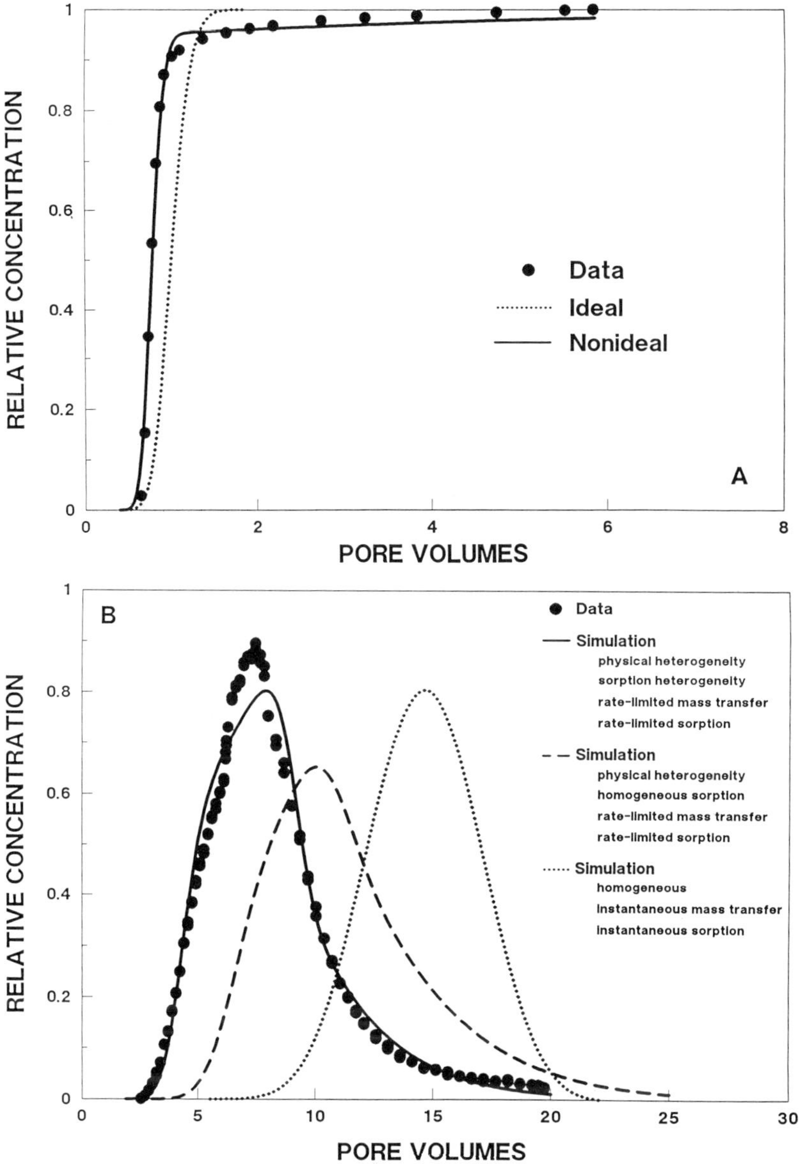

Figure 5. Illustration of the effect of porous-medium heterogeneity and rate-limited sorption/desorption on solute transport; A) Pentafluorobenzoate, B) Cobalt. Adapted from Brusseau and Zachara (1993).

shifted to the left of a simulated curve obtained for ideal conditions (homogeneous, instantaneous sorption) and exhibited tailing (see Figure 5B). The comparison reveals that transport of Co^{2+} through the heterogeneous porous medium was significantly nonideal. The optimised curve obtained for the case of hydraulic conductivity variability, sorption-capacity variability, rate-limited mass transfer between the two layers, and rate-limited sorption/desorption matched the experimental data quite well. In contrast, a simulated curve for the case of homogeneous sorption, hydraulic conductivity variability, and rate-limited mass transfer and sorption did not match the data well. The influence of sorption variability had to be considered to accurately simulate the transport of Co^{2+}. The impact of spatially variable sorption on solute transport was shown also by Brusseau (1991c) in an analysis of data reported by Starr et al. (1985), who investigated the transport of Sr in a flow cell containing layers of two media with different conductivity and sorption properties.

2.8 *Other factors*

Factors other than those discussed above may also contribute to observations of nonideal transport. Spatial or temporal changes in the magnitude and/or direction of the hydraulic gradient can influence transport. Another group of factors involves the use of incorrect models when analysing data. Sampling techniques can lead to observations of enhanced dispersion and, thus, influence observed behaviour.

3. Field experiments for subsurface transport of organic compounds

A multitude of field-scale experiments have been performed with nonreactive tracers to investigate solute transport in the subsurface (see Jury, 1985 and Gelhar et al., 1992 for recent reviews). Conversely, relatively few controlled field experiments have been performed to evaluate the transport of organic solutes under natural or induced fluid potential gradients. These few experiments have yielded results that are inconsistent with the simplifying assumptions of porous media homogeneity and instantaneous phase transfer or reaction that are often still used in the development and application of solute transport models. A detailed review of field-scale transport experiments for reactive solutes is provided by Brusseau (1994).

3.1 *Borden experiments*

An extensive field experiment was conducted at the Canadian Air Forces Base Borden, Ontario to investigate the transport of industrial-related organic solutes in a shallow, unconfined sand aquifer under natural gradient conditions (Curtis et al., 1986; Freyberg, 1986; Mackay et al., 1986b; Roberts et al., 1986; Sudicky, 1986). Well defined initial conditions were achieved by the injection of a pulse of solution ($12\,m^3$) containing two inorganic tracers (Cl^- and Br^-) and five organic chemicals (bromoform, tetrachloromethane, tetrachloroethene, 1,2-dichlorobenzene, and hexachloroethane). A dense, three dimensional array of sampling points was used to obtain time series data at selected points (i.e. breakthrough curves) as well as synoptic data on plume movement. During the course of the 2-yr experiment the plumes of the various solutes travelled 10 to 60 m.

Significant nonideal transport was observed for the organic solutes, wherein the

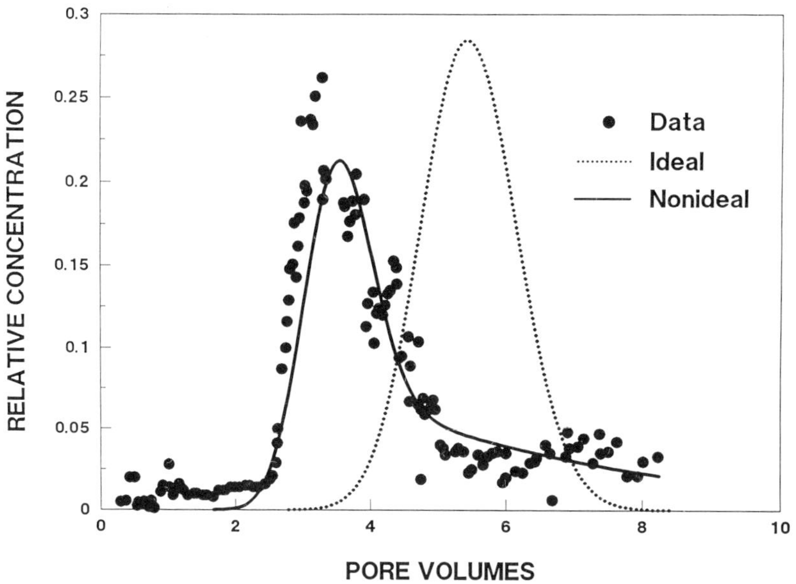

Figure 6. Measured breakthrough curves compared to simulated breakthrough curves obtained for ideal and nonideal conditions. The data represent results of tetrachloroethene transport during the Borden natural-gradient field experiment (data from Roberts et al., 1986). The nonideal curve accounts for spatially variable hydraulic conductivity, spatially variable sorption, and rate-limited sorption/desorption, whereas the ideal curve is based on assumptions of a homogeneous porous medium and instantaneous sorption (simulations from Brusseau, 1992b).

breakthrough curves were asymmetrical and the plumes decelerated with time (i.e. retardation factors were temporally nonuniform). This is illustrated in Figures 6 and 7, where measured data for transport of tetrachloroethene is compared to that expected for ideal conditions. The nonideal transport has been attributed to rate-limited sorption and/or diffusive mass transfer (Goltz and Roberts, 1986; Roberts et al., 1986; Ball et al., 1990; Quinodoz and Valocchi, 1993). It has also been postulated that more than one factor was responsible for the nonideal transport observed during the Borden experiments (Brusseau, 1989; Brusseau, 1992b).

Recent analyses have indicated that sorption heterogeneity and nonlinear sorption, combined with hydraulic conductivity variability, were the major factors causing the observed nonideal transport of the organic compounds (Brusseau and Srivastava, 1995). For illustration, the predicted simulation of the displacement of the centre of mass of the tetrachloroethene plume matched the measured field data quite well, as shown in Figure 7. This simulation was obtained with a model that accounted for hydraulic conductivity variability, sorption variability, rate-limited diffusive mass transfer, and nonlinear, rate-limited sorption.

Interestingly, it was concluded by Brusseau and Srivastava that rate-limited sorption/mass transfer played a relatively minor role in the transport of the organic compounds. This contradicts the conclusions reported by previous investigators

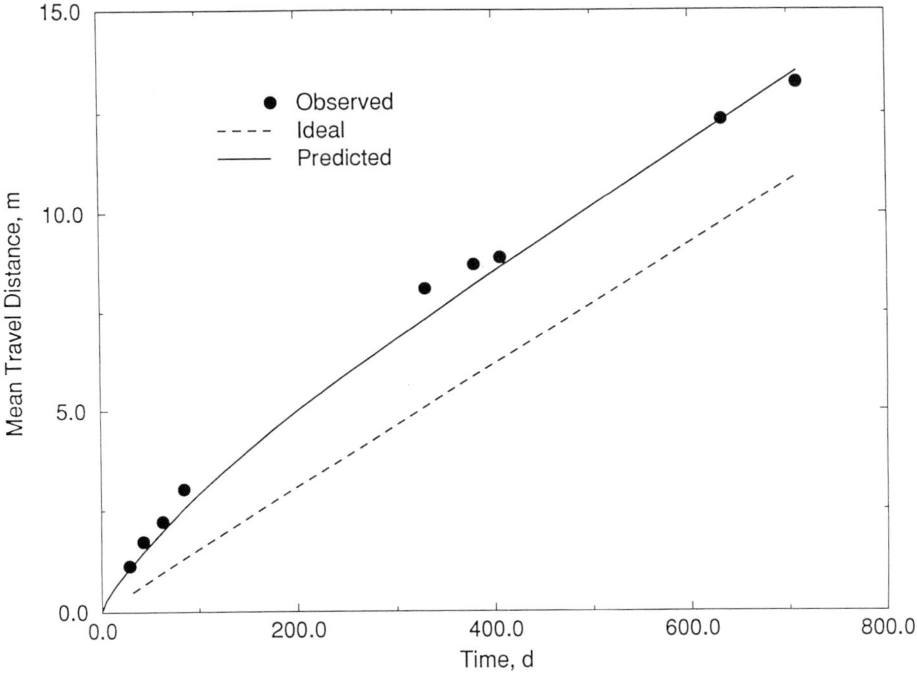

Figure 7. Displacement of the centre of mass of the tetrachloroethene plume measured during the Borden natural-gradient experiment. Data are from Roberts et al. (1986). The predicted simulation is from Brusseau and Srivastava (1995). The ideal curve represents transport under ideal conditions (homogeneous porous medium and linear, instantaneous sorption).

(Goltz and Roberts, 1986; Ball et al., 1990; Quinodoz and Valocchi, 1993) that rate-limited sorption/mass transfer was the primary causative factor. The results obtained in those analyses are due to the fact that they were based on curve fitting of mathematical models to the field data, with an a priori assumption that rate-limited sorption/mass transfer was the controlling factor.

3.2 Transport of organic compounds in the vadose zone

Innumerable field experiments have been performed to investigate various aspects of solute transport from the surface through the vadose zone. Much of this research has focused on agricultural related problems. An observation associated with many of these field experiments is transport of the solute to depths greater than those expected based on ideal transport theory. Jury (1985) summarises several of these field experiments. Relatively few experiments have been performed in such a manner that allows detailed analysis of the factors causing nonideal transport. This is especially true for organic compounds.

An early detailed field study of pesticide transport in the vadose zone was reported by Rao et al. (1974), who investigated the transport of picloram in a highly structured soil. They observed transport of a significant fraction of pesticide mass to depths approximately three times greater than those expected based on ideal water flow and instantaneous sorption. They attributed this and other nonideal behaviour to the impact of the porous medium structure on solute transport. This 'preferential flow' of organic pesticides has since been observed in several other well controlled field studies (Bowman and Rice, 1986; Jury et al., 1986; Ghodrati and Jury, 1992).

While many studies on transport of organic contaminants in the subsurface have been performed in North America and Europe, relatively few published studies on this topic are available for the Australasia-Pacific region. Most work on the fate and behaviour of organic compounds in soil environments focuses on the persistence and leaching of pesticides, and analyses of incidences of effluent disposal or leakage from storage facilities. A few of these case studies are briefly described below.

4. Case studies of transport and fate of organics in Australasia

4.1 Industrial contaminants in Australia

Several instances of groundwater contamination with organic compounds such as pesticides and hydrocarbons have been reported (Knight, 1993). Two such case studies are described below. Some other case studies on groundwater contamination by organic compounds and their remediation are discussed in Chapter 12.

4.1.1 Herbicides

A case study of severe groundwater contamination of a Western Australian shallow sand aquifer caused by herbicides originating from an agrochemical manufacturing plant has been reported by Appleyard (1993). The source of contamination was the disposal of the manufacturing effluent, which contained high concentrations of the herbicides (2,4-D and 2,4,5-T) and chlorinated phenols, into an onsite unlined pond. This method of effluent disposal was used from the 1960's to 1984. This industrial area near the coast is underlain by two aquifers; an upper, unconfined aquifer of sand and silt with moderate permeability, and a deeper, semiconfined limestone and sand aquifer of high permeability. The sand aquifer has a shallow water table (<5 m). Investigations carried out by the Geological Survey in 1982 revealed a contamination plume extending over an area of 0.7 km^2 near the agrochemical plant. Concentrations of 2,4-D and 2,4,5-T up to 180 mg L^{-1} and 60 mg L^{-1}, respectively, were noted in the plume within the sandy aquifer. Although samples from the aquifer have been found to contain microorganisms capable of degrading phenolic compounds under laboratory conditions, it has not been possible to establish whether biodegradation of phenols and herbicides is occurring at the site. A modelling study indicated that contaminants at concentrations high enough to potentially harm marine life could reach the nearby Cockburn Sound in 60–70 years (Appleyard, 1993).

Table 1. Sites with some probable organic chemicals in groundwaters of various states of Australia (Knight, 1993).

State/Territory	Sites with organics in groundwater	Provisional no. of total sites (some documentation)
Victoria	83	269–330
Western Australia	53	1080
New South Wales	16	1651*
Australian Capital Territory	4**	7
Queensland	?	1500**
South Australia	22	300
Northern Territory	2	?
Tasmania	3	143@
Total	>183	4950–5011#

*Numerous cattle dip chemicals sites, mainly soil contamination.
**Includes three petrol stations, figures for other sites do not include petrol stations.
@Consists of 140 landfills (domestic mainly).
#Excludes about 90 gasworks sites that are not documented.

4.1.2 Chlorinated organic compounds

A case study of groundwater contamination caused by chlorinated organic compounds originating from a pulp and paper mill in South Australia (SA) has been reported by Emmett (1993). A pulp and paper plant employing chlorine bleaching technology from 1966 until 1990 has been discharging its effluent to a local lake by use of a drain system. In 1989, identification of contamination in a private well close to the site led to a 2.5 year study, which revealed extensive contamination of the upper 10 m of the local aquifer. The contaminants were those typical of pulp mill effluents including chloroform, trichloroethene (TCE) and tetrachloroethene (PCE). Highly contaminated groundwater was found extending to a distance of at least 900 m parallel with the drain and 250 m away from the drain. In the 'hot spots', concentrations of total absorbable halogens of up to 8.1 mg L^{-1} and of volatile chlorinated hydrocarbons (chloroform, TCE, and PCE) up to 0.8 mg L^{-1} were observed. The local aquifer is shallow and unconfined, and is comprised predominantly of limestone (Gambier Limestone) with moderate to high permeability. The areas where groundwater contamination was found were those with the highest pollution potential based on analysis of the hydrological conditions (Emmett, 1993).

An overview of organic chemical contamination of groundwater in Australia has been provided recently by Knight (1993). The report lists several point-source incidences of groundwater contamination originating from a range of industrial, manufacturing, and service activities (Table 1). It is estimated that there are more than 183 sites (excluding gasoline service stations) in Australia where there is some probability that the groundwater is contaminated with organic compounds (Knight, 1993). Some case studies on contamination of groundwater with industrial organics

Table 2. Mean leaching depth (MLD) for pesticides observed under field conditions (Kookana et al., 1995).

Pesticides	Days after application	MLD (cm)	K_{oc}[#] (cm^3g^{-1})
Chlorpyriphos	148	<5	6070
Chlorthal dimethyl	142	<5	5000
Fenamiphos + metabolite	152	28*	100
Linuron	142	6	400
Metalaxyl	142	18	50
Metribuzin	142	12	60
Prometryne	152	9	400
Propyzamide	144	8	800
Simazine	142	6	130

[#]K_{oc} values as reported by Wauchope et al. (1992).
*Fenamiphos and metabolites might have leached below the maximum sampling depth (50 cm). The mean leaching depth is therefore likely to be deeper than 28 cm calculated on the basis of concentrations above 50 cm.

such as BTEX (benzene, toluene, ethylbenzene and xylene) due to leaking of underground storage tanks have been discussed in another chapter (see Chapter 12).

4.2 Agrochemical contaminants in Australia and New Zealand

Recent assessments of groundwater quality in shallow aquifers underlying irrigated agricultural areas of Australia have demonstrated that both particulate and dissolved contaminants are reaching aquifers from diffuse sources (Bauld et al., 1992). For example, triazine herbicides have been detected in the groundwaters of several states including New South Wales (NSW), Victoria and SA. Apart from these monitoring studies, several reports from the regions deal with degradation and leaching behaviour of pesticides through soils. An overview of studies on fate and behaviour of pesticides in the Oceania region has been provided recently by Theng et al. (1995). Some of the field leaching experiments on pesticides conducted in recent years are described below.

Briggs (1981) compared the vertical distribution of 8 herbicides in a podsolised sand after 10 cm of rain and noted that the distance travelled followed the order of octanol:water partition coefficients. Field experiments on leaching of nine different pesticides (including six herbicides, a fungicide, an insecticide, and a nematicide) through a coastal sand in Western Australia was conducted by Kookana et al. (1995). In this study, irrigation was applied daily on the basis of pan evaporation (@ 120% of PE) and distribution of pesticides in the soil profile was monitored at regular intervals to study both leaching and degradation behaviour. The mean leaching depths observed after a 5 month period followed the order of their sorption, i.e. chlorpyriphos and chlorthal dimethyl < linuron, simazine < propyzamide < prometryne < metribuzin < metalaxyl < fenamiphos (Table 2). While chlorpyriphos and chlorthal dimethyl showed essentially no movement and they degraded substantially during the study period, fenamiphos and its two metabolites showed relatively rapid leaching beyond 50 cm depth. Metalaxyl fungicide was also found to exhibit substantial leaching. The order of leaching for pesticides observed during these field experiments was in good agreement with the pollution potential assessment by Kookana and Aylmore (1994), carried out for several pesticides using the screening model of Jury et al. (1987).

Most studies on pesticide behaviour in Australian soil environments have dealt with persistence of herbicides, often in a qualitative way based on response of weeds and crops sown on treated areas. Such assessments obviously cannot discriminate between true degradation and losses due to various transport processes. Several studies on atrazine, chlorsulfuron, dichlobenil, diclofop methyl, 2,2-DPA, glyphosate, paraquat, picloram, simazine, and trifluralin have been reviewed by Jackson (1983), and more recently by Ferris and Haigh (1993). In Australian soils, degradation has been reported to follow first-order kinetics for atrazine (Bowmer, 1973; Haigh and Ferris, 1991), sulfonylureas (Blacklow and Pheloung, 1992) and trifluralin (Jolly and Johnstone, 1994) among others. Recently, Kookana et al. (1995) found that degradation of metalaxyl, linuron, metribuzin, and prometryn under laboratory conditions followed first-order kinetics with half-lives in the same order of magnitude as those noted for European and North American soils. However, data on pesticide behaviour are not always transferable between countries, mainly because of large variations in soil and climatic conditions.

Recent studies on sulfonylurea herbicides in the alkaline soils of Australia have shown rapid leaching and rather long persistence compared to what has been observed in the acid to neutral soils common to many parts of Europe and North America. This is because sulfonylureas, being acidic, dissociate to anionic forms in alkaline soils, and therefore are as susceptible to leaching as chloride and bromide anions. In fact, leaching beyond the root zone has been observed for these pesticides in experiments conducted in Victoria (Ferris et al., 1995). Although the major degradation pathways for sulfonylureas are acid hydrolysis and microbial degradation, only microbial degradation is possibly occurring in the highly alkaline soils (pH>8) of southern Australia. Furthermore, owing to their rapid leaching to subsurface horizons, little degradation occurs. Therefore, it is suspected that these pesticides may persist for years in alkaline Australian soils. The pH dependence of sorption and degradation of chlorsulfuron (a sulfonylurea herbicide) is shown in Figure 8. Long-term persistence for some other pesticides has also been noted in Australian soils. For example, atrazine has been found to have a half-life of more than 100 wk under certain field conditions in Australia, which is 13–20 times longer than what has been observed in soils of Romania and USA (Bowmer, 1991).

Long-term persistence of trifluralin has been observed by Jolly and Johnstone (1994). In New Zealand, several laboratory and field studies on leaching and persistence of a number of herbicides have also shown the important role that soil type plays in determining persistence and leaching potential (Rahman and Holland, 1985; James and Rahman, 1990; Baskaran, 1994). There are reports that pesticide use in New Zealand agriculture has resulted in contamination of groundwaters. Recent monitoring studies have detected seven different pesticides in groundwaters, mostly below the Health Advisory Limit established by USEPA (Close, 1993a)

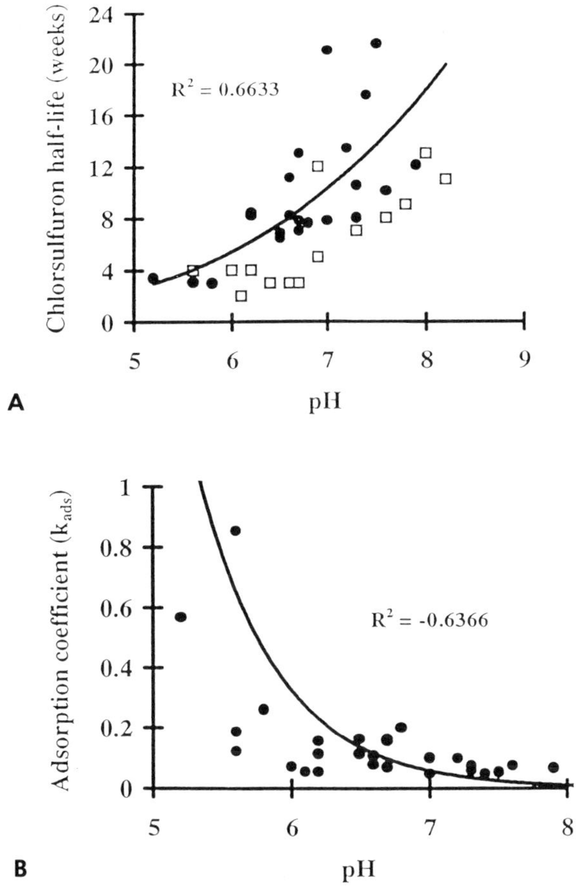

Figure 8. Effect of soil pH on half-life (A) and sorption (B) of chlorsulfuron. Based on data from several published studies from Europe and North America (A. Sarmah, Pers. Comm.).

Incidences of pesticide residues in groundwaters and surface waters of other countries in the Australasia-Pacific region have also been discussed in Chapters 14 to 25 of this book.

In recent years, several studies have been carried out in Australia and New Zealand to assess the vulnerability of groundwaters to pollution by pesticides and to evaluate the pollution potential of various pesticides. A risk assessment of groundwater pollution by pesticides for some regions of NSW was carried out using the pesticide root zone model (PRZM) by Milne-Home et al. (1991). On the basis of geological, hydrological, climatic, soil, agricultural, and pesticide (properties and use) factors, shallow aquifers of the Berriquin/Denimein Irrigation Area and the Richmond River Valley were assessed to have the highest risk of contamination of groundwater (high rainfall plus intensive agriculture: sugarcane, bananas, and vegetables). Recently, Kookana and Aylmore (1994) examined the groundwater

Table 3. Residual mass (as % of applied mass reaching 3 m depth in soil profile) of pesticides simulated (Jury et al., 1987) on the basis of measured K_{oc} and half-lives or those reported in literature* (Kookana and Aylmore, 1994).

Pesticide	K_{oc}, m^3kg^{-1} (measured)	K_{oc}, m^3kg^{-1} (literature)	Half-life, d (measured)	Half-life, d (literature)	Residual mass (measured)	Residual mass (literature)
Fenamiphos	0.55	0.17	30	10	6.2×10^{-9}	1.3×10^{-8}
Linuron	0.84	0.86	143	75	0.05	4.6×10^{-5}
Metribuzin	0.25	0.024	46	37	0.08	29.97
Prometryne	1.22	0.61	64	60	3.2×10^{-9}	2.2×10^{-4}
Simazine	0.15	0.14	75	75	6.69	7.91

*Literature values were taken from Jury et al., 1987.

pollution potential of some 40 different pesticides for a sandy soil profile in the Swan Coastal Plain, Western Australia. The mass of pesticides leaching past the zone of maximum sorption and biological activity was calculated using a simple screening model (Jury et al., 1987). They also examined the influence of OM distribution on prediction of pesticide pollution potential. Using local hydrological parameters and literature data on sorption (K_{oc}) and degradation (half-life) of pesticides, it was estimated that some 20 pesticides have the potential to leach beyond a 3 m depth in amounts greater than 0.1% of applied mass under the simulated conditions. However, when local K_{oc} and half-life values for five pesticides were used, the residual concentrations in some cases were found to be substantially different from those obtained on the basis of literature data (Table 3). This highlighted the need for site specific data for an accurate assessment.

Similarly in New Zealand, Close (1993b) developed an index of potential pollution of groundwater by pesticides and provided a ranking of regions for potential contamination. In the index, the factors such as amount of pesticides used, mobility and degradation characteristics of the pesticides, and vulnerability of an area to pollution were considered in conjunction with the DRASTIC ranking method. This method considers **D**epth to water table, (net) **R**echarge, **A**quifer media, **S**oil media, **T**opography (slope), **I**mpact of the vadose (unsaturated) zone and hydraulic **C**onductivity. Six out of 17 regions were identified as having a high pollution index ranking and three of these regions were selected for monitoring studies. For these regions, six of the 82 wells sampled were found to contain seven different pesticides, with concentrations mostly below USEPA Health Limits (Close, 1993a).

4.3 Organic contamination in Asia and the Pacific Islands

There are few readily available studies on fate and behaviour of pesticides and other organic compounds in soils of the Asia-Pacific region. However, during the last decade or so, the number of such studies from the Asia-Pacific region has certainly increased in the literature. Examples of some recent studies on the degradation, persistence, and movement of pesticides in the soil environment in the region are: China – Zhang et al. (1992), Japan and Thailand – Korpraditskul et al. (1993);

Kamiya et al. (1992), India – Bansal (1992); Bhuyan et al. (1993), Malaysia – Sahid and Teoh (1994); Ismail and Ahamad (1994), Bangladesh – Starr (1991), Taiwan – Wu et al. (1991). The fate, behaviour, and impact of pesticides in Indian soil environments have been discussed in detail in a recent book (Prasad and Gaur, 1994). Several country reports from Asia also discuss the issue of pesticides in the soil environment in other chapters of this volume.

Pesticide use in the Pacific Island countries of Oceania is small in comparison with that in Australia and New Zealand. However, a large number of active ingredients (nearly 300) are used in the region, some of which have been banned in the country of their origin (Mowbray, 1988; Watts, 1993). So far, very little work appears to have been undertaken in the region in terms of environmental fate and behaviour of these compounds.

5. Conclusion

The potential for human activities to adversely affect the environment has become of increasing concern during the past two decades. This has helped the transport and fate of contaminants in subsurface systems become one of the major research areas in the environmental/hydrological/earth sciences. An understanding of how contaminants move in the subsurface is required to address the problem of soil and groundwater contaminated by chemicals associated with industrial and commercial operations, service stations, waste disposal facilities, and agricultural production. Such knowledge is needed to evaluate the probability of contaminants associated with a chemical spill reaching an aquifer and contaminating groundwater. This knowledge is also required to develop and evaluate methods for cleaning up contaminated soils and aquifers. Just as importantly, knowledge of contaminant transport and fate is necessary to design 'pollution-prevention' strategies. In response to these needs, a tremendous body of literature on contaminant transport and fate – consisting primarily of results obtained by theoretical, experimental, and mathematical modelling based investigations, and to a much lesser extent, field experiments has been generated.

This brief review of some of the major aspects associated with the transport of organic contaminants in heterogeneous subsurface environments highlights the fact that the transport of organic contaminants at the field scale is characteristically non-ideal and many of the often used assumptions are not always valid. Recent research has demonstrated that phenomena such as diffusion constrained, rate-limited, sorption/desorption, sorption irreversibility and contaminant aging are the norm rather than exception for many organic contaminants. Although the effect of the presence of multi-component systems (e.g. co-existence of immiscible and miscible liquids, immiscible liquids), competitive or cooperative sorption between co-components and facilitated transport due to 'cosolvent' or 'colloid' effects have gained increased attention in recent years, their incorporation into transport models is at a preliminary state. The same is true for modelling of coupled processes such as sorption and transformation/degradation of organic chemicals in porous media.

Inherent heterogeneity of the soil subsurface and consequent spatial variability of physical, chemical and biological properties has been acknowledged to cause non-ideal transport of contaminants in porous media. However, relatively few experimental investigations examining the impact of spatial variability of chemical and biological properties on solute transport in porous media have so far been carried out.

Field experiments investigating various aspects of solute transport from surface through to vadose zone have largely been focussed on agrochemicals. While innumerable experiments have been performed in Europe and North America, relatively few have been reported from the Australasia-Pacific region. In the Australasia-Pacific region public awareness about contamination of the environment is on the increase and so is the monitoring of water, soil and air. The number of studies from the region dealing with the fate and behaviour of organic contaminants in the vadose zone is increasing. A sound knowledge base is crucial for designing effective preventative, management and cleanup strategies for organic contaminants in the soil environment of the region.

Acknowledgement

This research was supported in part by a grant from the National Institute of Environmental Health Sciences.

References

Abdul A S and Gibson T L 1986 Equilibrium batch experiments with six polycyclic aromatic hydrocarbons and two aquifer materials. Hazard. Waste Hazard. Mater. 3, 125-135.

Anderson M 1979 Using models to simulate the movement of contaminants through groundwater flow systems. CRC Crit. Rev. Environ. Cont. 9, 97-156.

Andricevic R and Foufoula-Georgiou E 1991 Modeling kinetic non-equilibrium using the first two moments of the residence time distribution. Stochastic Hydrol. Hydraul. 5, 155-171.

Angley J T, Brusseau M L, Miller W L and Delfino J J 1992 Nonequilibrium sorption and aerobic biodegradation of dissolved alkylbenzenes during transport in aquifer material: Column experiments and evaluation of a coupled-process model. Environ. Sci. Technol. 26, 1404-1410.

Appleyard S J 1993 Resourcing the investigation and management of groundwater contamination in Western Australia – a case study of contamination from herbicide manufacture near Perth. AGSO J. Aust. Geol. Geophysics. 14, 177-182.

Baehr A L 1987 Selective transport of hydrocarbons in the unsaturated zone due to aqueous and vapor phase partitioning. Water Resour. Res. 23, 1926-1938.

Baehr A L, Hoag G E and Marley M C 1989 Removing volatile contaminants from the unsaturated zone by inducing advective air-phase transport. J. Contam. Hydrol. 4, 126.

Bahr J M 1986 Applicability of the local equilibrium assumption in mathematical models for groundwater transport of reacting solutes, Ph.D dissertation, Stanford University, Palo Alto, Ca.

Ball W P and Roberts P V 1991a Long-term sorption of halogenated organic chemicals by aquifer material. 1. Equilibrium, Environ. Sci. Technol. 25, 1223-1236.

Ball W P and Roberts P V 1991b Long-term sorption of halogenated organic chemicals by aquifer material. 2. Intraparticle diffusion. Environ. Sci. Technol. 25, 1237-1249.

Ball W P, Goltz M N and Roberts P V 1990 Sorption rate studies with halogenated organic chemicals and sandy aquifer materials. Implications for solute transport and groundwater remediation. In Proc. 1990 Environmental Engineering Conference. pp. 307-313. Amer. Soc. Civil Engineers, New York, NY.

Bansal O P 1992 Effects of sewage sludge on oxamyl adsorption and degradation in soils. Indian J. Environ. Health. 34, 180-185.

Baskaran S 1994 Sorption and movement of ionic and non-ionic pesticides in selected soils of New Zealand. PhD thesis (unpublished). Massey University, Palmerston North, NZ.

Bauld J, Evans W R and Sandstrom M W 1992 Groundwater quality under irrigated agriculture: Murray Basin, southeastern Australia. In Groundwater and Environment. Proceedings of International Workshop on Groundwater and Environment, Beijing, China. Ed. Fei Jin. pp 447-457. Seismological Press, Beijing.

Bhuyan S, Sreedharan B, Adhya T K and Sethunathan N 1993 Enhanced biodegradation of g-hexachlorocyclohexane (g-HCH) in HCH (commercial) acclimatized flooded soil: factors affecting its development and persistence. Pest. Sci. 38, 49-55.

Bibby R 1981 Mass transport of solutes in dual-porosity media. Water Resour. Res. 17, 1075-1081.

Biggar J W and Nielsen D R 1962 Miscible displacement: II Behavior of tracers. Soil Sci. Soc. Am. Proc. 26, 125-128.

Blacklow W M and Pheloung P C 1992 Sulfonylurea herbicides applied to acidic sandy soils, movement, persistence, and activity within the growing season. Aust. J. Agric. Res. 43, 1157-1168.

Bowman R S and Rice R C 1986 Accelerated herbicide leaching resulting from preferential flow phenomena and its implications for ground water contamination. In Proc. Conf. on Southwestern Ground Water Issues, Phoenix, AZ, 20-22 Oct. pp. 413-425. National Water Well Assoc. Dublin, OH.

Bowmer K H 1973 The effect of temperature and soil type on atrazine persistence. Proc. Fourth Asian-Pacific Weed Science Society Conference, Rotorua, New Zealand, 242.

Bowmer K H 1991 Atrazine persistence and toxicity in two irrigated soils of Australia. Aust. J. Soil Res. 29, 339-350.

Briggs G G 1981 Adsorption of pesticides by some Australian soils. Aust. J. Soil Res. 19, 61-68.

Brown B D and Rolston, DE 1980 Transport and transformation of methyl bromide in soils. Soil Sci. 130, 68-75.

Brusseau M L 1989 Nonequilibrium sorption: Investigation at microscopic to megascopic scales, Ph.D Dissertation, University of Florida, Gainesville, FL.

Brusseau M L 1991a Transport of organic chemicals by gas advection in structured or heterogeneous porous media: Development of a model and application to column experiments. Water Resour. Res. 27, 3189-3199.

Brusseau M L 1991b Cooperative sorption of organic chemicals in systems composed of low organic carbon aquifer materials. Environ. Sci. Technol. 25, 1747-1752.

Brusseau M L 1991c Application of a multiprocess nonequilibrium sorption model to solute transport in a stratified porous medium. Water Resour. Res. 27, 589-595.

Brusseau M L 1992a Rate-limited mass transfer and transport of organic solutes in porous media that contain immobile immiscible organic liquid. Water Resour. Res. 28, 33-45.

Brusseau M L 1992b Transport of rate-limited sorbing solutes in heterogeneous porous media: Application of a one-dimensional multi-factor nonideality model to field data. Water Resour. Res. 28. 2485-2497.

Brusseau M L 1993a Using QSAR to evaluate phenomenological models for sorption of organic compounds by soil. Environ. Toxic. Chem. 12, 1835-1846.

Brusseau M L 1993b Complex Mixtures and Groundwater Quality. Environmental Research Brief, US Environmental Protection Agency, Wash. DC.

Brusseau M L 1994 Transport of reactive contaminants in heterogeneous porous media. Rev. Geophysics, 32, 285-313.

Brusseau M L 1995 The effect of nonlinear sorption on transformation of contaminants during transport in porous media. J. Contam. Hydrol. 17, 277-291.

Brusseau M L and Rao P S C 1989 Sorption nonideality during organic contaminant transport in porous media. CRC Critical Reviews in Environ. Control, 19, 33-99.

Brusseau M L and Rao P S C 1990 Modeling solute transport in structured soils: A review. Geoderma, 46, 169-192.

Brusseau M L and Rao P S C 1991 The influence of sorbate structure on nonequilibrium sorption of organic compounds. Environ. Sci. Technol. 25, 1501-1506.

Brusseau M L and Zachara J 1993 Transport of Co^{2+} in a physically and chemically heterogeneous porous medium. Environ. Sci. Technol. 27, 1937-1939.

Brusseau M L and Srivastava R 1995 Transport of Reactive Contaminants in Heterogeneous Porous Media: Reanalysis of the Borden natural-gradient field experiment. (in review).

Brusseau M L, Jessup R E and Rao P S C 1991a Nonequilibrium sorption of organic chemicals: Elucidation of rate-limiting processes. Environ. Sci. Technol. 25, 134-142.

Brusseau M L, Larsen T and Christensen T H 1991b Rate-limited sorption and nonequilibrium transport of organic chemicals in low organic carbon aquifer materials. Water Resour. Res. 27, 1137-1145.

Brusseau M L, Rao P S C and Bellin C A 1992a Modeling coupled processes in porous media: Sorption, transformation, and transport of organic solutes. In Interacting Processes in Soil Science, Advances in Soil Science, Eds. J Wagenet, P Baveye, and BA Stewart. pp 147-184. Lewis Pub. Ann Arbor, MI.

Brusseau M L, Jessup R E and Rao P S C 1992b Modeling solute transport influenced by multi-process nonequilibrium and transformation reactions. Water Resour. Res. 28, 175-182.

Bunzl K and Schimmack W 1988 Distribution coefficients of radionuclides in the soil: analysis of the field variability. Radiochimica Acta, 44/45, 355-360.

Chen Y M, Abriola L M, Alvarez P J J, Anid P J and Vogel T M 1992 Modeling transport and biodegradation of benzene and toluene in sandy aquifer material: Comparisons with experimental measurements. Water Resour. Res. 28, 1833-1847.

Chiou C T, Peters L J and Freed V H 1979 A physical concept of soil-water equilibria for non-ionic organic compounds. Science 206, 831-832.

Chiou C T, Porter P E and Schmedding D W 1983 Partition equilibria of nonionic organic compounds between soil organic matter and water. Environ. Sci. Technol. 17, 227-231.

Close M 1993a Assessment of pesticide contamination of groundwater in New Zealand. 2. Results of groundwater sampling. N. Z. J. Mar. Freshwater Res. 27, 267-273.

Close M 1993b Assessment of pesticide contamination of groundwater in New Zealand. 1. Ranking of regions for potential contamination. N. Z. J. Mar. Freshwater Res. 27, 257-266.

Coates J T and Elzerman A W 1986 Desorption kinetics for selected PCB congeners from river sediments. J. Contam. Hydrol. 1, 191-201.

Connaughton D F, Stedinger J R, Lion L W and Shuler M L 1993 Description of time-varying desorption kinetics: Release of naphthalene from contaminated soils. Environ. Sci. Technol. 27, 2397-2403.

Crittenden J C, Hutzler N J, Geyer D G, Oravitz J L and Friedman G 1986 Transport of organic compounds with saturated groundwater flow: Model development and parameter sensitivity. Water Resour. Res. 22, 271-284.

Curtis G P, Roberts P V and Reinhard M 1986 A natural gradient experiment on solute transport in a sand aquifer. 4. Sorption of organic solutes and its influence on mobility. Water Resour. Res. 22, 2059-2067.

Cvetkovic V D and Shapiro A M 1990 Mass arrival of sorptive solute in heterogeneous porous media. Water Resour. Res. 26, 2057-2067.

Elabd H, Jury W A and Cliath M M 1986 Spatial variability of pesticide adsorption parameters. Environ. Sci. Technol. 20, 256-260.

Emmett A J 1993 Organochlorine contamination of groundwater by pulp and paper mill effluent in South Australia. AGSO J. Aust. Geol. Geophysics. 14, 183-192.

Estrella M R, Brusseau M L, Maier R S, Pepper I L, Wierenga P J and Miller RM 1993 Biodegradation, sorption, and transport of 2,4-Dichlorophenoxyacetic Acid (2,4-D) in a saturated and unsaturated soil. App. Environ. Microbiol, 59, 4266-4273.

Falta RW, Javandel I, Pruess K and Witherspoon P A 1989 Density-driven flow of gas in the unsaturated zone due to the evaporation of volatile organic compounds. Water Resour. Res. 25, 2159-2169.

Ferris I G and Haigh B M 1993 Herbicide persistence and movement in Australian soils. Implications for Agriculture. In Pesticide Interactions in Crop Production. Ed. J Altman. pp 134-160. CRC Press, Boca Raton, Florida.

Ferris I G, Fawcet R, Stork P, Haigh B M, Pederson N and Rovira A 1995. Sulfonylurea behaviour in sodic soils. In Proceedings National Conference on Sodic Soils, Eds. R Naidu, M Sumner and P Rengasamy. pp 229-238. CSIRO Press, Melbourne.

Freyberg D L 1986 A natural gradient experiment on solute transport in a sand aquifer. 2. Spatial moments and the advection and dispersion of nonreactive tracers. Water Resour. Res. 22, 2031-2046.

Fried J J 1975 Groundwater Pollution. Elsevier, New York, NY. 2pp.

Fried J J, Muntzer P and Zilliox L 1979 Ground-water pollution by transfer of oil hydrocarbons. Groundwater, 17, 586-594.

Garabedian S P, Gelhar L W and Celia M A 1988 Large-scale dispersive transport in aquifers: field experiments and reactive transport theory. Report No. 315, Dept. of Civil Engin. Mass. Inst. Technol. Cambridge, MA.

Gelhar L W, Gutjahr A L and Naff R L 1979 Stochastic analysis of macrodispersion in a stratified aquifer. Water Resour. Res. 15, 1387-1397.

Gelhar L W, Welty C and Rehfeldt K R 1992 A critical review of data on field-scale dispersion in aquifers. Water Resour. Res. 28, 1955-1974.

Gerstl Z 1989 Predicting the mobility and availability of toxic organic chemicals. In Toxic organic chemicals in porous media: Eds. Z Gerstl, Y Chen, U Mingelgrin, and B Yaron. pp 151-162. Ecological Studies 73, Springer-Verlag, Berlin.

Ghodrati M and Jury W A 1992 A field study of the effects of soil structure and irrigation method on preferential flow of pesticides in unsaturated soil. J. Contam. Hydrol. 11, 101-125.

Gierke J S, Hutzler N J and McKenzie, D B 1992 Vapor transport in unsaturated soil columns: Implications for vapor extraction. Water Resour. Res. 28, 323-336.

Goltz M N and Roberts P V 1986 Interpreting organic solute data from a field experiment using physical nonequilibrium models. J. Contam. Hydrol. 1, 77-93.

Green R E and Karickhoff S W 1990 Sorption estimates for Modeling. In Pesticides in the Soil Environment, Processes Impacts and Modelling. Ed. H H Cheng. pp 79-101. Soil Sci. Soc. Am. Madison, WI.

Grisak G E, Pickens J F and Cherry J A 1980 Solute transport through fractured media 2. Column study of fractured till. Water Resour. Res. 16, 731-739.

Haigh B M and Ferris I G 1991. Predicting herbicide persistence using the CALF model. In Modelling the Fate of Chemicals in the Environment. Ed. I D Moore, pp 50-60. CRES, Australian National University, Canberra.

Hamaker J W and Thompson J M 1972 Adsorption. In Organic Chemicals in the Soil Environment. Eds C A I Goring and J W Hamaker, pp. 49-143. Wiley, NY.

Hunt J R, Sitar N and Udell K S 1988 Nonaqueous phase liquid transport and cleanup 1. Analysis of mechanisms. Water Resour. Res. 24, 1247-1258.

Ismail B S and Ahamad A R 1994 Attenuation of the herbicidal activities of glufosinate-ammonium and imazapyr in two soils. Agric. Eco. Environ. 47, 279-285.

Jackson R B 1983. Pesticide residue in soils. In Soils: an Australian Viewpoint. pp 825-842. CSIRO, Division of Soils, Melbourne/Academic Press, London.

James T K and Rahman A 1990 Soil mobility and moisture studies with some grass herbicides. Proc. 43rd New Zealand Weed and Pest Control Conf., 146-149.

Jaynes W F and Boyd, S A 1990 Trimethylphenylammonium-smectite as an effective adsorbent of water soluble aromatic hydrocarbons. J. Air Waste Manage. Assoc. 40, 1649-1653.

Johnson R L, Palmer C D and Keely J F 1987 Mass transfer of organics between soil, water and vapor phases: implication for monitoring, biodegradation and remediation. In Proc. of the NWWA/API Conf. on Petroleum Hydrocarbons and Organic Chemicals in Groundwater – Prevention, Detection and Restoration. pp 493-507. Nat. Water Well Assoc. Dublin, OH.

Jolly A and Johnstone P 1994. Degradation of trifluralin in three Victorian soils under field and laboratory conditions. Aust. J. Expt. Agric. 34, 57-65.

Jury W A 1985 Spatial variability of soil physical parameters in solute migration: A critical literature review. Report EA-4228. Electric Power Res. Inst. Palo Alto, CA.

Jury W A, Elabd H and Resketo M 1986 Field study of napropamide movement through unsaturated soil. Water Resour. Res. 22, 749-755.

Jury W A, Focth D D and Farmer W J 1987 Evaluation of pesticide groundwater pollution potential from standard indicies of soil-chemical adsorption and biodegradation. J. Environ. Qual. 16, 422-428.

Kabala Z J and Sposito G 1991 A stochastic model of reactive solute transport with time-varying velocity in a heterogeneous aquifer. Water Resour Res. 27, 341-350.

Kamiya M, Mitsuhashi S and Makino M 1992 Catalytic properties of cylcodextrins on the hydrolysis of parathion and paraxon in aquatic medium containing humic acids. Chemosphere. 25, 1783-1796.

Karickhoff S W, Brown D S and Scott T A 1979 Sorption of hydrophoxic pollutants on natural sediments. Water Res. 13, 241-248.

Karickhoff S W 1980 Sorption kinetics of hydrophobic pollutants in natural sediments. In Contaminants and Sediments. RA Baker, Ed. Ann Arbor Science, Ann Arbor, MI

Karickhoff S W, Brown D S and Scott T A 1979 Sorption of hydrophobic pollutants on natural sediments. Water Res. 13, 241-248.

Kinniburgh D G 1986 General purpose adsorption isotherms. Environ. Sci. Technol. 20, 895-904.

Knight M J 1993 Organic chemical contamination of groundwater in Australia – An overview to 1993. AGSO J Aust. Geol. Geophys. 14, 107-122.

Kookana R S, Di H J and Aylmore L A G 1995 A field study of leaching and degradation of nine pesticides in a sandy soil. Aust. J. Soil Res. 33.

Kookana R S and Aylmore L A G 1994 Estimating pollution potential of pesticides to ground water. Aust. J Soil Res. 32, 1141-55.

Kookana R S, Aylmore L A G and Gerritse R G 1992 Time-dependent sorption of pesticides during transport in soil. Soil Science 154, 214-225.

Kookana R S, Schuller R D and Aylmore L A G 1993 Simulation of simazine transport through soil columns using time-dependent sorption data measured under flow conditions. J. Contam. Hydrol. 14, 93-115.

Korpraditskul R, Katayama A and Kuwatsuka S 1993 Chemical and microbial degradation of atrazine in Japanese and Thai soils. J. Pest. Sci. 18, 77-83.

Koskinen W C and Harper S S 1990. The retention process, mechanisms. In Pesticides in the Soil Environment, Processes, Impacts and Modelling. Ed. H H Cheng. pp 51-78. Soil Sci. Soc. Am., Madison, WI.

Lee L S, Hagwell M, Delfino J J and Rao P S C 1992a Partitioning of polycyclic aromatic hydrocarbons from diesel fuel into water. Environ. Sci. Technol. 26, 2104-2110.

Lee L S, Rao P S C and Okuda, I 1992b Equilibrium partitioning of polycyclic aromatic hydrocarbons from coal tar into water. Environ. Sci. Technol. 26, 2110-2115.

Lee L S, Rao P S C, Brusseau M L and Ogwada R A 1988 Nonequilibrium sorption of organic contaminants during flow through columns of aquifer materials. Environ. Toxic. Chem. 7, 779-793.

Leonard R A 1990 Movement of pesticides into surface water. In Pesticides in the Soil Environment, Processes, Impacts and Modelling. Ed. H H Cheng. pp 303-349. Soil Sci. Soc. Am., Madison, WI

MacIntyre W G and deFur P O 1985 The effect of hydrocarbon mixtures on adsorption of substituted naphthalenes on clay and sediment from water. Chemosphere. 14, 103-111.

Mackay D M Ball W P and Durant M G 1986a Variability of aquifer sorption properties in a field experiment on groundwater transport of organic solutes: methods and preliminary results. J. Contam. Hydrol. 1, 119-132.

Mackay D M, Freyberg D L, Roberts P V and Cherry J A 1986b A natural gradient experiment on solute transport in a sand aquifer. 1. Approach and overview of plume movement. Water Resour. Res. 22, 2017-2029.

Massmann J W 1989 Applying groundwater flow models in vapor extraction system design. J Environ. Engin. 115, 129-149.

McCall P J and Agin G L 1985 Desorption kinetics of picloram as affected by residence time in soil. Environ. Toxic. Chem. 4, 37-45.

McCarthy J F and Zachara J M 1989 Subsurface transport of contaminants. Environ. Sci. Technol. 23, 496-502.

McGinley P M, Katz L E and Weber W J 1993 A distributed reactivity model for sorption by soils and sediments. 2. Multicomponent systems and competitive effects. Environ. Sci. Technol. 27, 1524-1531.

McMahon M A and Thomas G W 1974 Chloride and tritiated water flow in disturbed and undisturbed soil cores. Soil Sci. Soc. Amer. Proc. 38, 727-732.

Mendoza C A and Frind E O 1990 Advective-dispersive transport of dense organic vapors in the unsaturated zone 2. Sensitivity analysis. Water Resour. Res. 26, 388-398.

Miller C T, Poirier-McNeil M M and Mayer A S 1990 Dissolution of trapped nonaqueous phase liquids: Mass transfer constraints. Water Resour. Res. 26, 2783-2796.

Milne-Holme W A, Finlayson D and Yu X 1991 Generalised risk assessment of ground water contamination by pesticides, An expert system approach to PRZM. In Proc. Conf. Modelling the Fate of Chemicals in the Environment. Ed. I D Moore. pp 40-49. CRES, ANU, Canberra.

Mingelgrin U and Gerstl Z 1983 Reevaluation of partitioning as a mechanism of nonionic chemical adsorption in soils. J. Environ. Qual. 12, 1-11.

Molz F J, Guven O and Melville J G 1983 An examination of scale-dependent dispersion coefficients. Groundwater, 21, 715-725.

Mowbray D L 1988 Pesticide use in the South Pacific. UNEP Regional Seas Reports and Studies No. 89 and SPREP Topic Review No. 26.

Neretnieks I 1983 A note on fracture flow dispersion mechanisms in the ground. Water Resour. Res. 19, 364-370.

Nkedi-Kizza P, Biggar J W, van Genuchten M Th, Wierenga P J, Selim H M, Davidson J M and Nielsen D R 1983 Modeling tritium and chloride transport through an aggregated Oxisol. Water Resour. Res. 19, 691-700.

Nkedi-Kizza P, Brusseau M L, Rao P S C and Hornsby A G 1989 Nonequilibrium sorption during displacement of hydrophobic organic chemicals and ^{45}Ca through soil columns with aqueous and mixed solvents. Environ. Sci. Technol. 23, 814-820.

Pickens J F, Jackson R E, Inch K J and Merritt W F 1981 Measurements of distribution coefficients using a radial injection dual-tracer test. Water Resour. Res. 17, 529-544.

Pignatello J J 1990 Slowly reversible sorption of aliphatic hydrocarbons in soils. II Mechanistic aspects. Environ. Toxic. Chem. 9, 1117-1126.

Pignatello J J, Frink C R, Marin P A and Droste E X 1990 Field-observed ethylene dibromide in an aquifer after two decades. J Contam. Hydrol. 5, 195-210.

Powers S E, Abriola L M and Weber W J 1992 An experimental investigation of nonaqueous phase liquid dissolution in saturated subsurface systems: Steady state mass transfer rates. Water Resour. Res. 28, 2691-2705.

Powers S E, Loureiro C O, Abriola L M and Weber W J 1991 Theoretical study of the significance of nonequilibrium dissolution of nonaqueous phase liquids in subsurface systems. Water Resour. Res. 27, 463-477.

Prasad D and Gaur H S 1994 Soil Environement and Pesticides. Venus Publ. House, New Delhi, India.

Quinodoz H A and Valocchi A J 1993 Stochastic analysis of the transport of kinetically sorbing solutes in aquifers with randomly heterogeneous hydraulic conductivity. Water Resour. Res. 29, 3227-3240.

Rabideau A J and Miller C T 1994 Two-dimensional modeling of aquifer remediation influenced by sorption nonequilibrium and hydraulic conductivity heterogeneity. Water Resour. Res. 30, 1457-1470.

Rahman A and Holland P T 1985 Persistence and mobility of simazine in some New Zealand soils. NZ J. Expt. Agric. 13, 59-65.

Rao P S C and Davidson J M 1979 Adsorption and movement of selected pesticides at high concentrations in soils. Water Res. 13, 375-342.

Rao P S C, Evardsson K, Ou L T, Jessup R E, Nkedi-Kizza P and Hornsby A G 1986 Spatial variability of pesticide sorption and degradation. In Evaluation of Pesticides in Ground Water, ACS Symposium Series 315. pp. 100-115 Amer. Chemical Soc. Wash. DC

Rao P S C, Green R E, Balasubramanian V and Kanehiro Y 1974 Field study of solute movement in a highly aggregated oxisol with intermittent flooding: II Picloram, J. Environ. Qual. 3, 197-202.

Rao P S C, Rolston D E, Jessup R E and Davidson J M 1980 Solute transport in aggregated porous media: Theoretical and experimental evaluation, Soil Sci. Soc. Am. J. 44, 1139-1146.

Richter G and Jury W A 1986 A microlysimeter field study of solute transport through a structured sandy loam soil. Soil Sci. Soc. Am. J. 50, 863-868.

Rijnaarts H M, Bachmann A, Jumelet J C and Zehnder A J B 1990 Effect of desorption and intraparticle mass transfer on the aerobic biomineralization of a-hexachlorocyclohexane in a contaminated calcareous soil. Environ. Sci. Technol. 24, 1349-1354.

Roberts P V, Goltz M N and Mackay D M 1986 A natural gradient experiment on solute transport in a sand aquifer. 3. Retardation estimates and mass balances for organic solutes. Water Resour. Res. 22, 2047-2058.

Robin M J L, Sudicky E A, Gillham R W and Kachanoski R G 1991 Spatial variability of strontium distribution coefficients and their correlation with hydraulic conductivity in the Canadian forces base Borden aquifer. Water Resour. Res. 27, 2619-2632.

Robinson K G, Farmer W S and Novak J T 1990 Availability of sorbed toluene in soils for biodegradation by acclimated bacteria. Water Res. 24, 345-350.

Sabatini D A and Knox R C 1992 Transport and Remediation of Subsurface Contaminants. ACS Symp. Series 491, American Chemical Society, Washington, DC

Sahid I B and Teoh S S 1994 Persistence of terbuthylazine in soils. Bull. Environ. Contam. Toxicol. 52, 226-230.

Schwartz F W 1977 Macroscopic dispersion in porous media: Controlling factors. Water Resour. Res. 13, 743-752.

Schwille F 1988 Dense Chlorinated Solvents in Porous and Fractured Media. Lewis Publishers, Chelsea, MI

Scow K and Alexander M 1992 Effect of diffusion on the kinetics of biodegradation: Experimental results with synthetic aggregates. Soil Sci. Soc. Am. J. 56, 128-134.

Scribner S L, Benzing T R, Sun S and Boyd S A 1992 Desorption and bioavailability of aged simazine residues in soil from a continuous corn field. J. Environ. Qual. 21, 1115-1121.

Seyfried M S and Rao P S C 1987 Solute transport in undisturbed columns of an aggregated tropical soil: Preferential flow effects. Soil Sci. Soc. Am. J. 51, 1434-1443.

Sleep B E and Sykes J F 1989 Modeling the transport of volatile organics in variably saturated media. Water Resour. Res. 25, 81-92.

Smith J A, Chiou C T, Kammer J A and Kile D E 1990 Effect of soil moisture on the sorption of trichloroethene vapor to vadose-zone soil at Picatinny arsenal, New Jersey. Environ. Sci. Technol. 24, 676-682.

Smith L and Schwartz F W 1981 Mass Transport 2. Analysis of uncertainty in prediction. Water Resour. Res. 17, 351-369.

Smith L W and Bayer D E 1967 Soil adsorption of diuron as influenced by surfactants. Soil Sci. 103, 328-330.

Starr M A 1991 Degradation of diazinon in soils. Pak. J. Sci. Indust. Res. 34, 274-276.

Starr R C, Gillham R W and Sudicky E A 1985 Experimental investigation of solute transport in stratified porous media 2. The reactive case. Water Resour. Res. 21, 1043-1050.

Steinberg S M, Pignatello J J and Sawhney B L 1987 Persistence of 1,2-dibromoethane in soils: Entrapment in intraparticle micropores. Environ. Sci. Technol. 21, 1201-1210.

Stevenson F J 1982 Organic matter reactions involving pesticides in soil. In Humus Chemistry, Genesis Composition, Reactions. pp 403-419. Wiley-Interscience NY

Sudicky E A 1986 A natural gradient experiment on solute transport in a sand aquifer. Spatial variability of hydraulic conductivity and its role in the dispersion process. Water Resour. Res. 22, 2069-2082.

Taylor A W and Spencer W F 1990 Volatilization and vapour transport processes. In Pesticides in the Soil Environment, Processes, Impacts and Modelling. Ed. H H Cheng. pp 213-269 Soil Sci. Soc. Am. Madison, WI

Theis C V 1963 Hydrologic phenomena affecting the use of tracers in timing groundwater flow. In: Radioisotopes in Hydrology. pp. 193-207. Inter. Atomic Energy Agency, Vienna.

Theng B K G, Kookana R S and Rahman A 1995. Environmental concerns of pesticides in soil and groundwater and management strategies in Oceania. In Advances in the Environmental Science – Remediation of Degraded Soils and groundwaters, Volume 4: Worldwide Perspectives on Soil and Groundwater Pollution: Asia, Africa and Oceania. Ed. P M Huang. (in press) Science Reviews, London.

Travis C C and Etnier E L 1981 A survey of sorption relationships for reactive solutes in soil. J. Environ. Qual. 10, 8-17.

Valocchi A J 1989 Spatial moment analysis of the transport of kinetically adsorbing solutes through stratified aquifers. Water Resour. Res. 25, 273-279.

van Der Waarden M, Bridie A L A M and Groenewoud W M 1971 Transport of mineral oil components to groundwater I. Model experiments on the transfer of hydrocarbons from a residual oil zone to trickling water. Water Res. 5, 213-226.

van Genuchten M Th and Wierenga P J 1977 Mass transfer studies in sorbing porous media: Experimental evaluation with tritium. Soil Sci. Soc. Am. J. 41, 278-785.

Warren J E and Skiba F F 1964 Macroscopic dispersion. Soc. Pet. Eng. J. 215-230.

Watts M 1993 Poisons in Paradise – Pesticides in the Pacific. Greenpeace, Auckland.

Wauchope R D, Buttler T M, Hornsby A G, Augustijn-Beckers P W M and Burt J P 1992 The SCS/ARS/CES pesticide properties database for environmental decision-making. Rev. Environ. Contam. Toxicol. 123, 1-157.

Weber W J, McGinley P and Katz L E 1992 A distributed reactivity model for sorption by soils and sediments. 1. Conceptual basis and equilibrium assessments. Environ. Sci. Technol. 26, 1955-1962.

Willhite G P 1986 Waterflooding, Society of Petroleum Engineers, Richardson, TX

Wilson J L and Conrad S H 1984 Is physical displacement of residual hydrocarbons a realistic possibility in aquifer restoration? Paper presented at Petroleum Hydrocarbons and Organic Chemicals in Ground Water. pp. 274-298, Nat. Water Well Assoc. Dublin, OH

Wood L S, Scott H D, Marx D B and Lavy T L 1987 Variability in sorption coefficients of metolachlor on a captina silt loam. J. Environ. Qual. 16, 251-256.

Wu T C, Wang Y S and Chen Y L 1991 Degradation of isouron in three upland soils in Taiwan. J. Pest. Sci. 16, 195-200.

Zhang S M, Liu D S, Wang Z S and Ma X F 1992 A kinetic model describing the effect of temperature on the loss rate of pesticides in soil. Eco. Modelling 70, 115-125.

5
Adverse impacts of agriculture on soil, water and food quality

M.E. SUMNER AND M.J. McLAUGHLIN

1. Introduction

Modern agriculture has led to serious declines in soil and water quality as a result of practices which are not sustainable in the long-term. The greatest impact on soil quality has been in terms of soil loss, an irreversible process with 25, 8 and 12% of the crop-, pasture- and rangeland, respectively, in the United States of America (USA) eroding at rates faster than soil loss tolerance levels (National Research Council, 1993). Despite the ravages of erosion being known for a long time, only very small reductions in erosion rates have been recorded in these categories of land over the period 1982 to 1987. This possibly reflects the prediction that crop yields are only likely to decline by 10% over the next 100 years due to soil erosion at current high input levels (Putnam et al., 1988) which indicates that, to be effective in countering erosion, efforts should focus on offsite damage caused to rivers and lakes. In terms of water quality, agriculture has been the source of 64 and 57% of the non-point pollution of rivers and lakes in the US, respectively, with sediment and nutrients accounting for 60 and 81% of the pollution in affected rivers and lakes, respectively (Carey, 1991). Non-point source pollution from soil erosion impairs beneficial uses of water, fish spawning and rearing habitats and increases the cost of municipal water treatment, maintenance of navigational channels, irrigation systems and reservoir storage (Clark et al., 1985). In the USA, soil erosion costs $1.3 billion in lost productivity and $3 billion in off-site economic damage (Ribaudo, 1987).

In countries such as Australia, where water resources are scarce, protection of the quality of surface and groundwaters is critical for both agricultural and domestic users. Eutrophication and salinity are the two major water quality problems in Australia, with costs associated with eutrophication of water bodies estimated to be between A$10 and 50 million per annum (Cullen, 1996). The world's largest recorded riverine algal bloom was observed in 1991 in the Darling-Barwon river system in Australia (Verhoeven, 1993) where over 1000km of the river system was affected (Shafron, 1993). Considerable resources are now being directed to tackle this problem and determine the link between algal growth, water quality and land use.

Together the above examples reflect the position in many other countries practising intensive agriculture where the contributions to environmental degradation from agriculture are likely to be the greatest. Unfortunately not much information is available in the literature for the Asia-Pacific Region. Consequently, the major portion of this review will be devoted to an analysis of the impacts of agriculture on

R. Naidu et al. *(eds): Contaminants and the Soil Environment in the Australasia-Pacific Region, 125–181.*

soil quality in the remainder of the world and the consequent effects on water and food quality so that conclusions can be drawn for this region.

2. Adverse impacts of agriculture on soil quality

2.1 Definition

The National Research Council (1993) defined soil quality as 'the capacity of a soil to promote the growth of (healthy) plants (free from potential hazards to the food chain), protect watersheds by regulating the infiltration and partitioning of precipitation, and prevent water and air pollution by buffering potential pollutants such as agricultural chemicals, organic wastes, and industrial chemicals'. The authors' insertions in ( ) refine the above definition still further. Soil quality can be degraded by agricultural practices such as cultivation, pesticide application, irrigation, waste application, etc. resulting in erosion, compaction, salinisation, sodification, acidification, and pollutant toxicity, the impacts of which will be discussed in the sections to follow.

2.2 Acidity

Soil acidification is a natural process that occurs in most undisturbed ecosystems, particularly in high rainfall environments. Agricultural activities accelerate rates of acidification through interruption and modification of the C and N cycles in soil. The main factors responsible for soil acidification have been discussed in detail by Helyar and Porter (1989), but a brief discussion of the impact of agriculture on soil acidification through the C and N cycles is given here.

The major impact of agriculture on soil acidification stems from the use of both nitrogenous (ammoniacal-N) fertilizers and legumes (amino-N) to enhance the N nutrition of crops and pastures. Some soils having a high buffer capacity (high clay or organic matter contents) are better able to resist acidification than sandy soils. Amino- or ammoniacal-N forms in soil are readily oxidised by the biological process of nitrification to nitric acid (HNO_3) according to the following reactions:

(a) Organic N

$$CO(NH_2)_2 + 2\ H_2O + \text{urease} \Rightarrow (NH_4)_2CO_3 \qquad 1$$

$$(NH_4)_2CO_3 + 4\ O_2 \Rightarrow 2\ HNO_3 + CO_2 + 3\ H_2O \qquad 2$$

(b) Inorganic N

$$(NH_4)_2SO_4 + 4\ O_2 \Rightarrow 2\ HNO_3 + H_2SO_4 + 2\ H_2O \qquad 3$$

No matter what the source, all forms of reduced N added to soil are potentially acidifying depending on whether or not the nitrate produced is removed from the soil by plant uptake and removal or by leaching (see Chapter 2). Thus, a particular quantity of amino-N, whether from a legume source or urea, will generate on average half the acidity produced by ammoniacal-N. The reason for the difference lies in the dissipation of half the acidity produced from amino-N in the form of carbonic acid (H_2CO_3) as CO_2 and water (Equation 2) whereas, in the case of ammoniacal-N (Equation 3), the non-volatile anion (SO_4^{2-}, NO_3^-, Cl^-) conserves all the acidity produced in the soil.

In agricultural systems, the natural C cycle is interrupted by the harvesting and removal of plant and animal materials. These materials contain dissociated organic salts of Ca, Mg and K which are alkaline in nature and therefore their removal acidifies the soil. For example, the removal of 15 t lucerne hay with an organic anion concentration of 150 $cmol_c\ kg^{-1}$ has a net effect on acidification equivalent to just over 1 t lime (Chartres et al., 1992). Thus, most agricultural systems will have some requirement for inputs of lime to offset acidification generated by product removal.

Acidification of soil has several consequences as far as soil quality is concerned, most important amongst which is the production of elevated levels of soluble Al^{3+} and Mn^{2+} which are toxic to many crop plants according to the following reactions:

$$2\ Al(OH)_3 + 3\ H_2SO_4 — 2\ Al^{3+} + 3\ SO_4^{2-} + 6\ H_2O \qquad 4$$

$$MnO_2 + 4\ H^+ + 2\ e^- — Mn^{2+} + 2\ H_2O \qquad 5$$

The loss of soil quality in terms of crop production due to acidification results in decreased pH, elevated levels of Al^{3+} and loss of basic cations (Ca, Mg) which reflects the loss of cation exchange capacity (CEC). All of these changes take their toll on yield. In addition, soil acidification can have other deleterious effects on soil quality such as inhibition of nitrification, promotion of certain diseases and rendering certain elements (P, Mo) less available to plants. Acidification of polluted soils can result in increased availability of potentially toxic heavy metals such as Cd, Zn and Cu whose solubility increase with decreasing soil pH. In addition, the reduction in CEC increases the mobility and availability of these metals which is particularly important in variable charge soils common in tropical and subtropical regions (Naidu et al., 1996). This topic is discussed in Chapter 3.

In Australia, extensive areas of soil are naturally acidic and soil acidification has also occurred under legume-based pastures in high rainfall environments (Figure 1). Chartres et al. (1992) estimate that approximately 29 million hectares of land in Australia have topsoils (0–10 cm) with a pH in water less than 5.5.

The impacts of soil acidification on crop production can readily be neutralised by the use of lime with the amount required to maintain soil pH at the original level depending on source and amount of N applied to the soil.

2.3 Loss of organic matter and nutrients

Whenever virgin soils are cleared for cultivation of crops, the organic matter (OM) content invariably decreases as a result of the improved oxidative conditions and crop removal. This, in turn, leads to loss of soil quality manifested as degraded structure resulting from reduced interparticle bonding. This loss of structure has

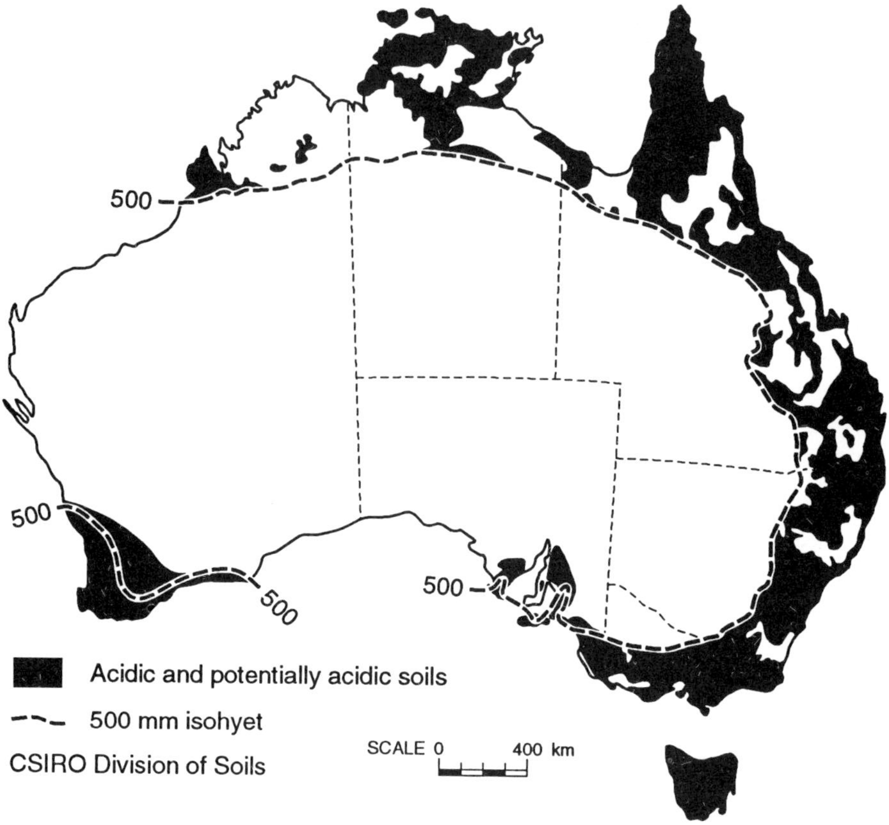

Figure 1. Areas of acidic or potentially acidic soils in Australia (CSIRO Division of Soils).

serious consequences for soil physical condition leading to increased compaction and reduced pore space when the soil is subject to loads from tillage machinery. This reduces the ability of the soil to readily conduct water and air within its mass leading to poor aeration and waterlogging. When exposed to the energy of impacting raindrops, crusts or seals can form at the soil surface reducing the infiltration rate of water into the soil which produces runoff and, consequently, erosion when rainfall intensity exceeds infiltration rate (Sumner and Stewart, 1992).

Reviews of the effects of agriculture on the turnover of OM in Australian soils have recently been published (Gifford et al., 1992; Grace et al., 1992). Significant losses of OM from soil have occurred in a number of regions of Australia due to clearing of native vegetation and cultivation of soils (Figure 2). Gifford et al. (1992) estimate that between 1880 and 1980 up to 39% of the organic C in Australian surface soils has been lost through both mineralisation and erosion. While some areas have noted significant losses of OM, large areas of permanent pasture have recorded significant gains in OM (Donald and Williams, 1954) due to increased C

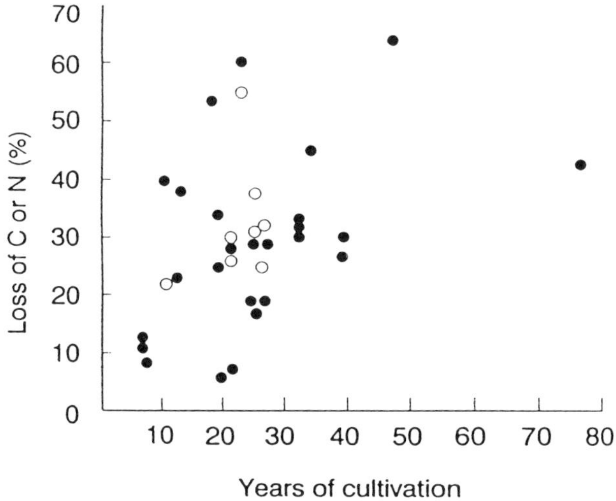

Figure 2. Percentage loss of organic carbon (open circles) and nitrogen (closed circles) in soil sown to wheat in relation to years of cultivation (from Gifford et al., 1992).

inputs from pasture growth stimulated by fertilisation. In these soils OM accumulates in the top few centimetres of soil and may form a dense mat containing significant CEC and nutrients (McLaughlin et al., 1990).

This negative train of events precipitated by cultivation can also be partially stemmed by the use of crop rotations involving reduced or no-tillage operations as a result of which soil disturbance is kept to a minimum and stable soil structure maintained or improved. Thus the closer the level of OM in a soil under a particular cropping system is to that in the virgin state, the greater is the potential sustainability. Comparisons of the impacts of maintaining OM at different levels on infiltration rate are presented in Figure 3.

In certain variable charge and sandy textured soils, loss of OM seriously reduces the cation exchange capacity and can increase anion exchange capacity thereby impairing the ability of the soil to retain nutrients such as Ca, Mg and K.

2.4 Erosion

2.4.1 Water erosion

Loss of OM under tillage results in structural degradation due to rupture of organic-inorganic bonds responsible for stable aggregation allowing clay to disperse under certain circumstances. Clay dispersion is promoted by (a) inputs of mechanical energy such as that from impacting raindrops; (b) elevated levels of exchangeable Na and Mg; and (c) low concentrations of electrolyte in the soil solution (Sumner, 1992; 1993). Obviously, bare soils are much more prone to dispersion by impacting

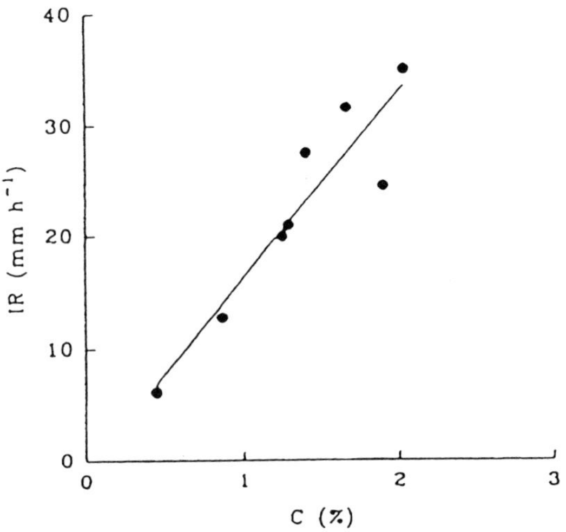

Figure 3. Effect of organic carbon content on the infiltration rate of Australian red-brown earths (Grierson et al., 1972).

raindrops than surfaces which are protected by vegetation or a mulch. Because the dispersed clay is mobile, clay transport blocks pores at the soil surface forming a seal which drastically reduces infiltration rate (IR). On drying, crusts with great strength can develop limiting seedling emergence. Consequently as IR is reduced, any excess precipitation must run off, resulting in soil loss, the magnitude of which depends on soil characteristics, slope, rainfall intensity and extent of soil cover (Shainberg, 1992; Sumner, 1992).

Because tillage in all its forms disturbs and exposes soil to impacting raindrops, some degree of erosion inevitably results. The great benefits which accrue to covering the soil surface at all times are illustrated in Figure 4 (Langdale et al., 1992) where replacing conventional by conservation tillage in 1975 had a profound effect on increasing residue cover and reducing soil loss. The buildup of crop residues over the 15 year period has resulted in marked improvements in soil quality in terms of total C and N as well as water stable aggregates resulting in a doubling of yields (Langdale et al., 1992). Similar results have been reported by McGregor and Greer (1982), and Andraski et al. (1985). Even in conventional tillage systems, residue retention can markedly reduce erosion losses which supply sediment to water bodies by intercepting and reducing raindrop energy (Gilley et al., 1986). The nature of the residue is less important than the amount. In comparison to Europe and North America, little information is available in the Asia-Pacific region on the link between land use, soil management and erosion. Olive and Walker (1982) found that in Australia soil loss can vary from <1 under pasture to 382 t ha^{-1} yr^{-1} under sugarcane, which represents an important mechanism for plant nutrient losses (Table 1). These nutrients ultimately contaminate water bodies.

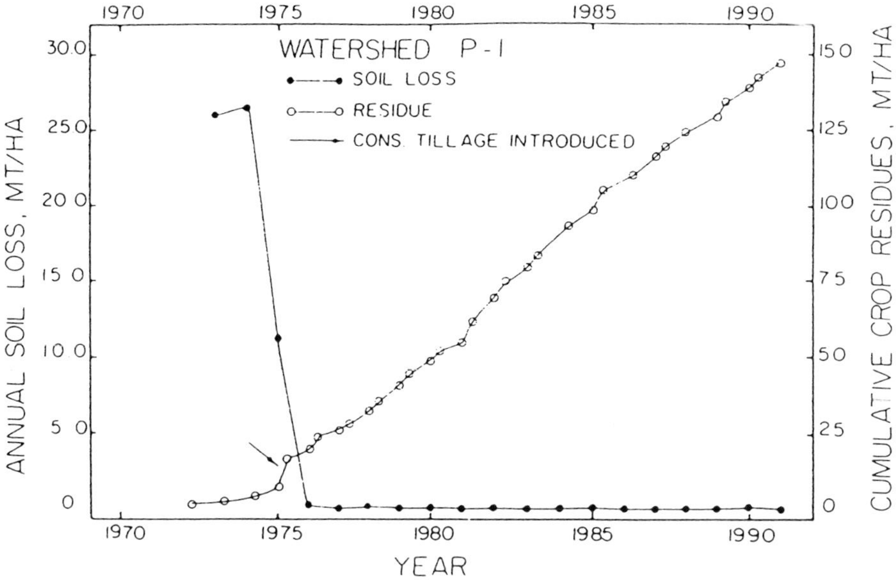

Figure 4. Changes in soil loss and cumulative crop residues as a result of the introduction of conservation tillage practices on a Typic Hapludult (Langdale et al., 1992).

Table 1. Estimated losses of nutrients through soil erosion from agricultural areas of Australia.

	Area $10^3\,km^2$	Soil loss $t\,km^{-2}\,yr^{-1}$	Nutrient lost in eroded soil (kt)		
			P	S	N
Crops	190	10–500	1.0–47.5	0.7–33.0	2.9–142.5
Pastures (unimproved)	160	5–50	0.3–3.0	0.2–2.0	2.8–28.0
Pastures (improved)	140	5–50	0.5–4.0	0.4–3.5	2.8–28.0
Forest/woodland	1050	1–10	0.3–3.0	0.2–1.6	1.1–10.0
Total			2.1–57.5	1.5–40.1	9.6–208.5

Ribaudo and Young (1989) made the first estimates of water quality benefits in terms of water storage, navigation, flooding, water conveyance, water treatment, municipal and industrial usage, etc. to be derived from soil erosion control on a regional basis in the US (Table 2). Benefits per unit of erosion reduction depend largely on demand for water resources and water quality rather than on the magnitude of erosion, which stresses the need to link erosion control to economic impacts.

Thus, the soil surface must be covered by vegetation or mulch for the maximum time possible in order to minimise runoff and erosion. This will also ensure that contaminants transported in association with sediment will be reduced to a minimum (Section 3 and Chapter 2).

Table 2. Estimated off-site benefits in 1983 arising from erosion reduction in various regions of the US (Ribaudo and Young, 1989).

Region	Off-site benefits (million \$) Range	Erosion reduction (million t)	Benefits per ton saved (\$/t)
Appalachia	21–52	66	0.53
Corn Belt	16–38	27	0.29
Delta States	18–42	27	1.07
Lake States	9–24	11	1.36
Mountain States	19–48	72	0.46
Northeast	7–18	8	1.50
Northern Plains	11–27	65	0.28
Pacific	14–36	31	0.77
Southern Plains	63–168	189	0.59
Southeast	23–54	48	0.79
Total	201–507	544	7.64

2.4.2 Wind erosion

In dryland agriculture and in semi-arid regions, large storm events have the potential to transport large quantities of soil by wind. For example, dust storms in Australia have been found to transport soil across the Tasman Sea to New Zealand (Mokma et al., 1972). Much wind erosion in Australia occurs in the arid interior and soil and nutrients are not 'lost' but merely redistributed within the landscape. Losses may occur however, from agricultural to non-agricultural regions. Indeed, many soils in south-eastern Australia owe their origin to wind blown deposits (Butler and Churchward, 1983).

2.5 Sodicity

Although large areas of soils are naturally sodic, particularly in Australia and Africa, the discussion to follow will be confined to the development of sodic soils as a result of agricultural practices. A sodic soil is one containing elevated levels of exchangeable Na, the exact level required being debated at present, but lying between an exchangeable sodium percentage (ESP) of 3 and 15% (Sumner, 1993; 1996). In most cases, this takes place by the removal of salts from saline soils by leaching as a result of the introduction of irrigation. During leaching, not all saline soils develop sodicity but only those containing sufficient levels of Na with little reserve of potentially soluble salts of divalent cations such as gypsum ($CaSO_4.2H_2O$) and lime ($CaCO_3$). These salts, when present, contribute to the electrolyte concentration and tend to maintain the hydraulic conductivity (HC) allowing the excess Na to be leached. Sodicity can also be developed by subjecting soil to

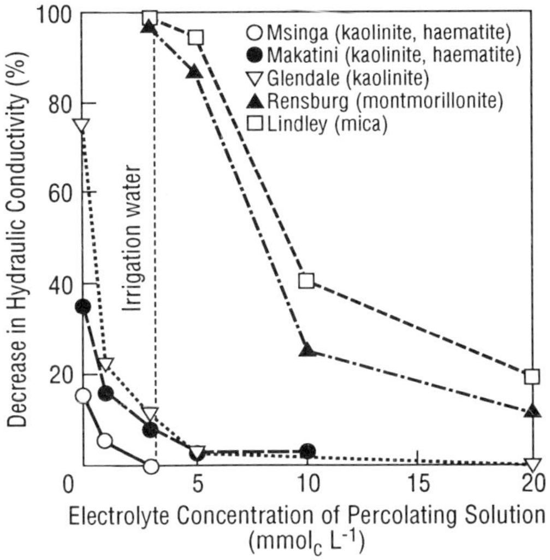

Figure 5. Effect of decreasing electrolyte concentration at a constant SAR = 20 on the hydraulic conductivity of South African irrigated soils (Msinga: Typic Haploxerult; Makatini: Oxic Rhodustaff; Glendale: Udic Paleustalf: Rensburg: Typic Pellustert; Lindley: Aridic Paleustalf) (Hensley, 1969).

water with a high sodium adsorption ratio (SAR). (A soil in equilibrium with a water of a given SAR will have an ESP almost numerically equal to the SAR value.) The development of sodicity in these cases is illustrated in the following equation:

$$\text{Soil-Ca} + 2\ NaHCO_3 \Rightarrow \text{Soil-2 Na} + CaCO_3\downarrow + CO_2 + H_2O \qquad 6$$

The removal of salts results in the dispersion of the clay and OM which become mobile and enter pores, causing blockage. As a result, infiltration, hydraulic conductivity and other physical properties are seriously impaired, greatly reducing soil quality. This is illustrated in Figure 5 for a number of soils in Africa in an area to be developed for irrigation. Both the Lindley and Rensburg soils which are quite permeable at high electrolyte concentrations of a solution of SAR = 20 become almost completely impermeable at the electrolyte concentration of the irrigation water. Thus the level of Na in the soil system must be balanced against the electrolyte concentration in order to maintain favourable physical properties. Similar demonstrations of the effect of salt removal from Na-affected soils on soil physical properties have been presented by other authors (McNeal et al., 1968; Pupisky and Shainberg, 1979; Cass and Sumner, 1982; Shainberg et al., 1991). On the other hand, in soils where the structure is stabilised by iron oxides (such as Msinga, Makatini, Figure 5), increasing levels of Na and decreasing electrolyte concentration do not have nearly the same effect because the clay particles are bonded together by the oxides. Once a soil has become sodic and physical properties have been seriously compromised, reclamation is difficult and costly. Thus the main management strategy should be one of prevention rather than cure, by utilising ameliorants such as gypsum to ensure that hydraulic conductivity is maintained while the deleterious Na is exchanged and removed.

2.6 *Salinity*

Vast areas of soils throughout the world are naturally saline, containing levels of salts which impair the ability of plants to obtain adequate water. However, the discussion here will focus only on the development of salinity in soils as a result of the agricultural pursuits of man, mainly irrigation causing severe declines in soil quality. Szabolcs (1989) has estimated that over 10 million hectares (ha) of irrigated land globally are abandoned annually as a result of salinisation and associated sodification and waterlogging. Globally, there is a direct relationship between irrigated and salinised areas (Szabolcs, 1994). Given that all natural waters contain salts, soil salinisation will occur under irrigation whenever insufficient water (rainfall plus irrigation) is applied to leach the salts accumulated in the soil as a result of evapotranspiration which removes only pure water. This usually occurs under poor water management or when natural or man-made drainage is inadequate or absent. Thus in arid areas with high evapotranspiration, irrigation schemes are short-lived unless leaching and drainage are practised. In some instances, natural drainage is sufficient for this purpose but in others, drainage systems must be installed. While drainage is required to keep salt levels in soils low enough for crop production, the drainage water poses a threat to both surface and groundwater quality at the point of discharge (see Section 3.2.4).

In addition, the clearing of natural vegetation for agricultural development often allows groundwater to rise which, if saline, can cause salinisation in the root zone. Flugel (1991) in South Africa has demonstrated increased river salinisation as a result of the development of dryland agriculture. Additional secondary sources of salts include atmospheric deposition and sea water intrusion. Salinity can also develop as a result of saline seeps which are of major concern in Australia (Sharma and Williamson, 1984) and North America (Halvorson, 1990). Saline seeps occur when water in excess of crop requirements moves laterally downhill due to an impervious layer, accumulating salts in transit and emerging in a seepage area. In the northern Great Plains, more than 1 million ha are affected (Brown and Miller 1981). Sometimes installation of an interceptor drain can prove effective but the need to dispose of the drainage water poses an additional problem which is also experienced in irrigated agriculture where the drainage water is several times more saline than the irrigation water. This presents the potential for off-site water quality degradation (Section 3.2.4).

An example of the extent of man-induced salinisation is presented for Australia in Figure 6 but this phenomenon is common throughout the world.

2.7 *Compaction*

Compaction of soil refers to the compression of soil which increases soil bulk densities and changes soil pore size volume and distribution. The consequences of compaction are changes in water retention characteristics, reductions in air filled porosity and changes in aggregate stability. As a result roots have great difficulty in

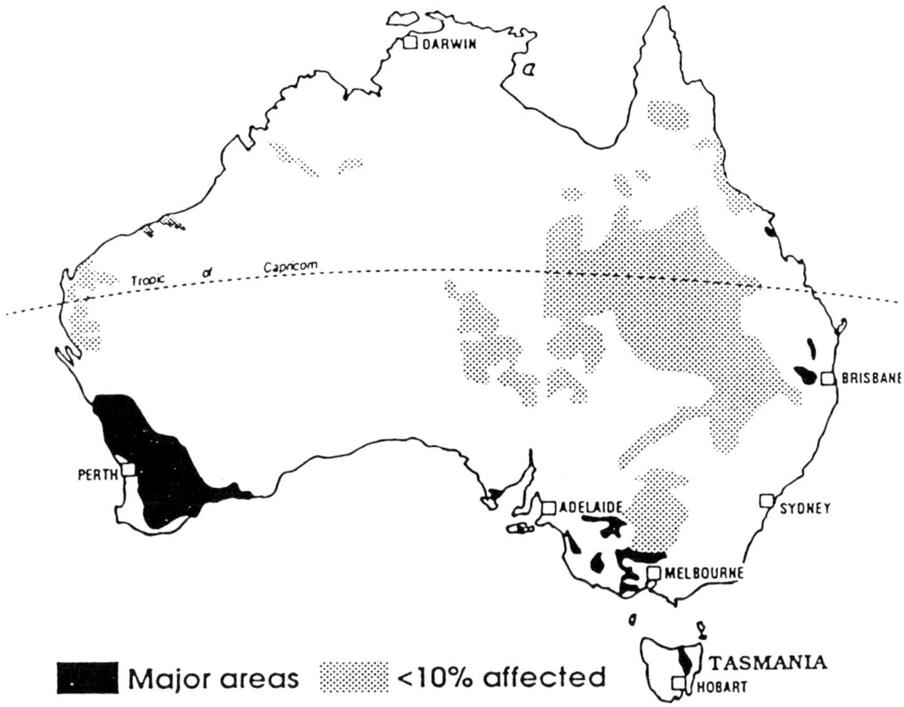

Figure 6. Areas of human-induced soil salinity in Australia (Chartres et al., 1992).

proliferating in compacted soil. Soils can become compacted by agricultural activities through use of heavy machinery (McGarry and Chan, 1984), trampling by stock (Burch, 1986) or by changes in soil structure induced by cultivation or irrigation practices. In general soils falling in the sandy to loam textural categories are most prone to compaction because of the enhanced packing ability of the nearly equal proportions of sand, silt and clay. Few studies on the extent and impact of compaction on soil quality have been carried out in Australia, except in cotton growing areas where compaction of Vertisols is common (Hodgson and MacLeod, 1989).

Solutions to soil compaction usually involve reduced animal and machine traffic, deep tillage and soil structure promotion, but little can be done to overcome the natural propensity for soils of appropriate texture to compact.

2.8 *Pollution*

Pollution of soils occurs due to inappropriate agrochemical applications, deliberate disposal of contaminated liquid and solid wastes and inadvertent addition of contaminants in amendments and fertilisers to soils. Gross contamination of soils rarely occurs in agricultural systems and is more common in close proximity to industrial complexes or in urban areas. Disposal of urban sewage waters and sludges on agricultural soils rather than discharge into water bodies is common in many countries and with growing concerns for pollution of surface and groundwaters, this practice is likely to increase in the future.

McLaughlin et al. (1995) recently reviewed the contamination of soils due to impurities in fertilisers. After consideration of contaminant concentrations in phosphate rocks and background concentrations in soils they concluded that the major elements of concern were As, Cd, Hg, Pb and F. Of these, Cd is the element of most concern due to its potential transfer to food crops (Section 4.1).

The major concerns related to soil pollution are due to the potential transfer of contaminants from soils to agricultural food crops which is discussed in more detail in Section 4. Other soil quality concerns in terms of pollution of agricultural soils relate to the survival and functioning of the soil biota in soils receiving significant loadings of pollutants, for example, sewage sludge disposal sites or sites contaminated by pesticides and other organic compounds. There is increasing evidence that metal pollutants in soil impair the functioning of the soil biota at concentrations well below those where phytotoxicity is first evident (Witter, 1993). Witter (1993) has reviewed a number of laboratory and field experiments dealing with impact of metals on microbial activity in order to suggest maximum allowable concentrations of metals for Swedish soils. If adopted, these limits, which are well below those recommended elsewhere, may well prove to be restrictive in terms of re-use of sewage sludges on land.

3. Adverse impacts of agriculture on water quality

To produce adequate food for the expanding world population, protection of water resources free from contamination by sediments, nutrients and pesticides from agriculture is absolutely essential. Unfortunately, the efficiency with which fertilisers and manures have been used has been low, resulting in losses of nutrients to surface and groundwaters while cultural practices have allowed the products of erosion carrying pollutants, to contaminate water supplies. In terms of rivers, the relative impact of the top ten pollutants in the USA is presented in Figure 7.

The major contributors from agriculture are sediment, nutrients and pesticides. A balanced approach to soil management resulting in economically sustainable productivity with minimum impact on water and food quality is required.

3.1 Surface water

3.1.1 Effect of tillage on sediment and nutrients

Tillage systems affect the partitioning of water falling on soil into runoff and infiltration but, once a crust has formed on conventionally tilled surfaces, the distribution increasingly favours runoff. Although conservation tillage reduces annual soil losses and sediment concentrations, the greater threat of nutrient loss immediately following surface application under no-till persists (McGregor and Greer, 1982; Gilley

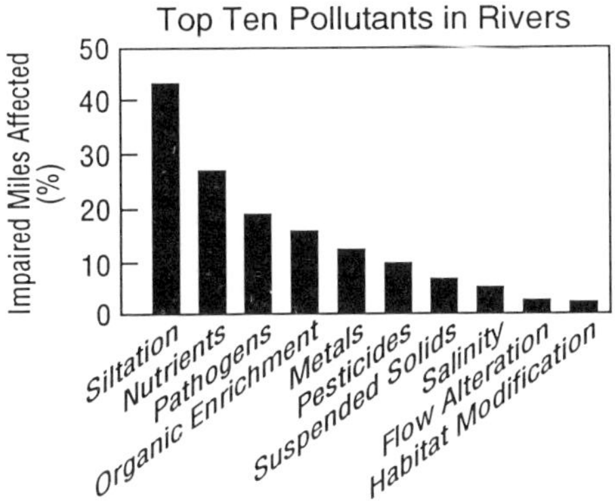

Figure 7. Ranking of river pollutants from the 1988 U.S. National Water Quality Inventory (CAST, 1992).

et al., 1986). However, substantial nutrient losses are associated with the sediment portion of runoff and most occur during seedbed preparation, planting and fertilising operations (Alberts et al., 1978). Reducing tillage intensity significantly reduced runoff rate and sediment concentration but NO_3-N, NH_4-N and PO_4-P concentrations in runoff were highest in the no-till system and decreased with tillage intensity (Seta et al., 1993). Similarly increasing surface residues had the same effect on solution P (McDowell and McGregor, 1984). Concentrations of these nutrients tended to be highest in the first rainfall event and decreased with subsequent events (data not presented). Chisel ploughing and no-tillage significantly reduced total sediment, NO_3-N and PO_4-P. Although the amounts of total nutrients lost were small, being less than 3% of the amount applied, concentrations of PO_4-P from all tillage greatly exceeded levels acceptable from a water quality point of view (0.01 mg L^{-1}) and NO_3-N concentrations were at the US Environmental Protection Agency human drinking water level (10 mg L^{-1}). Laflen and Tabatabai (1984) obtained similar results but because far more N and P were lost with sediment, and because erosion was greatest under conventional cultivation, the order of total nutrient loss was conventional > chisel plough > no-till (Table 3).

Thus, surface applied chemicals have less opportunity to interact with the undisturbed soil surface in the no-till treatments, rendering them more prone to movement with runoff water. On the other hand, Gold and Loudon (1986) found slightly higher soluble P concentrations in overland flow from conventional than conservation tillage areas which also exceeded the acceptable level. These results show that when nutrients are not incorporated, concentrations in runoff are increased substantially and may result in pollution of rivers and lakes. Similar results have been reported elsewhere in the US (McDowell and McGregor, 1980; 1984; Sharpley et al., 1991). Baker et al. (1987) confirmed that although conservation tillage practices in the Great Lakes Basin of North America were likely to increase, the impact on water quality was unlikely to be great except to the extent that changes in hydrology with tillage may change the relative losses of NO_3^- to surface versus ground-

Table 3. Effect of tillage system on sediment and nutrient concentrations in water and total losses of sediment and nutrients from two Hapludolls in a maize/soybean rotation* (Laflen and Tabatabai, 1984).

Tillage	Concentrations ($mg\,L^{-1}$)						Total losses ($kg\,ha^{-1}$)					
	Sediment	NH_4-N	NO_3-N	PO_4-P	Total N	Total P	Erosion	NH_4-N	NO_3-N	PO_4-P	Total N	Total P
					Clarion sandy loam							
Conventional	18 940a†	0.19a	0.18a	0.08a	2370a	728a	2380a	0.024a	0.027a	0.009a	5.21a	1.65a
Chisel plough	9710b	0.58a	0.21a	0.17a	2720ab	883ab	1250b	0.076a	0.027a	0.020a	3.01b	1.02b
No-till	4940b	1.23b	1.59b	0.60b	2940b	952b	820b	0.192b	0.208b	0.089b	2.28b	0.74b
					Monona silt loam							
Conventional	137 530a	0.23a	0.32a	0.16a	1620a	771a	46 620a	0.078a	0.107a	0.051a	75.43a	35.87a
Chisel plough	98 260b	0.64a	0.86a	0.32b	1770a	807a	31 520b	0.201a	0.275a	0.101a	55.17b	25.48b
No-till	33 870c	2.02b	1.78b	0.84c	2020b	915b	11 640c	0.691b	0.668b	0.289b	23.14c	10.54c

*Granular fertilizers were spread on all treatments before tillage at $26\,kg\,N\,ha^{-1}$ and $37\,kg\,P\,ha^{-1}$ (ammonium phosphate) for soybeans plus an additional $113\,kg\,N\,ha^{-1}$ (urea) for maize).

†Values in a column for a soil followed by the same letter are not significantly different at the 5% level.

Table 4. Effect of cover crops on nitrate, phosphate and sediment concentrations in runoff water for several cropping systems§.

Crop	Cover crop	State	Fertilizer		Mean annual concentration		
			N	P	Sediment	NO_3-N	PO_4-P
			kg ha^{-1} yr^{-1}			mg L^{-1}	
CT† Maize	None	GA	–	20	0.93	–	0.13 (4.1)*
	Rye		–	50	0.39	–	0.20 (1.4)
NT Cotton	None	AL	100	0	0.47	1.73 (3.1)	0.39 (0.4)
	Wheat		100	0	0.30	1.12 (0.9)	0.39 (0.2)
CT Peanuts 1985¤	None	OK	–	–	6.48	0.50 (4.4)	0.14 (1.4)
	Ryegrass		–	–	3.63	0.73 (1.5)	0.19 (0.5)
CT Peanuts 1988	None	OK	–	–	5.94	0.29 (20.9)	0.12 (5.9)
	Wheat		–	–	3.53	0.75 (3.3)	0.15 (0.9)
NT Soybean	None	MO	15	12	0.26	4.04	0.28
	Bluegrass		15	12	0.03	1.92	0.80

§Sources: GA=Georgia – Langdale et al. (1985); AL=Alabama – Yoo et al. (1988); MO=Montana – Zhu et al. (1989); OK=Oklahoma – Sharpley and Smith (1991).
†CT=Conventional tillage, NT=No-till.
*Values in ( ) are losses of total N and P in kg ha^{-1} yr^{-1}.
¤Values for peanuts are for a six month period.

water. They considered N management (rate, timing, placement) to be more crucial to NO_3^-losses than choice of tillage system.

Although conservation tillage and residue retention practices limit the transport of sediment and any associated pollutants to surface water bodies, the reverse is true for soluble nutrients. This is apparently due to the reduced opportunity which surface applied chemicals have in interacting with soil clay and OM with which they are often totally or partially immobilised.

3.1.2 Cropping systems

Arising from the current interest in sustainable agriculture, the use of cover crops to reduce erosion and retain nutrients, particularly N and P, during the winter season has received renewed attention and data for the Southern Plains are presented in Table 4.

Cover crops consistently decrease runoff, soil loss and total amounts of N and P transported but the effects on soluble NO_3-N and PO_4-P concentrations are not consistent. In the case of NO_3-N, cover crops both increased and decreased the concentration in runoff while for PO_4-P, the concentration was increased in all cases. Although cover crops generally reduce soil and total nutrient loss in runoff, the concentrations of soluble and bioavailable N and P, which are more important from the point of view of water quality, are often increased as a function of soil fertility, crop type and growth stage. In all cases (Table 4), PO_4-P concentrations exceed the water quality guidelines which means that cover crops will not prevent non-point pollution of water bodies by P from agricultural land. Both soluble and bioavailable P must be measured to accurately assess the impact of agricultural management practices on eutrophication of water bodies. Concentrations of NO_3-N were below the maximum potable levels for humans and livestock in USA (10 and 100 mg L^{-1}, respectively).

Comparisons of the inflow of nutrients to lakes from differently utilised catchments (arable land and forest immediately adjoining the lake) showed adverse impacts of agriculture on the lake water NO_3-N and PO_4-P concentrations (Misztal et al., 1992). On the other hand, conversion of natural vegetation to other uses such as pasture can greatly impact the NO_3-N concentrations in runoff water. In such a study in Arizona, Davis (1984) demonstrated that converting chaparral brush to pasture using herbicides increased streamflow and annual weighted mean NO_3-N concentrations from <1 to 16 mg L^{-1} in the first year. Over the 11 year period, NO_3-N from the treated watershed was 29-fold greater than that from the control, due to the continuing decomposition of the plant residues below the zone explored by the grass roots.

The greatest reductions in runoff, sediment and surface nutrient losses are obtained by combining structural best management practices (BMP) such as sediment basins, terraces and contours with improved nutrient management programmes (NMP) (optimum prescription manure and fertiliser rate and timing recommendations for the crop, soil and yield goal) and a cover crop (McIsaac and Hirschi, 1989; Hamlett and Brannan, 1991; Hamlett and Epp, 1994). Because a single BMP or NMP is not adequate to control pollutant losses from agricultural lands, programmes should emphasise the use of both BMPs and NMPs to minimise runoff, erosion, sediment and nutrient losses taking local field conditions (slope, soil type, cropping system, etc.) and economic factors into consideration.

In Australia, the limited data available indicate that soils producing irrigated high value crops or pastures, where rates of fertilisation are highest, have the greatest potential to export nutrients to surface waters (Campbell, 1978; Cullen and Rosich, 1979; Clark, 1988). In upland areas with Mediterranean climates where permanent or annual pastures predominate, losses of nutrients by surface runoff depend largely on stocking rate, which determines ground cover (Costin, 1980). In a study of soil and nutrient losses from a moderately to heavily grazed pasture in south-eastern Australia, Costin (1980) found a negative relationship between percentage ground cover and soil loss (Figure 8). Losses of soil, N and P by surface runoff over the five years of the experiment varied from 4–376, 0.02–1.23 and 0.003–0.22 kg ha^{-1}, respectively. While these figures are low, the impact on surface waters may be considerable.

Comparisons of cropped versus pasture land including pristine ecosystems show that concentrations of dissolved PO_4-P and NO_3-N in runoff water can be similar from both, but total concentrations are related to the quantity of particulate matter transported, which are likely to be greater in the former. In general, PO_4-P concentrations exceeded 0.15 mg P L^{-1} (Benoit, 1973; White and Williamson, 1973; Burwell et al., 1975; Beke et al., 1989).

The presence of livestock grazing on pasture can increase concentrations of

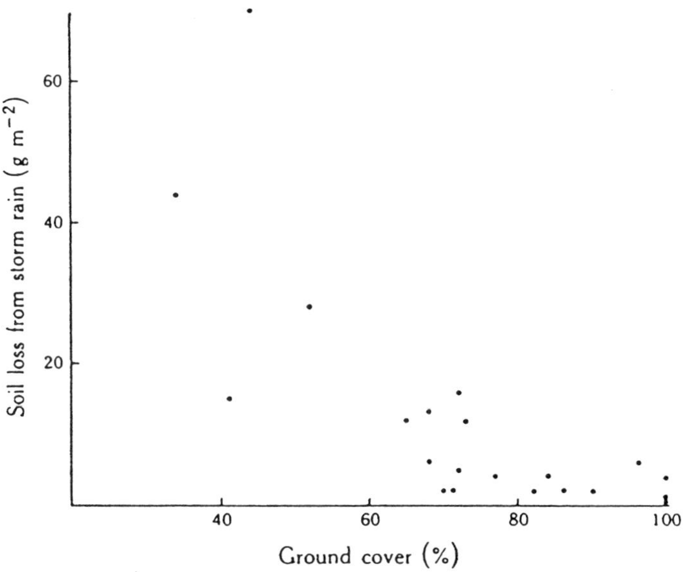

Figure 8. Relationship between soil loss under storm conditions and percentage ground cover in a moderately to heavily-grazed pasture (Costin, 1980).

NO_3-N, soluble and total P, chemical and total O_2 demand, Cl and total solids and decrease total N in surface runoff (Schepers and Francis, 1982). However, on an ungrazed control area (never grazed by cattle), the concentrations of these constituents were elevated to much higher levels (2- to 10-fold) which they attributed to wildlife activity and decomposition of plant material (Schepers et al., 1982). The concentrations of soluble P in runoff from all systems were high (0.54–4.1 mg L^{-1}). Feeding overwintering livestock on pasture drastically increases runoff and soil loss relative to the summer grazing when few losses occur (Owens et al., 1982). Under flood irrigation, return flows from grazed pastures can be enriched in biological oxygen demand (BOD), total P and PO_4-P, the latter being associated with fertilised pasture (Miller et al., 1984). Although these findings corroborate those of Schepers et al. (1982), the presence or absence of grazing animals and forage cropping pattern did not contribute as much as site characteristics, water management and hydrological variations, which agrees with the findings of Hollon et al. (1982).

3.1.3 Fertilisers and Manures

3.1.3.1 FERTILISERS

When applied on the surface as in pastures, fertilisers make a significant contribution to surface water contamination with NH_4-N and NO_3-N exceeding recommended levels for drinking water and accelerating eutrophication (Sharpley et al., 1983).

The processes associated with transformations of fertiliser N in soils, impacts of fertiliser type, fertiliser placement and timing are discussed in Chapter 2. Whether levels of N in runoff water from fertilised agricultural land can be maintained below recommended levels is questionable.

The effect of fertiliser P on the concentrations and losses of P in subsurface runoff is illustrated in Table 5. In all cases, addition of fertiliser increases both the concentrations and amounts of soluble and particulate P in the runoff water with the effect being much more pronounced in the case of soluble P. Although these losses account for a very small proportion of the applied fertiliser, the concentrations in the runoff water in nearly all cases exceed acceptable levels for discharge into surface water bodies (Sharpley and Withers, 1994). Losses of P from surface application are greater than when incorporated or injected but residue retention has little effect (Timmons et al., 1973).

Sharpley et al. (1987) found that the impact of fertiliser application on 20 watersheds in the Southern Plains was largely due to total and soluble P which both consistently exceeded the critical levels associated with accelerated eutrophication (0.01 and 0.02 mg L^{-1}) from both fertilised and unfertilised areas. While much of the soluble P stems from P leached from vegetation (Sharpley, 1981; Sharpley et al., 1981), the concentration in runoff will be determined by the extent to which this P is able to interact with the sorption capacity of the soil.

3.1.3.2 MANURES

Vast quantities of organic wastes such as animal manures and sewage sludge are being applied to agricultural land in general, and to the surface of pastures in particular, creating potential negative impacts for surface water quality. This has been investigated by a number of workers (McLeod and Hegg, 1984; Edwards and Daniel, 1992; 1993; 1994). Neither poultry litter nor fertiliser had any effect on runoff volume (data not presented). Concentrations of total N and P, NO_3-N, dissolved P (DP) and NH_3-N in runoff were highest for the fertiliser (McLeod and Hegg, 1984; Westerman et al., 1987) followed by the litter treatment, while total soluble salts (TSS) and chemical oxygen demand (COD) were greatest for the litter but, by the second simulated rainfall event, there were no significant differences between treatments in any of the variables measured. Thus most losses occur in the first rainfall event after application. Similar results have been obtained for other manures and sewage sludges with the concentrations of nutrients in runoff increasing more or less linearly with application rate but higher rainfall intensities tended to decrease runoff concentrations (Table 6) (Edwards and Daniel, 1993). In most cases concentrations in runoff exceeded permissible public water standards for P and sometimes for NO_3-N. Splitting manure applications had little effect on nutrient concentrations in runoff water (Quisenberry et al., 1980). Moore and Miller (1994) demonstrated that additions of substances such as gypsum, sodium aluminate and ferric salts markedly reduced levels of soluble P in poultry manure, suggesting that pretreatment of litter prior to field application might be the best management option in situations where

Table 5. Effect of fertilizer P on concentrations and losses of P in subsurface runoff (Sharpley and Withers, 1994).

Land use	P applied	Concentration		Amount		Fertilizer loss		Location
		Soluble	Particulate	Soluble	Particulate	Soluble	Particulate	
	$kg\,ha^{-1}\,yr^{-1}$	$mg\,L^{-1}$		$kg\,ha^{-1}\,yr^{-1}$		%		
Lucerne (tile drainage)	0	0.180	–	0.12	–			Canada
	29	0.210	–	0.19	–	1.0	–	
Continuous maize	40	0.007	–	0.03	–	–	–	Iowa, USA
	66	0.009	–	0.04	–	–	–	
Continuous maize (tile drainage)	0	0.20	0.100	0.13	0.29			Canada
	30	0.110	0.360	0.20	0.42	0.2	0.4	
Bluegrass sod (tile drainage)	0	0.02	0.15	0.06	0.09	–	–	Canada
	30	1.01	3.29	0.16	0.21	0.3	0.4	
Oats (tile drainage)	0	0.02	0.09	0.10	0.19			Canada
	30	0.42	1.10	0.20	0.30	0.3	0.4	
Grass	0	0.020	0.022	0.04	0.44			New Zealand
	50	0.033	0.019	0.12	0.07	0.2	+7.4[a]	
Grass (tile drainage)	0	0.064	0.072	0.08	0.09			New Zealand
	50	0.190	0.161	0.44	0.37	0.7	0.6	
Lucerne (tile drainage)	0	0.02	0.011	0.012	0.02			Canada
	30	0.37	1.03	0.20	0.31	0.3	0.3	

[a]Percent decrease in P loss from fertilizer compared to control treatment.

Table 6. Effect of pig manure rates and simulated rainfall intensities on runoff water quality from a fescuegrass sod on a Typic Fragiudult in Arkansas (Edwards and Daniel, 1993).

Treatment		NO_3-N	NH_4-N	Total N	DP§	Total P	COD★	TSS¤	EC
Manure†	Intensity ($cm\,h^{-1}$)				$mg\,L^{-1}$				$mS\,m^{-1}$
Control	5	0.1	1.3	5.7	0.8	1.1	78.3	6.6	58.0
Low	5	0.1	32.0	42.7	11.9	11.9	282.3	50.7	118.7
High	5	0.1	51.7	77.3	29.4	29.7	504.0	83.0	306.7
Control	10	0.1	2.3	4.7	0.9	1.0	82.7	7.8	51.3
Low	10	0.1	18.7	28.7	8.0	9.5	206.7	50.7	81.3
High	10	0.1	28.3	38.0	13.9	15.8	315.3	74.2	142.7

†Low and High represent applications supplying 214 and 435 $kg\,ha^{-1}$ total N and 7 and 14 $kg\,ha^{-1}$ DP, respectively.
§DP = dissolved phosphate.
★COD = chemical oxygen demand.
¤TSS = total soluble salts.

Table 7. Percentage of Vermont streams expected to have P concentrations exceeding a threshold of 20 $\mu g L^{-1}$ given one source per watershed (Clausen and Meals, 1989).

Source	Total P		Vermont streams exceeding 20 $\mu g P L^{-1}$ (%)
	Concentration ($mg L^{-1}$)	Daily export ($g d^{-1}$)	
Raw milkhouse waste	79.9	98.9	64
Milkhouse waste with filter strip	11.4	11.3	32
Raw barnyard runoff	20.1	79.7	58
Barnyard runoff with filter strip	18.8	44.0	41
Winter manure application	1.2	215.0	37

eutrophication of adjacent waters due to P runoff has been identified. Manure application on frozen ground causes major contamination of surface waters with nutrients (Young and Mutchler, 1976) but incorporation even to relatively shallow depths (15 cm) results in little nutrient runoff (Walter et al., 1987).

Vegetative filter strips, while effective in reducing the concentration of total P from milkhouse waste and barnyard runoff, were still not adequate to meet water quality standards (Table 7) (Dillaha, 1988; Clausen and Meals, 1989; Chaubey et al., 1994). Riparian buffer zones can have positive environmental impacts on sediment and P loads in surface runoff and on NO_3-N in diffuse subsurface flow (Muscutt et al., 1993). However on heavy soils with subsurface drains which provide major flow pathways through riparian areas, positive environmental effects would be minimal without additional measures. Fortunately many agricultural pollution sources are not adjacent to water bodies, affording additional opportunities to treat the runoff before it reaches the water.

These results indicate that water quality is impacted the most by runoff immediately following application of fertiliser or manure/sludge on the soil surface but the effect decays rapidly in subsequent runoff events. In addition, concentrations decrease with increasing rainfall intensity. The impact on water quality of manure disposal on soil is through a surface concentration of nutrients coupled with a decreased potential for the soil to sorb P applied subsequently (Sharpley and Halvorson, 1994). Incorporation of manures drastically reduces nutrient losses in runoff but may promote increased sediment transport depending on conditions. Nevertheless, levels of PO_4-P and COD which can enter surface water are of great concern from a eutrophication point of view. Thus management practices should focus on reducing runoff particularly during the period following surface application to pasture. Because of the pressures to dispose of these types of animal wastes in a relatively small area surrounding sites of production, continued applications to the same piece of land are likely, which will greatly exacerbate the effect by continuously increasing the levels of available N and P in the surface soil. With time,

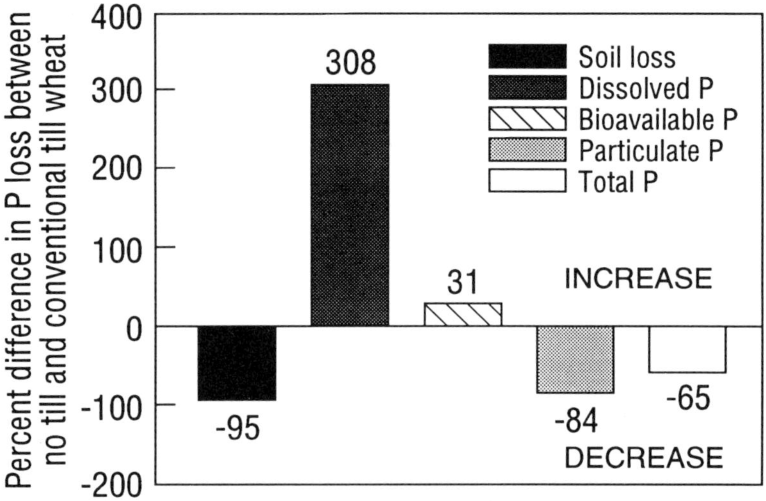

Figure 9. Differential effect of no-till wheat in the Southern Plains on soil loss and categories of P in runoff water (Sharpley et al., 1994).

overland flow is likely to be negatively impacted. Consequently, long-term assessments of the impacts of continued applications of animal wastes to land are required. In areas of concentrated production such as feedlots, filter strips offer only a partial solution to overland flow of nutrients to adjacent water bodies. Composting of animal wastes prior to application is a means of reducing the levels of soluble N and P which may enter runoff waters and is one management practice that will have beneficial effects on water quality. There are also indications that organic P applied in manures may leach more readily through soils than fertiliser P.

3.1.4 Managing soil phosphorus

From the previous discussion, P contamination of surface waters from agricultural runoff is clearly a major environmental concern. Williams et al. (1993), using the CREAMS model parameterised from 162 sites in 22 states, demonstrated that filter strips could potentially reduce N and P losses in surface runoff two-fold from a variety of row cropping systems in the Midwest, South and East of the USA. Some limitations of models are discussed in Chapter 2. The potential benefits of the use of filter strips has been recognised by Federal and State Governments who are encouraging the practice by financial incentives (Dillaha et al., 1989). Total P loss from land can be minimised by control of erosion utilising better vegetative cover to protect the soil surface (conservation tillage, buffer strips, riparian zones, cover crops, etc.) but this does not reduce the levels of soluble P in the runoff water. This effect is illustrated in Figure 9. In excellent reviews, Sharpley et al. (1994) and Davis (1994) addressed the issues and options in managing soil P which are summarised and discussed below (see also Chapter 2).

Incorporation of P fertilisers and manures reduces concentrations of P in runoff water. Furthermore basing P application rates on eutrophic rather than agronomic considerations will further reduce losses. To decrease potential losses, fertiliser inputs can easily be adjusted to balance crop P uptake. However current P soil tests do not adequately measure the availability of P in manures and are sometimes not particularly well calibrated. Fertiliser recommendations need to give credit for P and N in manures, which is seldom done. Manures should be appropriately analysed to determine N and P contents before application to increase effective nutrient use and reduce runoff losses. Application rates on new land should be based initially on N requirements of the crop but with continued use which results in accumulations of P, a change-over to management based on soil P should be introduced. Serious consideration should be given to pretreating manures with chemicals to reduce the levels of soluble P in the material. Manipulation of dietary intake of animals to reduce P concentrations in excreta can also reduce the adverse impacts of manure on water quality (Chapter 2). In addition, soil tests need to be interpreted from an environmental perspective to develop sound P management programmes which will reduce the potential for pollution to a minimum. Periodically no-till systems may have to be ploughed up to redistribute P accumulations from surface applications throughout the root zone. However the needs of conserving soil and preventing P pollution of surface waters are often in conflict. For example, BMPs may require the maintenance of 30% residue cover on the soil to prevent erosion while incorporating P sources may cause the cover to fall below this level. Thus BMPs should be flexible to maximise environmental benefit. Because soluble P concentrations in runoff from unfertilised native grass watersheds and in rainfall can exceed levels that promote eutrophication (0.05 mg L^{-1}), the above recommendations may not reduce P in agricultural runoff below critical values, emphasising the need to target source areas where potential P losses are greatest. Because P from the crop canopy is a major source of soluble P in runoff water, soil and plant should be considered as a continuum and appropriately assayed to assess the potential contributions of each to the final P concentration in solution.

Identification of sensitive or source areas in a watershed rather than the use of a general blanket treatment over a wide area will increase the effectiveness of strategies to minimise P in runoff water. This can be done using the P indexing system developed to identify soils vulnerable to runoff P loss (see Chapter 2). However, such a system may need recalibration in regions different from the area in which it was developed (USA). The site vulnerability index is obtained by summing the products of weight and value for each of the factors characteristic of a particular site. The index can then be used to identify situations which are likely to promote eutrophication and to develop management strategies to minimise runoff losses.

3.1.5 Irrigation

In areas where flood irrigation is practised, significant amounts of both N and P may be exported from soils in the irrigation tailwater or in runoff after rainfall. In a comparison of irrigated pastures versus cropped (sunflowers/wheat) soils in Victoria, Australia, Small (1985) found that pastures have a greater potential than crops for supplying P and N to runoff. In a study of nutrient runoff from fertilised dairy pastures, Mundy (1993) found that fertilisation with N and P increased nutrient runoff concentrations by 5 and 7 times, respectively. Approximately 2–3% and 3–4% of the applied N and P were lost in runoff following irrigation. More recently, Austin et al. (1994) demonstrated that typical irrigation practice on permanent pastures results in losses of 15% of the applied fertiliser P and 4% of the applied N. However, when irrigation is practised correctly, there should be little or no impact on surface runoff and consequently no further discussion will be offered here.

3.1.6 Best management practices for minimising soil, N and P losses in surface runoff

From the data presented above and the discussion of the processes in Chapter 2, the following guidelines which should be flexible to accommodate conflicting interests, will result in reduced levels of soil, N and P in surface runoff:

- Wherever possible avoid placement of fertilisers and animal manure slurries and sludges on the soil surface;
- Where surface placement is the only option, avoid wherever possible, applying fertilisers and manures immediately prior to rain or irrigation or in the autumn;
- Base fertiliser and manure application rates on sound principles designed to focus more on eutrophic than agronomic requirements;
- Use conservation tillage to minimise runoff and soil loss with occasional tillage to incorporate surface accumulations of P and residues;
- Use terraces, contours, cover crops, wetlands and riparian strips to intercept sediment and nutrients;
- To minimise P in runoff, use the vulnerability index to target other BMPs to sensitive or source areas rather than implementation over broad areas.

In a long-term (15 years) evaluation of BMPs to control non point source (NPS) pollution of a lake in Wisconsin, Garrison and Asplund (1993) found that even after the implementation of the NPS measures which involved improved animal waste (storage of manure for 6 to 12 months, redesign of feedlots to prevent clean water from entering, filter strips between feedlots and stream, manure only spread on unfrozen ground) and cropland (grassed waterways, less row and more hay crops, contour strips, settling ponds) management, the lake water quality with respect to P deteriorated. In another shorter-term Wisconsin study, Walker and Graczyk (1993) found that BMPs in one of two river catchments significantly reduced sediment and NH_4-N storm mass transport but had no effect on P in either catchment. The above information suggests that it might not be possible to prevent or even reduce the transfer of P from agricultural lands to water bodies.

3.1.7 Pesticides

Prior to about 1940, pesticides used in agriculture were derived mostly from natural products e.g. nicotine, pyrethrum, petroleum oils and inorganics such as S, As, Pb, Cu, Hg and lime (National Research Council, 1993). After this period, organochlorine and phenoxy herbicides were widely used up to the 1960s, when these were phased out in favour of triazine and amide herbicides and carbamates and organophosphates. The application of pesticides is increasing in many countries, particularly those developing countries in the Australasia-Pacific region where agriculture still represents the major economic activity. There are limited data on contamination of water supplies by pesticides in this region (see Chapters 14–25), but from experience in USA and Europe it is likely that contamination is widespread. Recent assessments of pesticide contamination of waters in USA by the US Environmental Protection Agency (US EPA, 1990) and Oceania (Theng et al., 1995) indicate that contamination is widespread, but often at low concentrations which do not exceed relevant water quality standards.

The fate and behaviour of pesticides in soils is described in Chapter 4. Contamination of surface waters by pesticides occurs either due to runoff of soluble pesticides, or transport of strongly sorbed pesticides by clays and organic colloids in suspension. Factors affecting surface runoff of pesticides are similar to those affecting P transport, namely climate, topography, soil structure and texture, soil dispersibility, pesticide properties and management factors – mode of application, application rate, etc. (Triegel and Guo, 1994). Pesticides that are strongly sorbed by soil will be particularly susceptible to losses in runoff rather than leaching. Pesticides that exhibit such behaviour include dichlorodiphenyltrichloroethane (DDT) and its metabolites, dieldrin, endosulfan, lindane, heptachlor and chlordane. Thus, erosion control practices, as discussed for non-point pollution due to P runoff above, will also generally help to limit runoff of these pesticides (see below). However, exceptions to this occur, where crop residues retained on soil to control soil erosion may actually lead to increased losses of applied pesticide, due to poor contact between pesticide and soil in such systems (Baker et al., 1978). Where rainfall (or irrigation) intensities are high following pesticide application, losses of soluble pesticides such as 2,4 dichlorophenoxyacetic acid (2,4 D), molinate or 4-chloro-2-methylphenoxyacetic acid (MCPA) may be high and unaffected by management to control soil erosion (Wauchope, 1978).

A list of the possible management options to reduce pesticide contamination of waters has been suggested (National Research Council, 1993):

- selection of proper pesticides and formulations;
- timing of and improvement in pesticide application methods to minimise drift and volatile losses;
- use of erosion and runoff control measures;
- use of non-chemical control measures such as crop rotations;
- integrated pest management, which embodies most of the above practices.

3.2 Groundwater

Groundwater pollution control is of growing concern worldwide. Among the nutrients, leaching of NO_3-N is of prime concern. In the US, more than 50% of community and rural wells are contaminated with NO_3-N with 2% exceeding the drinking water standard of 10 mg NO_3-N L^{-1}. Most of this contamination stems from production agriculture in the form of fertilisers and manures. Whenever the soil's capacity to retain water is exceeded, NO_3-N present will be subject to leaching and ultimately will affect stream concentrations. Thus precipitation patterns, spring thaws and irrigation have major influences on NO_3-N leaching with the effect increasing as the texture of the soil becomes more sandy. Tillage, cropping systems and nutrient applications also play major roles. However there are few long-term studies of the impacts of agriculture on groundwater quality. Because one of the foci of this paper is water quality, the remainder of the discussion in this section will deal mainly with NO_3-N concentrations in groundwater rather than losses of actual N which, in terms of the amounts of fertilisers and manures applied, are relatively small and of little significance from a crop production point of view.

High concentrations of NO_3-N (up to 80 mg L^{-1}) in groundwaters may occur naturally, particularly in arid regions (Heaton, 1964; Barnes et al., 1992). Barnes et al. (1992) suggest that these high NO_3-N concentrations are due to N fixed by bacteria associated with termite mounds. Thus the prime source of groundwater NO_3-N is not necessarily associated with agriculture.

3.2.1 Tillage

Baker (1987) concluded from a review of the literature that conservation tillage is likely to increase the proportion of water falling on a soil that ends up as subsurface flow. Thus relative to conventional tillage, leaching losses of NO_3-N and other contaminants not strongly sorbed by the soil are likely to increase, but this is not always the case. Early work by Thomas et al. (1973) and Tyler and Thomas (1977) in Kentucky showed that more NO_3-N leached below 90 cm in a killed sod no-till system than conventional tillage. On the other hand, no differences in and less leaching of NO_3-N have been observed in no-till than mouldboard plough treatments (Meek et al., 1995). These apparent differences were due to differences in macropore flow (Chapter 2) in the two systems studied, which highlights the importance of other variables such as fertilisation method, rainfall and soil type before any generalisations can be made. Considerably more information is needed before the long-term effects of conservation tillage on NO_3-N leaching can be predicted accurately.

Keeping land in fallow exacerbates NO_3-N leaching because of the absence of uptake. This increased leaching can be up to nine-fold more than from grassland and twice that of cropped land (Juergens-Gschwind, 1989). The practice of fallowing particularly in the growing season to control weeds or growth of a volunteer crop will increase the potential for NO_3-N leaching whereas fallowing to conserve moisture for the following season's crop is likely to present less of a problem in low rainfall areas. Nevertheless, even summer fallow practices to conserve water can contribute to shallow groundwater NO_3-N contamination as a result of the associated mineralisation of OM (Bauder et al., 1993). The payment-in-kind programme (PIK) in the USA in which farmers are required to idle 20% or more of their land while controlling weeds and providing residue cover creates ideal conditions for NO_3-N leaching in high rainfall areas.

3.2.2 Cropping Systems

Cover crops have a definite role to play in reducing erosion and NO_3-N leaching by incorporating the N into biomass provided that the reason for fallowing is not to conserve water. An ideal cover crop would be one which reduces NO_3-N leaching and, at the same time, supplies N to the following summer crop. Meisinger et al. (1991) summarised the effects of grass, brassicas and leguminous cover crops on subsurface water quality and only those studies dealing with lysimeters and field situations will be presented here (Table 8). Clearly non-leguminous crops are much more effective than legumes in preventing the leaching of NO_3-N to groundwater. The N stored in the winter cover crop will be mineralised in the growing season acting as a slow release fertiliser, but synchronisation of crop N requirements with the supply is difficult to achieve because the required cultural practices may not be physically possible.

In high production fertilised grazed pastures, the potential for NO_3-N leaching is greatly increased. On the other hand, losses from ungrazed pastures even at high rates of N (420 kg ha^{-1}) can be as low as 29 kg N ha^{-1} (Garwood and Ryden, 1986). As discussed in Section 3.2.3, grazing animals appear to be the key to NO_3-N leaching (Ball and Ryden, 1984; Ryden et al., 1984). Concentrations of NO_3-N in subsurface flow from summer and rotational winter grazing/feeding pastures in Ohio increased over a five year period with the winter pastures exceeding 10 mg L^{-1} for the entire period while summer grazing remained below this level (Owens et al., 1983).

Legumes are an alternative to fertiliser in supplying N to crops and pastures. Losses of 5–40 kg N ha^{-1} as NO_3-N have been reported in crop rotations involving legumes and small grains with most of the loss taking place immediately after residue incorporation (Bergström, 1987; Owens, 1990). Legumes therefore behave as slow release N sources for the associated crop and control leaching more effectively than when inorganic sources of N are used. On the other hand when the legume is ultimately killed, large quantities of N can be mineralised and re-enter the labile pool (Bergström, 1987). Thus the N fertiliser rate for subsequent crops must be adjusted accordingly to reduce leaching losses.

On the other hand legumes such as lucerne and grasses can be efficient scavengers for deep NO_3-N left over by previous crops or released under feedlots (Owens et al., 1994). In a long-term study in which grass watersheds were fertilised at 224 kg N ha^{-1} yr^{-1} for five years, NO_3-N concentrations increased throughout the period to levels exceeding the drinking water standard. In the sixth year, N applications were replaced by overseeding lucerne after which NO_3-N levels began to

Table 8. Effects of grass, brassicas and legume crops on subsurface water quality (Meisinger et al., 1991).

Cover crop	N leached		Reduction due to cover crop		Brief description of study
	N Mass $kg\,ha^{-1}$	NO_3 conc. $mg\,L^{-1}$	N Mass %	NO_3 conc. %	
				Grasses	
None	83	21	–	–	Lysimeter 75 cm deep, sandy loam, tobacco, $224\,kg\,N\,ha^{-1}$, 10 yr, Connecticut
Oats	36	11	57	48	
Rye	28	8	66	62	
None	103	48	–	–	Lysimeter 100 cm deep, silt loam, winter wheat, $200\,kg\,N\,ha^{-1}$, 1 yr, France
Ryegrass	40	25	63	48	
None		17			Groundwater, silt loam, maize, $336\,kg\,N\,ha^{-1}$, Maryland
Rye		12	–	29	
				Brassicas	
None	52	75	–	–	Lysimeter 1.2 m deep, loam, sudangrass unfertilized, cover crop with $110\,kg\,N\,ha^{-1}$, 5 yr, California
Mustard	10	15	80	80	
None	197	37	–	–	Lysimeter 80 cm deep, sand, $375\,kg\,N\,ha^{-1}$ as manure, 1 yr, Sweden
Rape	70	14	66	62	
				Legumes	
None		17		–	Groundwater, silt loam, maize, $330\,kg\,N\,ha^{-1}$, 1 yr, Maryland
Hairy vetch		18		6	

decline steadily until, in the tenth year, concentrations were at levels similar to those before fertilisation (Owens et al., 1994).

Field borders, filter strips and riparian zones if planted to deep rooted trees and grasses can serve as filters for subsurface flow (Cooper et al., 1994; O'Neill and Gordon, 1994) but in order for such systems to be sustainable, biomass would require periodic removal to prevent buildup adjacent to streams. Raising the level of the water table has also proved to be of benefit in reducing the concentrations of NO_3-N entering groundwater due to increased denitrification (Gilliam et al., 1979).

3.2.3 Fertilisers and Manures

3.2.3.1 EFFECT OF ROW CROPPING ON NITRATE MOVEMENT

Intensive crop production of non-legumes requires large inputs of N as fertiliser and manure (often >200 kg N ha^{-1}) to reach optimum yields, which creates an increased potential for NO_3-N leaching to groundwater. Timing and placement of N is crucial in minimising leaching losses (Burwell et al., 1976; Anderson and Randall, 1992; see also Chapter 2). Lysimeter measurements have shown that movement of applied fertiliser N can be rapid and losses of NO_3-N substantial, increasing with N fertiliser rate (Bergström and Johansson, 1991). Weil et al. (1990) in Maryland found that, on sandy Coastal Plain soils, N rates typical of irrigated maize production result in NO_3-N concentrations in groundwater in the range of 10–20 and 20–30 mg L^{-1} from fertiliser and manure, respectively. On the other hand, NO_3-N concentrations in groundwater under forest in watersheds dominated by crops were unaffected by the N applied to cropland. This suggests that forests may protect surface waters from eutrophication as a result of groundwater N discharges to the surface even when the forest is not riparian.

Under maize production in Nebraska, comprising over 84,000 ha of irrigated land, groundwater NO_3-N concentrations were positively correlated with the amount of NO_3-N remaining in the top 90 cm of soil from the previous season (Schepers et al., 1991); farmers were overestimating yield goals resulting in the application of excessive amounts of N fertiliser. Thus, protection of groundwater quality requires improved farmer education and monitoring and control of fertiliser practices (Rouvé and Bogacki, 1989).

While precision placement of N fertiliser might have a potential to reduce leaching losses (Timmons and Baker, 1991, 1992; Chapter 2), Baker and Timmons (1994) using ^{15}N-labelled fertiliser found that point-injection was better than knifing and surface banding (Table 9). Because maximum loss of N from fertiliser was only 20% of that applied, precision placement of N fertilisers is only likely to have a limited effect on NO_3-N leaching with N rate having a much greater impact. Even at economic optimum N rates (EON) of 192 and 67 kg N ha^{-1} for nonmanured and manured maize, respectively, concentrations in drainage water were 18.8 and 19.2 mg NO_3-N L^{-1}, respectively, twice the drinking water limit (Table 10). When no N was applied, NO_3-N concentrations were of the order of the drinking water

Table 9. Effect of method and rate of N fertilizer application on concentrations and losses of total and labelled ^{15}N as NO_3-N in lysimeters of an Aquic Hapludoll in Iowa (Baker and Timmons, 1994). Means followed by the same letter are not significantly different at the 5% level.

Method of N application	Concentrations ($mg\,L^{-1}$)						Losses ($kg\,ha^{-1}$)						Ratio of labelled $^{15}NO_3$-N to total NO_3-N		
	Total NO_3-N			Labelled $^{15}NO_3$-N			Total NO_3-N			Labelled $^{15}NO_3$-N					
	1984	1985	1986	1984	1985	1986	1984	1985	1986	1984	1985	1986	1984	1985	1986
Point injected (high/split)†	3.8a	23.0a	20.5a	1.3ab	5.1a	0.8a	8.4a	26.4a	76.8a	1.0bc	6.5a	3.1a	13.2ab	21.6a	4.2ab
Point-injected (low/split)*	4.0a	12.0b	8.1b	0.2b	1.0b	0.2b	7.5a	18.3a	36.0b	0.2c	1.7a	1.0b	2.6b	7.8b	2.4b
Knifed-in[¤]	5.9a	18.2ab	12.4b	1.9a	4.6a	0.6a	0.7a	26.8a	50.3ab	1.6ab	6.8a	2.4ab	17.7a	25.2a	4.9a
Surface-banded[§]	4.4a	16.4ab	10.4b	2.2a	3.5ab	0.5ab	14.5a	26.6a	48.7b	2.7a	5.7a	2.3ab	20.1a	20.9a	4.7ab

†50, 50, 100 $kg\,N\,ha^{-1}$ in three split applications.
*25, 50, 50 $kg\,N\,ha^{-1}$ in three split applications.
[¤]200 $kg\,N\,ha^{-1}$ at 0.15 m deep 0.3 m from row applied once.
[§]200 $kg\,N\,ha^{-1}$ at 0.3 m from row applied once.

Table 10. Effect of N rates and manure on flow-weighted NO_3-N concentrations and NO_3-N leached in zero tension lysimeters on a Typic Hapludalf in Pennsylvania (Jemison and Fox, 1994).

Treatment	Flow-weighted NO_3-N concentrations			NO_3-N mass leached	Fertilizer N leached at EON
	GS†	Period evaluated OW† mg NO_3-N L^{-1}	AN†	kg ha^{-1}	%
		Nonmanured			
0 N	9.2	6.2	7.7	37.1	
100 N	13.3	10.7	11.9	66.7	
200 N	21.6	18.2	19.8	109.7	
EON* (192)			18.8	106.7	35.8
$LSD_{0.05}$	3.7	2.5	1.7	18.4	
		Manured			
0 N	15.3	10.0	12.6	57.8	
50 N	15.8	11.5	13.5	67.5	
100 N	22.9	18.8	20.6	101.9	
EON (67)			19.2		
$LSD_{0.05}$	2.7	3.5	2.8	17.9	

*EON = economic optimum N.
†GS = growing season, OW = over-winter, AN = annual.

Table 11. Amount of N leached from lysimeters containing a Typic Paleudult in a cropped (1984) and uncropped (1983) season in Nigeria (Wong et al., 1987).

Treatment	Amount of N leached (kg ha^{-1})					
	1983			1984		
	Soil N	Fertilizer ^{15}N	Total N	Soil N	Fertilizer ^{15}N	Total N
1983 – 138 kg ^{15}N ha^{-1} 1984 – 138 kg N ha^{-1}	156	95	251	–	17	106
1983 – 0 N 1984 – 92 kg ^{15}N ha^{-1}	154	0	154	87	45	133

limit (Jemison and Fox, 1994). Under horticultural production about twice as much NO_3-N leaching occurs as with agronomic crops because there are long periods during which the soil is usually without cover, the soil is intensively cultivated and large amounts of crop residues are returned to the soil (Juergens-Gschwind, 1989). Under humid tropical conditions, great potential for NO_3-N leaching exists (Table 11) (Wong et al., 1987).

The long-term impact of row crop agriculture on groundwater quality in the Big

Spring basin (270 km^2) in Iowa over a period of 30 years is illustrated in Figure 10. The NO_3-N concentrations in the 1930s were <1 mg N L^{-1} but had increased three- and tenfold by 1950 and 1983, respectively, which directly paralleled the increases in manure and fertiliser applications of 30 and 300%, respectively. Notice the effect of the payment-in-kind (PIK) programme resulting in marked reductions in manure and fertiliser use, which were only reflected in lowered groundwater NO_3-N concentrations a few years later. Excesses of N in soils take long periods (>40 years) to appear in subjacent aquifers because of the slow transit time of NO_3-N which is determined by the thickness of the soil material overlying an aquifer. This means that reclamation time for such aquifers would be even longer.

The ability to use models to predict NO_3-N leaching through the soil profile (Chapter 2) in practice would be of great value in managing the quality of groundwater but that stage has not yet been reached (Jemison et al., 1994). Thus application of fertilisers and particularly manures to row crops is likely to cause contamination of groundwater by NO_3-N and, in some instances PO_4 as well. There are few control measures that can be adopted except the judicious use of these materials, such that they never exceed the N requirements of the crop being grown. Considerable carry-over of NO_3-N in the soil profile at the end of the maize season occurs in the US (Vanotti and Bundy, 1994a). Soil testing for NO_3-N prior to planting row crops can effectively predict N response and the amounts of supplemental N needed to produce optimum economic yields of crops following in the rotation (Bundy et al., 1992) which is an important strategy in minimising NO_3-N leaching losses from cropland (Vanotti and Bundy, 1994b). Use of rapid tissue assays has improved accuracy in determining the sidedress N requirements of rice, maize and cotton, thus reducing the potential for NO_3-N leaching (Edmisten et al., 1992). In addition, on-farm nutrient balance assessments provide an integrated approach to nutrient management which has the potential of reducing loadings on eutrophic farms (Lanyon and Beegle, 1989). The newly developed soil specific crop management technology (farming by soil) is an emerging, sophisticated technique for matching the field application of fertilisers and pesticides to soil types. This technology has been shown to be not only environmentally friendly, but also economically advantageous to farmers who practise it in terms of savings in fertilisers and pesticides (Marks and Ward, 1993). Young, technically well informed, educated and environmentally concerned producers are more likely to apply appropriate rates of N (Supalla et al., 1995), which indicates that education is the key and should address the linkages between N management and deteriorating water quality and also how recommendations are developed to increase confidence in following apparently low recommendations for N. Fertiliser N is not the main cause of NO_3-N leaching but rather arable agriculture is the key (Addiscott and Powlson, 1989). No matter how careful the farmer is in determining N requirements and following recommendations, a 'window' for NO_3-N to enter groundwater between tillage and the establishment of the next crop will always exist.

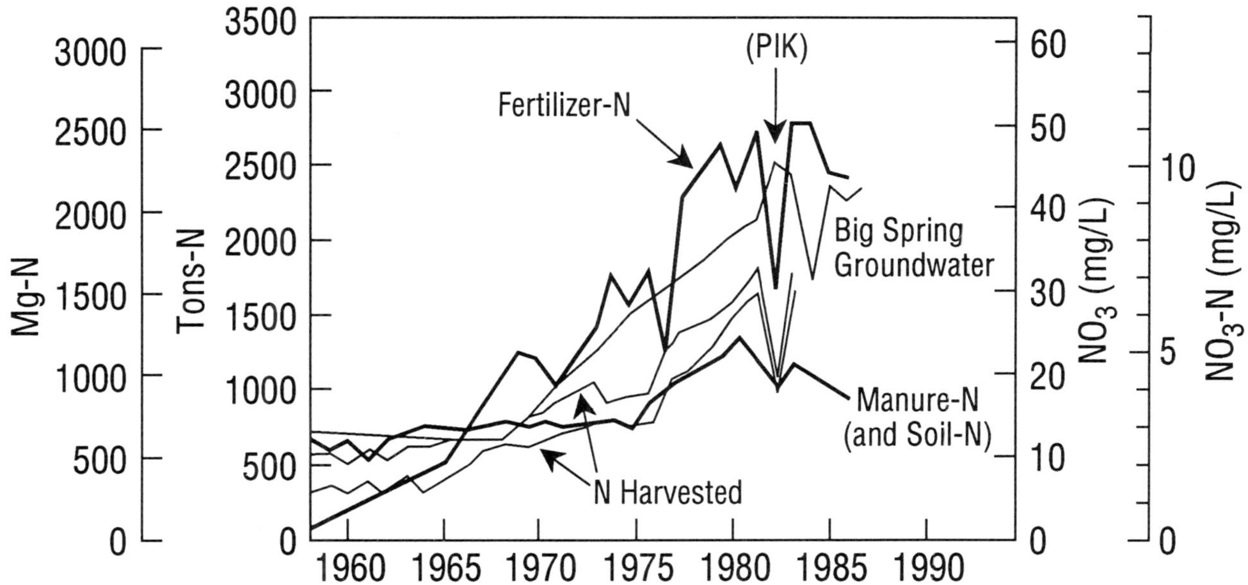

Figure 10. Changes in groundwater NO_3-N resulting from increases in manure and fertilizer usage in the Big Spring Basin in Iowa (Hallberg and Keeney, 1993).

3.2.3.2 EFFECT OF ROW CROPPING ON PHOSPHATE MOVEMENT

Histosols, when drained whether fertilised or not, are heavy contributors to both groundwater NO_3-N and PO_4-P with fertilisation exacerbating the problem (Hortenstine and Forbes, 1972). Draining a Lithic Medisaprist in Florida, resulted in high concentrations of soluble P (0.2–1.0 mg L^{-1}) in drainage water but sugar production had no measurable effect on groundwater quality (Coale et al., 1994). On a sandy Typic Paleudult receiving high rates of pig lagoon effluent (up to 1300 kg N ha^{-1} yr^{-1} for four years), flow weighted NO_3-N and PO_4-P concentrations in subsurface water increased with application rate to levels which would cause environmental damage to water bodies (Evans et al., 1984). Natural and accelerated (due to tile drainage) subsurface runoff contributes to water quality decline. When the water table approaches the plough layer (0–15 cm), elevated levels (0.12 mg L^{-1}) of dissolved P can enter tile drains (Sawhney, 1978). Gold and Loudon (1986) found flow-weighted mean concentrations over 22 rainfall events of 0.04 and 0.07 mg P L^{-1} and 11.7 and 10.5 mg NO_3-N L^{-1} for conventional and conservation maize tillage regimes, respectively. Normally P is sorbed from percolating water but artificial drainage accelerates movement into streams. Sharpley and Withers (1994) summarised the effects of fertiliser P on subsurface movement of P. Comparisons of drained and undrained fields demonstrate that P concentrations and losses are increased in the former because of a shorter contact time with the soil during which P could be sorbed. This phenomenon also accounts for the lower soluble P loads from a 1 m versus a 0.6 m deep tile drain under lucerne in Canada (Culley et al., 1983). Losses of P in subsurface flow are always insignificant in comparison to the amount added as fertiliser.

3.2.3.3 EFFECT OF GRASSLAND ON NITRATE AND PHOSPHATE MOVEMENT

An intensely grazed legume based pasture should be an ideal agroecosystem where fertiliser and fossil fuel inputs are minimal, soil OM is accumulating and the N cycle should prevent NO_3-N from leaching. However this has not proved to be the case. In New Zealand, Ryden et al. (1984) found 5.6 times as much NO_3-N below grazed than ungrazed ryegrass pastures while Baber and Wilson (1972) reported a mean NO_3-N concentration of 25 mg L^{-1} in groundwater under an intensive pastoral area on the North Island. Sharpley and Syers (1979) found that on permanent pasture receiving 60 kg N ha^{-1} as urea, grazing animals caused a dramatic increase in the concentrations of NO_3-N, dissolved inorganic P and particulate P (5-, 15- and 40-fold, respectively) in tile drainage water in New Zealand. The effect of urea fertiliser on NO_3-N concentrations was of the same magnitude as that for grazing. The increases in NO_3-N leaching due to grazing stem from urine affected areas where the conditions (high pH, high concentrations of urea) produce high concentrations of NO_3-N in excess of requirements (Ball et al., 1979). Groundwater under intensively fertilised grazing frequently exceeds water standards for NO_3-N (Roberts, 1987). The level of N fertilisation with inorganic and manure N relative to pasture requirements is directly related to losses of NO_3-N to groundwater, being highest under grazing (Owens et al., 1982, 1983; Sveda et al., 1992). Leaching losses of NO_3-N from pasture are often higher than from fallow or crop land (Goh et al., 1979). Even under legume-based pastures, leaching losses of N of the order of 20–30 kg N ha^{-1} yr^{-1} take place (Helyar and Porter, 1989). Indeed, high NO_3-N concentrations have been reported in 50% of samples of groundwater in the south-east of South Australia where the only major source of N added to soils is through N fixation by legumes (Scott and Dillon, 1989).

Use of slow release N sources (methylene urea) or nitrification inhibitors (nitrapyrin) merely delay leaching of N which in the long term is similar to that from NH_4NO_3 (Owens et al., 1992) but may reduce potential nonpoint source pollution if coupled with appropriate N rates and irrigation. However flow-weighted concentrations of NO_3-N in leachate do not meet drinking water standards most of the time irrespective of nitrification inhibitor treatment (Owens, 1987).

While grassland has the potential to limit the leaching of NO_3-N, ploughing old grassland under good drainage conditions releases large amounts of N which can result in a large pulse of NO_3-N entering groundwater (Juergens-Gschwind, 1989).

On light textured soils, leaching losses of P may be considerable, and in Australia there are significant areas of light textured soils where P movement presents agronomic problems for pasture nutrition (Lewis et al., 1981; Diggle and Bell, 1984). In Northern Ireland, McAllister (1976) found elevated concentrations of P in drainage water from areas treated with liquid livestock manures on cracking soils or where permeable backfill had been used over drains. Sharpley and Syers (1979) found rapid responses in dissolved P in drainage water under pasture when grazing commenced with a concentration of 0.25 mg L^{-1} being reached within one week. Losses of P through leaching on sandy soils can be as high as 100% of the amounts supplied in fertiliser (Lewis et al., 1981) and significant concerns have been raised with regard to eutrophication of streams and coastal waters (Weaver and Prout, 1993). This has led to the development of alternative fertilisers to superphosphate for topdressing of pastures (Yeates and Clarke, 1993), such as reactive rock phosphates and superphosphate/rock phosphate mixtures. Changing fertiliser type in this way resulted in significant reductions in P concentrations in rivers and streams in the affected catchments (Schofield and Birch, 1986). An alternative technique to reduce P leaching losses in light textured or organic soils is the addition of material with a high sorption capacity for P, either to the soil or to drains in the affected areas. In Germany and Australia, red mud residue from the alumina industry has been effectively used in this regard (Scheffer et al., 1986; Vlahos et al., 1989).

Most of the data for grassland and pastures suggest that fertilisation increases NO_3-N concentrations in groundwater which is further exacerbated by placing grazing animals on the pasture because most of the N is returned in the faeces and urine. Nitrification inhibitors merely delay the process. Composting of animal manures before application to land which reduces the potential for leaching, may be one of the most effective ways of reducing the levels of readily available N and P in the materials.

3.2.3.4 EFFECT OF FEEDLOTTING ON NITRATE AND PHOSPHATE MOVEMENT

With the increasing trend to feedlots and housing animals as a means of meat production in many countries, what was previously a diffuse source of pollution (manure on pastures) has become a point source of pollution. For example, in Australia more than 0.5 million cattle are now held in feedlots, and piggeries are expanding in size (Bowmer and Laut, 1992). Walker et al. (1979) estimate that in a typical feedlot, application rates of N and P to soil are approximately 37 and 9 t ha^{-1}, respectively per 150 day feeding period. However, provided the feeding pens and waste disposal systems are properly designed, movement of nutrients offsite can be minimised (Walker et al., 1979). However, some feedlots do not provide for adequate waste disposal and significant leaching of NO_3 in the area used for waste landspreading may be observed (Dillon, 1992).

3.2.4 Irrigation

3.2.4.1 NITRATES

Nitrate-N in a soil profile, being highly soluble, is readily leached with irrigation water and the high rates of leaching and nitrification common in irrigated soils together with relatively high fertiliser N rates combine to make NO_3-N losses to groundwater a serious problem. Extensive experimentation on irrigated maize (Timmons and Dylla, 1981; Hergert, 1986) and other crops (Hubbard et al., 1984) has shown that irrigation promotes the leaching of NO_3-N to groundwater, which is particularly true for sandy soils (Lembke and Thorn, 1980). The amount of nitrate moving through sandy soils is dependent on the concentration of NO_3-N and the flow of water. This makes N fertiliser management difficult but careful management of irrigation can reduce losses to a minimum (Smitka et al., 1977). Matching irrigation to evapotranspiration (ET) and N applied in fertiliser (rate and source) and irrigation water to crop requirements can reduce losses but on sandy soils, loss cannot be reduced to zero (Hergert, 1986). However even careful irrigation management will not always reduce NO_3-N losses to acceptable levels, mainly because of losses during winter and early spring caused by precipitation when the crop was not actively growing (Watts et al., 1994). Annual flow weighted concentrations of NO_3-N in leachate from irrigated maize can range as high as 70 mg N L^{-1} (Timmons, 1984) with losses in the range of 150–250 kg N ha^{-1} yr^{-1} being common (Timmons and Dylla, 1981). On fine textured soils, NO_3-N movement below the rooting depth can be more easily minimised than on sandy soils when irrigation and N are managed to meet crop needs (Edwards et al., 1972).

Losses of N in irrigated systems are related to the type of irrigation employed. Drip irrigation minimises N losses compared to furrow/flood irrigation (Weinbaum et al., 1992) due to better interception of applied nutrients in the wetted area under drip irrigation compared to furrow (Weinbaum et al., 1992). Low Energy Precision

Application and centre pivot irrigation systems to deliver N can reduce N losses by leaching relative to furrow irrigation (Geleta et al., 1994).

Irrigated high value horticultural crops often have rates of N and P application well in excess of crop demand (Nightingale, 1972). In the San Joaquin valley in USA, Miller and Smith (1976) estimated the removal of N in harvested crops as a percentage of N applied, with fruit tree production being the least efficient (<20% N applied removed in produce) followed by grapes (37%), vegetable crops (±50%), field crops (±55%) and grass for hay (>70%). In a review of nutrients in irrigation drainage in the Murray-Darling basin in Australia, Harrison (1994) found that drainage from irrigated horticulture was generally low in P and high in N, with the opposite being true for irrigated pasture soils. High nitrate N concentrations (>50 mg L^{-1}) were observed in many areas although losses of N were found to occur within the drainage system prior to entry into surface water supplies. This was attributed to uptake of N by aquatic vegetation/organisms and denitrification losses in drains.

Reducing NO_3-N losses to zero appears to be clearly impossible on sandy soils under irrigation. To minimise losses, applications of N prior to sowing should be limited to low levels to avoid leaching before the crop has established a sizeable root system, the bulk of the N fertiliser should be applied during the period of maximum uptake and irrigation scheduling should be practised.

3.2.4.2 SALTS

In general, drainage water from irrigation is often more saline than the irrigation water applied. Such saline water can have a great impact on water quality in some cases, rendering it unsuitable for further irrigation or domestic and stock uses. For example, on the Colorado River, 37% of the salt in the lower reaches stems from irrigated agriculture causing much greater impacts on industrial and municipal users (damage to plumbing and water heaters, increased costs of water treatment, increases in objectionable taste characteristics) (Holburt, 1984) than for agriculture (Jones, 1984). A major source of this water contamination with salts arises from drainage return flows and the use of unlined canals and ditches which not only lead to increased salt discharge due to seepage flow but contribute to low irrigation efficiency as well. Rhoades (1984) and Grattan and Rhoades (1990) have proposed strategies whereby drainage return flows would be intercepted before re-entering the river and be used for the irrigation of salt tolerant crops. When the reuse potential of drainage water is exhausted, it would be discharged to evaporation ponds. The volume of water required for irrigation can also be reduced by reusing shallow groundwater (Kruse et al., 1990).

However recently, an additional threat to water quality has arisen from the discovery of elevated (10- to 100-fold) levels of elements such as Se, B and Mo in reservoirs which have been used to store and evaporate drainage water in the western US (Deason, 1989). Reproductive failures in aquatic organisms and waterfowl due to high Se levels have been found in the Kesterson Reservoir in California (National

Research Council, 1989). The source of the Se is the land being irrigated from which it has been mobilised by the irrigation water. Thus this strategy may not always be valid but may be suitable in certain limited circumstances. There is little question that irrigated agriculture always degrades water quality. The National Research Council (1989) stated: 'irrigation.... over time cannot avoid causing an adverse off-site effect. This effect must be acknowledged: it can be minimised, internalised, or rejected, but it cannot be ignored. If irrigation is a desired use of water, then its waste waters must be treated and/or disposal provided for'.

3.2.5 Best Management Practices to minimise nitrate and phosphate leaching

Keeney (1986) has made a plea for more agroecosystem-oriented concepts and research in developing BMPs which can vary widely due to differences in soils, geology, crops, climate, irrigation, etc. In developing BMPs to reduce nitrate and phosphate losses to subsurface water, the following practices should be followed:

- Take N and P contributions from manures, soil OM, crop residues, stored inorganic N and available P into account accurately in N and P fertiliser recommendations, to prevent accumulations of N and P in excess of crop requirements;
- Tailor make fertiliser applications to supply the needs of the crop as it develops using split applications, slow release fertilisers and/or nitrification inhibitors;
- Use cover crops with deep roots to extract NO_3-N before it enters groundwater;
- Use more N efficient crops where possible;
- Manage irrigation in conjunction with rainfall to minimise leaching;
- Control and intercept runoff from feedlots and dairy barns;
- Where possible, recycle drainage waters through forest soils or natural grassland to reduce N and P levels;
- Pay particular attention to all of the above in sensitive areas (sandy soils and karst areas);
- Adopt a systems approach to solve the peculiar problems of a particular region.

Table 12 is a convenient summary guide to the best management practices for water quality protection from pollution by nutrients. However this is only a general guide and there may be situations where some BMPs will not perform as indicated.

3.2.6 Pesticides

One of the most detailed surveys of pesticide concentrations in groundwaters has been conducted by the US EPA (US EPA, 1990). A total of 101 pesticides and 25 pesticide by-products were monitored in 1292 community water supply or rural domestic wells. Based on the survey results, the US EPA estimates that 10.4% of the community water supplies and 4.2% of rural domestic wells contain detectable levels of one or more pesticides. The most commonly detected pesticides were atrazine and degradates of dacthal or DCPA (Briskin, 1994). In Australia and New Zealand, smaller surveys of pesticides in groundwaters have been conducted on a regional

Table 12. Best management practices for controlling soil (sediment), nitrate and adsorbed nutrients (e.g. phosphate) movement to water supplies.

BMP	Sediment	Soluble nutrients	Adsorbed nutrients	N Loss to groundwater
Conservation tillage	■†	□	■	★
Contour/terracing	■	△	■	★
Strip cropping	△	△	△	□
Filter strip	△	□	△	□
Cover crop	△	△	△	△
Crop rotation	■	△	■	□
Nutrient management	□	■	□	■
Irrigation management	△	△	△	△
Pasture management	△	□	△	□
Manure management	■	■	■	■
Grassed waterway	△	□	△	□
Critical area planting	■	△	△	□
Riperian forest	△	□	△	□
Composting	□	■	■	■
Streambank/waterbody protection	■	□	■	□

†□ No control to low effectiveness, △ Low to medium effectiveness.
■ Medium to high effectiveness, ★ May increase loading.

basis (for a review see Theng et al., 1995). As in the US, contamination is often found to be widespread, but below levels set for drinking water in each country. However, environmental levels (i.e. where impacts on aquatic biota are likely) are sometimes exceeded in localised areas, usually where agricultural activities are more intensive e.g. horticulture, cotton, etc.

The likelihood of pesticides leaching to groundwater depends on a number of factors – pesticide formulation, pesticide persistence, soil sorption and hydraulic characteristics, climate, management, etc. The processes controlling pesticide mobility through soil and modelling of transport to groundwater are discussed fully in Chapter 4.

4. Adverse impacts of agriculture on food quality

At present, a real issue of concern with regard to use of fertilisers and other agricultural chemicals in agriculture is food purity, specifically the introduction of unwanted elements such as heavy metals into foodstuffs. As many of the elements or compounds in question are only present in foods in trace quantities, their presence has only relatively recently been cause for concern as improvements in analytical techniques have indicated the level of contamination.

McLaughlin et al. (1996) recently reviewed the elements in fertilisers which could potentially cause concern from a human health perspective. They concluded that given concentrations of various elements in fertilisers, 'background' concentrations in soils and the potential transfer of contaminants from soils to plants and animals, only As, Cd, F, Hg and Pb need be considered as potential accumulators in soils and crops. Of these, Cd is the element of most concern as it has a relatively high transfer coefficient from soil to plant (compared to As, F, Hg or Pb) and human dietary intakes of this element are under scrutiny in many countries. Fluorine and Pb are strongly sorbed by soil components so their transfer from soil to plants is minimal (Stevens et al., 1995).

4.1 Cadmium

Cadmium is a widespread naturally occurring element, present in all soils, rocks, waters, plants and animals. Cadmium is used in the electroplating industries and is a by-product of smelting processes for other metals such as Zn. Much of the information on Cd toxicology comes from these industries where occupational exposure may be high.

Cadmium is absorbed by animals and humans through ingestion and/or inhalation. A large proportion of the Cd ingested is excreted by the body. However, a small proportion is retained, mostly in the kidneys and liver, so that it may accumulate in these organs. It is thought that above a level of 200 μg Cd g^{-1} of kidney (wet), pathological symptoms may appear (Kjellstrom et al., 1984; Friberg et al., 1985). However, many of the studies examining the effect of Cd on body functions have used short-term high dose techniques, often with pure Cd salts. The effect of prolonged exposure to low levels of Cd in foods (where Cd is often complexed by organic anions) has not been evaluated. To date the only reported human health problems directly attributed to Cd have been caused either by occupational exposure in the mining or metallurgical industries, or by the consumption by a malnourished population of food contaminated by industrial wastes (the famous outbreak of 'Itai-Itai' disease in Japan). Further details of this disease are given in Section 2.7, Chapter 11. Some workers argue that risks due to dietary intake of Cd are low (Ryan et al., 1982), although the World Health Organisation/Food and Agriculture Organisation Expert Committee on Food Additives proposed in 1972 a maximum tolerable level for intake by humans of 6.7–8.3 μg Cd kg^{-1} body weight wk^{-1}. This was revised in 1980 to 7 μg Cd kg^{-1} wk^{-1} and after review recently (1993) was not modified. In line with these regulations some countries have adopted food regulations controlling Cd concentrations (Walker, 1988). In Australia, the National Food Authority (NFA) has specified limits for Cd in certain foodstuffs under the Food Standards Code (NFA, 1993). These limits are detailed in Table 13.

Table 13. Maximum permissible concentrations (MPCs) for Cd in foodstuffs in Australia (Food Standards Code) (NFA, 1994).

Foodstuff	MPC for Cd ($mg\,kg^{-1}$)
Beverages and other liquid food	0.05
Crustaceans and the crustacean content of foods containing crustacean	0.2
Fish and fish content of products containing fish	0.2
Edible offal other than liver	2.5
Liver	1.25
Meat muscle	0.2
Molluscs and mollusc content of foods containing mollusc	2.0
Wheat germ	0.2
All other foods	0.05

There is considerable discussion of the appropriateness of these limits, particularly the broad blanket grouping 'all other foods', which can include both foods eaten in large amounts (staples such as rice, potatoes, bread, etc) and those eaten in small quantities (e.g. chocolate, herbs, etc). Indeed, food standards for Cd are currently under review in Australia.

4.1.1 Sources of Cd in agriculture

Sources of Cd in agricultural soils have been discussed by Bramley (1990), Jensen and Bro-Rasmussen (1992) and McLaughlin et al. (1996) for New Zealand, European and Australian soils, respectively. Atmospheric sources of Cd may be significant where there is significant industrial or urban activity (Jensen and Bro-Rasmussen, 1992) but phosphatic fertilisers are probably the largest single source of Cd additions to agricultural soils in many countries in the Asia-Pacific region (Bramley, 1990; McLaughlin et al., 1996). Inputs of Cd to soils in Australia from fertiliser use have been estimated to be approximately 1.6 g ha^{-1} yr^{-1}, which compares to data for UK of 4.3 g ha^{-1} yr^{-1} (Hutton and Symon, 1986), 3.5–4.3 g ha^{-1} yr^{-1} in Germany (Sauerbeck, 1982; Kloke et al., 1984), 3.0 g ha^{-1} yr^{-1} in Denmark (Hofmand, 1981), 8.9 g Cd ha^{-1} yr^{-1} in New Zealand (Bramley, 1990), 0.3–1.2 g Cd ha^{-1} yr^{-1} in USA (Mordvedt, 1987) and an average for EEC countries of 2.5 g ha^{-1} yr^{-1} (Biberacher and Shah, 1990). The amount of Cd present in phosphatic fertiliser is dependent on the quality of the rock used for fertiliser manufacture, with oceanic island sources of rock being higher in Cd than other sources (Table 14). Nitrogenous and potassic fertilisers normally have low concentrations of metal impurities.

4.1.2 Behaviour and plant availability of Cd in soils

Generally, Cd is strongly retained by soils but there is some evidence that in acidic Oxisols, retention may be weak due to net positive charge on oxide surfaces (Naidu

Table 14. Cadmium and phosphorus concentrations of some phosphate rocks (McLaughlin et al., 1996).

Phosphate rock	Cd mg kg^{-1}	P %	Cd mg Cd kg^{-1} P
USSR (Kola)	0.2	17.2	1
South Africa (Phalaborwa)	4	17.2	23
Chatham Rise phosphorite	2	8.9	23
China (Yunan)	5	14.4	35
Jordan	6	14.8	27
Australia (Duchess)	7	13.9	50
Mexico	8	14.0	57
Egypt (Quseir)	8	12.7	61
Makatea	10	13.0	77
Peru (Sechura)	11	13.1	84
Israel (Arad)	12	14.1	85
Tunisia (Gafsa)	38	13.4	108
Israel (Zin)	32	14.0	228
Morocco (Boucraa)	38	15.7	240
Christmas Island	43	15.3	275
North Carolina	47	15.1	311
Banaba (Ocean Island)	99	17.6	563
Nauru	100	15.6	641

et al., 1994). In most other soils, Cd retention decreases sharply with decreases in soil pH (Tiller et al., 1984). Cadmium is more strongly retained by soils that are finer textured i.e. clays compared to sands, and soils high in OM.

Availability of Cd to plants depends on a number of factors (Table 15). It is beyond the scope of this chapter to cover all these factors in detail but a number of reviews have been published in this area (Jackson and Alloway, 1992; Tiller et al., 1994). More recent research has also identified soil salinity as a major factor determining Cd uptake by both irrigated and dryland crops (McLaughlin et al., 1994; Li et al., 1994). In a survey of commercial potato crops in southern Australia, McLaughlin et al. (1994) found that Cl concentrations in soils were related (R^2=0.62) to Cd concentrations in tubers (Figure 11). The effect of salinity is through chloro-complexation of Cd in soil solution, allowing greater concentrations of Cd in solution (McLaughlin and Tiller, 1994) and enhanced uptake by plants due to the presence of the complexed ion (Smolders and McLaughlin, 1996).

4.1.3 Managing Cd in agriculture

Various farm management practices have been investigated to assess their effectiveness for decreasing Cd concentrations in food crops. The main practices studied are raising soil pH by liming; additions of macro- and micronutrients to the soil; cultivar selection; altering crop rotation sequences; and adjusting tillage practices. Further details of the effectiveness of these practices in minimising Cd accumulations by crops with the view of improving food quality are given in Section 6.3, Chapter 11.

Table 15. Factors affecting Cd uptake from soil (from Chaney and Hornick, 1978)

Any factor affecting Cd^{2+} activity
Soil
1. pH
2. amount of Cd present
3. metal sorption capacity of soil a. organic matter b. cation exchange capacity c. clay, Fe and Mn oxides
4. other micronutrients: Zn, Cu, Mn
5. macronutrients: NH_4, PO_4, K
6. temperature, moisture content, compaction
7. aeration; flooding = CdS
8. recurrent vs single application
Crop
1. species and cultivar
2. plant tissue: leaf > grain fruit and edible root
3. leaf age: older > younger
4. metal interactions

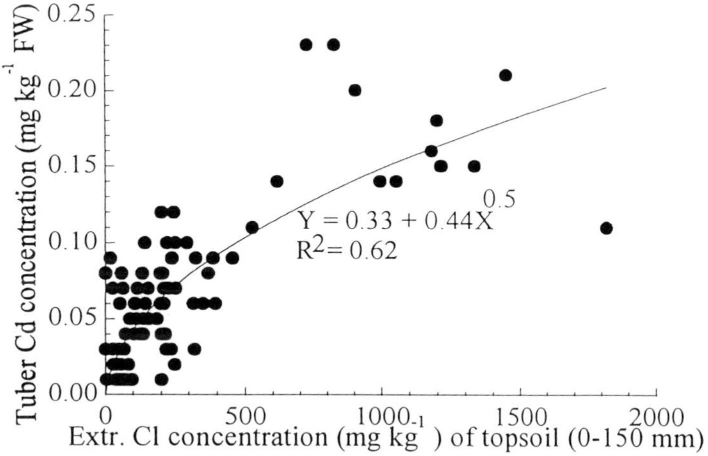

Figure 11. Relationship between soil Cl concentrations and Cd content of potato tubers (McLaughlin et al., 1994).

Results from surveys of Cd in agricultural commodities have found that most crops have low (<0.1 mg kg^{-1}) Cd concentrations (Table 16). Some areas close to industrial or urban centres, or where soils are naturally high in Cd, may produce foods with elevated Cd concentrations (Wiersma et al., 1986). Where elevated Cd

Table 16. Cadmium concentrations ($mg\,kg^{-1}$) in some agricultural produce.

Crop/Country	No. samples	Mean	Median	Range	Reference
Wheat					
Australia – Queensland	267	0.006	NR	<0.001–0.042	Best (1991)
Australia – SA	1,751	0.015	NR	<0.002–0.256	Richards et al. (pers. comm.)
USA – all states	288	0.043	0.030	<0.002–0.207	Wolnik et al. (1983)
Netherlands	84	0.070	0.060	0.02–0.35	Wiersma et al. (1986)
Germany	886	0.056	0.046	0.008–0.849	Weigert et al. (1984)
United Kingdom	20	<0.050	NR	<0.010–0.090	Anon (1983)
Barley					
Australia – all states	83	0.002–0.009*	NR	0.001–0.018	Anon (1990)
Australia – SA	1,557	0.01	NR	<0.002–0.073	Richards et al. (pers. comm.)
Netherlands	45	0.13	0.11	0.01–0.54	Wiersma et al. (1986)
United Kingdom	21	<0.030	NR	<0.010–0.050	Anon (1983)
Potatoes					
Australia – all states	32	0.040	0.030	NR–0.130	Stenhouse (1991)
Australia – all states	429	0.037	0.046	0.004–0.232	McLaughlin et al. (unpubl. data)
Australia – WA	116	0.035	0.030	0.005–0.120	Anon (1992)
USA – all states	297	0.031	0.028	0.002–0.181	Wolnik et al. (1983)
Netherlands	94	0.030	0.030	0.002–0.090	Wiersma et al. (1986)
Germany	133	0.047	0.030	0–0.320	Weigert et al. (1984)
Onions					
Australia – all states	24	0.040	0.020	NR–0.170	Stenhouse (1991)
Netherlands	83	0.013	0.011	0.004–0.032	Wiersma et al. (1986)
United Kingdom	11	0.040	NR	0.010–0.090	Thomas et al. (1972)
USA – all states	230	0.011	0.009	0.001–0.054	Wolnik et al. (1985)

*State means.
NR = not reported.

concentrations have been found in agricultural regions remote from point sources of pollution, this can usually be traced to soil acidity, Zn deficiency or soil salinity leading to enhanced Cd uptake by crops (Li et al., 1994; McLaughlin et al., 1994; Oliver et al., 1994a).

While most foods have low concentrations of Cd, the gradual accumulation of this element in most agricultural soils is of concern. Most fertilised agricultural systems import more Cd than is exported in agricultural produce (Table 17). As Cd is strongly retained in soils and is not easily leached, concentrations of Cd in soils are likely to continue to increase in future. The impact of this on food Cd concentrations is hard to predict, as the long-term availability of Cd to plants may change as soil conditions change over time; for example, build-up of OM may reduce phytoavailability of Cd (and vice versa) and acidification or salinisation may increase availability. Current research is focussing on the long-term behaviour of Cd in agricultural soils.

4.2 Nitrate toxicity

Nitrate toxicity is linked to the development of methaemoglobinaemia in mammals after consuming high concentrations of nitrite (NO_2^-), or NO_3-N which is subsequently reduced by bacteria to NO_2^-, in the stomach. The NO_2^- reacts with haemoglobin in the bloodstream and oxidises ferrous iron (Fe^{2+}) to ferric iron (Fe^{3+}), thus reducing the capacity of the haemoglobin molecule to carry oxygen. Numerous cases of acute intoxication in humans have been reported following consumption of well waters having high nitrate concentrations (WHO, 1978) with babies being particularly sensitive. Similar effects are evident in grazing animals, which appear to be more sensitive to high NO_3-N concentrations than non-ruminants (Walker, 1990). However, many of the reported cases are confounded by the presence of bacterial contamination in the water consumed. Much of the evidence for toxicity of NO_3-N stems from studies where contaminated waters were consumed, or from consumption of foods containing high concentrations of nitrate added as a preservative. There is no evidence that foods containing high concentrations of NO_3-N through uptake from soil have caused methaemoglobinaemia. In a review of the toxicity of NO_3-N in waters, Walker (1990) concludes that the acceptable daily intake recommended by the Joint FAO/WHO Expert Committtee on Food Additives (0.5 mg kg^{-1} body weight day^{-1} expressed as $NaNO_3$) could be increased up to five-fold.

Reviews of the carcinogenicity of NO_3-N^- have concluded there is little epidemiological evidence to suggest a significant risk at the concentrations found in most water supplies (Walker, 1990).

4.3 Pesticides

Concerns regarding pesticide residues in food crops have resulted in extensive testing of food commodities in both market basket type surveys and for regulation of imports and exports between countries. In Australia, the Australian Market Basket

Table 17. Estimated soil balance for contaminants in two agricultural systems in Australia – fertilized dryland wheat and irrigated potatoes (from McLaughlin et al., 1996).

Element	Input[a] ($g\,ha^{-1}$)	Crop harvest[b] ($g\,ha^{-1}$)	Net rate of addition ($g\,ha^{-1}\,yr^{-1}$)	'Background' soil concentration[c] ($mg\,kg^{-1}$)	Years to double soil concentration in 0–100 mm[d]	ANZECC Investigation level[e] ($mg\,kg^{-1}$)
Wheat						
As	1.0	0.30	0.70	4.0	7,500	20.0
Cd	6.0	0.12	5.88	0.2	45	3.0
Hg	0.1	0.03	0.07	<0.1	1,850[f]	1.0
Pb	4.0	0.30	3.70	21.0	7,500	300.0
F	4,000	3.0	3,997	300.0	100	400.0
Potatoes						
As	4.0	0.20	3.80	4	1,370	20.0
Cd	20.0	2.50	17.50	0.2	15	3.0
Hg	0.4	0.25	0.15	<0.1	870[f]	1.0
Pb	16.0	1.00	15.00	21.0	1,800	300.0
F	16,000	10.0	15,990	300.0	25	400.0

[a]Assumes 20 kg P ha^{-1} applied per wheat crop and 80 kg P ha^{-1} per potato crop and fertilizer contains (per kg P) 50 mg As, 300 mg Cd (250 mg Cd for potatoes), 5 mg Hg, 200 mg Pb and 200 g F. Element inputs in irrigation water assumed to be negligible, although F may be a significant impurity in some waters.

[b]Assumes 3 tonnes grain and 50 tonnes potatoes harvested per crop, all stubble or haulms returned to the soil. Metal concentrations in wheat grain assumed to be (per kg) 100 μg As, 40 μg Cd, 10 μg Hg, 100 μg Pb and 1 mg F and in potatoes (per kg fresh weight) 4 μg As, 50 μg Cd, 5 μg Hg, 20 μg Pb and 200 μg F. Data for element concentrations in crops taken from the AMBS (Stenhouse, 1991; Tiller et al., 1976; Anon, 1984; Wiersma et al., 1986).

[c]Data for 'rural' soils uncontaminated from industrial or urban sources taken from mean data for Cd and Pb in rural soils from Merry and Tiller (1991), Oliver et al. (1993, 1994a), data for As from Tiller (1992) and Olszowy et al. (1993), data for Hg from Olszowy et al. (1993) and data for F from McLaughlin et al. (unpublished data).

[d]Assumes a soil bulk density of 1300 kg m^{-3}.

[e]Guideline threshold values above which investigation of the contamination should take place, from ANZECC/NHMRC (1992) and Moen et al. (1988).

[f]Assumes 'background' soil concentration equals 0.1 mg kg^{-1} despite published analyses indicating values below this level (detection limit).

Survey (AMBS – Stenhouse, 1991) and the National Residue Survey (Bureau of Rural Resources, 1992) regularly monitor concentrations of pesticides in both meat and crop commodities (see Chapter 14). Compliance with Maximum Residue Levels is usually >99% for most commodities. Common residues in meat products were organochlorines, organophosphates and sulphonamides. Results from the AMBS suggest that there has been a gradual decrease with time in levels of organochlorines in foodstuffs and human milk. Organophosphates are found predominantly in foods with a cereal base, as these pesticides are widely used as protectants for grain during storage. In the AMBS, the most common organophosphates found were fenitrothion, chlorpyrifos methyl and pirimiphos methyl.

Where excessive concentrations of pesticide residues in food crops have been detected, this is usually as a result of improper use of the chemical in question, i.e. inappropriate application rates or insufficient withholding periods prior to crop harvest and marketing. There are few instances where pesticide residues in food crops are a result of plant uptake of the chemical from soil. For livestock however, ingestion of soil contaminated with pesticides may be a major route for exposure and contamination of the food chain (Harrison et al., 1970).

5. Discussion and conclusions

The adverse impacts of agriculture on soil water and food quality have been most significant in those countries where food production per unit area has increased to the greatest extent over the last 50 years. In these areas, the intensity of inputs and frequency of cultivation have had significant effects on the environment. Thus North American, European and some Asian countries have significant problems in terms of contamination of water supplies by sediment, nutrients and pesticides, while soil contamination by heavy metals has also been identified as a threat to long-term sustainability. At the other end of the scale, in low input agricultural systems such as in Australia and African countries, the soils may be less resilient to intensive agricultural management and hence soil erosion, compaction and soil structural degradation are often more significant issues.

Where nutrient inputs are made to soils, not all of the applied nutrients will appear in the produce exported from the field, with some remaining in the soil or being leached or emitted to the atmosphere. Under irrigation, the increased level of water makes it more difficult to prevent transport of nutrients, pesticides and sediment into surface and/or groundwaters. The inherent 'leakiness' of agriculture must be accepted and some balance struck between continued supply of high quality foods and some offsite impact of applied nutrients. Impacts of contaminants in fertilisers can also be controlled, but again a balance must be achieved between real health risks and the ability to source or manufacture fertilisers with low concentrations of contaminants. Where continued accumulation of metals in soils is identified, there is a real need to determine the long-term impacts on soil processes as remediation of metal contaminated soils is technically difficult and very costly.

Most of the management practices required to minimise adverse impacts of agriculture on soil, water and food quality are well known. What often minimises adoption of these practices is the altered economics of production. The challenge for policy makers is to formulate agricultural production systems where economic incentives are introduced to ensure continued food and fibre production with minimal impact on the environment.

References

Addiscott T and Rowlson D 1989 Laying the ground rules for nitrate. New Scientist. 122, 28-29.

Alberts E E, Schuman G E and Burwell R E 1978 Seasonal runoff losses of nitrogen and phosphorus from Missouri Valley loess watersheds. J. Environ. Qual. 7, 203-208.

Anderson B W and Randall G W 1992 Nitrate losses to tile drainage as affected by nitrogen fertilization of corn in a corn-soybean rotation. Agron. Abs. 30

Andersson A and Siman G 1991 Levels of cadmium and some other trace elements in soils and crops as influenced by lime and fertilizer level. Acta Agric. Scand. 41, 3-11.

Andraski B J, Daniel T C, Lowery B and Mueller D H 1985 Runoff results from natural and simulated rainfall for four tillage systems. Trans. Am. Soc. Ag. Eng. 28, 1219-1225.

Anon 1976 Application of sewage sludge to cropland: Appraisal of potential hazards of the heavy metals to plants and animals. Council for Agricultural Science and Technology Rep. 64, Ames, IA.

Anon 1983 Survey of cadmium in food: First supplementary report. Ministry of Agriculture, Fisheries and Food Surveillance Pap. 12, HMSO, London.

Anon 1984 Fluorine and fluorides. Environmental Health Criteria 36, World Health Organization, Geneva.

Anon 1989 Code of practice for the agricultural use of sewage sludge. UK Department of the Environment, HMSO, London.

Anon 1990 Cadmium and lead in Australian barley. In Final Report for Grains Research and Development Corporation Project "Screening of barley for cadmium and lead content at selected depots within Australia". Eds. anonymous. Queensland Department of Primary Industries, Indooroopilly, Queensland.

Anon 1992 Cadmium in Western Australian potatoes. West Australian Department of Agriculture/Health Department of Western Australia, Perth.

Austin N R, Prendergast J B, and Collins M D 1994 Reducing fertilizer nutrient losses through improved flood irrigation management. In Water Down Under 94. Eds. anonymous. pp 543-546. Groundwater/Surface Hydrology Common Interest Papers 1, Institute of Engineers, Canberra, Australia.

Baber H T and Wilson A T 1972 Nitrate pollution of groundwater in the Waikato region. J. N. Z. Inst. Chem. 56, 179-183.

Baker J L, Laflen J M and Johnson H P 1978 Effect of tillage systems on runoff losses of pesticides: A rainfall simulation study. Trans. Am. Soc. Ag. Eng. 21, 881-885.

Baker J L, Logan T J, Davidson J M, and Overcash M R 1987 Summary and conclusions. In Effects of Conservation Tillage on Groundwater Quality: Nitrates and Pesticides. Eds. T J Logan, J M Davidson, J L Baker and M R Overcash. Lewis Publishers, Chelsea, MI.

Baker J L and Timmons D R 1994 Fertilizer effects on leaching of labeled nitrogen for no till corn in field lysimeters. J. Environ. Qual. 23, 305-310.

Ball P R, Keeney D R, Theobald P W and Nes P 1979 Nitrogen balance in urine affected areas of a New Zealand pasture. Agron. J. 71, 309-314.

Ball P R and Ryden J C 1984 Nitrogen relationships in intensively managed temperate grasslands. Plant Soil. 76, 23-33.

Barnes C J, Jacobson G and Smith G D 1992 The origin of high nitrate groundwaters in the Australian arid zone. J. Hydrol. 137, 181-197.

Bauder J W, Sinclair K N and Lund R E 1993 Physiographic and land use characteristics associated with nitrate-nitrogen in Montana groundwater. J. Environ. Qual. 22, 255-262.

Beke G J, Lindwall C W, Entz T and Channappa T C 1989 Sediment and runoff water characteristics as influenced by cropping and tillage practices. Can. J. Soil Sci. 69, 639-647.

Benoit G R 1973 Effect of agricultural management of wet, sloping soil on nitrate and phosphorus in surface and subsurface water. Water Resour. Res. 9, 1296-1303.

Bergstrom L and Johansson R 1991 Leaching of nitrate from monolith lysimeters of different types of agricultural soils. J. Environ. Qual. 20, 801-807.
Best E K 1991 Cadmium and lead in Queensland wheat. In Final Report for Grains Research and Development Corporation Project "Screening of wheat for cadmium and lead content". Eds. anonymous. Queensland Department of Primary Industries, Indooroopilly, Queensland.
Biberacher G and Shah K D 1990 Fertilizer manufacture and EEC environmental activities. In Proceedings IFA Technical Conference, Venice, Italy. Eds. anonymous. pp 10(1)-10(14). International Fertilizer Industry Association, Paris, France.
Bowmer K H and Laut P 1992 Wastewater management and resource recovery in intensive rural industries in Australia. Water Res. 26, 201-208.
Bramley R G V 1990 Cadmium in New Zealand agriculture. N. Z. J. Agric. Res. 33, 505-519.
Briskin J S National Pesticide Survey: Methods, results, and policy implications. pp 143-154 In Mechanisms of Pesticide Movement into Ground Water. Eds. R C Honeycutt and D J Schabacker. Lewis Publishers, Boca Raton.
Brown P L and Miller M R 1978 Soil and crop management practices to control saline seeps in the U.S. northern Plains. Proc. 11th. Int. Soil Sci. Soc. Cong. (Edmonton). 7, 9-15.
Bundy L G, Schmitt M A, and Randall G W 1992 Predicting N fertilizer needs for corn in humid regions: Advances in the upper Midwest. In Predicting N Fertilizer Needs for Corn in Humid Regions. Eds. B R Bock and K R Kelley. pp 73-89. National Fertilizer and Environmental Research Center Bul. Y-226, Muscle Shoals, AL.
Burch G J 1986 Land clearing and vegetation disturbance. In Australian Soils: The Human Impact. Eds. J S Russell and R F Isbell. pp 169-184. University of Queensland Press, St. Lucia, Queensland.
Bureau of Rural Resources 1992 Report on the National Residue Survey 1989-1990 Results. Australian Government Printing Service, Canberra.
Burwell R E, Schuman G E, Saxton K E and Heineman H G 1976 Nitrogen in subsurface discharge from agricultural watersheds. J. Environ. Qual. 5, 325-329.
Burwell R E, Timmons D R and Holt R F 1975 Nutrient transport in surface runoff as influenced by soil cover and seasonal periods. Soil Sci. Soc. Am. Proc. 39, 523-528.
Butler B E and Churchward H M 1983 Aeolian processes. In Soils: An Australian Viewpoint. Ed. CSIRO. pp 91-99. CSIRO Publications/Academic Press, Melbourne/London.
Campbell I C 1978 Inputs and outputs of water and phosphorus from four Victorian catchments. Aust. J. Mar. Freshw. Res. 29, 557-584.
Carey A E 1991 Agriculture, agricultural chemicals, and water quality. In Agriculture and the Environment: The 1991 yearbook of agriculture. Ed. anonymous. pp 78-85. US Government Printing Office, Washington D.C.
Cass A and Sumner M E 1982 Soil pore structural stability and irrigation water quality: II. Sodium stability data. Soil Sci. Soc. Am. J. 46, 507-512.
CEC 1986 Commission of the European Communities Council directive of 12 June 1986 on protection of the environment, and in particular of the soil, when sewage sludge is used in agriculture. Official J. Europ. Comm. L181, 6-12.
Chartres C J, Helyar K R, Fitzpatrick R W, and Williams J 1992 Land degradation as a result of European settlement of Australia and its influence on soil properties. In Australia's Renewable Resources: Sustainability and Global Change. Eds. R M Gifford and M M Barson. pp 3-33. Bureau of Rural Resources Proceedings 14, Canberra, Australia.
Chaubey I, Edwards D R, Daniel T C, Moore P A and Nichols D J 1994 Effectiveness of vegetative filter strips in retaining surface-applied swine manure constituents. Trans. Am. Soc. Ag. Eng. 37, 845-850.
Clark E H, Haverkamp J A and Chapman W 1985 Eroding soils: The off-farm impacts. The Conservation Foundation, Washington D.C.
Clark R D S 1988 The export of nitrogen, phosphorus and turbidity from catchments with various land uses in the Mt. Lofty Ranges of South Australia. South Australian Engineering and Water Supply Department EWS Lib. Ref. 87/17, Adelaide, Australia.
Clausen J C and Meals D W 1989 Water quality achievable with agricultural best management practices. J. Soil Water Conserv. 44, 593-596.
Coale F J, Izuno F T and Bottcher A B 1994 Sugarcane production impact on nitrogen and phosphorus in drainage water from an Everglade Histosol. J. Environ. Qual. 23, 116-120.
Cooper J R, Gilliam J W, Daniels R B and Robarge W P 1995 Riperian areas as filters for agricultural sediment. Soil Sci. Soc. Am. J. 51, 416-420.
Costin A B 1980 Runoff and soil and nutrient losses from an improved pasture at Ginninderra, Southern Tablelands, New South Wales. Aust. J. Agric. Res. 31, 533-546.

Cullen P 1986 Managing nutrients in aquatic systems: The eutrophication problem. In Limnology in Australia. Eds. P De Deckker and W D Williams. pp 539-554. CSIRO, Melbourne, Vic.

Cullen P and Rosich R 1979 Effects of rural and urban sources of phosphorus on Lake Burley Griffin. Progress Water Tech. 11, 219-230.

Culley J L B, Bolton E F and Bernyk V 1984 Suspended solids and phosphorus loads from a clay soil: I. Plot studies. J. Environ. Qual. 12, 493-498.

Davis E A 1984 Conversion of Arizona chaparral to grass increases water yield and nitrate loss. Water Res. Research. 20, 1643-1649.

Davis J G 1994 Managing plant nutrients for optimum water use efficiency and water conservation. Adv. Agron. 53, 85-120.

Deason J P 1989 Irrigation-induced contamination: How real a problem. J. Irrig. Drain. Eng. 115, 9-20.

Diggle A J and Bell L C 1984 Movement of applied phosphorus following mining and revegetation of mineral sands on Australia's east coast. Aust. J. Soil Res. 22, 135-148.

Dillaha T A, Sherrard J H, Lee D, Mostaghimi S and Shanholtz VO 1988 Evaluation of vegetative filter strips as a best management practice for feed lots. J. Water Pollut. Cont. Fed. 60, 1231-1238.

Dillaha T A, Sherrard S H and Lee D 1989 Long-term effectiveness of vegetative filter strips. Water, Environ. Technol. 1, 419-421.

Dillon P 1992 Feedlots and nitrate contamination of groundwater. In Research Areas Pertinent to Intensive Rural Industry Waste Management.. Eds. K H Bowmer and P Laut. pp 53-60. CSIRO Division of Water Resources, Canberra, ACT.

Donal C M and Williams C H 1954 Fertility and productivity of a podsolic soil as influences by subterranean clover (*Trifolium subterraneaneum* L.) and superphosphate. Aust. J. Agric. Res. 5, 664-687.

Edmisten K L, Wood C W, Reeves D W and Tracy P W 1992 Determination of cotton nitrogen status with a hand held chlorophyll meter in Alabama and Missouri. Proc. Beltwide Cotton Conf. 3, 1101-1102.

Edwards D M, Fischbach P E and Young L L 1972 Movement of nitrates under irrigated agriculture. Trans. Am. Soc. Ag. Eng. 15, 73-75.

Edwards D R and Daniel T C 1992 Potential runoff quality effects of poultry manure slurry surface applied to fescue plots. Trans. Am. Soc. Ag. Eng. 35, 1827-1832.

Edwards D R and Daniel T C 1993 Effects of poultry litter application rate and rainfall intensity on quality of runoff from fescuegrass plots. J. Environ. Qual. 22, 361-365.

Edwards D R and Daniel T C 1993 Runoff quality impacts of swine manure applied to fescue plots. Trans. Am. Soc. Ag. Eng. 36, 81-86.

Edwards D R and Daniel T C 1994 Quality of runoff from fescuegrass plots treated with poultry litter and inorganic fertilizer. J. Environ. Qual. 23, 579-584.

Evans R O, Westerman P W and Overcash M R 1984 Subsurface drainage water quality from land application of swine lagoon effluent. Trans. Am. Soc. Ag. Eng. 27, 473-480.

Flugel W A 1991 River salinization due to dryland agriculture in the Western Cape Province, Republic of South Africa. In Sediment and Stream Water Quality in a Changing Environment: Trends and Explanation. Eds. N E Peters and D E Walling. pp 191-200. International Association of Hydrological Sciences Pub. 203, Wallingford, England.

Friberg L, Elinder C G, Kjellstrom T and Nordberg G F 1985 Cadmium and health: A toxicological and epidemiological appraisal. Effects and response. CRC Press, Boca Raton, FL.

Garrison P J and Asplund T R 1993 Long-term (15 years) results of NPS controls in an agricultural watershed upon a receiving lake's water quality. Water Sci. Technol. 28, 441-449.

Garwood E A and Ryden J C 1986 Nitrate loss through leaching and surface runoff from grassland: Effects of water supply soil type and management. In Nitrogen Fluxes in Intensive Grassland Systems. Ed. H G van der Meer. pp 99-113. Martinus Nijhoff Publishers, Dordrecht, Netherlands.

Gelata S, Sabbagh G J, Stone J F, Elliott R L, Mapp H P, Bernardo D J 1994 Importance of soil and cropping systems in the development of regional water quality policies. J. Environ. Qual. 23, 36-42.

Gifford R M, Cheney N P, Noble J C, Russell J S, Wellington A B, and Zammit C 1992 Australian land use, primary production of vegetation and carbon pools in relation to atmospheric carbon dioxide concentration. In Australia's Renewable Resources: Sustainability and Global Change. Eds. R M Gifford and M M Barson. pp 151-187. Bureau of Rural Resources Proceedings 14, Canberra, Australia.

Gilley J E, Finkner S C and Varvel G E 1986 Runoff and erosion as affected by sorghum and sorghum residue. Trans. Am. Soc. Ag. Eng. 29, 1605-1610.

Gilliam J W, Skaggs R W and Weed S B 1979 Drainage control to diminish nitrate loss from agricultural fields. J. Environ. Qual. 8, 137-142.

Goh K M, Edmeades D C and Hart P B S 1979 Direct field measurements of leaching losses of nitrogen in pasture and cropping soils using tension lysimeters. NZ J. Agric. Res. 22, 133-142.
Gold A J and Loudon T L 1986 Runoff water quality from conservation and conventional tillage. American Society of Agricultural Engineers Paper. 86-2540
Grace P J, Ladd J N, and Skjemstad J O 1994 The effect of management practices on soil organic matter dynamics. In Soil Biota: Management in Sustainable Farming Systems. Eds. C E Pankhurst, B M Doube, V V S R Gupta and P R Grade. pp 162-171. CSIRO Publications, Melbourne, Australia.
Grattan S R and Rhoades J D 1990 Irrigation with saline ground water and drainage water. In Management of Farm Irrigation Systems. Eds. G J Hoffman, T A Howell and K H Solomon. pp 432-449. American Society of Agricultural Engineers, St. Joseph, MI.
Grierson I T, Jijne J W and Greenland D J 1972 Changes in some physical properties of red-brown earths of different texture associated with increasing content of organic matter. Experimental Record. 6, 16-22.
Hallberg G R and Keeney D R 1993 Nitrate. In Regional Groundwater Quality. Ed. W M Alley. pp 297-322. Van Nostrand Reinhold, New York, NY.
Halvorson A D 1990 Management of dryland seeps. In Agricultural Salinity Assessment and Management. Ed. K K Tanji. pp 372-392. American Society of Civil Engineers, New York, NY.
Hamlett J M and Brannan K 1991 Water quality impacts of winter rye cover with selected best management practices in Pennsylvania. In Cover Crops for Clean Water. Ed. W L Hargrove. pp 53-55. Soil and Water Conservation Society of America, Ankeny, IA.
Hamlett J M and Epp D J 1994 Water quality impacts of conservation and nutrient management practices. J. Soil Water Conserv. 49, 59-66.
Harrison D L, Mol J C M and Healy W B 1970 DDT residues in sheep from ingestion of soil. NZ J. Agric. Res. 13, 664-672.
Harrison J 1994 Review of nutrients in irrigation drainage in the Murray Darling Basin. CSIRO Division of Water Resources Series 11, Canberra, Australia.
Helyar K R and Porter W M 1989 Soil acidification, its measurement and the processes involved. In Soil Acidity and Plant Growth. Ed. A D Robson. pp 61-101. Academic Press, Australia.
Hensley M 1969 Selected properties affecting the irrigable value of some Makatini soils. M.Sc. Agric. Thesis, University of Natal, Pietermaritzburg, South Africa.
Hergert G W 1986 Nitrate leaching through sandy soil as affected by sprinkler irrigation management. J. Environ. Qual. 15, 272-278.
Hodgson A S and MacLeod D A 1989 Oxygen flux, air-filled porosity and bulk density as indices of vertisol structure. Soil Sci. Soc. Am. J. 53, 540-543.
Hofmand M F 1981 Cirkulation af bly, cadmium, kobber, zink og nickl i dansk langbrug. In lammets jordbrugsandvendels. II. Fokusering. Ed. anonymous. pp 85-118. Polyteknisk Forlag, Lyngby, Denmark.
Holburt M B 1984 Colorado River salinity The user's perspective. In Salinity in Watercourses and Reservoirs. Ed R H French. pp 13-22. Butterworth Publishers, Boston, MA.
Hollon B F, Owen J R and Sewell J I 1982 Water quality in a stream receiving dairy feedlot effluent. J. Environ. Qual. 11, 59.
Hortenstine C C and Forbes R B 1972 Concentrations of nitrogen, phosphorus and potassium and total soluble salts in soil solution samples from fertilized and unfertilized Histosols. J. Environ. Qual. 1, 4-46
Hubbard R K, Assmussen L E and Allison H D 1984 Shallow groundwater quality beneath an intensive multiple cropping system using center pivot irrigation. J. Environ. Qual. 13, 156-161.
Humenik F J 1992 Water quality: Agriculture's role. CAST Task Force Rep. 120,
Hutton M and Symon C J 1986 The quantities of cadmium, lead, mercury and arsenic entering the UK environment from human activities. Sci. Tot. Environ. 57, 129-150.
Jackson A P and Alloway B J 1992 The transfer of cadmium from agricultural soils to the human food chain. In Biogeochemistry of Trace Metals. Ed. anonymous. pp 109-158. Lewis Publishers, London.
Jemison J M and Fox R H 1994 Nitrate leaching from nitrogen-fertilized and manured corn measured with zero-tension pan lysimeters. J. Environ. Qual. 23, 337-343.
Jemison J M, Jabro J D and Fox R H 1994 Evaluation of LEACHM: II. Simulation of nitrate leaching from nitrogen-fertilized and manured corn. Agron. J. 86, 852-859.
Jensen A and Bro-Rasmussen F 1992 Environmental cadmium in Europe. Environ. Contam. Toxicol. 125, 101-181.
Jones A R 1984 Controlling salinity in the Colorado River basin, the arid west. In Salinity in Watercourses and Reservoirs. Ed. R H French. pp 337-347. Butterworth, Boston, MA.

Juergens-Gschwind S 1989 Ground water nitrates in other developed countries (Europe). Relationships to land use patterns. In Nitrogen Management and Ground Water Protection. Ed. R F Follett. pp 75-138. Elsevier, Amsterdam, Netherlands.

Keeney D R 1986 Nitrate in groundwater: Agricultural contribution and control. In Agricultural Impacts on Groundwater. Ed. anonymous. pp 329-351. National Water Well Association, Worthington, OH.

Kjellstrom T, Elinder C G and Friberg L 1984 Conceptual problems in establishing the critical concentrations of Cd in human kidney cortex. Environ. Res. 33, 284-295.

Kloke A, Sauerbeck D R and Vetter H 1984 The contamination of plants and soils with heavy metals and the transport of metals in terrestrial food chains. In Changing Metal Cycles and Human Health. Ed. J O Nriagu. pp 131-141. Springer Verlag, Berlin.

Kruse E G, Willardson L and Ayars J 1990 On-farm irrigation and drainage practices. In Management of Farm Irrigation Systems. Eds. G J Hoffman, T A Howell and K H Solomon. pp 349-371. American Society of Agricultural Engineers, St. Joseph, MI.

Laflen J M and Tabatabai M A 1984 Nitrogen and phosphorus losses from corn-soybean rotations as affected by tillage practices. Trans. Am. Soc. Ag. Eng. 27, 58-63.

Langdale G W, Mills W C and Thomas A W 1992 Use of conservation tillage to retard erosive effects of large storms. J. Soil Water Conserv. 47, 257-260.

Lanyon L E and Beegle D B 1989 The role of on-farm nutrient balance assessments in an integrated approach to nutrient management. J. Soil Water Conserv. 44, 164-168.

Lembke W D and Thorn M D 1980 Nitrate leaching and irrigated corn production with organic and inorganic fertilizers on sandy soils. Trans. Am. Soc. Ag. Eng. 23, 1153-1156.

Lewis D C, Clarke A L and Hall W B 1981 Factors affecting the retention of phosphorus applied as superphosphate to the sandy soils of south eastern South Australia. Aust. J. Soil Res. 19, 167-174.

Li Y, Chaney R L and Schneiter A A 1994 Effect of soil chloride level on cadmium concentration in sunflower kernels. Plant Soil. 167, 275-280.

Marks R S and Ward J R 1993 Nutrient and pesticide threats to water quality. In Proceedings of Soil Specific Crop Management Conference. Eds. P C Robert, R H Rust and W E Larson. pp 293-299. American Society of Agronomy, Madison, WI.

McAllister J S V 1976 Studies in Northern Ireland on problems related to the disposal of slurry. In Agriculture and Water Quality. Ed. anonymous. pp 418-431. Ministry of Agriculture, Food and Fisheries Tech. Bul. 32, London, UK.

McDowell LL and McGregor KC 1980 Nitrogen and phosphorus losses in runoff from no-till soybeans. Trans. Am. Soc. Ag. Eng. 23, 643-648.

McDowell L L and McGregor K C 1984 Plant nutrient losses in runoff from conservation tillage corn. Soil Tillage Res. 4, 79-91.

McGregor K C and Greer J D 1982 Erosion control with no-till and reduced-till corn for silage and grain. Trans. Am. Soc. Ag. Eng. 25, 154-159.

McIsaac G F and Hirschi M C 1989 Nitrogen and phosphorus in eroded sediment from corn and soybean tillage systems. In Sediment and the Environment. Eds. R F Hadley and E D Ongley. pp 3-10. International Association of Hydrological Sciences Pub. 184, Baltimore, MD.

McLaughlin M J, Baker T, James T R and Rundle J A 1990 Distribution and forms of phosphorus and aluminium in acidic topsoils under pasture in southeastern Australia. Aust. J. Soil Res. 29, 371-385.

McLaughlin M J and Tiller K G 1994 Chloro-complexation of cadmium in soil solutions of saline/sodic soils increases phytoavailability of cadmium. Trans. 15th. Int. Cong. Soil Sci. 3b, 195-196.

McLaughlin M J, Tiller K G, Beech T A and Smart M K 1994 Soil salinity causes elevated cadmium concentrations in field-grown potato tubers. J. Environ. Qual. 34, 1013-1018.

McLaughlin M J, Tiller K G, Naidu R and Stevens D P 1996 The behaviour and environmental impact of contaminants in fertilizers. Aust. J. Soil Res. 34, 1-54.

McLeod R V and Hegg R O 1984 Pasture runoff water quality from application of inorganic and organic nitrogen sources. J. Environ. Qual. 13, 122-126.

McNeal B L, Layfield D A, Norvell W A and Rhoades J D 1968 Factors influencing hydraulic conductivity of soils in the presence of mixed-salt solutions. Soil Sci. Soc. Am. J. 32, 187-190.

Meek B D, Carter D L, Westermann D T, Wright J L and Peckenpaugh R E 1995 Nitrate leaching under furrow irrigation as affected by crop sequence and tillage. Soil Sci. Soc. Am. J. 59, 204-210.

Meisinger J J, Hargrove W L, Mikkelson R L, Williams J R, and Benson V W 1991 Effect of cover crops on groundwater quality. In Cover Crops for Clean Water. pp. 57-68. Ed. anonymous. Soil and Water Conservation Society of America, Ankeny, IA .

Merry R H and Tiller K G 1991 Distribution of cadmium and lead in an agricultural region near Adelaide, South Australia. Water Air and Soil Pollution. 57-58, 171-180.
Miller R J and Smith R B 1976 Nitrogen balance in the southern San Joaquin Valley. J. Environ. Qual. 5, 274-278.
Miller W W, Guitjens J C and Mahannah C N 1984 Water quality of irrigation and surface return flows from flood-irrigated pasture and alfalfa hay. J. Environ. Qual. 13, 543-548.
Misztal M, Smal H and Gorniak A 1992 Ground water inflow of nutrients to a lake from differently utilized catchments. Water Res. 26, 1237-1242.
Moen J E T, Cornet J P, and Evers C W A 1988 Soil protection and remedial actions: Criteria for decision making and standardization of requirements. In Assink J W and van den Brink W J. (Eds) Contaminated Soil "88". pp. 1495-1503. Kluwer Academic Publishers, Dordrecht, Netherlands.
Mokma D L, Syers J K, Jackson M L, Clayton R N and Rex R W 1972 Aeolian additions to soils and sediments in the South Pacific area. J. Soil Sci. 23, 141-162.
Moore P A and Miller D M 1994 Decreasing phosphorus solubility in poultry litter with aluminium, calcium and iron amendments. J. Environ. Qual. 23, 325-330.
Mordvedt J J 1987 Cadmium levels in soils and plants from some long-term soil fertility experiments in the United States of America. J. Environ. Qual. 16, 137-142.
Mundy G N 1993 Guidelines for the economic use of nitrogen fertilizer on dairy farms. Dairy Research and Development Corporation Final Rep. DAV 241, Glen Iris, Australia.
Muscutt A D, Harris G L, Bailey S W and Davies D B 1993 Buffer zones to improve water quality: A review of their potential use in UK agriculture. Agric. Ecosys. Environ. 45, 59-77.
MVP88 1991 Milieuprogramma voor 19881991. Voortgangsrapportage, Tweede Kamer, vergaderjaar 1987-1988. 20202, nrs 1-2.
Naidu R, Bolan N S, Kookana R S and Tiller K G 1994 Ionic strength and pH effects on the sorption of cadmium and the surface charge of soils. Europ. J. Soil Sci. 45, 419-429.
National Research Council 1989 Irrigation-induced water quality problems. National Academy Press, Washington D.C.
National Research Council 1993 Soil and water quality. National Academy Press, Washington D.C.
NFA 1994 Australian Food Standards Code, March 1993, National Food Authority. Australian Government Publication Service, Canberra, Australia.
Nightingale H I 1972 Nitrates in soil and groundwater beneath irrigated and fertilized crops. Soil Sci. 114, 300-311.
O'Neill G J and Gordon A M 1994 The nitrogen filtering capability of Carolina poplar in an artificial riparian zone. J. Environ. Qual. 23, 1218-1223.
Olive L J and Walker P H 1982 Processes in overland flow erosion and production of suspended material. In Prediction in Water Quality. Eds. E M O'Loughlin and P Cullen. pp 87-119. Australian Academy of Science, Canberra, Australia.
Oliver D P, Schulz J E, Tiller K G and Merry R H 1993 The effect of crop rotations and tillage practices on cadmium concentration in wheat grain. Aust. J. Soil Res. 44, 1221-1234.
Oliver D P, Hannam R, Tiller K G, Wilhelm N S, Merry H H and Cozens G D 1994a The effects of zinc fertilization on cadmium concentration in wheat grain. J. Environ. Qual. 23, 705-711.
Oliver D P, Tiller K G, Conyers M K, Slattery W J, Merry R H, and Alston A M 1994b The effects of soil pH on Cd concentration in wheat grain grown in southeastern Australia. In Third International Symposium on Plant Soil Interactions at Low pH. Ed. D G Edwards. In press.
Olness A, Smith S J, Rhodes E D and Menzel R G 1975 Nutrient and sediment discharge from agricultural watersheds in Oklahoma. J. Environ. Qual. 4, 331-336.
Olszowny H, Torr P, Imray P, Smith P, Hegarty J and Hastie G 1993 Studies of levels of trace elements in soils from rural and urban areas of Australia. Queensland Department of Health Report.
Omernik J M 1976 The influence of land use on stream nutrient levels. USEPA-600/3-76-014, Washington D.C.
Owens L B 1987 Nitrate leaching losses from monolith lysimeters as influenced by nitrapyrin. J. Environ. Qual. 16, 34-38.
Owens L B 1990 Nitrate-nitrogen concentrations in percolate from lysimeters planted to a legume-grass mixture. J. Environ. Qual. 19, 132-135.
Owens L B, Edwards W M and van Keuren R W 1992 Nitrate levels in shallow groundwater under pastures receiving ammonium nitrate or slow-release nitrogen fertilizer. J. Environ. Qual. 21, 607-613.
Owens L B, Edwards W M and van Keuren R W 1994 Groundwater nitrate levels under fertilized grass and grass-legume pastures. J. Environ. Qual. 23, 752-758.

Owens L B, van Keuren R W and Edwards W M 1982 Environmental effects of a medium fertility 12-month pasture program: II. Nitrogen. J. Environ. Qual. 11, 241-246.

Owens L B, van Keuren R W and Edwards W M 1983 Nitrogen loss from a high-fertility, rotational pasture program. J. Environ. Qual. 12, 346-350.

Pupisky H and Shainberg I 1979 Salt effects on the hydraulic conductivity of a sandy soil. Soil Sci. Soc. Am. J. 43, 429-433.

Putnam J, Williams J and Sawyer D 1988 Using the erosion-productivity impact calculator (EPIC) model to estimate the impact of soil erosion for the 1985 RCA appraisal. J. Soil Water Conserv. 43, 321-326.

Quisenberry V L, Hegg R O, Reese L E, Rice J S, and Torrence A K 1980 Management aspects of applying poultry or dairy manures to grasslands in the Piedmont region. In Livestock Waste: A Renewable Resource. Eds. anonymous. pp 170-173. American Society of Agricultural Engineers, St Joseph, MI.

Rhoades J D 1984 Reusing saline drainage waters for irrigation: A strategy to reduce salt loading of rivers. In Salinity in Watercourses and Reservoirs. Ed. R H French. pp 455-463. Butterworth Publishers, Boston, MA.

Ribaudo M O 1987 Agriculture's impact on water quality. In Agricultural Resources: Cropland, Water and Conservation Situation and Outlook Report. Eds. anonymous. USDA Econ. Res. Serv. AR8, Washington D.C.

Ribaudo M O and Young C E 1989 Estimating the water quality benefits from soil erosion control. Water Res. Bull. 25, 71-78.

Roberts G 1987 Nitrogen inputs and outputs in a small agricultural catchment in the eastern part of the United Kingdom. Soil Use Man. 3, 148-154.

Rouve G and Bogacki W 1989 Reduction of nitrate input into groundwater. In Protection of Water Quality from Harmful Emissions with Special Regard to Nitrate and Heavy Metals. Eds E Welte and I Szabolcs. pp 207-213. CIEC, Budapest, Hungary.

Ryan J A, Pahren H R and Lucas J B 1982 Controlling cadmium in the human food chain: A rationale based on health effects. Environ. Res. 28, 251-302.

Ryden J C, Ball P R and Garwood E A 1984 Nitrate leaching from grassland. Nature. 311, 50-53.

Sauerbeck D 1982 Zur Bedeutung des Cadmiums in Phosphatdungenmitteln. Landbauforsch. Volkenrode. 32, 192-197.

Sawhney B L 1978 Leaching of phosphorus from agricultural soils to groundwater. Water Air and Soil Pollution. 9, 499-505.

Scheffer B, Kuntze H and Bartels R 1986 Application of bauxite residues (red sludge) and industrial ferrous sulphate (green salt) on acid bog peat soils to reduce leaching of phosphates. Z. Kulturtechnik Flurbereinigung. 27, 76-82.

Schepers J S and Francis D D 1982 Chemical water quality of runoff from grazing land in Nebraska: Influence of grazing livestock. J. Environ. Qual. 11, 351-354.

Schepers J S, Hackes B L and Francis D D 1982 Chemical water quality of runoff from grazing land in Nebraska: II. Contributing factors. J. Environ. Qual. 11, 355-359.

Schepers J S, Moravek M G, Alberts E E and Frank K D 1991 Maize production impacts on groundwater quality. J. Environ. Qual. 20, 12-16.

Schofield N J and Birch P B 1986 Catchment management measures to reduce riverine phosphorus inputs to the eutrophic Peel-Harvey Estuary, Western Australia. In River Basin Management: Hydrology and Water Resources Symposium. Ed. anonymous. pp 113-118. Institution of Engineers National Conference publication 86/13, Barton, ACT, Australia.

Scott R I and Dillon P J 1989 Reconnaissance of soil water nitrate and related microorganisms near Mount Gambier, South Australia. Australian Centre for Groundwater Studies Summer Student project Report 20, Glen Osmond, SA, Australia.

Seta A K, Blevins R L, Frye W W and Barfield B J 1993 Reducing soil erosion and agricultural chemical losses with conservation tillage. J. Environ. Qual. 22, 661-665.

Shafron M 1993 A management strategy for blue-green algae in the Murray-Darling Basin. Aust. J. Soil Water Conserv. 6, 40-43.

Shainberg I 1992 Chemical and mineralogical components of crusting. In Soil Crusting: Chemical and Physical Processes. Eds. M E Sumner and B A Stewart. pp 33-53. Lewis Publishers, Boca Raton, Fla.

Shainberg I, Sumner M E, Miller W P, Farina M P W, Pavan M A and Fey M V 1989 Use of gypsum on soils: A review. Adv. Soil Sci. 9, 1-111.

Sharma M L and Williamson D R 1984 Secondary salinization of water resources in Southern Australia. In Salinity in Watercourses and Reservoirs. Ed. R H French. pp 571-582. Butterworth Publishers, Boston, MA.

Sharpley A N 1981 The contribution of phosphorus leached from crop canopy to losses in surface runoff. J. Environ. Qual. 10, 160-165.

Sharpley A N, Chapra S C, Wedephol R, Sims J T, Daniel T C and Reddy K R 1994 Managing agricultural phosphorus for protection of surface waters: Issues and options. J. Environ. Qual. 23, 437-451.

Sharpley A N and Halvorson A D 1995 The management of soil phosphorus availability and its impact on surface water quality. In Soil Processes and Water Quality. Eds. R Lal and B A Stewart. pp 7-90. Lewis Publishers, Chelsea, MI.

Sharpley A N, Menzel R G, Smith S J, Rhoades E D and Olness A E 1981 The sorption of soluble phosphorus by soil material during transport in runoff from cropped and grassed watersheds. J. Environ. Qual. 10, 211-215.

Sharpley A N and Smith S J 1991 Effect of cover crops on surface water quality. In Cover Crops for Clean Water. Eds. anonymous. pp 41-49. Soil and Water Conservation Society of America, Ankeny, IA.

Sharpley A N, Smith S J and Naney J W 1987 Environmental impact of agricultural nitrogen and phosphorus use. J. Agric. Food Chem. 35, 812-817.

Sharpley A N, Smith S J, Williams J R, Jones O R and Coleman G A 1991 Water quality impacts associated with sorghum culture in the Southern Plains. J. Environ. Qual. 20, 239-244.

Sharpley A N and Syers J K 1979 Loss of nitrogen and phosphorus in tile drainage as influenced by urea application and grazing animals. NZ J. Agric. Res. 22, 127-131.

Sharpley A N, Syers J K and Tillman R W 1983 Transport of ammonium and nitrate nitrogen in surface runoff from pasture as influenced by urea application. Water Air and Soil Pollution. 20, 425-430.

Sharpley A N and Withers P J A 1994 The environmentally-sound management of agricultural phosphorus. Fert. Res. 39, 133-146.

Small D R 1985 Effect of irrigation land use on nutrients in runoff. Aust. Water Res. Council Proj. Rep. 78/99

Smitka D E, Heermann D F, Duke H R and Batchelder A R 1977 Nitrate-N percolation through irrigated sandy soil as affected by water management. Agron. J. 69, 623-626.

Smolders E and McLaughlin M J 1996 Effect of chloride on Cd uptake by Swiss chard in nutrient solution. Plant Soil (in press).

Stenhouse F 1991 The 1990 Australian market basket survey report. National Health and Medical Research Council/National Food Authority, Australian Government Printing Service, Canberra, Australia.

Stevens D P, McLaughlin M J and Alston A M 1995 Speciation of fluoride and uptake of fluoride by plants. Proc. 3rd. Int. Conf. Biogeochem. Trace Elements. In press.

Sumner M E 1991 Soil acidity control under the impact of industrial society. In Interactions at the Soil Colloid Soil Solution Interface. Ed. G H Bolt. pp 517-541. Kluwer Academic, Netherlands.

Sumner M E 1992 The electrical double layer and clay dispersion. Adv. Soil Sci. (In press).

Sumner M E 1993 Sodic soils: New perspectives. Aust. J. Soil Res. 31, 683-750.

Sumner M E and Miller W P 1992 Soil crusting in relation to global soil degradation. Am. J. Alter. Agric. 7, 56-62.

Sumner M E, Rengasamy P, and Naidu R 1996 Sodic soils: A reappraisal. In Sodic Soils: The World Scene. Ed. M E Sumner and R Naidu. Oxford University Press, New York, NY.

Sumner M E and Stewart B A 1992 Soil crusting: Chemical and physical processes. Lewis Publishers, Boca Raton, FL.

Supalla R J, Selley R A, Bredeweg S and Watts D 1995 Adoption of nitrogen and water management practices to improve water quality. J. Soil Water Conserv. 50, 77-82

Sveda R, Rechcigl J E and Nkedi-Kizza P 1992 Evaluation of various nitrogen sources and rates on nitrogen movement, Pensacola bahiagrass production and water quality. Comm. Soil Sci. Plant Anal. 23, 2451-2478.

Szabolcs I 1989 Salt-affected soils. CRC Press Inc., Boca Raton, FL.

Szabolcs I 1994 Prospects of salinity for the 21st century. Trans. 15th. Int. Cong. Soil Sci. 1, 123-141.

Theng B K G, Kookana R S and Rahman A 1995 Environmental concerns of pesticides in soil and groundwater and management strategies in Oceania. Ed. P M Huang. Advances in Environmental Science – Remediation of Degraded Soils and Groundwaters Volume 4 – Worldwide Perspectives on Soil and Groundwater Pollution: Asia, Africa and Oceania. Science Reviews, London. (in press).

Thomas B, Roughan J A and Watters E D 1972 Lead and cadmium content of some vegetable foodstuffs. J. Sci. Food Agric. 23, 1493-1498.

Thomas G W, Blevins R L, Phillips R E and McMahan M A 1973 Effect of a killed sod mulch on nitrate movement and corn yield. Agron. J. 63, 736-739.

Tiller K G 1992 Urban soil contamination in Australia. Aust. J. Soil Res. 30, 937-957.

Tiller K G, de Vries M P C and Spouncer L R 1976 Some metals in selected vegetables purchased in Adelaide metropolitan area. CSIRO Division of Soils Rep. 13.

Tiller K G, Gerth J and Brummer G 1984 The relative affinities of Cd, Ni and Zn for different clay fractions and goethite. Geoderma. 34, 17-35.

Tiller K G, Oliver D P, McLaughlin M J, Merry R H and Naidu R 1994 Managing cadmium contamination of agricultural land. In Remediation of Soils Contaminated by Metals. Eds I K Isander and D C Adriano. Science and Technology Letters, Northwood, England.

Timmons D R 1984 Nitrate leaching as influenced by water application level and nitrification inhibitors. J. Environ. Qual. 13, 305-309.

Timmons D R and Baker J L 1991 Recovery of point-injected labeled nitrogen by continuous no-till corn. Agron. J. 83, 850-857.

Timmons D R and Baker J L 1992 Fertilizer management effect on recovery of labeled nitrogen by continuous no-till corn. Agron. J. 84, 490-496.

Timmons D R, Burwell R E and Holt R F 1973 Nitrogen and phosphorus losses in surface runoff from agricultural land as influenced by placement of broadcast fertilizer. Water Resour. Res. 9, 658-667.

Timmons D R and Dylla A S 1981 Nitrogen leaching as influenced by nitrogen management and supplemental irrigation level. J. Environ. Qual. 10, 421-426.

Triegel E K and Guo L 1994 Overview of the fate of pesticides in the environment, water balance; runoff vs. leaching. pp 1-13. In Mechanisms of Pesticide Movement into Ground Water. Ed. R C Honeycutt and D J Schabacker. Lewis Publishers, Boca Raton.

US Environmental Protection Agency 1990 National Pesticide Survey: Phase I Report. Report No. PB91-125765. US Department of Commerce, National Technical Information Service, Springfield, Virginia.

Vanotti M B and Bundy L G 1994 Frequency of nitrogen carryover in the humid Midwest. Agron. J. 86, 881-886.

Vanotti M B and Bundy L G 1994 An alternative rationale for corn nitrogen fertilizer recommendations. J. Prod. Agric. 7, 243-249.

Verhoeven J 1993 Implementing the New South Wales algal management strategy. Australian Journal of Soil and Water Conservation. 6, 30-34.

Vlahos S, Summers K J, Bell D T and Gilkes R J 1989 Reducing phosphorus leaching from sandy soils with red mud bauxite processing residues. Aust. J. Soil Res. 26, 177-190.

Walker J F and Graczyk D J 1993 Preliminary evaluation of effects of best management practices in the Black Earth Creek, Wisconsin, priority watershed. Water Sci. Technol. 28, 539-548.

Walker J M 1988 Regulation by other countries of cadmium in foods and the human environment. In Cadmium Accumulations in Australian Agriculture. Eds. J Simpson and B Curnow. pp 176-185. Bureau of Rural Resources Proc. 2, Australian Government Printing Service, Canberra, Australia.

Walker L W, Turner A K, and Vanselow D G 1979 Inflow to groundwater from a large cattle feedlot in Victoria. In Groundwater Pollution Conference Proceedings. Eds. C R Lawrence and R J Hughes. pp 375-390. Australian Water Resources Council Conference Series 1, Canberra, ACT, Australia.

Walker R 1990 Nitrate, nitrite and N-nitrosocompounds: A review of the occurrence in food and diet and the toxicological implications. Food Add. Contam. 7, 717-768.

Walter M F, Richard T L, Robillard P D, and Muck R 1987 Manure management with conservation tillage. In Effects of Conservation Tillage on Groundwater Quality: Nitrates and pesticides. Eds. T J Logan, J M Davidson, J L Baker and M R Overcash. pp 253-270. Lewis Publishers, Chelsea, MI.

Watts D G, Hergert G W and Nichols J T 1991 Nitrogen leaching losses from irrigated orchard-grass on sandy soils. J. Environ. Qual. 20, 355-362.

Wauchope R D 1978 The pesticide content of surface water draining from agricultural fields: A review. J. Environ. Qual. 7, 459-472.

Weaver D M and Prout A L 1993 Changing farm practice to meet environmental objectives of nutrient loss to Oyster Harbour. Fert. Res. 36, 177-184.

Weigert P, Muller J, Klein H, Zufelde K P and Hillebrand J 1984 Arsen, Blei, Cadmium und Quecksilber in und auf Lebensmittln. ZEBS Hefte 1, Bonn, Germany.

Weil R R, Weismiller R A and Turner R S 1990 Nitrate contamination of groundwater under irrigated Coastal Plain soils. J. Environ. Qual. 19, 441-448.

Weinbaum S A, Johnson R S and DeJong T M 1992 Causes and consequences of overfertilization in orchards. Hort. Tech. 2, 112-121.

Westerman P W and Overcash M R 1980 Short-term attenuation of runoff pollution potential for land-applied swine and poultry manure. In Livestock waste: A renewable resource. Ed. R J Smith. pp 289-292. Proc. 4th. Int. Symp. Livestock Wastes, Am. Soc. Agric. Eng., St. Joseph, MI.
White E M and Williamson E J 1973 Plant nutrient concentrations in runoff from fertilized cultivated erosion plants and prairie in eastern South Dakota. J. Environ. Qual. 2, 453-455.
WHO 1978 Nitrate, nitrite and N-nitrosocompounds. Envir. Health Crit. 5, World Health Organization, Geneva.
Wiersma S A, van Goor B J and can der Veen NG 1986 Cadmium, lead, mercury and arsenic concentrations in crops and corresponding soils in the Netherlands. J. Agric. Food Chem. 34, 1067-1074.
Williams R D, Nicks A D, Lewis J A, and Krider J N 1993 Impact of vegetative filter strips on soil loss and water quality. In Preserving our Environment: The Race is On. Ed. anonymous. pp 277-287. International Erosion Control Association, Steamboat Springs, CO, USA.
Witter E 1993 Heavy metal concentrations in agricultural soils critical to microorganisms. National Swedish Environmental Protection Board Rap. 40-79,
Wolnik K A, Fricke F L, Capar S G, Braude G L, Meyer M W, Satzger R D 1983 Elements in major raw agricultural crops in the United States. 1. Cadmium and lead in lettuce, peanuts, potatoes, soybeans, sweet corn and wheat. J. Agric. Food Chem. 31, 1240-1244.
Wong M T F, Wild A and Juo A S R 1987 Retarded leaching of nitrate measured in monolith lysimeters in southeast Nigeria. J. Soil Sci. 38, 511-518.
Yeates J S and Clarke M F 1993 Developing alternatives to phosphate fertilizers of high water solubility. Fert. Res. 36, 141-150.
Yoo K H, Touchton J T and Walker R H 1988 Runoff, sediment and nutrient losses from various tillage systems of cotton. Soil Tillage Res. 12, 13-24.
Young R A and Mutchler C K 1976 Pollution potential of manure spread on frozen ground. J. Environ. Qual. 5, 174-179.
Zhu J C, Gantzer S H, Anderson S H, Alberts E E and Beuselinck P R 1989 Runoff, soil, and dissolved nutrient losses from no-till soybean with winter cover crops. Soil Sci. Soc. Am. J. 53, 1210-1214.

6
Issues in the disposal of industrial and urban wastes

R.G. MCLAREN AND C.J. SMITH

1. Introduction

In many countries, the problems associated with the disposal of wastes generated by population centres, both large and small, together with associated industrial activity, have become significant environmental issues. The realisation that both controlled and uncontrolled disposal of wastes can result in the contamination of soils, surface waters and groundwaters has led to the development of a considerable assortment of regulatory measures and guidelines aimed at reducing or eliminating waste-related pollution. However, waste disposal and the associated environmental contamination from urban-industrial centres is an ongoing major issue, with the list of potential pollutants increasing as new chemicals and processes are developed. In addition, many sites contaminated in the past by waste disposal operations, or contaminated by a particular industrial process carried out at the site, continue to pose a serious threat to the environment.

Many different types of waste are produced by modern society as by-products of domestic and industrial life, each of which has its own set of potential contamination problems. Detailed information on the nature of wastes, their impact on the soil environment and regulatory measures adopted in the Australasia-Pacific region are addressed in Chapters 14 to 25. The aim of this chapter is to describe the various specific contamination issues relating to the disposal of some of the major industrial and urban wastes.

Irrespective of the source of the pollution, the main issues discussed in this chapter relate to potential contamination of the soil with both inorganic and organic pollutants and the subsequent effects on water and atmospheric pollution, plant growth, animal and human health and the overall fertility of the soil itself. In some cases, for example with sewage sludge and sewage effluent, an additional issue relates to the conflict between desires to re-use valuable waste constituents, while at the same time protecting the environment from some of the potentially harmful components.

2. Sewage sludge

One of the main waste products resulting from the biological treatment of human and industrial sewage, as carried out at sewage treatment works, is sewage sludge. Large amounts of sludge are continuously being produced and require disposal. For example in New South Wales, Australia approximately 5000 t of wet sludge are produced daily (Ross et al., 1991). Sewage sludge can have substantial problems

R. Naidu et al. *(eds): Contaminants and the Soil Environment in the Australasia-Pacific Region, 183–212.*

Table 1. Nutrient contents of sewage sludges applied to land in the Sydney region (1990–91)[A].

Nutrient	No. of samples	Nutrient content (% dry solids) Mean	Range
Total nitrogen	104	3.87	2.11–8.21
Ammonia-nitrogen	72	1.01	0.06–5.8
Nitrate-nitrogen	36	0.04	0.01–0.07
Total phosphorus	72	2.98	1.06–8.9
Sodium	72	0.17	0.05–0.53
Potassium	72	0.17	0.09–0.34
Calcium	72	1.87	0.95–4.33
Magnesium	72	0.37	0.18–0.68

[A]Adapted from Ross et al., 1991.

associated with its disposal and these are discussed below. Some of these problems arise from the fact that in most metropolitan areas there is no separation between domestic and industrial sewage and wastewater. In some situations, urban runoff and stormwater also find their way into the sewage works. However, although industrial contamination is a major problem with sewage sludge, domestic sewage alone is by no means completely free of a range of potential contaminants. For example, the heavy metals Cu and Zn, various pathogenic organisms and a whole range of organic chemicals can be present at concentrations of environmental concern.

The sewage sludge produced at municipal treatment plants is disposed of in several different ways. In some countries, large amounts of sludge have for many years been pumped out to sea, in other cases it has been landfilled, incinerated or applied to the land. Ocean discharge of sludge is generally no longer accepted by the public (Australian Water Resources Council, 1992) and has been, or is being, phased out in many countries. The costs of landfilling sludge, including transport and site costs, are increasing and, as discussed below, landfill sites themselves can be sources of environmental contamination. Incineration of sludge is not widespread even in highly industrialised countries, for instance in the European Community in 1984 less than 9% of the sewage sludge produced was incinerated (Van den Berg, 1993). In addition, the ash resulting from incineration still requires disposal and the incineration process itself can result in atmospheric pollution.

Against this background, there is increasing pressure for the land disposal of sewage sludge. Indeed in many countries a substantial proportion of the sludge produced has always been applied to the land, such as in many developing countries that do not have appropriate treatment facilities (also see Chapters 14 to 25). Sewage sludge is basically an organic waste material which contains significant amounts of N, P and other nutrients (see Table 1) and therefore has been recognised as having considerable potential as a fertiliser material and soil conditioner. However, the application of sewage sludge to land is not without its problems, mainly as a result of various contaminants present within the sludge. These contaminants can include the heavy metals (Table 2), pesticide residues and other organic compounds such as

Table 2. Mean metal concentrations in some sewage sludges from Australia and New Zealand (mg kg^{-1} dry solids).

Metal	Bolivar (Adelaide)[A]	Port Adelaide[A]	Glenfield/West Camden (Sydney)[B]	Auckland (NZ)[C]	Christchurch (NZ)[C]
Arsenic	3.6	5.2	5.5	21	11.9
Cadmium	2.8	20.7	3.9	9.2	6.2
Chromium	925	194	81	1209	1618
Copper	888	1120	1427	571	572
Lead	190	250	140	579	443
Mercury	1.7	2.1	5.1	1.8	5.2
Nickel	61	86	25	179	117
Selenium	0.5	0.5	10.6	–	–
Zinc	834	1350	839	1826	1648

[A]Personal communication; [B]Ross et al., 1991; [C]Department of Health (NZ), 1992

polychlorinated biphenyls (PCBs) together with a range of pathogenic organisms. Sewage sludges will vary in the degree of contamination. For example, sludges from industrialised cities will have higher metal loadings than those from small rural towns. Contamination with organic compounds and pathogens will also vary greatly between sludges. Hence not all of the issues discussed below will be relevant to every situation where sludge is applied to the land. Similarly the exact nature of the sewage treatment process will also effect the nature of the sludge produced and the types of problems likely to be encountered during land disposal.

2.1 Heavy metals

Sewage sludge contains a whole range of heavy metals, some of which can be present in very high concentrations (Table 2). Indeed, the bulk of the metals entering a sewage processing plant are retained in the sludge. The most common metals found are Cd, Cr, Cu, Ni, Pb and Zn. However other metals including As and Hg may also be present in appreciable amounts. These metals are known to be potentially toxic to plants and animals, so pose a risk when applying sludge to the land. Many experiments have shown that high levels of metals in soils can reduce plant growth and that the concentrations of some metals in plants can be increased substantially by applications of sludge to the soil. In several countries, including Australia and New Zealand, guidelines for sludge use, based on sludge and soil metal concentrations, have been developed with the aim of preventing metal toxicity to plants and to human and animal consumers of the plants (e.g. Ross et al., 1991; Department of Health (NZ), 1992). In some countries, application of sludge to forests is being used as a means of ensuring that metals do not enter the food chain.

One of the major problems with the application of metal-containing sludges to soils, particularly with repeated applications, is that the metals accumulate in the surface layers of the soil. In general, there appears to be little movement of most heavy metals below the zone of incorporation of the sludge (Emmerich et al., 1982;

Chang et al., 1984). Dowdy and Volk (1984), in an extensive review of heavy metal movement, concluded that movement was most likely to occur where heavy applications of sewage sludge are made to sandy, acid, low organic matter soils receiving high rainfall or irrigation. Some evidence for this appears in Figure 1, which shows the distribution of Ni and Zn down the profile of a sandy soil treated with sludge at intervals over a period of 25 years. Concentrations of the metals at depths well below the zone of incorporation are much higher than those in a nearby untreated soil. Direct leaching of metals has also been noted by Lamy et al. (1993) who observed increased leaching of Cd from a sludge-treated agricultural soil, and by Cameron et al. (1994) who demonstrated increased leaching of Zn from a sludge-treated forest soil. However, in both these studies concentrations of metals in drainage leachates from the sludge-treated soils remained below drinking water limits and thus would be unlikely to cause significant contamination of groundwater.

In addition to the potentially toxic effects of metals on plant growth and their introduction into the food chain, current concern is focusing on the effects of high soil metal concentrations on soil biological activity. Considerable evidence shows that, in the long-term, elevated metal concentrations can reduce soil microbial biomass levels, inhibit N_2 fixation by both free-living and symbiotic organisms and reduce certain enzyme activities such as urease and phosphatase (e.g. Tyler, 1981; Baath, 1989; McGrath, 1994). The implications of such effects on the long-term fertility and sustainability of the soil remain unclear and several important questions still require answers. In particular we need to understand any changes in metal bioavailability which may take place over long periods of time following sludge application to the soil. Berrow and Burridge (1980) have hypothesised that when the organic matter added in sludge breaks down, the metals held by it will be released and become more mobile and phytotoxic. However it has also been suggested that, with time, metals added to the soil in sludge will react with the soil and 'revert' to less mobile and bioavailable forms (Lewin and Beckett, 1980). Whatever the answers are, the most conservative approach would be to minimise metal inputs to the soil wherever possible.

2.2 *Nutrients*

Although sewage sludge can be a valuable source of plant nutrients, particularly N and P, care must be taken to ensure that these nutrients are not allowed to pollute the environment by contaminating surface or groundwater supplies. To help achieve this, the aim should be to match nutrient application rates to plant requirements. In the case of N, this may not be easy since a high proportion of the N in sludge will be present in organic forms and will only become available for plant uptake after mineralisation to NH_4^+-N or NO_3^--N. Rates of N mineralisation in the field are difficult to predict and if NO_3^- in particular is released from the sludge faster than it is taken up by plants, leaching of excess NO_3^- from the soil may result. Once leached from the soil profile, the NO_3^--N has the potential to contaminate groundwater and ultimately drinking water supplies (see Chapters 2 and 5). As discussed

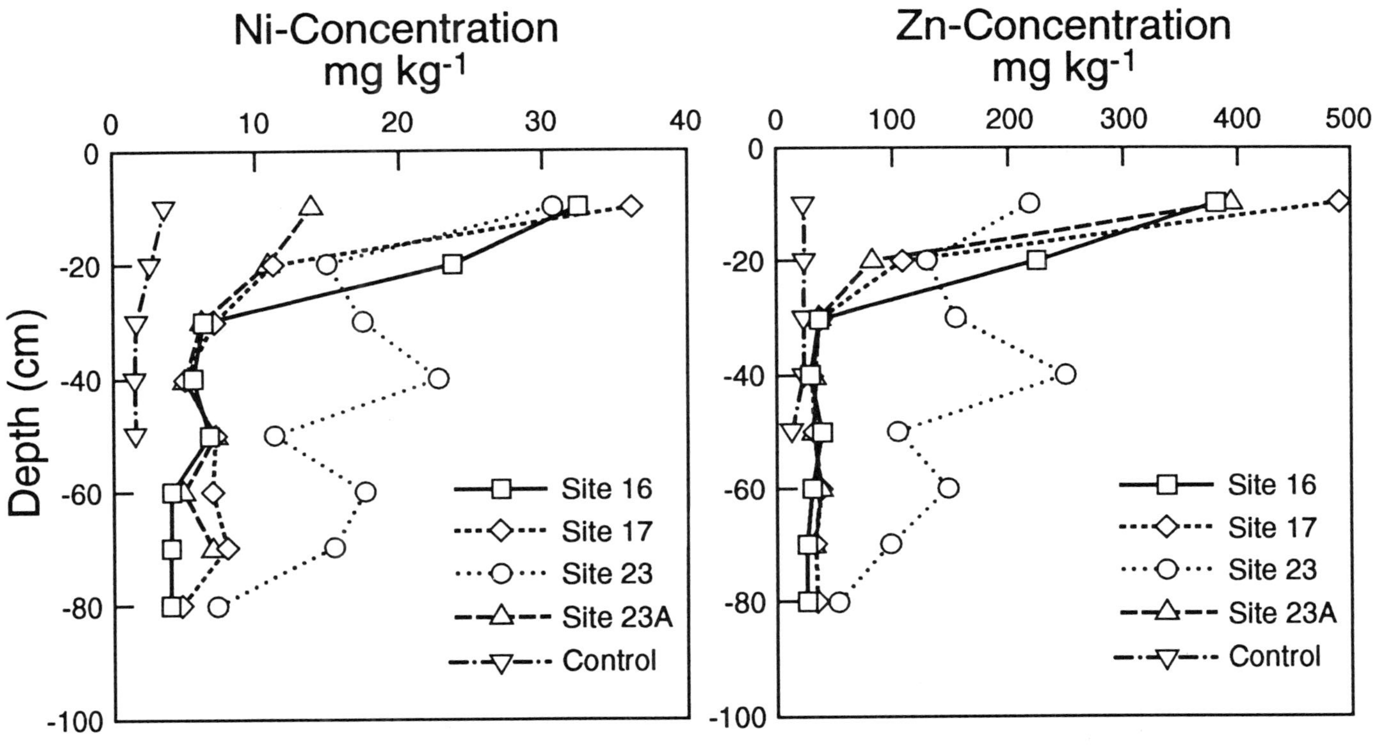

Figure 1. Distribution of Ni and Zn in soil profiles at sewage sludge disposal site.

in Chapters 2 and 5, the leached NO_3^--N may eventually find its way into open water bodies such as streams and lakes and cause eutrophication.

Compared to N, P is relatively immobile in soils and applications of sewage sludge are unlikely to result in leaching of P from the soil profile (Lindo et al., 1993). However with both P and N (and other sludge contaminants) there is a risk that if sewage sludge is surface applied in unsuitable locations, heavy rain may cause run-off leading to the contamination of nearby surface water. In such cases, transport of the contaminants may not necessarily take place in solution but as suspended solids. Clearly sludge application management practices must minimise such possibilities, usually by ensuring good buffer zones around bodies of surface water, and by not applying sludge during high rainfall periods (Ross et al., 1991). Incorporation of sludge into the soil will also minimise pollution of surface waters.

2.3 Pesticides and other organic chemicals

The wastewaters entering a sewage treatment works will contain a whole range of organic chemicals including pesticides and other toxic and potentially carcinogenic compounds. The types and amounts of compounds present vary with the sources of the wastewater (Table 3). The majority of these compounds are recalcitrant and lipophilic and therefore end up closely associated with the sewage solids sludge (Lester, 1982). As a result, concern has been expressed regarding the potential health hazards associated with the application of sludge-borne organic chemicals to the land. There are indeed several possible pathways of human contamination from these chemicals which include (O'Connor et al., 1991):

- Uptake by plants and consumption by humans;
- Direct application to edible plant parts and consumption by humans;
- Direct ingestion of sludge-contaminated soil by children;
- Uptake by plants grazed by animals, transfer to animal food products and consumed by humans;
- Direct ingestion of sludge-contaminated soil by grazing animals, transfer to animal food products and consumption by humans; and
- Leaching from the soil and contamination of water supplies.

In practice, some of these potential pathways would seem unlikely to be the cause of significant human exposure to organic chemicals. O'Connor et al. (1991) carried out a detailed review of the bioavailability to plants of sludge-borne toxic organics. They concluded that the vast majority of these chemicals in sludges occur initially in sludge-amended soils at low concentrations and are so strongly sorbed in the sludge-soil matrix as to have low bioavailabilities to plants. Alternatively they may be degraded or otherwise lost from the soil during the cropping season, and are accumulated, if at all, at very low concentrations in the edible portion of food chain crops. Direct contamination of plants is also unlikely since sludge application to standing food crops is not permitted.

The most likely pathway for human exposure arises from the ingestion of

Table 3. Major sources of organic contaminants in waste water (adapted from Lester, 1982).

Chemical class	Sources
Aliphatic and aromatic hydrocarbons (including benzenes and petroleum hydrocarbons)	Petrochemical industry wastes Heavy/fine chemicals industry wastes Industrial solvent wastes Plastics, resins, synthetic fibres, rubbers and paint production Coke oven and coal gasification plant effluents Urban run-off Disposal of oil and lubricating wastes
Polynuclear aromatic hydrocarbons (PAHs)	Urban run-off Petrochemical industry wastes Various high temperature pyrolytic processes Bitumen production Electrolytic aluminium smelting Coal-tar coated distribution pipes
Halogenated aliphatic and aromatic hydrocarbons	Disinfection of water and wastewater Heavy/fine chemicals industry wastes Industrial solvent wastes and dry cleaning wastes Plastics, resins, synthetic fibres, rubbers and paint production Heat-transfer agents Aerosol propellants Fumigants
Organochlorine pesticides	Agricultural/horticultural run-off Domestic usage Pesticide production Carpet mothproofing Timber treatment
Polychlorinated biphenyls (PCBs)	Capacitor and transformer manufacture Disposal of hydraulic fluids and lubricants Waste carbonless copy paper recycling Heat transfer fluids PCB production
Phthalate esters	Plastics, resins, synthetic fibres, rubbers and paint production Heavy/fine chemicals industry wastes Synthetic polymer distribution pipes

sludge-contaminated soil by grazing animals. Cattle diets can contain on average 6% soil (Healy, 1968) and sheep can also consume considerable amounts of soil. Organochlorine pesticides in particular are known to accumulate in cattle primarily as a result of soil ingestion (Lindsay, 1982). These chemicals are deposited in the body fat and can appear in meat products and milk. However where such accumulation has occurred, for example with DDT, it has been related to the past use of pesticides on pastures rather than as a result of sludge application.

There appears to be little data on the leaching of organic chemicals from sludge-treated soils, but in view of the strong adsorption of these compounds by the soil, it seems unlikely that this represents a major potential pathway for human exposure. As with nutrients, run-off from sludge-treated areas is an alternative pathway for the contamination of surface waters with toxic organics. This could certainly have implications for aquatic life.

2.4 Pathogens

Raw sewage can contain a whole range of pathogenic organisms, including viruses, bacteria, fungi and parasites such as tapeworms and liver flukes. Although sewage treatment processes are designed to eliminate or reduce pathogen numbers they are never 100% effective and substantial numbers of organisms can be present in sewage sludge. The pathogens within sewage sludge can cause diseases if they come in contact with humans. According to Ross et al. (1991), this contact can occur in several ways: by inhaling sludge aerosols or dust while spreading; by eating vegetables or fruits contaminated by sludge; by drinking water contaminated by run-off or percolation from sludge applied to land or by eating meat from livestock infected while grazing pastures or crops fertilised with sludge. Of the large range of potential hazards, it has been concluded (Ross et al., 1992) that salmonellosis and bovine cysticercosis are the two infectious diseases most likely to affect animals and humans as a result of sludge contamination. Outbreaks of both these diseases in animals in European countries have been associated with surface application of sewage sludge to pastures (Jones, 1986; Nansen and Henriksen, 1986). However, the risks can be minimised by incorporation of sludges into the soil and by using suitable withholding periods before allowing animals to graze sludge treated land.

3. Wastewater/sewage effluent

Wastewater is the general term applied to the liquid waste collected in sewers and treated in wastewater treatment plants. Domestic wastewater is generally high volume with low to medium concentrations of nutrients. The average rate of generation varies from about 200 to 600 litres per day per person (Asano et al., 1985; Breen 1992). Total dissolved salt (TDS) range from 250–850 mg L^{-1}; total N: 15–90 mg L^{-1}; and total P: 5–20 mg L^{-1} (Asano et al., 1985; Breen, 1992). In contrast, wastewater from intensive agricultural industries (feedlots, piggeries) has a much higher chemical (COD) and biological oxygen demand (BOD), a higher total solid content (2990–18000 mg L^{-1}) and contains more N (85–1080 mg L^{-1}) and P (40–440 mg L^{-1}) than domestic wastewater (Breen, 1992). Similarly rural industry wastewaters (abattoirs, canneries) have two to five times more total solids, dissolved salts and nutrients than domestic wastewater. For example, TDS of cannery wastewater can range from 22000–72000 mg L^{-1}. The volume of wastewater generated will depend on the type and size of the industry (Breen, 1992).

The quality of treated urban effluent clearly depends on the quality of the water supplied to the treatment works and the degree of treatment of the wastewater.

Treated sewage effluent differs from regular water in the following ways (Feigin et al., 1991; Westcot and Ayres, 1985):

- Biological and chemical oxygen demand of the water is increased through municipal and industrial use. Treatment of the wastewater may reduce the BOD and COD but the level of reduction depends on the intensity of the treatment;
- Inorganic salts levels are increased in wastewater as a result of its use. Major increases are seen in sodium, chloride and bicarbonates. These ions cause an increase in the TDS (salinity) and the sodicity (sodium content) of the water and are generally not removed during treatment, except for some precipitation with bicarbonates;
- Effluent contains different concentrations of macronutrients, especially N and P;
- The concentration of trace elements may also increase as a result of the addition of industrial wastewater to the sewer. Trace elements required by the plant (B, Cu, Mn, Mo and Zn) may be present in excessive levels. Other elements such as Cd, Pb, Ni and Hg may also contribute to the toxic hazard of the effluent;
- Pathogenic microorganisms (bacteria and viruses) are present, although their concentration is reduced during the treatment process.

The quality of effluent may be improved in some cases by storage in treatment ponds before final release or by some form of chemical treatment carried out in the sewage-works. For example, N may be removed by increasing the effluent pH and stripping out the ammonia, or by biological nitrification and denitrification aided by the addition of an energy source (Lance, 1986). Phosphorus removal can be achieved by chemical precipitation using lime at high pH, or Al or Fe salts at neutral pH.

In many countries, including Australia, the effluent produced at municipal treatment plants had traditionally been disposed by pumping into inland rivers or water bodies or release into oceans. Direct discharge of wastewaters into water bodies is generally no longer acceptable by the public because this practice has contributed to outbreaks of toxic blue-green algae in many Australian rivers (Harvey, 1993; Chapter 2). In arid and semi-arid environments, where supplies of fresh water are limited, there is increasing pressure for urban authorities to investigate alternative water supplies, including the re-use of effluents. The greatest motivation for reclamation and recycling of water is the lack of fresh water supplies, and an inability to develop new fresh water supplies (Hartling, 1994). Although effluents can be re-used in several different ways, crop irrigation is probably one of the methods for which there is a high potential application.

3.1 Effluent irrigation

Correctly designed and managed effluent irrigation schemes have the ability to remove many contaminants from the wastewater, so that their impact on the environment is reduced, while at the same time providing a valuable source of water and nutrients to enhance plant growth (Feigin et al., 1991; Pettygrove and Asano, 1985). In addition to other considerations that apply to all irrigation water, irrigation with effluent must take into account the special composition of the effluent. Since effluents contain impurities, attention must be given to the water quality in order to evaluate the possible long-term effects on the soil/plant ecosystem that arise from the accumulation of salt, nutrients and other trace elements added during its use. Incorrect management of irrigation schemes may lead to the contamination of groundwater with salts or NO_3^-. Human health risks may result if drinking water is pumped from the contaminated groundwater. The main issues to be considered in the management of effluent irrigation schemes are discussed below.

3.1.1 Crop type

The range of plants which can be irrigated with effluent include woodlots, pastures, arable crops, and recreation areas. Irrigation has considerable potential in arid and semi-arid regions such as the inland areas of Australia where there is a large deficit between evapotranspiration and rainfall. The plant species chosen for an irrigation scheme depends on the efficiency in taking up and removing nutrients and water from the selected site. Crops or pastures whose biomass is regularly removed from the site, have the capacity to harvest and remove large amounts of nutrients over a long period of time. However, if the plant is harvested by grazing there will be little export of nutrients from the site. The export of nutrient is maximised when the biomass is mechanically harvested and moved offsite. In contrast, trees are effective in nutrient removal during the early growth stages before the canopy has closed (Stewart et al., 1990; Cromer et al., 1983). From this time onwards, most of the additional nutrients applied in the effluent is in excess of the tree's capacity to accumulate them and may be leached to groundwater. Nutrient uptake rates are high, but much of this nutrient is returned to the soil in litterfall (Attiwill and Leeper, 1987). Consequently, the storage of nutrient by the tree will be low. The nutrients in the litter rapidly decompose, causing the release of nutrients back to the soil. Thus the potential for leaching of NO_3^--N to groundwater is great.

3.1.2 Hydraulic loading rate

The hydraulic loading rate is defined as the amount of water used by the plant under non-restrictive conditions to achieve full production potential. For an effluent irrigation system to be effective it is essential that the irrigation water be applied at the correct time to meet the plants' requirement. Determination of the 'correct' amount of water to apply may be achieved with a water balance model (e.g. Myers, 1992; Gardner and Littleboy, 1994). Such models consider rainfall (R_f), effluent applied (I_e), evapotranspiration (ET), the accretion of water in the root zone (ΔS) and deep drainage (D). The water balance equation is expressed as:

$$I_e = ET + D + \Delta S - R_f \quad (1)$$

The models based on the above equation are generally calculated on a monthly, weekly, or daily time step. Comparison of the irrigation requirement with the rate of effluent production determines the area of land required for effluent irrigation. The method of water application, and specific application rate needs to be matched to the soil. If effluent re-use schemes are to be sustainable, it is essential that well-established irrigation and soil monitoring procedures are followed (Pruitt and Snyder, 1985).

3.1.3 Salinity

Salt is continually added to the soil with the irrigation water and since most plants take up limited amounts of salt, as water is taken up by the plant or evaporates from the soil surface, most of the salt remains in the soil. Unless leached from the plant rootzone the salt will accumulate to levels that impact on the growth of the plant (Rhoades and Loveday, 1990). The rate of accumulation depends on the concentration of salt in the irrigation water, irrigation application and the rate at which salt is removed by leaching (salt out). When the system comes to equilibrium, the amount of salt being leached equals the annual increment of salt applied. Applying water in excess of what can be used by the plant ensures that salt is removed from the soil profile by leaching (Feigin et al., 1991; Oster and Rhoades, 1985). Thus, establishing a net downward flow of water and salt through the root zone is the only way to manage the salinity problem. Good drainage is essential for this to happen, but the leachate may have an adverse impact on groundwater resources.

The above discussion assumes that it is permissible to have leaching and that the required leaching rate can be achieved in the soil. If drainage is not adequate, shallow water tables develop and further contribute to the salinity hazard. Clearly, long term re-use of wastewater is only possible if there is adequate leaching and drainage at the site.

3.1.4 Soil sodicity

The accumulation of Na in the soil may result in the deterioration of soil physical conditions through clay particle swelling and dispersion. In turn, this causes a reduction in soil porosity, hydraulic permeability and infiltration rates, and destruction of the soil structure (Rhoades and Loveday, 1991; Feigin et al., 1991; Naidu et al., 1995). The soils are less suitable for irrigation and the growing of crops and are thus less efficient in the treatment of effluent. When the clay particles disperse and move to deeper horizons in the soil, the damage is irreversible, as is the case for some sandy soils. Leaching by winter rains will dilute the saline soil solution, and consequently the soils become more sensitive to sodicity effects of rapid particle dispersion and soil crusting. Shainberg and Oster (1978) found that this process increased run-off and erosion during the winter, and if the infiltration capacity was not restored, difficulties may be experienced in supplying adequate water to satisfy the plants' demand.

Table 4. Trace element water quality criteria for irrigation used continuously on all soil types.

Element	Concentration (NSW†) mg L^{-1}	Concentration (Feigin et al., 1991) mg L^{-1}	Concentration (Canberra, Feb. 94) mg L^{-1}
Aluminium	5	5	–
Arsenic	0.1	0.1	–
Beryllium	0.1	0.1	–
Boron	–	0.75	–
Cadmium	0.01	0.01	0.006
Chromium	0.1	0.1	0.009
Cobalt	0.05	0.05	–
Copper	0.2	0.2	0.1
Fluoride	1	1	1.08
Iron	1	5	2.1
Lead	0.2	5	–
Manganese	0.2	0.2	–
Molybdenum	0.01	0.01	–
Nickel	0.2	0.2	0.003
Selenium	0.02	0.02	–
Vanadium	–	0.1	–
Zinc	2	2	0.082

†NSW Recycled Water Coordination Committee (1993). Guidelines for Urban and Residential Use of Reclined Water.

3.1.5 Heavy metals

As discussed above, the bulk of the metals entering a sewage processing plant are retained in the sludge. As a result, metal concentrations in domestic or tertiary-treated sewage effluent are very low, although higher concentrations may be found in industrial sewage effluent and especially in raw sewage (Feigin et al., 1991).

The concentration of heavy metals in the treated effluent for Canberra (a non industrialised city) on 11 February, 1994 and the permissible concentration in irrigation water (Feigin et al., 1991) and effluent used for urban and residential reuse (State Pollution Control Commission, 1979; NSW Recycled Water Coordination Committee, 1993) are summarised in Table 4. The concentration of heavy metals in the non-industrial effluent is within the range of values reported by Feigin et al. (1991) for secondary effluents. Furthermore, the data show that the concentrations are below the recommended maxima for effluent and irrigation water used continuously on soil (State Pollution Control Commission, 1979; Feigin et al., 1991; NSW Environmental Protection Agency, 1989).

A way of evaluating the potential hazard of heavy metals is to determine the cumulative total loading of Cd, Cu, Ni, Zn and Pb against NSW Agriculture and Fisheries and the US Environmental Protection Agency (USEPA) recommendations for heavy metals contained in sludge (Page and Chang, 1985; NSW Agriculture and Fisheries, 1989; Feigin et al., 1991). Given the trace-element concentration in the effluent and the annual application rate, one can calculate the time required for the soil to reach the loading limit of the metal (Table 5).

The time required for the heavy metals to reach the limits proposed by the NSW

Table 5. Calculated length of time for effluent-irrigated agriculture soil to reach the heavy metal limit.

Element	Concentration $mg\,L^{-1}$	Annual input (@60 mm yr^{-1}) $kg\,ha^{-1}$	Suggested loading at soil CEC† 5–15‡ $kg\,ha^{-1}$	Time to reach soil loading years
Cd	0.006	0.04	2.55††–10	62–250
Cu	0.1	0.6	100–250	166–416
Ni	0.003	0.02	50–250	>2,500
Zn	0.082	0.49	250–500	>510
Pb	0.019	0.11	50–100	>450

†CEC of the soil at Canberra in the upper 30 cm ranged from 5 to 12 $cmol_c\,kg^{-1}$.
††Lower loading limit based on NSW Agriculture and Fisheries guidelines for the use of sewage sludge on agricultural land.
‡CEC range for soil in $cmol_c\,kg^{-1}$ soil.

Agriculture and Fisheries (1989) or the USEPA (Page and Chang, 1985) varies from 62 years to more than 2500 years. The proposed USEPA limits for metal inputs are specified for soil pH>6.5. The pH of the metal-affected soil should be maintained above this value to ensure the low mobility and availability of trace metals in the soil. One major problem with the irrigation of metal-containing effluent is that the metals accumulate in the surface layer of soil. Although it may take a long time for heavy metal concentrations in effluent irrigated soils to reach the currently proposed limits, the potentially harmful effects that may arise as a result of future land use should not be ignored (Page and Chang, 1985).

3.1.6 Phosphorus

Application of P through effluent irrigation will usually not be excessive, especially if the effluent is applied at the water use rate of the crop. Assuming a median P concentration of 10 mg L^{-1}, an annual application of 1000 mm applies 100 kg P ha^{-1}. This is considerably above the fertiliser application rate for most agricultural and pasture crops grown in Australia, and greater than the amount removed in the above-ground portion of the plant (NSW Agriculture and Fisheries, 1989; NSW Environment Protection Agency, 1995). However, as a result of reactions of added P with the soil, the concentration of soluble P in the soil solution generally remains low. For sandy soils that have a minimal P sorption capacity, P may be transported through the soil profile into groundwater after several years of effluent irrigation (Iskander and Syers, 1980; Latterell et al., 1982; Mansell et al., 1992). Therefore soils need to be carefully evaluated for their capacity to sorb P before effluent is irrigated. Several models have been proposed to estimate P sorption/transport in soil, e.g. Ryden and Pratt (1980), Ryden et al. (1981) and Mansell et al. (1992). However, because of the uncertainties involved in the estimation of P sorption, effluent irrigated sites should be frequently monitored, especially as the cumulative P loading approaches the estimated P retention capacity of the system.

3.1.7 Nitrogen

Nitrogen is an essential element for plant growth and is usually added in large quantities in agricultural lands in order to optimise the crop yield. Irrigation with sewage effluent can add considerable quantities of N to the soil depending on the irrigation rate (hydraulic loading) and the N concentration in the effluent. The N enters the soil-plant N cycle, details of which can be found in reviews by Keeney (1981) and Stevenson (1982).

Irrigation management determines the distribution of N application over time, since N is added in every irrigation. The N may be either taken up by the plant, lost as gaseous N (NH_3 volatilisation and/or biological denitrification) or leached below the root zone. Some of the N in the effluent will be present in the organic form and will only become available for plant uptake after mineralisation. When the N loading exceeds the capacity of the plants to accumulate the N, the NO_3^--N concentration rapidly increases in the soil solution and significant leaching occurs (Smith et al., 1994). Once leached below the root zone, the NO_3^--N has the potential to contaminate groundwater supplies. Breuer et al. (1979) and Sopper and Kerr (1979) have shown the importance of active N uptake on reducing the leaching of NO_3^--N. Leaching of N may also be reduced as a result of biological denitrification.

The fraction lost by denitrification can be variable and is difficult to predict under field conditions. In soils that are well drained, the fraction of the NO_3^--N likely to be lost by denitrification is low, whereas leaching losses are likely to be the dominant loss mechanism. Although the basic N cycle processes are well understood, the ability to accurately predict the process requires investigation. Gaseous N loss processes are the most difficult to predict and need to be quantified and predicted under conditions of flowing soil solution. Our understanding of the interaction between leaching and biological denitrification needs to be improved in order to devise management methods to enhance gaseous N loss. For example, the addition of organic carbon (OC) (in this case sewage sludge) increased N removal from soil columns flooded with secondary treated effluent (Lance, 1986). The increased N removal was attributed to decomposition processes in the soil providing a constant source of readily available C for the denitrifiying bacteria. Perhaps the approach of adding a readily available C supply can be applied to other effluent irrigation schemes. The development of management options to enhance N removal by biological denitrification requires further investigation.

Given the uncertainty in gaseous N losses, the total N loading to a site from effluent and mineralisation should balance the net N uptake of the above-ground vegetation. This will give a conservative N loading to the site, and the processes such as NH_3 volatilisation and denitrification ($NO_3^- \Rightarrow N_2$) may provide a safety margin against excessive N loading (NSW Environment Protection Agency, 1995).

3.2 Effluent treatment by wetlands

In addition to the reuse of effluent by irrigation, in recent years there has been considerable interest in the possibility of using wetlands to treat effluents. Either modified natural wetlands or constructed wetlands can be used for this purpose (Venus, 1994). In the wetlands, the effluent water evaporates or percolates through the soil and/or sediment, while the aquatic plants remove the nutrients. The oxidised surface soil layer, oxidised rhizosphere of the plants and the slime layer on the plant stems and in the litter act in a manner analogous to fixed growth biological filter systems. Treatment also occurs by plant filtration, which filters suspended particles from the water stream. The sediments are characterised by a high organic content, are anaerobic and are a sink for the accumulation of N, P and S (Millin and Heritage, 1992; Breen, 1992). The aquatic plants are adapted to growing under waterlogged conditions and have a well-developed system for gaseous exchange between the roots and the tops.

The formation of aerobic zones within the flooded sediment favours the production of NO_3^-. The NO_3^- then diffuses into anaerobic zones where it is converted to di-nitrogen or nitrous oxide gas that is released to the atmosphere. If the objective of the scheme is to remove nutrients, the wetland must be managed to maximise dry matter productivity and the accumulation of N and P in the plants. Furthermore, it is important that the biomass be harvested and exported, rather than allowing the wetland plants to die *in situ* and return nutrients (recycle) to the ecosystem.

Wetland systems are probably not recommended for use where soil and groundwater conditions are suitable for irrigation of the effluent. Wetlands may have applications where agricultural, pasture or tree production is not of a sufficient scale to cover the cost of operating a land based irrigation scheme. Advantages of constructed wetlands include low to moderate cost of construction, low energy requirement, simple operation, good hydraulic and load buffering capacity, good pathogen removal and good nutrient removal (Breen, 1992). Disadvantages are high land requirement, performance may be seasonally dependent, may result in contamination of groundwater or cause the development of a perched watertable and salinisation of surface soils, and are potential sites for insect pests (mosquitoes). Effluents from wetlands often have low dissolved oxygen levels, which may pose a problem if the water is being discharged into sensitive water bodies. However, dissolved oxygen levels can be easily increased by using a cascading discharge structure in the outfall design (Venus, 1994).

4. Landfills

Disposal of solid wastes in landfill sites is a common disposal option worldwide. For example, Australia disposed of about 12.8×10^6 t of solid wastes in 1989, or just over 776 kg per person per year (van den Broek, 1994). By comparison 2.3×10^6 t are disposed of each year in New Zealand, which equates to about 733 kg per person per year. Domestic garbage, mainly paper, cardboard and vegetable matter, formed the major portion of the waste (49–55%) deposited in landfills (van den Broek, 1994; Willmot, 1994a). The other major contributor was industrial waste. The shortage of landfill sites, and community opposition, is forcing authorities to seek to adopt integrated waste management options and alternative disposal methods. Part of the approach includes waste minimisation, waste avoidance, recycling and resource recovery before resorting to disposal. Within the landfill, uncontrolled fermentation reactions break down the solid wastes. The major environmental problems associated with landfills are the impact of leachate on surface and groundwaters and the production of landfill gas (Willmot, 1994b).

The chemical composition of the leachate is highly variable and site specific due to the diverse chemical composition of the material deposited in the landfill. However, the major determinant of the leachate quality is the stage of fermentation of the material. In general, the leachates have a high organic strength, ie. high BOD and COD, and in many ways resemble sewage. Contamination of groundwater by landfill leachate is an extensive and increasing problem. For example, data from Finland show detrimental changes in groundwater quality down gradient of the landfills (Assmuth and Strandberg, 1993). Common contaminants in the leachate were heavy metals and chlorinated hydrocarbons such as dichloromethane and 1,2-dichloroethane. In addition, the concentration of Fe, Mn and Cl in the leachate exceeded drinking water standards. Not all of the organic pollutants deposited in the solid waste will be leached. Data from landfills in Florida would suggest that some pollutants are metabolised by microorganisms, or otherwise degraded, sorbed onto particulate matter, or diluted to such an extent as to be non-detectable in the leachate (Hallbourg et al., 1992).

One method of leachate control is to use a low permeability top cover to minimise water entry into the landfill (Willmot, 1994b; Weeks et al., 1992). Possible choices include clay or manufactured liners. The containment of toxic material, salts and other pollutants is achieved through the use of an underlying liner. This liner collects and directs the leachate generated as a result of fermentation to drainage points. Clays are frequently chosen because of their low permeability. However, high concentrations of some soluble organic material can alter the permeability of water through the clay therefore limiting their usefulness as a liner material. Furthermore, the liners may fail as a result of waste subsidence, and consequently the topcover may serve as the principal pathway for water entry into the landfill (Schulz et al., 1992). The permeability of liners, possible effect of chemicals on their permeability and the ability to predict the changes over a 100 year period require further investigation.

Problems associated with landfill gas are its accumulation to explosive quantities, smell, groundwater acidification and reduced growth following final restoration of the site (Willmot, 1994b). The net emission of CH_4 is a function of the biodegradable OC content, the proportion of the OC that degrades and the CH_4 component of the landfill gas. The average CH_4 emission rate from municipal solid wastes in Canberra was $49\,g\ m^{-2}\ d^{-1}$, corresponding to an annual emission of $10\,kg\ CH_4\ t^{-1}$ of municipal solid waste (Denmead, 1995, unpublished data). Methane emissions have been found to be seasonal, and are dependent on water content (Jones and Nedwell, 1993). Control of gas migration may be achieved by the use of French drains, whereas gas extraction or air injection systems may be required for deeper sites (Willmot, 1994b).

5. Coal combustion residues

Combustion of coal, predominantly for the generation of electricity, produces large amounts of residues which require disposal and which have the potential to contaminate soils, ground and surface waters. The main material requiring disposal is fly ash which is the residue from coal combustion that enters the flue gas stream. It is composed predominantly of fine particles, and is collected in emission control devices, such as electrostatic or mechanical filters (Carlson and Adriano, 1993). Fly ash is sometimes mixed with 'bottom ash' which is the residue from coal combustion that remains in the boiler that is generally a mixture of ash and slag. In addition, other residues such as flue gas desulphurisation waste or fluidised bed boiler ash are sometimes produced. However, compared to the information available on fly ash, these latter residues have been the subject of comparatively little study (Carlson and Adriano, 1993).

Currently Australia produces in excess of 7.7×10^6 t of fly ash annually (Beretka and Nelson, 1994); the corresponding figure for the USA is over 60×10^6 tonnes (Carlson and Adriano, 1993). The physical, mineralogical and chemical properties of fly ash depend on the composition of the parent coal, conditions during coal combustion and the efficiency of emission control devices. In addition, the chemical properties, in particular, can be affected by storage and handling of the material (Adriano et al., 1980). Particle sizes are commonly between 0.01 and 100 μm. In a study of 13 fly ashes from Australian power stations, particles were found predominantly (67–98%) in the silt plus fine sand fraction (Aitken et al., 1984). Fly ash tends to be dominated by Si, Al and Fe which are present in the minerals quartz (SiO_2) and mullite ($3Al_2O_3.2SiO_2$) together with smaller amounts of haematite (Fe_2O_3) and magnetite (Fe_3O_4) (Adriano et al., 1980; Aitken et al., 1984). They also tend to be enriched in As, B, Ca, Mo, S, Se and Sr (Page et al., 1979) and in addition contain a wide range of other trace elements including Cd, Cu, Mn and Zn (Table 6). Many fly ashes have a high alkalinity (e.g. Aitken et al., 1984) but acidic ashes are produced by high S coals (Adriano et al., 1980). Fresh (unweathered) ash is often characterised by high concentrations of soluble salts (Carlson and Adriano, 1993). However in a study of Australian fly ashes, Aitken et al. (1984) found that the level of soluble salts was extremely variable, with electrical conductivities of saturation extracts ranging from 0.63–55 mS cm^{-1}.

The main environmental contamination problems resulting from the disposal or utilisation of fly ash relate to its high soluble salts content and potential toxicity of some individual elements, particularly B. Table 7 indicates the potential toxicity of some of the trace elements found in fly ash to both terrestrial and aquatic life and Figure 2 outlines the various potential impacts of fly ash on the environment. Fly ash is commonly deposited in water suspensions in ponds adjacent to the electricity-generating stations and the ash particles allowed to settle out on the bottom of the pond. The ponds themselves are a potential source of pollution since the effluent water is sometimes discharged or overflows into natural water courses causing damage to aquatic life. Also, leachate from the ponds has the ability to contaminate

Table 6. Elemental composition of fly ash.

Element	Range of values from literature[1]
Si	1.41 – 28.6%
Al	0.1 – 17.3%
Fe	1 – 29%
Ca	0.11 – 22.2%
Mg	0.04 – 7.6%
Na	0.01 – 2.03%
K	0.15 – 3.5%
S	0.1 – 1.5%
P	0.04 – 0.8%
N	negligible
As	2.3 – 6300 $\mu g\,g^{-1}$
B	10 – 618 $\mu g\,g^{-1}$
Cd	0.7 – 130 $\mu g\,g^{-1}$
Co	7 – 520 $\mu g\,g^{-1}$
Cr	10 – 1000 $\mu g\,g^{-1}$
Cu	14 – 2800 $\mu g\,g^{-1}$
Mn	58 – 3000 $\mu g\,g^{-1}$
Mo	7 – 160 $\mu g\,g^{-1}$
Ni	6.3 – 4300 $\mu g\,g^{-1}$
Pb	3.1 – 5000 $\mu g\,g^{-1}$
Se	0.2 – 134 $\mu g\,g^{-1}$
Zn	50 – 1286 $\mu g\,g^{-1}$

[1]Adapted from Page et al. (1979).

Table 7. Potential toxicity of elements in fly ash (adapted from Van Hook and Shults, 1976).

Element	Terrestrial		Aquatic
	Plant	Animal	
As	Low	Low	Low
B	High	Medium	Low‡
Cd	High	High	High
Co	High	Medium	High
Cr	High	Medium	Medium
Cu	High	Medium	High
F	High	High	Low
Hg	Medium	High	High
Mn	Low	Low	Low
Mo	Low	Medium	Low
Ni	High	High	Medium
Pb	Low	Medium	Medium
Se	Medium	High	Low‡
Zn	Low	Medium	Medium

‡Uncertain.

the groundwater. Ultimately the ponds fill up with ash and become a source of air pollution, since dry ash is extremely susceptible to wind erosion (Campbell et al., 1983). At this stage, attempts are made to stabilise the ash *in situ* by establishing a vegetative cover, or the ash is removed to landfill sites. In some cases, fly ash from power stations is landfilled directly without prior ponding. Once in a landfill, the fly

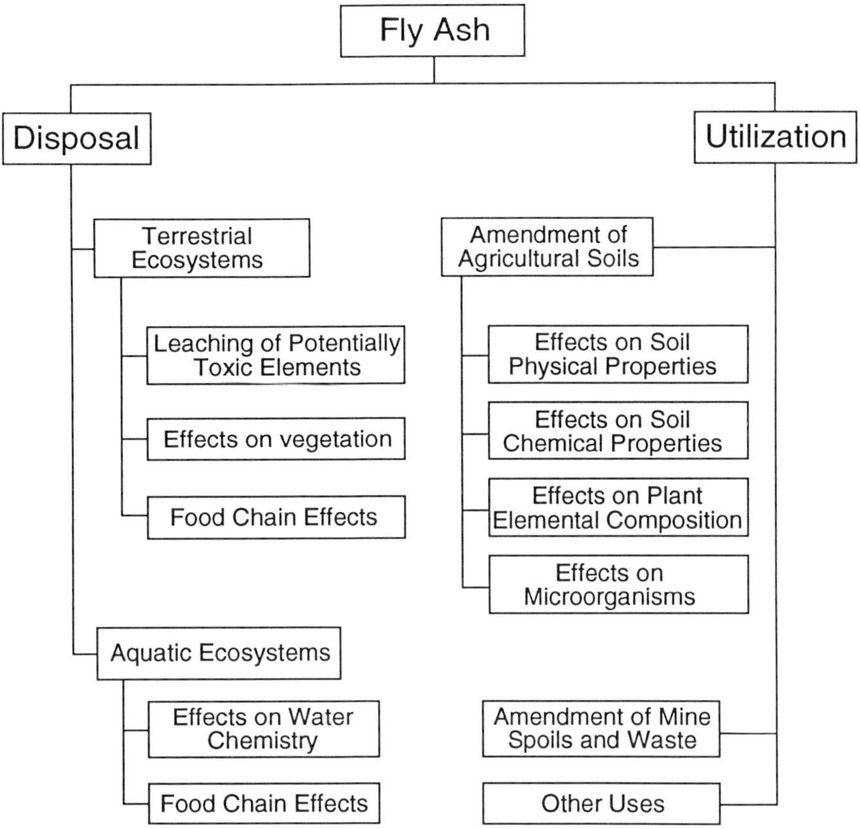

Figure 2. Schematic diagram of the potential environmental impacts of fly ash (adapted from Carlson and Adriano, 1993)

ash will again be subject to leaching, resulting in potential groundwater contamination.

The revegetation of ash ponds or landfill sites is often slow as vegetation establishment may be limited by: (a) lack of essential nutrients, usually N and P; (b) toxicity caused by high pH and/or high soluble salt concentrations, high B and other toxic trace elements; and (c) the presence of compacted and/or cemented layers in the ash (Carlson and Adriano, 1993). Revegetation may be speeded up by covering the ash deposits with a layer of soil, but plants whose root systems penetrate the underlying ash may still accumulate high concentrations of toxic elements (Weinstein et al., 1989).

As a result of decreasing available space and increasing costs associated with ponding and landfilling fly ash, other methods of disposal and/or utilisation have been examined. Some ash is used in the manufacture of building products such as concrete and bricks, and ceases to be a potential contaminant. Approximately 10%

of the ash produced in Australia is used in this way (Beretka and Nelson, 1994). Many studies in North America and Europe have investigated the agronomic utilisation of fly ash with a view to improving the physical composition of soils, neutralising soil acidity and applying nutrients. However, detailed reviews of this work have concluded that although some liming and fertilising effects are observed, the accumulation of soluble salts, B and other toxic elements appears to be the most serious constraint associated with land application of fly ash (Adriano et al., 1980; Carlson and Adriano, 1993). Similarly, a study by Aitken and Bell (1985) demonstrated that B toxicity can limit plant growth on a range of Australian fly ashes although B uptake was dependent on plant species and rate of fly ash application. Their data implied that caution should be exercised in utilising fly ashes, and the use of fly ash to improve soil characteristics should be undertaken with a knowledge of the B status.

6. Wastes containing hydrocarbons and other organic chemicals

Although some of the wastes already discussed in this chapter may contain a range of organic chemicals, they are not generally regarded as a major problem in the disposal of those wastes. However, there are other materials where this is clearly not the case, and the presence of potentially toxic organic chemicals represents a serious threat to the environment and/or human health. In this section we deal with three such wastes, fuel hydrocarbons, gasworks wastes and pulp mill effluents. It is realised that the types of waste and range of organic chemicals covered in this section are by no means comprehensive. However they do serve as useful examples of the main types of problems likely to be encountered. Unfortunately, new chemicals are continuously appearing on the market and will undoubtedly also find their way into these and other waste products.

6.1 Fuel hydrocarbons

Although not strictly wastes, fuel hydrocarbons have been included here because in recent years there have been many examples of large-scale pollution of the environment by oil spills of one type or another. For example, the Exxon Valdez disaster in Alaska and the rupturing of the Siberian oil pipeline. However, smaller-scale pollution of soil and water bodies with a range of hydrocarbons would appear (see Chapter 12) to be a relatively widespread occurrence. Leaks from underground fuel storage tanks and pipes, above-ground spills and discharges from tanks and delivery vehicles can result in significant quantities of oil, petroleum and other fuels entering the soil (Bridges, 1991).

Petroleum contaminated sites may pose significant risks to the public because of potential health and environmental effects (Kofi Asante-Duah, 1991). In the United States, the failure of underground storage systems for hydrocarbon fuels has led to an environmental contamination and public health risk problem of considerable proportions (Lindon, 1993). All commercial fuels consist of complex mixtures of

Table 8. Seed germination after 10 days in fuel contaminated lysimeter soil, 4 and 14 weeks after the spill (adapted from Wang and Bartha, 1990).

Fuel product	Germination at 4 weeks (%) soybean	ryegrass	Germination at 14 weeks (%) soybean	ryegrass
none	80	80	70	87
jet fuel[A]	66	20	78	95
heating oil[A]	12	2	65	88
diesel oil[A]	6	0	68	55

[A]no mixing or bioremediation.

Table 9. Plant dry weight yield after 30 days of growth in fuel contaminated lysimeter soil, 4 and 14 weeks after the spill (adapted from Wang and Bartha, 1990).

Fuel product	Plant growth at 4 weeks (g) soybean	ryegrass	Plant growth at 14 weeks (g) soybean	ryegrass
none	14.53	2.70	9.82	2.00
jet fuel[A]	8.17	0.07	10.32	0.30
heating oil[A]	0.00	0.00	15.96	0.30
diesel oil[A]	0.00	0.00	9.67	0.10

[A]no mixing or bioremediation.

hydrocarbons, including a range of alkanes, cycloalkanes, alkenes and various mono- and poly-aromatic compounds. In addition, some fuels such as petrol also contain additives such as the Pb tetraalkyls and the Pb scavenger 1,2-dibromoethane (DBE). Hydrocarbon fuels are known to have toxic effects on humans, sometimes causing intoxication and irritation of mucous membranes, whether brought about by inhalation or ingestion (Lindon, 1993). Some individual compounds in hydrocarbon fuels also have proven (e.g. benzene and DBE) or suspected (e.g. cresols and phenols) carcinogenic effects. Once soil is contaminated with hydrocarbon fuels there are several potential pathways whereby humans can be exposed to their toxic effects. These include inhalation of volatile compounds, dermal contact with the contaminated soil and soil ingestion (a potential problem with children). In addition, depending on the viscosity of the contaminant fuel, the permeability of the soil and the quantity spilt, hydrocarbons can leach through the soil and lead to the contamination of groundwater supplies (Stone, 1991 – also see Chapter 12).

In addition to its threat to human health by contamination of soil by hydrocarbon fuels, oil pollution can result in changes in soil physical and chemical properties which affect the growth of plants (Udo and Fayemi, 1975). Although in the long-term, many hydrocarbons will be decomposed biologically in the soil, the immediate effects may include the creation of anaerobic conditions, the death of existing vegetation and soil fauna and the prevention of re-vegetation. For example, Wang and Bartha (1990) have shown that soil contamination with three hydrocarbon fuels (jet fuel, heating oil and diesel oil) can inhibit seed germination and growth of soybean and ryegrass (Tables 8 and 9). Similarly, Udo and Fayemi (1975) observed a drastic reduction in germination and yield of corn grown in a soil contaminated with crude oil.

Table 10. Contaminants likely to be present at former gasworks sites (Nakles et al., 1993).

Inorganics	Metals	Volatile Aromatics	Phenolics	Polynuclear Aromatic Hydrocarbons
Ammonia	Aluminium	Benzene	Phenol	Acenaphthene
Cyanide	Antimony	Ethylbenzene	2-Methylphenol	Acenaphthylene
Nitrate	Arsenic	Toluene	4-Methylphenol	Anthracene
Sulphate	Barium	Total Xylenes	2,4-Dimethylphenol	Benzo(a)anthracene
Sulphide	Cadmium			Benzo(a)pyrene
Thiocyanates	Chromium(VI)			Benzo(b)fluoranthene
	Copper			Benzo(g,h,i)perylene
	Iron			Benzo(k)fluoranthene
	Lead			Chrysene
	Manganese			Dibenzo(a,h)anthracene
	Mercury			Dibenzofuran
	Nickel			Fluoranthene
	Selenium			Fluorene
	Silver			Indenol(1,2,3-cd)pyrene
	Vanadium			Naphthalene
	Zinc			Phenanthrene
				Pyrene
				2-Methylnaphthalene

6.2 *Gasworks wastes*

In many countries, the manufacture of town gas by coal carbonisation has been replaced over the past 30 years by the introduction of natural gas. Pressure for land in urban areas has often led to the redevelopment of many former gasworks sites, some of which may have been in use for periods of over 100 years. The redevelopment of such sites is often hampered by severe contamination problems. In the UK it has been estimated that there were between 3000 and 4000 gas production sites and in the US between 100 and 1500. In Australia it has been estimated that there are only between 100 and 150 such sites (Turczynowicz, 1993), however the contamination problems are common to all sites.

The process of gas manufacture resulted in a considerable range of byproducts and solid and liquid wastes which were often disposed of on site. Table 10 lists some of the major contaminants found at gasworks sites although not all are necessarily present at all sites. The list includes inorganics such as cyanide and ammonia, heavy metals, organics such as volatile aromatic compounds, phenolics and a number of polynuclear aromatic hydrocarbons (Nakles et al., 1993). Many of the chemicals listed are potentially toxic to humans and, as for fuel hydrocarbons, exposure may result from inhalation of volatile compounds, dermal contact or ingestion via soil or contaminated water. Some of these compounds have had so long to seep through the soil that contamination may be found to depths of 10 m or more, and seepage may extend well beyond the gasworks site (Bridges, 1991).

One of the major problems confronting redevelopers of gasworks sites is the extremely variable distribution of contaminants over the site, which will depend on the layout of the original plant and the location of waste dumps. This is a particular problem at older sites where the gas plant may have been dismantled many years ago and site records are virtually non-existent. Clearly, detailed sampling of such sites is required before any clean-up can be undertaken. Site remediation itself involves a whole series of issues which are dealt with in another chapter (see Chapter 13). However, rather than *in situ* remediation, redevelopment of former gasworks sites can sometimes involve the complete removal of contaminated soil and disposal in a landfill. This, of course, only transfers potential problems with contaminants from one site to another. The issues associated with landfills are covered in Section 4.

6.3 Pulp mill effluent

Paper pulp mills can represent a significant threat to the environment because of the large volumes of potentially toxic waste effluent produced during the chemical processing of pulp (Johnson and Ryder, 1988). In particular, use of the common Kraft process, which involves extensive chlorine bleaching, produces a wide range of organic chemicals many of which have known carcinogenic or mutagenic properties. Although most modern mills employ some form of effluent treatment prior to its discharge, the final effluent can still contain considerable concentrations of potentially toxic chemicals. In the past, effluents have traditionally been discharged into aquatic environments with adverse effects on both marine and freshwater biota (Higashi et al., 1992; Hodson et al., 1992). Partly as a result of such adverse effects, there has been increasing interest in land disposal of pulp mill effluent.

One of the potential benefits of land application is that, in addition to toxic compounds, pulp mill effluents contain substantial amounts of nutrients (including N, P, Ca, Mg and trace elements) and organic matter, and can therefore act as fertilisers. Indeed crop responses to pulp mill effluents have been recorded in New Zealand (Johnson and Ryder, 1988), India (Lobo and Kettar, 1980) and the USA (Narum et al., 1979). However, unless the potentially toxic compounds in the effluent are ameliorated by soil processes, such as degradation, volatilisation and sorption, land treatment may result in adverse environmental effects. These may include contamination of ground and surface waters, plant uptake and phytotoxicity, entrance to the human food chain and toxicity to soil organisms. Unfortunately, as noted by Kookana and Rogers (1995), the behaviour of toxic organic compounds from pulp mill effluent in the soil environment has received relatively little attention.

The issue is a complex one because of the many different types of compounds that can occur in pulp mill effluents. These include halogenated alkanes and alkenes, chlorinated acids, chlorophenols, chlorobenzenes, dioxins and furans, chloroligno compounds, terpenes, sterols, and some nonchlorinated organic compounds such as toluene and naphthalene. In a recent comprehensive review, Kookana and Rogers (1995) have examined the likely fate and behaviour of these compounds when applied to the soil in pulp effluents. They reached the conclusion that, considering potential toxicity, bioavailability, mobility and groundwater contamination, the two most important groups of compounds in pulp mill effluents are chlorinated alkanes and alkenes and chlorophenols. Chlorinated alkanes and alkenes, being highly soluble and strongly acidic, are generally more mobile in soils than other chlorine-substituted organic compounds. Chlorophenols tend to be resistant to microbial degradation, their environmental persistence and toxicity increasing with increasing Cl substitution (Muller, 1992). Although more toxic compounds do occur in pulp effluent, for example dioxins and furans, given their relatively low concentrations and interactions with soils, their environmental impact is likely to be minimal (Kookana and Rogers, 1995). However these authors note that their conclusions are based on studies with individual toxic chemicals, often conducted *in vitro*, rather than on studies carried out under realistic effluent disposal conditions. They (Kookana and Rogers, 1995) stress the need for fieldbased investigations of pulp mill effluent in soils to be carried out before disposal of pulp effluent on land is accepted as a reuse option.

7. Contamination at timber treatment sites

The timber industry uses chemicals such as copper chrome arsenate (CCA), creosote and pentachlorophenol in order to preserve timber from biological decay. As might be expected from their use in this way, these chemicals are generally toxic to living organisms and are long-lasting (Soong and Emmett, 1993). Timber treatment sites can provide a point of entry for the preservative chemicals into the general environment by way of spills and leaks and run-off from treated wood, with resultant contamination of surrounding soil and, in some cases, surface and groundwaters.

CCA is currently the most widely used preservative for wood in Australia (Soong and Emmett, 1993) and New Zealand. Unfortunately in both countries there are examples where poor management practices and accidents at treatment sites have resulted in contamination of soil and water with Cu, Cr and As. Figure 3 shows the potentially contaminated areas at CCA treatment sites. Under dry conditions, contaminated soil particles can become airborne dusts and in addition sprays and mists produced by pumps at the sites can also lead to air pollution. Burning of waste CCA-treated products can also release contaminants into the atmosphere (Soong and Emmett, 1993). Thus pollution resulting from CCA treatment may extend well beyond the treatment site.

Copper, Cr and As are all known to be toxic to humans, therefore contamination of water supplies with these elements represents a considerable threat to human health. Although all three elements are adsorbed by soil, Cr and As are present in CCA solution as the dichromate ($Cr_2O_7^{2-}$) and arsenate ($H_2AsO_4^-$) anions and are therefore not as strongly retained by soil as copper which is present as the Cu^{2+} cation (Carey et al., 1995). In a study of the leaching of CCA solution through undisturbed soil cores, McLaren et al. (1994) showed that of the three elements, Cr was most readily leached, followed by As, with relatively little leaching of Cu (Table 11). This study also showed that preferential flow of metals in structured soils could be of particular importance in terms of speeding up their movement down through the soil towards groundwater. The above adsorption and leaching studies

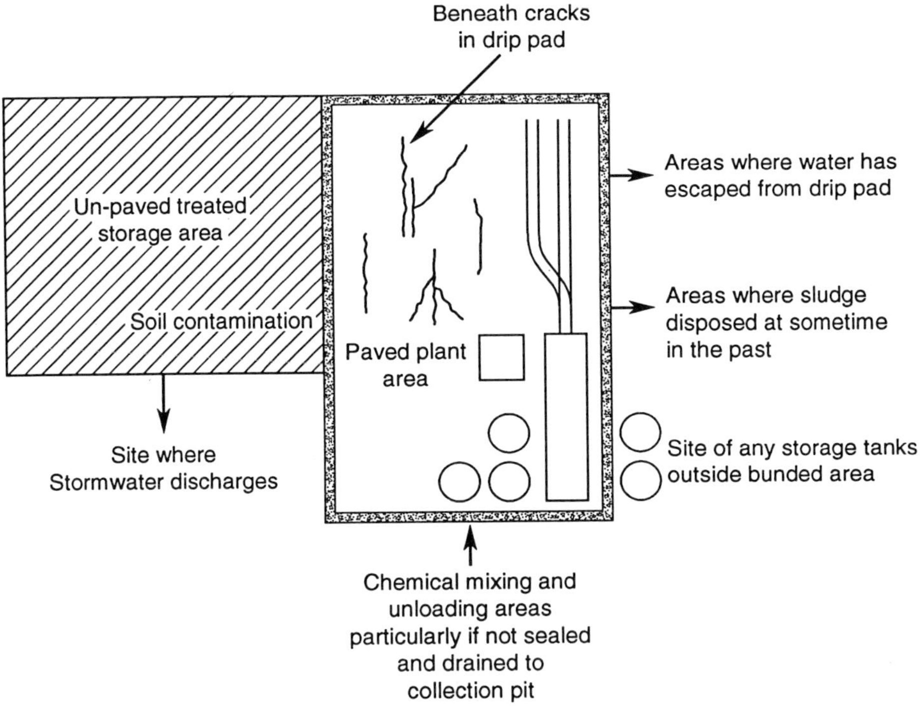

Figure 3. Potentially contaminated areas at CCA treatment sites (Soong and Emmett, 1993).

Table 11. Leaching of Cu, Cr and As from CCA solution through intact soil cores.

	% Cu, Cr and As leached through soils after 2.5 pore volumes of drainage[A]		
Soil	Cu	Cr	As
Templeton A horizon	1.0	53.3	15.2
Templeton B horizon	3.6	92.4	4.1
Selwyn A horizon	0.6	56.8	18.5
Selwyn C horizon	0.2	95.5	28.3

[A]Data from McLaren et al., 1994.

help to explain the observation that in waters contaminated by CCA in South Australia, concentrations of the elements are in the order Cr>As>Cu (Soong and Emmett, 1993). In some situations, in addition to leaching of CCA through the soil to groundwater, surface runoff may lead to contamination of surface waters. It is known that Cu, Cr (particularly as the dichromate ion) and As are all potentially toxic to many forms of aquatic life.

Table 12. Effect of CCA contamination on pasture growth and herbage composition[A].

Level of Contamination	Soil heavy metal content (0–5 cm, mg kg^{-1})			Herbage DM yield (g m^{-2} 35 d^{-1})	Herbage metal concentrations (mg kg^{-1})		
	Cu	Cr	As		Cu	Cr	As
Control	19.3	47.3	12.3	180.6	10.5	2.4	<1
Low	109	148	161	182.9	14.6	2.8	3.5
Medium	425	382	469	85.3	18.4	5.2	5.7
High	835	739	790	18.6	23.9	7.9	11.1

[A]Data from Yeates et al., 1994.

In addition to potential leaching to groundwater, Cu, Cr and As contamination of soil also creates problems for plant growth and soil biological activity. Grant and Dobbs (1977) showed in greenhouse pot experiments that growth of a range of crops (dwarf French beans, carrots, tomatoes, perennial ryegrass, bent grass) was inhibited in CCA-contaminated soil. Comparing the levels of CCA used in their experiments with the levels of soil contamination found at timber treatment sites, indicated that many contaminated timber treatment sites would not support plant growth. Less heavily contaminated soil would support plant growth but might give rise to crops containing twice the recommended limit for As in food. A field study by Yeates et al. (1994) showed that effects on plant growth and soil biological activity depended on the level of contamination with CCA. Whereas contamination by 100 mg kg^{-1} of Cu, Cr and As caused little depression of soil biological activity, there was some suppression at 400 kg ha^{-1}, and at 800 mg kg^{-1} normal processes were inhibited and pasture herbage production was negligible. Herbage Cu, Cr and As concentrations were also significantly increased in the highest contaminated soil (Table 12). Indicators of soil biological activity that were inhibited by CCA contamination included: decomposition of buried cotton strips, earthworm numbers, nitrification rates and enzyme activities.

8. Conclusion

In a review of this nature, it is impossible to cover all potential urban and industrial wastes and other forms of soil contamination. However, we have attempted to cover the major types of wastes currently of concern in the Australasia-Pacific Region. Although other types of wastes, for example, wastes from food processing industries are of importance in some areas, many of the problems discussed in this chapter relating to contaminants such as metals, nutrients or organic chemicals, are common to these other materials. It is clear that many wastes have the potential to seriously damage the environment and have adverse effects on human health. Other chapters (see Chapters 9–13) in this book examine remediation and rehabilitation of contaminated sites and the regulations and legislation relating to contamination issues.

References

Adriano D C, Page A L, Elseewi A A, Chang A C and Straughan I 1980 Ultilization and disposal of fly ash and other coal residues in terrestrial ecosystems: a review. J. Environ. Qual. 9, 333-344.

Aitken R L and Bell L C 1985 Plant uptake and phytotoxicity of boron in Australian fly ashes. Plant Soil 84, 245-257.

Aitken R L, Campbell D J and Bell L C 1984 Properties of Australian fly ashes relevant to their agronomic utilization. Aust. J. Soil Res. 22, 443-453.

Asano T, Smith R G and Tchobanoglous G 1985 Municipal wastewater: treatment and reclaimed water characteristic. In Irrigation with Reclaimed Municipal Wastewater: A Guidance Manual. Eds. G S Pettygrove and T Asano. pp 2-1-2-26. Lewis Publishers, Inc., Chelsea, MI.

Assmuth T W and Strandberg T 1993 Ground water contamination in Finnish landfills. Water Air Soil Pollut. 69, 179-199.

Attiwill P M and Leeper G W 1987 Forest Soils and Nutrients. Melbourne University Press.

Australian Water Resources Council 1992 Guidelines for Sewerage Systems; sludge management (draft). AWRC. 34 p.

Baath E 1989 Effects of heavy metals in soil on microbial processes and populations (a review). Water Air Soil Pollut. 47, 335-379.

Beretka J and Nelson P 1994 The current state of utilisation of fly ash in Australia. In Proc. 2nd Int. Symp. Ash - A Valuable Resource, Vol 1. pp 51-63. South African Coal Ash Association, Johannesburg.

Berrow M L and Burridge J C 1980 Trace element levels in soils: effects of sewage sludge. In Inorganic Pollution and Agriculture. pp 159-183. MAFF Reference Book 326, HMSO, London, UK.

Breen P F 1992 Artificial wetlands for wastewater treatments: A review with particular reference to rural areas. In CSIRO Division of Water Resources: Research Areas Pertinent to Intensive Rural Industry Waste Management. Eds. K H Bowmer and P Laut. pp 34-46. Divisional Report 92/4.

Breuer D W, Dale W C and Schiess P 1979 Nitrogen transformation and leaching associated with waste-water irrigation in Douglas Fir, Poplar, grass and unvegetated systems. In Utilization of Municipal Sewage Effluent and Sludge on Forest and Disturbed Land. Eds. W E Sopper and S N Kerr. pp 19-33. Pennsylvania State University, Pennsylvania.

Bridges E M 1991 Waste materials in urban soils. In Soils in the Urban Environment. Eds. P Bullock and P J Gregory. pp 28-46. Blackwell Scientific Publications, Oxford.

Cameron K C, McLaren R G and Adams J A 1994 Application of municipal sewage sludge to low fertility forest soils: The fate of nitrogen and heavy metals. Transactions of 15th World Congress of Soil Science, Vol. 3a, 467-482.

Campbell D J, Fox W E, Aitken R L and Bell L C 1983 Physical properties of sands amended with fly ash. Aust. J. Soil Res. 21, 147-154.

Carey P L, McLaren R G and Adams J.A. 1995. Sorption of cupric, dichromate and arsenate ions in some New Zealand soils, Water Air Soil Pollut. in press.

Carlson C L and Adriano D C 1993 Environmental impacts of coal combustion residues. J. Environ. Qual. 22, 227-247.

Chang A C, Warneke J E, Page A L and Lund L J 1984 Accumulation of heavy metals in sludge treated soils. J. Environ. Qual. 13, 87-91.

Cromer R N, Tompkins D and Barr N J 1983 Irrigation of Pinus radiata with waste water: tree growth in response to treatment. Aust. Forest. Res. 13, 57-65.

Department of Health (NZ) 1992 Public health guidelines for the safe use of sewage effluent and sewage sludge on land. Department of Health, Wellington, New Zealand. 66 p.

Dowdy R H and Volk V V 1984 Movement of heavy metals in soils. In Chemical Mobility and Reactivity in Soil Systems. pp 229-240. SSSA, Madison, USA..

Emmerich W E, Lund L J, Chang A C and Chang A L 1982 Movement of heavy metals in sewage sludge treated soils. J. Environ. Qual. 11, 174-178.

Feigin A, Ravina I and Shalhevet J 1991 Irrigation with Treated Sewage Effluent. Management for Environmental Protection. Springer-Verlag, Berlin, Heidelberg. 224.

Gardener E and Littleboy M 1994 Using water balance models to size effluent reuse schemes. In Proceedings of Recycled Water Seminar. 19-20 May 1994. Newcastle.

Grant C and Dobbs A J 1977 The growth and metal content of plants grown in soil contaminated by a copper/chrome/arsenic wood preservative. Environ. Pollut. 14, 213-226

Hallbourg R R, Delfino J J and Miller W L 1992 Organic priority pollutants in groundwater and surface water at three landfills in north central Florida. Water, Air, Soil Pollut. 65, 307-322.

Hartling E C 1994 Water recycling in Los Angeles County. In Proceedings of Recycled Water Seminar. 19-20 May 1994. pp 13-20. Newcastle.

Harvey P 1993 Control of Nutrient Inputs. In Algal Management Strategy Background Papers. pp 9-20. Algal Management Working Group, Murray Darling Basin Commission.

Healy W B 1968 Ingestion of soil by dairy cows. N.Z. J. Agric. Res. 13, 664-672.

Higashi R M, Cherr G N, Shenker J M, Macdonald J M and Crosby D G 1992 A polar high molecular mass constituent of bleached Kraft mill effluent is toxic to marine organisms. Environ. Sci. Technol. 26, 2413-2420.

Hodson P V, McWhirter M, Ralph K, Gray B, Thivierge D, Carey J H, Van Der Kraak G, Whittle D M and Levesque M 1992 Effects of bleached Kraft mill effluent on fish in the St Maurice River, Quebec. Environ. Toxicol. Chem. 11, 1635-1651.

Iskander I K and Syers J K 1980 Effectiveness of land application for phosphorus removal from municipal wastewater at Manteca, California. J. Environ. Qual. 9, 616-621.

Johnson B and Ryder I 1988 The disposal of pulp and paper mill effluents by spray irrigation onto farmland. In Alternative Waste Treatment Systems. Ed. R Bhamidimarri. pp 55-65. Elsevier Applied Science, London.

Jones H A and Nedwell D B 1993 Methane emission and methane oxidation in land-fill cover soil. Micro. Ecol. 102, 185-195.

Jones P W 1986 Sewage sludge as a vector of salmonellosis. In Epidemiological Studies of Risks Associated with the Agricultural Use of Sewage Sludge. Ed. J C Block, A H Havelaar and P L'Hermite. pp 21-33. Elsevier Applied Science Publishers, London.

Keeney D R 1981 Soil nitrogen chemistry and biochemistry. In Modelling Wastewater Renovation, Land Treatment. Ed. I K Iskander. pp 259-276. John Wiley & Son, New York.

Kofi Asante-Duah D 1991 Using a risk assessment protocol to develop soil cleanup criteria for petroleum contaminated sites. In Hydrocarbon Contaminated Soils. Volume 1. Eds. E J Calabrese and P T Kostecki. pp 493-523. Lewis Publishers, Chelsea, Michigan.

Kookana R S and Rogers S L 1995 Effects of pulp mill effluent disposal on soil. Rev. Environ. Contam. Toxicol. 142, 13-63.

Lamy I, Bourgeois S and Bermond A 1993 Soil cadmium mobility as a consequence of sewage sludge disposal. J. Environ. Qual. 22, 731-737.

Lance JC 1986 Effect of sludge additions on nitrogen removal in soil columns flooded with secondary effluent. J. Environ. Qual. 15, 298-301.

Latterell J J, Dowdy R H, Clapp C E, Larson W E and Linden D R 1982 Distribution of phosphorus in soil irrigated with municipal wastewater effluent: 5 year study. J. Environ. Qual. 11, 124-128.

Lester J N 1982 Occurrence, behaviour and fate of organic micropollutants during waste water and sludge treatment processes. In Environmental Effects of Organic and Inorganic Contaminants in Sewage Sludge. Eds. R D Davis, G Hucker and P L Hermite. pp 3-18. Reidel, Dordrecht, The Netherlands.

Lewin V H and Beckett P H T 1980 Monitoring heavy metal accumulation in agricultural soils treated with sewage sludge. Effluent Water Treat. J. 71, 205-208.

Lindo P V, Taylor, R W and Shuford J W 1993 Accumulation and movement of residual phosphorus in sludge-treated decatur silty clay loam soil. Commun. Soil Sci. Plant Anal. 24, 1805-1816.

Lindon P 1993 A health risk assessment for soils contaminated with fuel hydrocarbons: petrol. In The Health Risk Assessment and Management of Contaminated Sites. Contaminated Sites Monograph Series No. 2. Ed. A Langley and M van Alphen. pp 313-334. South Australia Health Commission, Adelaide.

Lindsay D G 1982 Effects arising from the presence of persistent organic compounds in sludge. In Environmental Effects of Organic and Inorganic Contaminants in Sewage Sludge. Eds. R D Davis, G Hucker and L Hermite. pp 19-26. Reidel, Dordrecht, The Netherlands

Lobo A J and Kettar C M 1980 Treatment and disposal of effluent at the Pudumjee Pulp and Paper Mills Ltd. Ippta. 17, 23-25.

Mansell R S, Bloom S A, Burgoa B, Nkedi-Kizza P and Chen J S 1992 Experimental and simulated P transport in soils using a multireaction model. Soil Sci. 153,185-194.

McGrath S P 1994 Effects of heavy metals from sewage sludge on soil microbes in agricultural ecosystems. In Toxic Metals in Soil-Plant Systems. Ed. S M Ross. pp 247-274. John Wiley & Sons, Chichester, UK.

McLaren R G, Carey P L, Cameron K C, Adams J A and Sedcole J R 1994 Effect of soil properties and contact period on the leaching of copper, chromium and arsenic through undisturbed soils. Trans. 15th Int. Cong. Soil Sci., Vol. 3a: 156-169.

Millin P and Heritage A 1992 Application of constructed wetlands for the treatment of domestic sewage in rural Australia. In CSIRO Division of Water Resources: Research Areas Pertinent to Intensive Rural Industry Waste Management. Eds. K H Bowmer and P Laut. pp 25-33. Divisional Report 92/4.

Muller R 1992 Bacterial degradation of xenobiotics. In Microbial Control of Pollution. Eds. J C Fry, G M Gadd, R A Herbert, C W Jones and I A Watson-Craik. pp 35-57. Cambridge University Press, Cambridge, UK.

Myers B J 1992 Effluent loading rates for irrigated plantations - a water balance model. Aust. For. 55, 39-47.

Naidu R, Sumner M E and Rengasamy P (Eds) 1995 Australian Sodic Soils: Distribution, properties and Management. pp 351. CSIRO Publications, Melbourne, Australia.

Nakles D V, Linz D G and Hayes T D 1993 Wastes and chemicals of interest at MGP sites. In Management of Manufactured Gas Plant Sites Technology Seminar. pp 32-36. Remediation Technologies, Inc, Arlington, VA.

Nansen P and Henriksen Sv Aa 1986 The epidemiology of bovine cysticercosis (*C. bovis*) in relation to sewage and sludge application on farmland. In Epidemiological Studies of Risks Associated with the Agricultural Use of Sewage Sludge. Eds. J C Block, A H Havelaar and P L'Hermite. pp 76-82. Elsevier Applied Science Publishers, London.

Narum Q A, Mickelson D P and Roehne N 1979 Disposal of an integrated pulp-paper mill effluent by irrigation. EPA-600/2-79-033. USEPA, Washington, DC.

NSW Agriculture and Fisheries 1989 Guidelines for the use of sewage sludge on agricultural land.

NSW Agriculture 1991 The Feedlot Manual. NSW Agriculture, Sydney, Australia.

NSW Environmental Protection Agency 1989 Design Guide for the Land Application of Wastewaters, NSW EPA, Publication WP-8, 46p.

NSW Environmental Protection Agency 1995 The Utilisation of Treated Effluent by Irrigation. NSW EPA, Sydney. 92p.

NSW Recycled Water Coordination Committee 1993 NSW guidelines for urban and residential use of reclaimed water. 1st edition. 21p.

O'Connor G A, Chaney R L and Ryan J A 1991 Bioavailability to plants of sludge-borne toxic organics. Rev. Environ. Contam Toxicol. 121, 129-155.

Oster J B and Rhoades J D 1985 Water management for salinity and sodicity control. In Irrigation with Reclaimed Municipal Wastewater: A Guidance Manual. Eds. G S Pettygrove and T Asano. pp 7-1-7-20. Lewis Publishers, Inc., Chelsea, MI.

Page A L and Chang A.C 1985 Fate of wastewater constituents in soil and groundwater: trace elements. In Irrigation with Reclaimed Municipal Wastewater - A Guidance Manual. Eds. G S Pettygrove and T Asano. pp 13-1–13.16, Chelsea.

Page A L, Elseewi A A and Straughan I R 1979 Physical and chemical properties of fly ash from coal-fire power plants with reference to environmental impacts. Residue Rev. 71, 83-120.

Pettygrove G S and Asano T 1985 Irrigation with Reclaimed Municipal Wastewater: A Guidance Manual. Lewis Publishers, Inc., Chelsea, MI. 518p

Pruitt W O and Snyder R L 1985 Crop water use. In Irrigation with Reclaimed Municipal Wastewater: A guidance Manual. Eds. G S Pettygrove and T Asano. pp. 5-1–5-48. Lewis Publishers, Inc., Chelsea, MI

Rhoades J D and Loveday J 1990 Salinity in irrigated agriculture. In Irrigation of Agricultural Crops. Agronomy Monograph No. 30. Eds. B A Stewart and D R Nielson. pp 1089-1142. ASA-CSSA-SSSA, Madison, WI.

Ross A D, Lawrie R A, Keneally J P and Whatmuff M S 1992 Risk characterisation and management of sewage sludge on agricultural land - implications for the environment and the food chain. Aust. Vet. J. 69, 177-181.

Ross A D, Lawrie, R A, Whatmuff M S, Keneally J P and Awad A S 1991 Guidelines for the use of Sewage Sludge on Agricultural Land. NSW Agriculture, Sydney, Australia. 17 p.

Ryden J C and Pratt P F 1980 Phosphorus removal from wastewater applied to land. Hilgardia. 48, 1-36.

Ryden J C, Syers J K and Iskandar I K 1981 Evaluation of a simple model for predicting phosphorus removal by soils during land treatment of wastewater. In Modelling Wastewater Renovation, Land Treatment. Ed. I K Iskander. pp 647-667. John Wiley & Son, New York.

Schulz R K, Ridky R W and O'Donnell E 1992 Control of Water Infiltration Into Near Surface LLW Disposal Units. Progress Report on Field Experiments at a Humid Regional Site, Beltsville, Maryland. NUREG/CR-4918. Vol. 6.

Shainberg I and Oster J D 1978 Quality of irrigation water. Int. Irrigation Inf. Cent (IIIC), Bet Dagan, Israel Publ. 2. 65p.

Smith C J, Bond W J and Polglase P J 1994 Estimating gaseous and leaching losses of nitrogen under effluent-irrigated plantations. In Measurement and management of nitrogen losses for groundwater protection in agricultural production systems. Proc. of Water Down Under 94 Workshop. Ed. W J Bond. pp 53-57. Adelaide, November 1994, Land and Water Resources R & D Corporation.

Soong F S and Emmett A J 1993 Assessment and management of CCA timber preservation plants. In The Health Risk Assessment and Management of Contaminated Sites. Contaminated Sites Monograph Series No. 2. Eds. A Langley and M van Alphen. pp 229-246. South Australia Health Commission, Adelaide.

Sopper W E and Kerr S J 1979 Renovation of municipal wastewater in eastern forest ecosystems. In Utilization of municipal sewage effluent and sludge on forest and disturbed land. Eds. W E Sopper and S N Kerr. pp 61-76. Pennsylvania State University, Pennsylvania.

State Pollution Control Commission 1979 Design guide for the disposal of wastewaters by land application. Publication WP-6. 41p.

Stevenson F J 1982 Nitrogen in Agricultural Soils. ASA-CSSA-SSSA, Madison, WI. 940p

Stewart H T L, Hopmans P and Flinn D W 1990 Nutrient accumulation in trees and soil following irrigation with municipal effluent in Australia. Environ. Pollut. 63, 155-177.

Stone W A Jr 1991 Assessing health risks associated with diesel contaminated soils and groundwater. In Hydrocarbon Contaminated Soils. Volume 1. Eds. E J Calabrese and P T Kostecki. pp 167-179. Lewis Publishers, Chelsea, Michigan.

Turczynowicz L 1993 The assessment and management of gasworks sites. In The Health Risk Assessment and Management of Contaminated Sites. Contaminated Sites Monograph Series No. 2. Eds. A Langley and M van Alphen. pp 261-312. South Australia Health Commission, Adelaide.

Tyler G 1981 Heavy metals in soil biology and biochemistry. In Soil Biochemistry, Vol. 5. Eds. E A Paul and J N Ladd. pp 371-414. Marcel Dekker, New York.

Udo E J and Fayemi A A A 1975 The effect of oil pollution of soil on germination, growth and nutrient uptake of corn. J. Environ. Qual. 4, 537-540.

van den Berg J J 1993 Effects of sewage sludge disposal. Land Degrad. Rehab. 4, 407-413.

van den Broek B 1994 Solid waste management in Australia. In The Interdata Environmental Resource Management Handbook. Ed. N A M Eastwood. pp 27-33. I.D.P. Inter Data Pty. Ltd.

van Hook R I and Shults W D 1976 Effects of trace contaminants from coal combustion. ERDA 77-64. Nat. Tech. Inf. Serv., U.S. Dep. Commerce, Springfield, Va.

Venus G C 1994 Wetlands wastewater treatment in New Zealand. In The Interdata Environmental Resource Management Handbook. Ed. N A M Eastwood. pp 92-95. I.D.P. Inter Data Pty. Ltd.

Wang X and Bartha R 1990 Effects of bioremediation on residues, activity and toxicity in soil contaminated by fuel spills. Soil Biol. Biochem. 22, 501-505.

Weeks O L, Mansell R S and McCallister S W 1992 Evaluation of soil top-cover systems to minimize infiltration into a sanitary landfill: A case study. Environ. Geol. and Water Sci. 20, 139-151.

Weinstein L H, Osmeloski J F, Rutzke M, Beers A O, McCahan J B, Bache C A and Lisk D J 1989 Elemental analysis of grasses and legumes growing on soil covering fly ash landfill sites. J. Food Safety 9, 291-300.

Westcot D W and Ayers R 1985 Irrigation water quality. In Irrigation with Reclaimed Municipal Wastewater: A Guidance Manual. Eds. G S Pettygrove and T Asano. pp 3-1–3-37. Lewis Publishers, Inc., Chelsea, MI.

Willmot C D 1994a Refuse data - New Zealand. Solid wastes. In The Interdata Environmental Resource Management Handbook. Ed. N A M Eastwood. pp 96-101. I.D.P. Inter Data Pty. Ltd.

Willmot C D 1994b The environmentally sound management of landfill sites. In The Interdata Environmental Resource Management Handbook. Ed. N A M Eastwood. pp 160-163. I.D.P. Inter Data Pty. Ltd.

Yeates G W, Orchard V A, Speir T W, Hunt J L and Hermans M C C 1994 Impact of pasture contamination by copper, chromium, arsenic timber preservative on soil biological activity. Biol. Fertil. Soils 18, 200-208.

7
Exploration, mining and mineral processing

GRAHAM F. TAYLOR

1. Introduction

Mining is of economic importance to many countries in the Australasia-Pacific region because in addition to providing export income, it is also the source of fuel, construction materials and raw materials for many manufactured goods. However, this chapter is essentially based on Australian experiences, unless otherwise stated, and therefore may not truly represent conditions in all countries of the Australasia-Pacific region.

In Australia, mining and mineral processing are mature industries with many Australian companies now operating overseas and their concepts and technologies being applied throughout the world. From the first extraction of coal at Newcastle, New South Wales (NSW) in 1796 and export in 1800, it has expanded to become Australia's biggest export income earner (minerals including coal represent 40% of export income). Developing countries, such as Papua New Guinea, depend even more heavily on the mining industry for income, whilst many of the smaller island nations of the region have no mineral resources.

Australia's continued economic well-being demands a competitive mining industry, able to explore for and develop new mines and undertake down-stream processing. Each of these activities impacts on the environment and the minerals industry has readily accepted the need for environmental management to meet community expectations and regulatory authorities' requirements. Thus, many of the badly contaminated sites are the result of early mining and mineral processing activities, although there is some contamination derived from existing activities.

2. The minerals industry

In a country the size of Australia there are many different types of both primary and secondary mineral deposits which can be broadly grouped as:

- metallic minerals;
- non-metallic minerals;
- construction materials;
- fuels

These groupings, and commodities within them, have a different origin or are associated with different host rocks which require different exploration, mining and processing technologies. These, in turn, lead to quite different environmental impacts, nature of contamination and methods for remediation/rehabilitation/amelioration,

R. Naidu et al. *(eds): Contaminants and the Soil Environment in the Australasia-Pacific Region, 213–266.*

which are further complicated by the effects of climate, topography, weathering history and present use of adjacent lands.

2.1 Exploration

Regional exploration for any commodity is largely dependent on geological factors such as lithology, tectonics, structure and topography, as mineral deposits commonly have a known or recognisable association (e.g. hydrocarbons are associated with sedimentary rocks). With modern remote sensing techniques and new interpretive tools, regional exploration has become less invasive with resultant diminished effect on the environment.

From the regional, through prospect to deposit/mine scale exploration, geophysical and geochemical exploration techniques are more invasive as they commonly depend on measurement of properties of host rocks, or the deposits themselves. The most obvious effect upon the environment is the extensive 'cut lines' used for seismic and other geophysical traverses and for geochemical sampling using drilling techniques. Such 'cut lines' remain visible for decades and may well contribute to contamination through erosion. Additional contamination may result from costeans (trenches) and drilling spoil, particularly where the latter is left in plastic bags around the countryside.

Much of the contamination resulting from exploration programmes is due to subsequent mining activities. More recent and future exploration will have far less impact on the environment, as restoration is generally undertaken immediately after exploration ceases. With the ever increasing environmental awareness and reduced land access through native titles, national parks, heritage lands etc., new non-invasive (airborne) exploration techniques are being developed.

2.2 Mining

Australia has an abundant diversity of mineral commodities which require varying mining techniques. Table 1 lists a range of commodities, the most commonly used mining technique(s), possible contaminants associated with each commodity and their impact on the environment.

Each of the various mining techniques poses a set of unique environmental problems, e.g. in quarrying operations there is usually little waste rock and all that remains is a topographic change whilst at the far end of the spectrum strip mining produces huge tonnages of waste rock and spoil and commonly an extensive final void (e.g. coal mines in the Hunter Valley and Bowen Basin).

The process of mining shifts around the near surface materials, resulting in holes in the ground. Such activities, whether at or below the surface, may result in large waste rock or spoil dumps, possibly massive changes to the topography, and the removal or disruption of flora and fauna. In some instances the process creates dust and noise, but subsequent contamination of the minesite and downstream areas is often due to natural processes acting upon the mined material.

Commodity	Mining technique	Nature of contaminants	Impact on environment
1) Metallic minerals			
Copper	u/g, o/p (leaching)	heavy metals, AMD, tailings	water, flora, fauna, humans
Lead/zinc/silver	u/g	heavy metals, AMD, tailings	water, flora, fauna, humans
Gold	o/p, u/g, dredging	heavy metals, AMD, tailings	water, flora, fauna, humans
Nickel (PGE's)	u/g, o/p	heavy metals, AMD, tailings	water, flora, fauna, humans
Aluminium	strip	disturbed regolith profile	flora, fauna
Iron	o/p	dust, spontaneous combustion, dust	atmosphere, human health
Manganese	strip	waste rock, disturbed profile, dust	flora, fauna
Tin	dredging, sluicing, u/g	waste rock, AMD, tailings	water, flora, fauna, humans
Molybdenum	o/p	AMD, waste rock	water, flora, fauna
Titanium/zirconium	dredging	disturbed land, increased radioactivity	vegetation
b) Non-metallic minerals			
Diamonds	o/p, dredging	waste rock	little
Opals	u/g, o/p	disturbed profile, waste rock, dust	human health
Other gemstones	dredging	waste rock, siltation	little
Magnesium	strip	disturbed profile	vegetation
Silica	quarrying	dust	little
Phosphate	strip	waste rock, disturbed profile	possible human health from radioactivity
Feldspar	quarrying	dust, waste rock	little
Kaolin	strip	disturbed regolith profile	little
Talc	quarrying	dust, disturbed profile	little
c) Construction materials			
Gypsum	quarrying	few	little
Limestone	quarrying	few	little
Aggregate	quarrying	siltation	rivers
Dimension stone	quarrying	few	little
Asbestos	u/g	fibrous particles	human health (mesothelioma)
d) Fuels			
Coal	u/g, strip	waste rock, AMD, dust	water, vegetation, silicosis
Oil-shale	strip	waste rock, AMD, heavy metals	water
Uranium	o/p	radioactivity, waste rock	human health
Oil/gas	drilling	condensate flaring	human health atmosphere

u/g = underground; o/p = open pit; AMD = Acid mine drainage.
Note that in some instances e.g. uranium, a commodity could have been assigned to another group.

2.3 Mineral Processing

Just as exploration and mining of mineral commodities require many differing techniques, so does mineral processing. Initially there is a requirement to separate the commodity from waste rock, gangue or low grade mineralisation. Most simply this is done by washing in the presence of various additives to ensure maximum separation. Other techniques utilised are heavy liquids, gravity, screening, magnetics and other physical properties. For sulphide minerals, fine grinding followed by froth flotation separates the ore minerals from waste which is then disposed of in tailings/ slimes dams.

Subsequent processing may involve roasting, sintering, smelting, calcining and electrolytic refining, all of which produce by-products which may be potentially harmful to the environment (Table 2). In many instances, these metallurgical processes occur in remote arid parts of Australia (e.g. Kalgoorlie, Tanami, Mount Isa) and may appear to have minimal effect on the environment. Other processing plants are close to the coast in heavily populated regions (e.g. Cockle Creek, Newcastle, Wollongong, Port Pirie, Whyalla, Hobart, Kwinana) and may have significant impact on the environment.

3. Contaminants

The nature of contaminants derived from exploration, mining and processing activities (Table 1) can be divided into:

- solid wastes;
- water contaminants;
- air contaminants.

3.1 Solid wastes

Overburden, spoil and waste rock from open pit, strip mining and quarrying stockpiled within close proximity to the mine may contribute dust (Broken Hill, NSW) or siltation of drainage channels (Mount Morgan, Mount Lyell). If this material is benign, resultant contamination is more of nuisance value. However, where these waste materials contain significant concentrations of heavy metals and/or sulphides, the problems are exacerbated. Dust laden with heavy metals can readily enter the food chain and be directly ingested, particularly by young children (e.g. Broken Hill). Sulfide rich waste rock has a high acid generating potential resulting in acid mine drainage (AMD), acid soils and groundwaters and secondary dispersion of toxic elements. Although Australian coals are generally low in sulfur, some coal washery wastes have a relatively high acid generating potential.

Beneficiation processes produce tailings and residues which potentially pose an even greater threat to the environment. Sulfide-rich tailings associated with base metal, Cu and Au mineralisation may result in AMD and secondary dispersion of toxic elements into soils and groundwaters. These processes are hastened by the fine

Table 2. Mineral processing by-products.

Base metals: (Cu, Pb, Zn, Ni)
- Mixed oxide and sulfide ores in leach dumps and tailings leading to AMD and dispersion of heavy metals;
- Slags from smelting;
- Baghouse dusts and fumes;
- Jarosite residue and roaster dust from zinc refining
- Refinery sludges and impurity bleeds from refining
- Base metal contaminated ammoniacal sulfate or carbonate tailings from concentrate or laterite nickel processing;
- Arsenical solid wastes;
- Ferrite wastes.

Steel making:
- slags from furnaces;
- baghouse dusts;
- aromatics and phenols in coke oven gas residues.

Aluminium:
- red mud from alumina refineries;
- aluminium fluoride from scrubbers at aluminium smelters;
- sodium CN and other metal impurities in disused carbon anode and liner material from smelters.

Precious metals:
- cyanide in leach residues;
- roaster gases bearing arsenic oxides and sulphur dioxide;
- carbon residues fouled with base metals;
- mercury amalgam in old placer projects.

Uranium processing:
- acid, ammonium sulfate and radioactive wastes in tailings dams;
- acid formation in old sulfide dumps;
- possible scrubber dust dump areas.

Phosphate processing:
- acid and low levels of U_3O_8 in tailings dams of phosphoric acid production plants.

Ferroalloys:
- slag.

Secondary lead smelters:
- slag and flue dusts.

Titanium dioxide:
- sulfate or chloride leach residues.

Molybdenum disulfide concentrates:
- sulfate leach residues.

Beach sands:
- monazite in tailings.

Tin and tungsten:
- heavy media, gravity and magnetic tailings.

Asbestos:
- tailings and dusts.

grain size needed to liberate the ore minerals. Although the residue from the concentration of mineral sands is inert to oxidation, it may be slightly enriched in radioactive minerals.

Some metals, particularly gold, copper, uranium, titanium dioxide and aluminium, are liberated by chemical leaching with the resultant residue contaminated with the chemicals (e.g. CN, sulfuric acid, caustic soda). However, because of the initial cost of these chemicals, every effort is made to recover them before the residue is disposed of. Older gold mines utilised mercury which may be retained in the soil over very long periods.

Residues from high temperature processes (sintering, smelting, calcining, roasting, electric furnaces and molten salt electrolysis) such as slag and sinter are also potential sources of contaminants although release of heavy metals and other contaminants may be much slower because of the refractory nature of the material. Another contaminant from zinc processing is jarosite containing concentrations of heavy metals. This material is presently dumped at sea south of Tasmania.

A source of pollution often overlooked is that due to the infrastructure. Metal, chemicals, oil, grease and abandoned equipment, if not properly disposed of, may significantly contribute to the total contaminant level.

3.2 Water contaminants

In the mining industry, heavy metals (Cu, Pb, Zn, Cd, Hg, As, Sb, Bi) are the chief contaminants of surface and groundwaters. The most significant source is AMD derived from tailings, waste rocks, spoil, rejects and mine adits. However, some heavy metals may be derived from burial of slag/sinter etc. and leaching by saline-rich groundwaters which are common in many parts of Australia including the coastal fringe. Such contaminants will have a long-term impact upon the environment. Other water contaminants include CN (from the processing of gold ores), other chemical leachates, and the complex organic chemicals used in the flotation process.

The mining industry contributes significantly to the sulfate content of Australian water but the majority of this load is derived from marine aerosols and evaporation. Little, if any, salt is derived from the mining industry.

The petroleum industry in Australia has had a remarkably good record with minimum spillage either on land or at sea. By far the greatest source of oil/grease is the poor house-keeping at minesites and mills where lubricants, cleansers and fuels can enter groundwaters.

3.3 Air contaminants

Although mining operations, disturbed lands, waste rock and tailings, and stockpiles of mineral products are potential sources of dust, suppression of dust by watering and revegetation is common. Of great concern are the fibrous particulates such as asbestos and silica which pose grave environmental health problems.

The principal output from the mining industry which affects air quality is sulfur dioxide together with the volatile elements (As, Sb, Hg, Bi) derived from base metal smelters, and gold roasting. Modern technology ensures that much of SO_2 is scrubbed from the stack emissions. Therefore the amount of acid rain in Australia is minimal due also in part to the arid climate of much of the country. This was not always the case, with the 'moon-scape' (area of hills depleted of vegetation due to SO_2) at Queenstown being a classic example of the effects of SO_2 on the environment. Modern technology has also ensured that fumes no longer contain significant concentrations of heavy metals.

Uranium mining poses two possible environmental hazards: external exposure to gamma radiation and internal exposure to alpha radiation from inhaled radon gas and ore dust containing uranium and its decay products. Regulation of the industry ensures that there is minimal occupational health risk and old mines have lower radiation than many granites and other igneous rocks.

Another gaseous contaminant is fluorine derived from the aluminium and ceramic industries. Generally the levels are sufficiently low to be of little concern.

4. Environmental concerns

In recognising the need for improved environmental management, the Australian Mining Industry Council (AMIC) commissioned a review of the needs and priorities for environmental research (AMIC, 1994). From the 27 strategic issues identified, 7 were selected by a scoring/ranking system as the main strategic issues faced by industry:

- ecosystem establishment and resilience;
- final voids;
- biodiversity;
- acid drainage;
- greenhouse;
- tailings management;
- waste management.

with the next five issues in order of ranking being:

- completion criteria;
- water management systems;
- water treatment processes;
- alternative land uses;
- contaminated lands.

The key strategic issues of biodiversity and greenhouse were not further addressed as it was deemed that biodiversity is more of a policy issue whilst greenhouse is being addressed on a national basis. Whilst contaminated lands ranked only twelfth amongst environmental issues, each of the other issues is a contributing factor to contamination or its remediation.

The investigation of contaminated sites and assessment of the data has in large part been by regulatory authorities or consultants employed by them. Generally, there have been no guidelines or standards by which such investigations/assessments were conducted, and there is much anecdotal evidence of substandard practice. To overcome such situations the Australian and New Zealand Environment and Conservation Council (ANZECC) and the National Health and Medical Research Council (NH&MRC) have jointly developed Australian and New Zealand Guidelines for the Assessment and Management of Contaminated Sites (see Chapter 14). However these guidelines are not specific to the mineral industry and therefore do not recognise all the issues raised in previous sections.

To this end, the mining industry, government authorities, tertiary institutions and consultants have initiated guidelines and practices for the mineral industry. For instance, the Commonwealth EPA, in conjunction with the Australian mining industry, has produced a series of modules on Best Practice Environmental Management in Mining. At the same time, for example, the Queensland Department of Mines and Energy has produced Guidelines for Environmental Management for the Mining Industry and all States have legislation concerning the quality of the environment. The Australian Conservation Foundation has recently released a discussion paper on principles for Australian mining companies in PNG.

Of significance is the role played by the mining industry in environmental management. The AMIC Environment Committee has been instrumental in initiating projects, workshops and symposia. It has produced a Mine Rehabilitation Handbook and a Water Management Handbook for use by farming, land-care, government, community and mining groups. AMIC also publishes Landline, a magazine with a considerable amount of information for environmental professionals. Each year, AMIC sponsors an Environmental Workshop and has also sponsored two workshops on acid mine drainage (AMD).

The Australian Minerals and Energy Environment Foundation was established in 1991 to promote the implementation of sustainable development through excellence in environmental science and management in the mineral and energy industries. It has compiled two volumes of a Bibliography – Mining and the Environment, and each year awards 5 prizes for environmental excellence. There are also a number of research groups throughout the country which provide assistance to the mining industry: CSIRO Minesite Rehabilitation Research Program; the University of Queensland Centre for Mined Land Rehabilitation, Australian Nuclear Science and Technology Organisation and the Australian Centre for Minesite Rehabilitation Research. In addition, the mining industry is well-serviced by many consultants.

5. Contaminated mine sites

All sites subjected to exploration, mining (and mineral processing) activities are potentially sources of contamination. Indeed, the US Geological Survey has started an inventory of all sites 'disturbed' by mining and exploration. Some of the states have initiated similar programs, e.g. the Abandoned Mine Land Division of the Wyoming Department of Environmental Quality (Richmond, 1995) which receives federal funding to eliminate hazards to public health and safety and to correct environmental degradation of off-site areas resulting from mining activities abandoned prior to reclamation laws. In Canada, CANMET Mining Research Laboratories maintains the Inactive Mines Database and the Province of Ontario has undertaken a comprehensive program of identification and remediation of abandoned mine hazards (Mitchell and Mackasey, 1995).

Australia and PNG, like the United States and Canada, have many minesites (Figures 1 and 2). There are, however, some significant differences which directly influence the extent and impact of contamination (Table 3). Generally, only those mines along the highly populated Eastern seaboard, Tasmania and adjacent to the major rivers will have widespread contamination resulting in severe impact upon flora, fauna and humans. In PNG and other countries of tropical SE Asia and Oceania, high relief and excessive precipitation result in different impacts upon the environment (Table 3), particularly siltation of rivers and instability of constructed landforms. However, the long-term consequence of mining in semi-arid to arid regions of Australia has not been established.

In Australia, there are a number of both abandoned and existing mines which are the subject of extensive studies and remediation. Some of the problems and remedial activities are discussed below.

5.1 Base metal mines

5.1.1 Captains Flat, NSW

This Cu, Au, Pb, Ag, Zn and pyrite deposit, about 50 km SE of Canberra was mined from 1882 to 1962. Processing waste water and mine drainage water polluted the Molonglo River upstream of Canberra. In the 1940s, 2×10^6 L d^{-1} of mill waste water and mine water containing approximately 300 mg L^{-1} Zn was continuously discharged directly into the river. This represents 600 kg d^{-1} Zn load in addition to unknown quantities of other metals, acid, cyanides, sulfides, thiocyanates, cresols and phenols. Over the complete period of mining, contaminated material was eroded from the tailings dumps and contaminated leachate seeped into the river. Significant problems arose on several occasions when the tailings dams collapsed and erosion and leaching of the minesite added to contamination of the river bed, floodplain sediments and pastures, resulting in landholders suing the mining company for destruction of pastures.

Because of this, and its proximity to the Australian capital, Canberra, substantial rehabilitation of the site was undertaken in 1976. This work effectively stabilised the tailings dams, eliminating risk of failure, and greatly reduced erosion and failure. However, despite these measures, some pollution of the Molonglo River continues to occur because of AMD generation and dispersion of soluble elements (Table 4). This serves to show that a combination of poor mine planning and local geological

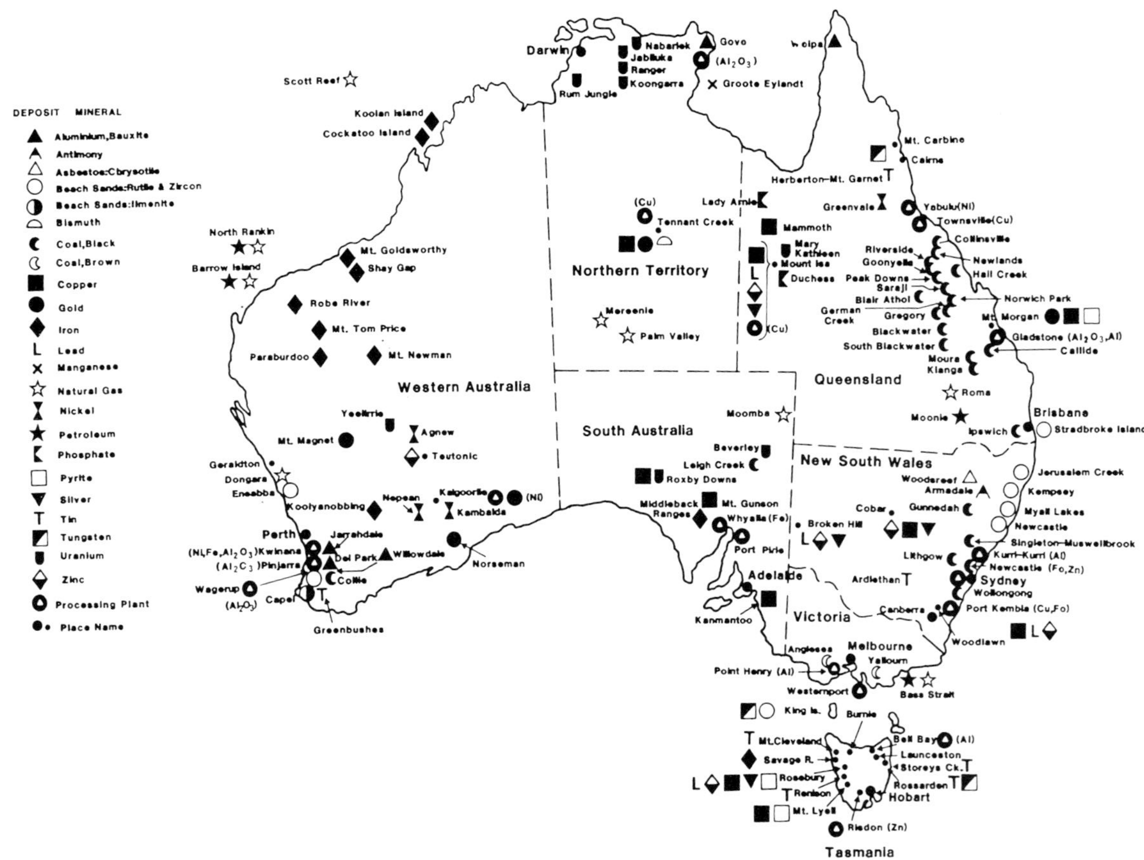

Figure 1. Major mining and mineral-processing operations in Australia.

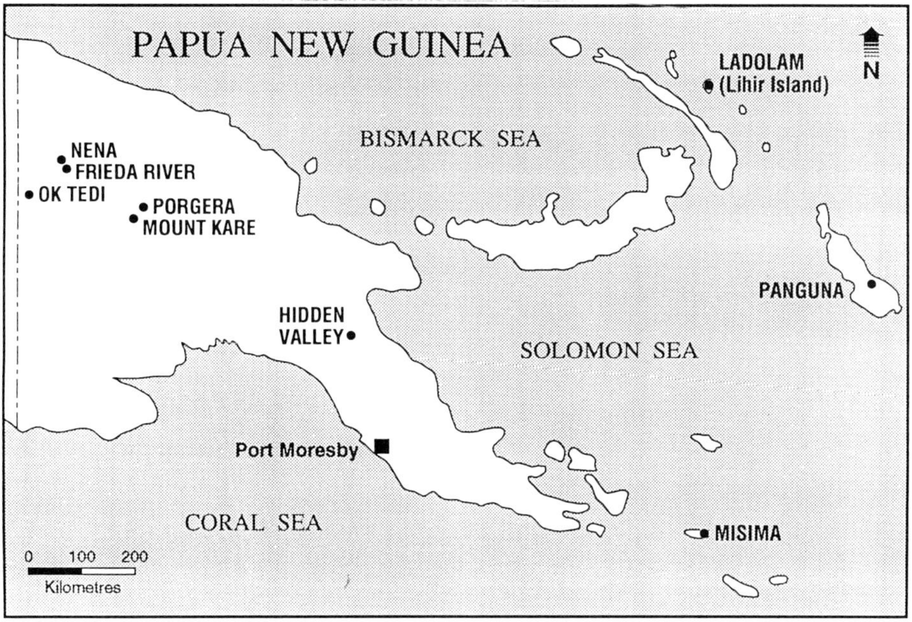

Figure 2. Major mining and mineral-processing operations in Papua New Guinea.

structure make it nearly impossible to prevent water and oxygen entering the mine and waste dumps. Consultants (Dames and Moore, 1993) have indicated that unless further remedial action is undertaken, the present situation will remain indefinitely. Any additional remediation would be very costly and could not hope to prevent all future contamination.

5.1.2 Woodlawn, NSW

The Woodlawn deposit occurs some 70 km NE of Canberra in a similar geological environment to Captains Flat. Mining commenced in 1978 to produce Cu, Pb and Zn concentrates (with Au, Ag having been leached from the stock-piled gossan early in the operation) and is continuing today. The mine occurs in a sensitive area adjacent to Lake George and is visible from the Federal Highway, the main route from Sydney to Canberra, so the monitoring and remediation are especially important.

Although not suffering from the catastrophic failure experienced at Captains Flat, several periods of high intensity rainfall have caused problems of excess water in the tailings and evaporation dams and waste rock dumps. Carefully designed water management techniques and constant monitoring ensure that offsite discharges meet acceptable criteria. The now disused open pit produced 60–70 million

Table 3. Comparison of factors influencing extent and impact of contamination from minesites in North America, Australia, PNG, SE Asia and Oceania.

	North America	Australia	Impact in Australia	PNG and SE Asia/Oceania	Impact in PNG SE Asia/Oceania
Topography	Commonly very rugged but with large areas of subdued relief in central region	Moderate to low relief even in upland areas	Restricted area of contamination	High relief	Siltation of rivers, failure and erosion of constructed landforms
Tectonics	Subject to earthquakes and volcanic eruptions	Relatively stable over long period	Damage to infrastructure and constructed landforms unlikely	Subject to earthquakes and volcanic eruptions	Damage to infrastructure and constructed landforms likely
Climate	Extremes in precipitation, temperature and wind speed	Less extreme and mainly semi-arid to arid climate	Movement of contaminants more restricted	Tropical rainfalls of > 3000 mm and hot weather	Widespread movement of contaminants by erosion
Surface water	Abundant freshwater lakes and rivers	Scarce and often saline	Limited	Large rivers but few lakes	Siltation of rivers for long distances
Population	US densely populated; Canada has ribbon development along US border	Low population concentrated around coast	Limited	Highly variable but many villages and some major centres of population	All villagers exposed to potential contamination
Agriculture	Intense over wide area	Intense in restricted area	Limited	Highly variable	Highly subject to contamination

Table 4. North Adit Spring, Captains Flat chemistry summary (post rehabilitation data).

Study	Statistic	pH	Zn mg L^{-1}	Zn load kg d^{-1}	Flow L s^{-1}
Brooks (1980)	Max	3.4	315	55.5	3.1
Jan 77–Oct 78	Min	2.7	170	3.2	0.2
7 samples pH & conc	Median	3.2	230	34.3	1.9
6 samples load & flow					
Brown & Train (undated)	Max	3.7	450	151	5.4
Dec 76–Aug 78	Min	2.4	195	3	0.1
60 samples	Median	2.7	240.5	19	1.0
Sparksman (1982)	Max	3.2	180		
Jun 82–Nov 82	Min	2.8	153		
4 samples	Median	3.0	160		
NSW Dept Min Res*	Februrary 1992	3.4	120		
	August 1992	3.1	105		

*New South Wales Department of Mineral Resources (unpublished), by kind permission of Kerry Brooks.

Table 5. Typical Woodlawn waste rock dump leachate analyses.

Date	pH	Conductivity (mS cm^{-1})	Cu	Zn	Fe mg L^{-1}	Cd	SO_4
03 Jan 92	2.9	2670	230	7650	350	45	39000
29 Apr 93	3.1	1860	125	6500	350	30	32800
27 Feb 95	3.0	2060	88	4980	310	24	27900

tonnes of waste rock which covers an area of 100 ha and is constructed in five levels to a total height of 80 m. This has been rehabilitated by compaction of the surface, covering with weathered rock and soil and revegetation with grasses, shrubs and trees. Despite a sulfide content of 6–7%, the vegetation is healthy with the roots seldom penetrating the compacted layer. There have been some small sections of acid burning. Leachate flow has reduced from 3 L s^{-1} to 0.4 L s^{-1} over a four year period but still has a relatively high contaminant load (Table 5).

The AMD generating potential is high because of the high pyrite content of the tailings. Seepage associated with two sheared contacts between massive sandstone and a banded siltstone in the southern wall of the southern tailings dam is highly contaminated (Table 6; Niven, 1990). The tailings in two dams are presently being reprocessed, but this will not reduce the heavy metal content (except for Zn) or the AMD potential. As the mine has only 4–5 years to run, methods are now being sought to ensure its environmental integrity.

Table 6. Woodlawn south dam seepage and piezometer analyses.

Location	pH	Eh (mV)	Cond ($\mu S\ cm^{-1}$)	Cu ($\mu g\ ml^{-1}$)	Pb ($\mu g\ ml^{-1}$)	Zn ($\mu g\ ml^{-1}$)	Cd ($\mu g\ ml^{-1}$)	Al ($\mu g\ ml^{-1}$)	Mn ($\mu g\ ml^{-1}$)	Ca ($\mu g\ ml^{-1}$)	Mg ($\mu g\ ml^{-1}$)	SO_4 ($\mu g\ ml^{-1}$)
South Dam	3.2		7 600	40	2.1	1 100	5.8	140	68	510	640	7 300
SS-1	6.00	16	4 730	0.87	0.74	600	1.55	6.25	66	615	720	5 420
SS-2	4.91	80	4 380	5.39	2.7	780	3.16	120	65	620	755	6 180
SS-3	4.56			110	3.79	2 660	6.92	1 300	260	555	2 510	21 090
SS-3	4.56			120	3.8	2 270	6.40	1 200	240	600	2 600	20 000
SS-3R			8 400	100	3.8	2 100	6.00	990	250	580	2 450	18 000
SS-5A	4.87			15	6.15	1 025	2.40	150	120	585	1 060	8 490
SS-5B	4.89			15	5.97	1 075	2.44	135	130	535	100	8 070
SS-5C	4.88			17	6.6	1 075	2.64	140	130	550	1 075	8 640
SS-7	4.91			20	3.4	925	2.58	120	110	535	910	7 480
SE-1	4.18	119	4 750	2.73	2.95	700	3.02	11	55	580	615	5 750
SE-2-Rut	6.17	6	1 707	<0.01	0.06	1.94	0.002	0.24	0.61	25	45	135
SE-2-Pond	8.57	-130	4 290	<0.01	0.03	0.75	<0.001	0.30	0.30	35	260	280
SE-3	7.33	-60	4 140	0.04	0.08	195	0.88	0.24	35	350	550	3 410
SE-4	6.09	11	4 090	1.06	1.18	390	1.21	0.73	60	410	725	4 870
SE-5	6.53	-13	4 060	1.96	0.06	855	2.99	1.94	85	525	800	5 750
X1	7.62	-69	3 230	ND	ND	0.33	0.004	0.34	0.61	3.18	159	307
X2	7.47	-53	2 060	ND	ND	0.15	0.003	0.68	0.00	0.80	51	68
Y1	6.94	-23	4 540	ND	0.05	1.02	0.002	ND	0.02	17	344	318
Y2	7.18	-36	4 770	0.02	0.03	1.09	0.007	0.06	0.02	30	319	200
Z1	6.97	-24	2 070	0.01	0.02	0.77	0.004	0.06	0.72	25	221	691
Z2	7.27	-40	1 501	0.01	0.03	0.51	0.005	0.07	0.03	76	145	479

Notes: 'South Dam' is a sample of surface south dam water;
'SS' = seepage from beneath the south wall, south dam;
'SE' = seepage from beneath the east wall, south dam;
X1 to Z2 are piezometers.

5.1.3 Sunny Corner, NSW

A series of small mines developed on faulted lenses of Pb-Zn-Cu-Ag-pyrite-arsenopyrite mineralisation at Sunny Corner 20 km NNW of Lithgow were abandoned last century. Effluents from the adits and numerous spoil heaps contaminate Daylight Creek resulting in a pH of 2.5-3.5 and high concentrations of Fe, Al and Zn. The creek is devoid of life except for algal growths and bacterial slimes on thick hard iron hydroxide concentrations (Chapman et al., 1983). The chemistry of water at 12 sites in Daylight Creek is shown in Figure 3.

Although the abandoned mines are in a heavily wooded, high rainfall, moderate-high relief environment, there has been little concerted effort at rehabilitation. This has, in part, been due to the proposed end use for the site, with the local community accepting its small tourist potential (a similar 'moon-scape' to that at Mount Lyell). Because of fracturing of the rock and the positioning of adits, any rehabilitation to prevent AMD and formation of iron hydroxides would be extremely difficult in this relatively high rainfall environment.

5.1.4 Mount Lyell, Tasmania

The Mount Lyell mineral field on the west coast of Tasmania was initially worked for Au following its discovery in 1883. Copper mineralisation was subsequently found in 1893 and the company developed a process for pyrite smelting in 1896. These smelting operations combined with timber cutting, frequent bushfires and high annual rainfall resulted in extensive loss of vegetation and considerable erosion. In addition, waste rock dumps and abandoned mine workings are significant sources of AMD. Tailings have been consistently discharged into the local river system and in excess of 90 million tonnes have been discharged into the Queen/King River systems and Macquarie Harbour to date (Wood, 1991).

Some natural revegetation of the estimated 1500 ha of completely denuded and 2500 ha of substantially denuded countryside has occurred, and research into revegetation of the remaining area is continuing. Although welcomed by regulatory authorities, there have been some objections to the revegetation program by members of the local community who want the bare hills to remain as a 'tourist attraction'.

Again, AMD and dissolved metals derived from waste rock dumps (50 million tonnes containing 10% pyrite), mine dewatering, underground drainage and pyritic outcrops creates the greatest environmental problems, particularly for aquatic ecosystems. The nature of leachates from three sources is shown in Table 7. A constant problem is rain falling into the old West Lyell Open Cut which then percolates through 350 m of fractured and caving pyritic rock before it is pumped out of the mine. Discharge from the abandoned adit shows that high levels of contaminants are still present after 50 years when these underground workings were abandoned. AMD from tailings is reduced because of the practice of sub-aqueous disposal (a method recommended in Canada – see later). However, some tailings have been

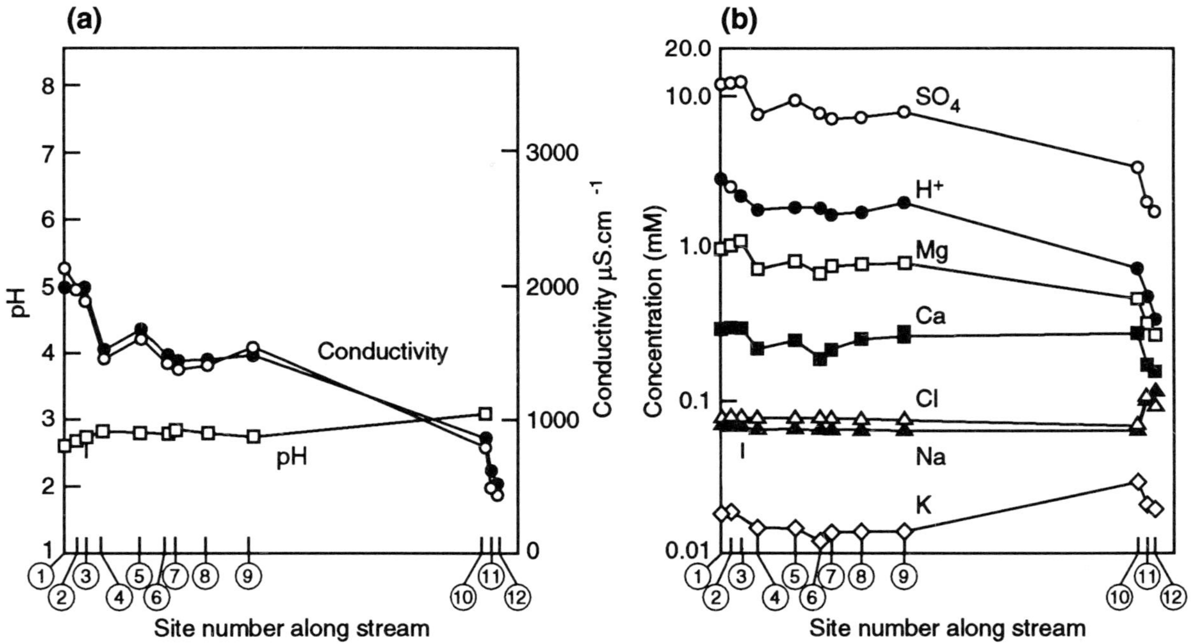

Figure 3. For the scan at Daylight Creek carried out on 27th October 1981 is shown (a) pH and conductivity, (b) concentrations of major cations and anions, (c) concentrations of toxic metals and (d) map of Daylight Creek study area at Sunny Corner indicating the position of sampling sites referred to in the text. The main flow and contamination arises from the collapsed adit marked.

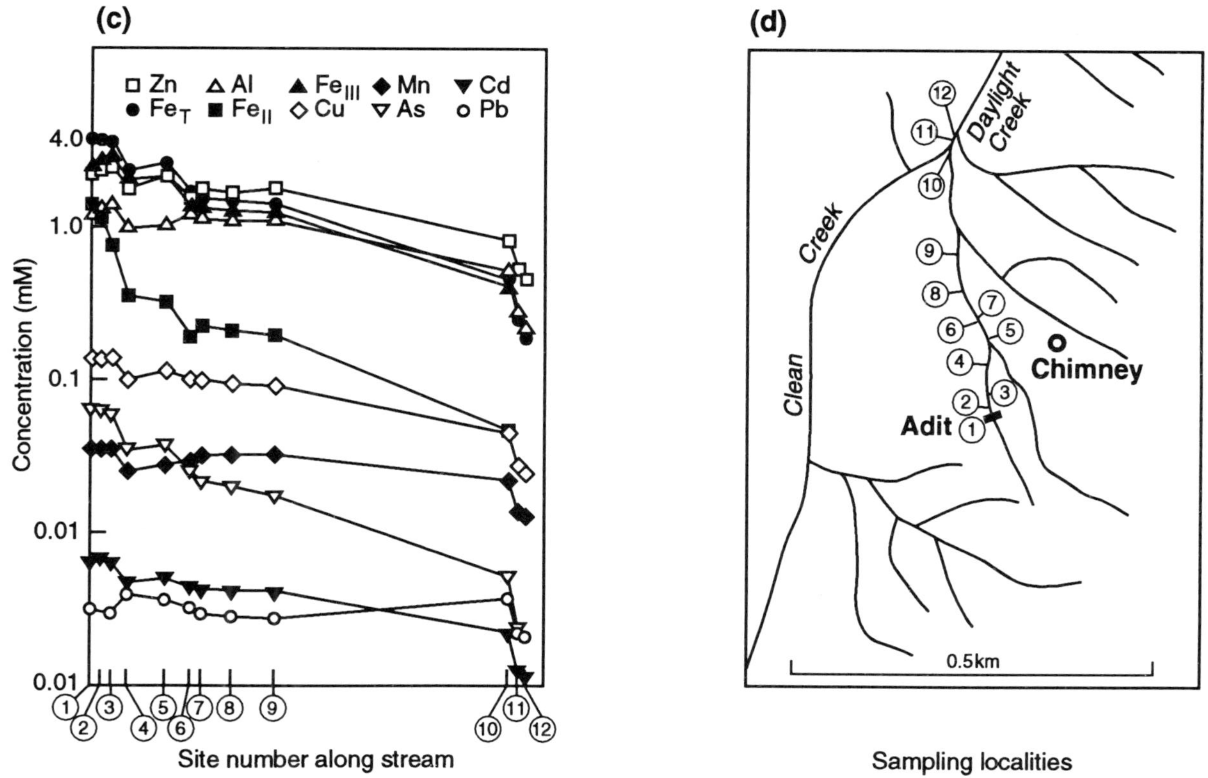

Figure 3. Continued

Table 7. Nature and composition of leachates from Mount Lyell abandoned Au mine.

	pH	Fe	Mn	Cu mg L^{-1}	Zn	SO_4	Flow m^3 d^{-1}
Waste rock	2.5	1 160	140	130	13	8 200	4 700
Mine water	2.9	520	40	170	7	3 800	4 300
Abandoned adit	2.8	138	174	28	45	1 100	400

deposited on the banks of the King River and over a considerable period have been intermittent sources of AMD over a distance of 15 km down river (Wood, 1991). The contamination at Mount Lyell has recently attracted state and federal funding and will be the subject of a number of research projects. There is considerable doubt as to whether contaminant flow into Macquarie Harbour or Lake Burbury will ever be eliminated.

5.1.5 Brukunga, South Australia

The Brukunga Pyrite Mine, 50 km ESE of Adelaide, was mined for pyrite over the period 1952 to 1972 for the manufacture of sulfuric acid. Since abandonment, highly contaminated AMD from waste rocks and tailings has been polluting Dawesley Creek (Table 8). Measures have been taken to minimise the effect on the environment by installation of a water treatment plant, seepage collection sumps and pump-back systems (Smith and Hancock, 1992). Additionally, covering and revegetation of the tailings dam and minor work on the waste dumps have been undertaken and rehabilitation research is continuing (Agnew, 1994). The estimated cost of water treatment funded by the South Australian Government is in excess of A$500 000 annually.

5.1.6 Mount Morgan, Queensland

Gold, Cu and Ag were mined at Mount Morgan, 35 km SSW of Rockhampton, between 1882 and 1991. As a result of this long period of mining there is now a very large open cut filled with low quality water, numerous waste rock dumps, tailings dams and smelter material. There have also been discharges of tailings, waste rock, mineral processing solutions as well as AMD from the tailings and waste rock into the nearby Dee River resulting in considerable contamination and a consequent massive drop in plant and microinvertebrate species. Poor quality water from the open cut is also entering the groundwater zone.

A recent study of the mine and drainage systems including 21 sites up to 50 km downstream on the Dee River (Figure 4) shows high levels of contamination (Table 9) and precipitates of ferric hydroxides, aluminium hydroxides, gypsum, jarosite and other hydrated sulfate minerals. Concentrations of metals in the sediments are also high (Table 10). As the Dee River is used as a water supply and passes through agricultural and grazing country the impact of contaminants is widespread. The Queensland Department of Minerals and Energy are presently funding a detailed

Table 8. Representative water quality from mine sources at Brukunga.

	pH	Acidity†	SO_4	Fe	Al	Ca	Mg	Na	Cd	Mn	Zn
		mg L^{-1}									
Mine cuts	2.7	10000	12000	2500	500	400	400	200		50	20
Tailings	2.3	7000	8000	4000	50	450	300	200	0.005	150	15
Dawesleys u.s*	7.5	0	100	2	<1	40	30	400	<0.2	<0.1	<0.2
Dawesley d.s* (KAN 2)	3.5	2000	2500	200	200	100	130	350	0.11	20	25
Waste dumps	2.5	64000	35000	2000				450	15	500	
Mine benches	3	4000									

†as $CaCO_3$.
*Winter 1973.

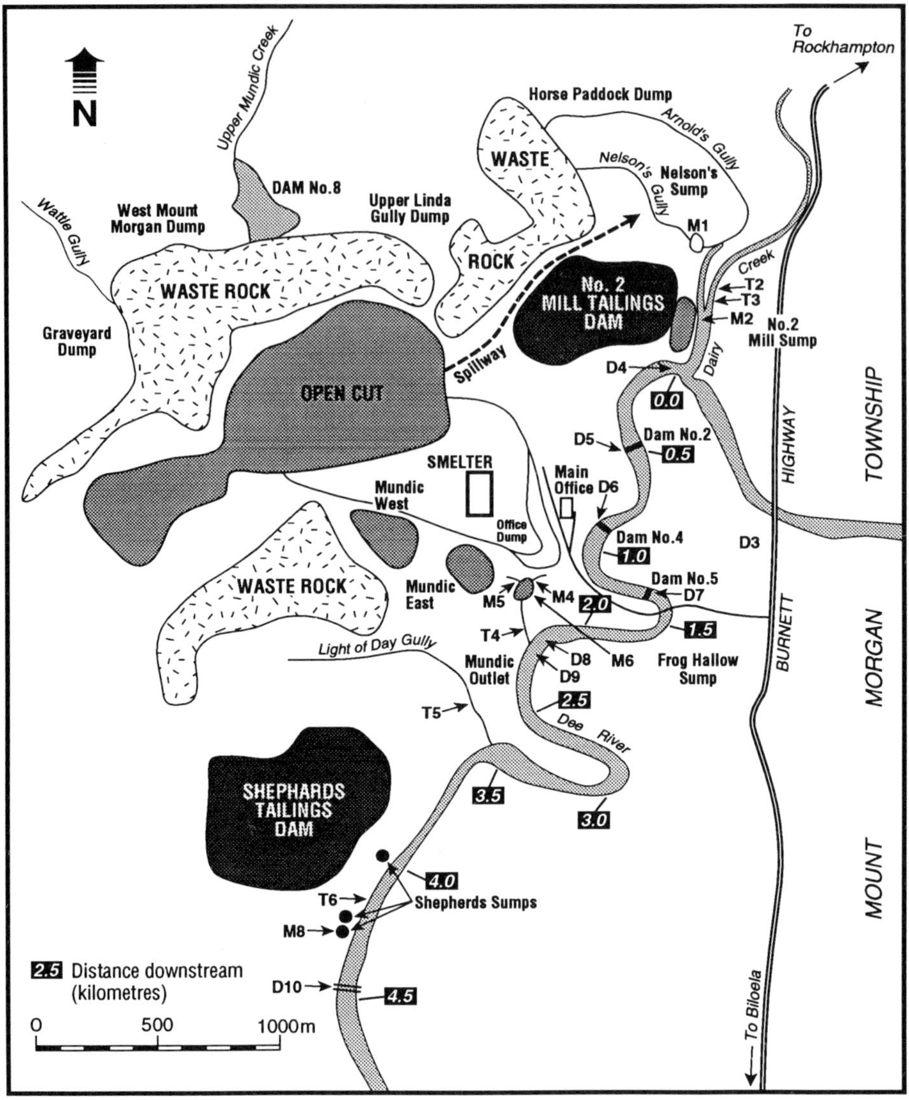

Figure 4. Map of Mount Morgan minesite showing sampling sites.

study to determine: the locations and relative contributions of various sources of pollution in the Dee River; the potential for remobilisation into the water column of contaminants currently held within the river sediment; and the natural processes leading to removal of contaminants from the water column. Rehabilitation of the mine and dumps are considered to involve a considerable capital investment and continuing outlays for maintenance.

Table 9 Water quality in sumps and seeps on the Mt. Morgan minesite and in tributaries of the Dee River.

Analyte	Sumps							Tributary				
	M1 Nelsons	M2 No 2 Mill	M3 Open Cut	M4 Frog Hollow Creek	M5 Frog Hollow Pipe	M6 Frog Hollow Sump	M8 Shepherds	T2 Dairy Ck	T3 Dairy Ck	T4 Mundic Ck	T6 Shepherds	T7 Fletchers Ck
Al mg L^{-1}	610	3 020	330	803	461	628	640	0.17	0.093	1 090	340	<0.005
Ca	478	509	550	437	466	475	450	482	428	493	446	36
Cd	0.14	0.34	0.09	0.36	0.48	0.49	0.05	<0.01	<0.01	0.74	0.05	<0.01
Co	1.80	5.8	1.23	2.50	2.74	2.72	1.37	0.56	0.54	3.04	1.82	<0.01
Cu	20	121	20	106	39	102	26	0.11	0.46	196	16	<0.005
Fe	3.56	554	104	1 460	1 920	1 730	51	0.13	46	1 420	2.25	0.043
Mn	136	389	46	131	153	152	90	33	48	124	152	0.10
Na	236	306	792	213	325	260	277	731	388	172	266	50
Ni	0.47	1.60	0.49	0.65	0.71	0.69	0.28	0.13	0.07	0.72	0.28	<0.01
Zn	41	104	11.4	45	59	60	15.4	1.42	2.30	55	11.3	<0.01
SO_4 mg L^{-1}	13 500	43 600	8 240	16 100	15 000	15 800	11 700	7 402	9 375	17 940	10 770	12
pH	3.66	2.24	2.84	3.29	3.72	3.33	2.94	7.03	6.20	2.76	3.62	7.15
E.C. $\mu S\ cm^{-1}$	12 460	11 200	10 420	13 600	13 400	13 800	-	10 350	10 580	14 900	-	590

Table 10. Concentrations of metals in sediments from the Dee River (mg kg^{-1} dry weight unless otherwise indicated).

Element	Site											
	T2 Dairy Ck fines	T3 Dairy Ck coarse[b]	T3 Dairy Ck fine	D9 Dee at Mundic Ck coarse	D9 Dee at Mundic Ck fine	D11 1st ford fines[a]	D13 coarse[b]	D13 fines	D14 Walmul coarse	D15 Dee u/s Fletchers coarse	D16 Dee d/s Fletchers coarse	D16 Dee d/s Fletchers white ppt
Ag	<5	-	-	<5	<5	<5		<5	<5	<5		<5
Al %	5.06	2.40	6.38	0.74	1.81	4.32	0.80	1.37	0.97	0.77	0.59	15.83
As	-	-	-	32.8	42.7							
Ba	26	33	20	49	162	28	29	52	30	26	13	9
Ca %	0.86	0.10	0.11	0.454	0.255	0.14	0.09	0.08	0.126	0.079	0.162	0.60
Cd	<5	-	-			10.4		<5	<5	<5		<5
Co	30	-	-	7	10	19		<	6	10		38
Cr	5	17	-	14	18	17	10	12	9	9	8	<5
Cu	448	784	1 306	945	1 187	903	489	784	296	370	263	11 945
Fe %	3.96	15.21	20.62	9.08	11.58	17.58	3.45	8.33	4.75	6.85	3.91	0.22
K	208	806	782	1 114	1 857	688	600	781	613	614	519	446
Mg %	1.58	0.52	0.53	0.439	0.798	1.72	0.24	0.34	0.328	0.283	0.258	0.36
Mn	321	147	205	191	307	1 480	116	178	242	228	185	1 215
Na %	0.36	0.19	0.21	0.14	0.29	0.89	0.10	0.05	0.04	0.03	0.05	0.21
Ni	<10	-	-			<10		<10	<10	<10		<10
Pb	<50	-	-	74	148	<50		<50	<50	<50		<50
Ti	46	197	81	263	273	99	203	70	380	435	398	11
V	42	159	153	79	112	59	70	95	79	87	67	11
Zn	67	76	137	112	205	144	47	138	62	74	60	955

[a] <63 μm,

5.1.7 Broken Hill, NSW

The Broken Hill Lode has been mined since 1883 and has generated open cuts, waste rock dumps and tailings dams as well as smelting residues from the older activities. Pasminco and Normandy, the present lease-holders, have developed an extensive 'Rehabilitation and Environmental Management Plan' which recognises:

- the arid, semi-desert nature of the region;
- the concentration of activity and long history of mining at Broken Hill; and
- the tourist, educational, scientific and historical significance of geological features and mining structures.

Because of the low pyrite content of the ore and country rocks, and the arid climate, AMD is not a problem, particularly as a considerable proportion of waste rock and tailings are returned underground as fill (McIlveen and Lewis, 1990). The main contaminant has been dust and there has been a program, initiated in 1936, of regeneration of native flora on dumps and bare areas. Present management practices include stabilisation of tailings dams, continued revegetation, and monitoring of water run-off into Stephens Creek Reservoir. A new open-cut mine will generate large volumes of waste rock, which has the potential to increase the dust problem. Indications are that all mining will cease early in the next Century, with the resultant need for extensive rehabilitation.

5.2 Gold Mines

Gold is presently mined either from large open-cuts (e.g. Telfer, Boddington, Mount Leyshon, Kidston, Kalgoorlie Superpit) or from underground (e.g. Mount Charlotte, Norseman, St Ives). Waste rock dumps and tailings dams are potentially the source of similar contamination to that derived from base metal mines. The one significant difference is the comparatively high levels of residual CN used to extract gold from the ore. Examples of environmental conditions at a representative range of gold mining sites in Australia and PNG follow:

5.2.1 Boddington, Western Australia

The Boddington Gold Mine is located in jarrah forests about 100 km SSE of Perth. While bauxite mining has been a recognised land use in the jarrah forests, the uniqueness of gold mining in this area resulted in different and elevated levels of environmental scrutiny, protection and management (John, 1988). The fundamental difference between bauxite- and gold-mining is the management of residue from carbonic leach (CIL) processing plant (CN, saline and alkaline). A detailed study (Miller et al., 1991) showed that free CN was rapidly volatilised from the decant and residue surface and that weak acid dissociable (W.A.D.) CN forms (mainly copper CN complexes) were transformed to stable non-toxic iron CN complexes within the residues dam. Soluble CN and Cu concentrations decreased sharply with depth within the first 4 m from the surface and were negligible at a depth of greater than 8 m. Seepage water entering local catchment has very low CN concentration, and is monitored in 29 bores. Thus, the main source of potential contamination is dispersed by natural processes.

5.2.2 Horn Island, Queensland

Horn Island, off the tip of Cape York in Torres Strait, has been mined for gold since 1894 with most intense production occurring in the late 1980's. The mine closed suddenly in December 1989 due to the operating company being placed in receivership. The Queensland Minister for Resource Industries instructed the receivers to rehabilitate the mine with the provision of A$2.2 million for the task (Frankcombe, 1993). The key environmental issues were identified as:

- probability and quality of pit water discharge;
- contamination of groundwater through leaching of heavy metals from tailings; and
- acid mine drainage from mineralised rocks and its impact on fringing reefs and benthic fauna.

Initial monitoring of the plant site, tailings dams, process water dam, mine pits, waste rock dumps, low grade ore stockpile and other facilities showed that the mine was more environmentally benign than originally anticipated with pH in open pit water currently averaging 6.2. Because of the location of the mine in an environmentally and politically sensitive location between Australia and PNG, and the possibility of contamination of the surrounding reefs and marine life, this mine is an example of the need for proper environmental management on small tropical islands.

5.2.3 Kidston, Queensland

The Kidston Gold Mine is located 280 km WNW of Townsville in gently undulating country subjected to distinct wet and dry seasons. It lies within the Copperfield River catchment which eventually enters the Gulf of Carpentaria through the Gilbert River. The area is mainly used for cattle raising. Gold production from a low grade porphyry system is approximately 200 000 ounces per annum, resulting in large waste rock dumps and tailings dams both of which have AMD generating potential. In addition, the tailings have low residual CN levels, with the majority being complexed and insoluble (Table 11). The operators of the mine have already planned decommissioning and rehabilitation of the mine (Ritchie, 1991) and have undertaken considerable research into regeneration of native flora on waste rock and tailings. Planning has already commenced on water management once the mine is decommissioned. Seepage and run-off from the tailings and waste rock will be diverted into the 252 m deep open pit where, because evaporation exceeds rainfall by 1612 mm per annum, it will take 20 years to equilibrate with the regional aquifer

Table 11. Cyanide content and forms in tailings core samples at Kidston, Queensland.

Sample Core Number	Age of the Sediment	Core Paste pH	Residual Total CN ($mg\,kg^{-1}$)	Soluble Total CN ($mg\,kg^{-1}$)	Soluble W.A.D. CN ($mg\,kg^{-1}$)	Soluble Free CN ($mg\,kg^{-1}$)	Soluble/ Total (%)	W.A.D./ Total (%)	W.A.D./ Soluble (%)	Soluble SCN $mg\,kg^{-1}$
1A	1.5	8.3	1.3	0.18	0.08	0.08	12.2	5.4	44	0.20
1B		8.0	1.6	0.04	0.04	0.04	2.4	2.4	100	0.20
1C		8.1	1.2	0.08	0.02	0.02	6.3	1.6	25	1.20
1D		8.1	1.2	0.04	0.02	0.02	3.2	1.6	50	1.20
2A	1.5	8.3	1.1	0.06	0.04	0.02	5.2	3.5	67	0.20
2B		8.4	1.6	0.50	0.32	0.02	24	15	64	1.72
2C		8.7	1.1	0.06	0.04	0.04	5.2	3.5	67	0.20
2D		8.4	0.6	1.10	1.10	0.02	65	65	100	13.5
3A	0.5	8.1	1.9	0.10	0.04	0.02	5.0	2.0	40	57
3B		7.9	1.9	0.08	0.04	0.02	4.0	2.0	50	49
3C		8.0	6.2	0.12	0.04	0.02	1.9	0.63	33	52
3D		8.1	0.6	0.10	0.02	0.02	14	2.9	20	37
5A	2.0	8.5	1.3	2.14	1.04	0.02	62	30	49	2.96
5B		8.1	1.8	0.38	0.36	0.02	17	17	95	2.08
5C		7.9	1.8	0.22	0.22	0.02	11	11	100	5.60
5D		8.1	1.6	0.86	0.78	0.02	35	32	91	20
6A	1.5	2.7	1.2	0.28	0.20	0.02	19	14	71	2.04
6B		3.0	1.1	0.48	0.02	0.02	30	13	4.2	2.08
6C		4.2	0.6	0.12	0.02	0.02	17	28	17	0.70
6D		6.0	1.2	0.06	0.02	0.02	4.8	16	33	4.60
7A	5.5	7.5	9.6	0.34	0.14	0.14	3.4	1.4	41	1.24
7B		8.3	6.2	0.22	0.04	0.04	3.4	0.62	18	0.20
7C		8.5	1.3	0.32	0.16	0.02	20	9.9	50	0.32
7D		8.0	1.2	0.24	0.02	0.02	17	1.4	8.3	2.16

Table 12. Cyanide forms in fresh tailings solution

Constituent	Concentration (mg L^{-1})
Total CN	72
W.A.D. CN	68
SCN	130
CNO	44
$Cu(CN)_4^{3-}$	37
$Fe(CN)_6^{4-}$	15
$Ni(CN)_4^{2-}$	2.8

(Ritchie, 1991). Analysis of the residual CN in fresh tailings solutions (Table 12) suggests that CN will not pose a problem, although it is planned to study CN degradation kinetics.

5.2.4 Pine Creek, Northern Territory

Gold was discovered at Pine Creek, 220 km south of Darwin, in the 1870's. Present mining operations began in 1985 and are due to cease in late 1995. The mine is subject to monsoonal rainfall during November to March with an average annual rainfall of 1147 mm.

Much of the initial mining was of oxidised ore from the Enterprise Pit (Figure 5), but recently has consisted of both oxidised and unweathered rock. Tailings are therefore a mixture of both oxidised and fresh material, the latter with an AMD producing potential. Both South and North Waste Dumps are producing AMD (Table 13) which is discharged into the process water dam. Table 14 demonstrates that since capping of the dumps began in 1991 in conjunction with a catchment management plan (Fawcett, 1995) there has been a significant reduction in the amount of AMD reaching the process water dam.

Decommissioning of the mine will leave a number of open pits, waste rock dumps and a tailings dam in the immediate vicinity of the town of Pine Creek. It is presently proposed to divert water from Pine Creek into the Enterprise Pit which will provide a large lake for recreational uses, reduce risk to public safety by reducing exposed height of pit walls and control local groundwater levels.

5.2.5 Kalgoorlie, Western Australia

Gold was discovered around Kalgoorlie in the early 1890's and the Golden Mile is still a significant gold producer. In 1985 a working party established to examine conservation and rehabilitation found the environmental record of the gold mining industry to be poor (Howard et al., 1991). Since then the Department of Mines has appointed two environmental officers at Kalgoorlie and each of the producers has hired environmental specialists. The two most critical environmental issues are sulfur dioxide from smelting or roasting of concentrates (see under processing) and dust. However, in this arid environment 'background' dust levels are high and pre-

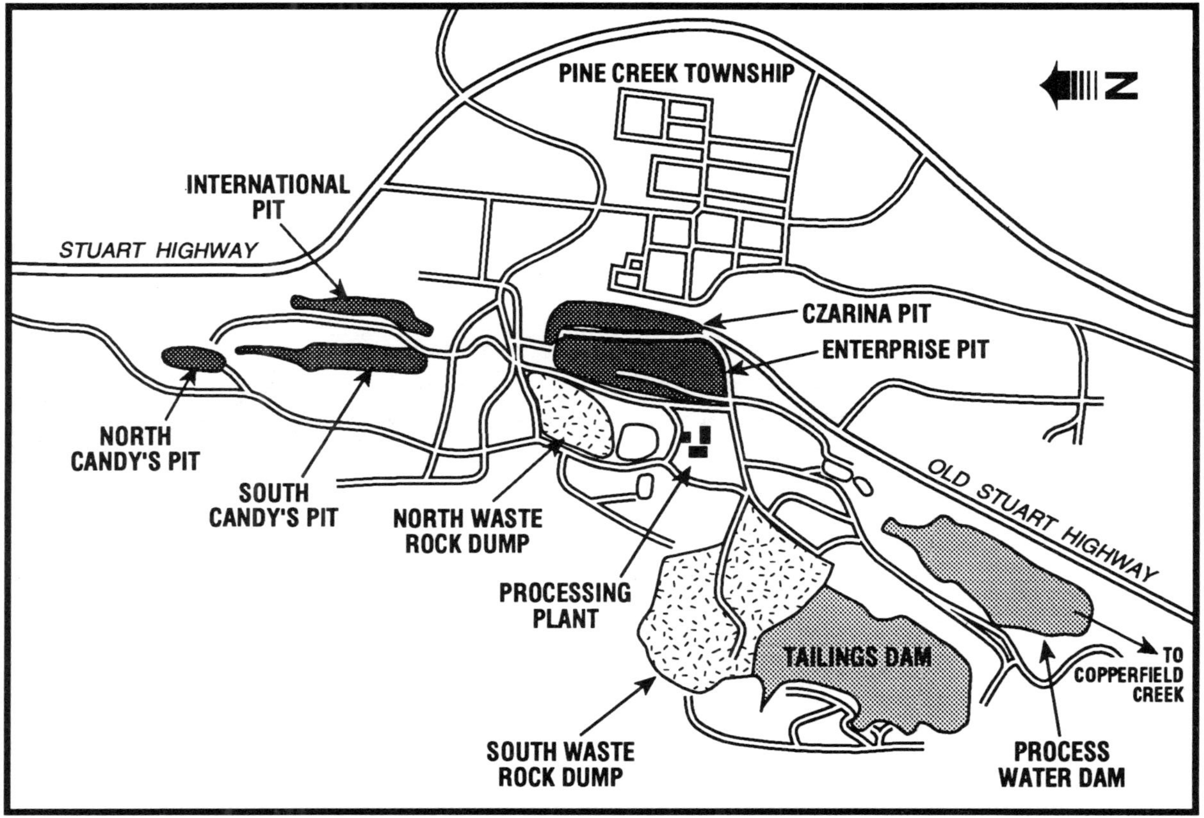

Figure 5. Site plan of the Pine Creek Goldfields Minesite.

Table 13. Water analysis results from monitoring site CK7, east of north waste dump, Pine Creek.

Attribute	Date sampled 21.1.91	13.2.92	20.12.94
pH	2.9	3.9	5.1
EC ($\mu S\,cm^{-1}$)	10890	850	400
Mg ($mg\,L^{-1}$)	1976	60	21
SO_4 ($mg\,L^{-1}$)	14920	410	56
Mn ($mg\,L^{-1}$)	262	5.4	1.3
Al ($mg\,L^{-1}$)	649	9.5	0.26
As^A ($\mu g\,L^{-1}$)	882	8	15
Cd^A ($\mu g\,L^{-1}$)	1960	74	7
Cu^A ($\mu g\,L^{-1}$)	48800	1200	32
Fe^A ($\mu g\,L^{-1}$)	53000	250	30
Pb^A ($\mu g\,L^{-1}$)	264	41	10
Zn^A ($\mu g\,L^{-1}$)	879000	15000	1400

[A]Measured in filtrate.

Table 14. Water analysis results from the process water dam, Pine Creek

Attribute	Date sampled 23.4.91	18.3.92	3.2.95
pH	3.6	4.2	4.5
EC ($\mu S\,cm^{-1}$)	2400	1700	1200
Mg ($mg\,L^{-1}$)	240	110	51
SO_4 ($mg\,L^{-1}$)	1800	1000	530
Mn ($mg\,L^{-1}$)	27	12	5.7
Al ($mg\,L^{-1}$)	70	19	3.7
As^A ($\mu g\,L^{-1}$)	12	10	20
Cd^A ($\mu g\,L^{-1}$)	300	10	19
Cu^A ($\mu g\,L^{-1}$)	4700	3500	680
Fe^A ($\mu g\,L^{-1}$)	4800	3600	280
Pb^A ($\mu g\,L^{-1}$)	280	10	17
Zn^A ($\mu g\,L^{-1}$)	102000	30000	6900

[A]Measured in filtrate

liminary monitoring suggests that more dust is generated in the town than at the mines (Howard et al., 1991). It is assumed that this will be alleviated by revegetation with native flora. However, rainfall is low, droughts are frequent and summer temperatures commonly exceed 40°C, so that rehabilitation techniques used in less extreme environments are not applicable. Companies have supplemented their conventional wisdom with local pastoral rehabilitation experience and some internal research.

5.2.6 Porgera and Ok Tedi, PNG

Although not in Australia, Porgera and Ok Tedi are both operated by Australian mining companies, and illustrate totally different environmental problems to those experienced in much of Australia. Both mines occur at high altitude in the central mountainous region of PNG with an annual rainfall in excess of 3500 mm and are based on low grade porphyry systems resulting in high waste rock and tailings pro duction. Under these conditions it is not possible to construct safe permanent storage for tailings, although the waste rock is dumped in erodible storage areas. At both mines, the tailings are disposed of in the local rivers and eventually make their way to the delta of the Fly River. The potential for contamination of the river systems with CN, acid, alkalis, dissolved and particulate heavy metals and silt is high. Although there has been considerable research and monitoring (Eagle and Higgins, 1990; Ross, 1991) it is extremely difficult to extrapolate into the future to determine what effect the mining will have on the flora and fauna.

The CN leach plant at Ok Tedi has been closed down for several years with the result that CN base metal complexes are no longer discharged to the Fly River. However, concentrator tailings continue to be discharged to the river, and in the case of Ok Tedi has led to a compensation claim by the native peoples. Porgera tailings from the concentrator and from the neutralised CN residue of the pressure oxidation plant are still discharged to the Fly River.

5.3 Asbestos

Contamination derived from asbestos mining is the most insidious form of mine contamination, often leading to the incurable lung disease mesothelioma and other forms of cancer, especially in those engaged in mining and processing the ore (de Klerk et al., 1989).

At Wittenoom in the Pilbara region of Western Australia the highly fibrous 'blue asbestos' crocidolite was mined from thin veins for many years. Because of deteriorating health of the workforce, the mine was closed in 1966 and has since been the subject of extensive litigation. Out of a workforce of 6500 men and 400 women, there had been 32 deaths due to mesothelioma, 91 from cancer of trachea, bronchus and lung and 34 due to pneumoconiosis. To prevent further problems resulting from the ingestion of asbestos fibres, access to these sites has been prohibited, which is not difficult in this rugged and sparsely populated terrain. Because of its rugged beauty, there has been some opposition to this prohibition as Wittenoom townspeople see it as a source of tourist income.

At Woods Reef in northern NSW, mining of 'white asbestos' chrysotile (which is less of a health risk) has left a large open pit surrounded by exposed heaps of waste rock and tailings. Some remedial work and natural processes have lowered the airborne threat although there are still problems of silt being washed into waterways. A recent estimate suggested that A$15 million may be required to clean up the mine with A$80 million being required to fully rehabilitate the site. Contamination risks would be considerably lowered if the proposed wet-milling of the tailings ever occurs, but this would seem to be economically unviable.

5.4 Bauxite

Australia produces approximately 40% of the world's bauxite from 3 major areas: the Darling Ranges, Western Australia; Gove, Northern Territory and Weipa,

Queensland. Although contamination from bauxite mining is relatively benign, the companies involved have been at the forefront of environmental management and rehabilitation of the large mined and disturbed areas.

5.4.1 Darling Range, Western Australia

Alcoa of Australia Ltd. mines bauxite from four separate mines in the environmentally sensitive jarrah forests of the Darling Ranges SE of Perth. This area is particularly sensitive because of the nutrient-poor lateritic soils, high evaporation, summer drought and high temperatures, dieback disease of the jarrah forest and a popular tourist area close to a major population centre. The main potential problem is erosion and siltation of local water courses and catchment areas during mining. Alcoa has developed sophisticated techniques for rehabilitation of the mined areas to the highest standards expected by regulatory authorities and the public (Slessar, 1988).

5.4.2 Weipa, Queensland

Comalco Aluminium Ltd. has operated an extensive open cut bauxite mine at Weipa on the western coast of Cape York Peninsula, Queensland since 1966. The climate is tropical with hot wet summers and dry mild winters. Bauxite mining significantly alters the landscape and soil profile so that techniques were developed to revegetate disturbed areas with self sustaining and maintenance free native vegetation after an initial attempt at growing commercial timbers. The company continues to explore improved methods of establishing native vegetation and is utilising remote sensing techniques to monitor regrowth and sustainability.

5.5 Coal Mining

5.5.1 Bowen Basin, Queensland

Coal is mined from four major measures within the Permian Bowen Basin of central Queensland by both open-cut and underground techniques. Open-cut techniques result in extensive voids (up to 25 km long and 100 m deep) and huge piles of overburden. This overburden varies considerably with respect to sulfide content, sodicity and mineralogy, and is both weathered and fresh. There is considerable research to determine the most effective means of rehabilitation of these waste rock dumps to ensure that ground and surface water quality is maintained and that dust is minimised. However, because of the variability and mixing of the strata in the waste rock dumps and soils derived from them, revegetation is not simple. Stability of both these waste rock dumps and the final voids impacts on the rehabilitation measures and is again dependent on site structure and hydrology, as well as mineralogy, geochemistry of the sediments.

Some of the mines have a potential AMD problem as the washery rejects and waste rocks contain substantial pyrite. However, this is not always immediately evident as any acid produced in the washery rejects may be buffered by carbonate-rich mine water which is used in the milling operation. Any contaminant elements are initially retained within the rejects. Once deposition of rejects has ceased and the carbonate disappears or is used up, AMD will occur with possible transport of toxic elements into surface and groundwaters. As the mines are surrounded by intense agriculture, rehabilitation measures are aimed at preventing this happening.

5.5.2 Hunter Valley

Coal mining in the Hunter Valley poses similar contaminant problems to that in the Bowen Basin with large overburden dumps, reject piles and potential for AMD and release of heavy metals. Two additional sources of contamination add to the environmental problems in the Hunter Valley.

- contribution to the salinity of the Hunter River by interception of saline groundwaters in Permian strata; and
- spontaneous combustion of waste material adding to soil and atmospheric contamination.

In a region of intense agriculture and large population, both of which are dependent on Hunter River water, the coal mining industry is constantly monitoring the quality of run-off water and developing appropriate management techniques. Many of the products of spontaneous combustion are carcinogenic and are difficult to remove from soils. A similar problem occurs at the Leigh Creek coal mine in South Australia.

5.6 Uranium Mines

Australia's two operating uranium mines are subject to the strictest regulations with respect to contamination off the mining lease (i.e. mining leased area), which is in marked contrast to the abandoned Rum Jungle mine which ceased operation in 1971.

5.6.1 Ranger, Northern Territory

Uranium mining at Ranger has been from the large No. 1 open pit and has resulted in extensive waste rock dumps, low-grade ore stockpiles, and a massive tailings dam (Figure 6). The mine is surrounded by Kakadu National Park, and during the summer months is subjected to monsoonal rains. In addition to contaminants normally associated with mining (dust, silt, AMD, heavy metals) there is potential for radioactive elements to enter the surrounding wetlands. All run-off from the minesite is channeled into a number of retention ponds (RP 1-3 in Figure 6). Because of the high rainfall during the wet season, some of this water is used to irrigate surrounding bushland during the dry season. Table 15 shows the composition of water from RP2 and the annual loadings based upon 1800 mm of water applied to the bushland (Willett et al., 1993).

Tailings from the mine are produced after the extraction of U by sulfuric acid.

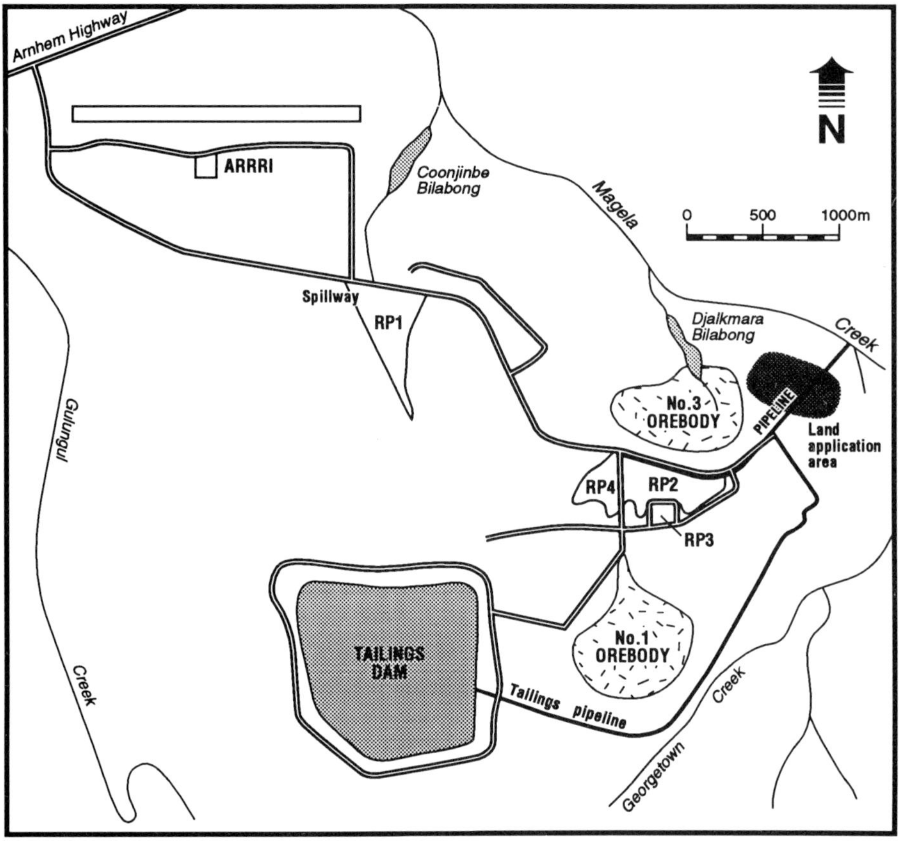

Figure 6. Location of the land application site at the Ranger Uranium Mine.

Manganese dioxide is added as an oxidant for the extraction process and the tailings are neutralised by addition of lime prior to discharge to the tailings dam. Composition of the tailings is shown in Table 16 (Willett et al., 1994) to which a high $MgSO_4$ content must be added due to reaction between acids and the high Mg-chlorite host rocks. ERA, the operators, undertake constant monitoring of surface and groundwaters and have undertaken extensive environmental management and rehabilitation measures. These are subject to constant scrutiny by CEPA. Future development of the No 3 orebody will see the tailings deposited in the disused No. 1 open pit.

5.6.2 Olympic Dam, South Australia

The Olympic Dam Cu-U-Au mine 500 km N of Adelaide occurs in arid, sparsely-populated poor pastoral country which bears little resemblance to the country surrounding Ranger. Although the mining is underground, there has been considerable disturbance of the surface. Many of the potential contaminants derive from the

Table 15. Approximate annual loadings of solutes from RP2 water, Ranger.

		Estimates based on 1 800 mm of applied water	
		mg L^{-1}	g m^{2}
(a) Mean concentrations derived from	Na	35	66.6
	K	3.0	5.3
	Mg	88	158.4
	Ca	18	32.4
	Cl	4.8	8.6
	SO_4	437	786.6
		μg L^{-1}	mg m^{-2}
	Cu	4	7.2
	Pb	1.5	2.7
	Mn	2 068	3 722
	Zn	12.4	22.3
	U	1 190	2 142
		Bq L^{-1}	kBq m^{-2}
	^{226}RA	1.22	2.2
		mBq L^{-1}	Bq m^{-2}
(b) Alligator Rivers Region Research Institute data	^{238}U	(1–40).10^{3}	(2–70).10^{3}
	^{234}U	(1–40).10^{3}	(2–70).10^{3}
	^{230}Th	30–100	50–206
	^{226}Ra	(0.5–3).10^{3}	(1–5).10^{3}
	^{210}Po	150	270
	^{235}U	50–2 000	(0.1–4).10^{3}
	^{231}Pa	1–5	2–9
	^{237}Ac	15	30
	^{232}Th	2	4
	^{228}Ra	5–30	9–54
	^{228}Th	5–30	9–54

Source: Willett et al. (1994).

milling process used to extract the ore elements. Unlike normal tailings which are usually deposited with alkaline mill water, the tailings at Olympic Dam are pumped out at a pH of 2-3 as a result of leaching of U with sulfuric acid. Under such conditions, the potential exists for heavy elements to enter groundwaters, unless the tailings dam integrity is assured.

5.6.3 Rum Jungle, Northern Territory

The Rum Jungle uranium mine 80 km south of Darwin in the Northern Territory was operated from 1954 and continued until shutdown in 1971. Open cut mining of White's orebody (Figure 7) required diversion of the Finniss River with overburden being deposited in a dump on the south bank of the diversion channel (Harries and Ritchie, 1988). Dyson's, another uranium orebody, was mined in 1957–8. In 1963

Table 16. Total concentrations and fractionation of metals in uranium mine tailings, Ranger. Means of 5 analyses (s.d. where available).

	Cu	Fe	Mn	Pb	U	Zn
Exchangeable						
$mg\,kg^{-1}$	<0.4	<1	654(9.3)	13(0.8)	<0.01	0.5
% of total	<0.1	<0.003	21.3	1.1	<0.01	<11
Reducible						
$mg\,kg^{-1}$	47.8(6.1)	3 396(169)	1 084(30)	802(23.3)	18(0.8)	6.1(0.61)
% of total	15.3	9.4	35.3	70.1	10	11.1
Oxidizable						
$mg\,kg^{-1}$	114(15.0)	471(62)	13.6(0.96)	29(2.2)	23(1.4)	36.6(3.7)
% of total	36.5	1.3	0.4	2.5	13	66.6
Residual						
$mg\,kg^{-1}$	150	32 333	1 318	300	135	12.3
% of total	48.1	89.3	42.9	26.2	77	22.4
Total[A] $mg\,kg^{-1}$	312	36 200	3 070	1 140	176	55

[A]Determined by XRF.

the Intermediate copper orebody was mined with the low-grade sulfide and oxide ores being put in a pile where an attempt was made to extract the Cu by heap leaching. Until 1961, tailings were pumped to a disposal area, with supernatant water flowing into the Finniss River. After 1961, tailings were pumped into Dyson's and White's open-cuts. Initially, highly acidic (pH 1.5) waste liquor from the treatment plant was also discharged to the tailings area, but after 1961 it was pumped straight into a holding dam in the bed of the Finniss River and was released downstream in the wet season. From 1967, part of the waste liquor was pumped into White's open cut.

When the mine closed in 1971, there remained three waste rock dumps, three open cut pits, a tailings dam and a copper heap leach pad (Figure 7). All of these contributed significantly to pollution of the Finniss River with AMD and heavy metals (Cu, Mn, Zn), resulting in greatly reduced aquatic flora and fauna. An estimated A$16 million was spent in the early 1980's to rehabilitate the mine by putting all the tailings in one of the open cuts and revegetating the area; the heap of copper ore was removed and the area revegetated; and the two pyritic waste dumps were reshaped and capped with a three-layer cover and sown with native species. Although there has been a marked improvement in the quality of water in the Finniss River (Table 17) (Bennett and Lawton, 1995) the two pyritic waste rock dumps continue to generate acid and release heavy metals. The tailings are no longer a source of contamination, but low pH (~2.5), high Cu water from the two remaining open cuts continues to enter groundwaters.

The Rum Jungle Creek South uranium mine several km south of the main Rum Jungle mine was operated from 1961–63 resulting in an open pit and a slightly radioactive overburden heap of $2 \times 10^6\,m^3$ of chloritic schists and slates. In 1990 for

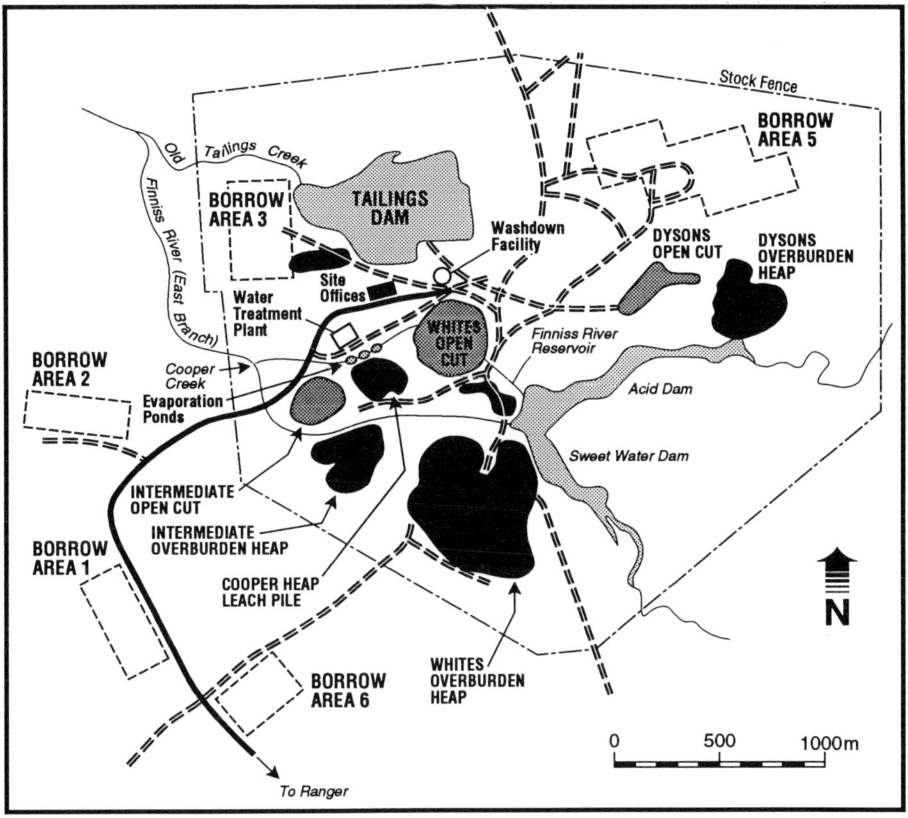

Figure 7. Rum Jungle mine site plan prior to rehabilitation.

a cost of A$1.9 million the area was rehabilitated by burying spilled radioactive material from around the site and replacement with imported non-radioactive material and top soil, reshaping and stabilising the waste rock dump, and revegetating all disturbed areas. A number of facilities were constructed and the site is now used as a public recreational area including swimming in the old open pit.

5.7 Heavy mineral sands

Australia is the world's foremost source of heavy mineral sands (ilmenite, rutile, zircon) from deposits which most commonly occur along the coast (Queensland, New South Wales, Western Australia). In eastern Australia, mining has been of frontal dunes, high dunes and old strand lines with a variety of vegetation and fauna. In WA much of the present mining is on the northern sandplains (Brooks, 1987). Disturbance of the land may result in erosion and siltation, dust and a soil profile vastly different from that pre-mining. As a result of the beneficiation process, the waste material may be radioactive due to the presence of xenotine and monazite.

Table 17. Historical load data (tonnes) in the Finniss River at a site 5.6km downstream from Rum Jungle together with a record of flow and rainfall.

Year	Flow ($\times 10^6 m^3$)	Rainfall (mm)	Copper total	Zinc total	Manganese total	Sulfate
1969/70	7	896	44	–	46	3300
1970/71	33	1611	77	24	110	12000
1971/72	31	1542	77	24	84	6600
1972/73	22	1545	67	22	77	5500
1973/74	69	2000	106	30	87	13000
1982/83	9.5	1121	23	5	6	1520
1983/84	48	1704	28	9	21	3600
1984/85	11.7	1136	9.1	4.1	7.2	1600
1985/86	11.4	1185	3.7	2.7	8.2	4400
1986/87	13.2	1222	5.6	2.7	8.6	2870
1987/88	6.3	1064	3.2	2	5.4	1230
1988/89	35	1600	5.4	4.4	19.2	3940
1989/90	3.1	900	1.8	1.6	3.9	760
1990/91	40.5	1590	14.9	7.4	30.5	4000
1991/92	7.1	1002	3.8	2.7	9.1	1260
1992/93	29.9	1421	11.9	3.9	24.7	2696

Source: Bennett and Lawton (1995).

5.8 Other potential contaminants

All open-cut and strip mining results in complete or partial loss of the regolith profile. At many mines the soil is removed and stockpiled for later use in covering and revegetating tailings and waste rock. At bauxite mines, the underlying lateritic material (bauxite and sometimes kaolinite) are removed, thus completely changing the character of the profile. At mines where a considerable depth of overburden is dumped the various strata are mixed and soils formed on them may be saline, sodic, alkaline or acid and totally different from the original. Revegetation of such soils may be difficult.

Soils conserved for later use may result in either loss or reduction in biological productivity (Bell, 1986). He notes that over long periods, there may be physical, chemical and microbiological degradation of stockpiled soil. Other problems may arise through compaction by earth moving machinery, changes to the water regime, increased erosion, reduction in fertility through horizon mixing, contamination with salts and/or toxic elements, reduction in soil microbes and fauna and reduction in native seed reserves.

Another impact upon the environment is transport of mineral resources to the processing plant, port facilities for export or to the user. Disruption to the environment is caused by the burying of gas and oil pipe lines, railways for the haulage of coal, iron ore and base metals to loading facilities for export or to power stations and smelters in Australia, and extensive conveyor belt systems, in use at some mines. In PNG some of the concentrates are moved by barge along the larger river systems. Not only do these and the use of road transport pose potential dust and contamination problems through spillage, but they are also disruptive of the native flora and fauna.

6. Contamination of soils from mineral processing operating

The separation of ore minerals from waste by flotation, gravity, leaching, screening etc. to leave fine-grained tailings is part of mineral processing and is included in Table 2, which lists the contaminants of processing of various ores and metals. However, as this separation usually takes place at the mine-site, contaminants related to tailings have been discussed above.

6.1 Aluminium refining, Western Australia

Alcoa of Australia refines bauxite by the Bayer process at Kwinana, Pinjarra and Wagerup on the coast south of Perth. The bauxite has a low alumina content and for each one tonne of alumina extracted there remains two tonnes of residue. At full capacity, the three refineries could generate 12 Mtonnes of alkaline residue each year (Ward, 1987). These residues consist of Fe, Al and Si minerals and are separated into two fractions; residue sand at $>150\ \mu m$ and red mud at $<150\ \mu m$, the latter causing some problems for reclamation and vegetation. The main contaminant from these residues is dust, although some caustic solutions may enter groundwaters over a long period as dewatering of the red muds is a very slow process.

6.2 Smelter slags

Slags derived from ferrous and non-ferrous smelters are recycled or used (for instance as road base material) whenever possible, but there is still a considerable amount of slag put to land fill. Again, there are marked variations in the nature of the slag and its properties in relation to rehabilitation.

6.2.1 Pt Kembla Steelworks, NSW

In the steel industry the disposal of blast furnace and steel furnace slags and reject shale from the washing of coking coal is of major concern. Blast furnace slag consists mainly of Ca and Al silicates with minor Mg and Fe depending on the composition of the feed and a pH of 8.3. The slag is cooled with seawater and therefore also has a high Na and Cl content. Slag from the steel works has a much higher Fe content and an initial pH of 11.6. These slags are disposed of in an environment with temperate climate, high rainfall and acidic soils. Research has been directed at investigating the potential of using the wastes as substitutes for soils in the rehabilitation of landfill emplacements (Hodges et al., 1987). In trials, the slag is being mixed with other waste materials to improve its potential to sustain plant growth.

6.2.2 Port Pirie, South Australia

Smelting and refining of Pb has been carried out at Port Pirie since 1889 with production of large amounts of slag consisting of Fe, Ca, Mn, Al silicates with relatively high levels of trace elements including nutrient elements required for plant growth. This material is dumped in an area with a semi-arid climate, low rainfall and high pH soils. There is little problem in revegetating the slag.

6.3 Gaseous and particulate emissions

The calcining of ores or concentrates and smelting of concentrates to produce metals or metal oxides produces gases and particulates which in modern processing operations are collected in scrubbers, electrostatic precipitators and baghouses to reduce contamination of the surrounding areas. There is substantial evidence that there has been considerable contamination in the past and that much of that contamination is still present.

6.3.1 Port Pirie, South Australia

Studies of Pb concentrations in soils around the BHAS Pb/Zn smelter at Port Pirie by Tiller et al. (1975) showed that a 270 km^2 area contained 5–10 times background EDTA-extractable Pb and that small areas close to the smelter had Pb levels exceeding 100 times background (Figure 8). The Pb concentration decreased exponentially with distance from the smelter but was detectable up to 65 km away. The pattern of contamination for Zn and Cd was similar to that of Pb and these elements were taken up by vegetables grown in the area. It is believed that much of this contamination occurred prior to 1925–30 when measures to upgrade emission control were improved.

6.3.2 Sulfide, NSW

A sulfide plant at the northern end of Lake Macquarie has had Pb smelting, concentrate roasting and fertiliser and cement manufacturing over the last century. Current operations include a sinter plant, Imperial Smelting furnace and zinc refinery, producing Pb, Zn, sulfuric acid, $CuSO_4$ and Cd. In recent years there has been considerable focus on increasing production, improving occupational health and safety and environmental performance. However there has been considerable community concern and adverse media coverage concerning elevated children's blood lead levels. As at Port Pirie, the soils at this site are contaminated with Pb and Cd.

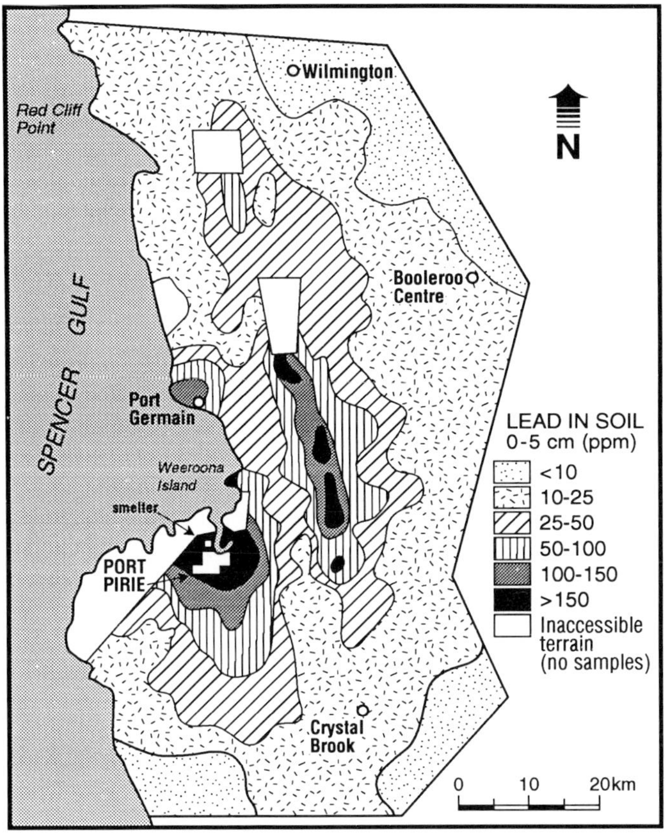

Figure 8. The regional distribution of EDTA-extractable Pb in the 0–5 cm layer of soils surrounding the Port Pirie smelter in South Australia. The concentration of Pb to the north and northeast of the plant is related to both low- and high-level wind direction and is associated with a mountain range rising steeply from the coastal plain bordering Spencer's Gulf (Tiller et al., 1975).

6.3.3 Kalgoorlie, Western Australia

One of the major contaminants from the roasting of sulfide ore is the production of sulfur dioxide. This is most evident at Queenstown, Tasmania where the combined effects of SO_2, timber filling and erosion have combined to form the existing 'moonscape'. The impact of SO_2 on the environment around Kalgoorlie is low due to low rainfall, sparse vegetation and air quality management. Sulfur dioxide is produced from the Kalgoorlie Nickel Smelter and the Gidgi Gold Roaster (which replaced a number of smaller roasters within the city limits in 1989). Air quality management systems consisting of eight SO_2 monitors, two wind monitoring stations and two weather stations allow both reactive and predictive control of air quality in Kalgoorlie-Boulder (Spencer et al., 1991). Unlike at Queenstown, the main impact is not in the form of acid rain, but adsorption by vegetation and inhalation by humans. Other elements which may be released by these roasting processes are As, Sb and Te.

6.3.5 Aluminium Smelters

Australia has aluminium smelters located at Gladstone, Queensland; Tomago and Kurri Kurri, NSW, Portland and Geelong, Victoria and Bell Bay, Tasmania. The process involves passing electricity through a molten mixture of cryolite salts and alumina within a carbon-lined reaction cell resulting in gaseous and particulate fluoride production which is treated in scrubbers to minimise fluoride contamination (Moller and Forth, 1991). However, there is little published data on the fallout of fluoride on soils surrounding the Australian smelters. Some research indicates that the major effect of these emissions is on plant growth through direct contact, with very little uptake through soils. There have been studies of the impact of F on animal and human health in both Australia and New Zealand.

6.4 Chemical leachates

Many mineral extraction processes involve leaching with corrosive or toxic chemicals such as sulfuric acid and CN. As a result both soils and water in the immediate vicinity of tailings dams at these operations can become contaminated not only with the leachate but also with heavy metals liberated by the process. The latter is similar to the effect of AMD at both abandoned and operating mines.

6.4.1 Australind, Western Australia

A sulfate pigment plant commenced operation in Australind, 150 km south of Perth, in 1964. A by product of the process is Fe-rich acidic effluent and the State agreed to be responsible for its disposal. It was pumped to the Leschenault Peninsula and discharged into the 40 m high quartz and calcareous sand dunes. It was initially discharged onto the ocean beach from whence it flowed directly into the ocean. In 1968, in response to public concern, the beach discharge was discontinued and the effluent was disposed instead onto the lime-rich sands. The effluent seeped into the sands where the acid was neutralised and Fe precipitated. The neutralised liquid leached into the underlying groundwater and thus ultimately reached the ocean. However solids carried in the effluent commonly sealed off the floor of lagoons preventing further infiltration of effluent. These solids were removed by dragline and dumped in large mounds by the lagoons. This process stopped in 1990, and rehabilitation of the peninsula for use by the public as a conservation and recreation park is underway.

6.4.2 Townsville, Queensland

A problem faced by many companies is the disposal of waste water, particularly in regions of high rainfall (e.g. Ranger Uranium Mine) or in populated or agricultural regions. At Townsville, waste water containing Cu, Ni and As from the copper refinery, was used to irrigate grass pasture for hay production and short-term fattening of cattle. These elements are concentrated in the top 10 cm of a yellow podzolic soil with little movement occurring below 30 cm. Adsorption was greater onto a clay soil within the refinery site, with As being adsorbed onto Fe oxides in both soils. Although there is presently little effect upon the grass, there is need to assess the long-term impact of sustained irrigation. Similar problems are faced wherever excess water is disposed of, although the nature of the contaminant may vary. In Western Australia where the mill waters are commonly very saline, their disposal may provide a sol-ution to dust suppression with the formation of crusts on the surface of tailings dams.

6.5 Jarosite

About 80% of the Zn used in the Western World is produced by the electrolytic smelting process with the remainder produced by the Imperial Smelting Process (e.g. Sulphide near Newcastle – see Section 4.3.2). Zinc concentrates contain a wide range of contaminants depending on their origin, with Fe being the major elemental impurity. Various methods are used to extract the Fe from the electrolyte solution with the process at Risdon in Tasmania since 1973 producing jarosite $KFe_3(SO_4)_2(OH)_6$ as a waste. The jarosite also contains 4.3% Zn, 1.1% water-soluble Zn, 0.026% Cd, 0.11% water-soluble Cd, 1.3% Pb, 0.16% Cu, 0.43% As and 0.34 mg Hg kg^{-1} (Sanderson and Wilson, 1994) and at present is disposed of at sea 100 km south of Hobart in 2000 m of water. Because of international treaties, ocean disposal must cease and an alternative found. Land disposal is considered to be inappropriate as jarosite will not form a self-supporting stockpile and under certain conditions the heavy elements may be released. The operators have undertaken considerable research to develop alternative processing techniques which will produce a more benign residue in the form of inert vitreous slag. In developing a new technique, account has to be taken of up to 25 associated elements which often create specific problems that require unique solutions.

Processing of other ores produces a wide range of contaminants some of which are relatively benign whilst others are toxic to flora and fauna. No simple rehabilitation/remediation techniques can be expected to solve all contamination problems although modern processing and environmental management techniques are reducing hazards to within internationally acceptable limits.

7. Assessment

The Australian and New Zealand Guidelines for the Assessment and Management of Contaminated Sites (ANZECC/NHMRC, 1992) provide a framework for the proper assessment and management of contaminated sites. Detailed information on ANZECC guidelines has been reviewed in Chapters 9 and 14. Although not specific to mines and mineral processing sites, Sections 4.3, 4.4, 5.1, 5.2 of these Guidelines are particularly relevant in the assessment of the degree of contamination of a particular site. Emphasis on some particular aspects is however warranted.

Table 18. Sampling media for assessment of contaminated mine and mineral processing sites.

Biological	Gases	Liquids	Solids	Vegetation
Aquatic algae	Soil gas	Surface water	Soil (*sensuo lato*)	Trees/shrubs
Bacteria	Atmosphere	Groundwater	Stream sediments	Grasses
	Stack emanations	Seepage waters	Lake sediments	
		Effluent including chemicals	Waste rock	
			Tailings	
			Ore	
			Other waste materials	

7.1 Sampling

Too much emphasis cannot be placed on the need for proper sampling procedures. Often the selection of an appropriate sample and the sampling technique is incorrect, leading to an incorrect assessment of the nature and extent of contamination.

Sampling media which may be used in assessment of contamination are listed in Table 18. Experience shows that a specialist in each of these general areas is most able to select the appropriate sample. For instance in sampling trees in the Australian environment, only a specialist can hope to distinguish between the various eucalyptus species, and to tell the difference between second and third year growth. Similarly, in sampling soils, it is often difficult to distinguish between residual and transported material and, in Australia, whether the transported material is aeolian or not. Because it is recognised that correct sampling is the key to any geochemical studies, there are many references to sampling techniques e.g. Description of Sampling of Contaminated Soils – A Field Guide (Boulding, 1994); Field Sampling Manual for Reactive Sulphide Tailings (Mend, 1989); Handbook of Exploration Geochemistry Volume 4 (Butt and Zeegers, 1992). These references also provide guidelines to the correct handling and preparation of samples prior to analysis. Again, incorrect procedures, resulting in cross-contamination, can lead to erroneous results.

7.2 Analysis

Standard analytical procedures for assessing contaminated sites are available through such organisations as US EPA and Standards Australia. Recent advances in analytical techniques in relation to geological samples (Hall, 1992) indicate that detection limits in the ppb to ppt range are possible using inductively coupled plasma mass spectrometry (ICPMS) and ppm-ppb range using instrumental neutron activation analysis (INAA). It must be remembered that no one analytical technique is capable of analysis of all elements at such low levels, and that a combination of analytical techniques is both desirable and necessary.

Having selected an analytical technique appropriate for the nature of the sample, it is equally important to ensure that the analytical data are correct. Errors are most commonly introduced by contamination, as well as instrumental errors. Both may be eliminated by thc usc of standards, duplicate samples and accredited standard check samples. If all normal sampling, sample preparation and analytical techniques are adopted the chemical data will accurately reflect the environment surrounding the mine or processing plant.

Techniques other than chemical analysis are commonly required to ascertain the nature of contamination. In many instances it may be necessary to determine the mineralogy using optical microscopy, X-ray diffraction (XRD), scanning electron microscopy (SEM), electron microprobe (EMP) and, in extreme cases, ion microprobe. Such techniques are used, for example, in the detection and characterisation of asbesti-form minerals, speciation of minerals containing heavy metals (as many such minerals are extremely stable and have extremely low bioavailability) or simply for the characterisation of secondary minerals and dusts.

The application of isotope analyses to environmental problems is well documented (e.g. Gulson et al., 1994) and is discussed in more detail in the next section. A variety of instruments are available to characterise and quantify a wide range of gases, airborne particulates, organic contaminants and radioactivity. However, the location of such monitoring instruments is as equally significant as appropriate sampling for chemical analyses.

7.3 Data interpretation

Correct interpretation of the data is of extreme importance for the appropriate assessment of the nature and extent of contamination. Common problems which arise in data interpretation are:

- is the anomaly anthropogenic or natural;
- what is the source of the anomaly; and
- does the anomaly pose a threat.

7.3.1 Anthropogenic or natural?

Most mines and mineral processing plants will have an associated geochemical anomaly in soil, water or atmosphere. Unless situated on a minesite, the anomalies associated with mineral processing plants are probably anthropogenic. However in the vicinity of the minesite, such anomalies may well be natural. Many mineral deposits are formed by the passage of mineral-laden solutions through rocks, until such time as an appropriate location for the minerals to be precipitated in sufficient concentration to be mined is found. Thus, the surrounding rocks may have levels of heavy and toxic elements in excess of those recommended by the regulatory authorities. As examples, the recently opened McArthur Pb-Zn-Ag mine in Northern Territory has anomalies of 5000 mg kg^{-1} Zn, 400 mg kg^{-1} Hg, and 270 mg kg^{-1} As extending 20 km. Many granites and some shales have uranium contents sufficient for the natural radiation to exceed presently defined safety limits. When such rocks weather, the contaminant that enters the surrounding soils can eventually pollute streams or underground water.

As shown in section 3.1, acid mine drainage (AMD) is the greatest single contaminant associated with the mining industry. It derives from the oxidation of sulfides, principally pyrite and pyrrhotite, in waste rock, tailings and from mines. However, there is abundant evidence from the many gossans which occur throughout Australia, that sulfide oxidation has been occurring for millions of years and that AMD and heavy metal dispersion are both natural and anthropogenic. Such naturally occurring sulfide oxidation and dispersion is utilised in mineral exploration. In PNG, where heavy rainfall and mountainous terrain result in high erosion rates and constant exposure of fresh sulfides, naturally-occurring AMD is common (and is used as an exploration tool).

Several countries have embarked upon extensive geochemical surveys to determine the 'background' or 'baseline' values for a wide range of elements. Such a database is of use to environmentalists, for mineral exploration, agriculture, forestry etc. In Queensland, the Australian Geological Survey Organisation has recently completed a geochemical survey of Cape Yorke Peninsula, Queensland, to add to an older survey of the Georgetown, Queensland area. In NSW, a geochemical survey of the Clarence-Moreton Basin in the northeast of the state has been completed and a compilation of exploration company data has been commissioned by NSW Department of Minerals and Energy. Until such time as the whole of Australia has been covered by a detailed survey, the data acquired by exploration companies are generally all that is available to provide a 'background' in any given area.

7.3.2 Source of the anomaly (contamination)

Sometimes contamination is incorrectly assigned to mining and processing activities. The most recent example is that of Broken Hill, NSW, where high blood Pb levels in children were attributed to the mining operation. Detailed Pb isotope studies by Gulson et al. (1994) have shown that although some of this Pb is derived from the mine, approximately half is derived from leaded petrol and leaded paint (used at the turn of the century to paint houses and now flaking due to age and high temperatures). Even the Pb from the orebody may not all be due to mining activities as a considerable amount of the orebody has been eroded into the surrounding soils. In the absence of 'base-line' data, it is not now possible to determine the Pb content of soils due to weathering and erosion.

8.3.3 Toxicology

Although generally outside the scope of this chapter (see Chapter 9), the potential toxicity of metal contamination should be considered when the data are interpreted. As noted in Chapter 1 the impact of metal contaminants on agroecosystems has been well studied but less is known of their bioavailability in the human food chain. Generally, contamination is measured as a total value with little or no effort at speciation (see Chapter 3) in aqueous soil phase. At some mines and mineral processing plants, the contaminant(s) may be in a form which has very low bioavailability, i.e. it will not be taken up by plants, animals or humans and therefore poses little concern. Considerable research is needed to ascertain the bioavailability of these various species, but if it will not dissolve in 0.1 M HCl (to replicate the digestive system) then there is diminished possibility of it being toxic. It is recommended that solution phase speciation of the contaminant be undertaken, at least in the preliminary phases of investigation.

8. Remediation and rehabilitation

8.1 Acid Mine Drainage

There is little doubt that the greatest source of contamination from mines and some mineral processing plants is AMD and the associated secondary dispersion of heavy metals and other toxic elements. Sulfides, and particularly pyrite, are associated with many types of metalliferous mines, and occur in coal measures, oil-shales and as H_2S in natural gas reservoirs. Upon exposure to water and oxygen, they are readily oxidised to form low pH, sulfate-rich solutions in which many of the heavy metals are readily soluble. Other complex reactions (Thornber and Taylor, 1992) may ensure that the pH remains low and that the heavy metals remain in solution for long distances from the source ensuring that the contamination is widespread. Additional sources of acid are the SO_2 derived from roasting and smelting of sulfidic ores which results in acid rain (which is killing many forests in the Northern Hemisphere and was partially responsible for the barren hillsides at Queenstown, Tasmania) and from sulfides entrapped in slags.

In Canada, problems associated with AMD have been addressed by the Mine Environment Neutral Drainage (MEND) Program. The current recommended method of preventing AMD is to dispose of sulfidic waste rocks and tailings sub-aqueously to prevent oxidation. In a country with high precipitation, low temperatures, and an abundance of lakes, this is possible, but as yet has not received environmentalists' acceptance. However, many tailings dams are kept constantly covered by water to prevent oxygen ingress.

In Australia, with low precipitation, high evaporation and a lack of freshwater lakes (except in Tasmania), this method of disposal cannot be considered. Many lakes in Australia are ephemeral and highly saline, characteristics which enhance AMD and secondary dispersion. It has therefore been necessary to utilise other means to remediate the AMD problem and to rehabilitate waste rock dumps and disused tailings dams. However, no successful means have been found to prevent AMD generation within abandoned mines such as at Sunny Corner and Captains Flat. Generally, this is a problem in higher rainfall areas only, and is best ameliorated by natural or constructed wetlands. The processes which occur in wetlands that reduce metal concentra-tions and acidity of solutions are outlined in Table 19 (Jones and Chapman, 1995).

Table 19. Processes in wetlands that reduce metal concentrations and acidity.

Process	Nature	Controlling Variables
Dilution	Physical	Volume of water
Dispersion	Physical	Flow velocity, channel geometry and roughness
Oxidation	Chemical Microbiological	Concentrations of oxygen and organic carbon
Precipitation/co-precipitation	Chemical	Concentration of components, pH redox potential
Adsorption on precipitates	Chemical	Amount of precipitate, concentrations of major cations and anions, pH
Adsorption/ion exchange on suspended and bed sediments	Physical Chemical	Concentration of suspended sediment, particle size and mineralogy, pH, concentrations of major cations and anions
Sulfate reduction	Microbiological	pH, organic carbon, concentrations of oxygen and sulfate
Uptake by biofilms, algae, and aquatic macrophytes	Chemical Biological	Density of plants, temperature, light intensity, availability of nutrients

8.1.1 Waste Rock Dumps

The most commonly used technique in Australia to prevent AMD is to encase the waste rock with benign material such as barren rock or soil. Because of costs it is preferable to use local material, so that it is common practice to stock-pile all top-soil at the beginning of mining operations. Oxide material, because of the absence of sulfides and the high clay content, is commonly used, although at Mount Leyshon in Queensland, 1 m thickness of fresh porphyritic rock from the open pit is being compacted over the waste rock dumps (Orr, 1995a;b). The main purpose of encapsulation is to prevent O_2 and water infiltration, and is designed to form a capillary break to minimise salt rise or root penetration, to allow for surface water runoff, and ensure long-term stability in relation to desiccation/shrinkage cracking, weathering and erosion (Watson, 1995). Often, the surface is revegetated with native grasses, shrubs and trees.

In some older mines where such remediation/rehabilitation measures have been undertaken, AMD generation has continued. This is probably due to the difficulty in designing adequate encapsulation once mining operations are complete or well under way. In such circumstances, the acidic solutions may be recovered using containment ponds, subsurface seepage drains and groundwater recovery bores and then neutralised.

Another possibility is the use of in-pit disposal provided that a void of sufficient volume is available. Such a technique is used in the open cut coal mines where the overburden is placed back into the void from which the coal has been mined. Such a technique does not prevent AMD generation *per se*, and some sealing of the void is necessary to prevent contaminants entering the groundwater. There is also the possibility of acidic solutions reacting with the wall rocks of the void. The huge volumes of waste rock generated by open-cut coal mining are impossible to encapsulate and other methods to remediate contamination due to AMD are required.

Rehabilitation of waste rock dumps at Woodlawn, NSW; Mount Leyshon, Queensland; Pine Creek, Northern Territory and Pajingo, Queensland have yet to stand the test of time, although constant monitoring suggests that the measures are continuing to be successful.

8.1.2 Tailings

Typical tailings storage in Australia comprises above ground containment in a constructed dam or behind an embankment in a valley. Coarser tailings may be returned underground as filling for open slopes. Because of the fine grain size sulfidic tailings are more reactive than waste rock in which sulfide minerals may be armoured against oxidation. Considerable research has been conducted on the rate of oxidation within waste rock dumps and tailings by Ritchie (1994).

Again, prevention of AMD by encapsulating the tailings is the most commonly used technique. This includes designing the embankment with an impervious and benign core, partial or full-storage floor lining (typically earthfill construction) and a low permeability cover over the final tailings surface. An alternative is to dispose of the tailings within any voids, such as open pits, a method being considered at Ranger Uranium Mine.

Design issues relate to the nature of the floor lining and the surface cover. The use of locally derived clay dominant soils is commonly used as a floor liner and when not available a synthetic lining such as high density polyethylene (HDPE) has been used. The cost and uncertain design life of this alternative is a significant factor. An additional consideration in designing a cover is the reduction of capillary rise which brings dissolved salts to the surface which may hinder vegetation growth. Similarly, root penetration may well allow water and oxygen to migrate down, which would recharge the AMD mechanism.

Other covers have been tried including organic blankets (sewage sludge), cementitious covers, hard pan formation or chemicals such as phosphoric acid to armour the sulfides. The various techniques for tailings disposal are dealt with by Chalkley et al. (1989) and Gadsby et al. (1990).

Undoubtedly the key to preventing AMD-generation is planning during the initial stages of mine development. This will include the siting and construction of waste rock dumps and tailings dams, stockpiling of sufficient quantities of benign material for encapsulation and water management. There is no doubt that problems associated with AMD are of greater significance in older and abandoned mine sites.

Consideration is presently being given to the technique of codisposal of tailings with waste rock. The technique is used in the coal mining industry in USA and has been adapted by at least one underground mine in the Bowen Basin. Advantages are less land disturbance, better return of processing water and reduced rate of sulfide oxidation. As yet this technique has not been applied to metalliferous mines.

8.1.3 Effluent treatment

In addition to the measures described above, which are seldom, if ever, 100% effective, acidic runoff can be collected and treated prior to entering drainage systems. As already mentioned above constructed or natural wetlands have been used to treat mine effluents. However, these effluents such as at Ranger, Northern Territory; Tom's Gully gold mine, Northern Territory; Woodcutters base metal mine, Northern Territory; Hilton base metal mine, Queensland; the Hellyer mine, Tas., and the Westralian Sands synthetic rutile plant, Western Australia, are near-neutral pH and low sulfate content. Systems to treat moderate strength, low pH AMD are still in the development stage.

A technique which is commonly used is to neutralise with lime effluent that has been collected in containment ponds. This has several disadvantages, being labour intensive, costly, and producing a sludge of $CaSO_4.2H_2O$ (gypsum) which contains precipitated contaminant elements and which has to be disposed of. This method is presently used at Brukunga, South Australia at a cost of A$500 000 per annum and at Mount Leyshon, Queensland.

8.2 Cyanide and other chemicals

A variety of toxic chemicals are used in mineral processing. Prominent amongst them is NaCN which is used for both Au exploration and processing. Somewhere in the order of 80 000 t of NaCN is used each year in Australia of which 30 000 t is 'lost', usually with the tailings. This may constitute an environmental risk as evidenced by the recent death of a large number of birds at the Northparkes Mine.

Normally CN levels within tailings are approximately 30 mg L^{-1} with some in the form of insoluble complexes. The soluble and volatile CN breaks down under ultraviolet light or by the reaction of cyanophilic bacteria, and is volatilised and lost to the atmosphere. In assessing the CN content of tailings water, it is therefore of importance to determine its various forms (speciation).

There are many complex organic compounds used in mineral processing particularly in the flotation process. Considerable emphasis is now being placed on bioremediation to break down these organic compounds and to complex or accumulate toxic elements (see for example Mining Environment Management, September, 1995).

8.3 Landform stability

Mining activities result in new landforms such as waste rock dumps, tailings dams, open pits (including strip mines) and truncated regolith profiles. Contamination from these constructed landforms may occur through massive failure (e.g. of a tailings dam wall), erosion, weathering resulting in dust, siltation of drainage channels, contamination by toxic elements etc.

Because of differences in rock type, mining and processing methods, topography, hydrology, and climate the stability of landforms is unique to each mine. In some States, the actual design is encapsulated within legislation, but it is more appropriate that the design of stable constructed landforms by geotechnical engineers should be in accordance with local landform characteristics. Any such design should be in collaboration with geomorphologists and ecologists to ensure that ecosystem reconstruction is in accord with stability design.

8.4 Development of sustainable ecosystems

Revegetation of disturbed lands is the simplest form of reconstruction undertaken by the mining industry. However, this does not necessarily coincide with ecosystem development or sustainability. As with the development of stable landforms, the construction of a sustainable ecosystem is dependent on a variety of factors many of which are common to both.

Perhaps the most significant is the proposed final use of the site. In the Bowen Basin, the present use of land surrounding the coal mines is agriculture and grazing. If the land disturbed by coal mining is to be rehabilitated for this use, the measures taken will be totally different than if it was returned to its original brigalow scrubland or an appropriate native forest equivalent. In some instances, rehabilitation may be minimal to attract tourists as at Queenstown, Tasmania, and Broken Hill, New South Wales. On the other hand, extensive remediation/rehabilitation was necessary to restore the Rum Jungle South mine for recreational use.

Two sectors of the mining industry have made significant advances in re-establishing sustainable ecosystems, viz. bauxite and beach sand mining. Because of their location in the jarrah forests of southwest Western Australia, Alcoa of Australia have invested considerable time and effort in establishing a stable, self-regenerating forest ecosystem which is in accordance with a detailed 'prescription' developed jointly with Department of Conservation and Land Management. This includes enhancing or maintaining water, timber, recreation, conservation and other nominated forest values. Similar efforts have been made to restore the totally different ecosystems at Weipa, Queensland and Gove, Northern Territory.

At Capel, Western Australia mining of beach sands occurred in a State pine forest plantation of low productivity with some private grazing land. Rather than returning the area to its previous land use or to a maintenance-free bushland, the opportunity was taken to substantially improve the nature conservation values of the area. With help from the Royal Australian Ornithologists' Union, the area was developed as a self-sustaining wetland for waterbird conservation.

Thus in reconstructing an appropriate landform and ecosystem many factors have to be considered including:

- final use of the site;
- characteristics such as climate, soil and water chemistry, hydrology, stability of landforms;
- native flora and fauna;

- cost-effectiveness of revegetation;
- sustainability in relation to fire, drought, flood, grazing etc.;
- low maintenance.

Reconstruction is therefore site specific and should be planned in collaboration with relevant authorities during initial stages of mining.

9. Indicative costs

The cost of rehabilitation of abandoned mines will vary according to the nature and extent of contamination. In USA it has been estimated the total cost of cleaning up the Summitville, Colorado gold mine to be US$100–120 million. Similarly, the Iron Mountain, California Cu mine which has mine drainage pH of −2 will cost somewhere in the order of US$150 million for the most effective and politically negotiable combination of treatments. Both of these mines are in high rainfall, mountainous areas and adjacent to major water supplies. In Wyoming US$253 million has been spent on abandoned mine projects since 1977. This includes about $3 million per hectare for subsidence; US$22,000 per hectare for reclaiming abandoned surface coal mines; an average of $3500 to close each underground coal mine; and US$32,000 per hectare to reclaim surface uranium mines (Richmond, 1995).

Costs in Australia are comparable. Estimated cost for complete rehabilitation of the Brukunga Pyrite Mine is A$50 million with present annual cost of treating AMD with lime, and covering and revegetating the tailings being A$500,000. A detailed costing for the rehabilitation work at Rum Jungle uranium mine (Bennett and Lawton, 1995) is given in Table 20. It is interesting to note that the cost of rehabilitating the waste rock dumps, including reshaping before covering, was A$68500 per hectare. Despite this expenditure, rehabilitation has not been completely effective, although the quality of water in the Finniss River is much improved.

At Pajingo, Queensland. capping of the waste rock dump with three 0.5 m thick compacted layers and a loose rock mulch layer cost approx. A$50,000 per hectare (Tredinnick and Cornwell, 1995). Revegetation, consisting of minor earthworks, seeding with natives and pasture cost an additional $1500 per hectare. At Mount Leyshon, Queensland the cost of building a runoff control scheme was $1 million with annual extensions costing around A$40,000 (Orr, 1995b). To construct the porphyry seal (1 m thickness) on the waste rock dumps costs A$4100 per hectare and this does not include the cost of selectively handling the porphyry or shaping of the dump, both of which are integral parts of the mining operations. The total cost of dump rehabilitation for flat surfaces is about A$12,000 per hectare.

Despite the considerable sum spent on rehabilitation (A$2.3 million in 1976), pollution problems still exist at Captains Flat. Dames and Moore have estimated that the only feasible way of overcoming these problems is by treatment of the point sources with chemical and/or biological processes. The cost of a chemical treatment plant is estimated at A$641,000 with annual operating costs of A$81,000. Combined chemical and biological treatment would cost an estimated A$679,500 with annual

Table 20. Cost of components of rehabilitation work at Rum Jungle.

Component	Expenditure 1986 A$
Copper heap leach pile/tailings dam/Dyson's open cut	3685925
Waste rock dumps	2826997
Open cut water treatment	6234221
Other areas	904720
Management	4666040
Monitoring	322346
Total	18640249

operating costs of A$30,000. Broad scale rehabilitation of the diffuse sources is estimated to cost over $1.5 million.

Finally, the costs of constructing a wetland can vary enormously and depend particularly on the topography of existing landforms, whether natural or synthetic liners are required, and the composition of the effluent. Typical price ranges quoted in the United States are US$10–100 m^{-2}.

10. Conclusions

Metalliferous and coal mining and mineral processing are potential contaminators of the environment. The greatest sources of contamination are old and abandoned mines developed when the necessity for and techniques of environmental management were poorly understood. In recent years, community expectations for the environment have risen, and the mineral industry has responded accordingly. However, this has not prevented recent widespread contamination from some mines (e.g. in PNG and Irian Jaya) where the climate, topography and seismic activity require entirely different environmental management techniques from those which, for example, are used successfully at the majority of Australian mines.

The major source of contamination is acid mine drainage (AMD) caused by oxidation of sulfides in waste rock, low grade ore and tailings. In the more humid coastal zones much of this contamination finds its way into surface and groundwaters and is known to have detrimental impact on aquatic ecosystems, and, during periods of flood, on the surrounding soils. However, many of Australia's mines occur in semi-arid to arid environments and soil contamination results from aeolian processes. Similarly, contamination from processing plants is caused by airborne particulate matter.

The extent of soil contamination from mines and mineral processing is not well documented in Australia. One reason is the lack of baseline data which would provide a measure of the contribution by natural processes (ore genesis and weathering). There is, however, substantial data on contaminant levels in surface and groundwaters. Recent geochemical mapping by government instrumentalities will provide the necessary information for some restricted areas.

Successful remediation and rehabilitation techniques are very much dependent on the climate, topography, soil characteristics and proposed end use. Present techniques can be extremely costly and are seldom totally successful. Little is known of their long-term viability. Industry-funded research is continuing to develop more cost-effective techniques of, for example, dewatering tailings, prevention of AMD, stabilizing constructed landforms and developing sustainable ecosystems. In the Australasian–Pacific region where many countries are dependent on the minerals industry, better environmental management is the key to meeting the community expectations of a high standard of living.

References

Agnew M 1994 Weathering products and geochemistry of waste residues at the Brukunga pyrite mine, Adelaide Hills, South Australia, in relation to environmental impacts. Unpublished Honours thesis, Univ. of Adelaide.

ANZECC/NHMRC 1992 Australian and New Zealand guidelines for the assessment of contaminated sites. 57pp.

Australian Mining Industry Council 1994 Environmental research in the Australian mining industry. Priorities and needs. 50pp.

Bell L C 1986. Mining. In Australian Soils – The Human Impact. Eds. J S Russell and R F Isbell. pp 444-466. Univ. Queensland Press, St. Lucia.

Bennett J W and Lawton M D 1995 Assessment of the Rum Jungle strategy for acid mine drainage control. In 2nd Australian Acid Mine Drainage Workshop. Eds. N J Grundon and L C Bell. pp 179-190.

Boulding R 1994 Description and sampling of contaminated soils: a field guide. 2nd edition. Boca Raton, Lewis.

Brooks D 1987 Rehabilitation in Australia's mineral sands industry. Mining Rehabilitation '87' AMIC Canberra. pp 24-34.

Brooks K A 1980 Mine waste pollution of the Molonglo River: An investigation into the effect of remedial measures carried out in 1976 at Captains Flat. Chemical Laboratories N.S.W. Dept of Mineral Resources.

Butt C R M and Zeegers H 1992 Sample media used in geochemical exploration in tropically weathered environments: definitions and use. In Regolith Exploration Geochemistry in Tropical and Subtropical Terrains. Eds. C R M Butt and H Zeegers. Elsevier, Holland pp 513-523.

Chalkley M E, Conard B R, Lakshmanan V Z and Wheeland I C G (eds) 1989 Tailings and effluent management. Pergamon, New York. 336pp.

Chapman B M, Jones D R and Jung R F 1983 Processes controlling metal ion attenuation in acid mine drainage streams. Geochim. Cosmochim. Acta 47, 1957-1973.

Dames and Moore 1993 Final report, Captains Flat mine site. Assessment of options for further remediation for NSW Environment Protection Authority.

de Klerk N H, Armstrong B K, Musk A W and Hobbs M S T 1989 Cancer mortality in relation to measures of occupational exposure to crocidolite at Wittenoom Gorge in Western Australia. British J. Indus. Medicine. 46, 529-536.

Eagle M and Higgins R 1990 The establishment of environmental management criteria for the Ok Tedi project. AMIC Environmental Workshop, Wollongong. AMIC, Canberra. pp 112-121.

Fawcett M N R 1995 Management of AMD at the Pine Creek Mine. In 2nd Australian Acid Mine Drainage Workshop, Charters, Towers. Eds. N J Grundon and L C Bell. pp 83-90.

Frankcombe M 1993 Rehabilitation strategy for the derelict Horn Island gold mine. AMIC 18th Annual Environmental Workshop, Burnie. AMIC, Canberra. pp 268-284.

Gadsby J W, Malick J A and Day S J (eds) 1990 Acid Mine Drainage – Design for Closure. Bitech Vancouver BC. 518pp.

Gulson B L, Mizon K J, Law A J, Korsch M J, Davis J J and Howarth D 1994 Source and pathways of

lead in humans from the Broken Hill mining community – an alternative use of exploration methods. Econ. Geol. 89, 889-908.
Hall G E M 1992 Geoanalysis. J. Geochem Explor. 44, 1-349.
Harries J R and Ritchie A I M 1988 Rehabilitation measures at the Rum Jungle mine site. In Environmental management of solid wastes. Eds. W Salomons and U Förstner. pp 131-151. Springer-Verlag, Berlin.
Hodges D, Thompson S and Tucker J 1987 Rehabilitation of smelter slags. Mining Rehabilitation '87'. AMIC Canberra. pp 103-119.
Howard R, Wilcox D and Cahill M 1991 Environmental management in the W.A. goldfields: examples from the Kalgoorlie area. AMIC Environmental Workshop, Perth. AMIC, Canberra 1, 188-202.
John C 1988 Residue and cyanide management at the Boddington gold mine. AMIC Environmental Workshop, Darwin. AMIC, Canberra 2, 214-221.
Jones D R and Chapman B M 1995 Wetlands to treat AMD – facts and fallacies. In 2nd Australian Acid Mine Drainage Workshop, Charters Towers. Eds. N J Grundon and L C Bell. pp 127-145.
Mcnd 1989 Field sampling manual for reactive sulfide tailings. 153pp.
Miller S D, Murray G S, Jeffrey J J, Songberg T A and Wong J W C 1991 The nature and fate of cyanide in gold processing residue and implications for future residue disposal at the Boddington gold mine. AMIC Environmental Workshop, Perth. AMIC Canberra 2, 147-164.
McIlveen G and Lewis G 1990 The preparation of a rehabilitation and environmental management plan for the operations of Pasminco Mining at Broken Hill NSW. AMIC Environmental Workshop, Wollongong. AMIC, Canberra 2, 19-54.
Mining Environment Management. Sept 1995. Mining Journal Ltd., London. pp 3-27.
Mitchell A and Mackasey W O 1995 The identification, location and assessment of abandoned mine hazards – the first step towards rehabilitation. Sudbury '95 Conference on Mining and the Environment. Sudbury Ontario. pp 289-298.
Moller I and Forth J 1991 Environmental management at Boyne smelters. AMIC Environmental Workshop, Perth AMIC Canberra 1, 161-171.
Niven R 1990 The hydrogeology and geochemistry of seepage beneath the south tailings dam, Woodlawn mine, NSW. AMIC Environmental Workshop, Wollongong. AMIC, Canberra 1, 172-206.
Orr M S 1995a Control of acid mine drainage through water management at Mount Leyshon gold mine. In Grunden N J and Bell L C (eds) 2nd Australian Acid Mine Drainage Workshop, Charters Towers. pp 67-73.
Orr M S 1995b Development of a waste dump management strategy at Mount Leyshon gold mine. In 2nd Australian Acid Mine Drainage Workshop, Charters Towers. Eds. N J Grundon and L C Bell. pp 91-100.
Richmond T C 1995 Abandoned mine land reclamation costs in Wyoming. Sudbury '95 Conference on Mining and the Environment. Sudbury, Ontario. pp 1221-1230.
Ritchie A I M 1994 Sulfide oxidation mechanisms: controls and rates of oxygen transport. In The environmental geochemistry of sulfide in mine waste. MAC Shortcourse Handbook. Eds. D W Blowes and J L Jambor. pp 201-244. Nepean, Ontario.
Ritchie P 1991 Decommissioning and rehabilitation planning, Kidston gold mine, north Queensland. AMIC Environmental Workshop, Perth. AMIC, Canberra 2, 121-146.
Ross C 1991 Environmental planning and development of the Porgera gold mine, Papua New Guinea. AMIC Environmental Workshop, Perth. AMIC, Canberra. 2, 98-108.
Sanderson I and Wilson C W 1994 Solving the jarosite issue. AMIC Environmental Workshop, Karratha. AMIC, Canberra. pp 311-325.
Slessar G C 1988 Management of strategic environmental issues – Alcoa's bauxite mining operation in Western Australia. AMIC Environmental Workshop, Darwin. AMIC, Canberra 1, 85-100.
Smith P and Hancock S 1992 Brukunga – the acid test. Waste Disposal and Water Management in Australia. pp 3-12.
Spencer R E, Howard R K and Riddle J C C 1991 The strategy for air quality management in the Kalgoorlie-Boulder region. AMIC Environmental Workshop, Perth. AMIC, Canberra 2, 67-78.
Thornber M J and Taylor G F 1992 The mechanisms of sulphide oxidation and gossan formation. In Regolith Exploration Geochemistry in Tropical and Subtropical Terrains. Eds. C R M Butt and H Zeegers. pp 119-138. Elsevier, Holland.
Tiller K G, Merry R H, Cartwright B and Bartlett N R 1975 Dispersal of lead emissions from an isolated lead smelter within an agricultural region of South Australia. Search, 6, 437-439.
Tredinnick I A and Cornwell P J 1995 A manager's operational perspective of AMD management at the

Pajingo gold mine. In 2nd Australian Acid Mine Drainage Workshop, Charters Towers. Eds. N J Grundon and L C Bell. pp 43-51.

Ward S C 1987 Reclaiming bauxite residual disposal areas in South-West Australia. Mining Rehabilitation '87. AMIC, Canberra. pp 61-70.

Watson A 1995 Practical engineering options to minimize AMD potential. In 2nd Australian Acid Mine Drainage Workshop, Charters Towers. Eds. N J Grundon and L C Bell. pp 53-65.

Willett I R, Bond WJ, Akber R A, Lynch D J and Campbell G D 1993 The fate of water and solutes following irrigation with retention pond water at Ranger Uranium Mine. Supervising Scientist for the Alligator Rivers Region Research Report 10. 122pp.

Willett I R, Noller B N and Beech T A 1994 Mobility of radium and heavy metals from uranium mine tailings in acid sulfate soils. Aust. J. Soil Res. 32, 335–355.

Wood I 1991 The Mount Lyell Mining and Railway Company Limited - an environmental case study. AMIC Environmental Workshop, Perth. AMIC, Canberra 1, 203-220.

8
Legal responses to the problem of soil contamination – the Australian experience

ROBERT J. FOWLER

1. Introduction

'A look at the current state of practice in contaminated site management cannot help but arouse some grave doubts as to whether we are any more able to deal with the complexity of evaluating environmental impacts of chemical contaminants theory than we were thirty years ago'.

(Robinson and Shoebridge, 1993)

This observation, which was made by the Director of the Victorian Environment Protection Authority, suggests that there is a need for better responses to the problem of land contamination in Australia. Although the problem now is widely regarded as a matter of serious environmental concern, effective legal responses are still being developed. The Federal constitutional system in Australia, under which environmental management functions are performed primarily by State and Territory governments (Chapter 14), has produced a variety of approaches to the management of contaminated land. The role of the Commonwealth government has been limited to a contribution to the production of National Guidelines for the Assessment and Management of Contaminated Sites and a Position Paper on Financial Liability for Contaminated Site Remediation. Only three States (New South Wales, Victoria and Queensland) have developed legislation which is directed specifically at the subject of contaminated land. For the other States and Territories, the problem is addressed to varying degrees through their general environment protection legislation.

Thus, Australia occupies a rather uncertain middle ground between those jurisdictions which have not yet developed any legal or policy arrangements for dealing with land contamination, and those which have in place extensive arrangements. In the latter category, the United States provides the best example, in the form of the scheme known as Superfund (the legislation for which is formally entitled the Comprehensive Environmental Response, Compensation and Liability Club 1980). Other examples may be found in Europe in countries such as the Netherlands, Denmark and Germany. The last-mentioned country provides an interesting parallel with Australia insofar as it is also a federation in which the national government has not yet developed a comprehensive, uniform approach (Rehbinder, 1993).

Despite the lack of appropriate legal responses in all jurisdictions in Australia, there has been considerable attention devoted to the land contamination problem, much of which is described in Rowe and Seidler (1993). Overseas experience has been evaluated by Australian environmental and health authorities and has been taken into consideration in proposing a nationally consistent approach. In particular,

R. Naidu et al. *(eds): Contaminants and the Soil Environment in the Australasia-Pacific Region, 267–280.*

there has been substantial wariness about the United States Superfund model, due to the high levels of litigation which have been associated with its implementation. Also, the use of clean-up standards such as those developed in the Netherlands has been the subject of much debate. As a result, the general approach favoured in Australia, and reflected in the National Guidelines for the Assessment and Management of Contaminated Sites, is distinctive and worthy of consideration, particularly by other countries where similar measures are being contemplated.

This chapter will examine why there is a need for a legislative approach to land contamination which is over and above common law principles and the dictates of general environmental legislation. It will then consider the crucial question of where and how legal responsibility should be imposed for funding the remediation of contaminated land, looking in particular at the extent to which the regulatory schemes in various countries have imposed liability on private parties or, on the other hand, relied upon public funds for clean-up operations. The Australia Position Paper on Financial Liability for Contaminated Site Remediation also will be considered in this context. Finally, the paper outlines the most important elements of a regulatory process for the remediation of contaminated land and explores some of the key issues which must be addressed in developing such a process.

2. The need for a legislative approach

Land contamination presents challenges which do not appear to be capable of being addressed by traditional common law actions brought in the courts or through the operation of general environmental legislation. In each context, there are significant limitations which can only be overcome by specific legislation directed to the problem of land contamination.

2.1 Common law liability

The emphasis of common law tortious remedies is on the protection of private rights, in particular where damage is caused to property or an injury is experienced by persons. This means that common law remedies are unlikely to lead to orders for environmental remediation, the focus rather being placed upon the payment of monetary compensation. Nevertheless, it is interesting to note some recent developments in common law principles concerning liability for environmental damage, including in the specific context of land contamination.

The most significant question to arise in relation to common law liability for environmental damage is whether to impose liability on the basis of fault or, on the other hand, to impose a system of strict liability. A fault-based system will result in liability being imposed on parties only where they can be shown to have been negligent in their actions, in particular by allowing or causing harm which they should have foreseen. Strict liability, which has been considered a part of the common law for over one hundred years as a result of the rule in the *Rylands v Fletcher*

(1868), renders parties liable for loss they have caused even where they have taken all due care to avoid it.

In cases where land contamination gives rise to damage to persons or their property, it now seems clear that courts in Australia and England will approach the question of liability on the basis of fault rather than apply a principle of strict liability. This was made clear by the British House of Lords in *Cambridge Water Company v Easter Counties Leather* (1994), where it was held that liability for historic pollution should not be imposed upon a party who could not have foreseen the consequences of their activities at the time of undertaking them. It was concluded that the development of a strict liability rule for environmental damage should be left to the legislature:

> ...I incline to the opinion that, as a general rule, it is more appropriate for strict liability in respect of operations of high risk to be imposed by Parliament, than by the courts. If such liability is imposed by statute, the relevant activities can be identified, and those concerned can know where they stand. Furthermore, statute can where appropriate lay down precise criteria establishing the incidence and scope of such liability.

The High Court recently decided, in *Burnie Port Authority v General Jones Pty Ltd* (1994) that the rule in *Rylands v Fletcher* should be considered to have been absorbed by the principles of ordinary negligence for the purposes of the common law of Australia. However, the High Court also qualified its support for the principles of ordinary negligence by suggesting that where a person uses land to carry on a dangerous activity or to introduce a dangerous substance, the duty of care in such circumstances may involve a degree of diligence so stringent as to amount practically to a guarantee of safety. This statement clearly has some potential relevance to cases involving land contamination. Although it has been suggested that this qualification 'imposes a standard of care so high that it comes very close to the strict liability it purports to replace' (McDonald, 1994), there is still the difficulty that an activity or substance may not be regarded as 'dangerous' if the particular damage to which it gives rise was not foreseeable at the time. This kind of legal issue has arisen repeatedly under the rule in *Rylands v Fletcher*. The operation of a fault-based approach to liability for land contamination in Australia is reflected in the observation that the outcome of the *Cambridge Water Company* case would probably have been the same were it dealt with under the principles laid down by the High Court in the *Burnie Port Authority* case.

To the extent that common law liability for land contamination is dependent upon fault, and in particular upon foreseeability of the harm caused, a recent decision of Burchett, J in the Federal Court of Australia in *Alec Finlayson Pty Ltd v Armidale City Council* (1994) is of considerable significance. The defendant Council was held liable in negligence for having approved a residential subdivision in 1984 without taking steps to discover or impose conditions in relation to land contamination from a known prior industrial use. The Council was required to pay substantial damages to a builder-developer who had innocently purchased allotments in the subdivision only to discover that they were chemically contaminated. Burchett, J placed considerable reliance on the evidence of an expert witness who had enjoyed a lengthy career as a town planner and senior assessor in the Land and Environment Court in New South Wales. This witness testified that questions of land contamination were well known to and had been considered by council planning officers in that State from the 1960s onwards.

If the operators of facilities which have given rise to land contamination are attributed with a similar knowledge and understanding of the likely consequences of allowing contamination to occur, unlike the defendant in the *Cambridge Water Company* case, then liability in negligence could be imposed on them. However, as noted previously, the focus of such claims will be the award of damages for loss incurred as a result of land contamination rather than remediation of the land in question. This limitation means that it is necessary to turn to environmental legislation to deal with the fundamental tasks of identification, assessment and remediation of contaminated land. As the House of Lords observed in the *Cambridge Water Company* case, such legislation also may establish criteria for the imposition of strict liability on the relevant parties.

2.2 General environmental legislation

Only three jurisdictions in Australia have adopted legislation which deals specifically with land contamination. In New South Wales, there is the *Environmentally Hazardous Chemicals Act 1985* and in Queensland, the *Contaminated Land Act 1991*. In Victoria, amendments to the *Environment Protection Act 1970* were adopted in the late 1980's which enable clean-ups to be directed by the Environment Protection Authority where land or water has been 'polluted', and for the issue of clearance certificates by environmental auditors where land has been remediated (Robinson and Shoebridge, 1993). Other jurisdictions rely entirely on powers contained in their general environment protection legislation to deal with pollution in order to address problems of land contamination at the present time.

Increasingly, it is being recognised that these general measures are not adequate to address the complex scientific, administrative and legal issues which are raised by the problem of land contamination. In South Australia, for example, the former South Australian Waste Management Commission (SAWMC) and the Department of Environment and Planning (SADEP) released a detailed discussion paper in September 1991 authorising a proposed legislative approach to the contaminated land problem (SAWMC/SADEP, 1991). However, the new *Environment Protection Act 1993*(SA) did not attempt to implement these proposals or to address the issue of land contamination, because of ongoing uncertainties about the question of liability for remediation. Instead, the South Australia Environment Protection Authority (SAEPA), which was established by the Act, is preparing proposals to amend the legislation during 1996 so as to incorporate specific measures on land contamination. The Western Australian government is also expected to release a discussion paper by the end of 1995 outlining more specific measures for the imposition of liability for contaminated land in that State.

Through the observation of various land contamination schemes overseas and existing measures in Australia, it is possible to discern two key areas of regulatory activity which specific legislation in this area needs to address:

- first, the identification and assessment of contaminated sites; and
- second, procedures for the remediation of such sites, including appropriate management and oversight arrangements.

In addition to these two matters, there is the fundamental question of funding site remediation, which involves quite difficult issues concerning the nature and extent of liability on the part of the private sector, and the related subject of how far the public purse should contribute to the costs of clean-ups. The following sections will address the subject of funding first, and then return to explore briefly the key elements of the regulatory process for remediation of contaminated land.

3. Funding the remediation of contaminated land

A recent survey of clean-up programmes in Europe and the United States by Church and Nakamura (1994) noted that the initial assumption in each country surveyed was that clean-up costs were capable of being met primarily from public funds, with some supplementation from private funds where possible. Underlying this approach in each instance was a belief that land contamination was confined to a fairly limited number of sites. Experience has dictated that these assumptions were significantly wrong:

> Yet, in each country came the subsequent identification of many more sites than originally expected and recognition that remediation costs at those sites can be very high indeed. These factors put the cumulative costs of hazardous waste clean-up well beyond the willingness of legislators and the public to pay with general tax revenues. Thus, all these governments turned to devices to augment the role of private parties in the clean-up process. (Church and Nakamura, 1994)

This is a very significant observation in connection with Australia also, where recognition of the magnitude of the land contamination problem and development of legal response mechanisms is still at a relatively early stage. Based on overseas experience, it is reasonable to assume that the extent of land contamination will prove to be greater than anticipated when liability schemes are put in place. As will be seen below, the Position Paper on Financial Liability for Contaminated Site Remediation favours placing the major financial responsibility upon private parties and, correspondingly, places less emphasis upon the use of public funds for environmental remediation. In this respect, Australia may ‘leapfrog’ the experience of overseas jurisdictions by moving more immediately to a liability regime focused on private parties. Before considering the Position Paper more fully, it is helpful to understand how the issue of financial liability for site remediation has been approached in other jurisdictions overseas and, in particular, to compare the United States Superfund scheme with the approaches of several European countries.

3.1 The United States – Superfund

The *Comprehensive Environmental Response, Compensation and Liability Act* 1980 (CERCLA), or Superfund, empowers the United States Environmental Protection Agency (USEPA) to adopt a highly coercive approach to the clean-up of contaminated sites. Liability under CERCLA is strict, or in other words, does not depend upon establishing that land contamination was caused by negligence or that the resulting harm was foreseeable. A crucial element of the liability scheme under CERCLA is that it operates retrospectively, so that parties can be made responsible for activities which may have been undertaken twenty or thirty years ago.

Another significant aspect of Superfund liability is that it is 'joint and several', which means that any single responsible party can be held liable for the full cost of a clean-up, irrespective of how much, or how little, it has contributed to the overall contamination. The USEPA has focused its recovery efforts in most instances on those parties which have substantial resources to pay for clean-ups – what has been termed the 'deep pocket' strategy. The pursuit of this strategy has been facilitated by the wide definition of potentially responsible parties (PRPs) provided by CERCLA, which includes landowners and generators, transporters and disposers of wastes. The net cost of liability has even extended to lenders as a result of the controversial *Fleet Factors* decision in 1990. In that case, the court held that lenders may be liable under Superfund if they participate in the financial management of a facility to a degree indicating a capacity to influence the treatment of hazardous waste at the facility.

The coercive approach adopted under Superfund has had several, significant consequences. First, as noted above, it has placed much greater emphasis on private funding of environmental remediation and resulted in very few sites being funded exclusively from public funds (the so-called 'orphan' sites). Second, it has prompted PRPs to seek frequently to delay the imposition of liability and to reduce the level of their liability by suing as many other parties as possible. A related development has been a significant increase in litigation between PRPs and their insurers. Delays of up to a decade have arisen in Superfund clean-ups whilst PRPs engage in litigation and negotiations concerning liability and the extent of clean-up required. The result is costly legal fees, or as they are often termed, 'transaction costs'.

Whilst no more than a dozen Superfund cases have actually gone to a full trial, many hundreds more have bogged down in expensive and time-consuming preliminary legal proceedings. This adversarial system has produced transaction costs which range from 12% to 40% of total costs of a clean-up, with the average being about 19% for large industrial firms (Church and Nakamura, 1994). The sums involved in this respect are very considerable, given that the average cost of a Superfund clean-up is A$33 million.

Another significant feature of the Superfund scheme has been the statutory clean-up standard, which requires remediation to background levels of pollution for virgin land. There is no legal requirement to address current or future land uses when determining the level of clean-up to be required, although in practice the negotiated arrangements which are reached by the USEPA with PRPs do take such matters into account to some extent. Relatively few Superfund clean-ups have been completed (approximately 220 by 1994), with some 1100 others in various stages of completion. Just as significant is the fact that CERCLA applies only to those sites designated as having national priority, and that many more sites will require remediation under State laws independently of the Superfund scheme. It is estimated that there are as many as 70 000 other sites where clean-ups could be ordered (Quint, 1994).

The proposed reauthorisation of CERCLA by Congress in 1994 provided an opportunity for significant reform of the existing system. Among the matters under consideration in the re-authorisation debate were proposals to require the USEPA to revise its approach to the selection of remedies so as to reduce the coercive focus and to replace joint and several liability with a 'fair share' allocation scheme (Johnston, 1994). The latter proposal included a requirement for insurers to pay taxes totalling $8.1 billion over ten years into a special fund, and at a lower rate thereafter. It was estimated that this tax would contribute about 40% of the clean-up costs for Superfund sites and that polluters would pay the rest. The tax was not seen as a major burden for the insurance industry insofar as it constituted less than one percent of business insurance premiums.

However, the political machinations which occurred in the lead up to the Congressional elections in late 1994 included the CERCLA re-authorisation as one of their victims. In the wake of the Republican ascendancy in the Congress and the emergence of the Newt Gringrich 'Contract with America', the future of the Superfund scheme is clouded with doubt. Nevertheless, the lessons of its past history are of considerable relevance to Australia and indeed appear to have influenced the development of a national position with respect to financial liability.

3.2 European experience

In May 1993, the European Community issued a draft Directive on Remedying Environmental Damage which advocates a stringent system of strict liability for future environmental damage. However, liability for historic pollution has not yet been addressed comprehensively by the European Union (as it is now known, following the Treaty of Maastricht) and remains essentially a matter for individual member countries to tackle.

In their survey of European approaches to environmental remediation, Church and Nakamura (1994) found that countries such as the Netherlands, Denmark and Germany have placed substantial reliance on public funds, principally from general tax revenue, to address land contamination problems. There has been considerably less confrontation between governments and private industry in these countries over clean-ups, but a number of shortcomings have been overlooked by those who have praised this aspect of the European approach to environmental remediation.

First and foremost, even though governments have realised that the public purse is not sufficient to address the problem of land contamination, they have been relatively unsuccessful in pursuing private contributions. Governments have been unable to get judicial support for the recovery of clean-up costs from private parties to the same extent as in the United States. Courts in the Netherlands and Denmark, for example, have declined to impose liability on private parties to reimburse governments for the costs of clean-ups in numerous circumstances.

Second, where industry has contributed to clean-ups, this has been largely on a voluntary basis in situations where the costs of clean-up have been relatively low and a market incentive has existed in terms of the future use of the land concerned. Voluntary clean-ups have been facilitated also by the application of lesser standards for clean-up than in the United States. Even in the Netherlands, where some of the strictest standards are considered to have been developed, these standards have altered over time, in particular to allow for the consideration of present and future land-uses. Thus, whilst the existence of health or environmental concerns is the common trigger for clean-up action, containment has often been the final step in Europe, whereas it is simply the first step in the United States.

The overall effect of these various factors has been that many seriously polluted sites in Europe remain unaddressed, due either to a lack of sufficient market incentives to encourage a voluntary, private clean-up or the absence of sufficient public funds. This contrasts sharply with the United States, where the emphasis in designating national priority sites has been placed on the remediation of the most seriously polluted sites.

Finally, it is interesting to note that in the Federal constitutional system in Germany, the absence of national laws or standards concerning environmental remediation has left up to 70000 sites largely untreated. Only in the more affluent Western states of Germany has some remediation action been pursued, under a variety of approaches which generally have led, as in other European countries, to a combination of a limited number of public clean-ups and a larger number of voluntary private clean-ups of low-cost sites.

A comparison of the United States and European experience with environmental remediation presents two sharply contrasting systems. The benefits of the European approach, in terms of fewer transaction costs and potentially faster clean-ups, must be balanced against the lack of action at major polluted sites as a result of the substantial reliance upon public funds to deal with such sites. The United States approach, which places considerable emphasis upon the coercion of private industry to meet the costs of clean-ups, inevitably produces substantial acrimony and litigation, with attendant transaction costs. Clearly, Australian industry would prefer to see the European approach adopted in this country. However, State governments are particularly reluctant to draw on limited public funds to undertake environmental remediation and it can be expected therefore that the focus will be placed primarily upon recovery from private parties.

3.3 Australia: the position paper on Financial Liability for contaminated site remediation

Awareness within Australia of the need to put in place clear and consistent principles concerning liability for contaminated land is very high. The Australian and New

Zealand Environment and Conservation Council (ANZECC), a body comprised of the Commonwealth, State and Territory environment Ministers, released a Position Paper in April 1994 which proposes a consistent approach across jurisdictions with respect to financial liability for contaminated site remediation (ANZECC, 1994). The Position Paper sets out agreed national principles for the imposition of financial liability which are to be adopted by the various jurisdictions through their own administrative and legislative frameworks. It is reasonable to assume therefore that liability for environmental remediation will expand quite rapidly in Australia and New Zealand over the rest of this decade.

The liability scheme proposed by ANZECC is much closer to the Superfund model than the European approach. Where parties are directed to take remedial action in relation to risk sites, the Position Paper suggests they should be strictly liable. Liability is also intended to be retrospective. Most significantly, the principal responsibility for the costs of remediation is intended to be private rather than public. The basic approach is to ensure that polluters bear the cost of remediation where they are solvent and identifiable. Where the polluter cannot be made to pay, the Position Paper recommends that the owner or current occupier should be made liable. The Position Paper also recommends that Federal, State and local government should be subject to the same liability rules as private parties.

The use of public funds for the clean-up of sites contaminated by private parties is addressed only peripherally in the Position Paper by way of a vague recommendation that the means of funding the remediation of orphan sites needs to be determined by the governments concerned. The idea of a national fund along the lines of Superfund has been advanced in Australia (Osmers, 1993) but is explicitly rejected by ANZECC on the dubious ground that it would lead to inequities in the distribution of funds between jurisdictions. What really lies behind this conclusion is a firm resistance on the part of most State and Territory governments to Commonwealth involvement in the remediation of contaminated sites. This reflects a broad resistance by the States to an expanded presence of the Commonwealth in the field of environmental management more generally.

It will be interesting to see whether, if the Position Paper recommendations are implemented in each State and Territory, there will follow a similar degree of acrimony and litigation (with attendant transaction costs) to that which has been experienced in the United States. The Position Paper supports the apportionment of liability between multiple polluters on the basis of the contribution each has made and suggests that this task should be undertaken by an appropriate fact finding tribunal. However, the American experience of PRPs seeking to join as many other parties as possible and thereby to deliberately delay the clean-up process has clearly prompted ANZECC to suggest an approach which avoids such problems. Paragraph 6.12 of the Position Paper provides that:

> The resolution of apportionment, however, must not be allowed to delay the primary objective of abatement of the environmental or human health risk associated with the site. Consequently, any joining of, or contribution from, other liable parties should be subject to the necessary remedial action first being undertaken.

Interestingly, this is exactly the approach recommended for Australia by an American critic of Superfund prior to the release of the Discussion Paper (Latin, 1993). Thus, it would seem to be envisaged that liability could be imposed on a joint and several basis at first instance, thereby facilitating the same ‘deep pocket’ approach as has been pursued by the USEPA. If apportionment proceedings are precluded until remediation has been completed, the initial transaction costs may be considerably less than in the United States, and clean-ups may be conducted more speedily. Of course, it remains to be seen whether, and how quickly, these various proposals are adopted in State and Territory legislation.

4. The regulatory process for the remediation of contaminated land

Whilst there are many administrative procedures which can be incorporated within a regulatory scheme for the management of contaminated land, these can be grouped conveniently within the two areas identified earlier in this paper, namely identification and assessment of contaminated sites and remediation procedures.

4.1 Identification and assessment

As discussed in Chapters 9 and 14, the identification of contaminated sites is a two-stage process, with the first stage involving some form of preliminary investigation to determine whether particular sites may warrant further attention and the second stage constituting a risk assessment to determine whether a site is sufficiently contaminated to warrant remediation.

At the first stage, a number of mechanisms may be employed in order to trigger a preliminary investigation (Fowler, 1993). These can include: the use of the planning system wherever re-zoning or development of land used previously for industrial purposes is proposed; or a system developed in New Jersey in the United States, whereby a prohibition is imposed on commercial transactions involving industrial land (for example, sale, lease, or even simply the cessation of operations) pending an audit of the land; or the imposition of disclosure requirements on owners and occupiers of land who know or ought to know that their land is contaminated. In addition, environmental authorities may endeavour to seek out contaminated sites, for example by working in close cooperation with local government authorities who often hold extensive historical records concerning past land-use.

The National Guidelines for the Assessment and Management of Contaminated Sites, published by ANZECC and the National Health and Medical Research Council (NHMRC) in January 1992 (ANZECC/NHMRC, 1992) propose investigation level guidelines based on both health and environmental soil capability parameters. These guidelines are to be used to determine whether a problem may exist, or, as the document itself states, ‘as starting points in the investigation process’.

Where preliminary investigations indicate a potential land contamination problem, the Guidelines envisage that the next step will be to commission a risk assessment. The purposes of risk assessment are described as including the following:

- to establish whether site remediation is necessary on health grounds alone; and
- to determine a tolerable level of contaminants that can remain on site with adequate probition of public health (assuming no unacceptable environmental impacts are involved)

The Guidelines suggest that 'a site-specific approach to risk assessment will often prove to be the best approach to ensuring the most effective management strategy is adopted, ie, one based on its particular characteristics, history and proposed land use'. Reliance on numerical assessments is regarded as 'not feasible because of limitations in toxicological and exposure data'. Chapter 9 presents a detailed review of the ANZECC/NHMRC - national guidelines for the assessment and management of contaminated sites.

A related administrative process in relation to identification and assessment is to establish a Register on which suspected and known sites can be recorded, as can sites which have been remediated and cleared for future uses. Such a Register has been established in Queensland pursuant to the *Contaminated Land Act 1991* (Granger and Imry, 1993), and has led to the listing of several thousand sites in that State. Public access is guaranteed to information on all sites recorded in the Register except those which are classed as 'possible' contaminated land.

4.2 Remediation procedures

The decision to request remediation, and the actual form of remediation to be required, will involve an evaluation of the technical or scientific advice contained in a health risk assessment alongside economic, political and social considerations (Langley, 1993). This approach deliberately avoids that developed some years ago in the Netherlands whereby remediation is required once levels of a substance within soil exceed certain numerical values.

Where, as is intended in Australia, the remediation of contaminated land is funded primarily from the private rather than the public purse, environmental authorities will require appropriate powers to direct remediation activity under the relevant legislation. As noted above, the National Guidelines do not favour an 'absolute' approach to the development of remediation criteria, that is, the use of fixed concentrations of given contaminants for all sites below which no adverse health impacts are expected. Rather, criteria are to be developed on a site specific basis, drawing on the health risk assessment procedure, so as to take into account the many variables which exist, including:

- the toxicology of the contaminants involved;
- background concentrations;
- the amount and type of contaminants;
- contaminant mobility and migration pathways;
- synergistic effects of multiple contaminants;
- sensitivity of the surrounding environments;
- phytotoxicity;

- land use;
- available treatment technologies; and
- economic considerations (SAWMC, 1991)

The preference for a site-specific approach to remediation in Australia poses difficulties in drafting the relevant legislation. It is widely considered that one of the most significant deficiencies of the Superfund legislation is that the USEPA frequently compromises clean-up standards in the process of negotiating settlements with PRPs (Office of Technology Assessment, 1989; Latin, 1993). One means of avoiding this would be to preclude responsible parties from conducting site assessments or preparing remediation plans and to request instead that the relevant environmental authority perform these tasks. By this means, the 'negotiation' element would be removed. However, the costs of such an approach are considerable and a compromise might be to assign the responsibility for the performance of such tasks to PRPs and to specify by legislation the minimum requirements for site assessments and the preparation of management plans. However, the site-specific approach favoured in Australia means that it will be very difficult to produce such requirements in a clear and accountable manner by legislation.

Another corollary of the site specific approach favoured in Australia is that public consultation and involvement in the remediation process is highly desirable. The South Australian Discussion Paper released in 1991 acknowledged that 'public involvement in a remediation programme is essential if public confidence is to be engendered'. It proposes that the public have access to the site assessment and be entitled to comment on the proposed remediation plan. It also recommends that public communication should continue during the remediation operations.

Management and oversight of remediation operations is an essential element of any regulatory scheme. The role of the administering environmental authority has been the subject of some debate in this context also. Authorities are reluctant to become too heavily involved in the oversight function or, in particular, to issue clearance certificates upon the completion of remediation operations because of a concern that they may attract legal liability subsequently. Hence, in jurisdictions such as Victoria, a system of private environmental auditors has been instituted, whereby the Environment Protection Authority can approve Environmental Auditors who are able to issue certificates which attest that a site is free from contamination.

5. Conclusions

Australian jurisdictions have taken a cautious approach to the development of legal responses to the problem of land contamination. Observation of the difficulties experienced with regulatory systems overseas has led to proposals for a pragmatic approach to the imposition of financial liability for remediation. Whilst many of the elements of the United States Superfund scheme are supported (for example, strict and retrospective liability), a clear aim has been to avoid the high incidence of litigation which has attended that scheme. Thus, apportionment of liability amongst private parties is intended to be the subject of legal proceedings only after remediation has been completed. Private parties rather than the public purse seem likely to be the major source of funding for site remediation in Australia.

Pragmatism also is the hallmark of the National Guidelines for the Assessment and Management of Contaminated Sites. Rather than relying upon absolute or fixed standards, the determination of whether a site requires remediation and the development of remediation measures are intended to be pursued on a site-specific basis. It is envisaged that this will allow the consideration of a wide range of matters, including future land-use proposals and economic factors. As a corollary, it should be accepted that a high level of public involvement in the regulatory process for remediation is desirable.

The role of environmental and health authorities in the administration of this process is still the subject of some debate, particularly in connection with the conduct of site assessments, the preparation of remediation plans, and the issue of clearance certificates for remediated sites. Authorities are reluctant to assume full responsibility for these tasks, preferring instead to see them performed by independent experts.

Finally, it must be emphasised that only a few jurisdictions in Australia have developed specific legal responses to the problem of land contamination at the present time. A great deal of work remains to be done in this regard in most States and Territories, particularly in relation to the development of principles concerning financial liability. As in other areas of environmental management such as environmental impact assessment, pollution control and nature conservation, there is a very distinct possibility that significantly different regulatory schemes will emerge from one jurisdiction to another, despite the adoption of National Guidelines. This raises the issue whether a national approach could be better developed through Commonwealth legislation than by seeking the cooperation of all of the States and Territories.

References

Australian and New Zealand Environment and Conservation Council/National Health and Medical Research Council 1992 Guidelines for the Assessment and Management of Contaminated Sites. Canberra, Australia. 57p.

Australian and New Zealand Environment and Conservation Council 1994 Position Paper on Financial Liability for Contaminated Site Remediation. Canberra, Australia. 35p.

Church T W and Nakamura R T 1994 Beyond Superfund: Hazardous Waste Clean-up in Europe and the United States. Georgetown International Environmental Law Review: 7: 15-57. Allen and Unwin, St Leonards, New South Wales, Australia.

Fowler R J 1993 Identification of Sites : The South Australian Proposals. In Contaminated Sites in Australia - Challenges for Law and Public Policy. Eds. GC Rowe and S Seidler, pp111-115. Allen and Unwin, St Leonards, New South Wales, Australia.

Granger K and Imray F P 1993 Contaminated Land Registration in Queensland. In Contaminated Sites in Australia - Challenges for Law and Public Policy. Eds. GC Rowe and S Seidler, pp136-139. Allen and Unwin, St Leonards, New South Wales, Australia.

Johnston C 1994 Superfund Reauthorization: Who Decides Who's Liable Under CERCLA? EPA Slips a Bombshell into the CERCLA Reauthorization Process. Environmental Law: 24: 1045-1054.

Langley A J 1993 Improving Health Risk Assessment and Management of Contaminated Sites. In Contaminated Sites in Australia - Challenges for Law and Public Policy. Eds. GC Rowe and S Seidler, pp19-24. Allen and Unwin, St Leonards, New South Wales, Australia.

Latin H 1993 Why Agencies Seldom Do What Regulatory Laws Require: Some Lessons from the Superfund Fiasco. In Contaminated Sites in Australia - Challenges for Law and Public Policy. Eds. GC Rowe and S Seidler, pp 66-76. Allen and Unwin, St Leonards, New South Wales, Australia.

McDonald J 1994 Financial Liability for Contaminated Site Cleanups - The Increasingly Strict Standard of Fault-based Liability. Environ. and Plan. Law J: 11, 516-522.

Office of Technology Assessment, United States States Congress, 1989 Coming Clean: Superfund's Problems Can be Solved. US Government Printing Office, Washington DC, USA.

Osmers C 1993 Cleaning of contaminated land; Who pays? In Contaminated Sites in Australia - Challenges for Law and Public Policy. Eds. GC Rowe and S Seidler, pp 181-180. Allen and Unwin, St Leonards, New South Wales, Australia.

Quint M 1994 For Insurers, a 'Black Hole' Looms. New York Times, 22 September 1994: C1, C17.

Rehbinder E 1993 Contaminated Sites in Germany. In Contaminated Sites in Australia - Challenges for Law and Public Policy. Eds. GC Rowe and S Seidler, pp 43-55. Allen and Unwin, St Leonards, New South Wales, Australia.

Robinson B and Shoebridge L 1993 Contaminated Site Management in Victoria : A Strategic Overview. In Contaminated Sites in Australia - Challenges for Law and Public Policy. Eds. GC Rowe and S Seidler, pp 37-42. Allen and Unwin, St Leonards, New South Wales, Australia.

Rowe G C and Seidler S Eds 1993 Contaminated Sites in Australia - Challenges for Law and Public Policy. Allen and Unwin, St Leonards, New South Wales, Australia. 228p.

South Australian Waste Management Commission and Department of Environment and Planning, 1991 Contaminated Land - A South Australian Legislative Approach. Adelaide, Australia. 77p.

Office of Technology Assessment, United States States Congress, 1989 Coming Clean : Superfund's Problems Can be Solved. US Government Printing Office, Washington DC, USA.

9
Health risk assessment and management of contaminated sites in Australia

A. LANGLEY

1. Introduction

Chapters 1 to 8 emphasize the adverse effects of contaminants on the soil and water environment. In this chapter the health risk assessment and management of contaminated sites are discussed with particular emphasis on Australia. There has been increasing concern over the assessment and management of contaminated sites in the Australasia-Pacific region over the last 10 years. This arises from community concerns about potential health risks, increasing urban redevelopment where old industrial sites may be reused for residential developments, and an increasing need by administrative authorities to ensure effective and cost conscious remediation of considerable numbers of sites affected usually by anthropogenic activities.

The scale of the problem is difficult to assess as the definition of 'contaminated' has often been subjective or poorly defined and only limited efforts have been made to develop an inventory of contaminated sites.

Sites range from very localised point sources of contamination (e.g. a single leaking underground storage tank) to large industrial or defence sites to regional examples of contamination (e.g. areas surrounding smelters). The nature of the assessment and management of sites is dependent upon the proposed land use, community and governmental expectations, and the resources available for the task. The massive costs incurred under the USEPA assessment and management processes have been of concern. Currently many of the costs in the US have been incurred in assessment and litigation and the number of sites remediated has been limited although increasing. Within the Australasia-Pacific region there has been a strong attempt to define appropriate models for assessment and management that incorporate adequate safety margins and which avoid excessively conservative assessment and management methodologies.

The Australasia-Pacific region has the opportunity to select the best aspects of the various assessment and management methodologies that are available. To date a variety of methodologies have been influential in Australia. The USEPA methodology has been influential because of its systematic and well-documented methods, the links of many local consultants to parent companies in the USA, and its use of a quantitative risk assessment endpoint. However, the model has been criticised because of its conservatism so that risk may be overestimated by orders of magnitude, leading to remediation for extremely low actual risk levels and where there is minimal contamination. The initial Dutch Soil Criteria (A, B and C levels) were also widely used in Australia. An extensive list of soil criteria was available and these

R. Naidu et al. *(eds): Contaminants and the Soil Environment in the Australasia-Pacific Region, 281–307.*

criteria were generally regarded as conservative and therefore perceived to be of low risk by both consultants and the community. The Dutch B values were provided as default values for the Environmental Soil Quality Guidelines in the Australian and New Zealand Guidelines for the Assessment and Management of Contaminated Sites. Within the Netherlands, the A, B and C criteria, which had limited rationales, have since been superseded by a more rigorously modelled and documented approach.

After considerable discussion between health agencies and environmental agencies a set of Australian and New Zealand National Guidelines for the Assessment and Management of Contaminated Sites (ANZECC/NHMRC) was published in 1992. A relatively brief document, it established the basic principles for the assessment and management of sites. It provided a range of environmentally-based investigation levels and a limited number of health-based investigation levels. It detailed a process of health risk assessment.

A key principle of the Guidelines is that 'a site-specific approach to risk assessment will often prove to be the best approach to ensuring the most appropriate management strategy is adopted' (ANZECC/NHMRC, 1992). The site-specific approach is based on the 'particular characteristics, history and proposed land use of a site' (ANZECC/NHMRC, 1992).

The Guidelines also provide both environmental and health soil quality guidelines to assist in the site assessment process (see chapter 14). The environmental soil quality guidelines cover background levels and investigation levels. It is important to note that the background levels are essentially background pristine levels rather than 'background' urban levels, and the investigation levels are not the levels at which clean-up is required but rather 'assist in judging whether a detailed investigation of site is necessary' (ANZECC/NHMRC, 1992).

Levels at which remediation is required were not provided in the Guidelines. Site-specific appraisal was generally proposed for determining the need for, and nature of, further investigation and remediation. This appears to be the basis of legislation or administration in each state. A formal framework of risk assessment provides a sound basis for such site-specific appraisal and the Guidelines provide assistance in undertaking such site-specific risk assessments.

A quantitative estimate of risk was considered not feasible 'because of limitations of toxicological and exposure data' and a semi-quantitative estimate was preferred if there were sufficient data available. The process calls for uncertainties to be detailed for each stage of the risk assessment process and for these uncertainties to be considered in the overall management of a site.

2. The process of risk assessment

Definitions of risk assessment vary slightly but a well-accepted one is that of the United States National Academy of Science (1983), which recommended:

'*risk assessment* ... mean(s) the characterisation of the potential adverse health effects of human exposures to environmental hazards. Risk assessments include several elements: description of the potential adverse health effects based on an evaluation of results of epidemiological, clinical, toxicological, and environmental research [*hazard identification*]; extrapolation from those results to predict the type and estimate the extent of health effects in humans under given conditions of exposure [*dose response assessment*]; judgments on the number and characteristics of persons exposed at various intensities and durations [*exposure assessment*]; and summary judgments on the existence and overall magnitude of the public-health problem [*risk characterisation*]. Risk assessment also includes characterisation of the uncertainties inherent in the process of inferring risk.'

The process of health risk assessment detailed in the Guideline is intended to fulfil the following objectives:

- 'to establish baseline risks and whether site remediation or other action is necessary on health grounds alone;
- to determine a tolerable level of contaminants that can remain *in situ* with adequate protection of public health (assuming no unacceptable environmental impacts are involved);
- to enable comparison of potential health impacts of various remediation techniques;
- to provide a consistent method for appraising and recording public health risks...' (ANZECC/NHMRC, 1992)

3. Prerequisites for risk assessment

Prior to undertaking any form of risk assessment, there are several key issues to be considered:

- What is the purpose of the risk assessment?;
- Who (or what) are you trying to protect?;
- What is the expected end point of the risk assessment? Is the concern for cancer and/or noncancer outcome?;
- Is it to be a quantitative or qualitative assessment of risk? Is it the actual, calculated or perceived risk (McKone & Bogen, 1991)? There are likely to be considerable differences between the three types and an appreciation of each is necessary. If it is for quantitative risk assessment there is a need to determine what quantitative level of risk is acceptable or tolerable (unless several risks are being compared in order to rank them rather than to consider the absolute level of risk). If a qualitative risk assessment process is used, what criteria will be used to categorise or interpret the results?;
- How is the risk assessment information going to be used? for regulatory purposes? for standard setting? for education? to provide advice?;
- Is there concern for current exposures, or past and future exposures?;
- Is there concern for the risks to individuals, groups of people or for the population as a whole? If for individuals, are those typical individuals or those with 'worst case' exposures?;
- Are there mixtures with the possible interactions between substances?;
- What information is needed for sensible decision making?

(Langley, 1993)

4. Use of models

Exposure assessments are comprised of permutations of modelling, sampling, and modelling/sampling combinations (McKone, 1994). Even the use of complex models still provides a static picture of a dynamic world albeit a more elaborate representation of reality (McKone, 1994) and such a picture must be placed within a sound theoretical framework (Langley, 1995).

The use of more complex models 'does not necessarily increase precision' (McKone, 1994). The costs of collecting and analysing data, and constructing new models 'must be balanced by the value of the information obtained' (McKone, 1994). There is a need to appraise the value of information along with its uncertainties in 'defining the capabilities and limits of exposure models' (McKone, 1994).

If a model is to be used, both the model and the reason for modelling must be justified. The ANZECC/NHMRC (1992) proposes a semiquantitative model for health risk assessment and use a form of semiquantitative risk assessment for establishing health-based investigation levels.

A model must be validated, i.e. it must provide accurate, reliable, reproducible, sensitive and specific measurements of risk. The reliability will relate to the ability of a model to consistently produce similar results for similar scenarios; the reproducibility to the ability for different operators to produce similar results; the sensitivity for low levels of risk to be measured; and the specificity for the ability to relate risks to a particular agent (Langley, 1995).

Both the construction of a model and the data used with it should be judged. Any model is only as good as the data upon which the model is based (USEPA, 1993) and the data used with it (Langley, 1993). For practical purposes, a model needs to be able to:

- determine if the risks are acceptable or unacceptable or to accurately rank or categorise the level of risk;
- select the most cost effective method to prevent or reduce unacceptable risk;
- evaluate the success of risk mitigation efforts (Sexton et al., 1992).

To provide useful and accurate information about the level of risk the model must ideally be able to:

- 'identify the type of data required for the risk assessment;
- incorporate only appropriate and accurate data;
- determine exposures from all possible routes of exposure;
- identify individuals and population groups which are particularly susceptible in the particular exposure situation;
- be sufficiently flexible to be able to incorporate new exposure data, new toxicological findings or new innovations in risk assessment;
- quantify the contribution to risk from each exposure pathway and route (Meyninger and Molof, 1993);
- provide a record of how the risk assessment is performed so that the decision making process can be understood and audited at any future time.'

(Langley, 1993)

The qualitative considerations in appraising a risk assessment model include:

- how extensive the supporting database is;
- whether human epidemiological data as well as experimental data are included;
- whether laboratory data are for single or multiple species;
- if multiple species were used in toxicological testing, whether all species responded in a similar way;
- how the model extrapolates laboratory animal data to humans;
- the nature and extent of data gaps in the model;
- the scientific uncertainties relating to toxicology and exposure assessment;
- how the uncertainties in the model were addressed;
- what are the assumptions underlying the model;
- what is the level of confidence in the risk assessment (USEPA, 1993 cited in Langley, 1993).

5. The Australian model of risk assessment for contaminated sites

Health risk assessment is defined as 'the process of estimating the potential impact of a chemical or physical agent on a specified human population system under a specific set of conditions' (Paustenbach, 1989).

The risk assessment process detailed in the ANZECC/NHMRC uses as a skeleton the USEPA risk assessment process. This comprises four steps:

- data collection and evaluation;
- exposure assessment;
- toxicity assessment;
- risk characterisation.

These steps are represented in Figure 1.

This approach is not markedly different from the US National Academy of Sciences (NAS) approach, the four steps of which are:

- hazard identification;
- dose-response assessment;
- exposure assessment; and
- risk characterisation.

The risk assessment guidelines are intended for chemical contamination but could also be used for physical (e.g. radiation) and microbiological hazards.

5.1 Data collection

Data collection entails the acquisition and analysis of information about chemicals on a site that may affect human health and which will be the focus for the particular risk assessment (USEPA, 1989).

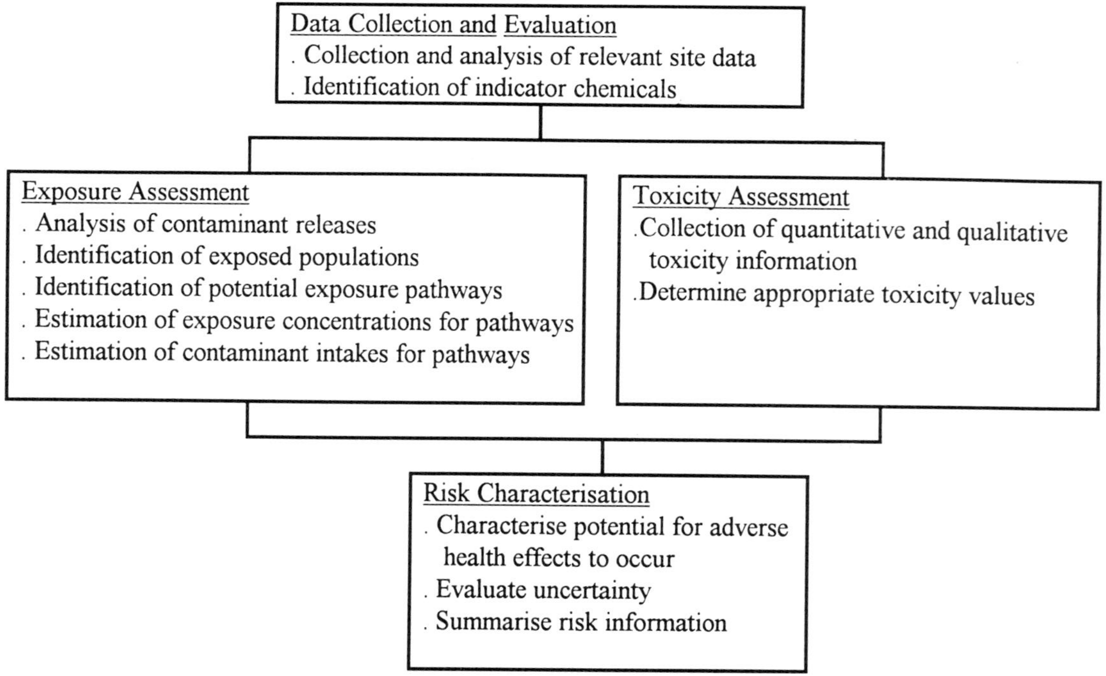

adapted from USEPA, 1989.

Figure 1. Risk assessment model.

5.2 Toxicity assesment

Toxicity assessment involves (USEPA, 1989):

- the nature of adverse effects related to the exposure;
- the dose-response relationship for various effects;
- the weight of evidence for effects such as carcinogenicity.

5.3 Exposure assessment

Exposure assessment is the determination of the frequency, extent and duration of past, current and future exposures. It also identifies exposed populations (including particularly susceptible sub-populations) and potential exposure pathways. Environmental monitoring and predictive models can be used to determine the levels of exposure at particular points on the exposure pathways. The contaminant intakes from the various pathways under a range of scenarios, including worst case situations, can then be estimated (USEPA, 1989).

5.4 Risk characterisation

Given the above information, risk characterisation details the potential type and incidence of effects for the exposure conditions described in the exposure assessment. An integral part of this stage is to evaluate the uncertainties and assumptions (for the individual steps and cumulatively) (USEPA, 1989). Consideration must always be given to the differences between calculated, actual and perceived levels of risk in the process of risk characterisation (McKone and Bogen, 1991).

6. Issues of data collection and evaluation

The most efficient way of data collection is to use judgemental sampling based on a thorough understanding of the chemicals on a site and their likely distribution. For judgemental sampling to be effective there is a need for a comprehensive and validated site history.

6.1 Site histories

A sufficient site history will include adequate details about the following (Edwards et al., 1994):

- site definition and description;
- present owners and occupiers and current use of the site;
- previous occupiers;
- previous industry/uses;
- previous/present building structures;
- products manufactured;

- raw materials used;
- intermediate products;
- wastes produced;
- disposal locations;
- discharges to land and water;
- product spills, losses, incidence and accidents (including fire);
- sewer and underground service plans;
- chemical storage and transfer areas;
- history of adjacent land uses to enable consideration of the effects of off-site activities causing contamination.

To complement judgemental sampling based on site history (or if there is an inadequate site history), grid sampling is required although this is recognised to be less efficient. Set protocols for grid sampling have not been established in Australia but generally a square grid is used. Triangular and herringbone (Ferguson, 1992) grids are more efficient but may be more difficult to survey than square grids. Randomised patterns of sampling have rarely been used, probably for the same reasons related to difficulties in surveying. The New South Wales Environmental Protection Agency (NSW EPA) has recently released a detailed protocol for sampling locations for a specific use, petroleum filling stations (NSW EPA, 1994a). The NSW EPA has also released a document (NSW EPA,1994b) detailing methods of statistically evaluating sampling patterns. It is much more difficult to interpret judgemental sampling because of the biased nature of the sampling. This biased sampling creates problems in statistically evaluating the size of a patch that may be missed by sampling and for the summary statistics to be used in the evaluation of the data. Data is very rarely normally distributed and a geometric mean is likely to be a better measure of central tendency (but not necessarily of exposure). There is no single perfect summary statistic and it is been considered appropriate to call for a range of summary statistics including:

- arithmetic mean (and standard deviation);
- geometric mean (and geometric standard deviation);
- range of values;
- frequency distribution of values.

The data may need to be evaluated for particular strata so that a three dimensional conceptualisation of the contamination becomes possible.

The amount of data generated can be difficult to interpret and ways of tabulating the data and mapping it are being devised. Table 1 gives an indication of the presentation of numerous parameters related to data in a single table.

Software graphics packages are available to map data so that the location of sampling points; depth of a borehole; depths of sampling and results can all be represented. This visual characterisation of the data enables more accurate and efficient appraisal of data by owners, consultants and regulatory authorities and is being strongly encouraged in Australia.

Table 1. Quick Scan Presentation of results.

Project No. : 88.74.08.01	*Lot No.* :	2 of Section 306 South St	7th May 1991
Suburb : Adelaide	*Analysed by* :	Grand Lab.	
Site Address : 212 South Street	*Analysis No.* :	8874 Grand	

Selection >>	Locations 7–12										
Sample	CHEMICAL SCREENING RESULTS										
Depth (mm)	Pb	As	Cd	Cr	Co	Ni	Zn	Cu	Hg	S(%)	pH
Loc 7											
0–150	200	3	−1	100	4	14	210	28	0.25	0.040	8.6
150–300	170	3	−1	80	6	15	220	100*	0.25	0.030	8.7
300–450	10	−1	−1	60	8	20	34	20	−0.05	0.030	8.6
Loc 8											
0–150	36	2	−1	90	18	75	24	8.0	0.50	0.030	8.0
150–300	−1	2	−1	110	12	28	46	28	0.05	0.030	7.6
Loc 9											
0.15–	250	3	−1	90	4	15	310	50	0.55	0.030	8.8
150–300	160	2	−1	85	5	13	240	60	0.40	0.040	8.4
750–900	4	−1	95	11	22	44	26	7.6	−0.05	0.030	7.6
Loc 10											
0–150	10	−1	−1	70	−1	8	16	10	−0.05	0.060	8.3
150–300	24	5	−1	85	5	13	34	18	0.05	0.030	8.1
300–450	12	3	−1	90	7	15	30	18	−0.05	0.030	8.1
750–900	4	−1	−1	50	6	14	22	15	−0.05	0.020	8.4
Loc 11											
0–150	290	5	−1	80	4	11	540*	24	0.10	0.050	8.3
150–300	450*	10	−1	85	5	15	760*	1750***	0.70	−0.010	8.1
300–450	90	5	−1	110	9	17	30	19	0.05	0.010	7.8
	12	2	−1	110	9	17	30	19	0.05	0.010	7.8
Loc 12											
0–150	100	3	2	85	6	15	80	28	0.25	0.020	8.4
150–300	940*	5	−1	130	7	18	190	60	2.70*	0.010	8.4
300–450	46	1	−1	110	12	24	46	26	0.20	−0.010	7.8
I.L.	300	100	20	600[a]	100[a]	100[a]	500[a]	100[a]	2.00[a]	0.500[a]	<5 or >9[a]

(1) I.L. = Investigation Level[b]; (2) All units are in ppm except where shown; (3) Negative value indicates <Detection Level[b].
*[c]denotes>=and <2×I.L., **denotes >=2 and <5×I.L., ***denotes >=5 and <10×I.L., ****denotes >=10×I.L.
[a] Provisional Investigation Levels used by South Australian Health Commission in 1991.
[b] Health Investigation Levels.
[c] Arbitrary method of categorising data (from Langley and El Saadi, 1991, p. 33 and adapted from SA Housing Trust/SA Health Commission format).

Table 2. Primary screen of indicator analytes

Metals:	As Cd Co Cr III Cr VI Cu Hg Ni Pb Zn
Inorganic Non-Metallics:	CN (total cyanide) S (sulphur) Sulphides Sulphates Thiocyanates
Organics:	*Total phenolic substances* (as phenol) *Total hydrocarbons* *Total polycyclic aromatic hydrocarbons* benzo[a]pyrene benzo[a]anthracene dibenz[a,h]anthracene benzo[k,b]fluoranthene indeno[1,2,3-c,d]pyrene acenaphthalene acenaphthene anthracene benzo[g,h,i]perylene chrysene fluoranthene fluorene naphthalene phenanthrene pyrene *Pesticides:* chlorinated cyclodienes *Polychlorinated biphenyls*
Other:	pH

(from Turczynowicz, 1991)

6.2 Chemical analysis

Problems have been noted to arise in risk assessment due to a lack of quality assurance and analysis. Instances of the use of inadequate extraction methods have been encountered leading to underestimates of the burden of contamination. While the USEPA has a requirement for routine testing of more than 100 substances, in Australia it has been considered that a more targeted approach based on site history will usually suffice. Substances usually analysed are given in Table 2.

	↗	Air	→	Personal air	→	Inhalation
	→	Surface water		Tap water	→	
Contaminant				→		Ingestion
	→	Ground water		Food	→	
	↘	Soil	→	Household soil	→	Dermal contact

(adapted from McKone & Daniels, 1991)

Figure 2. Exposure pathways.

7. Issues in exposure assessment

7.1 General exposure issues

The ANZECC/NHMRC considered the following points are important considerations in exposure assessment:

- 'Children usually receive a higher exposure to soil contaminants per unit body weight than adults;
- Soil ingestion by small children is usually by far the most important exposure route;
- One exposure route will normally predominate;
- The inhalation route will be important for highly volatile contaminants but, as they rapidly evaporate, they will rapidly disappear from a site unless new sources are added;
- In large-scale contamination (i.e. regional, such as near a smelter) more exposure pathways will be involved than in small-scale (very localised) contamination;
- All exposure pathways (see Figure 2) must be considered for health risk assessment (e.g. con-taminant levels in food, water and air) as these will contribute to the level of risk.'

(ANZECC/NHMRC, 1992)

When appraising the exposure assessment as part of a risk assessment, some of the key issues are (Langley, 1993):

- Are all exposure pathways considered? Are any important exposure pathways omitted?;
- Is the contribution of each pathway quantified?;
- How accurately is exposure determined?; and
- What uncertainties exist for the exposure assessment?.

7.2 Exposure data

A vast amount of data is required about a large number of parameters to be able to accurately calculate the dose to which a person (or population) is exposed. An effective exposure assessment will need to determine the relevant exposure pathways and provide sufficient information about each of these pathways.

As an example of the scale of data potentially required, Lioy (1990, cited in

National Research Council, 1991) states that information is required about the following parameters to determine exposure to only a soil-borne contaminant:

'Soil and Sediment

1. Concentrations (µg g^{-1})
   - A. Dusts
     1. Outdoor
     2. Indoor
   - B. Contaminated soil
     1. Uncontrolled dumps
     2. Airborne deposition
     3. Landfills
     4. Resuspension

II Patterns of exposure
   - A. Frequency and duration
   - B. Intensity of contact

III. Percolation rate
   - A. Soil composition
   - B. Water table
   - C. Solubility
   - D. Transport

IV Volatilisation
   - A. Contaminant
   - B. Soil composition
   - C. Top soil and cover

V Contact rate via exposure route
   - A. Dermal deposition and permeability
   - B. Lung
   - C. Gastrointestinal tract (pica)
     1. Normal population
     2. Abnormal ingestion behaviour

VI Body parameter
   - A. Lung volume
   - B. Exposed skin surface (condition of skin)

VII Absorption
   - A. Soil composition
   - B. Contact and absorption rates'

(Lioy, 1990 cited in National Research Council, 1991)

To be able to determine total exposures from all exposure pathways and routes similar amounts of information are also required about water contaminants, air and food (commercial and home-grown produce).

The diversity of exposure pathways which need to be represented in a model is shown by the example of the number of pathways considered in assessing the risks of hazardous agents in sewage sludge. For an accurate assessment all pathways need to be considered and their contributions to total exposure assessed. The 15 pathways for sludge are:

'a. Land Application (Beneficial Use)
- Sludge-soil-plant-human
- Sludge-soil-plant-home gardener
- Sludge-soil-child
- Sludge-soil-plant-animal-human
- Sludge-soil-plant-animal toxicity
- Sludge-soil-animal toxicity
- Sludge-soil-plant toxicity
- Sludge-soil-soil biota-predator of soil biota toxicity
- Sludge-soil-airborne dust-human
- Sludge-soil-surface water-contaminated water-fish toxicity-human toxicity
- Sludge-soil-air-human
- Sludge-soil-ground water-human

b. Surface Disposal
- Sludge-soil-air-human
- Sludge-soil-groundwater-human

c. Incineration
- Sludge-incineration particulate-air-human.'

(USEPA, 1992)

7.3 Point estimates vs probability distributions

Point estimates rather than probability distributions (e.g. Monte Carlo approaches) have been adopted in assessing exposures in Australia. It has generally been considered that the methods of exposure will take into account a very high proportion of the population. Where possible, authorities have required that Australian data are used. This has provided difficulties for some of the most important exposure pathways. For instance, it is considered that soil ingestion is a critical exposure pathway but there are extremely limited Australian data and even the international data are very contentious. The 'critical receptor' is considered to be a child aged between two and three years. This is because of that child's hand-to-mouth activities leading to potentially significant soil ingestion, and aspects of the child's physiology and development that may make children particularly susceptible to soil contaminants.

Recently there have been exhortations to undertake stochastic (e.g. Monte Carlo type) assessments. Such assessments require probability distributions of data and Australian data are even more lacking in this regard. There is very limited data on important variables such as soil ingestion, dietary intakes and children's behaviour.

7.3.1 Monte Carlo analysis

Previously, USEPA risk estimates used the concept of a maximally exposed individual (MEI) or a reasonably maximally exposed individual (RMEI) based on the use of mathematical models of a series of 95th percentile values. The result of this layer upon layer of conservative assumptions leads to extremely conservative assessments which may verge on the implausible.

Recent USEPA documents have supported the use of Monte Carlo analysis (Alsop et al., 1993). At a meeting of the Australian Society of Clinical and Experimental Pharmacology and Toxicology in February 1995, there was enthusiasm for Monte Carlo methods, especially for use in the field of exposure assessment. At that time no Monte Carlo-based assessments had been completed but one was being undertaken for a major site in Sydney.

Monte Carlo analysis provides a range of estimates of exposures and risks and should enhance the process of risk assessment and management if done properly (Langley, 1995). The technique enables the display of the range of predicted potential risks including the average and upper 95th percentile (Alsop et al., 1993).

In Monte Carlo analysis a 'distribution of possible values for each of the parameters (is) described .... along with the probability of occurrence of each value' (Alsop et al., 1993). Using standard mathematical formulae several thousand iterations of a mock mathematical model are performed. For each iteration, values for each parameter are selected randomly from each distribution. The estimated values are combined to provide a frequency distribution of possible exposures or risk (Alsop et al., 1993).

USEPA (1992a) states that simulations such as that found with the Monte Carlo model often 'include low probability estimates at the upper end that are higher than those actually experienced in a given population, due to the improbability of finding these exposures or doses in a specific population of limited size, or due to non-obvious correlations among parameters at the high ends of their ranges'.

7.3.2 Limitations of Monte Carlo-type assessments

With Monte Carlo-type assessments there is still a need to determine what is an acceptable level of exposure and this has always been a difficult and contentious issue (Langley, 1995). Smith (1994) considers that 'the level of exposure exceeded by 1 in 20 exposed persons would seem to be an appropriate reasonable maximum'. However, this would allow 5% of the population not to be included in the exposure assessment and may exclude their consideration in the risk management process (Langley, 1995).

Another problem is that Monte Carlo assessment does not indicate 'which variables are the most important contributors to output uncertainty' (USEPA, 1992b).

Limited information is available about many variables for the exposure assessments for Australia. As a consequence of this many input variables are described as triangular distributions with some or all of the parameters for the distributions being only estimates. Fundamental exposure factors such as soil ingestion rates are poorly described. There are significant limitations to all the soil ingestion studies undertaken to date and no detailed studies have occurred in Australia (Langley, 1995). Important data such as duration of residence are, surprisingly, unavailable from institutions such as the Australian Bureau of Statistics. Shower duration is one area where there are extensive Australian data (James and Knulman, 1987 cited in AIHC, 1994) and this data is used in the USEPA exposure model and in American Industrial Health Council (1994) to provide probability distributions but even this data is not stratified by gender or age (Langley, 1995).

Smith (1994) stresses the need 'to collect and verify distributions from many currently undescribed input assumptions' to improve accuracy. He also highlights the need to distinguish between 'variability' (measurable factors that differ across populations such as height) and 'uncertainty' (unknown, difficult to measure factors such as frequency of trespassing on a site) – something that currently available software packages do not do, but which a reviewer of a Monte Carlo risk assessment must be able to assess (Smith, 1994).

The use of Monte Carlo methods may be inappropriate where the predictions of exposure are very uncertain. McKone (1994) gives the example of benzo(α)pyrene where information on benzo(α)pyrene exposure is 'not readily available' so that the use of Monte Carlo methods to assess variability in population exposures is somewhat inappropriate (Langley, 1995).

Burmaster and Anderson (1994) detail fourteen principles for good practice in Monte Carlo assessments and Langley (1995) proposes requirements that regulatory authorities in Australia will set for assessments using Monte Carlo methods:

- Meeting the 14 principles of good practice due;
- The provision of adequate information to the authority to enable review of the assessment; This may require the provision of the software (and underlying formulae) and data;
- A demonstration of the relevance of the exposure data to the site – data from other countries or cultural backgrounds may not be relevant;
- An explanation of the data and model which will be able to be understood by the relevant community;
- The use of data that account for age and gender differences and take into account susceptible populations.

(Langley, 1995)

8. Issues in toxicological assessment

8.1 Data sources

In the first instance toxicological data from the World Health Organisation (WHO) are recommended. The principal data relate to Acceptable Daily Intakes (ADIs) and Provisional Tolerable Weekly Intakes (PTWI). The carcinogenicity evaluations of the International Agency for Research on Cancer (IARC) are also utilised in the determination of the presence of a potential carcinogenic effect. There has been considerable support for the principle of substance-specific evaluation but, particularly in the case of carcinogens, data and/or the toxicological resources are limited in

Australia. In other Australian administrative jurisdictions (e.g. the assessment of pharmaceuticals and food) a substance-specific approach has been used rather than the utilisation of a quantitative risk assessment approach for carcinogens.

Even for common chemicals there is often a paucity of toxicological data. The lack of data is even more pronounced for mixtures. Toxicological data has usually been developed using pure chemicals or solutions of chemicals.

8.2 Bioavailability

It is important to note that the absorbed dose is not equivalent to the intake as absorption from the lungs, gastrointestinal tract and skin may be substantially affected by numerous biological and physico-chemical factors. This difference between intake and absorbed dose is described by the term 'bioavailability' which indicates the overall proportion of a substance that is absorbed into the body. Information on bioavailability for the relevant substances is required for exposure assessment. The bioavailability of substances when adsorbed to soil has been only considered in a limited way and yet may be particularly important (e.g. the inclusion of contaminant in a vitreous slag).

8.3 Carcinogens

Australia has been heavily influenced by toxicological data from the USEPA. The USEPA has slightly different methods of classifying carcinogens to the IARC and its methods of assessing cancer slope factors are conservative and tend to lead to overestimates of cancer risk. The USEPA methods do not accept threshold concepts and their use of linearised multi-stage methods causes an extrapolation of high dose, usually animal, data back to low dose human exposures. There are many controversies related to the extrapolation of high dose animal data to humans. Unlike laboratory rats, humans have a diverse range of behaviours and genetic heritage and this must be taken into account in risk assessments. The method of developing a cancer slope factor from studies involving small numbers of animals may lead to a significant degree of conservatism in its own right. Increasingly, information about substance-specific mechanisms of carcinogenicity will be important in risk assessment (Stone, 1995).

There are continuing debates as to whether there are threshold doses below which carcinogenesis will not occur or whether there will be risks arising from the slightest exposures, and whether there are linear or some other forms of relationships between dose and carcinogenic risk.

8.4 Quality of toxicological data

The quality of toxicological data used in an assessment must be reviewed. The following are key factors in reviewing toxicological studies to determine which study's conclusions will prevail where there are conflicting studies:

- quality of data;
- resolving power of the studies (significance of the studies as a function of the number of animals or subjects);
- relevance of route and timing of exposure;
- appropriateness of dose selection;
- replication of effects;
- number of species examined;
- availability of human epidemiologic study data (USEPA, 1992a).

9. Issues relevant to risk characterisation

9.1 General

- Are any individuals or sub-populations at risk from exposure to soil contaminants?
- What are the highest levels of exposure and risk? These may not be synonymous as children may be at higher risk than adults with a lesser exposure. This is seen with lead where the susceptibility of children is greater because of their sensitivity during neurobehavioural development.
- What is the high exposure, high risk population and what factors describe it?
- Can sensitive populations (e.g. people with unusual degrees of susceptibility to the hazard) be identified?
- What is the typical level of individual risk?

(Langley, 1993a)

The best assessments will provide a range of estimates for different individuals and population subgroups. The confidence intervals around the risk estimates should be specified. In the USEPA method, estimates have been upper bound estimates: more recently moves have occurred for 'best' or 'most reasonable' estimates of risk to be also provided.

9.2 Quantitative risk assessment

The USEPA has devised elaborate methods of quantitative risk assessment but these contain numerous conservative assumptions, the cumulative effect of which tends to be an estimate of risk far exceeding the typical level of risk. The ANZECC/NHMRC (1992) state 'in most situations, a numerical estimate of risk is not feasible because of limitations in toxicological and exposure data' and 'risk assessment is based on probabilities rather than absolutes and this should be reflected in decision-making'.

Quantitative risk assessments attempt to provide a numerical estimate of risk and are to be contrasted with qualitative risk assessments which categorise risk (e.g. IARC appraisals which assess chemicals and place them into certain pre-classified categories of carcinogenicity).

9.3 Uncertainties in site-specific risk assessment

The assessment of uncertainty is a fundamental part of the process. The risk assessment 'can be judged by how well uncertainties are determined, delineated and addressed. It is therefore important that the level of uncertainty can be assessed for results arising from any model and that the uncertainty is taken into account in preparing site specific risk analyses' (Langley, 1993).

The sources and magnitude of uncertainties must be identified for each stage in the risk assessment process (i.e. data collection, toxicological assessment, exposure assessment) and the cumulative effect of these various sources of uncertainty in the risk characterisation determined (Langley, 1993).

Reasons for addressing uncertainties in exposure assessments include (USEPA, 1992b, detailed in Langley, 1993):

- Uncertain information from various sources will have been combined for use in the model;
- The need to make decisions about whether further resources should be used on seeking further information and data to reduce uncertainty;
- To highlight biases that may have crept into the process;
- As risk assessment is an iterative process, uncertainty analysis may enhance the outcome of the process;
- Being able to characterise the uncertainty will assist the decision-makers and ultimately improve the decision making;
- Exposure assessors have a responsibility to present not just numbers but also a clear and explicit explanation of the implications and limitations of their analyses; Uncertainty characterisation helps carry out this responsibility (USEPA, 1992a).

9.3.1 Data collection

Contamination is rarely evenly distributed across a site, e.g. landfill operations may have different materials deposited at different times. There is a need to be aware that relevant contaminants have been investigated and large areas of high concentrations have not been missed.

9.3.2 Exposure assessment

The highest soil (and contaminant) intakes tend to be in toddlers who swallow soil because of their hand-to-mouth activities. It is assumed that toddlers are unlikely to swallow more than 100 mg soil d^{-1}. However a small number of children may have a soil-eating disorder and may eat considerably more than this. Have high risk individuals and sub-populations been identified?

9.3.3 Toxicological assessment

There is often limited or controversial information about the toxicity of a substance. The weight of evidence may need to be considered where conflicting toxicological evidence is available.

9.4 Managing uncertainty

The most important method of dealing with uncertainty is 'to incorporate conservative assumptions' (e.g. assume toddlers on the site will 'eat' soil at higher than average levels when assessing exposure) into the risk assessment process (Langley, 1993b). This provides safety factors to 'allow for unusual scenarios or developments in knowledge' (e.g. the contaminant may be found to be more toxic in the future) (Langley, 1993b).

10. Requirements for site-specific risk assessment in Australia

- There must be adequate data to characterise the nature and distribution of contamination on a site. For a large site where differing exposure scenarios may be in place, sufficient data will be required for each of these exposure scenarios with subsequent separate risk assessments.
- The exposure estimates must include realistic typical exposure assessments but may also include an estimate of 99th percentile exposures.
- The underlying toxicological data must be appraised with details given of its derivation and the level of confidence.
- The uncertainties for each step of the risk assessment as well as the overall uncertainty must be adequately discussed.
- Any assumptions used in the risk assessment must be thoroughly detailed and references provided as to their source. Assumptions must be justified. This will include a justification of the relevance of the data to the Australian environment.

11. Development of Australian health-based soil criteria

There has been considerable emphasis on developing soil criteria for Australia to assist in the assessment and management of contaminated sites. However, 'there are two prerequisites for comparison of soil test results with defined soil criteria. The first prerequisite is a uniform soil sampling methodology which provides an appropriate amount of information about the distribution and level of contaminants on a piece of land. The second is a uniform approach to data analysis to enable a meaningful interpretation of sampling results' (ANZECC/NHMRC, 1992).

In some quarters there were calls to use the Dutch A, B and C criteria as they covered an extensive range of compounds and appeared conservative. There was, however, limited rationale available for their derivation, their use was qualified, and they reflected different environmental and soil conditions to those existing in Australia. They were used as default values (if a value for a particular substance was not provided) for the 'environmental soil quality guidelines' in the ANZECC/NHMRC (1992). The Dutch criteria have since been extensively reviewed, a much more detailed rationale is now provided for their use, and the values and application of soil criteria have changed.

Ultimately, 'Investigation Levels' were established and detailed, and the concept of site-specific 'Response Levels' noted.

Contaminated Site

a 1 2 3 4 b

Background Level	Investigation Level	Response Levels (single or multiple)	Possible overt health effects

Proposed Land Use:
1. Residential
2. Horticultural
3. Recreational
4. Commercial/Industrial

Environmental Factors:
a. Groundwater protection
b. Plant protection

(Figure not to scale)

(ANZECC/NHMRC, 1992)

Figure 3. The relationship of soil criteria levels for substance, X.

Separate health and environmental investigation levels were established to take into account the different sensitivity of human health or environmental systems. 'Site-specific decisions will be required to determine whether the health or environmental investigation level (or both) will be applied to that particular site' (ANZECC/NHMRC, 1992). In practice, health-based criteria have tended to be used where the site is used for purely residential purposes. The choice of a health- or environmentally-based investigation level will depend on the particular circumstances of a site, e.g. environmentally-based levels may be of paramount importance where a 'natural' wetland is to be established.

11.1 Australian soil criteria

Australian health-based soil criteria are intended to protect people living on a site for a lifetime with few exceptions. Safety factors are considered to have been built into them. Details of the relationship of the soil criteria to the possibility of overt health effects are provided in Figure 3. Investigation levels have been set. These are intended to prompt site specific risk assessment if contaminant levels are found above the investigation levels during a site investigation. They have been misapplied as triggers for clean-up which was not intended when they were drafted. More reasonably they have been used as the targets for clean-up when remediation has been warranted. Investigation levels have been based around the child as a critical receptor and the possibility of a lifetime occupation of a site. Toxicologically, ADIs or PTWIs of the WHO have been used when available. When these have not been available recourse to basic toxicological data has occurred.

Currently there is a process to develop investigation levels for a variety of different exposure scenarios (e.g. residential, recreational, industrial). Given that site-specific assessments may not be practicable ultimately, there may be a set of soil criteria which will prompt some form of response related to site management should a site-specific risk assessment not be undertaken.

11.1.1 Investigation levels

The definition provided was 'an investigation level is the concentration of a contaminant above which further appropriate investigation and evaluation will be required' (ANZECC/NHMRC, 1992).

Investigation levels are intended 'to ensure a detailed investigation of a site is done whenever necessary' (ANZECC/NHMRC, 1992).

The investigation and evaluation is intended to ascertain:

- the typical and extreme concentrations of the contaminant(s) on the site;
- the horizontal and vertical distribution(s) of the contaminant(s) on the site;
- the physico-chemical form(s) of the contaminant(s);
- the bioavailability of the contaminant(s).

(Langley and El Saadi, 1991).

Where soil concentrations exceed investigation levels, factors which need consideration are: 'potential ground water contamination; land use; the history and nature of the contamination; the local background levels; and the size of the site. Exposure pathways will be more diverse for a larger site' (ANZECC/NHMRC, 1992). They were not intended to be less than the natural background variations unless it is considered that the natural levels within the soil may present a health or environmental problem.

Directions for the use of Health-based Investigation Levels included: 'site-specific evaluation of the available data and proposed land use will be required to determine whether single, occasional or typical values in excess of the investigation level will prompt further investigation' and 'levels slightly in excess of the investigation levels do not imply unacceptability or levels likely to pose a significant health risk' (ANZECC/NHMRC, 1992).

11.1.2 Response levels

Response levels were considered to 'apply to a specific site and site assessment and are levels at which some form of response to protect public health and/or the environment with a wide margin of safety is required' (ANZECC/NHMRC, 1992). They were conservative and 'overt health effects will not be expected to occur until well in excess of response levels' (ANZECC/NHMRC, 1992). The nature of the response may range from informing occupants of the site so that they are aware of hazards arising from unusual behaviours (such as pica) where lower levels of contamination exist, to complex soil treatment at higher levels of contamination. There may need to be a series of response levels representing a stepwise progression of actions along the range of appropriate responses. A site-specific assessment may be required to enable consideration of mixtures of contaminants where no single contaminant exceeds response levels. The site-specific assessment must comply with a uniform approach (Langley and El Saadi, 1991).

Different response levels were intended to be used for different exposure situations (e.g. residential, recreational, or commercial/industrial land uses). There may be situations, such as with Cd, where plant uptake rather than direct human exposures is the limiting factor and the order in Figure 3 may be 2,1,3,4.

Factors affecting the nature of the response were stated to include:

'• Land use e.g. residential, agricultural/horticultural, recreational or commercial/ industrial;
- Potential child occupancy;
- Potential environmental effects including leaching into groundwater;
- Single or multiple contaminants;
- Depth of contamination;
- Level and distribution of contamination;
- Bioavailability of the contaminant(s);
- Toxicological assessment of the contaminant(s) e.g. toxicokinetics, carcinogenicity, acute and chronic toxicity;
- Physico-chemical properties of the contaminant(s);
- State of the site surface e.g. paved, grassed or exposed;
- Potential exposure pathways;
- Uncertainties with the sampling methodology and toxicological assessment.'

(ANZECC/NHMRC, 1992, p 34)

11.2 Determination of health-based investigation levels

Health-based investigation levels were determined taking into account:

- The bioavailability of a substance. The bioavailability should be assumed to be 100% if specific information is not available;
- The PTWI or ADI as determined by the World Health Organisation/Food and Agricultural Organisation (WHO, 1987; WHO, 1989);
- Other potential sources of the substances that comprise a proportion of the PTWI or ADI (e.g. background levels of the substance in food, water, air; and the amount of exposure through these routes) (ANZECC/NHMRC, 1992).

Exposures were represented by the equation:

Exposure to substance X = Background exposures (e.g. from food and water) + Exposures from contaminated soil by ingestion, inhalation and skin absorption.

= Background exposures + amount of substance absorbed from soil.

$$= BE + (S_{ing} \times C_{ing} \times B_{ing} + S_{inh} \times C_{inh} \times B_{inh} + S_{skin} \times C_{skin} \times B_{skin})$$

$$= BE + SE_{soil}$$

BE	=	Background Exposures (e.g. from food and water).
S_{ing}	=	Amount of soil ingested.
S_{inh}	=	Amount of soil/dust inhaled and retained.
S_{skin}	=	Amount of soil on skin.
C_{ing}	=	Concentration of substance in soil ingested.
C_{inh}	=	Concentration of substance in soil/dust inhaled and retained.
C_{skin}	=	Concentration of substance in soil on skin.
B_{ing}	=	Bioavailability, i.e. percentage absorbed, of substance when ingested.
B_{inh}	=	Bioavailability of substance when inhaled.
B_{skin}	=	Bioavailability of substance when on skin.
SE_{soil}	=	Substance exposure from soil (ANZECC/NHMRC, 1992).

Qualifications to setting the Health-based Investigation Levels were:

- In setting an investigation level guideline, total exposure to substance X, the sum of the background exposure and the substance exposure from soil is not intended to exceed the ADI or PTWI, i.e., $BE+SE_{soil} < ADI$ or PTWI;
- The degree to which exposures at a proposed investigation level guideline are below the ADI or PTWI will be set by NHMRC and will depend on factors such as: the nature of the threshold effects, the completeness of toxicological data, exposure variability within a population and the relative sizes of BE and SE_{soil};
- It should be recognised that short-term exposure to levels exceeding the PTWI is not a cause for concern provided the individual's intake averaged over longer periods of time does not exceed the level set (WHO, 1989);
- If no PTWI or ADI is available a tolerable level of risk will need to be determined (e.g. for carcinogens) and used for calculations (ANZECC/NHMRC, 1992).

It was 'considered that this way of determining investigation levels would protect the entire population with few exceptions. People who may have unusual sensitivity to contaminants may need to be considered in a site assessment' (ANZECC/NHMRC, 1992).

Health-based Investigation Levels have been established for Pb, Cd, As and benzo(a)pyrene. There is sufficient toxicological information to propose investigation level guidelines for these substances. The substances are also frequent and toxicologically important contaminants.

Further investigation level guidelines will be proposed as sufficient toxicological appraisals become available. Values for a range of metals, organochlorine pesticides and organic compounds are under consideration.

As guidelines they will be subject to change as more information becomes available. The proposed land use; distribution of contaminants; and the frequency distribution of elevated levels will all be very important in interpreting the results for a site.

11.3 Assumptions in the Australian standards-setting model

Key assumptions used in the setting of Australian health-based investigation levels are:

11.3.1 Inhalation-related factors

- Inspirable particulates inside a house will be 75% of the level of inspirable particulates outdoors (Hawley, 1985). USEPA (USEPA, 1989) found indoor airborne Pb levels were 0.3 to 0.8% of outdoor levels for houses without air-conditioning.
- 75% of the inhaled dust will be retained in the respiratory tract and 25% will be exhaled (Hawley, 1985).
- Half the inspirable dust will be sufficiently small to reach the pulmonary alveoli. This will be the respirable dust fraction and will be considered to have a diameter of less than 10 μm.

11.3.2 Ingestion-related factors

- Where bioavailability data for ingested soil contaminants are unknown the value of 100% absorption will be used. If bioavailability data are available they can be used providing the values are able to be justified.
- Ingested soil is considered to comprise one part indoors and five parts from outdoors (Hawley, 1985).
- The concentration of contaminants in indoor soil will be considered to be 120% of that for outdoor soil (South Australian Health Commission estimate from Port Pirie data).

For the purpose of exposure assessment the ANZECC/NHMRC adopted the soil ingestion estimates in Table 3. These incorporate a wide margin of safety (i.e. they are conservative estimates which overestimate typical exposures). Inhalation and skin routes of exposure are considered relatively insignificant in most circumstances. These figures are used for preliminary risk assessment but with an awareness of the very small number of children who have a soil eating habit and who may ingest more than this level.

In most situations, the person most exposed to contaminated soil will be a child aged 2 to 3 years because of their hand-to-mouth activities. Background exposures from food and water will also need to be considered for exposure assessment.

11.3.3 Dermal absorption-related factors

- Where available, substance specific data should be used.
- A child's soil contact area will be equivalent to the area of both hands, both legs and both feet (Hawley, 1985). This area of skin will be taken as 0.21 m^2 (Hawley, 1985).
- The child will wash once each day.
- The soil adherence factor will be 11 mg per 21.5 cm^2 (Hawley, 1985) i.e. a total of 1074 mg of soil on the exposed skin

Minor variations of these have occurred in standards setting. Variation has also occurred in the extent of the percentage allowed for soil exposures from the remnant after background exposures from food, water and air have been subtracted from the ADI or PTWI.

Table 3. Soil Intake Estimates

Age (years)	Soil Intake (mg d^{-1})
0–1	negligible
1–5	100*
5–15	50*
Adult	25*

*conservative estimates (from ANZECC/NHMRC).

11.4 Australian health-based investigation levels

Table 4. Health Investigation Level Guidelines.

Substance	Health level, mg kg^{-1}
Pb	300
As (Total)	100
Cd	20
Benzo(a)pyrene	1
(PAHs)[b]	(20)

[a]These levels should be compared with the appropriate summary statistic(s).
[b]The level for PAHs is based on 5% of the PAHs being present as benzo(a)pyrene.

These investigation level guidelines 'relate to specific extraction and analytical techniques' and 'they should not be compared to other tables of values which do not use similar techniques' (ANZECC/NHMRC, 1992).

12. South Australian experience at Port Pirie

In the late 1980s the legacy of industrial sites became apparent in several places in Australia. Investigations were prompted by concerns about health or environmental effects as well as the need to rehabilitate land, often as part of urban consolidation processes.

Much of the Australian health risk assessment work has been influenced by the assessment and remediation work done in Port Pirie, South Australia. This town of 15 000 people has the world's largest Pb smelter. The smelter was commissioned in 1889 and has operated continuously since then. It has been estimated that 150 000 t of Pb were lost to the environment in and around Port Pirie (Body et al., 1988). Assessment of this regional contamination commenced in the mid 1970s and a remediation program commenced in 1984 (Maynard et al., 1994). Investigations in Port Pirie have focussed on: the environmental burden of Pb (and also Cd, As, Zn and other heavy metals); blood Pb monitoring of children and neurobehavioural assessments. The research has provided epidemiological information on the relationships between Pb exposure and neurobehavioural effects (e.g. McMichael et al., 1988). Further research is being undertaken to assess the environmental behaviour of Pb in Port Pirie so as to determine the principal exposure pathways and the sources of contamination and re-contamination after remediation.

Information about examples of health effects in Australia from contaminated sites has been limited. The Port Pirie Cohort Study of approximately 540 children from pre-natal to ten years of age has provided valuable epidemiological information about the subtle neurobehavioural affects of Pb (McMichael et al., 1988).

13. Outstanding issues in risk asssessment in Australia

Major issues still to be considered in risk assessment in Australia include:

- Development of an Australian model for the toxicological evaluation of carcinogens;
- The incorporation of bioavailability data for the assessment of particular materials;
- Greater information about key exposure pathways. For example Australian data on soil ingestion and the rates of pica are almost nonexistent;
- Better sampling strategies;
- Better laboratory practices and quality assurance;
- Better descriptions of uncertainty and better narratives of risk incorporating uncertainty to enable improved risk management.

References

Alsop W R, Ruffle B, Frohmberg E J and Anderson P D 1993 Development of risk-based clean-up standards using Monte Carlo analysis. In Developing clean-up standards for contaminated soil, sediment, and groundwater. How clean is clean?. Washington: ATSDR/USEPA. 407-416.

American Industrial Health Council 1984 Exposure Factors Sourcebook. Washington, DC: American Industrial Health Council Environmental Health Risk Assessment Sub-committee. pp 144.

Australian and New Zealand Environment and Conservation Council 1992 National Health and Medical Research Council. Australian and New Zealand Guidelines for the Assessment and Management of Contaminated Sites, Canberra: ANZECC, NHMRC. 57p.

Body P E, Inglis G R and Mulcahy D E 1988 Lead Contamination in Port Pirie, South Australia. South Australian Department of Environment and Planning, Adelaide. 87p.

Edwards J W, van Alphen M and Langley A 1994 Identification and assessment of contaminated land. Improving site history appraisal. (Contaminated Sites Monograph Series No. 2) Adelaide, South Australian Health Commission. pp 196.

Ferguson C C 1992 The statistical basis for spatial sampling of contaminated land. Ground Engineer. June, 34-38.

Hawley J K 1985 Assessment of health risks from exposure to contaminated soil. Risk Anal. 5, 289-302.

Langley A J and El Saadi O 1991 Protocol for the Health Risk Assessment and Management of Contaminated Sites. Summary of a National Workshop on the Health Risk Assessment and Management of Contaminated Sites. Adelaide, SA Health Commission.

Langley A J 1993 Finding your way through the maze of health risk assessment methodologies used in Australia. In Proceedings of a conference. Risk Assessment: Avoiding the Risks. Sydney, Australian Water and Wastewater Association. Ed. W Symonds. Australian Waste Water Association.

Langley A J 1993a Refining exposure assessment. In The health risk assessment and management of contaminated sites. Proceedings of the Second National Workshop on the Health Risk Assessment and Management of Contaminated Sites. (Contaminated Sites Monograph Series No. 2). Eds. A J Langley and M Van Alphen. pp 89-117. Adelaide, South Australian Health Commission.

Langley A J 1993b Improving health risk assessment and management of contaminated sites. In Fulbright Symposium. Contaminated Sites in Australia: Challenges for Law and Public Policy. Eds. G C Rowe and S Seidler. pp 19-24. Sydney: Allen and Unwin.

Langley AJ 1995 Exposure factors in risk assessment. In The health risk assessment and management of contaminated sites. Proceedings of the Third National Workshop on the Health Risk Assessment and

Management of Contaminated Sites. (Contaminated Sites Monograph Series No. 4). Eds. B Markey and A J Langley. Adelaide, South Australian Health Commission.
McKone T E 1994 Uncertainty in Variability in Human Exposures to Soil Contaminants Through Home Grown Food: A Monte Carlo Assessment. Risk Anal. 14, 449-463.
McKone T E and Bogen K T 1991 Predicting the uncertainties in risk assessment. Environ. Sci. and Technol. 25, 1674-1682.
McKone T T and Daniels J I 1991 Estimating human exposure through multiple pathways from air, water and soil. Regulat. Toxicol. Pharmac. 13, 36-61.
McMichael A J, Baghurst P A, Wigg N R, Vimpari GV, Robertson E F and Roberts R J 1988 Port Pirie Cohort Study: environmental exposure to lead and children's abilities at the age of four years. N. Engl. J. Medicine. 319, 468-475.
Maynard E J, Calder I C and Phipps C V 1993 The Port Pirie Implementation Program. Review of Progress and Consideration of Future Directions (1984-1993). Adelaide: South Australian Health Commission. (A detailed review of regional contamination and remediation options).
Meyninger R and Molof A H 1993 Not how clean is clean? - but who is safe? How safe? In Developing clean up standards for contaminated soil, sediment, and groundwater. How clean is clean? ATSDR/ USEPA, speciality conference series proceedings. 429-439.
National Academy of Sciences 1983 Risk Assessment in the Federal Government: Managing the Process. Washington,DC: National Academy Press. pp 191.
National Research Council 1991 Environmental Epidemiology: Public Health and Hazardous Wastes. Volume l. National Research Council, Washington. pp 296.
NSW EPA 1994a Guidelines for Assessing Service Stations. Sydney, New South Wales Environment Protection Authority, 1994. pp 31.
NSW EPA 1994b Draft Soil Sampling Guidelines for Contaminated Site Investigation. Sydney, New South Wales Environment Protection Authority, 1994. pp 41.
Paustenbach D J 1989 A Survey of Health Risk Assessment. In The Risk Assessment of Environmental and Human Health Hazards: A Textbook of Case Studies. Ed. D J Paustenbach. 27-124. John Wiley & Son, New York.
Sexton K, Seleran S G, Wagener D K and Lybarger J A 1992 Estimating human exposures to environmental pollutants: availability and utility of existing databases. Arch. of Environ. Hlth. 47, 398-407.
Smith RL 1994 Monte Carlo simulation for human exposure assessment at a superfund site. Risk Anal. 14, 433-439.
Stone R 1995 A molecular approach to cancer risk. Science. 268, 356-357.
Turczynowicz L 1991 Approaches to Analyte Selection. In The health risk assessment and management of contaminated sites. Proceedings of the First National Workshop on the Health Risk Assessment and Management of Contaminated Sites. Canberra (Contaminated Sites Monograph Series No. 1). Eds. O El Saadi and A J Langley. pp 31-51. South Australian Health Commission, Adelaide.
U S Environmental Protection Agency 1989 Risk Assessment Guidance for Superfund, Volume 1. Human Health Evaluation Manual, (Part A). Interim final. Washington DC: US Environmental Protection Agency, Office of Emergency and Remedial Response. EPA/540/1-89/002.
U S Environmental Protection Agency 1992 Standards for the use of disposal of sewage sludge: Final. 40 CFR Parts 257, 403 and 503.
U S Environmental Protection Agency 29 May 1992a. Guidelines for Exposure Assessment; Notice. Part VI. Federal Register 57(104): 22888-22938.
U S Environmental Protection Agency 1993 What's in a Number? [Editorial]. EPA Journal January/February/March : 11.

10
Listen and learn – developing effective communications strategies to deal with communities affected by contaminated sites

TONI MEEK

1. Introduction

As discussed in this book, there are thousands of contaminated sites in prime residential areas throughout the world. With the increased publicity on the negative impacts of contaminants on the soil environment and on human health, the social impacts of contaminated sites are becoming a serious issue within the community and regulatory bodies. Thus the management of a contaminated site can be challenging for organisations charged with the responsibility for dealing with it. There are often many diverse interests, particularly if the site is located close to a residential area. Frequently these interests may be at odds with each other but each position held may, in its own right, have some validity and the problems are never clear cut. People who discover that they are living on or near a contaminated area of land generally react strongly to such news. Their reactions vary, and to a large extent are controlled by their existing knowledge of the environmental and health impact of the particular contaminant(s) and their lack of knowledge or access to information and decision makers.

Organisations responsible for dealing with these problems can gain significant advantages if they carefully plan and time their communications with affected or potentially affected communities. These advantages include:

- enhanced credibility with the community;
- increased trust that the problem will be adequately dealt with;
- less resistance and greater co-operation with measures proposed to deal with the contamination (provided they are reasonable); and
- less negative media interest.

More successful attempts at effective communications with the community are growing in number. There are still many instances, however, where better strategic planning in communications could lead to more constructive outcomes for the people involved.

2. Discussion

Less stringent planning controls in the past, and the need for workers to live close to their workplace, have meant that many industrial sites are likely to be physically located close to residential areas. Whereas once the community was more accepting

R. Naidu et al. *(eds): Contaminants and the Soil Environment in the Australasia-Pacific Region, 309–322.*

of this, today it is often no longer the case, particularly with the increasing concern about the use of hazardous chemicals. As a consequence, some industries have recognised the need to relocate to a more suitable site. Sometimes when this happens, they leave behind a 'legacy' for nearby residents – a contaminated site.

Because many former industrial sites are close to residential areas, it not uncommon for them to be re-zoned to other more sensitive uses like residential ones. Usually a site assessment is done when this is proposed. This assessment will include a risk assessment to determine the nature and extent of the contaminants present and an evaluation of the potential risks to public health and the environment.

Any good risk assessment should also give some attention to the physical location of the particular site and explore the broader community context, if relevant. This involves determining: (a) whether the contaminated site is situated close to residential areas; (b) the impact on community resources such as water supply, food resources etc; and (c) the level of concern displayed by the community about the possible consequences of this contamination. The community needs to be satisfied with the clean-up levels and techniques proposed. If these needs are not satisfied, the organisations responsible will lose credibility and could meet with great resistance when attempting to implement policies.

How then can those responsible for the remediation of a contaminated site avoid these problems? The answer lies in being able to ask some fundamental questions when formulating management plans for a site. For example: is the site located close to a residential area? are the contaminants likely to be ones in which the local community might take an interest? If so, designing a community relations strategy in the formative stages of the development of the site management plan can be a wise and profitable move.

What then might a community relations strategy look like?

Obviously there cannot be one set 'blueprint'. Every situation will be different. Building on, and learning from, past experiences, can provide a useful starting point. The following strategy which has a practical rather than a theoretical emphasis is based upon the experience of the Victorian Environment Protection Authority (Vic. EPA).

2.1 Developing a community relations strategy

Having a clear operating philosophy is going to result in better management and actions for different situations. This is fundamentally good business practice. Contaminated site management is no different. Provided below is a set of key principles to consider when working with many of these community issues.

2.1.1 Defining a set of key principles

A useful first step in formulating a community relations strategy is to define a set of key principles upon which such a strategy is to be developed. These principles ultimately need to inform and guide any decisions and hence the behaviour of those involved in the management of the problem. They are not meant to be exhaustive but rather, provide a starting point for thinking about how the problem will be managed. The key principles are listed below:

- People have a right to be informed of all the issues that may be involved in proposals that could potentially affect them.
- People need to be able to inform and influence the decision-making process.
- A community relations strategy needs to be integrated into established decision-making structures in organisations.
- Information needs to be communicated in ways that people can easily comprehend. This includes transparency about the rationale for deciding what kinds of information to convey, and also includes the need for information to be available in other languages if necessary.
- Early planning and timing of any community relations program is critical to its success.
- Adequate timelines, flexibility and appropriate resources need to be available to all parties.

The following discussion elaborates upon these key principles.

2.1.1.1 PEOPLE HAVE A RIGHT TO BE INFORMED OF ALL THE ISSUES THAT MAY BE INVOLVED IN PROPOSALS THAT COULD POTENTIALLY AFFECT THEM

An increasingly debated concept that has had varied acceptance in the past is the notion of the community's 'right to know'. Such a concept accepts that people have a right to be informed of all the issues involved in proposals or activities that might affect them. The timing of when this information is provided is also critical.

Rydant (1984) suggests that all people potentially affected by a proposal need to have early and complete access to corporate information and decision-makers, as well as a real opportunity to assist in project planning. Identification of the individual groups who have a right to, or want to, know is the responsibility of those managing a particular issue or problem.

Recognition of this right has not always been given a high priority, and affected communites have often heard of proposals after the project design has been completed. An example in recent years in Melbourne was the construction of the Brunswick to Richmond power line. The then State Electricity Commission of Victoria decided an above-ground power line was the best option. Such was the extent of community objections to this that a panel was appointed to review options, and the power line was ultimately placed underground. This was what the community preferred. More recently however, McNeil et al. (1992) suggest that this is changing. The Vic. EPA experience suggests that the timing of involvement is still a contentious area.

2.1.1.2 PEOPLE NEED TO BE ABLE TO INFORM AND INFLUENCE THE DECISION-MAKING PROCESS

The idea of the community participating in decision-making about issues that might affect them is often seen as verging on the unacceptable, particularly by some people in executive positions. Concern is frequently expressed about the community 'taking over' and forcing proponents or existing industries to move out of the area. However, when open dialogue between industry, EPA and the community has been possible, more mutually agreeable decisions have been reached. To understand why, it is once again useful to attempt to appreciate and understand the situation from a community perspective.

When people have no reason to trust, then it is not unreasonable to expect them to behave in a cynical or disbelieving way. It has been the Vic. EPA's experience in dealing with environmental problems causing concern, that the community can be very vocal and intimidating early on in the consultation process. Traditionally, industry has not been required to be open with the community about their practices. From the community's perspective there may be no reason to believe that industry is wanting to become a better neighbour and reduce the impact of its activities.

People often believe that they have only one opportunity to get their views heard, hence the loud voices and angry words. However, it is important for managers of these problems to recognise that whilst total trust may never be achievable, building whatever trust is possible will take time.

Governments are often similarly viewed with suspicion by the community and have had a checkered history when it comes to undertaking effective community consultation. Two such examples in Australia include the siting of the Brunswick to Richmond power line mentioned earlier, and the attempt to locate a high temperature incinerator in Corowa, New South Wales. As a result, it is also not uncommon to hear strong words from affected communities when it comes to government agencies managing contaminated site remediation.

As more serious consultative efforts have been undertaken there have been increasing numbers of examples in which the Vic. EPA has been involved where local community representatives have been heard to comment that they do not see it as their role to be making decisions in some circumstances. The public do, however, want to have their views heard and to have them seriously considered in any decision that might affect them. As trust builds up and the dialogue becomes more transparent and open, the community is often more willing to compromise on certain issues and become more supportive of some decisions.

Providing opportunities for people to inform and influence the decision-making process also confronts the issue of the community's right to know. Those responsible for managing problems such as contaminated sites need to be aware that there may be different information requirements for different sections of the community. For example, local residents are frequently concerned about the potential health risk and so want access to the information but this is also tempered with concerns about the potential for land values to decline and social stigma. There may also be local environment action groups that have a broader agenda of publicising the environmentally harmful activities of certain industries, and there may be interest shown from larger environmental organisations involved in campaigns to stop the production of certain products because of their environmental impact. Environmental problems like a con-taminated site issue can be seen by them to be an opportunity to further their ends.

Meeting the information needs of those vested interests can prove challenging for agencies managing the problem because of the different sensitivities to certain information being disclosed. Tension can quickly develop between the rights of the individual and the broader community's 'right to know'. Chances are, opinions are likely to vary considerably and there will be no clear cut answers. The consequences of this can be seen in the following example.

In the Melbourne suburb of Yarraville, there is a large vacant allotment (a former quarry/landfill) surrounded by houses. The site has been known to have been contaminated with As and other heavy metals for some thirty years. The Vic. EPA monitors the site to check for any off-site contamination. In 1993 as part of this activity, Vic. EPA undertook a sampling program around the site to assess the spread of contaminants. This involved soil sampling of many individual residential properties and gaining residents' co-operation to do this.

The Vic. EPA developed a communications plan, based upon previous experiences, that attempted to deal with these issues as sensitively as possible. An important consideration was determining how much information to release for wider public consumption versus the issue of protecting an individual property owner's rights.

The plan therefore took into account the views of individual property owners but also allowed for the wider community to know what the sampling program had found. Regular liaison with, and briefing of, the local media was undertaken and general trends in results were provided. Property owners were individually advised of results and given an undertaking that they would not be publicly released. A broad summary of results was also made available to satisfy the needs of the wider community. Residents were happy with this but there were still some others in the community who nevertheless believed that all the results should be made public. The Vic. EPA ultimately decided to respect residents' wishes and individual results were not publicly released.

2.1.1.3 INTEGRATING A COMMUNITY RELATIONS STRATEGY INTO EXISTING DECISION-MAKING PROCESSES IN AN ORGANISATION

For any community relations strategy to have meaning and impact, it needs acceptance at all levels in organisations and to be integrated into organisational procedures and culture. For example, customer service is being increasingly promoted and emphasised as an integral part of a successful organisation. Whilst definitions of who are customers and stakeholders may differ across organisations, each group is still deserving of a certain level of service. The community is no exception. Without such acceptance it is questionable as to how effective that organisation will be in dealing with community concerns.

A good place to start in addressing this issue is for the organisation to ask:

'Where does the community fit into our work?'

In dealing with this question, there must be deliberation about the planning, development and implementation of suitable training programs that will assist in developing and building upon the skills required to enhance and maintain positive relationships with the community. Clear messages and expectations from senior members of the organisation will help. A work performance appraisal system with individual performance criteria reflecting these expectations is another way of incorporating these desired behaviours into organisational norms and culture.

Understanding this question from a risk perception viewpoint will also shed some light on how to deal with it. Sandman (1987) for example, argues that organisations need to re-think their views about how they assume the public understands risk:

'... just as the public pays too much attention to outrage and ignores hazard, industry and regulatory agencies pay too much attention to hazard and ignore outrage. In the process they may make the outrage worse.'

This is a clear message for risk assessors and managers. Things may not in reality be hazardous, but if the contaminants in question are seen to be hazardous in the public's eye, then processes must be put in place to deal with any potential outrage. The Vic. EPA was confronted with this problem when dealing with land contaminated with dioxin and chlorophenols in the Melbourne suburb of Fawkner. Whilst Vic. EPA had assessed the risk of any off-site contamination to be minimal and had discussed this with people living near the site, a media campaign by an environmental action group, warning of the risks of dioxin, was very successful in increasing local community outrage about health concerns.

2.1.1.4 INFORMATION NEEDS TO BE COMMUNICATED IN WAYS THAT PEOPLE CAN EASILY COMPREHEND

The way information is conveyed needs to be carefully considered. It is important to recognise that not everyone has a technical background so information must be targeted to the level of comprehension of the consumer. Early on in consultative processes, community participants may appear to accept the volumes and complexities of information being conveyed regardless of whether or not the information is relevant or in fact understandable. However, a lack of comprehension about the problem can lead to negativity and mistrust at later stages: as a local resident once said to industry and government members of a consultative group, 'Don't "expert out" people'. This may be because they feel intimidated and unable to speak out. In the interests of maintaining open communication links, responsible agencies should never assume too much and should ask people what they would like to know. Equally important is the need to realise that we live in a multi-lingual society with various levels of literacy. The challenge is to be able to communicate in simple English and consider whether information in other languages is required.

Another important factor to consider is the need to examine reasons behind providing information in the first place. Pollak (1985) suggests that attempts made by governments and large companies to provide information to the public as a way of allaying fears, have often been viewed suspiciously. Contrary to what might have been the motives of well-intentioned governments and proponents, the public often label such actions as 'paper warfare' and public relations exercises. Such responses highlight the need for responsible agencies to evaluate more carefully the purpose of providing information. If this does not happen, Pollak (1985) suggests that providing more information can more frequently lead to what he calls 'anti-technology' sentiments.

O'Hare et al. (1983) argue that what might be more useful is to distinguish the reason(s) information is required. They believe that little is achieved if information simply reconfirms prior beliefs. O'Hare et al. (1983) also suggest that the value of information is difficult to determine prior to its consumption, and decision-makers need to draw upon past experiences to assess the value of the type of information to be provided.

Finally, O'Hare et al. (1983) argue that it is important for responsible agencies to develop a profile of information users. In that way, they suggest that it is going to be more likely that decision-makers will have a clearer picture of the information requirements of potentially affected parties. For example, the information needs of local residents are going to be different to those of the local media.

2.1.1.5 EARLY TIMING AND PLANNING OF ANY COMMUNITY RELATIONS STRATEGY IS CRITICAL TO ITS SUCCESS

Careful planning needs to be undertaken early, to clearly define the goals and timing of any community relations program. As discussed above, in the past there has been a tendency to think about a community relations strategy at the end stages of program planning. This has sometimes resulted in dealing with crises on the run, not having anticipated emerging issues in a timely way. Hence a fundamental requirement for achieving a successful community relations program is the need for responsible agencies not to go too far down the track of planning a consultation process in isolation from those who need to be involved.

Successful outcomes are possible with joint planning involving the community. If the goal is to have an effective community relations program, certainly have some tentative ideas and strategies in mind, but do not simply impose them on the community. See the community as a partner. Find out what people think. Do some preliminary work to find out who in the community might be interested in being involved. This action by those managing the process, recognises the value of the community as a resource. Again, it is possible that initially the community may be suspicious, but the Vic. EPA's experience has shown that much goodwill can be developed with early community involvement, leading to more constructive outcomes for all concerned.

2.1.1.6 ADEQUATE TIMELINES, FLEXIBILITY AND APPROPRIATE RESOURCES NEED TO BE AVAILABLE TO ALL PARTIES

Attempting to involve the community early on in the consultation process can mean that it may take more time than expected to plan things. The process of seeking out views and reaching consensus about actions is time consuming, particularly as involving the community will frequently mean meeting people out of normal working hours. Not everyone will have similar views on how to go about things. Equally, not everyone will come to the process with the same level of knowledge and understanding of the issues. The community will need time to assess the credibility of the people involved and to begin to trust what the so-called experts say.

Those managing the process could also consider making the experts available to the community on an 'as needed' basis. Alternatively, they may offer the community resources to seek out their own experts. This activity may sometimes lead to opposite views being given, but if this arises, then perhaps there is a need to question the reliability of the expert advice given in the first place.

The reality is that community consultation can be messy and imprecise. Working with people is not always straightforward. Administrative systems and processes in which the decision-makers operate may not always be suitable to transpose, without modification, into community-based activities.

3. Who is the community?

So far in this discussion, words like the 'community' and the 'public' have been used freely and as if they are a homogeneous entity. This is rarely the case. The public is in fact made up of a variety of vested interests, including what Macgill (1989) describes as government, industry, active environmental lobbies, the media and powerless communities. Mostly there are a variety of interests and these interests might have different goals. The key is to be able to define who the public is and how it thinks.

Once having defined who the public is, it is important to:

- examine and attempt to understand the motivations of the various vested interests involved;
- be able to understand the ways in which people process information to arrive at a judgement about what is or is not, an acceptable degree of risk; and
- to be able to understand why people selectively use information from the various sources to which they have access, and how they can at times transcend their own rules of reasoning (Covello, 1984). An example of this is how people accept the risks involved in driving their motor car. The risks of having an accident may be far greater than the risk to health of exposure to a certain pollutant being discharged from a factory chimney stack nearby.

Such considerations ultimately mean that it is a challenge to arrive at a representative view of a heterogeneous public (Macgill, 1989). However, attempts might be made, so that those in power can enhance their understanding of the issues.

As well as coming to an understanding of what the public thinks, Macgill (1989) also suggests that there is a certain inevitability of tensions and contradictions, and of the need to expose the reality of the impossibility that everyone's needs can be met. This means that compromise will become an inevitable part of the process. Macgill (1989) suggests that how the compromises are justified is reflected in the motives for wanting to understand public perceptions. A well planned and considered community relations program that seriously attempts to understand community needs and issues has the potential to more effectively manage and address the various concerns that may be expressed.

Having discussed the key principles involved, the next section proposes some key behaviours to be adopted by agencies responsible for contaminated sites management, or indeed for anyone intent upon undertaking a successful community relations program.

4. Key behaviours to adopt when undertaking a community relations program

4.1 Act as a good neighbour

It is important for those managing a contaminated site problem to be mindful of the activities or non-activities being proposed for the site. This aspect will be discussed in two ways: managing a vacant or non-operational site and managing any site remediation.

If no clean-up activities are planned then the site will still require a degree of 'management' so that it does not become a problem for the people living nearby. This might mean regular maintenance of the site if it is a vacant lot, keeping weeds down, or ensuring that the site is grassed to minimise dust blowing around.

If clean-up activities are planned, then all aspects of the clean-up should be assessed for their potential impact on the surrounding neighbourhood and these factors should be included in any community relations plan that is developed.

The ability of site managers to sensitise themselves and see things as if they were neighbours places them in a strong position to begin to appreciate the problem from the community's point of view.

4.2 Accept the perception of the community

Community concerns can build up as a result of misinformation, having little or no information, or because there is really something to be concerned about. It is important for those managing these problems to be able to accept that the community may have a different perception of the problem and the risks posed.

How the public perceives risks can vary. It is important for those communicating about risk to be aware of this and this understanding will ultimately serve to better inform any actions planned. As Covello (1984) states:

'Policy analysts and decision-makers can benefit from a better understanding of how the public thinks and how it makes decisions about technological risks.'

Sandman (1987) and Macgill (1989) provide some useful ways to think about public perceptions of risk. Sandman (1987) suggests that there is a difference of opinion between public and management agencies, perception of risk is often viewed by the management agency as a public misunderstanding of the issue. Sandman also argues that this may be an oversimplification by scientists and managers. He believes more can be gained from the 'experts' questioning their own actions to see if what they are doing may be problematic.

Macgill (1989) also lends support for this assertion when she argues that it is important to be aware of the variable realities we all experience. She suggests that:

'...to speak in terms of irrationality and misperception, among other elements of self-blinding vocabulary, is plainly to misunderstand how people, all of us, cope with conflicting realities.'

Macgill (1989) believes that it is important to understand that people's perceptions are complex and ambiguous and are never as straightforward as we may think.

Those managing problems could do well to keep an open mind about why and how the public develops the perceptions they do. A significant challenge for them becomes one of being able to accept that, right or wrong, these perceptions are real for the people involved. The key is to be able to understand and work with these perceptions.

4.3 Develop effective listening skills

Listen and learn. This is a simple message but frequently overlooked by many of those dealing with community issues. It is very easy when consulting with communities to assume that telling people what is to be done, will alleviate their concerns. There may be a time and a place for this but with issues generating a great deal of community concern, it is almost certainly not early on in the process.

One of the greatest challenges for those who want to do community consultation well is to listen. This can be extremely difficult to do at times, particularly if one believes they are being misrepresented. It is easy to feel aggrieved and become defensive. Again, there is a time and a place to correct these impressions. It may not be in the early stages of a consultative process, and certainly how it is done is important. The important thing is to try to clarify what the concerns are. Some of the most significant actions that can be demonstrated are to show a willingness to understand the views of the community. This does not mean that you have to agree with them.

4.4 Do not assume all the knowledge about the problem rests with those managing it

It is important for those managing the problem to give an honest appraisal to the community about what they do and do not know. With their colleagues there may often be more openness in discussing uncertainties. If there are uncertainties and incomplete information (and there always will be), explain them to the community. Explain how these uncertainties will be managed. This is one of the fundamental rules of good risk communication.

It is also important to be very wary of discussing risk comparisons in the consultation process. This is not meant to suggest that one avoids this, because sometimes giving concrete examples to which people can relate can be very effective. The timing of such comparisons is probably more critical. For example, it may be the case that the actual risk data about the problem indicates that other more everyday risks are greater than those from the contaminants on the site. People do not necessarily want to hear this when they are being told about risks associated with activities over which they have no control.

Even though the community may be exposed to less risks from exposure to certain pollutants, people may find this hard to accept because they do not have control over exposure to that risk. They may see themselves at the 'mercy' of others to manage that risk, hence the mistrust. Generally, people are more likely to accept such risk comparisons when they have begun to develop some trust in those doing the consultation or when they feel the risk comparisons are not being used to dismiss summarily their concerns.

Another significant action for those managing the problem would be to have their claims verified by another independent source in which the community has some trust. However, finding someone in whom everyone has some confidence could prove challenging.

4.5 View conflict as positive

Few would disagree with the notion that conflict can be unpleasant. It is not very enjoyable to be subjected to a group of angry people. At the height of some conflicts, it could soon become easy to imagine that the community is made up of unreasonable people. It therefore becomes easy sometimes to fall into an 'us versus them' situation. The skill is to be able to avoid doing this and not become defensive. Again the key to managing this is to understand why this anger may have emerged in the first place.

One simple action to deal with this is to acknowledge that everyone has a right to their opinion and to be heard. Early on in the consultation process, listening will be one of the most effective actions that can be done by those managing the problem. This means that there may need to be several opportunities for people to express their concerns. It is important not to rush this.

5. Key tasks to adopt when managing a community relations program

The key tasks outlined below flow from the principles and behaviours identified earlier. Again it is important to stress that there are no right and wrong answers and there may be more than one way of dealing with the problem. The tasks address 'who, what, when, how and why' questions and are designed to provide at least an overview to consider.

5.1 Undertake an assessment of the presentation problem

Identify the key players who may have an interest or have a role in solving the problem. They may not only include local people and groups, but other government agencies, for example, local government. The media can also be a key player in conveying information. The important thing to do is to understand how the media works and target information accordingly.

Attempt to define the interests of each player. Do not assume. Go and ask. Assumptions can sometimes be wrong.

5.2 Endeavour to keep the process as clear as possible

People are more likely to accept decisions if they have participated as fully as possible in the consultation process and can see how their views have been considered. These need to be reported as fully as possible and thought needs to be given by managers as to how this will occur.

5.3 Develop structures in which the community can easily participate

These will tend to range from the more formal, for example, meetings and information sessions, through to informal one to one contacts as work progresses. When choosing the structures, take time to ask people in the local community what they think might work best. Working in this way has the potential for greater acceptance of solutions.

5.4 Develop mechanisms for how the wider community will be kept informed

It is very easy when dealing with localised problems to assume that those residents involved are the voice for everyone else. Certainly they may have expressed an interest in participating but there are others who may also have an interest who do not present themselves in the more 'formal' venues. It is therefore important to look at how the wider community will be kept informed. Media releases to local newspapers could be one way. Letter box drops, information sessions and door knocks could also have a place.

5.5 Develop conflict management structures and processes

It is unrealistic to assume that any consultative process will be devoid of conflict. What is essential to do early on in the process is to negotiate with all parties, mechanisms or 'ground rules' for dealing with any potential conflicts. If expectations are clear and 'up front' then it can often be easier to manage any problems that emerge.

5.6 Plan for regular evaluation of the program

It is important to regularly review efforts and to develop a critically reflective approach to examine how the program could be enhanced. Critical reflection refers to the need to regularly question what is being done, why, and whether or not it is the best way.

Ideally evaluation should not take place as an 'event' at the completion of activities, although there will be benefits in a debriefing of the overall program upon its conclusion. Instead, evaluation should occur as a continuous process. The program of evaluation should be flexible enough to change and adapt as needs and issues change.

Evaluation need not be cumbersome. It may involve using a variety of formal and informal mechanisms such as regular debriefs with staff, through to surveys and feedback sheets (Social Impact Unit, 1991).

5.7 Factoring in the costs of a community relations program

From the discussion so far, it will have become increasingly apparent that there are potentially financial and other costs involved with undertaking a successful community relations program. Activities like these are frequently resource intensive and this needs to be recognised from the outset.

Depending on the size of the problem several people may be required. For example, personal contact with affected residents is essential, particularly with some of the sensitive issues that may emerge, such as possible health risks, the potential for property values to be affected and any clean-up activity planned.

Meeting with communities also frequently involves out of hours work and for many of these meetings it will be preferable that senior members of the organisation managing the problem, attend. This gives a clear message to the community about how seriously the problem is being taken.

There are also likely to be other financial costs. These include the production and distribution of any written material and other media produced.

Whilst these costs may seem expensive, they are a sound investment, when compared to the costs involved from loss of credibility. Costs involved from loss of credibility run high. These costs for example, could be a loss of shareholder confidence and ultimately profits if a company has been identified as a poor environmental performer. For example, Vic. EPA prosecuted a public company for breaking environmental laws, after shareholders indicated their displeasure to the company, threatening to sell off their shares because they did not want to be associated with a company that did not take its environmental obligations seriously.

6. Conclusions

Community consultation is invariably never straightforward or precise. There is often a myriad of vested interests and it becomes a real challenge for agencies charged with the responsibility of consulting to come up with a process that everyone accepts. No community relations strategy will ever be 'ideal' and not everyone will be pleased, but that is the nature of the world in which we all live.

Recognising that conflict will invariably be an inherent part of the process and identifying some simple principles and practices to manage this conflict can ultimately lead to better outcomes overall. This chapter provides a snapshot glimpse of what some of these are.

References

Covello V T 1984 Overview and Definition of Risk: Its Assessment and Management. Environ Int. 10, 359-366.

Macgill S 1989 Risk Perception and the Public: Insights from Research around Sellafield. In Environmental Threats: Perception, Analysis and Management. Ed. J Brown. pp 48-66. Belhaven Press, Great Britain.

McNeil, Barker V and Sinclair, M 1992 Best Overseas Practice In Relation To Community Education About Hazardous Chemicals And Relative Health Risk. Report prepared for the Public Health Displan Management Group. Department of Social and Preventive Medicine, Monash University, Melbourne. 30 p.

O'Hare M, Bacow L and Sanderson D 1983 Facility Siting and Public Opposition. Van Nostrand Reinhold Company Inc, New York. 223 p.

Pollak M 1985 Public Participation. In Regulating Industrial Risks. Ed. H Otway and M Peltu. pp 76-93. Butterworth & Co Ltd, London.

Rydant A L 1984 A Methodology for Successful Public Involvement. Social Impact Assessment. 96-98, 4-9.

Sandman P M 1987 Risk Communication in the Chemical Industry: Explaining Hazard or Coping with Outrage. Paper given at Exxon Chemical Americas, Houston, Texas, September 29, 1987. 12p.

Social Impact Unit 1991 Working with Communities. A Guide for Proponents. Perth, Western Australia. 22p.

11

Sources, potential adverse effects and remediation of agricultural soil contaminants

RUFUS L. CHANEY AND DANIELLE P. OLIVER

1. Introduction

The need for increased food production and the desire to improve the economic conditions have led to rapid industrialisation in many Asian countries. At present the impact of industrialisation on the soil environment in many countries in the Australasia-Pacific region is not clear although reports discussed in this book (Chapters 14 to 25) suggest that increasing rcliance on agrichemicals and industrial effluents for irrigation purposes has resulted in extensive contamination of agricultural land. The lack of industrial zoning legislation in these countries has led to expansion of industries into regions traditionally used for agricultural purposes. This expansion together with the disposal of industrial wastes, fallout from industrial atmospheric emissions, and the use of polluted waters for irrigation of crops have resulted in contamination of valuable agricultural land in many countries in the Asia-Pacific region.

An additional problem in this region, and in particular Australia and New Zealand, has arisen from greater public awareness of environmental and food quality issues and the emphasis by marketing authorities for 'cleaner, greener' food for both local consumption and export. The appropriateness of food quality regulations and the practicality of achieving certain regulatory guidelines is under review in many countries. Thus many agricultural products which would not be deemed contaminated in terms of adverse effects on health are now being required to meet more stringent food quality guidelines to be acceptable for sale. Given that many Australasia-Pacific countries are becoming increasingly dependent on agricultural commodities, there is a need for scientists and regulators to develop management strategies that help minimise soil-plant transfer of contaminants. Such an approach will ensure that developing countries maintain their traditional export market.

In this chapter we review the sources and nature of agricultural contaminants, the pathways of contaminant transfer, and outlined remediation processes applicable to both severely contaminated sites and regions where agricultural quality may be affected by historical agricultural practices. Since there are only limited examples available for most countries in the Australasian-Pacific region we have mainly used examples from European and North American countries to illustrate points.

R. Naidu et al. *(eds): Contaminants and the Soil Environment in the Australasia-Pacific Region, 323–359.*

2. Sources and nature of agricultural contaminants

Contaminants enter the soil environment through both natural and anthropogenic activities. All elements are a natural part of the environment, present in all rocks, soils, and plants. Background levels of elements in soils depend upon both the nature of the parent materials, and on the soil texture since many elements are associated with the fine particles which have the ability to adsorb or form solid solutions with the element of interest. The natural occurrence or 'background' concentrations of soil elements and their variability is often overlooked in considerations of pollution problems. Anthropogenic contamination of soils results both from residential processes, agricultural processes, and industrial emissions (see Chapters 14 to 25), although the latter is usually the focus of concern. During the last decade, the potential for soil contamination by agricultural practices has become more generally recognised, and includes contamination by fertilisers, manures, pesticides, and even limestone and gypsum. In many cases, factors other than the contamination event itself are as important or even more important in making the contamination become a large enough problem to require remediation (e.g. increased availability of the contaminant due to soil pH decline; in the case of Cd, increased availability due to the use of irrigation water with high concentrations of chloride; increased evidence of the contaminant by growing more sensitive crops). In each case, a better understanding of the sources and nature of contamination hastens the identification of the level of hazard and possible strategies to reduce the potential for problems from resulting contamination. Examples of sources and the nature of agricultural contaminants are given in the following section.

2.1 Fertilisers

Phosphatic fertilisers are used extensively in Australia and New Zealand and to a lesser degree in many Asia-Pacific countries. However, in many of these countries the use of P fertilisers is expected to increase particularly since yield increases from P application to crops have been significant and are economically attractive (Karim, 1993). Therefore, concerns raised about contaminants in fertilisers in industrialised countries where use is widespread are pertinent to developing countries.

Schroeder and Balassa (1963) were the first to raise concerns about Cd concentrations in fertilisers in relation to the accumulation of Cd in food crops. Cadmium is the main element of concern in P fertilisers and is discussed in detail in Section 4.1.1, Chapter 5. Other elements of concern are As, Cr, F, Ni, Sr, Th, U and Zn (Mortvedt and Sikora, 1992; Sauerbeck, 1993). The concentration of contaminants in the fertilisers is dependent upon its concentration in the parent rock (Williams and David, 1973; Mortvedt and Sikora, 1992), and the processing technologies used to manufacture the product. In Australia, the main rock phosphate sources for the manufacture of P fertilisers until the last decade were from oceanic sedimentary, guano-based deposits with Cd concentrations ranging from 42 to 99 $mg\,kg^{-1}$ (McLaughlin et al., 1995).

Table 1 shows the levels of Cd and Zn in high analysis P fertiliser materials prepared from four rock phosphate sources in the US. Cadmium and P concentrations in phosphate rocks from several sources around the world are listed in Table 14, Chapter 5. Several nations have expressed concern about Cd in phosphate fertilisers, and Sweden established a limit of 150 mg Cd kg^{-1} P, which was later decreased to 100 mg Cd kg^{-1} P. Since nearly all Cd in commercial fertilisers was derived from the phosphate component the expression of the limit on the basis of P content allows normalisation across different fertiliser products. Because different ore sources differ so widely in Cd concentration (e.g. Florida, Kola, and South Africa are very low in Cd, while Idaho, Morocco, and Senegal are quite high in Cd) and because some nations have begun imposing strong limitations on Cd in fertiliser products, commerce has been affected. There is currently little agreement between countries regarding regulations of Cd in fertiliser products. For example, the Netherlands has imposed a substantially lower fertiliser Cd limit (25 mg Cd kg^{-1} P_2O_4) than that proposed by Sweden (Landner et al., 1995). Current, existing or proposed legislation regarding Cd contents of fertilisers in several countries is given in Table 2. Technologies to remove Cd during phosphate fertiliser manufacture have been evaluated and different technologies may be required depending upon the Cd limitation imposed on the product (Landner et al., 1995).

Radionucleide impurities in P fertilisers are also of concern. Florida phosphate rock ranks high in its content of U, Ra and Th relative to other major phosphate producing areas of the world (Menzel, 1968). Rothbaum et al. (1979) found most of the U in P fertilisers applied in four long-term experiments remained in the surface 23 cm layer of soil. Several studies have shown that annual applications of P fertil isers at the recommended field application rates do not appear to result in increased levels of radionucleides in field crop plants, even at three sites where triple superphosphate had been applied annually for 50 years (Mays and Mortvedt, 1986; Mortvedt, 1986).

Heavily P fertilised soils have also been found to have elevated levels of F. After seven years of P application, water-soluble F in soil increased from 0.29 to 0.49 mg kg^{-1} (Larsen and Widdowson, 1971). Fluoride is strongly adsorbed by amorphous aluminium oxyhydroxides prevalent in acid soils and can precipitate with other elements to form various solid phases. Common F concentrations in plants grown in uncontaminated soils range from 2 to 20 mg kg^{-1} (Brewer, 1966) but Kabata-Pendias and Pendias (1984) found much higher F concentrations in plants surrounding Al smelters where airborne HF is considered much more toxic to plants than soil F. Most studies have found F additions to soil via fertilisers should not cause a hazard through plant uptake (MacIntire et al., 1942; Braen and Weinstein, 1985; Singh, 1990; Sikora et al., 1992). However the solubilisation of Al through addition of F to soil may result in Al phytotoxicity.

2.2 *Limestone*

Limestone is another source of agricultural contamination. The Cd concentration in natural agricultural limestone, gypsum and commercial nitrogenous and potassic

Table 1. Cd and Zn concentrations in US phosphate fertiliser production (Wakefield, 1980).

	Rock Feed		Gypsum		CSP		54% WPA			
Ore Source	Cd	Zn	Cd	Zn	Cd	Zn	Cd	Zn	Cd:Zn in CSP	mg Cd kg^{-1}P
	mg kg^{-1} dry									
Central FL	5	61	1	19	8	89	6	70	0.089	39
North FL	8	50	2	6	9	99	8	98	0.091	43
NC	38	303	14	45	66	603	49	500	0.109	140
Western	122	871	12	48	83	973	128	1 202	0.085	436
(Swedish suggested limit)										150

FL – Florida; NC – North Carolina.
CSP – concentrated superphosphate.
54% WPA – wet process acid.

Table 2. Current existing or proposed legislation regarding Cd contents of fertilisers in Australia and in Australia's trading partners (from McLaughlin et al., 1995).

	Cd Level $mg\,Cd\,kg^{-1}$ P
Austria	275
Denmark	150 from July 1992 110 from July 1995
France	None
Germany	200 voluntary
Japan	340
Netherlands	40
Norway	100
Sweden	over 5 = tax of 30 Swedish krona mg^{-1} Cd; 100 maximum
Switzerland	50 from 1992
UK	None
USA	None
Australia	
Tasmania	
Phosphatic fertilisers	450
Trace element fertilisers	80 (per kg product)
Soil amendments	10 (per kg product)
Victoria (proposed)	
Phosphatic fertilisers	350
Other fertilisers	10 (per kg product)

fertilisers is generally quite low (<$1.0\,mg\,kg^{-1}$) (Stenstrom and Vahter, 1974; Zarcinas and Nable, 1992). However the Cd concentrations of by-products of industrial processing (e.g. phosphogypsum) may be quite high depending on the industrial processes involved (Zarcinas and Nable, 1992). In Australia the inputs of Cd to soils from non-phosphatic fertilisers are negligible compared to those from phosphatic fertilisers (McLaughlin et al., 1995).

In other nations, limestone produced during processing of some metal ores has been marketed as agricultural limestone. Where very high Pb or other elements were found in some limestone products, regulations were developed to limit metal concentrations, but few countries have such regulations. In the US, Chaney et al. (1988) noted that dolomitic limestone produced as a by-product with Zn ore in

Pennsylvania and Tennessee contained 1 500 to 3 000 mg Zn kg^{-1} and 5-15 mg Cd kg^{-1}. The Cd:Zn ratio found in the dolomitic limestone was similar to that of most Zn ore bodies, which reduces the potential for the Cd to be transferred to food chains.

2.3 Organic contaminants

Just as with fertilisers organic contaminants including pesticides, herbicides and fungicides have been used extensively in Australia and New Zealand (see Chapters 14 and 20). Their use in other countries in the Asia-Pacific region is increasing as the requirements for greater food production in these countries increase.

An example of the use of organic chemicals is DDT which was introduced in the early 1950s in many countries to control pests in agricultural situations. In Australia DDT and later a wide range of organophosphates and carbamates were used from 1955–1962 to control cattle tick in eastern Australia (Beard, 1993). In Australia DDT from cattle tick dips was generally disposed by subsurface burial alongside the dips. This has led to extensive localised contamination of soils throughout Australia (see Section 5.2.2, Chapter 14). Particular concern has arisen over these sites because they have become disused and difficult to identify, and are now being used for residential development (Beard, 1993). Contamination by DDT at the cattle dip sites can be as high as 100 000 mg kg^{-1} soil (Beard, 1993).

DDT was also used extensively in Australia on cotton (*Gossypium hirsutum*) crops. In the Namoi Valley, New South Wales, data showed that although DDT contamination was essentially confined to the cotton growing area, land surface contamination extended to at least 30 km from the cotton perimeter. Cattle grazing in areas adjoining DDT-sprayed cotton fields were found to be contaminated (Gilbert et al., 1990).

In New Zealand DDT was used extensively from 1950–1970 on pastures (Chapters 20) to control grass grub (*Costelytra zealandica* White) and porina caterpillar (*Wiseana* spp.) (Hume et al., 1989). In Taiwan DDT was used (1953–1957) to eradicate *Anopheles minimus* and *A. sinensis*, the principal vectors of malaria. In India OC insecticides constitute more than 70% of the pesticide used in the mid 1980s (Chapter 17; Orio et al., 1986). This is reflected in the high concentrations of DDT and hexachloro-cyclohexane (HCH) found in a wide range of agricultural commodities in India (Kawano et al., 1992).

The predominant process which allows transfer of soil OCs to livestock is soil ingestion. This was identified by researchers who were studying impacts of soil ingestion on the wear of sheep's teeth. Residues of DDT and other OCs were absorbed from soil fed to sheep under controlled conditions, while forages had little or no detectable DDT contamination (Harrison et al., 1970). Because DDT and other OCs react strongly with the soil particles they will remain at the soil surface for prolonged periods. Thus management of such soils may involve tilling pastures to move the contaminant below the depth of soil ingested by grazing livestock in order to reduce contaminant residues in livestock.

In Australia sulphonylurea herbicides are used to control broad-leaved and grass weeds in cereal crops. Current farming practices in Australia, including the use of minimum tillage, rely heavily on this group of herbicides. However concern has arisen about the persistence of sulphonylureas in the highly alkaline soils that are characteristic of parts of the wheat growing regions. In some areas the rate of breakdown is also hindered by low rainfall. Concerns have arisen that the persistence of sulphonylureas and their movement down the soil profile may lead to contamination of groundwater. This is of particular concern in areas where groundwater is used for crop irrigation such as in parts of south-eastern Australia. See Chapter 4 for other examples of organic contaminants polluting groundwater.

In Asia organic compounds, such as PCBs, have also contaminated rural soils when released from different munitions during the Second Indochina war (Thao et al., 1993a). Thao et al. (1993b) found PCBs ($>150\,\mathrm{ng\,g^{-1}}$) in non-cultivated soils in south-western Vietnam where armaments and solvents had been stored on former USAF air bases during the war. Further, PCBs released to the air entered the global circulation and all soils on earth are contaminated to some extent. Soils near cities, or at locations where PCBs were used commercially or dispersed to the environment, often have over 1 mg g^{-1}. In the UK, a survey of PCB levels in 'background' soils found concentrations ranging from 1.7 to $1200\,\mathrm{ng\,g^{-1}}$ (Creaser et al., 1989). In these background soils, no source is suspected other than general environmental dispersal during use or disposal of PCBs at distant locations. Fortunately, PCBs are not absorbed from soils into plants except into the skin of root vegetables such as carrot (*Daucus carota* L.). Consequently, food chains are well protected from soil PCBs except for the ingestion of soil by grazing livestock (Chaney et al., 1995).

2.4 Swine and poultry manures

In areas where access to fertilisers is limited either due to availability or cost, the use of animal manures is an attractive alternative for increasing soil fertility. However manures must be disposed of safely. Since manures are bulky commodities with low nutrient concentrations, higher application rates often occur 'nearer the barn' with concomitant increases in the concentration of NO_3^- in groundwater. Manures of some monogastric livestock often contain high levels of Zn and Cu, which were added to the diets as growth promoters. Kao (1991) evaluated sawdust swine waste and conventional pig compost on an acidic and neutral loam. One year after the swine waste application the Zn and Cu concentrations in Chinese cabbage (*Brassica chinensis* L.) leaves were 28–43% greater with compost application than with no compost in the control soil. In the neutral soil there was no significant difference in leaf concentrations of Zn and Cu between no compost and compost treatments, a result commonly observed in many studies (see also Zhu et al., 1991). Many scientists have expressed concern about Cu and Zn in swine and poultry manures, especially where swine manures could be surface applied on pastures grazed by ruminant livestock including sheep – the livestock most sensitive to excessive dietary Cu. After a number of field studies of sheep grazing on pastures to which high Cu swine manures were applied, it has become clear that manure with concentrations of 1000 mg Cu kg^{-1} or higher does not threaten the health of sheep in full season grazing programs (e.g. Poole et al., 1983; Bremner, 1981). Interactions between manure Cu and other elements in the manure, metal binding components of the manure, and nutrients in the pasture plants reduce the potential for toxicity to sheep of the manure Cu.

2.5 Biosolids, effluent and composted urban solid waste

Although land disposal of biosolids is widespread in North America and Europe, biosolids have not been applied in appreciable amounts to agricultural soils in Australia. This situation is expected to change since the government is reviewing the discharge of sewage into the sea. The main restriction to the use of sewage sludge on land has been the presence of high concentrations of contaminants, which is dependent upon the level of industrial effluent in the sewage. Although both organic and metals contaminants in biosolids are of concern the latter has been the subject of much research due to its greater impact on human health and food quality. Substantial concern was expressed about the metal content in biosolids in the period before much research was completed on this subject (e.g. Chaney, 1973; Chaney and Giordano, 1977) but the potential for the adverse effects expected from the metals has seldom been seen unless the materials were highly contaminated by industrial discharges.

The main metals of concern in biosolids in North America are Zn, Cu and Ni because they commonly occur in sewage sludge and effluents and can be toxic to plants. In Australia however, the main metal of concern in biosolids is Cd (S. Rogers, pers. comm.). This concern has arisen from a marketing viewpoint because the current maximum permissible concentration for Cd in a range of foodstuffs, including cereals and agricultural products, is 0.05 mg kg^{-1} fresh weight (FW). This limit is set by the Australian National Food Authority (NFA, 1993). Discussion of the appropriateness of this limit is given briefly in Section 4.1, Chapter 5. Other metals of concern in sludge include Pb, Hg, Se, and Mo which, in excess, threaten animals, livestock, or wildlife.

By contrast with the water-borne sewage systems used in most Western countries human waste in Asian countries has been collected for thousands of years and spread on crops after little if any treatment (see Chapters 14 to 25). Therefore the problem of contamination with industrial waste is not as great as in most Western cities where industrial wastes are commingled with domestic wastes and street runoff (Taiganides, 1976). However, in Japan although human waste was used on land until the end of World War 2 this ceased when chemical fertilisers became abundant at low cost and the high cost of labour prevented the collection of human waste (Takahashi, 1976). Newly constructed sewerage treatment works in Japan produce biosolids, much of which are land applied as fertiliser. As in most Western cities the most serious concern in Japan regarding use of sewage sludge on farmland is the need to enforce limits on heavy metals.

2.6 *Aerosol contamination of farmland near smelters*

Heavy metal contaminants of soils and plants in the vicinity of ore smelters is dependent upon the technology used in the smelting process, climatic factors and distance from the smelter. An example of contamination associated with a smelter is the Pb smelter at Port Pirie, South Australia, which is an isolated plant located in an otherwise rural region. A survey of metal contamination in the surrounding soil found the EDTA-extractable concentrations ($mg\,kg^{-1}$) of Zn, Cd and Cu ranged from <5 to >40, <0.15 to >1.0, and <2 to 6–10, respectively (Cartwright et al., 1977). Merry et al. (1981) investigated the Pb, Cd and Zn in wheat (*Triticum aestivum* L.) plant tops and grain. They found the concentrations ($mg\,kg^{-1}$) of Pb, Cd and Zn in wheat plant tops ranged from >120 to <1.5, >0.2 to <0.075 and >40 to <20, respectively, and in grain ranged from >0.6 to <0.15, >0.10 to <0.03, and >40 to <30, respectively. The concentrations of metals in grain decreased with distance away from the smelter and at a distance of 30 km metal concentrations had reached 'background' values. Readers are directed to Chapters 1 and 14 for a detailed review of the environmental impact of emissions arising from this smelter.

Similarly at Palmerton, Pennsylvania, US, soils and plants in gardens and farms located in close proximity to a Zn smelter had elevated concentrations of Zn and Cd. Chaney et al. (1988) found that the average concentration of Zn and Cd in garden soils was $5830\,mg\,kg^{-1}$ and $76\,mg\,kg^{-1}$ respectively. Potatoes (*Solanum tuberosum* L.) from home gardens were found to contain up to $6\,mg\ Cd\,kg^{-1}$ dry weight. Applications of limestone and organic fertilisers were used to counteract yield reduction and chlorosis due to excessive soil Zn. The Zn and Cd concentrations in the liver and kidney of cattle grazing near the smelter were also found to be elevated indicating that part of the Zn and Cd in forages was bioavailable (Chaney et al., 1988). Interestingly, a study of the human population in the vicinity of the smelter found no evidence of Cd toxicity (Sarasua et al., 1995), even though the soil contained 50 times higher Cd concentrations than that which caused a high incidence of Cd disease in humans in Japan and China (see below). The families in Palmerton had home gardens, and some ingested a large percentage of their diet as garden-produced crops. Similar findings were reached in Shipham, UK (Strehlow and Barltrop, 1988) and in Stolberg, Germany (Ewers et al., 1993) where smelter or mining wastes were redeveloped for housing after World War 2. In both these communities, many persons consumed home grown garden foods as a significant part of their diets. The reasons for the discrepancies between the communities in the occurrence of Cd toxicity symptoms are discussed further in section 2.7.

2.7 *Contamination of farmland by dispersal of mining and smelting residues by water and wind erosion*

Cadmium contamination of rice soils has been extensively studied in several countries following detection of severe human disease (renal tubular dysfunction and osteomalacia) induced in farm families dependent upon rice as their staple diet for many decades. In most cases, Zn, Zn–Pb, or Cu mining or smelting caused contamination of a waterway which was then used for irrigation of rice paddies. In these cases, as little as 2 mg Cd kg^{-1} soil caused Cd disease in humans (Kobayashi, 1978; Nogawa et al., 1987; Shiwen et al., 1990). Further details of Cd toxicology are given in Section 4.1, Chapter 5.

In Japan, $CdCl_2.2.5H_2O$ was a common contaminant in wastewater effluent from a number of factories and CdO was emitted as a vapour from Zn refineries. In areas with Zn smelters or refineries in close proximity to agriculture, the dominant form of Cd is CdS in paddy fields and CdO in other fields (Muramoto, 1990). The toxicity of different forms of Cd to glutinous and non-glutinous rice (*Oryza sativa* L.) was found to be in the order $CdCl_2 2.5H_2O > CdO > CdS$. The Cd concentration in unpolished rice in a polluted versus a non-polluted area in this study ranged from 0.5–5.0 and 0.1–0.5 mg kg^{-1} dry weight, respectively. In soil the concentration of Cd ranged from 1–60 and 0.5–1.0 mg kg^{-1}, respectively (Muramoto, 1990).

Hsieh (1990) studied the distribution of Cd in soils of rice fields in Taiwan which were contaminated from Cd-containing water discharged from Cd-plating factories. Cadmium concentrations in brown rice ranged from 0.5–2 mg kg^{-1}. Although Cd did not affect rice growth or yield, the concentration in brown rice where soil concentrations were 1–2 mg kg^{-1} was still above the critical level of 0.5 mg kg^{-1} stipulated by the Environmental Protection Bureau (Taiwan). As noted in Chapter 25, Chen (1991) investigated Cd and Pb contamination in paddy soils and rice plants of northern Taiwan associated with irrigation with wastewater from two chemical plants producing Cd-based plastic stabilisers. The Cd concentration of paddy soils and brown rice was 0.22–1486 and 0.51–7.94 mg kg^{-1}, respectively. The Pb concentration of paddy soils and brown rice was 6.3–12 740 and 0.10–58.27 mg kg^{-1}, respectively. This case is extremely unusual because Cd contamination occurred independent of Zn, in contrast with most paddy soil Cd contamination (e.g. Kobayashi, 1978; Shiwen et al., 1990).

Chaney and Ryan (1995) summarised research findings which now explain why disease occurred in subsistence farm families in Japan and China who grew rice on contaminated soils, but similar harm has not befallen farm families in western countries from either cropland or vegetable garden contamination. They concluded that rice culture separates Zn and Cd such that ZnS remains in soil, or Zn remains in leaves, while soil CdS is oxidised and readily absorbed and translocated to the grain during the grain filling period following paddy drainage. Further, primary subsistence on rice causes humans to experience Fe, Zn, and Ca deficiencies, which in turn substantially increases Cd absorption and retention by humans. Thus soil chemistry, agronomy and human nutrition each played important roles in causing renal tubular dysfunction in Japan and China.

3. Pathways of contaminant transfer

Public concern about contaminants in soils is based on their potential for adverse effects on agriculture, humans, or the environment. The pathways by which soil contaminants may cause harm are principally related to:

- plant uptake moving the contaminant into agricultural or ecosystem food chains, or causing injury to sensitive plants;
- ingestion of soils by grazing livestock or wildlife, causing metals and organics to move into human food chain or harm grazing livestock;
- ingestion of soil or housedust by children;
- leaching of soil contaminants into groundwater, into surface waters, or erosion of soil materials into surface waters where contaminants are released to cause ecosystem harm through food web 'bio-magnification';
- volatilisation of the contaminant into the air; or
- movement of soil particles into air for inhalation by humans.

For each contaminant, one pathway dominates in causing the highest probability of adverse effects to some environmental receptor. The extent of contamination, the characteristics of the receptor, the nature of the soil and the contaminant, and the nature of agriculture or the ecosystem strongly affect the potential for any adverse effects. For example, although water soluble contaminants can leach to groundwater, most of the contaminants discussed in this paper are of concern because they persist in surface soils for centuries to millennia in the absence of erosive loss. The persistence in surface soils is the very reason why soil contaminants are of such concern to agriculture. Thus, although some pesticide residues, especially the relatively persistent but water soluble compounds such as dibromochloropropane (DBCP), ethylene dibromide (EDB), etc. can move to groundwater by either movement with leaching water or colloidal transfer down the profile, groundwater contamination is of minor significance. One exception is the persistence of sulphonylureas in highly alkaline soils in Australia and their subsequent movement to the groundwater which is used for irrigation.

3.1 Plant uptake of soil contaminants and the 'Soil-Plant Barrier' concept

Plants are both a mechanism of transfer of soil contaminants to higher trophic levels, and an important limit to transfer. It is well known that some elements are readily absorbed by plant roots and translocated into the shoots. Poisoning of livestock that ingest plants containing Se and Mo has been known for centuries. On the other hand, plants have limited uptake of most elements from soil such that humans, livestock, and wildlife are not at any chronic risk from that metal in soils. This body of knowledge was integrated by Chaney (1980) and named the 'Soil-Plant Barrier' concept. Different groups of elements are well known to fit one of several categories, which are defined as follows. In Group 1 are those metals which are so insoluble in the soil, or in plant fibrous roots, that the plant shoots are not a significant source of element transfer even when the soils are highly contaminated. These include elements used as 'tracers' for soil contamination of plants: Ti, Cr, Zr, Y, Ag, Sn. Group 2 includes those elements which can be absorbed into roots, but are not translocated to shoots at high enough rates to comprise risk: Hg, Pb, As. Group 3 includes those elements which are readily absorbed by roots and translocated to shoots, but the plant suffers severe yield reduction before humans, livestock, or wildlife are at any risk from chronic lifetime consumption of the crop. Thus phytotoxicity protects the food-chain. Metals in Group 3 include Zn, Cu, Ni, B, Mn and those elements for which plants do not provide food-chain protection. Group 4 includes Se and Mo, the well known food-chain toxic elements, and Cd in rice (*O. sativa* L.) when it is grown under flood irrigation systems and is a staple food in the diet. Cobalt is also in Group 4 because it can harm ruminant livestock at plant concentrations of 5–10 mg kg^{-1}, which is well below phytotoxic levels (25–100 mg kg^{-1} DW). Fortunately, few Co contamination episodes for soil are known to have occurred. High Co contamination has resulted from direct industrial waste disposal on soils as 'land treatment' before some wastes were regulated as hazardous (Pinkerton et al., 1981; Pinkerton and Brown, 1985).

Development of the 'Soil-Plant Barrier' concept provides structure to the wide body of knowledge which has been generated by scientists. The usual focus is on fields with high enough levels of required nutrients to support normal yields of the crops in question. Thus, the ability of P to inhibit Pb release from the roots is part of the protection. Similarly, when soils have adequate Cl for plant growth, the Cl level is high enough to precipitate Ag to prevent transfer into plants. This contrasts with Cl effects on Cd uptake by potatoes (see below). A comprehensive or holistic view is required, and exceptions must be sought. This concept or model for the role of plants in transfer of soil contaminants is critical to efficient and cost-effective solutions for contaminated soils. Focus of public concern and regulatory effort on the validated pathways for transfer of metal contaminants from soil can provide great public savings of anguish and money.

A general summary of the foliar concentrations of elements in plants required for normal growth and development, and those high enough to cause phytotoxicity, is shown in Table 3. Knowledge of phytotoxicity remains imperfect, but there have been sufficient studies of major crop plants and vegetable crops that we are comfortable that the foliar concentrations shown in Table 3 can be used as indicators of phytotoxicity. The third column shows concentrations above which further evaluation should be undertaken.

Research in the last few years has indicated some of the most important aspects of plant uptake of metals and organics from soils. In the case of those metals which are strongly precipitated, adsorbed or chelated in soils but can be absorbed by plants under 'worst-case' conditions, movement from the soil particles to the root membrane is usually the factor which limits plant uptake. Only the free ion (e.g. Zn^{2+}) can cross the plasmalemma and enter the cytoplasm. Metal ions chelated with organic ligands or complexed with inorganic ligands in the soil solution cannot be absorbed directly, but must be separated from the ligands by exchange processes. The root carriers which use energy to cause uptake may out-compete the ligands in soil solution and carry the ion across the membrane. This understanding has been improved substantially with the introduction of practical chelator buffers. Chelating

Table 3. Foliar analysis (mg kg^{-1} dw) for diagnosis of heavy metal problems for plant productivity or for food chain safety.

Element	Deficient	Ordinary	Evaluate further	Phytotoxic diagnostic
Zn	< 15	20–100	≥ 300	> 500
Cu	< 5	5–20	≥ 20	> 30
Ni	< 0.1	< 1–10	≥ 50	> 100
Mn	< 15	15–200	> 300	> 500
Pb*	–	< 1–3	> 10	–
Cr**	–	< 1–5	> 10	–
Mo***	< 0.1	< 1–5	> 5	> 10

*Pb risk is predominantly through soil ingestion.
**Soil Cr^{3+} is not phytotoxic; foliar analysis is useless.
***Mo is risk to ruminant livestock, not plants.

agents can buffer metal ion activities just as pH buffers can buffer solution pH. Different pH buffers can be used to provide particular solution pH, and different metal chelator buffers can be used to provide desired levels of Zn^{2+} activity (see Parker et al., 1995, for review). Application of chelator buffers allowed researchers to impose Zn deficiency in nutrient solutions much more easily than previous approaches, and the requirement for Zn^{2+} activity by most plants has been found to be about 10–10.6 mol L^{-1}, or pZn^{2+} = 10.6. Roots can obtain adequate Zn for growth if this level is provided, but lower Zn activities cause yield reduction, and leaves have characteristic deficient levels.

Since adsorption, precipitation or chelation processes normally allow slow diffusion of metal ions from the soil to the roots, processes which increase the concentration of ions in solution will increase uptake. In the case of Cd, Bingham et al. (1983; 1984; 1986) showed that addition of chloride to soils increased the uptake of Cd by plants, but equivalent addition of sulfate had little effect on Cd uptake. Their research indicated that while Cl decreased metal ion activity in soil solution, formation of weak chloro-Cd complexes increased the solubility of Cd in soils. These observations were recently substantiated by McLaughlin et al. (1994) who found that high soil chloride (from high Cl irrigation waters) caused greater uptake of Cd by potatoes, *S. tuberosum* L. The impact of Cl and salinity on Cd uptake by potatoes has been discussed in Section 4.1.2, Chapter 5. Li et al. (1995a) also found that high soil Cl facilitated Cd uptake into the shoots and kernels of non-oilseed sunflower. Similarly, the addition of chelating agents to soils can increase metal uptake even though activity of the free ion is decreased (Wallace et al., 1977). In the same way, an artifact in the study of the fate of metals applied to soils as part of biodegradable organic matter was found to be the formation of chelating agents during periods of anaerobic biodegradation. Immediately after the addition of the organic waste to soil, the production of biodegradation byproducts which chelate metals caused increased Cu and Zn uptake by radish, *Raphanus sativa* L. (see Chaney and Ryan, 1994). When high rates of biosolid application are used to reach high cumulative metal applications, data from the period during organic matter biodegradation are likely to be useless due to the biodegradation error. Although even normal fertiliser rates of application of biosolids, composts, or manures may cause detectable increased uptake of some elements, the 'rapid biodegradation error' is seldom important in risk assessment.

Organic contaminants can be evaluated in the same general way as metals. Many organics (e.g. chloroform) are so volatile that they do not remain in soils long enough to provide soil burdens which could be absorbed and translocated to plant shoots. Other compounds are water soluble and are leached from the surface soil and do not comprise risk through plant uptake. Other compounds are persistent in the soil, bound to the organic matter of the soil, and are not water soluble. These compounds are of much greater risk through soil ingestion than plant uptake because plants cannot translocate the compounds from the soil into edible plant parts. Root crops such as carrot (*D. carota* L.) can accumulate PCBs into their 'peel' layer, a lipophilic layer on the outside of the root. Shoots are contaminated mostly through soil adherence, or by volatile contaminant transfer from the soil surface. For these organic compounds which persist long enough in soils, plant uptake is not an important transfer pathway – rather, soil ingestion greatly dominates over plant uptake (see O'Connor et al., 1991; Chaney et al., 1991; 1995). Post-harvest application of organics to prolong storage of horticultural commodities has also resulted in organic contaminants violating food quality guidelines.

The brief discussion above indicates that plants allow transport of a few elements up the food-chain, but no organics of consequence. For nearly all elements, food-chains are well protected from soil contaminants. The exceptions are well known: Se and Mo, and Cd in rice (*O. sativa* L.) diets or in the absence of co-contaminating Zn. Cobalt appears to offer potential risk through uptake by forage crops, but no cases have been observed due to the rarity of Co contamination of farm soils.

3.1.1 Plant uptake of contaminants in the field compared with greenhouse studies

Information on both crops and soils is required for the estimation of soil-food chain transfer of a contaminant. Both field and glasshouse studies have been used to investigate the plant uptake of contaminants. However consideration must be given to the method adopted for the estimation of soil-food chain transfer of a contaminant. For instance, in the field, roots are not constrained to only the contaminated soil as they are in pots. Soil and plant water status, and soil temperature, are different in the field than that found in the greenhouse. Rooting depth changes during the growing season, and is affected by available soil water and rainfall. Greenhouse studies usually involve harvesting seedlings as opposed to growing crops to maturity, so deeper rooted crops that require a longer season, are likely to have substantial differences between uptake slopes derived from greenhouse pots compared with those derived from field studies.

Such an effect has been demonstrated by deVries and Tiller (1978) who found that the soil to plant transfer slopes for Cd, Zn, Cu, etc. were much higher when estimated in greenhouse pots using biosolids as the metal source than when the same biosolid was applied to lysimeter units (incorporated to plough layer depth) or to farmers' fields. A similar result has been found for Cu (Davis, 1981) and toxic organics such as PCBs (O'Connor et al., 1990). However, in the case of PCBs the source of the compounds was found to have a higher impact (O'Connor et al., 1990) (see Section 3.1.2).

3.1.2 Effect of source of contaminant on plant uptake

Phytoavailability of contaminants is controlled by both its interaction with the soil environment and the source of contaminants. The simplistic view that soluble metal salts, or pure PCBs added to soils would have the same soil chemistry and phytoavailability as the contaminant, is wrong. The other components of a contaminant source may have very important effects on the phytoavailability of the contaminant, either by adsorption to constituents of the source, or due to the presence of another element which interacts with the contaminant of concern.

Source differences are especially important in the case of land-applied biosolids where the metals and organics in biosolids can minimise the presumed risk from applied contaminants in at least three ways. The first is the absorbent aspect of the carrier organic material in biosolids. The second is the competition from co-contaminants i.e. Zn competing with Cd for uptake. The third is the time dependent increased adsorption such as that seen with PCB contamination. The common misunderstandings about whether soluble salts of metals added to soils can provide useful estimates of metal uptake were recently demonstrated in a review by McBride (1995). Until researchers become aware of the role of metal adsorbing components of biosolids, and the overriding interactions between Zn and Cd in Cd transfer in food-chains, they will continue to reach invalid conclusions about risks from soil contaminants. McBride presented a model graph for plant:soil relationships for a metal which started the confusion process (his Figure 1).

3.1.2.1 SALT-Cd VS BIOSOLID-Cd

The common response curves for applied salt-Cd or salt-Cd+Zn, compared with applied biosolid-Cd are shown in Figure 1. Part of this figure is based on the results in Figure 2 which illustrates the usual response of metal concentration in plants to added soluble salts of Cd and Zn. The adsorption strength of the Pocomoke sandy loam (Typic Umbraqualt), which is high in organic matter (OM), is strong enough to cause a linear response of plant-Cd vs. soil-Cd. However, in the Sassafras sandy loam (Typic Hapludult), which is low in OM, the specific metal binding sites are more limited. Thus the Sassafras soil shows a linear response for salt-Cd alone, but an upward curving response for salt Cd and Zn. In the absence of the added Zn, a simple linear response is seen until plant Cd approaches phytotoxic levels, and it then either curves upward, or the slope declines (Figure 1). Model A in Figure 1 is the pattern for Cd phytoavailability with either Cd alone, or Cd and Zn in a

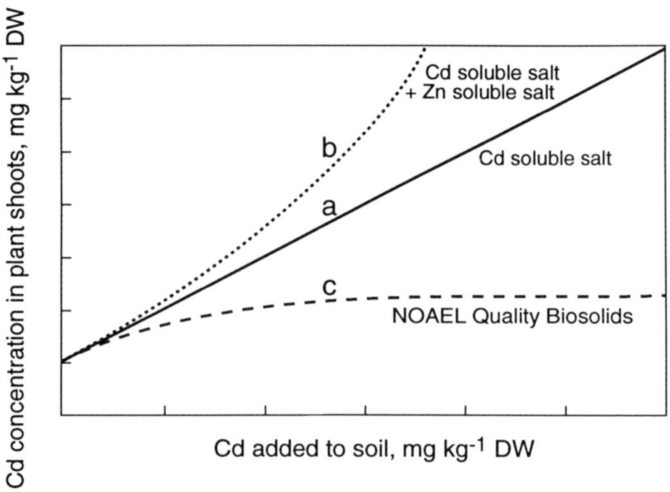

Figure 1. Hypothetical models corrected from McBride (1995) of plant Cd response to total soil Cd concentrations: a) from addition of soluble Cd salt; b) from addition of a Cd soluble salt with 100 times more Zn as a soluble salt; and c) from addition of NOAEL quality biosolids.

NOAEL = no observed adverse effects limit.

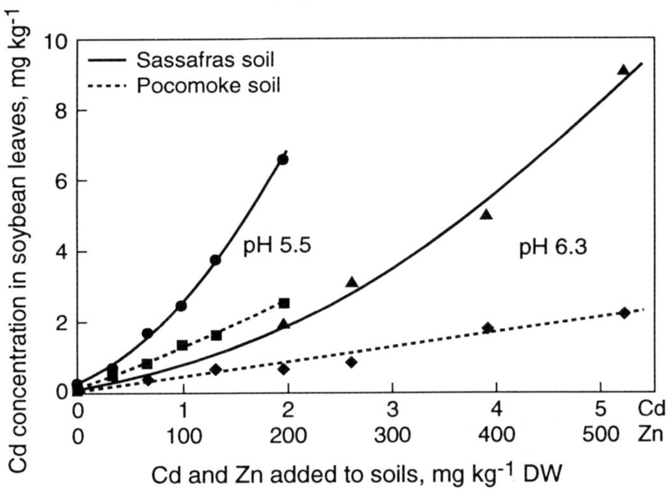

Figure 2. Influence of Zn and Cd addition to Sassafras and Pocomoke sandy loam soils, and soil pH (adjusted to pH_w 5.5 or 6.3) on the Cd concentration in soybean trifoliate leaves (from White and Chaney, 1980).

strongly adsorbing soil, while Model B is the pattern for Cd phytoavailability when Cd and Zn are added to a light textured soil. Model C in Figure 1, however, represents the response pattern commonly seen with high quality biosolids which

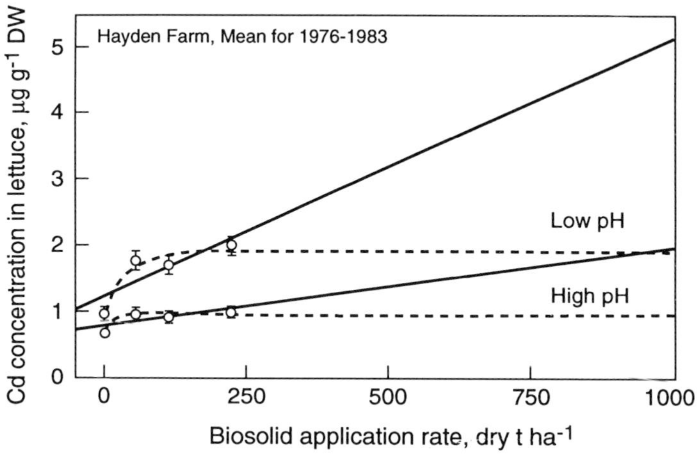

Figure 3. Linear vs plateau regression analysis of lettuce (*Lactuca sativa* L.) uptake of Cd from Christiana fine sandy loam amended with 0, 56, 112 or 224 Mg dry heat-treated biosolid ha^{-1}. Results are averages for 1976 to 1983 (from Chaney et al., 1982).

apply adsorbing materials (e.g. Fe, Mn, Al hydrous oxides) as well as Cd and Zn to soils. Rather than having a linear response or an increasing slope as seen with salt additions, the response is a plateau due to the simultaneous addition of Cd and Cd-adsorbing materials. This response is also shown in Figure 3.

The long-term response to sludge applied Cd in field plots at Beltsville is shown in Figure 3. Dr. R.B. Corey (R.B. Corey, 1980 personal communication) had predicted the plateau response in 1980, based on theoretical consideration of adsorption and chelation in soils. This pattern was verified by Chaney et al. (1982). Subsequent research on these same plots shows that this pattern has been retained through to 1994. Corey et al. (1987) interpret these responses in terms of the specific adsorption (adsorption of trace levels of soluble metals in the presence of mM levels of Ca) characteristics of the applied biosolid, based on hydrous oxides of Fe, Mn and Al, and possibly humic materials in the sludge. First, the metals must be bound strongly by the adsorption surfaces in the sludge to be retained through wastewater treatment and biosolids processing. The plateau observed in Figure 3 can be explained if soil specific adsorption sites control phytoavailability at low additions of sludge, but sludge specific adsorption controls phytoavailability after some amount of biosolids has been incorporated. Actually, some sludges have such low Cd, and such high Fe, Mn, and P, that the plant uptake of Cd declines from the control treatment. This has been observed in a few cases for field studies. The results from long-term biosolids plots on Christiana fine sandy loam (Typic Paleudult) at Beltsville, Maryland, US (Chaney et al., 1982; Chaney and Ryan, 1994) are illustrated in Figure 3. Both the plateau regression results and simple linear regression results for the same data are shown to illustrate that when a linear regression is used to derive plant-Cd:soil-Cd relationships or slopes for the USA rules for land application of sewage sludge, the estimated Cd transfer to plants is significantly exaggerated.

3.1.2.2 EFFECTS OF Cd:Zn RATIO ON PLANT UPTAKE

Part of the effect of other sludge constituents on Cd uptake by plants is due to the presence of 100 to 200 times more Zn than Cd in the sludge. Zinc is a competitive inhibitor of the uptake of Cd and the interaction is reciprocal. This case is illustrated in Table 4 for a study of Cd uptake by lettuce (*L. sativa* L.) grown on a soil which had been amended ten years ago with a sludge which had equal concentrations of Cd and Zn – approximately 1000 mg kg^{-1}. Limestone was added to raise soil pH to decrease Cd uptake by plants, but no reduction in uptake was found which is in contrast with the usual findings (e.g. Figure 3) (Chaney et al., 1982). A pot experiment was conducted to study the ability of different amendments to decrease Cd uptake and it was found that when Cd and Zn concentrations in the soil were equal and the pH_w raised to 7.5 by the addition of 2% $CaCO_3$ the Cd contributed to induced Zn deficiency. Basic studies using chelator-buffered nutrient solutions have confirmed that Zn deficient plants have no reduction in Cd uptake, but a reduction in yield resulted in higher plant concentrations. Oliver et al. (1994a) demonstrated in field studies that the amelioration of Zn deficiency decreased Cd concentration in grain. The decrease in Cd accumulation could not be explained by yield dilution effects. There is some evidence that plants respond to Zn deficiency by increasing the instantaneous rate of Zn uptake by the roots, and this can cause higher Cd uptake under the conditions of our experiment. Thus, Cd soluble salts added to soils can never model the responses of Cd contamination sources other than Cd-Ni battery production wastes, PVC stabilisers with Cd, and Cd pigments which have no Zn present. Geologic Cd is usually about 0.5% of the Zn present. Purified Zn of commerce usually has <0.01% Cd for commercial reasons as well as to produce Cd for markets; nearly all Cd in commerce is derived from ores which had 100–200 times higher Zn. Phosphate fertilisers, on the other hand, have a much higher Cd : Zn ratio, on the order of 10%, such that fertilisers rich in Cd can substantially increase the phytoavailable Cd in soils. Some marine shale deposits also have a 10% Cd : Zn ratio, which allows much greater food-chain mobility of the soil Cd (e.g. Lund et al, 1981; Burau et al., 1981).

When Cd contamination is accompanied by 100 : 1 higher Zn concentration (the usual relationship in common Zn contamination sources such as smelters, mine wastes, urban aerosol, and Zn-fertilisers derived from waste products), another protection is provided by the natural processes of soils and plants. Excessive uptake of Zn can cause phytotoxicity, and under acidic soil conditions the increased uptake of Cd and Zn to leaves is in the same proportion, such that Zn phytotoxicity is an absolute protection against excessive bioavailable Cd in leaves and storage tissues. This is illustrated in a study in Palmerton, Pennsylvania, USA (Baker and Bowers, 1988) and is discussed in Section 2.6 and 2.7. Briefly, vegetable garden soils in the area were contaminated by a nearby smelter which resulted in severe yield reduc-

Table 4. Effect of amendments with biosolid composts on soil pH, yield and Cd and Zn concentrations in lettuce (*Lactuca sativa L.*).

Treatments	pH_w[1]	Yield	Metal concentrations in lettuce Cd	Zn
		$g\,pot^{-1}$	$mg\,kg^{-1}$	
Control	5.6	5.9 c**	14.3 b	12.8 e
$CaCO_3$	7.5	1.6 d	19.5 a	9.9 f
$CaCO_3$+Zn	7.4	6.8 abc	1.9 g	53.7 c
$CaCO_3$+Peat	7.3	1.2 e	12.0 bc	9.3 f
$CaCO_3$+Peat+Zn	7.3	7.9 ab	2.2 f	62.8 b
$CaCO_3$+Compost	7.3	8.4 a	9.1 de	37.3 d
$CaCO_3$+Compost+Fe	7.1	8.4 a	10.1 cd	40.3 d
$CaCO_3$+Compost+Mn	7.3	7.8 ab	7.7 e	35.6 d
$CaCO_3$+Compost+Zn+Mn	7.2	8.4 a	7.6 e	38.5 d
$CaCO_3$+Compost+Zn	7.2	6.6 bc	3.9 f	64.7 b
$CaCO_3$+Compost+Zn+Fe	7.1	7.8 ab	3.8 f	63.1 b
$CaCO_3$+Compost+Zn+Mn	7.3	7.6 ab	3.7 f	67.8 ab
$CaCO_3$+Compost+Zn+Fe+Mn	7.0	7.0 abc	3.9 f	75.5 a

**Geometric means followed by the same letter are not significantly different ($p<0.05$) according to the Waller-Duncan K-ratio t test.

[1]pH_w = 1 : 1 soil : water.

Factorial additions of Zn, Fe and Mn to a compost with basal levels of Zn, Fe, Mn and compost to a soil (pH_w 6.1) which had been historically contaminated by Cd-rich sewage sludge (soil contained 6.8 mg $Cd\,kg^{-1}$ and 89 mg $Zn\,kg^{-1}$). Composted biosolid was amended with the different combinations and incubated for one week, dried and mixed. The compost (10% dry weight) was then mixed with the test soil; 2% reagent $CaCO_3$ was added to all but an acidic control, incubated for one week, leached to remove soluble salts, fertilised and remixed, and seeded. Romaine lettuce was grown in a growth chamber for five weeks, and analysed for metals of interest.

tions. No individual would have increased their consumption of Cd in the order of 10 mg day^{-1} despite eating 50% of their annual consumption of garden crops grown on soils with Cd concentrations up to 100 mg $Cd\,kg^{-1}$. Either they added limestone which decreased lettuce (*L. sativa* L.) uptake of Cd, or the plant suffered Zn phytotoxicity and decreased the amount of food available for consumption.

3.1.2.3 PCBS

In the case of PCBs, research by O'Connor et al. (1990) has shown that PCBs added to soils in biosolid amendments are transferred to carrot roots at a much lower rate than chemical PCBs added to the same soil. Two processes play an important role. First, the addition of humic materials in the biosolid can increase the organic matter in the soil and hence the adsorption capacity of the soil for PCBs. Secondly, and more important is the process of pre-equilibration of PCBs with the organic matter in the biosolids. Evidence has been obtained over many years which shows that potentially toxic organic compounds become more strongly adsorbed over time (O'Connor et al., 1990). Some view it as dissolving in more lipophilic sites in the soil or biosolid organic matter. This time dependent increased adsorption greatly reduces PCB transfer to carrot roots, forage crops, or the general environment. Predictions of lipophilic compound transfer to foods and wildlife can only be valid when biosolid applications made a long time ago are assessed now in field studies. Freshly added potentially toxic organics can readily volatilise and contaminate above ground plant tissues, but the same amounts of PCBs well equilibrated with the soil have non-detectable volatilisation. This phenomenon can be described as 'hyst-eresis', but it is more complex than inferred by this term. Thus risk assessment for food-chain transfer of both metals and organics can only be responsibly conducted using field studies of long-term amendments from the source which causes contamination of concern.

Thus, the nature of the source of contaminants can have a very strong influence on the plant uptake and food-chain transfer of Cd, PCBs, and many other contaminants.

3.1.3 Bio-magnification of contaminants by terrestrial vs aquatic food chains

Concern has been expressed about 'bio-magnification' of contaminants because lipophilic contaminants are effectively absorbed and retained in animal tissue. This can lead to increases in the concentration of the contaminant as one moves up the trophic level. The concept of 'bio-magnification' arose from studies of DDT movement in wildlife food chains, and it is now well known that only one limited case allowed increase in concentration of a metal (methyl-Hg) in terrestrial food webs. Elements of wide concern such as Cd, As, and Pb are actually 'bio-minified' in food webs because each organism absorbs and retains only a small portion of the element ingested in their food (Beyer, 1986). In the case of cattle or goats which ingested biosolids or forage crops grown on soils with high cumulative rates of biosolids, less than 1% of ingested Cd and Pb was retained in the bodies of the animals. Homeostasis for most elements prevents build-up in the body of high levels which could comprise risk to the next highest trophic level.

Methyl-Hg is lipophilic compared to other natural metal species in the environment. This compound is stored in body fat, while ingested inorganic Hg is stored in the liver and kidney. Terrestrial food-chains have not been shown to cause Hg transfer at levels which comprise risk to humans, livestock, or wildlife, but aquatic food chains allow many more levels of 'bio-magnification' than do terrestrial food chains. Thus, the critical pathway for excessive soil Hg is erosion to rivers and lakes where methyl-Hg can be formed much more readily, and be accumulated and 'bio-magnified' through many layers of the food web. Selenium eroded or released to runoff water can cause similar bio-accumulation and 'bio-magnification' in aquatic food webs where the incorporation of Se in proteins allows it to be largely retained upon ingestion.

3.1.4 Ingestion of contaminants by grazing livestock and wildlife

In general, livestock are more exposed than humans to plant-absorbed soil contaminants because they ingest locally grown plant materials as nearly 100% of their diet for their lifetime. Soil can be splashed onto the basal parts of plants, or plants can be stomped into the mud by livestock or farm equipment. Over-grazing of pastures causes livestock to ingest a higher amount of soil. And some plants are pulled up by livestock, and the soil-covered roots are also ingested. The maximum tolerable levels of dietary minerals for domestic livestock are given in Table 5. By comparison humans have lowered exposures to plants grown on a particular soil. Most aspects of risk through feed and forage crops were discussed above in Section 2.2. The 'Soil-Plant Barrier' is applicable to livestock protection.

The phenomenon of soil ingestion was first well described by Healy of New Zealand (see Harrison et al., 1970), who found that ingested soil contributed to wear of sheep's teeth resulting in shortened life cycles and lower profitability. Because elements such as Pb, As, F, Fe, Hg, etc. are not accumulated into the normally edible parts of plants, ingestion of soil is the dominant transfer pathway for soil elements to enter the digestive systems of ruminant or non-ruminant animals (Abrahams and Thornton, 1994). Direct ingestion of high F soils by livestock is expected to cause adverse effects before plant uptake of F would harm plants or animals. Calcium fluoride is often formed in non-acidic soils and is readily dissolved in the gastro-intestinal system. From there F is absorbed, moves to the bones causing them to become brittle. Aerosol contamination of F around Al smelters has caused severe toxicity to cattle in many locations.

In Australia concerns have arisen about the accumulation of Cd in a common pasture weed, capeweed (*Arctotheca calendula* L.) and the accumulation of Cd in the liver and kidneys of sheep over two years of age. Merry (1988) found the highest Cd concentrations of pasture plants in capeweed, ranging from $0.38-4.04\,mg\,kg^{-1}$. A common management practice to control capeweed in Australian pastures is to spray the pasture with a weedicide which increases the palatability of the weed. Sheep are then allowed to graze on the pasture with the aim of the sheep removing the capeweed. This however, results in the sheep getting a much higher concentration of Cd in their diet than they would otherwise receive. The concern was that since Cd accumulates in the liver and kidneys the concentrations of Cd in these sheep organs may exceed the maximum permissible limit for kidney and liver, 2.5 and $1.25\,mg\,Cd\,kg^{-1}$, respectively.

4. Risk assessment

In order to conduct a comprehensive analysis of the potential for soil contaminants to be transferred to plants, humans, livestock, and wildlife where they might cause injury or a reduction in crop quality which may affect trade, the United States Environment Protection Agency (USEPA) developed a group of quantitative path-

Table 5. Maximum tolerable levels of dietary minerals for domestic livestock in comparison with levels in forages (based on literature summarised in Chaney, 1980; Chaney, 1983; and Chaney and Ryan, 1994).

Element	'Soil–Plant Barrier'	Concentration in plant foliage[A]		Maximum concentrations chronically tolerated[B]			
		Normal	Phytotoxic	Cattle	Sheep	Swine	Chicken
		mg kg^{-1} dry foliage		mg kg^{-1} dry diet			
As_{INORG}	Yes	0.01–1	3–10	50	50	50	50
B	Yes	7–75	75	150	(150)	(150)	(150)
Cd[C]	No	0.1–1	5–700	0.5	0.5	0.5	0.5
Cr^{3+}	Yes	0.1–1	20	(3000)	(3000)	(3000)	3000
Co	No	0.01–0.3	25 –100	10	10	10	10
Cu	Yes	3–20	25–40	100	25	250	300
F	Yes?	1–5	–	40	60	150	200
Fe^{3+}	Yes	30–300	–	1000	500	3000	1000
Mn	?	15–150	400–2000	1000	1000	400	2000
Mo	Fails	0.1–3.0	100	10	10	20	100
Ni	Yes	0.1–5	50–100	50	(50)	(100)	(300)
Pb[c]	Yes	2–5	–	30	30	30	30
Se	No	0.1–2	100	(2)	(2)	2	2
V	Yes?	0.1–1	10	50	50	(10)	10
Zn	Yes	15–150	500–1500	500	300	1000	1000

[A]Based on experience as summarised in Chaney (1983).
[B]Based on chronic animal tolerance of elements summarised by NRC (1980).
Continuous long term feeding of minerals at the maximum tolerable levels may cause adverse effects.
Levels in parentheses were estimated (by NRC) by extrapolating between animal species.
[C]Maximum levels tolerated were based on Cd or Pb in liver, kidney, and bone in foods for humans rather than simple tolerance by the animals; ignored role of Zn in inhibition of Cd uptake by plants or animals.

way assessment protocols for the land application of biosolids (Table 6). For each pathway, and for each principal contaminant in biosolids, field-derived data from long term biosolid-amended soils were used to estimate transfer coefficients. Toxic thresholds were derived from feeding studies and grazing studies with livestock on metal rich soils or biosolid-amended soils or diets. The full logical analyses of Cd, Pb and As risk assessment are shown in a monograph by Chaney and Ryan (1994). The original description of the risk assessment pathways was published by USEPA (1989a), but the data selected by USEPA staff and contractors to develop a national regulation for land application of sewage sludge was found to be seriously flawed during the public review period. The technical basis for the Rule was then corrected by a team led by Dr. J.A. Ryan of USEPA, and the details are reported in the Technical Support Document (USEPA, 1992) for the 40 Codes of Federal Regulations 503 Rule (USEPA, 1993b).

The most important aspects of the quantitative risk assessments have identified alternative methods for contaminated soils and biosolids and are described in the following sections. Improved understanding of the soil and other factors which control transfer of contaminants from soils has allowed improved risk assessments, and better assessments have identified alternative methods to treat soils *in situ* to reduce the risk from soil contaminants. In many cases, improved understanding has lowered the estimate of risk per unit soil contaminant concentration. Reduction in uncertainty usually shows that higher levels could have been tolerated had we only known. In other cases, such as the occurrence of toxicity symptoms in communities eating rice contaminated with Cd and Zn in Japan and China, a fuller understanding has shown why this one food production pathway allowed much higher transfer of soil Cd to humans than found possible with wheat (*T. aestivum* L.), soybean (*Glycine max* L.), or lettuce (*Lactuca sativa* L.). Thus, the evaluation of contaminant transfer is an important issue where improved scientific methods and more comprehensive research have provided substantial public benefit by reducing the unnecessary fear of contaminated soils, and reduced the cost of remediation of these soils.

5. Remediation of contaminated agricultural soils

Some agricultural soils are so contaminated that calculations support the conclusion that humans, livestock, plants, or wildlife are at risk from the soil contaminants. When this is found, many different options are available to society and the land owner. The Pathway Risk Assessment allows one to calculate the allowable level for each pathway, and the estimates for different pathways can differ by at least 10^4 due to different properties of contaminants and pathways processes.

For example, sites where children might be at risk from soil Pb, As, etc. might not have any children visiting the site. At industrial sites, or remote agricultural areas where children are prevented from playing the exposure to the Highly Exposed Individual does not occur. However, for society to be assured that children

Table 6. Pathways for risk assessment for potential transfer of trace contaminants in biosolids to humans, livestock, or the environment, and the highly exposed individuals to be protected by a regulation based on the Pathway Analysis (USEPA, 1989a; 1993a; 1993b; Chaney and Ryan, 1994). Each pathway presumes 1 000 t dry biosolids ha^{-1} and/or annual application of biosolids as N fertiliser.

Pathway	Definition of highly exposed individuals
1 Biosolids → Soil → Plant → Human	Individuals whose diets contain 2.5% of all food produced on biosolids-amended soils.
2 Biosolids → Soil → Plant → Human	Home gardeners who apply 1000 t ha^{-1} to strongly acidic soils and consume 60% 'home garden' foods for 70 years.
3 Biosolids → Human	Children aged 1–6 years who ingest biosolid products at 200 mg d^{-1} for 5 years.
4 Biosolids → Soil → Plant → Animal → Human	Farm family members whose diets include meat and dairy products, 45% of which are produced using crops grown on soils with 1000 t ha^{-1}.
5 Biosolids → Soil → Animal → Human	Same as Pathway 4, except annual surface applications allow soil ingestion.
6 Biosolids → Soil → Plant → Animal	Livestock which consume feedstuffs which are grown on soils with 1000 t biosolids ha^{-1}.
7 Biosolids → Soil → Animal	Livestock which graze 1/3 of year on pastures with surface applied or incorporated (1000 t ha^{-1}) biosolids.
8 Biosolids → Soil → Plant	'Crops' grown on strongly acidic soils amended with 1000 t biosolids ha^{-1}.
9 Biosolids → Soil → Soil biota	Earthworms, microbes, etc. in soil with 1000 t ha^{-1}.
10 Biosolids → Soil → Soil Biota → Predator	Shrews or other small mammals which live on site, and consume 33% earthworms in diet for whole life.
11 Biosolids → Soil → Airborne dust → Human	Tractor operator for normal farm tillage.
12 Biosolids → Soil → Surface water → Human	Individuals whose diets consist mainly of fish.
13 Biosolids → Soil → Air → Human	Farm households.
14 Biosolids → Soil → Groundwater → Human	Well water on farms where individuals rely solely on this source for their water supply.

would not be exposed in the future, land deeds or government records must carry the constraint for the contaminated land that it may not be developed in such a way that children might be exposed. This requires the land owner to relinquish some value that the land might have had without this constraint.

Where contaminated land would cause risk through soil ingestion, through the garden foods pathway, the soil ingestion by livestock pathway, etc. specific remedies may be possible. As discussed in Section 2.3 tillage can greatly reduce the exposure of OC compounds in the surface organic layer of a pasture, without transferring the risk to other parts of the ecosystem. But when soils are highly contaminated, some remediation act may be needed to protect humans or wildlife.

The remediation processes associated with heavily contaminated sites are discussed in Chapters 12 and 13. The remediation processes discussed in the following section are unique to agricultural situations and include management practices used to minimise plant uptake. The latter processes are used where contamination occurs on a broad scale and is of a low level such that the site would not be deemed 'contaminated' but the produce grown at this site may not meet the stringent food quality guidelines imposed by regulatory authorities.

Two new approaches for remediation of moderately contaminated soils have come from agricultural research: Phytoremediation, and remediation by treatment with 'Tailor-made' composts *in situ.*

5.1 Phytoremediation

Phytoremediation is thc use of plants to remove contaminants from a soil. Plants now called 'Hyperaccumulators' have evolved on geochemically enriched sites around the world. The name was assigned because these plants tolerate about 10-100 times higher metal concentrations in their shoots than agronomic species. For example, the definition for a Ni hyperaccumulator is the occurrence of $>1000\,mg\ Ni\ kg^{-1}$ in dry leaves collected at the native habitat of the plant, while normal plants start to suffer Ni phytotoxicity at $50-100\,mg\,kg^{-1}$. Plants are known which readily hyperaccumulate Zn, Cd, Ni, Co, Cu, As, and some other elements in nature (Table 7). For Zn and Mn, which do not ordinarily cause phytotoxicity until concentrations in the shoots are of the order of $500-1000\,mg\,kg^{-1}$, the level required to achieve the 'hyperaccumulator' label is 1% ($10\,000\,mg\,kg^{-1}$). Three characteristics have been found in these species: 1) hyper-tolerance of soil metals and shoot metals; 2) extreme uptake of the metal from soils, and 3) hyper translocation of metals from roots to shoots. Although most metal tolerant plant species exclude metals from the roots and shoots, hyperaccumulator plants apparently evolved this ability to limit predation by chewing insects and gain resistance to plant diseases caused by microbial pathogens (Boyd and Martens, 1994; Boyd et al., 1994).

Although the existence of these crops was known to some researchers before 1970 (see Ernst, 1975), the intense efforts of Brooks and co-workers brought hyperaccumulator plants to general attention in the late 1970s (see Baker and Brooks, 1989). Brooks et al. (1977) considered the plants to be bio-geo-prospecting plants, or botanical prospecting plants since the plant occurred on a soil that contained high concentrations of the metal of concern. Most hyperaccumulators have been found to be 'endemic metallophytes', or plants which occur only on soils with very high metal concentrations. By analysis of 1 cm^2 samples of leaves of herbarium specimens, Brookes et al. (1977) could look for economic mine deposits without going to the field. They have also found this a powerful method in evaluating phylogeny.

Table 7. Examples of plant species which hyperaccumulate Zn, Ni, Se, Cu, Co, or Mn to over 1% of their shoot dry matter, far above levels tolerated by normal crop plants.

Element	Plant species	Max. metal concentration in leaves $mg\,kg^{-1}$	Location	Reference
Zn	*Thlaspi calaminare*	39600	Germany	Reeves (1983)
Cu	*Aeolanthus biformifollus*	13700	Zaire	Brooks (1978)
Ni	*Phyllanthus serpentinus*	38100	N. Caledonia	Kersten (1979)
Co	*Haumaniastrum robertil*	10200	Zaire	Brooks (1977)
Se	*Astragalus sp.*	>10000	S. Dakota	Rosenfield (1964)
Mn	*Alyxia rubricaulis*	11500	N. Caledonia	Brooks (1981)

Since Chaney (1983) could see the same hyperaccumulating process as a method to remove soil contaminants introduced the concept of developing a phytoremediation crop to decontaminate polluted soils. The value of metals in the biomass might offset part or all of the cost of cleaning up the site, or even be used for 'phytomining' low grade ore materials. In the case of Zn, if the shoots contain 2.5% Zn in the dry matter, the ash may contain 20–40% Zn, which is equivalent to high grade ores. Although it has taken over ten years for this idea to become of greater interest to the remediation community, many researchers have not entered this field of research and development (Moffat, 1995)

Unfortunately, no plant has been found to hyperaccumulate Pb or Cr under the conditions which would produce meaningful biomass yields. If enough phosphate is present to support high yields, Pb is retained in the roots (Berti and Cunningham, 1995). Progress has been made in field evaluation of the Zn and Cd hyperaccumulator, *Thlaspi caerulescens*, a Brassica species which can accumulate up to 2.7% Zn and over $1000\,mg\,Cd\,kg^{-1}$ without yield reduction (Baker et al., 1994; Brown et al., 1994; 1995). Although this plant can hyperaccumulate Zn and Cd in the field, it has a small biomass, and a rosette growth form which prevents efficient mechanical harvesting. Because many of the metal tolerant species occur in very infertile soils and have also evolved slow growth, it appears that it will usually be necessary to breed improved hyperaccumulator crops, or to use biotechnology to move the genes required for hyperaccumulation of metals into agronomical adapted plants with high biomass production and a growth habit better adapted for harvesting. Table 8 shows estimates for Zn and Cd concentration in normal high yielding forage crops (at the point of 50% yield reduction due to Zn toxicity), in the native hyperaccumulator plants (at the near maximum levels without phytotoxicity), and in a phytoremediation crop developed by breeding or biotechnology. The higher the biomass, and the higher the concentration of a metal of value in the biomass ash, the higher the economic value of the ash may be.

5.2 *'Tailor-made' composts and biosolids*

The failure of hyperaccumulation to deal with Pb and Cr, and the demonstration that soil properties strongly affect the bioavailability of Pb in soil ingested by mammals

Table 8. Selected hyperaccumulators of zinc (≥1% dry weight) with or without Cd hyperaccumulation.

Species	Max. Zn %	Max. Cd $mg\,kg^{-1}$	Reference
Cardaminopsis halleri	1.4		Ernst, 1968
Haumaniastrun katangense	2.0		Baker & Brooks, 1989
Thlaspi alpestre	2.5		Reeves & Brooks, 1983
Thlaspi calaminare	3.9		Reeves & Brooks, 1983
Thlaspi caerulescens	2.7	1020.	Brown et al., 1994
Thlaspi cepaelfollum	2.1		Reeves & Brooks, 1983
Thlaspi praecox	2.1		Reeves & Brooks, 1983
Thlaspi tatraense	2.7		Reeves & Brooks, 1983
Dichapetalum geloniodes	3.0	2.1	Homer et al., 1991
Athyrium yokoscense	0.64	165.	Nishizono et al., 1987
Arenaria patula	1.31	238.	Chaney et al., 1988

(Freeman et al., 1992), suggested to Chaney and colleagues that addition of biosolid composts rich in Fe, P, and $CaCO_3$ might substantially reduce the bioavailability of soil Pb. Composts would also correct the severe infertility of many metal contaminated sites, correct phytotoxicity caused by co-contaminants with the Pb such as Zn and Cd, mix surface deposited Pb with some depth of soil which would dilute the Pb concentration at the surface, and provide improved soil physical properties and organic-N which would facilitate development of a remediated ecosystem at a contaminated site.

In order to determine whether biosolids might also reduce soil Pb bioavailability, feeding studies were conducted by Heneghan et al. (see Chaney and Ryan, 1994). The incorporation of a 10% biosolid/soil mixture into soil at the rate of $225\,t\,ha^{-1}$ significantly reduced soil Pb bioavailability for an urban soil from New Orleans, LA, which contained $1\,600\,mg\,Pb\,kg^{-1}$. The most effective product reduced soil Pb bioavailability by 65%, with only 10% of the Pb as bioavailable as Pb-acetate added to the diet. Lead acetate is generally absorbed very efficiently by test animals, and it is considered as bioavailable as the Pb in drinking water (50% absorbed on fasting). Many risk assessment models for Pb presume that the soil Pb is 60% as bioavailable as water Pb, so to find a reduction to 10% is an important result.

This demonstration of effective reduction in soil Pb bioavailability offers promise of a method to treat both hazardous waste site soils with intermediate Pb levels, and urban soils rich in Pb, in a cost-effective manner. The next stage in development of this work requires field validation of the effectiveness and persistence of the reduction in soil Pb bioavailability. In a field test of this technology at Palmerton, Pennsylvania, USA, on Zn smelter contaminated soils with $30\,000\,mg\,Zn\,kg^{-1}$ and over $1\,000\,mg\,Cd\,kg^{-1}$, addition of limestone and fertilisers did not allow lawn grasses to persist after seeding or sodding. The application of compost provided a 'one-shot' remediation aid because of several factors. Firstly the high Fe levels in the compost favoured Fe uptake by grasses. The Fe also counteracted the Zn displacement of Fe from phytosiderophores used by grasses to obtain Fe from soils. The compost raised soil pH and adsorbed the high soil Zn and Cd very effectively. 'Tailor-made' composts rich in the chemical factors which can inactivate soil metals and reduce potential for phytotoxicity and mammalian toxicity offer great potential for solving soil remediation needs in many nations.

At the same time, chemical extraction methods or *in vitro* extractability methods are being developed which should be well correlated with the results of mammalian bioassay of soil Pb bioavailability (e.g. Ruby et al., 1993). Ingestion of soil by children is widely recognised as an important pathway for risk from some soil contaminants, but the amounts of soil ingested daily remain uncertain. Methods to estimate soil ingestion based on analysis of elements not absorbed by plants or added to foods, in both diets and faeces of young children provide quantitative information (Calabrese et al., 1989). Recent studies of soil ingestion by children by Calabrese et al. and other research groups have shown that the policy decision by USEPA to presume 200 mg soil per day is at about the 95th percentile of soil ingestion by 1 to 4 year old children who participated at an university day care center.

5.3 Management practices that can be manipulated to improve food quality

An increasing awareness of food quality issues by consumers, increasing emphasis by marketing authorities on the sale of 'clean' food and improvement in detection levels in analytical techniques has resulted in increasing concern about the concentration of contaminants in food. As mentioned earlier the main source of contaminants in agricultural soils is from phosphatic fertilisers (see Section 2.1). The main element of concern in phosphatic fertilisers is Cd (McLaughlin et al., 1995). In Australia the specified limits for Cd in certain foodstuffs are given in Table 13 in Chapter 5. The effectiveness of various management practices for decreasing contaminant uptake by food crops is discussed in the following section.

5.3.1 Liming

The most commonly recommended management practice for minimising metal uptake, and in particular Cd uptake, by crops is raising soil pH (Page et al., 1981; Alloway, 1990). The influence of pH on minimising metal uptake can be attributed to the role that ionic speciation in soil solution and surface charge characteristics of soil particles play in Cd adsorption (Naidu et al., 1994). The effectiveness of raising soil pH to minimise metal, and in particular Cd, uptake by crops has been variable. While some studies under glasshouse conditions have found raising soil pH decreased Cd uptake (Andersson and Nilsson, 1974; Street et al., 1978) others have found no decrease in both biosolid-amended (Pepper et al., 1983) and fertilised soils (Oliver et al., 1994b; Maier and McLaughlin, 1995). Maier and McLaughlin (1995) even found liming increased Cd concentrations in potatoes (*S. tuberosum* L.).

Thus the effectiveness of raising soil pH to minimise metal uptake by crops appears limited under field conditions and requires further investigation to be validated.

5.3.2 Addition of macro- and micronutrients

The similarity in chemical behaviour between Cd and Zn has led many researchers to evaluate the interaction between Cd and Zn and their effect on plant uptake of each metal. The Cd content of lettuce (*Lactuca sativa* L.) (Maclean, 1976), barley (*Hordeum vulgare* L.) (Singh and Steinnes, 1976) and soybean (*G. max.* L.) (White and Chaney, 1980) was found to be independent of applied Zn treatments. By contrast, Zn applications (0–50 mg kg^{-1}) decreased Cd concentrations in Swiss chard (*Beta vulgaris* L.) and maize (*Zea mays* L.) grown in the glasshouse (Abdel-Sabour et al., 1988). Oliver et al. (1994a) found in several field-based experiments that Cd concentrations in wheat (*T. aestivum* L.) grain can be decreased by up to 50% by the addition of 2.5 to 5.0 kg Zn ha^{-1} to soils that were marginally or severely Zn deficient (Figure 4). In Australia the application of Zn is a relatively inexpensive management practice that may have a major impact on grain quality by decreasing the Cd concentration to below the current Australian limit of 0.05 mg kg^{-1} (NFA, 1993). The amelioration of Cu and Mn deficiency under field conditions has not been effective in decreasing Cd concentration in wheat (*T. aestivum* L.) grain (D. Oliver, unpublished data).

Soil N status and the rate and type of N fertiliser applied have also been found to affect Cd accumulation by crops (Willaert and Verloo, 1992; Oliver et al., 1993). The elevated Cd concentrations in crops have been attributed to the acidifying effect of the N fertilisers used.

5.3.3 Choice of cultivar

There are two aspects to the selection of cultivar: i) the selection of species or cultivars which do not accumulate high concentrations of the contaminants of concern; and ii) breeding low contaminant-accumulating cultivars. The first method involves growing existing low-accumulating cultivars in regions where contamination is a concern. This is one of the easiest management practices available for minimising contaminant uptake of crops. However, the availability of suitable cultivars and their adaptability to the soil and environment in the area of concern limits the success of this approach. McLaughlin et al. (1994) found the choice of an appropriate potato (*S. tuberosum* L.) cultivar often decreased Cd concentrations by up to 75%. Differences in Cd accumulation have also been found in different cultivars of soybeans (*G. max* L.) and corn (*Zea mays* L.) grown on sludge-treated soils (Bogess et al., 1978; Hinesly et al., 1978) and field grown wheat (*T. aestivum* L.) (Oliver et al., 1995).

The second option is more applicable to areas where the concentrations found in crops affect human health or where the concentrations are so high as to affect trade. This option would only be chosen when no other management practices had

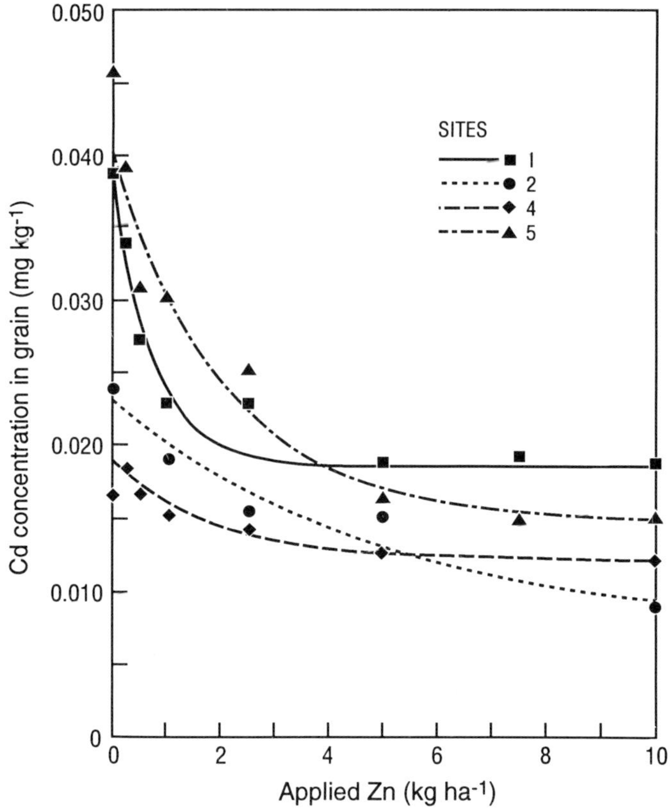

Figure 4. Effects of Zn fertilisation on Cd concentration in wheat (*Triticum aestivum* L.) grain.

succeeded in minimising uptake and the area of concern was so extensive that mixing the commodity from different sources, and effectively diluting the contaminant, was not possible. This is an expensive option which may not always be feasible since breeders must select for genetic characters that favour low accumulation of the contaminant, while at the same time combine all other required characteristics (e.g. yield, disease resistance etc.). Some examples of successful breeding programmes are given below.

Non-oilseed sunflower kernels from the US are marketed in Germany and other countries as a speciality food but examination of Cd concentration in the kernels showed considerably higher concentrations than found in other grain crops (Ocker et al., 1991). The high concentrations could not be traced to soil contamination (Chaney et al., 1993) and the sale of US sunflower kernels in Germany was threatened since the concentrations exceeded the limit of 0.60 mg Cd kg^{-1} for this product. Using germplasm identified to have lower kernel Cd concentrations the industry breeders have developed hybrid cultivars with low Cd concentrations for commercial production in 1996 (Li et al., 1995b). This is the first example of specific commercial breeding of lower grain Cd genotypes with the purpose of meeting commercial limitations. Canadian breeders have undertaken a similar breeding program with durum wheat since the Cd concentrations in durum wheat are not meeting the regulatory guidelines required by their trading partners. In this case they have been able to identify a marker gene which allows inexpensive identification of the lower Cd progenies, substantially decreasing the cost of the breeding and selection process (Penner et al., 1995).

5.3.4 Sequence of crop rotation and changes to tillage practices

The importance of crop rotation is well known in countries where shifting agriculture is practised and has only been recognised in many developed countries as soil fertility declines. The need to minimise disease, increase weed control, improve soil structure and fix N has led to the inclusion of legumes in rotation sequences in many parts of Australia and changes to tillage practices (Pratley and Rowell, 1987). As mentioned in Section 2.3 the incorporation of PCBs into the 'plough layer' is necessary to minimise the availability of OCs to livestock by placing them below the depth of soil ingested. Crop rotations were found to effect Cd concentrations in wheat (*T. aestivum* L.) grain with higher concentrations in wheat grown after legumes, and in particular lupins (*Lupinus angustifolius* L.), compared with wheat (*T. aestivum* L.) grown after barley (*H. vulgare* L.) or wheat (Oliver et al., 1993). Thus the choice and order of crops grown in a rotation sequence can have significant effects on the Cd concentration in grain.

6. Summary and conclusions

Soil contamination has been a universal process both in urban and in agricultural areas. Industrial discharges have caused extreme contamination of soils in many locations throughout the world. Both historic and recent mining activities have dispersed many contaminants, as has smelting of most metals. Historically, certain agricultural pesticides and fertilisers caused elemental contamination of soils that require evaluation today.

Because of increasing concern about the adverse impacts of contaminants on animal and human health extensive research to evaluate contaminant transfer to plants or livestock has been undertaken to conduct risk assessments. In the last five years, the Pathway Risk Assessment methods have been rigorously evaluated. The fundamental approach is usually comfortable to scientists but data were not always readily available. As more effort was taken to estimate the number of persons who comprised the Highly Exposed Individuals for particular contaminants from particular sources of contaminants, it has usually become clear that risks were overestimated.

Many of the remediation processes applicable to urban contaminated sites can be transferred to agricultural situations. However, the cost-effectiveness of such approaches may be limited owing to the broad-scale nature of agriculture and in many countries the distance of agricultural sites from urban centres where the necessary engineering facilities may be located.

More recently concern has arisen in developed countries about food quality and the concentration of contaminants in many food crops. This has led to research into management practices that can be manipulated to minimise the accumulation of contaminants in produce.

The issues raised in many developed countries are now becoming concerns in many countries in the Asia-Pacific region as these countries become increasingly industrialised and the population in these countries increases. Many of the lessons learnt in other countries are applicable in this region and if heeded early enough may circumvent the environmental issues now facing developed countries.

References

Abdel-Sabour M F, Mortvedt J J and Kelsoe J J 1988 Cadmium-zinc interactions in plants and extractable cadmium and zinc fractions in soil. Soil Sci. 145, 424-431.

Abrahams P W and Thornton I 1994 The contamination of agricultural land in the metalliferous province of southwest England: Implications to livestock. Agric. Acosys. Environ. 48, 125-137.

Alloway B J 1990 Cadmium In Heavy Metals in Soils. Ed. B J Alloway. pp 100-124. John Wiley and Sons, NY, USA.

Andersson A and Nilsson K O 1974 Influence of lime and soil pH on cadmium availability to plants. Ambio 3, 198-200.

Andresen A M, Johnson A H and Siccama T G 1980 Levels of lead, copper and zinc in the forest floor of the northeastern United States. J. Environ. Qual. 9, 293-296.

Baker A J M, McGrath S P, Sidoli C M D and Reeves R D 1994 The possibility of in situ heavy metal decontamination of polluted soils using crops of metal-accumulating crops. Resources, Conservation and Recycling 11, 41-49.

Baker A J M and Brooks R R 1989 Terrestrial higher plants which hyperaccumulate metal elements - A review of their distribution, ecology and phytochemistry. Biorecovery 1, 81-126.

Baker D E and Bowers M E 1988 Health effects of cadmium predicted from growth and composition of lettuce grown in gardens contaminated by emissions from zinc smelters. Trace Subst. Environ. Health 22, 281-295.

Beard J 1993 The evaluation of DDT contaminated soil associated with cattle tick dip sites. In Proceedings of the Second National Workshop on the Health Risk Assessment and Management of Contaminated Sites. Eds. A. Langley and M. van Alphen. pp. 119-133. South Australian Health Commission, Adelaide, South Australia.

Berti W R and Cunningham S D 1995 Remediating soil lead with green plants. In Trace Substances, Environment and Health. Ed. C R Cothern. pp 43-51. Science Reviews, Northwood, UK.

Beyer W N 1986 A re-examination of biomagnification of metals in terrestrial food chains. Environ. Toxicol. Chem. 5, 863-864.

Bingham F T, Sposito G and Strong J E 1984 The effect of chloride on the availability of cadmium. J. Environ. Qual. 13:71-74.

Bingham F T, Strong J E and Sposito G 1983 Influence of chloride salinity on cadmium uptake by Swiss chard. Soil Sci. 135:160-165.

Bingham F T, Sposito G and Strong J E 1986 The effect of sulfate on the availability of cadmium. Soil Sci. 141:172-177.

Bogess S F, Willavize S, Koeppe D E 1978. Differential response of soybean varieties to soil cadmium. Agron. J. 70:756-760.

Boyd R S and Martens S N 1994 Nickel hyperaccumulated by *Thlaspi montanum* var montanum is acutely toxic to an insect herbivore. Oikos 70, 21-25.

Boyd R S, Shaw J J and Martens S N 1994 Nickel hyperaccumulation defends *Streptanthus polygaloides* (Brassicaceae) against pathogens. Am. J. Bot. 81, 294-300.

Braen S N and Weinstein L H 1985 Uptake of fluoride and aluminium by plants grown in contaminated soils. Water, Air Soil Pollut. 24, 215-223.

Bremner I 1981 Effects of the disposal of copper rich slurry on the health of grazing animals. pp. 245-260. In Copper in Animal Wastes and Sewage Sludge. Eds. P L'Hermite and J Dehandtschutter. Reidel Publ., Dordrecht, Neth.

Brewer R F 1966 Fluorine. In Diagnostic Criteria for Plants and Soils. Ed. H D Chapman. Univ of Calif, Div. of Agric. Sci.

Brooks R R, Lee J, Reeves, R D and Jaffr T 1977 Detection of nickeliferous rocks by analysis of herbarium specimens of indicator plants. J. Geochem. Explor. 7, 49-57.

Brown S L, Chaney R L, Angle J S and Baker A J M 1994 Zinc and cadmium uptake by *Thlaspi caerulescens* and *Silene cucubalis* in relation to soil metals and soil pH. J. Environ. Qual. 23, 1151-1157.

Brown S L, Angle J S, Chaney R L and Baker A J M 1995 Zinc and cadmium uptake by *Thlaspi caerulescens* and *Silene cucubalis* grown on sludge-amended soils in relation to total soil metals and soil pH. Environ. Sci. Technol. 29, 1581-1585.

Burau R G, Jopling W F, Martin C V and Snow G F 1981 Monterey Basin Pilot Monitoring Project. Volume 2. Appendix L. Market basket survey of cadmium and zinc in the soils and produce of Salinas Valley. Appendix M. Cadmium, zinc, and phosphorus in Salinas Valley Soils. State of California.

Calabrese E J, Barnes R, Stanek E J III, Pastides H, Gilbert H E, Veneman P, Wang H, Lasztity A and Kostecki P T 1989 How much soil do young children ingest: An epidemiological study. Regulat. Toxicol. Pharmacol. 10, 123-137.

Cartwright B, Merry R H, and Tiller K G 1977 Heavy metal contamination of soils around a lead smelter at Port Pirie, South Australia. Aust. J. Soil Res. 15, 69-81.

Chaney R L 1973 Crop and food chain effects of toxic elements in sludges and effluents. In Proc. Joint Conf on Recycling Municipal Sludges and Effluents on Land. Nat. Assoc. St. Univ. and Land-Grant Coll., Washington, DC. pp 129-141.

Chaney R L 1980 Health risks associated with toxic metals in municipal sludge. In Sludge – Health risks of Land Application. Eds. G Bitton, B L Damro, G T and J M Davidson. pp 59-83. Ann Arbor Sci. Publ. Inc., Ann Arbor, Michigan, USA.

Chaney R L 1983 Potential effects of waste constituents on the food chain. In Land Treatment of Hazardous Wastes. Eds. J F Parr, P B Marsh and J M Kla. pp 152-240. Noyes Data Corp., Park Ridge, New Jersey, USA

Chaney R L 1983 Plant uptake of inorganic waste constituents. In Land Treatment of Hazardous Wastes. Eds. J F Parr, P B Marsh and J M Kla. pp 50-76. Noyes Data Corp, Park Ridge, New Jersey, USA.

Chaney R L and Giordano P M 1977 Microelements as related to plant deficiencies and toxicities. In Soils for Management and Utilization of Organic Wastes and Wastes and Wastewaters. Eds. L F Elliott and F J Stevenson. pp 235-279 Soil Sci. Soc. Amer. Inc., Madison, Wisc..

Chaney R L, Beyer W N, Gifford C and Silco L 1988 Effects of zinc smelter emissions on farms and gardens at Palmerton, PA. In Trace Substances in Environmental Health XX11. Ed. D D Hemphill. pp 263-280.

Chaney R L and Ryan J A 1994 Risks Based Standards for Arsenic, Lead and Cadmium in Urban Soils. (ISBN 3-926959-63-0) DECHEMA, Frankfurt. 130 pp.

Chaney R L and Ryan J A 1995 Guidelines for organics utilization and future developments in land applications. In Proc. Int. Symp. Land Application of Organics - 1995. July 28, 1995, Tokyo, Japan. Council for the Promotion of Utilization of Organic Materials (in press).

Chaney R L, Ryan J A and O'Connor G A 1995 Organic contaminants in municipal biosolids: Risk assessment, quantitative pathways analysis, and current research priorities. Sci. Total Environ. (in press).

Chaney R L, Ryan J A and O'Connor G A 1991 Risk assessment for organic micropollutants: US point of view. In Treatment and Use of Sewage Sludge and Liquid Agricultural Wastes. Ed. P L'Hermite. pp 141-158. Elsevier Applied Science, New York.

Chaney R L, Sterrett S B, Morella M C and Lloyd C A 1982 Effect of sludge quality and rate, soil pH, and time on heavy metal residues in leafy vegetables. In Proc. Fifth Ann. Madison Conf. Appl. Res. Pract. Municipal and Industrial Waste. pp 444-458. Univ. Wisconsin-Extension, Madison, Wisconsin.

Chaney R L, Li Y -M, Schneiter A A, Green C E, Miller J F and Hopkins D G 1993 Progress in developing technologies to produce low Cd concentration sunflower kernels. In Proc. 15th Sunflower Research Workshop. Eds. A A Schneiter, M L Straley and B D Vick pp 80-92. National Sunflower Association, Bismark, ND.

Chen Z S 1991 Cadmium and lead contamination of soils near plastic stabilizing materials producing plants in northern Taiwan. Water, Air Soil Pollut. 57-58, 745-754.

Corey R B, King L D, Lue-Hing C, Fanning D S, Street J J and Walker J M 1987 Effects of sludge properties on accumulation of trace elements by crops. In Land Application of Sludge. Eds. A L Page, T J Logan and J A Ryan. pp 25-51. Lewis Publishers Inc., Ann Arbor, MI.

Creaser C S, Fernades A R, Harrad S J, Hurst T and Cox E A 1989 Background levels of polychlorinated biphenyls in British soils A II. Chemosphere 19, 1457-1466.

Davis R D 1981 Copper uptake from soil treated with sewage sludge and its implications for plant and animal health. In Copper in Animal Wastes and Sewage Sludge. Eds. P L'Hermite and J Dehandtschutter. pp. 223-241. Reidel Publ., Boston, MA.

deVries M P C and Tiller K G 1978 Sewage sludge as a soil amendment, with special reference to Cd, Cu, Mn, Ni, Pb and Zn - comparison of results from experiments conducted inside and outside a greenhouse. Environ. Pollut. 16, 213-240.

EPAWA 1989 Monitoring Pesticides - A Review. The Environment Protection Authority, Perth, Western Australia, Bulletin No. 407.

Ernst W H C 1975 Physiology of heavy metal resistance in plants. In Proc. Symp. Int. Conf. Heavy Metals in the Environment vol II (Part I). Eds. T C Hutchinson, S Epstein, A L Page, J Van Leen and T Davey. pp. 121-136. University of Toronto, Toronto, Canada.

Ewers U, Freier I, Turfeld M, Brockhaus A, Hofstetter I, Konig W, Leisner-Saaber J and Delschen T 1993 Heavy metals in garden soil and vegetables from private gardens located in lead/zinc smelter area and exposure of gardeners to lead and cadmium (in German). Gesundheitswesen 55, 318-325.

Freeman G B, Johnson J D, Killinger J M, Liao S C, Feder P I, Davis A O, Ruby M V, Chaney R L, Lovre S C and Bergstrom P D 1992 Relative bioavailability of lead from mining waste soil in rats. Fundam. Appl. Toxicol. 19, 388-398.

Gilbert W S, Singh G and Ahmad N 1990 Pesticides residue studies in the Gwydir and Namoi cotton growing environments, 1981 and 1983. Technical Bulletin - NSW Agriculture & Fisheries, 68 pp.

Harrison D L, Mol J C M and Healy W B 1970. DDT residues in sheep from the ingestion of soil. N. Z. J. Agr. Res. 13, 664-672.

Hinesly T D, Alexander D E, Ziegler E L and Barrett G L 1978 Zinc and cadmium accumulations by corn inbreds grown on sludge amended soil. Agron J. 70, 425-428.

Hornbuckle K C, Achman D R and Eisenreich S J 1993 Over-water and overland polychlorinated biphenyls in Green Bay, Lake Michigan. Environ. Sci. Technol. 27, 87-98.

Hsieh C F 1990 Cadmium content in the soil and its uptake and distribution in the different parts of rice plants. Bulletin of Taichung District Agricultural Improvement Station. No. 29, 11-27.

Hume T M, Fox M E and Wilcock R J 1989 Use of organochlorine contaminants to measure sedimentation rates in estuaries : a case study from the Manukau Harbour. J. Royal Soc. of New Zealand 19, 305-317.

Kabata-Pendias A and Pendias H 1984 Trace Elements in Soils and Plants. CRC Press, Inc., Boca Raton, Florida, USA. 315 pp.

Kao M H 1991 The evaluation of sawdust swine waste compost on the soil ecosystem, pollution and vegetable production. Soils Fertil. Taiwan, pp 31-47.

Karim Z 1993 Phosphatic fertilizers and the environment. In Proceedings of the Regional FADINAP seminar on fertilization and the environment, Chiang Mai, Thailand, 7-11 September 1992. pp 103-111. New York, USA. United Nations.

Kawano M, Ramesh A, Thao V D and Tatsukawa R 1992 Persistent organochlorine insecticide residues in some paddy, upland and urban soils of India. Int. J. Environ. Anal. Chem. 48, 163-174.

Kobayashi J 1978 Pollution by cadmium and the *itai-itai* disease in Japan. In Toxicity of Heavy Metals in the Environment. Ed. F W Oehme. pp. 199-260. Marcel Dekker, Inc., New York.

Landner L, Oberg M O and Aringberg-Laanatza M 1995 Cadmium in Fertilizers (Preparation of an OECD Workshop). National Chemicals Inspectorate of Sweden, Stockholm.

Larsen S and Widdowson A E 1971 Soil fluorine. J. Soil Sci 22, 2210-221.

Li Y-M, Chaney R L and Schneiter A A 1995 Effect of soil chloride level on cadmium concentration in sunflower kernels. Plant Soil 167, 275-280.

Li Y-M, Chaney R L, Schneiter A A and Miller J F 1995b Combining ability and heteosis estimates for kernel cadmium level in sunflower. Crop Sci. 35, 1015-1019.

Lund L J, Betty E E, Page A L and Elliott R A 1981 Occurrence of naturally high cadmium levels in soils and its accumulation by vegetation. J. Environ. Qual. 10, 551-556.

Ma W-C 1989 Effect of soil pollution with metallic lead pellets on lead bioaccumulation and organ/body weight alterations in small mammals. Arch. Environ. Contam. Toxicol. 18, 617-622.

MacIntire W H, Winterberg S H, Thompson J G and Hatcher B W (1942) Fluorine content of plants, fertilized with phosphates and slags carrying fluorides. Ind. Eng. Chem. 34, 1469-1479.

Maclean A J 1976 Cadmium in different plant species and its availability in soils as influenced by organic matter and additions of lime, phosphorus, cadmium and zinc. Can. J. Soil Sci. 56, 129-138.

Maier N A and McLaughlin M J 1995 Agricultural limestone is ineffective in controlling Cd uptake by commercial potato crops. Fert. Res. (in review)

Mays D A and Mortvedt J J 1986 Crop response to soil applications of phosphogypsum. J. Environ. Qual. 15, 78-81.

McBride M B 1995 Toxic metal accumulation from agricultural use of sludge: Are USEPA regulations protective? J. Environ. Qual. 24, 5-18.

McLaughlin M J, Palmer L T, Tiller K G, Beech T W and Smart M K 1994 Increasing soil salinity causes elevated cadmium concentrations in field-grown potato tubers. J. Environ. Qual. 23, 1013-1018.

McLaughlin M J, Tiller K G, Naidu R and Stevens D P (1995) Review: The behaviour and environmental impact of contaminants in fertilizers. Aust. J. Soil Res. 34, 1-54.

Menzel R G 1968 Uranium, radium, and thorium content in phosphate rocks and their possible radiation hazard. J. Agr. Food Chem. 16, 231-234.

Merry R H 1988 Investigations on cadmium in South Australia: rainfall, soils, cereals, pastures and soil-plant relations. In Cadmium accumulations in Australian agriculture: National Symposium. Canberra 1-2 March 1988. Eds. J Simpson and B Curnow. pp 62-79. Aust. Gov. Publ. Serv., Canberra, Australia.

Merry R H, Tiller K G, De Vries M P C and Cartwright B 1981 Contamination of wheat crops around a lead-zinc smelter. Environ. Pollut. Ser. B 2, 37-48.

Moffat A S 1995 Plants proving their worth in toxic metal cleanup. Science 269, 302-303.

Mortvedt J J 1986 Effect of calcium silicate slag application on radium-226 concentrations in plant tissues. Comm. Soil Sci. Plant Anal. 17, 75-84.

Mortvedt J J and Sikora F J 1992 Heavy metals, radionuclides and fluorides in phosphorus fertilizers. In Future Directions for Agricultural Phosphorus Research. Ed. F J Sikora. Agricultural Research Department, National Fertilizer and Environmental Research Centre, Tennessee Valley Authority Bulletin Y-224. TVA, Muscle Shoals, Alabama, USA.

Muramoto S 1990 Comparison of metal uptake between glutinous and non-glutinous rice of cadmium chloride, oxide and sulfide at the critical levels. Bull. Environ. Contam. Toxicol. 45, 415-421.

Naidu R, Bolan N S, Kookana R S and Tiller K G 1994 Ionic strength and pH effects on the sorption of cadmium and the surface charge of soils. Europ. J. Soil Sci. 45, 419-429.

National Research Council 1980 Mineral tolerance of domestic animals. National Academy of Sciences, Washington DC, USA. 577 pp.

NFA (1993) Australian Food Standards Code. National Food Authority. Aust. Govt. Publ. Service, Canberra, Australia.

Nogawa K, Honda R, Kido T, Tsuritani I and Yamada Y 1987 Limits to protect people eating cadmium in rice, based on epidemiological studies. Trace Subst. Environ. Health 21, 431-439.

Ocker H -D, Bruggemann J, Ruhl Ch -St and Klein H 1991 Cadmium in sunflower kernels, poppy and sesame (in German). Bundesgesundheitsblatt 34, 556.

O'Connor G A, Kiehl D, Eiceman G A and Ryan J A 1990 Plant uptake of sludge-borne PCBs. J. Environ. Qual. 19, 113-118.

O'Connor G A, Chaney R L and Ryan J A 1991 Bioavailability to plants of sludge-borne toxic organics. Rev. Environ. Contam. Toxicol. 121, 129-155.

Oliver D P, Schulz J E, Tiller K G and Merry R H 1993 The effect of crop rotations and tillage practices on cadmium concentration in wheat grain. Aust. J. Soil Res. 44, 1221-1234.

Oliver D P, Hannam R, Tiller K G, Wilhelm N S, Merry R H and Cozens G D 1994a The effects of zinc fertilization on cadmium concentration in wheat grain. J. Environ. Qual. 23, 705-711.

Oliver D P, Tiller K G, Conyers M K, Slattery W J, Merry R H and Alston A M 1994b The effects of soil pH on Cd concentration in wheat grain grown in south-eastern Australia. In Third International Symposium on Plant-Soil Interactions at Low pH. Ed. D G Edwards. Kluwer Academic Publishers, Dordrecht, Germany (in press).

Oliver D P, Gartrell J W, Tiller K G, Correll R, Cozens G D and Youngberg B L 1995 Differential responses of Australian wheat cultivars to cadmium concentration in wheat grain. Aust. J. Agric. Res. 46, 873-886.

Orio A A, Botkin D B and Pillai M K K (1986) Pesticide pollution of soil, water and air in Delhi area, India. Sci. Total Environ. 55, 321-327.

Page A L, Bingham F T and Chang A C 1981 Cadmium. In Effects of Heavy Metal Pollution on Plants vol. 1: Metals in the Environment. Ed. N W Lepp. pp 77-103. Appl. Sci. Pub., London, England.

Parker D R, Chaney R L and Norvell W A 1995 Equilibrium computer models: Applications to plant nutrition research. pp. 163-200. In Chemical Equilibrium and Reaction Models. Eds. R H Loeppert, A P Schwab and S Goldberg. Soil Science Society of America Special Publ. No. 42. Soil Sci. Soc. Am./Am. Soc. Agron., Madison, WI.

Penner G A, Clarke J, Bezte L J and Leisle D 1995 Identification of RAPD markers linked to a gene governing cadmium uptake in durum wheat. Genome 38, 543-547.

Pepper I L, Bezdicek D F, Baker A S and Sims J M 1983 Silage corn uptake of sludge-applied zinc and cadmium as affected by soil pH. J. Environ. Qual. 12, 270-275.

Pinkerton B W and Brown K W 1985 Plant accumulation and soil sorption of cobalt from cobalt-amended soils. Agron. J. 77, 634-638.

Pinkerton B W, Brown K W, Thomas J C and Zabcik D 1981 Plant uptake of cobalt from a petroleum sludge affected soil. Agron. Abstr. 1981, 32.

Poole D B R, McGrath D, Fleming G A and Sinnott J 1983 Effects of applying copper rich pig slurry to grassland. 3. Grazing trials: Stocking rate and slurry treatment. Ir. J. Agric. Res. 22:1-10.

Pratley J E and Rowell D L 1987 From the first fleet - evolution of Australian farming systems. In Tillage - New Directions in Australian Agriculture. Eds. P S Cornish and J E Pratley. pp 2-24. Inkata Press, Melbourne, Australia.

Ruby M V, Davis A, Link T E, Schoof R, Chaney R L, Freeman G B and Bergstrom P 1993 Development of an *in vitro* screening test to evaluate *in vivo* bioaccessibility of ingested mine-waste lead. Environ. Sci. Technol. 27, 2870-2877.

Sarasua S M, McGeehin M A, Stallings F L, Terracciano G J, Amler R W, Logue J N and Fox J M 1995 Final Report. Technical Assistance to the Pennsylvania Department of Health. Biological indicators of exposure to cadmium and lead. Palmerton, Pennsylvania. Part II (May 1995). Agency for toxic substances and diseases registry, US-DHHS, Atlanta, Georgia. 57 pp.

Sauerbeck D 1993 Conditions controlling the bioavailability of trace elements and heavy metals derived from phosphate fertilizers in soils. In Proceedings of the International IMPHOS Conference - Phosphorus, Life and Environment. Institute Mondial du Phosphate, Casablanca. (in press).

Schroeder H A and Balassa J J 1963 Cadmium: uptake by vegetables from superphosphate and soil. Science 140, 819-820.

Shiwen C, Lin Y, Zhineng H, Xianzu Z, Zhaolu Y, Huidong X, Yuanrong L, Rongdi J, Wenhau Z and Fangyuan Z 1990 Cadmium exposure and health effects among residents in an irrigation area with ore dressing wastewater. Sci. Total Environ. 90, 67-73.

Sikora F J, Copeland J P, Dillard E F and Burnell J R 1992 Fluorosilicic acid in suspension fertilizers: Effects on maize growth and soil chemistry. Soil Sci. Soc. Am. J. 56

Singh B R 1990 Cadmium and fluoride uptake by oats and rape from phosphate fertilizers in two different soils. Norw. J. Agri. Sci. 4, 239-249.

Singh B R and Steinnes E 1976 Uptake of trace elements by barley in zinc polluted soils. 2. Lead, mercury, selenium, arsenic, chromium and vanadium in barley. Soil Sci. 121, 38-43.

Smith W H and Siccama T G 1981 The Hubbard Brook ecosystem study: Biogeochemistry of lead in the northern hardwood forest. J. Environ. Qual. 10, 323-333.

Street J J, Sabey B R and Lindsay W L 1978 Influence of soil pH, phosphorus, cadmium, sewage sludge and incubation time on the solubility and plant uptake of cadmium. J. Environ. Qual. 7, 286-290.

Strehlow C D and Barltrop D 1988 The Shipham Report -- An investigation into cadmium concentrations and its implications for human health: 6. Health studies. Sci. Total Environ. 75, 101-133.

Stenstrom T and Vahter M 1974 Cadmium and lead in Swedish commercial fertilizers. Ambio 3, 91-92.

Taiganides E P 1976 Principles and techniques of animal wastes management and utilization in Asia. In Organic recycling in Asia. FAO Soils Bulletin 36. pp 341-362.

Takahashi J 1976 Role of night soil in Japanese agriculture. In Organic recycling in Asia. FAO Soils Bulletin pp. 363-364.

Thao V D, Kawano M and Tatsukawa R 1993a. Persistent organochlorine residues in soils from tropical and sub-tropical Asian countries. Environmental Pollution 81, 61-71.

Thao V D, Kawano M, Matsuda M, Wakimoto T, Cau H D, Quynh H T and Tatsukawa R (1993b) Chlorinated hydrocarbon insecticide and polychlorinated residues in soils from the southern provinces of Vietnam. Intern. J. Environ. Anal. Chem. 50, 147-159.

U.S. Environmental Protection Agency. 1989a. Development of risk assessment methodology for land application and distribution and marketing of municipal sludge. EPA/600/6-89/001.

U.S. Environmental Protection Agency. 1993a. 40 CFR Part 257. Standards for the Use or Disposal of Sewage Sludge; Final Rules. Fed. Reg. 58,9248-9415.

U.S. Environmental Protection Agency. 1993b Technical support document for land application of sewage sludge. Vol. I. PB93-110575, Vol. II PB3-110583. NTIS

Wakefield Z T 1980 Distribution of cadmium and selected heavy metals in phosphate fertilizer processing. Bull. Y-159. National Fertilizer Development Centre, Tennessee Valley Authority, Muscle Shoals, AL.

Wallace A., Romney, Alexander G V, Soufi S M and Patel P M 1977 Some interactions in plants among cadmium, other heavy metals, and chelating agents. Agron. J. 69,18-20.

Weitzman M, Aschengrau A, Bellinger D, Jones R, Hamlin J S and Beiser A 1993 Lead-contaminated soil abatement and urban children's blood lead levels. J. Am. Med. Assoc. 269, 1647-1654.

White M C and Chaney R L 1980 Zinc, cadmium, and manganese uptake by soybean from two zinc and cadmium-amended Coastal Plain soils. Soil Sci. Soc. Am. J. 44, 308-313.

Willaert G and Verloo M 1992 Effects of various nitrogen fertilizers on the chemical and biological activity of major and trace element in a cadmium contaminated soil. Pedologie 43, 83-91.

Williams C H and David D J 1973 The effect of superphosphate on the cadmium content of soils and plants. Aust. J. Soil Res. 11, 43-56.

Zarcinas B and Nable R O 1992 Boron and other impurities in South Australian fertilizers and soil amendments. CSIRO Division of Soils Divisional Report 118, 10 pp.

Zhu Y M, Berry D F and Martens D C 1991 Copper availability in two soils amended with eleven annual applications of copper-enriched hog manure. Commun. Soil Sci. Plant Anal. 22: 769-783.

12
Technologies for enhanced remediation of contaminated soils and aquifers: overview, analysis, and case studies

P. S. C. RAO, G. B. DAVIS AND C. D. JOHNSTON

1. Introduction

In many industrialised countries, contamination of soils and groundwater from improper disposal of hazardous industrial wastes is an environmental issue of increasing public concern, regulatory activity, and scientific investigation. In the United States (USA) alone, there are estimated to be about 400 000 waste disposal sites where soil and groundwater contamination is deemed to be of sufficient extent and magnitude, that some type of remedial action is warranted to protect public health or to minimise adverse environmental and ecological impacts. The costs of cleaning up such sites over the next 20–30 years may easily exceed one trillion USA dollars. Experiences of environmental scientists and engineers in the USA over the past decade have shown, however, that the required regulatory levels may not be achieved at most sites, even after several decades of clean-up efforts using the existing technologies.

The experiences in other industrialised countries are quite similar to those in the USA, although the differences in regulatory process and socio-economic factors that drive the remediation attempts must be considered in any such comparison. For example, the remediation endpoints required by regulations, the scope of monitoring programs, and the assignment of legal/financial liability for remediation efforts all may vary greatly from country to country. The environmental literacy and the public perception of the relative risks of soil and groundwater contamination (in comparison to other hazards) can influence the regulatory policy and levels of allowed contamination as well as the expectations of the required clean-up of contaminated sites. Such factors play a dominant role in determining the range of technologies used for monitoring to identify soil and groundwater contamination problems, and the range of technologies used to remediate contaminated sites. It is entirely possible that a large number of waste disposal sites remain yet to be identified in many countries. Equally likely is the scenario that existing groundwater contamination, and efforts to clean-up such contamination, may be entirely undocumented or that such information is either inaccessible or inadequate for scientific analyses.

2. Soil and groundwater contamination

A study of the release of chemicals from a contaminated site is concerned with two aspects: (a) the *amounts or concentrations* of chemicals being released, and (b) the

R. Naidu et al. *(eds): Contaminants and the Soil Environment in the Australasia-Pacific Region, 361–410.*

rate of release of these chemicals. These two aspects are addressed by two types of quantitative approaches: thermodynamics (equilibrium theory) and kinetics. Thermodynamic calculations provide information on the maximum concentrations at which a given contaminant may be present, as well as on the equilibrium distribution coefficients of the contaminants when they are in contact with more than one phase (e.g. solution, solid, and gas phases). Approaches using kinetics provide information on chemicals. Approaches using kinetics provide information on the rates at which contaminant concentrations change, and permit estimation of the time required to attain equilibrium concentrations predicted from thermodynamic considerations.

The release and migration of organic contaminants from a waste disposal/spill site will produce a contaminant plume, whose composition will vary with time and distance as the plume size increases. Because of the nonuniformity in the plume composition, we find it convenient to separate the problem into three regions: a *near-field or source region*; a *transition zone*; and a *far-field or dissolved plume region*. Rather than distance from the contaminant source, the criterion employed to designate these regions is the chemistry of the contaminant mixture.

The near-field region corresponds to the source itself and its immediate vicinity. In this region, the composition and concentrations of most waste constituents are similar to that in the waste itself. There are usually two, possibly three, liquid phases in this region. This would be the case, for example, at a fuel spill site, where we may find nonaqueous phase liquids (NAPLs) and an aqueous phase as well as a vapour phase. Based on their density, NAPLs are usually referred to as Dense (or DNAPLs) when they are denser than water, and as Light (LNAPLs) when their density is less than water. This distinction is particularly important because of the dramatic differences in retention and transport of these water-immiscible fluids in the vadose zone and aquifers (Gillham and Rao, 1990). Examples of NAPLs of environmental concern include: chlorinated solvents used as industrial solvents and cleaning agents (e.g. trichloroethene, TCE, and perchloroethene, PCE); various automobile and aviation fuels (gasoline, diesel, JP-4); lubricating oils (motor oil, transmission fluids); and industrial processing wastes (coal tar, creosote).

In the transition zone, the solution phase is likely to be predominantly a single-phase, homogeneous liquid made up of water and varying amounts of cosolvents or surfactants or 'facilitators' (e.g. dissolved organic carbon or other colloidal matter) if they were present in the source region. The concentration of one or more contaminants may be sufficiently high to make these contaminants behave differently than they would in dilute aqueous solutions.

The far-field region corresponds to the dissolved phase plume in which contaminants are present in the aqueous phase well below their aqueous solubility limits. During migration of the plume through the vadose zone and the saturated zone, chromatographic separation of the constituents occurs due to their different mobilities. Furthermore, dilution resulting from hydrodynamic dispersion and attenuation resulting from abiotic/biotic transformations decreases contaminant concentrations. Thus, high concentrations of multiple contaminants are not likely to be found as the distance from the source increases. Nevertheless, it is likely that these contaminant concentrations may be higher than the standards set by the regulatory agencies, and thus remediation may be required.

2.1 Migration and retention of NAPLs

The migration of NAPLs is important in understanding the final distribution of complex waste mixtures in the subsurface. When a NAPL is introduced into the subsurface, it will migrate downward through the vadose zone as an immiscible liquid due to gravitational and hydraulic potential gradients. Lateral spreading also occurs due to the effect of physical heterogeneity (e.g. layering) and capillary forces (Schwille, 1984). As the NAPL progresses downward, part of it is trapped in the pore spaces, and will become immobilised due to capillary forces. The immobilised portion that is retained by the soil is referred to as the residual saturation of NAPL. Whether the NAPL reaches the saturated zone as a free-moving liquid depends on the amount introduced, the area of infiltration, and the residual saturation in the vadose zone.

Once the NAPL reaches the saturated zone, further migration is determined by its density. LNAPLs will float on top of the capillary fringe and spread laterally in the direction of the declining water table. DNAPLs will stay on top of the capillary fringe until a sufficient pressure is developed to overcome the surface tension of the capillary fringe, and then penetrate into the saturated zone. When groundwater is displaced by the pressure of the moving NAPL, the NAPL-water interface may become unstable, as manifested by 'fingers' of NAPL penetrating through the aquifer (Schwille, 1988). As the DNAPL migrates downward through the saturated zone, it leaves behind a residual saturation of DNAPL entrapped in the pore space. This residual NAPL saturation is generally much higher than the residual saturation in the vadose zone (Wilson et al., 1990). The DNAPL migrates downward until it encounters a less permeable layer, and then spreads laterally in accord with the slope of this less-permeable stratification, forming pools of DNAPL. This lateral spreading could be in a direction different than that of regional groundwater flow. Accumulation of the NAPL may occur at any location in the porous medium (saturated or unsaturated zone) where the NAPL encounters zones of lower permeability. The distribution of NAPL becomes more complicated when the soil or aquifer consists of discontinuous laminae of different permeabilities.

In contrast to a DNAPL which may sink through the saturated zone, a LNAPL accumulates at the top of the capillary fringe. If the mass of the LNAPL becomes sufficiently large, a depression in the water table may be created. Due to seasonal fluctuations or pumping activities the elevation of the water table moves up and down. This periodic vertical displacement results in trapping of the LNAPL in capillary spaces below the average depth of the groundwater table (Pfannkuch, 1984; Hunt et al., 1988). The LNAPL may only periodically be remobilised and appear in monitoring boreholes (Steffy et al., 1995).

The NAPL phase, present as residual saturation or as pools, forms a source of soluble contaminants. Groundwater that comes in contact with the NAPL will dissolve components from the NAPL. The dissolved contaminants will be carried away from the source in the direction of the groundwater flow, thus creating a dissolved-phase contaminant plume. The contaminants dissolved in the aqueous phase may be retained by the solid phase, reducing their rate of transport. In the vadose zone, an envelope of organic vapors can develop around the source when compounds volatilise from the NAPL or soil water. These vapors may diffuse through the vadose zone and escape at the ground surface or contaminate upstream regions of pristine groundwater. Water infiltrating through the vadose zone will dissolve some of the NAPL constituents, and cause a periodic influx of contaminants into the saturated zone. Figures 1 and 2 show schematic representations of the distribution of a NAPL spill and an DNAPL leak at hypothetical sites.

Transport of NAPLs in subsurface systems can be described by multi-phase flow models. A growing body of literature is presently available, in which various mathematical models are presented that describe the migration of NAPLs in saturated and unsaturated media (Corey, 1986; Abriola, 1989). These transport models range from those based on simple sharp-interface or piston-flow assumptions (e.g. Van Dam, 1967) to more sophisticated models in which the flow equations for each of the fluid phases are simultaneously solved, including hysteresis in displacing patterns (Lenhard et al., 1989). Although these multi-phase transport models can be used to get a general idea of how NAPLs migrate through porous media, their practicality may be limited. In general, transport of NAPLs is limited to a relatively small volume of the total contaminated zone. In most cases, the wetting history and soil heterogeneity are unknown, making prediction of the transport and distribution of NAPLs very difficult. Furthermore, depending on the characteristics of the porous medium and NAPL, the transport of the NAPL as a separate fluid phase may take place in a relatively short time period. It is the release of constituents from the residual NAPL and pools of NAPL that determines the long-term (over several decades or centuries) contamination problem.

2.2 Release and transport of contaminants

The release and transport of organic chemicals in NAPL-contaminated soils is controlled by four major processes: dissolution, volatilisation, sorption, and biodegradation. The rates at which these processes occur, and the manner in which these coupled processes impact each other, will determine the spatial extent of the source zone and the dissolved phase and vapor phase contaminant plumes. Furthermore, the ability to sustain or enhance some of these processes will determine the extent and rate at which site remediation can be achieved.

Augustijn (1993) suggested that it is useful to distinguish five different zones in which each of these processes play a different role. Zone I is the region in which the NAPL is a free moving liquid; Zones II and III represent the regions containing residual NAPL in the saturated and unsaturated zone, respectively; Zone IV represents the dissolved-phase contaminant plume in the saturated zone; and Zone V

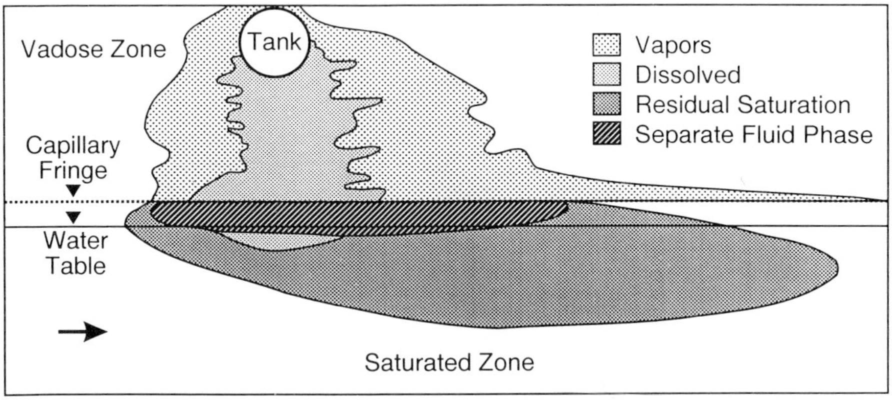

Figure 1. Schematic representation of LNAPL distribution in the vadose and saturated zones at a hypothetical contaminated site with an unconfined aquifer. The source of the LNAPL is a leaking underground storage tank.

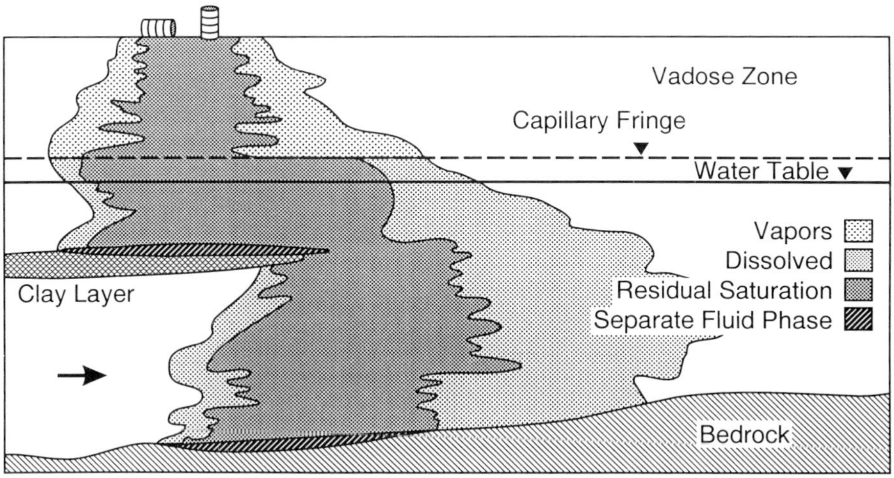

Figure 2. Schematic representation of DNAPL distribution in the vadose and saturated zones at a hypothetical contaminated site with an unconfined aquifer underlain by bedrock. The source of the DNAPL is a spill at the ground surface.

represents the vapor phase contaminant plume in the vadose zone. Zones I through III are considered as the source (near-field) region of the contaminated site, whereas Zones IV and V are referred to as the plume (far-field) regions. Dissolution and volatilisation are the primary processes controlling the release of organic contaminants from the source; sorption is the dominant process in the far-field; volatilisation is only important in the vadose zone; whereas biodegradation plays a significant role in the vadose and the saturated regions of Zones IV and V. Figure 3 shows

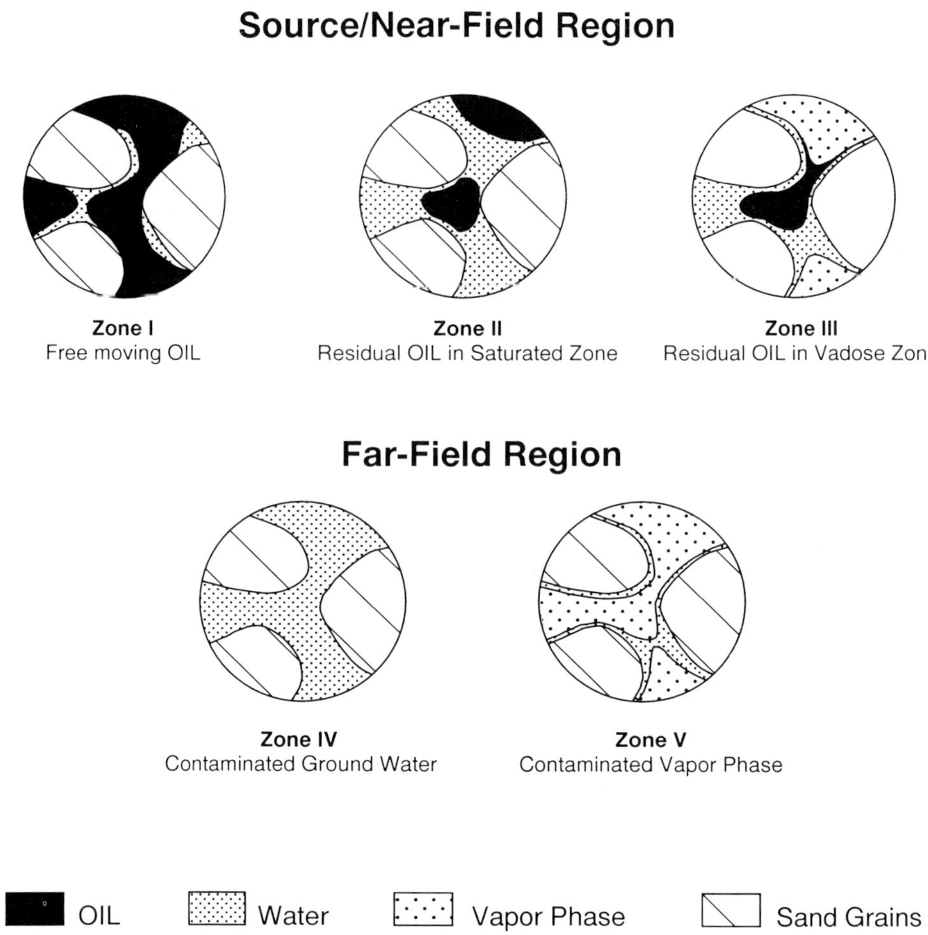

Figure 3. Schematic representation of the fluid phase distribution in five different zones at a contaminated site [from, Augustijn, 1993]

schematics of the distribution of phases in each of these zones. Augustijn (1993) presented the following description of each of those five regions.

Zone I is the region in which free moving NAPL is present, and occurs where the NAPL has accumulated due to a reduced permeability for the NAPL, such as at the water table or at a low-permeable stratification. In this zone, the NAPL content exceeds the residual saturation, but will almost never equal porosity. The contaminant concentrations in the air and water phases in this zone, are most likely in equilibrium with the NAPL and sorbed phases. It is, however, the contact between this region and the flowing water that determines the release of organic contaminants. This contact area is often very limited compared to the total volume of the NAPL pool. A review of the specific problems associated with dense NAPLs is given by Huling and Weaver (1991).

The region in which the NAPL is trapped as a residual fluid phase in the saturated zone is indicated by Zone II. The entrapment of an immiscible liquid occurs when the capillary pressure exceeds other forces acting on the liquid. The capillary pressure is the result of a combination of adhesive forces between the liquid and solid phases, and cohesive forces within the fluid itself. The capillary force is proportional to the interfacial tension and the strength of fluid wetting to the solid surface, and inversely proportional to the pore size. Hence, the residual NAPL saturation depends on several factors such as the media's pore-size distribution; wettability; fluid viscosity and density ratios; interfacial surface tension; gravity and buoyancy forces; and hydraulic gradients (Mercer and Cohen, 1990).

The residual saturation is expressed as the fraction of pore spaces occupied by the NAPL. At the stage of residual saturation, the NAPL phase is discontinuous and may occur as single droplets or ganglia (Chatzis et al., 1983; Schwille, 1988; Wilson et al., 1990). The residual saturation can vary widely. Values have been reported from less than a percent for low viscosity fluids and highly permeable media (Schwille, 1984), to as high as 60% for gasoline in an initially dry, fine sand (Hoag and Marley, 1986). The residual saturation is highest in heterogeneous media (Wilson et al., 1990).

For most minerals, water is the wetting fluid, i.e. water preferentially covers the predominantly hydrophilic mineral surfaces. Soil organic matter, however, prefers contact with the organic liquid due to hydrophobic forces. Therefore, in surface soils, the solid matrix may have an intermediate preference for wetting by water or NAPL. For this reason, the distribution of residual NAPL in subsurface systems may be different from the distributions observed for experimental media in which the solid surfaces are preferably wetted by water. Furthermore, the NAPL mixture may also be sufficiently polar to interact with the mineral phases. Thus, the NAPL may be present not only as ganglia, but also as films.

Zone III represents the regions in the unsaturated zone that contain residual NAPL. Air is the least-preferential wetting fluid for mineral surfaces; hence, the most likely sequence of phase distributions in the vadose zone is mineral/water/NAPL/air. Since the NAPL is a discontinuous phase at residual saturation, both NAPL and water may contact the air phase.

The unsaturated zone is a very dynamic component of the subsurface hydrology. During wetting and drying cycles, the air and water contents and fluxes can vary considerably, changing contact areas between the different phases almost continuously. Dissolution of the residual NAPL will be important when water percolates through the vadose zone. Sorption and desorption will determine the rate at which the NAPL constituents move down once they are dissolved. The predominant pathway for degradation in the vadose zone would be aerobic biodegradation. Water infiltration at the surface causes periodic inputs of contaminants into the saturated zone. If the site is covered by structures that prevent any infiltration of water, this pathway for groundwater contamination is essentially eliminated.

Zone IV represents the contaminant plume in the saturated zone, originating from the NAPL phases present in the zones described previously. The size and shape of the dissolved phase plume depends on the time, the compositional complexity of the NAPL that was introduced in the soil, the groundwater flow rate, heterogeneity of the soil, and sorption dynamics. Sorption is the primary process that controls the relative rate at which the contaminants are being transported. Different chemicals experience different affinities for the solid phase, which leads to chromatographic separ-ation of the chemicals in the contaminant plume. Rates of biodegradation via aerobic or anaerobic pathways play a major role in the attenuation of contaminants in Zone IV (e.g. Thierrin et al., 1995).

The contaminated vadose zone free of NAPL is indicated by Zone V. As was discussed under Zone III, the vadose zone is a very dynamic system in which the phase distribution frequently changes. During the periods that air exists as a continuous phase, transport through the vapor phase may be important for volatile organic compounds. Transport of organic vapors is controlled by vapor diffusion, advection due to density differences (Falta et al., 1989; Mendoza and Frind, 1990), or advection due to pressure gradients caused, for example, by infiltrating water. In these cases, the vapor phase acts as a medium in which contaminants are slowly being transported from NAPL contaminated regions to the atmosphere, groundwater, or other regions of the unsaturated zone. At low water contents, sorption of organic vapors at the air/water interface (Pennell et al., 1992), and onto mineral surfaces becomes more important (Rhue et al., 1989; Chiou, 1990; Pennell et al., 1992), affecting the rate of transport through the vapor phase.

2.3 The importance of nonequilibrium mass transfer

Nonequilibrium mass transfer conditions have an important impact on contaminant concentrations, the dispersion and attenuation of contaminants, and the efficiency of remediation techniques. In many cases, some type of nonequilibrium constraint for inter-phase mass transfer exists, caused by either large-scale heterogeneity or rate-limiting processes at the local scale, or both (see Chapter 4).

The heterogeneous character of most porous media causes the hydraulic conductivity to be spatially variable, which results in a heterogeneous distribution of soil water or groundwater velocities. The concentration gradients created by the non-uniform velocity distribution can result in rate-limited diffusion between different flow domains, causing nonequilibrium conditions. The presence of a NAPL phase also reduces the hydraulic conductivity for water and enhances this type of nonequilibrium.

At the local scale (pore- or grain-scale), nonequilibrium conditions occur when the contact time between phases is not sufficient to complete the process that leads to an equilibrium distribution of the chemicals in all phases. The time required by a process to reach equilibrium can be determined by a chemical reaction rate or diffusion process. Chemical processes such as dissolution, volatilisation, sorption, and degradation, are generally considered rapid, but may require an extended time for completion. The rate of many processes, however, is controlled by diffusion. Dissolution of multi-component NAPLs, for example, involves diffusion of the constituents through the NAPL phase toward the NAPL-water interface, dissolution at the interface, and diffusion away from the NAPL-water interface. Sorption is also considered to be controlled by diffusion. Two mechanisms have been proposed to describe rate-limited sorption: intra-sorbent diffusion, and retarded intra-particle diffusion. The first mechanism views sorption as diffusion of solutes into the sorbent matrix, the second mechanism views sorption as diffusion into the micropores of porous particles in which the solutes are instantaneously sorbed (see Brusseau and Rao, 1989; and Chapter 4 this volume). Sorption dynamics can also significantly limit bioavailability and biodegradation rates of contaminants in contaminated soils and aquifers (e.g. Rao et al., 1993; 1995; Alexander, 1994).

The contact area between phases has an important impact on nonequilibrium mass transfer. Small contact areas between phases result in short contact times and smaller mass flux across the interface which increases the potential for non-equilibrium conditions. For this reason, the distribution of the NAPL phase is very important, since it determines the contact area between NAPL-water, NAPL-air, and air-water. Large groundwater or soil water velocities also reduce the contact time, and therefore may enhance nonequilibrium conditions. This may be the case not only in regions of large intrinsic permeability for fluid flow under natural hydraulic gradients, but also the case for fluid flow under induced gradients during pumping.

In NAPL-contaminated soils and aquifers, conditions observed at the larger scale may be different from the conditions at the local scale. For example, if dissolution of the NAPL is instantaneous, but dilution of contaminated water with fresh water leads to concentrations well below expected equilibrium, the observations at a larger scale may be misinterpreted as nonequilibrium conditions. On the other hand, when dissolution is rate limited at the local scale, but contact between NAPL and water extends over a large distance, apparent equilibrium conditions may be reached at the whole-plume scale. In these cases, it is again important to know the distribution of the NAPL phase to be able to assess the current situation and predict further plume migration and attenuation.

3. Overview of remediation strategies

The need to invoke any remediation strategy at a given contaminated site is based on risk analysis, which determines that the level of exposure to contaminants via various exposure pathways is unacceptable in terms of either human health considerations or ecological impacts. Thus, the magnitude and type of remediation required is determined by three primary factors: the source; the path(s) of exposure; and the impacted target(s) or receptor(s). This approach is referred to as the 'Source-Path-Target' methodology (Loxham, 1992). The required levels of clean-up at a site may include restoration to pre-contamination or 'pristine' levels. More likely, however, regulatory requirements may specify clean-up to achieve some target value (e.g. maximum contaminant level, or MCL) that reduces the expected risk to some acceptable level (e.g. 1 in 10^6).

The available remediation technologies may be grouped in two categories: (1)

ex-situ techniques which require removal of the contaminated soil or groundwater, for treatment either on site or off site, and in some cases returning the treated water/soil to the restored site; and (2) *in-situ* methods, which attempt remediation without excavation of contaminated soils or pumping of contaminated groundwater. *In-situ* techniques may be favored over the *ex-situ* techniques for the following reasons: (1) cost reduction by eliminating or minimising site excavation, transportation, and treatment/disposal costs; (2) reduced health impacts on the public or remediation crews by minimising exposure to contaminants during excavation/ treatment processes; and (3) remediation of inaccessible sites located at greater depths or located under or around existing superstructures (e.g. buildings) such that excavation is not an option. In this paper, we will focus on the *in-situ* remediation techniques for soil and groundwater clean-up. However, one case study of *ex-situ* treatment using similar methods to those used *in-situ* is also presented (see also Chapter 13 for more detail).

In-situ remediation technologies may be grouped into three broad categories (Figure 4) based on whether the clean-up goals are achieved by containment/immobilisation of the source or by treatment to reduce/remove the source amount or to decrease exposure. Containment technologies are based on the placement of barriers that prevent contaminant transport beyond the source zone, and are used to provide a short-term solution while the contaminant plume (far-field regions) is being remediated using other technologies. Containment technologies are also useful for long-term isolation of source regions when no immediate danger or threat is perceived. An active area of development is technologies based on permeable or semi-permeable reactive barriers which attenuate contaminant concentrations without influencing groundwater flow patterns. Here, we will only examine *in-situ* technologies based on treatment (e.g. extraction) or process manipulation (e.g. biodegradation). For further information on containment technologies refer to Chapter 13.

4. Overview of pump-and-treat technologies

The Pump-and-Treat (PAT) method is the most common groundwater remediation technique attempted to date in the USA (EPA, 1993). At contaminated sites, several groundwater extraction wells are placed within the contaminant plume at strategic locations. Large volumes of contaminated water are extracted, usually at rates greater than the regional groundwater flow, over extended periods of time. The contaminated water is then treated *ex-situ* to remove the suite of contaminants that are of specific concern. For this purpose, variations of well-known water treatment technologies (e.g. activated carbon or air-stripping towers) have been adapted for above-ground treatment of contaminated groundwater. The treated water may be used for various purposes depending on the level of clean-up achieved, or simply returned to the aquifers using injection wells located at up-gradient locations.

It has been observed that at most sites contaminant concentrations in extracted groundwater decrease dramatically during the early stages of pumping. However,

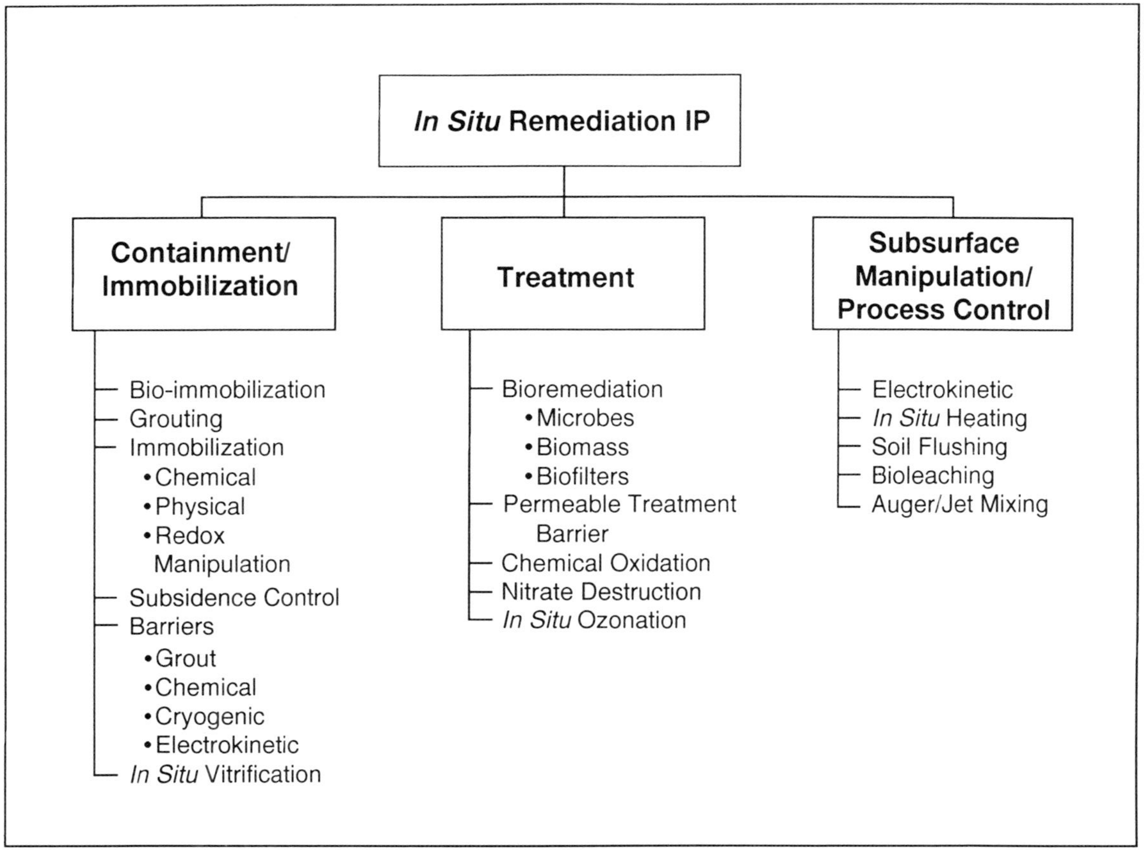

Figure 4. Categorisation of *in-situ* remediation technologies being developed and tested by the USA Department of Energy [from, DOE, 1994].

low levels of contaminants may persist for extended periods of time after the initial drop, and may even rebound significantly, usually above the acceptable levels, soon after pumping has been terminated (Palmer and Fish, 1992). The processes responsible for such observations include (EPA, 1992a): (1) diffusion of contaminants from low-permeability zones; (2) hydraulic isolation of certain regions within the well field; (3) slow desorption of sorbed-phase contaminants from the aquifer solid matrix; and (4) slow dissolution of contaminants from distributed sources. Variations to the traditional PAT technology have been attempted to overcome the limitations described above. Such improvements include: (1) reconfiguration of the well fields, based on the hydrogeology of the site, to minimise hydraulic and hydrodynamic isolation; (2) pulsed pumping to overcome the concentration rebound effect; and (3) integration of PAT with other remediation options.

Evidence from numerous sites where the traditional and modified PAT technology has been attempted suggests that clean-up of groundwater to levels prescribed by health-based standards cannot be achieved (Mackay and Cherry, 1989; NRC, 1994). The volumes of water required to be pumped and the time required for such pumping usually have been seriously underestimated, and the claims for clean-up levels have often been greatly overestimated. Even where such low levels are claimed to have been attained, it is often difficult to establish unequivocally that the contaminated aquifer has indeed been restored.

The two major limitations of past applications of PAT for aquifer remediation are: the uncertainties or inadequacies in site characterisation; and contaminant source characterisation. The geohydrochemical complexity of the aquifers was often underestimated or poorly understood at most sites. Site characterisation was often side-stepped in the public/regulatory pressures to get something started. Uncertainties in determining the contaminant source(s) and contaminant concentrations also played an equally important role in past failures of PAT applications. The analytical challenges in characterising complex, multi-component wastes, and poor records of waste disposal at most sites added to such uncertainties. The aforementioned complexities were often not considered in locating or operating the wells for extraction of contaminated groundwater. The implicit assumptions of homogeneous aquifers and known contaminant source(s) are now known to be too simplistic and somewhat naive representations of vastly complex and diverse subsurface systems. It must be recognised that the conventional PAT technology is designed to deal with the symptoms, i.e. the contaminant plume, and fails to consider the causes, i.e. the contaminant sources, which may be located both within the vadose and saturated zones.

The criteria to be used for site characterisation and in evaluating performance of PAT and its variations have been presented in a recent report (EPA, 1992a). Increased efficiency in aquifer remediation may be achieved by better site/source characterisation. This, at the minimum, must include: (1) detailed history of waste disposal at the site; (2) characterisation of the hydrogeochemistry of the vadose and saturated zones; (3) contaminant source identification and quantification; (4) contaminant plume delineation; (5) estimation of parameters for various physical/

Table 1. Criteria used for categorisation of contaminated aquifers in terms of clean-up using PAT technologies.

Category*	Potential for Aquifer Clean up
1	PAT systems may be used to achieve health-based clean up levels in a reasonable period of time.
2	Similar to Category 1, but greater uncertainties in achieving or in establishing clean-up levels.
3	Only partial clean up, to levels much greater than regulatory standards, may be possible, which is also subject to significant uncertainties.
4	Clean up to health-based regulatory standards is unlikely.

*Category 1 is the easiest to clean up, while Category 4 is the most difficult scenario. This grouping is entirely subjective.

chemical/biological processes governing contaminant behavior in aquifers; and (6) uncertainty analyses based on flow and transport simulation models for site-specific conditions. To do all this is neither an easy task nor an inexpensive effort. Even if all of the above specified information were to be collected at a site, certain fundamental limitations of the PAT technology may limit its usefulness. A recent report, released by the USA National Research Council (NRC, 1994), has ranked the likelihood of success of PAT based on the relative ease with which contaminated aquifers may be cleaned up. The subjective criteria by which the sites are grouped are explained in Table 1. The rank matrix, reproduced here as Table 2, is based on the complexity of the aquifer hydrogeochemistry and the contaminant chemistry.

As might be expected from earlier discussions, increases in the complexity of either the aquifer or the contaminant characteristics contribute to increasing difficulties in remediation, and to the greater likelihood of failure of conventional PAT. The easiest cases for remediation (grouped in Category 1) are homogeneous aquifers, either single- and multi-layered, with the contaminant existing only in the aqueous phase. The only challenge in such cases is to design the location and operation of the extraction wells to simply pump out the contaminated groundwater plume in an efficient manner for above-ground treatment. The existence of various sinks (e.g. biodegradation) for contaminants only makes it easier to achieve faster clean-up.

Heterogeneous/fractured aquifers with contamination limited to the aqueous phase are grouped either in Category 2 or 3, reflecting increased problems in clean-up (Table 2). Sorption of the contaminants by the aquifer matrix only increases the difficulties, primarily due to slow desorption kinetics. Such effects are especially pronounced in heterogeneous or fractured media, where release of the contaminants from the sorbed phase can be severely constrained by diffusion-controlled mass transfer processes (Brusseau and Rao, 1989; 1991). Thus, such contaminated aquifers are grouped in Category 3 in Table 2.

The most difficult cases for remediation are heterogeneous and fractured media

Table 2. Relative ease of clean-up of contaminated aquifers as a function of contaminant chemistry and hydrogeology. (Adapted from NRC, 1994).

	Contaminant Chemistry					
Hydrogeology	Mobile, Dissolved (degrades/ volatilizes)	Mobile, Dissolved	Strongly Sorbed Dissolved (degrades/ volatilizes)	Strongly Sorbed, Dissolved	Separate Phase LNAPL	Separate phase DNAPL
Homogeneous, single- or multi-layered	1[A]	1	2	2	2–3	3
Heterogeneous, single- or multi-layered	2	2	3	3	3	4
Fractured	3	3	3	3	4	4

[A]Relative ease of clean-up, where 1 is easiest and 4 is most difficult

contaminated with NAPLs. The complexities of flow/transport processes in heterogeneous or fractured media are only compounded by overlaying the dynamics of NAPL retention patterns, accessibility, and dissolution rates. Thus, these types of cases receive the highest ranking in terms of difficulty (grouped in Category 4 in Table 2).

The NRC study (NRC, 1994) also evaluated evidence from 80 sites in the USA where PAT technology had been attempted. Only 2 of the 80 sites were placed in Category 1, 14 sites in Category 2, 29 sites in Category 3, and the remainder in Category 4. It is clear that in less than 20% of the sites evaluated (Categories 1 and 2) PAT technologies are likely to succeed, while at a vast majority of the sites (about 80%) only a partial clean-up might be feasible.

5. Enhanced remediation technologies

5.1 Alternate and innovative technologies

With mounting evidence for failure of PAT at most sites, many alternative technologies have been developed (EPA, 1992b; NRC, 1994). Two distinct types of approaches are being considered for guiding regulatory policy and the required remedial actions at contaminated sites. The first approach has been to seek less-stringent regulatory standards for remediation goals based on human-health concerns. Convincing arguments for this approach can be made, especially in cases where the groundwater is not likely to be used for human consumption and land use scenarios do not include human exposure (e.g. remote sites) or when the site is designated for industrial uses (e.g. brownfield sites). Discussions of risk based or cost benefit analyses that are used in arguing for less stringent remediation endpoints are beyond the scope of this paper. Here, our analysis will be limited to a hydro-geochemical perspective in examining several variations to the conventional PAT that have been developed. The motivation for all alternate technologies is that they offer enhanced rates of *in-situ* clean-up of contaminated soils and aquifers.

USA EPA (1994) distinguishes between 'traditional,' 'alternative,' and 'innovative' treatment technologies as follows:

Alternative technologies are alternatives to land disposal. The most frequently used alternative technologies are incineration and solidification or stabilisation. Innovative technologies are alternative treatment technologies for which applications at superfund and other similar sites are inhibited by lack of data on performance and cost. In general, a treatment technology is considered innovative if it has had limited full-scale application. Often, it is the application of a technology or a process to soils, sediments, sludge, and solid-matrix waste (e.g. mining slag) that is innovative. Groundwater treatment after the water has been pumped to the surface often resembles traditional water treatment technologies; thus, in general, pump-and-treat or *ex-situ* groundwater remedies are considered established. *In-situ* bioremediation and other *in-situ* treatment of groundwater, however, are considered innovative technologies.

Thus, land disposal and pump-and-treat are considered as established, traditional technologies, while a number of *in-situ* and *ex-situ* treatment approaches based on a novel application that has not been field-tested or used at full-scale would be considered as being alternate, innovative technologies. Here, we will examine some examples of such technologies in the developmental stages.

5.2 Barriers to adoption of alternate technologies

The selection of remediation technologies for use at waste disposal/spill sites is a long and complex process, at least in the USA, involving social, legal, financial, and economic factors, in addition to the scientific and engineering considerations. Various local, state, and national regulatory approvals and permits must be obtained as dictated by the applicable regulatory codes (Herndon, 1987). Competing interests of many stakeholders must be considered including those of the site owners (the 'potentially-responsible parties'), the regulatory agency enforcement personnel, and public interest groups. Site-specific monitoring and characterisation data must also be collected, evaluated, and interpreted by a team of technical experts. Conflict resolution in any of these phases may be a time-consuming activity. (For further discussion of this see Chapter 10). Even if there is a general consensus that the site in question needs to be cleaned up in the shortest time period and at least cost, it may not be possible for all parties involved to agree upon one or more remediation technologies to be used at the site to achieve the clean-up goals. Economic drivers, uncertainty in success of a given untested innovative technology, and a history of regulatory agency acceptance all contribute to the selection of traditional, but not necessarily equally or more effective, remediation technologies. Data compiled by the USEPA for remedial actions taken at 728 superfund sites showed pump-and-treat was the choice technology in 72% of the cases, while only at 2% of the sites some type of innovative technology for groundwater remediation was attempted. At other sites, either non-treatment remedies had been adopted, or a treatment technology had not yet been selected.

The remediation technology market place in the USA is regulatory-driven, rather than being client or site driven. That is, the demand for and acceptance of any remediation technology is primarily a function of the likelihood of regulatory agency acceptance. Thus, any uncertainty in the regulatory process (expected changes in regulations) leads to the selection of an established technology that has been considered the 'best available' by the state or federal regulatory agency. The site owners and the technical experts/consultants involved in site management and remediation are usually risk averse, unwilling to take chances on unproven or inadequately tested technologies, and concerned about additional legal and financial liabilities of having to pay enforcement penalties if the technology fails to meet the specified regulatory goals. This then presents a 'catch-22' problem for the technology developers and vendors: a technology cannot be field tested because it is difficult to obtain regulatory permits to conduct the test(s), and the regulatory permits cannot be obtained because there is only limited field data on technology performance. As a result, the lag time for adoption of alternate and innovative technologies can be a slow and arduous process. Major changes are being put in place in the USA to overcome these limitations for remediation technology development. Co-ordinated field tests of innovative approaches to soil and aquifer remediation are presently being conducted at several locations in the USA. Adoption of similar testing protocols in other countries would be most conducive to rapid adoption of innovative (and, hopefully, more efficient and cost-effective) remediation technologies.

Various alternate remediation technologies available for soil and groundwater remediation are summarised in Table 3. They are broadly categorised as being 'demonstrated' and 'developing'. Technologies listed under the first category have been field-tested at many locations, and full-scale systems have been used, and commercialisation is well underway. Technologies under the second category are the innovative technologies that have had laboratory-scale and limited field-scale testing. In the following sections, we will present an overview of the proposed alternative technologies, and examine two of them in greater detail: (1) *in-situ* flushing techniques, which use chemical adjuvants (e.g. surfactants and cosolvents) that greatly facilitate dissolution and desorption of contaminants, thus improving the efficiency of aquifer remediation (Palmer and Fish, 1992); and (2) *in-situ* bioremediation techniques, that are based on enhanced biodegradation of target contaminants to innocuous metabolites (NRC, 1993). These two groups of techniques were selected for further discussion because the first is representative of physical approaches, while the latter is based on microbiological approaches. Such a contrast might be helpful in understanding the diverse challenges facing environmental scientists and engineers involved in soil and aquifer remediation.

5.3 In-situ aquifer flushing

Among the new aquifer remediation technologies being developed is the use of various chemical adjuvants, such as surfactants and cosolvents, that enhance the release and mobility of contaminants. In the 1950's and 1960's, both cosolvents and

Table 3. Enhanced Pump-and-Treat and alternate technologies for soil and aquifer remediation. (Adapted from NRC, 1994).

Technology	Description	Application	Limitations
Demonstrated Technologies			
Soil Vapor Extraction	Flushes air through soil above water table	Extracts volatile contaminants above the water table	Difficulty flushing zones of low permeability and removing contaminants bound to soils
In-Situ Bioremediation – Hydrocarbons	Pumps materials through the subsurface to stimulate growth of organisms that biodegrade contaminants	Removes petroleum products and derivatives above and/or below the water table	Difficulty delivering growth-stimulating materials to zones of low permeability; slowed by presence of NAPLs; difficulty delivering adequate oxygen to the organisms
Bioventing	Pumps air and nutrients through soil above water table to stimulate growth of organisms that biodegrade contaminants	Removes petroleum products and derivatives above the water table	Similar to those for soil vapor extraction; also, adding nutrients in aqueous solution may inhibit air movement and affect soil's load-bearing capacity
Developing Technologies			
Pulsed or Variable Pumping	Varies the pumping rate to allow contaminants to dissolve, desorb, and/or diffuse from stagnant regions	Improves removal efficiency for sites with NAPLs and other residual contaminants	Increases clean up time because of reduced pumping rate; other limitations similar to those of conventional pump-and-treat systems
In Situ Bioremediation – Metals	Pumps materials through the subsurface to stimulate growth of organisms that change chemical form of metals	Dissolves metals to facilitate extraction or immobilizes them to prevent spreading	Similar to those for the other forms of *in situ* bioremediation
In Situ Bioremediation – Chlorinated Solvents	Pumps materials through the subsurface to stimulate growth of organisms that biodegrade contaminants	Removes chlorinated solvents above and/or below the water table	Similar to those for in situ bioremediation of petroleum hydrocarbons; also possible accumulation of hazardous intermediate compounds

Table 3. (continued)

Technology	Description	Application	Limitations
Developing Technologies			
Air Sparging	Injects air below the water table and captures it above the water table	Extracts volatile contaminants; promotes bioremediation	Difficulty flushing low-permeability zones; difficulty operating at depths below 30 ft; difficulty extracting multi-component mixtures
Steam-Enhanced Extraction	Injects steam above the water table	Volatilizes contaminants above the water table	Difficulty volatilizing contaminants from low-permeability zones
In Situ Thermal Desorption	Adds heat above the water table by Joule heating, radio frequency heating, or other means	Improves removal of contaminants with low volatility above the water table	Difficulty distributing heat evenly throughout the contaminated zone
In Situ Chemical Treatment	Injects chemicals to transform contaminants in place	Chemically transforms contaminants into less hazardous products	Difficulty delivering chemicals to low-permeability zones; possibility that chemical reactions may adversely affect the subsurface; slow reaction rates
In Situ Soil Flushing	Flushes surfactants or cosolvents below the water table	Enhances recovery of contaminants with low water solubility	Similar to those of conventional pump-and-treat systems; possibility that chemical reactions may adversely affect the subsurface; slow reaction rates
Natural Assimilative Capacity	Allows natural contaminant biodegradation to occur without human intervention other than careful monitoring	May remove easily degraded contaminants such as petroleum products under proper conditions	Lack of operating history for judging effectiveness
Physical Containment	Contains contamination with cutoff walls, caps or liners or by solidifying soil	Can prevent off-site contaminant migration and improve control of pump-and-treat systems	Long-term reliability uncertain
In Situ Reactive Barriers	Treat contaminated water as it passes through a physical barrier containing reactive chemicals, organisms or activated carbon	Can treat a variety of contaminants by chemical, physical or biological removal	Not field tested

For more detail refer to Chapter 13.

surfactants were used in the petroleum industry for enhanced oil recovery. However, unfavorable economic factors limited its routine use at the full scale operations. Now there is a renewed interest in solvent and surfactant *in-situ* flushing for *in-situ* aquifer and soil remediation. Several commercial vendors in the USA and Europe are already using *ex-situ* solvent washing of excavated contaminated soils (EPA, 1992a; 1993). The use of organic cosolvents and surfactants for *in-situ* aquifer remediation is based on four principles. First, cosolvents and surfactants decrease the interfacial tension between the aqueous phase and nonaqueous phase liquid (NAPL), inducing mobilisation of the residual NAPL phase. Second, cosolvents and surfactants increase the solubility of nonpolar organic chemicals, enhancing the release of organic constituents of an immobile NAPL phase. Third, cosolvents and surfactants reduce sorption, and the resulting decrease in retardation facilitates faster advective transport of dissolved contaminants. Fourth, mass-transfer rates generally increase in the presence of cosolvents and surfactants. (See Chapter 4).

The use of cosolvents to recover sorbed-phase contaminants has been demonstrated in laboratory column studies reported by Augustijn et al. (1994), while enhanced dissolution of residual NAPLs in packed sand or glassbead columns was investigated by several workers (Boyd, 1991; Roy, 1992; Boyd and Farley, 1992; Augustijn, 1993; Farley et al., 1993; Imhoff et al., 1995). The immiscible fluid sources studied have included single-component NAPLs (TCE, PCE) and multi-component NAPLs (gasoline, coal tar). NAPL dissolution during aqueous displacement has been examined by several researchers (e.g. Hunt et al., 1988; Borden and Kao, 1992; Imhoff et al., 1994) to examine the significance of mass-transfer limitations. Lab-scale column studies have also been conducted on surfactant flushing for removal of sorbed-phase contaminants (e.g. Gannon et al., 1989; Abdul and Gibson, 1991; Edwards et al., 1991). More recent studies have evaluated the pore-scale processes governing nonequilibrium NAPL dissolution during surfactant flushing (Pennell et al., 1993; 1994).

To date, there have been only two field-scale evaluations of *in-situ* cosolvent flushing to remediate contaminated aquifers. Broholm and Cherry (1994) reported the results from the first pilot-scale field testing of *in-situ* cosolvent flushing for enhanced recovery of residual NAPL. They constructed a hydraulically-isolated test cell ($4.5 \times 5.5 \times 2.3$ m) in a shallow sandy aquifer at Borden, Ontario, Canada, and introduced a small volume (about 5 litres) of a ternary mixture of NAPLs (10% chloroform, 40% TCE, and 50% PCE) at 5 cm below the water table. About 220 days after releasing the NAPL mixture, the test cell was flushed for 5.5 days with 30% methanol at a pore-velocity of 13 cm day^{-1}. This was followed by another 100 days of flushing with water. During this entire period about 30% of the introduced NAPL mixture was recovered as determined by post-flushing soil sampling in the test cell. Maximum solubility enhancement for TCE and PCE in the pore fluids sampled near the extraction wells ranged from 2 to 5. This increase was larger than that expected based on the maximum methanol concentrations likely to be present in the NAPL-contaminated zone. Broholm and Cherry (1994) suggested that some NAPL mobilisation may have occurred.

A team of researchers from the University of Florida (UF) have recently completed another field study to evaluate *in-situ* cosolvent flushing technology (Rao et al., 1995). The study was conducted at a site on the Hill Air Force Base near Salt Lake City, Utah. The shallow, sand-gravel, unconfined aquifer (located at about 6.1 m below ground surface) had become contaminated with jet fuel (JP-4) and various chlorinated organic solvents (including chlorobenzenes and chloroalkenes) disposed at the site. Note that while the Borden study involved the controlled release of a small amount of a three-component DNAPL into an uncontaminated aquifer, the UF study at the Utah site was conducted in an aquifer that became contaminated as a result of long-term waste disposal and spillage of a mixture of LNAPLs and DNAPLs. The site had been used over a 20 yr period for fire training activities and contained two chemical disposal pits; the fuels and solvents infiltrated through the vadose zone and contaminated the shallow aquifer at the site. The mixed waste LNAPL contamin-ation currently exists at the site as a residual phase in the shallow aquifer, over a smear zone of about 2 m. A hydraulically-isolated test cell (50 m × 4 m), similar to that used in the Borden study discussed earlier, was built at the site by driving 10 m long, interlocking sheet piles that were grouted at the joints. The sheet piles penetrated about 3 m into the clay confining unit. The cell was instrumented with multi-level samplers and injection and extraction wells. A binary alcohol mixture (70% ethanol, 10% n-pentanol, 20% water) was pumped through four injection wells into the test cell over a three week period. The dissolved waste constituents in the fluids collected in three extraction wells were monitored. The objective here was to cause enhanced solubilisation of the LNAPL constituents, but not ganglia mobilisation. Soil core samples were taken before and after cosolvent flushing to quantify the amount of LNAPL that had been removed by the flushing event. Preliminary analysis of the data from these studies indicates that the flushing had efficiently removed about 70–80% of the LNAPL from aquifer zones that had been flushed with the cosolvent mixture. Further comments and claims about these data must await complete and careful analysis.

Several field-scale studies have been conducted to examine the efficiency of surfactant flushing of contaminated aquifers. Fountain (1992) described the results of two such studies. In one field test, PCE introduced into a hydraulically-isolated test cell placed in a shallow, sandy aquifer was recovered successfully by flushing with about 15 pore volumes of a surfactant solution. In a second field test, attempts are under way to remediate an aquifer contaminated with carbon tetrachloride. Other such tests are under way in the USA and Canada.

Although cosolvents and surfactants may have potential application for *in-situ* remediation, several problems still need careful consideration before implementation in the field. Such problems include: (1) delivery of the additive to the source; (2) reduced efficiency of the flushing treatment due to soil heterogeneities and non-uniform distribution of the contaminants; (3) transport of dissolved contaminants off site; (4) interactions of the adjuvant with soil/aquifer matrix; (5) toxicity of the adjuvants; (6) recovery and waste management; and (7) costs. The choice of a suitable solvent depends on the extraction capacity of the solvent or surfactant (cosolvency power or MSR), biodegradability, toxicity, and cost. The remedial fluid must have favorable density, viscosity, and interfacial tension properties. In some cases, a combination of cosolvents, or surfactants, or both may be considered to optimise the properties of the solvent mixture. Cosolvent or surfactant *in-situ* flushing treatment will most likely be suitable for hydrocarbons with low vapor pressure, since less costly remediation techniques are available for volatile compounds.

5.4 In-situ bioremediation

A host of well-accepted technologies that depend on the use of native or introduced microorganisms to detoxify or transform contaminants in municipal wastewaters and industrial waste streams has been practiced with considerable success over the past several decades. The use of microbial transformations for *ex-situ* land treatment of soils contaminated with petroleum hydrocarbon residues has also been demonstrated in recent years (e.g. Pope and Mathews, 1993). *In-situ* bioremediation of organic contaminants in the vadose zone has also been demonstrated (e.g. Sims et al., 1993). However, employing bioremediation for *in-situ* restoration of contaminated aquifers has a much shorter history, but has generated considerable controversy and scientific debate (NRC, 1993). Yet, the use of this technology holds continuing interest because of its potential for cost-effectiveness. Another appealing aspect of *in-situ* bioremediation is that it is a natural approach to aquifer restoration that does not require the addition of chemical agents other than microbial nutrients (see also Chapter 12).

Existing and developing *in-situ* bioremediation technologies may be grouped into the following two broad categories (NRC, 1993):

(1) Intrinsic bioremediation, where the essential materials required to sustain microbial activity exist in sufficient concentrations and amounts that naturally occurring microbial communities are able to degrade the target contaminants without the need for human intervention. This technique is the more likely choice for restoring sites with low levels of contamination over an extensive area (e.g. far-field regions within the contaminant plume).
(2) Engineered bioremediation, which relies on various approaches to accelerate *in-situ* microbial degradation rates. This is accomplished by optimising the environmental conditions by adding nutrients and/or electron acceptors, thus promoting the proliferation and activity of existing microbial consortia; this approach is referred to as bioenhancement. In some cases, specific degraders may need to be introduced (referred to as bioaugmentation) to achieve the required biodegradation rates. Engineered *in-situ* bioremediation is favored at highly contaminated sites, where significant reductions in contaminants may need to be achieved over a short period of time.

The site characteristics that lead to successful application of these two types of *in-situ* bioremediation technologies are summarised in Table 4. To date, the second group of approaches has been the focus of research, although there is growing inter-

Table 4. Site characteristics that favor in situ bioremediation. (Adapted from NRC, 1994).

ENGINEERED BIOREMEDIATION

Engineered bioremediation requires installing wells and other engineering systems to circulate electron acceptors and nutrients that stimulate microbial growth. Key site characteristics for engineered bioremediation are:

(1) Adequate transmissivity of the subsurface to fluids, hydraulic conductivity greater than 10^{-4} cm s^{-1} if the system circulates water, intrinsic permeablity greater than 10^{-9} cm if the system circulates air;
(2) Relatively uniform subsurface medium (common in river delta deposits, flood plains, or large rivers, and glacial outwash aquifers);
(3) Residual concentration of nonaqueous-phase contaminants of less than 10 000 mg kg^{-1} of subsurface solids.

INTRINSIC BIOREMEDIATION

Intrinsic bioremediation destroys contaminants without human intervention, as the population of native microbes capable of degrading the contaminant increases naturally. The process requires thorough site monitoring to demonstrate that contaminant removal is occurring. Key characteristics of sites amenable to intrinsic bioremediation are:

(1) Consistent groundwater flow (speed and direction) throughout the seasons, seasonal variation in depth of water table less than 1 m, seasonal variation in regional flow trajectory less than 25 degrees;
(2) Presence of carbonate minerals (e.g. limestone, dolomite, shell material) to buffer pH;
(3) High concentrations of electron acceptors, such as oxygen, nitrate, sulfate, or ferric iron;
(4) Presence of elemental nutrients, especially nitrogen and phosphorus.

See also Chapter 13.

est in using intrinsic bioremediation since it offers a less costly alternative to most other remediation technologies. The major limitation to this technique however, is that the rate at which a site is restored may be too slow and monitoring costs could be considerable. The likely success of restoring a site to regulated levels is also uncertain and intrinsic bioremediation is currently less likely to be accepted as an alternative to more aggressive approaches.

Alexander (1994) lists the following criteria in judging *in-situ* bioremediation to be a practical alternative: (1) existence of microbial consortia possessing the requisite catabolic activity; (2) ability to sustain adequate microbial transformation rates for decreasing contaminant concentrations to regulated levels within an acceptable period of time; (3) absence of metabolites that are either toxic to the microbial communities, or the absence of by-products that are themselves of environmental concern; (4) target contaminants are bioavailable; (5) presence of environmental conditions that are conducive to microbial activity, including adequate concentrations of nutrients, electron acceptors, and energy sources; and (6) cost-effective when compared to other competing technologies that may be used at the site.

In addition to the microbial factors listed above, we must also consider the hydrogeologic characteristics of the contaminated site designated for *in-situ* bioremediation. As discussed earlier (see Table 3), hydrogeologic heterogeneity can severely restrict the ability to deliver various remedial adjuvants (e.g. nutrients, electron acceptors, energy sources) to aquifer locations where contaminants and microorganisms are present. For example, when the contaminants are trapped within immobile zones (e.g. clay lenses, intra-matrix micro-pores) that are only accessible by diffusion-controlled mass transfer, the rate at which the adjuvants can reach these regions can be much smaller than that required to sustain high biodegradation rates. Even in homogeneous aquifers, the groundwater flow rates may be just too slow to deliver the adjuvants to all regions within the contaminant plume. Slow groundwater flow rates, on the other hand, provide more favorable conditions for intrinsic bioremediation, because much longer residence times are required to compensate for the slower intrinsic biodegradation rates. Note that slower groundwater flow rates also ensure that mass-transfer constraints are not likely to be a limitation for bioavailability of contaminants.

The efficiency of *in-situ* bioremediation is controlled by bioavailability, a term used to represent the ease with which the target contaminants are accessible to the microorganisms. In many cases, the contaminants degrade in soils and aquifers at much slower rates than that expected based on in-vitro pure culture studies. This might often be the case even when appropriate microbial consortia are present, and the extant environmental conditions are optimal for biodegradation. The limited biodegradation in such cases may be because the contaminants are present in a form, or in a location, that is not readily accessible for microbial uptake and metabolism. Factors contributing to limited bioavailability include (Alexander, 1994): (1) contaminant sorption by the aquifer/soil solid matrix; (2) presence of residual NAPLs; and (3) physical entrapment of the contaminant in intra-matrix micro-pores or immobile zones. The rate of desorption, dissolution, or diffusion may be very slow in heterogeneous aquifers. Thus, the rate at which the microbial consortia can potentially degrade the contaminants far exceeds the rate of release from inaccessible regions or unavailable forms.

Efforts to bioremediate sites contaminated with a variety of organic contaminants are summarised in Table 5. Much of the current controversy and debate surrounding *in-situ* bioremediation stems from the difficulties in unequivocally establishing that microbial processes were in fact involved in the observed loss of contaminants (NRC, 1993). In most field studies, sufficient care is not taken to adequately characterise the site, or the necessary evidence to prove *in-situ* bioremediation is not collected. The question is not about whether the specific microbial consortia (whether native or introduced) have the capability to biodegrade the contaminants, as usually demonstrated in *in-vitro* studies. It must be established that *in-situ* biodegradation leads to contaminant detoxification and loss. As pointed out by Alexander (1994):

'It is thus essential to distinguish between biodegraDABLE and biodegraDED. The former term indicates a susceptibility (of the contaminant) to destruction, while the latter describes what actually occurs in a particular set of conditions.'

In heterogeneous aquifers, proving that *in-situ* biodegradation is occurring is a daunting task. The hydrogeologic complexity of the aquifers, the uncertainties regarding contaminant chemistry, and the vagaries of microbial communities, all

Table 5. Contaminant susceptibility to bioremediation. (Adapted from NRC, 1993).

Contaminant Chemical Class	Frequency of Occurrence	Current Status	Evidence of Future Success	Limitations
Hydrocarbons and Derivatives				
Gasoline, fuel oil	Very frequent	Established		Forms nonaqueous phase liquid
Polycyclic aromatic hydrocarbons	Common	Emerging	Aerobically biodegradable under aerobic conditions	Sorbs strongly to subsurface solids
Creosote	Infrequent	Emerging	Readily biodegradable under aerobic conditions	Sorbs strongly to subsurface solids; forms nonaqueous phase liquid
Alcohols, ketones, esters	Common	Established		
Ethers	Common	Emerging	Biodegradable under a narrow range of conditions using aerobic or nitrate-reducing microbes	
Halogenated Aliphatics				
Highly chlorinated	Very frequent	Emerging	Cometabolized by anaerobic microbes; cometabolized by aerobes in special cases	Forms nonaqueous phase liquid
Less chlorinated	Very frequent	Emerging	Aerobically biodegradable under a narrow range of conditions; cometabolized by anaerobic microbes	Forms nonaqueous phase liquid
Halogenated Aromatics				
Highly chlorinated	Common	Emerging	Aerobically biodegradable under a narrow range of conditions; cometabolized by anaerobic microbes	Sorbs strongly to subsurface solids; forms nonaqueous phase solid or liquid
Less chlorinated	Common	Emerging	Readily biodegradable under aerobic conditions	Forms nonaqueous phase solid or liquid
Polychlorinated Biphenyls				
Highly chlorinated	Infrequent	Emerging	Cometabolized by anaerobic microbes	Sorbs strongly to subsurface solids
Nitroaromatics	Common	Emerging	Aerobically biodegradable; converted to innocuous volatile organic acids under anaerobic conditions	

confound efforts to assign fractional losses to multiple, coupled processes operating in aquifers.

Much research needs to be done to better understand the advantages and limitations of *in-situ* bioremediation technology. The USA National Research Council report (NRC, 1993) offers three strategies for such research planning: (1) development of innovative and inexpensive site characterisation techniques that permit a better understanding of a site's hydrogeochemistry and contaminant distributions; (2) developing comprehensive evaluation protocols for establishing the occurrence of *in-situ* biodegradation; and (3) development and validation of improved simulation models that appropriately represent the many coupled processes in heterogeneous aquifers. This indeed is the research agenda currently supported by funding agencies and the scientific community in the USA

6. Bioremediation case studies in Australia

Three case studies are presented below which illustrate application of some of these remediation technologies in Australia: (1) intrinsic bioremediation at a service station site where gasoline leaked into an anaerobic aquifer; (2) *in-situ* bioventing of a diesel fuel spill, and (3) *ex-situ* treatment of hydrocarbon contaminated soil material in biotreatment piles.

6.1 Case Study 1: Intrinsic bioremediation at a site of gasoline leakage to groundwater

As indicated in the discussion above, there are potentially several levels of detail at which intrinsic bioremediation may be shown to be occurring. First, the contaminant plume is shown to be static (at some acceptable concentration) via monitoring and perhaps modelling. Second, estimated degradation rates are high enough to reduce contaminant plume concentrations below action levels at receptor locations (i.e. at the boundary of a property, at a well head, etc.) or to maintain the plume at a static state. Third, the extent of bioremediation is estimated through assessment of the limitations to biodegradation and investigation of the microbial processes and communities that play a role in the degradation process. Each of these aspects requires an increasingly greater amount of research effort, but none of the options is a do nothing option - a planned and sometimes intensive monitoring program is at least required. Below we examine the plausibility of intrinsic bioremediation at two levels of detail with respect to a BTEX (benzene, toluene, ethylbenzene, xylene) plume in a shallow unconfined aquifer.

6.1.1 Site background

The site is situated in an urban area 15 km north-east of Perth, Western Australia on the Quaternary Bassendean Sand Formation of the Swan Coastal Plain. Leakage of gasoline from an underground storage tank at a petroleum service station has contaminated anaerobic, sulphate-rich groundwater in the aquifer beneath the service station and further down the hydraulic gradient. More detail of the investigation at the site and distribution of contaminants can be found in Barber et al. (1991) and Davis et al. (1993).

The aquifer at the site is comprised of medium to fine grained dune sand (7–12 m thick) and overlies a clay aquitard of the Guildford Formation. The depth to the water table is approximately 4 m below ground surface, the saturated aquifer thickness is approximately 6 m and water table fluctuations over a year can be greater than 1.6 m. Sampling of groundwater (see Davis et al., 1992) has been carried out from 11 stainless steel multiport boreholes (denoted MP in subsequent text and figures) with 12 ports (internal diameter of 2.5 mm, slotted over a 0.2 m depth) placed at 0.25 m and at 0.5 m depth intervals over the saturated aquifer thickness. Groundwater at the site is anaerobic, slightly acidic (pH 5–6) and sulphate-rich, with background concentrations of sulphate in the range 20–50 $mg\,L^{-1}$. Iron oxide/hydroxide precipitates are also present on the aquifer sediments.

6.1.2 Steady-state (static) plumes?

Groundwater sampling from the multiport boreholes and organic analyses allow mapping of long (greater than 400 m), thin (less than 50 m wide and 3–4 m thick) plumes (see Figures 5 to 8), extending down hydraulic gradient in groundwater towards the south-east. For example, the benzene plume for April 1991 has a minimum length of 420 m, width 20 to 50 m (Figure 5) and thickness 0.5 to 2 m. The concentrations of the organic compounds in the plume are high, especially in multiport boreholes MP10/MP12 and MP11. Typical benzene and toluene concentrations in MP10/MP12 (and often MP11) are 30–50 $mg\,L^{-1}$ and 40–70 $mg\,L^{-1}$, respectively. These concentrations are orders of magnitude above environmental and drinking water criteria, and are comparable to concentrations in equilibrium with NAPL gasoline (Cline et al., 1991).

Groundwater data contoured for August 1994 (Figures 6 and 8), when compared to data from April 1991 (Figures 5 and 7), indicate that some of the individual organic plumes may display steady-state (static) behavior. The benzene plume for August 1994 is similar in extent to the plume in April 1991, although no extra boreholes have been installed down-gradient of MP8 to observe further movement of the plume to the south-east. The toluene plume for August 1994 (Figures 6 and 8) is again truncated (less than 250 m long) with respect to the benzene plume and shows a similar shape to that of the April 1991 plume (Figures 5 and 7). The apparent increase in width of the plumes for August 1994 compared to April 1991, may be due to contouring uncertainties in locating the plume boundaries. Therefore for toluene and some of the other compounds (not shown here) it appears that the plumes may be static, or intrinsically attenuated.

Despite this comparison, it is clear from time trends for organic compounds in selected boreholes along the plume (Figure 9) that the plume is at best pseudo-steady state. The significant variations in organic concentrations at the MP locations

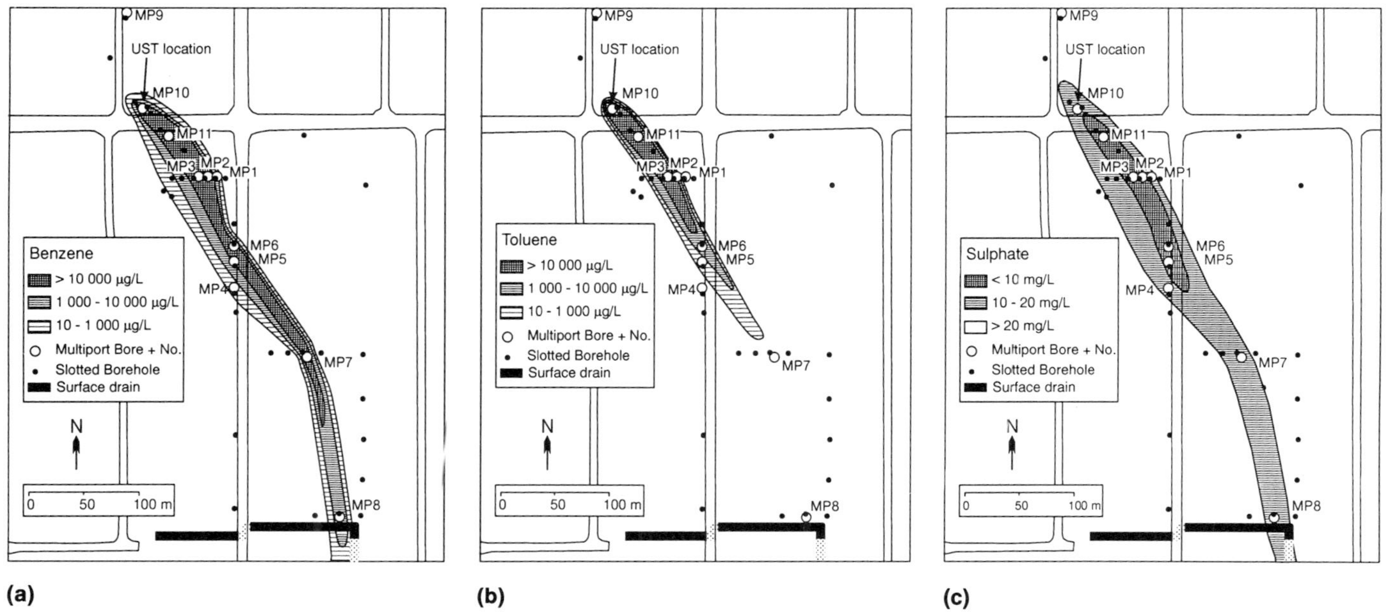

Figure 5. Contoured plumes for (a) benzene, (b) toluene and (c) sulphate in groundwater for April 1991, also showing locations of multiports and other boreholes

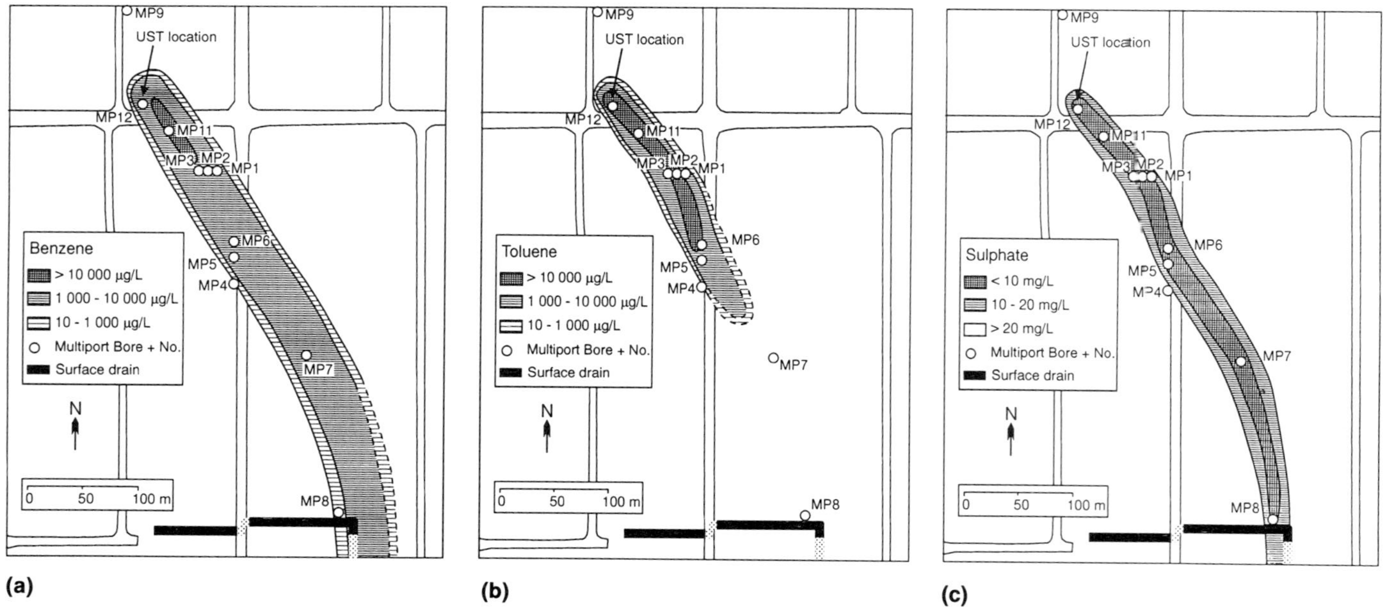

Figure 6. Contoured plumes for (a) benzene, (b) toluene and (c) sulphate in groundwater for August 1994

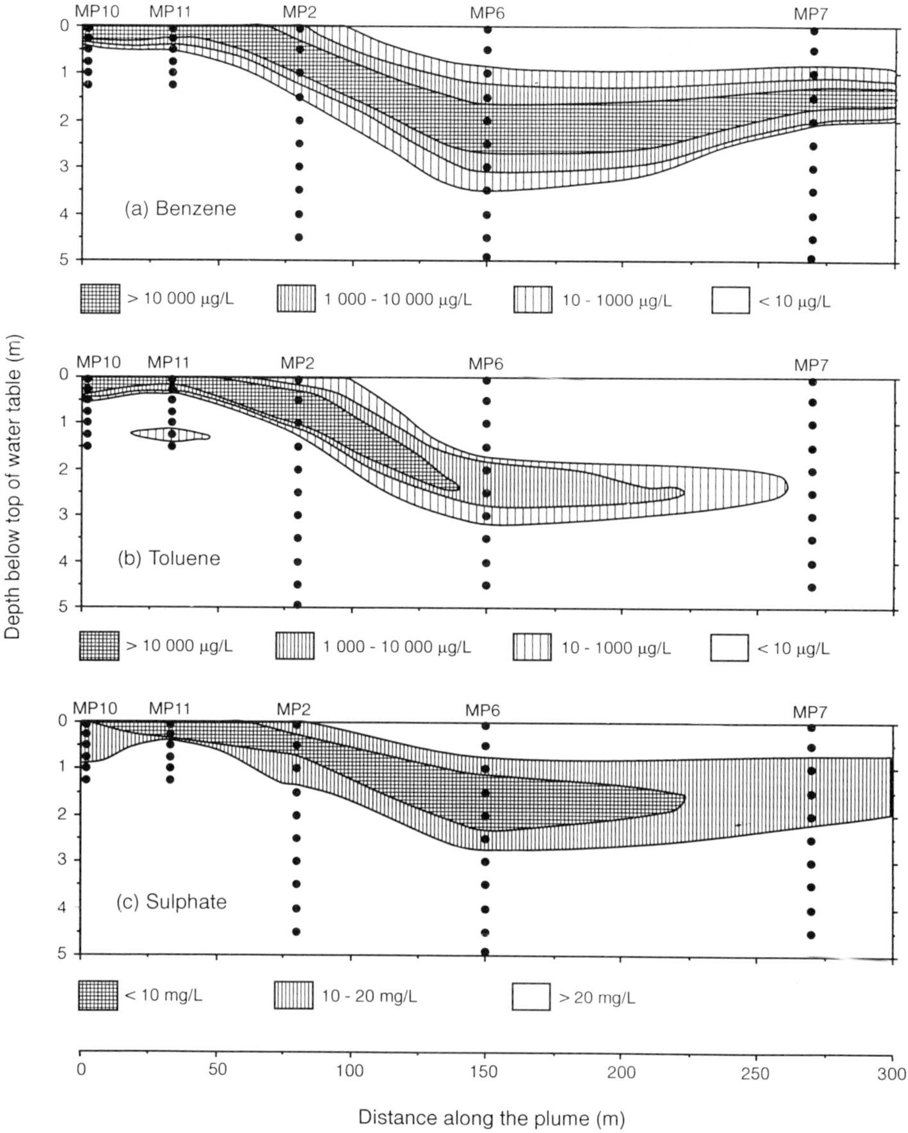

Figure 7. Cross-sections along the plume showing (a) benzene, (b) toluene and (c) sulphate concentrations in groundwater for April 1991. Dots are locations of monitoring ports (MP).

over time, especially 30 m (at MP 11) and 80 m (at MP3) down-gradient of the source location, indicate seasonal variations and perhaps other episodic up-gradient inflows of water with higher and lower hydrocarbon concentrations. This may be due to seasonal inundation of residual gasoline near the source of the leakage to groundwater, or due to changes in groundwater flow directions. The latter explanation is

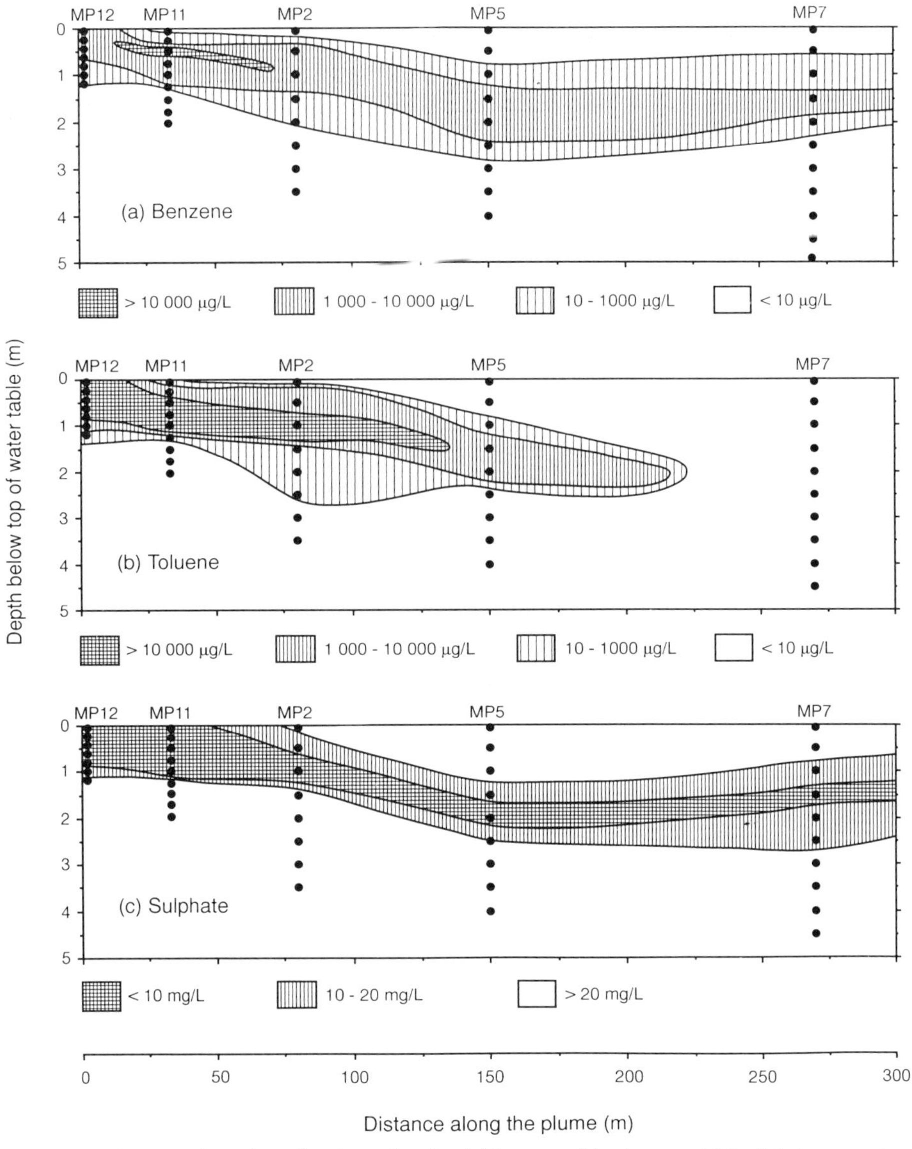

Figure 8. Cross-sections along the plume showing (a) benzene, (b) toluene and (c) sulphate concentrations in groundwater for August 1994. Dots are locations of monitoring ports (MP)

supported by groundwater table data (not shown) and time trends for organic compound concentrations, depicted as cross-sections across the plume in Figure 10. These transient features have significant implications for reliable monitoring of groundwater plumes, and in assessment of plume stability, and therefore for assessment of the effectiveness of intrinsic attenuation/biodegradation in minimising plume migration.

From monitoring the plume it can be concluded that the toluene plume is at a pseudo-steady state, and is substantially bioattenuated, despite migrating significantly off-site. However, it is unclear from the data whether the benzene plume is stable. From preliminary modeling of the plume (Thierrin et al., 1993), benzene was found not to be at steady state until the plume had extended to a distance beyond 4km. This conclusion is largely based on the small dispersivities observed in the sand aquifer, and an assumption of no biodegradation of the benzene in the anaerobic aquifer.

6.1.3 Biodegradation

The significant attenuation of toluene and other BTEX compounds with respect to benzene within the plume is likely to be due to preferential biodegradation. The depletion of electron acceptors and the production of mineralisation by-products would also be indicators of intrinsic bioremediation, as would direct or indirect measures of biodegradation rates.

At the site, ferric iron and SO_4^{2-} are significant potential electron acceptors (there is no dissolved O_2 or NO_3^-). No systematic measurement of ferric iron (Fe^{3+}) has been carried out; however, concentrations of ferrous iron (Fe^{2+}) are variable and tend to become depleted where dissolved sulphide is detectable. Systematic decreases of SO_4^{2-} concentrations (see Figures 5 and 6), and increases in hydrogen sulphide concentrations (up to 4.0 $mg\,L^{-1}$ within the plume) and bicarbonate concentrations within the plume are strong indicators of biodegradation. The presence of phenols and cresols which are known degradation products (Grbic-Galic and Vogel, 1987) in regions of the BTEX plume also supports this proposition. More direct measures of biodegradation have been obtained from modeling of the plume dimensions to give average plume-scale degradation rates for some of the BTEX compounds (Thierrin et al., 1993), and from a small-scale tracer test (Thierrin et al., 1995) to give *in-situ* estimates of the degradation rates.

Modeling of the hydrocarbon plume was performed with the three-dimensional analytical transport model SOLUTE (Sellmeijer, 1990) which accounts for uniform advective flow, dispersion, diffusion, retardation and first-order degradation (see Thierrin et al., 1993 for details). The groundwater flow velocity (150 $m\,yr^{-1}$), the effective porosity (0.28) and the saturated aquifer thickness (6 m) were kept constant in the model. Retardation coefficients estimated by Thierrin et al. (1992), and given in Table 6, were used. The model was calibrated to benzene data for April 1991, since benzene showed least evidence of attenuation due to degradation. The calibration indicated that the plume in April 1991 had taken approximately 4 years to develop. The model estimates of the first order degradation rates (half-lives) for each of these compounds are given in Table 6, and range from the most rapid degradation for toluene (120 ± 25 days) to the slowest degradation rates for ethylbenzene (230 ± 30 days) and benzene (>800 days). These rates are in the approximate ranges quoted by Howard et al. (1991) and others for anaerobic environments, apart from the xylene isomers (see Table 6).

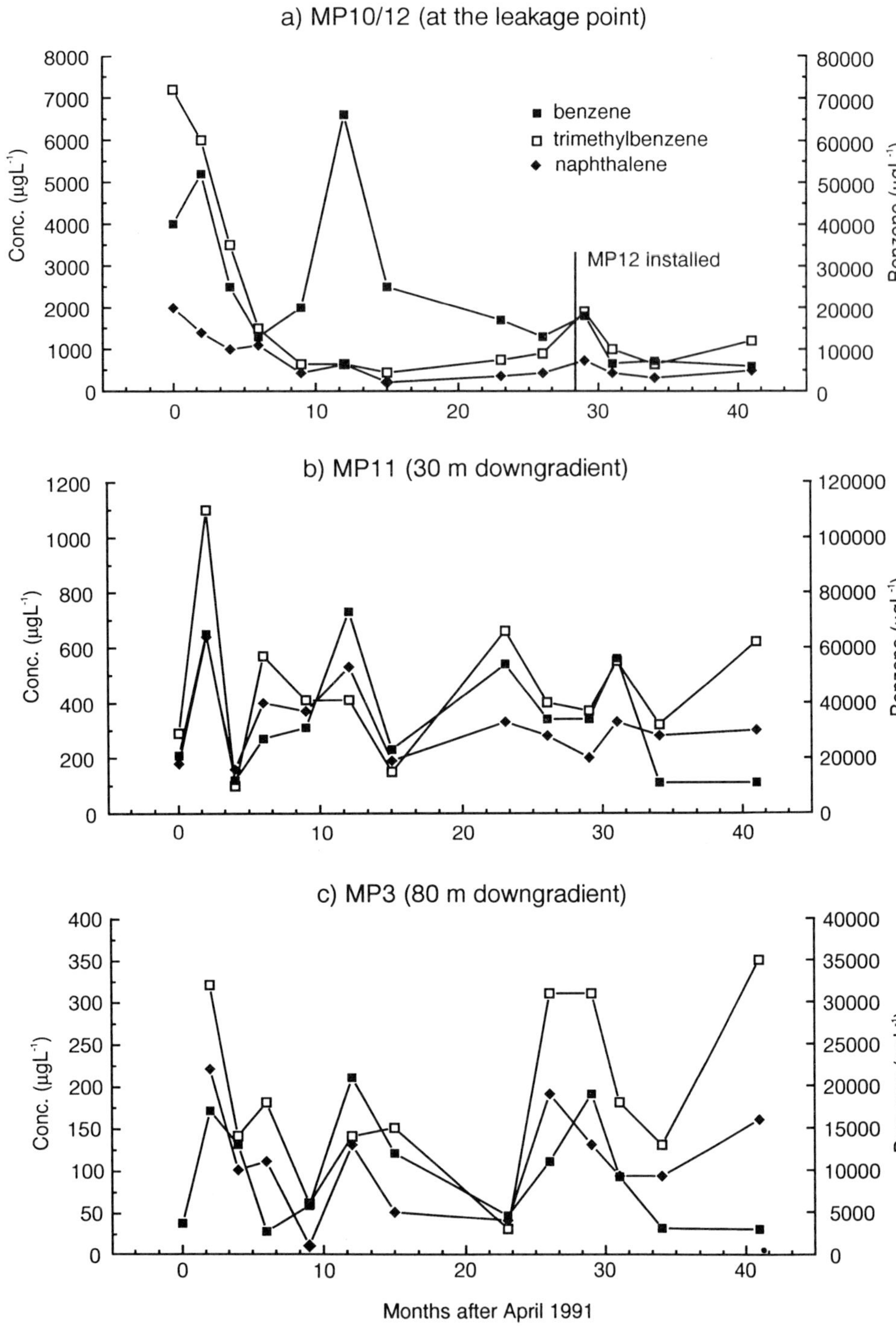

Figure 9. Peak concentrations of benzene, trimethylbenzene and naphthalene in selected multiport piezometers between April 1991 and August 1994.

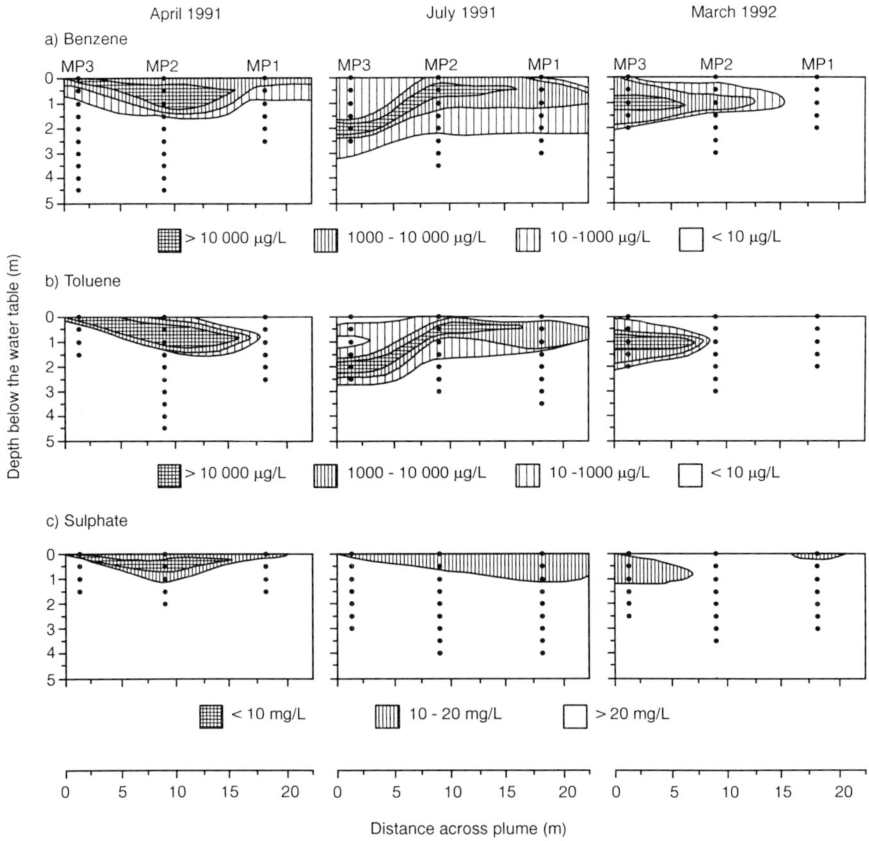

Figure 10. Cross-sections across the plume showing (a) benzene, (b) toluene and (c) sulphate concentrations in groundwater for April 1991, July 1991 and March 1992. Dots are locations of monitoring ports (MP).

A small-scale tracer test was carried out immediately down-gradient of MP3 (Thierrin et al., 1995). Bromide and selected deuterated hydrocarbon compounds were injected into the plume and monitored in groundwater over a 17 m travel distance. From breakthrough data, benzene showed little evidence of degradation (Table 7) within the anaerobic plume (half-life greater than 800 days). However, near the water table some degradation was observed, perhaps due to contact with oxygenated recharge water at that depth. This is consistent with previous laboratory studies (Patterson et al., 1993b). Naphthalene had the lowest half-life, although the estimate may be affected by strong sorption of naphthalene.

The half-life for toluene was also quite small (100 days) and comparable to those estimated at plume scale (120 days). Good agreement between the tracer test and plume estimates was also found for p-xylene (tracer test: 225 days) and m- and p-xylene (plume scale: 170 days). All half-lives estimated at the two different scales are larger than those observed by Acton and Barker (1992), Beller et al. (1992 a, b),

Table 6. Retardation coefficients (R) and half-life (days) estimates for hydrocarbon compounds from plume modeling.

Compound	R*	Half-life (days) Model	Half-life (days) Howard et al. (1991)
Benzene	1.05	>800	112–720
Toluene	1.05	120±25	56–210
Ethylbenzene	1.4	230±30	176–228
m-& p-Xylene	1.12	170±10	28–112
o-Xylene	1.12	125±10	180–360
1,3,5-Trimethylbenzene	1.4	180	–
Naphthalene	1.32	160±20	25–258

*Values taken from Thierrin et al. (1992).

Table 7. Half-lives (days) estimated from the tracer test data, assuming first-order degradation rates.

Parameter	Half-life (days)
Benzene	>800
Toluene	100±40
p-Xylene	225±75
Naphthalene	33±6

Edwards and Grbic-Galic (1992), or Edwards et al. (1992) in laboratory microcosm and column experiments under sulphate reducing conditions. Half-life estimates up to 100 days were obtained from laboratory experiments by Edwards et al. (1992) for o- and p-xylene.

6.1.4 Summary and implications

Hydraulic data indicate substantial seasonal and other transient changes in groundwater flow directions and water table elevations. These are reflected by changes in hydrocarbon concentrations within individual multiport boreholes and by movement of the plume. Despite this, intrinsic bioremediation of some of the components of the hydrocarbon plume (e.g. toluene) can be inferred from frequent monitoring of groundwater chemistry at carefully placed boreholes, and reaffirms the criteria listed in Table 4 for determining sites that are suitable for application of intrinsic bioremediation principles. Trends in inorganic compounds and degradation by-products in groundwater and field estimates of degradation rates indicate that significant intrinsic biodegradation is occurring within the plume of hydrocarbons. Rates differ depending on the hydrocarbon compound of concern. In particular, benzene which is likely to be of most concern, is not significantly bioremediated. This is shown both by the extent of the benzene plume and the large half-life estimates. The overall attenuation of the organic species intrinsically bioremediated may make the impact on groundwater acceptable, if some off-site migration is permissible. Benzene, however, is not intrinsically bioremediated at this site and is insufficiently attenuated to prevent an extensive off-site plume.

Further work could be carried out to determine the principal hydrocarbon degrading bacteria at the site and perhaps to determine limitations to degradation rates. Other studies could further investigate the biochemical pathways for the compounds to degrade. These tasks would broaden understanding of the underlying processes and aid in determining the controls on biodegradation of the hydrocarbon compounds. Unfortunately, current microbiological assessments are extremely expensive to carry out and may not fit a remediation time frame. Additional tests and studies are more likely to be carried out where active/enhanced remediation may be proposed.

6.2 Case Study 2: Bioventing to degrade diesel NAPL

6.2.1 Background and methodology

A six month bioventing field evaluation trial was carried out to bioremediate residual diesel fuel NAPL that had contaminated the vadose zone of the shallow unconfined sand aquifer at a site south of Perth, Western Australia. Contamination of the site resulted from spills of diesel from above ground storage tanks. The diesel was weathered due to degradation and water washing which has depleted the diesel NAPL of readily soluble and volatile compounds. As a result, groundwater contamination is limited to low concentrations of naphthalene and methylnaphthalenes. Coring at the site has shown the NAPL to have migrated 40 m westward and to extend 100 m north/south (Figure 11).

The stratigraphy at the field site is variable, with thin layers of inter-bedded fine- and coarse-grained sand in the interval of NAPL contamination and water table fluctuation. Peaks (5 to 26% by weight – Figure 11) in the NAPL distributions were shown by Johnston and Patterson (1994) to be located at or immediately above the water table. The interval of NAPL contamination which typically is 3.9 to 4.5 m below ground, coincided with the interval of water table fluctuation (about 0.5 m annually).

The bioventing trial consisted of nutrient addition and aeration of a 100 m^2 plot to induce increased microbial degradation of the weathered diesel. During aeration, the water table was drawn down to approximately 4.5 m below ground to expose the NAPL-contaminated zone to air. Nutrients were added to a major portion of the trial plot; however, one region was retained as an aerated control (i.e. not amended with nutrients). The site was instrumented with bundled multilevel boreholes (denoted MI), along with recently developed oxygen sensors (Patterson et al., 1994), placed at multiple depths across the water table and the NAPL-contaminated zone. These installations allowed monitoring of groundwater chemistry and semi-continuous

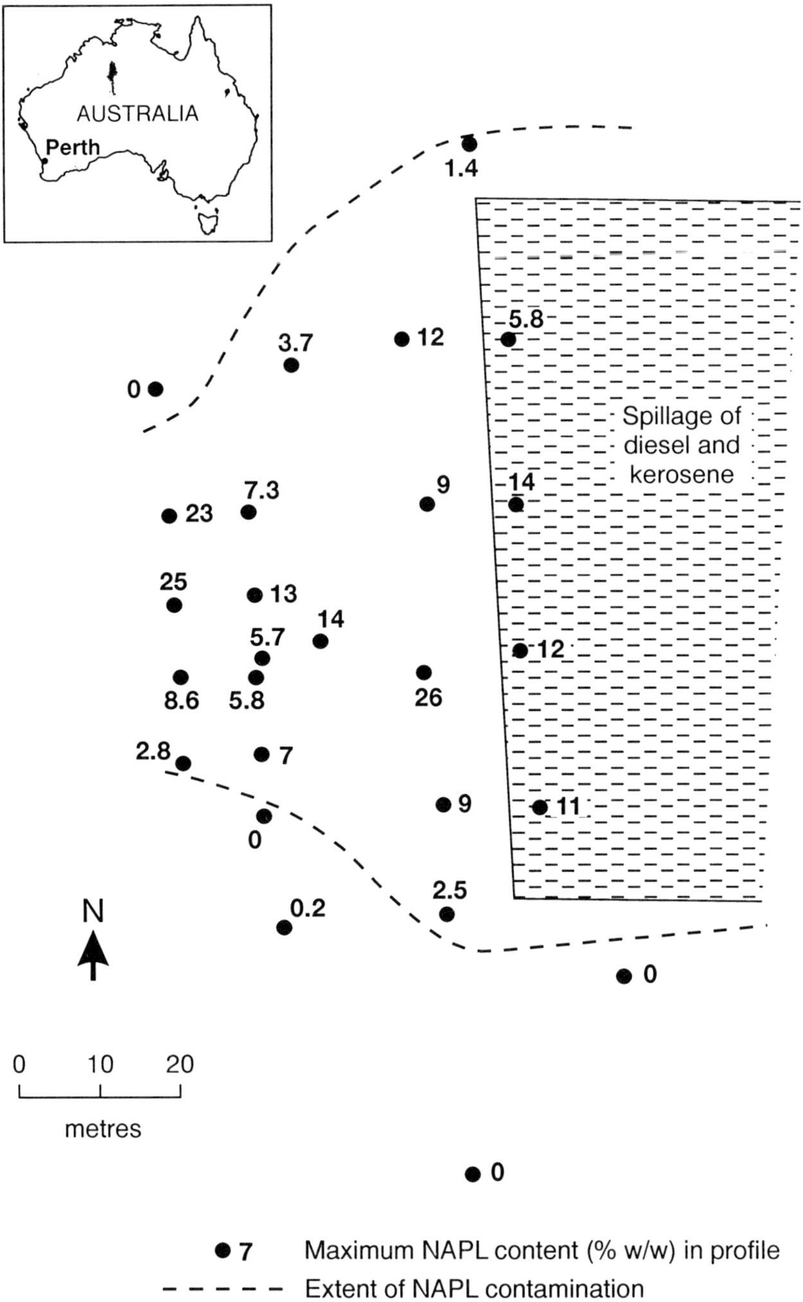

Figure 11. Maximum diesel NAPL content encountered at core sampling sites showing distribution of NAPL contamination away from spill location.

logging of O_2 concentrations. The air-filled porosity was also estimated from bulk density measurements and neutron logging when the water table was lowered. Further details of the procedure can be found in Davis et al. (1995).

Several indicators can be used to judge the effectiveness of bioventing, including monitoring of nutrient depletion and microbiological population changes, respirometry tests (Hinchee and Ong, 1992) to determine degradation rates, and repeated coring to quantify the loss of NAPL. Of these indicators, coring is the most direct measure but entails destructive sampling, and variability of NAPL contents within the target plot may make small losses difficult to quantify. Of the remaining indicators, respirometry tests, where oxygen concentration decreases enable calculation of NAPL degradation rates, are the most reliable, giving at least an indirect measure of degradation during the progress of bioventing. Results of a respirometry test carried out during the bioventing trial are given below.

During bioventing, respirometry tests were carried out by halting aeration periodically to measure oxygen concentration decreases over time. Using these data, oxygen usage rate coefficients were estimated (assuming zero order usage), and with air-filled porosity estimates, can be used to calculate NAPL degradation rates (Γ_h – mass NAPL degraded per mass of soil per day) via the equation (Hinchee and Ong, 1992):

$$\Gamma_h = [(O_c\ P\ \theta_a)/(100\ R\ T\ \rho_b)]\ .\ [(MW_h)/(9.5)] \qquad (1)$$

where O_c is zero-order oxygen usage rate coefficient (%O_2 day^{-1}), P is soil gas pressure (atm), R is the universal gas constant ($8.206\times10^{-5}\,m^3\,K^{-1}\,mole^{-1}$), T is temperature (K), θ_a is air-filled porosity ($m^3\,m^{-3}$), ρ_b is soil bulk density ($kg\,m^{-3}$), and MW_h is molar mass of hexane ($kg\,mole^{-1}$). Note that in Equation (1) parameters that are critical to a reliable estimate of NAPL degradation are O_c, ρ_b and θ_a. In particular, the oxygen usage rate coefficient and air-filled porosity (O_c and θ_a) need to be monitored each time a respirometry test is conducted, since soil-water contents and/or NAPL contents may change during bioventing.

6.2.2 Results

Two locations within the trial area were chosen for monitoring oxygen concentrations for the respirometry tests; one in an area of the trial plot that was amended with nutrients (MI1); and the other in a control area where no nutrients were added (MI3). Cores were collected from each of these locations prior to the bioventing trail and enabled correlation of NAPL depth distributions with the depths of the oxygen sensors, and hence locations for degradation rate estimates. NAPL distributions at location MI1 and MI3 (Figure 12) show the bulk of the NAPL situated above the water table at the time of coring. NAPL is distributed over similar depth intervals at locations MI1 and MI3, with the first occurrence of NAPL contamination 3.86 m below ground. Peak NAPL contents differ at the two locations; however, the total NAPL mass in the soil cores is similar.

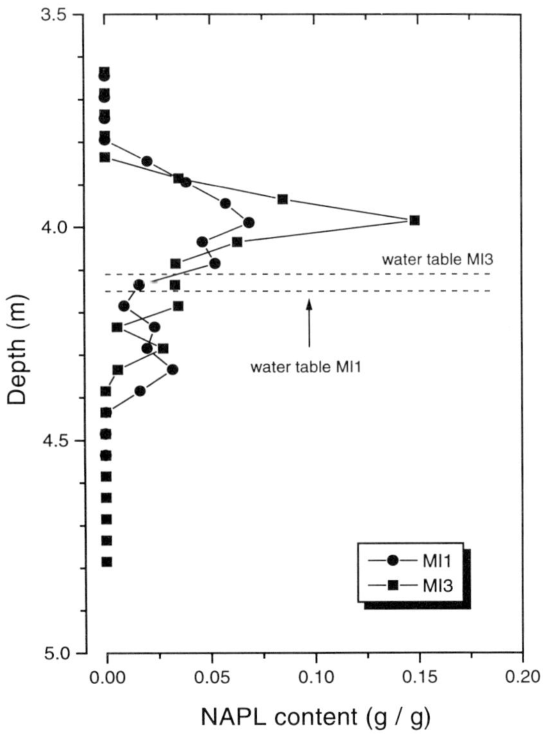

Figure 12. Vertical distribution of diesel NAPL at locations MI1 and MI3.

Air-filled porosity, θ_a, estimates (Figure 13) decrease sharply with depth immediately above the water table. Due to layering in the profile and hysteresis, θ_a profiles do not remain invariant with respect to the water table position (see for instance, Figure 13). Direct measurements are therefore critical to a good estimate of θ_a and, hence, NAPL degradation rate estimates. Fluctuating water tables, in particular, will impact on θ_a (and oxygen concentration) measurements and, therefore, calculations of NAPL degradation rates. During drawdown of water tables, a sharp decrease in θ_a to $0.16\,m^3\,m^{-3}$ is observed at 3.75 m below ground (Figure 13), coinciding with a finer sand layer at that depth.

Oxygen concentrations were measured hourly over a period of six months. Typical oxygen depletion profiles measured with the oxygen sensor installations are shown in Figure 14. Note that although water tables were lowered to approximately 4.5 m below ground, the oxygen sensors at 4.4 m showed no response to aeration, indicating the presence of a capillary fringe. Zero-order oxygen usage rate coefficients at the 4.15 m depth are greater than at the 3.9-m depth, which may be expected if θ_a were lower at deeper depths (i.e. nearer the water table). However, the layering in the profile causes θ_a to be slightly higher at the deeper depth. Increased rate coefficients at the 4.15 m depth could also be due to increased microbial activ-

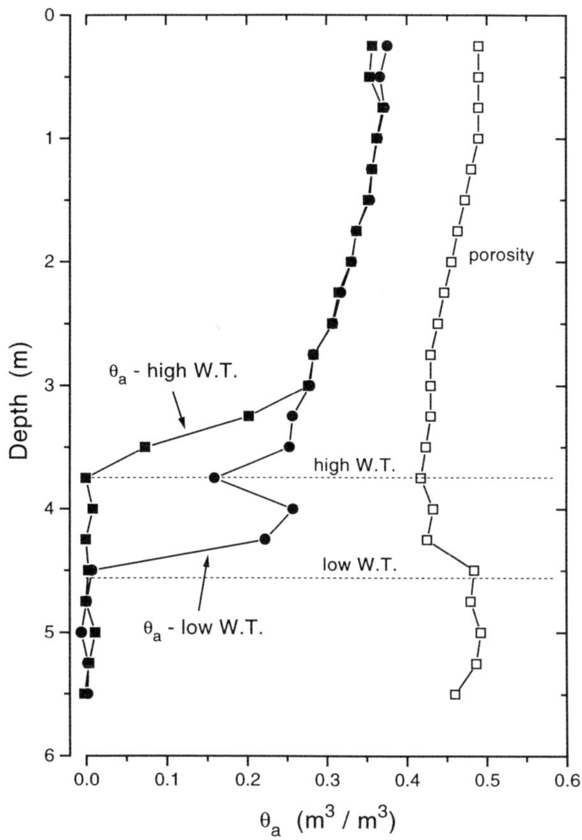

Figure 13. Vertical distribution of air content, θ_a, under high and low water table (W.T.) conditions.

ity within the NAPL-contaminated zone, or be influenced by diffusion of oxygen from the ground surface.

NAPL degradation rates between 7.8 and 40 mg hexane kg soil^{-1} day^{-1} were calculated from Equation (1) using the zero-order rate coefficients (Table 8). These rates are higher than, but comparable to, rates reported by Hinchee and Ong (1992) and Miller et al. (1993), and substantially higher than rates estimated from data of Pijls et al. (1994) for fresh diesel contamination. The higher rates may be due to higher background rates of oxygen utilisation or higher abiotic oxygen usage although recent data from the site suggest that abiotic oxygen usage is negligible. Also, NAPL degradation rates determined for the nutrient-amended area of the trial plot were two to three times those in the unamended area, consistent with laboratory observation that addition of nutrients enhanced biodegradation of the weathered diesel NAPL at the site. This is also consistent with data in Pijls et al. (1994) who reported a contrast of between 5 and 8 for nutrient-amended and unamended sites of fresh diesel contamination.

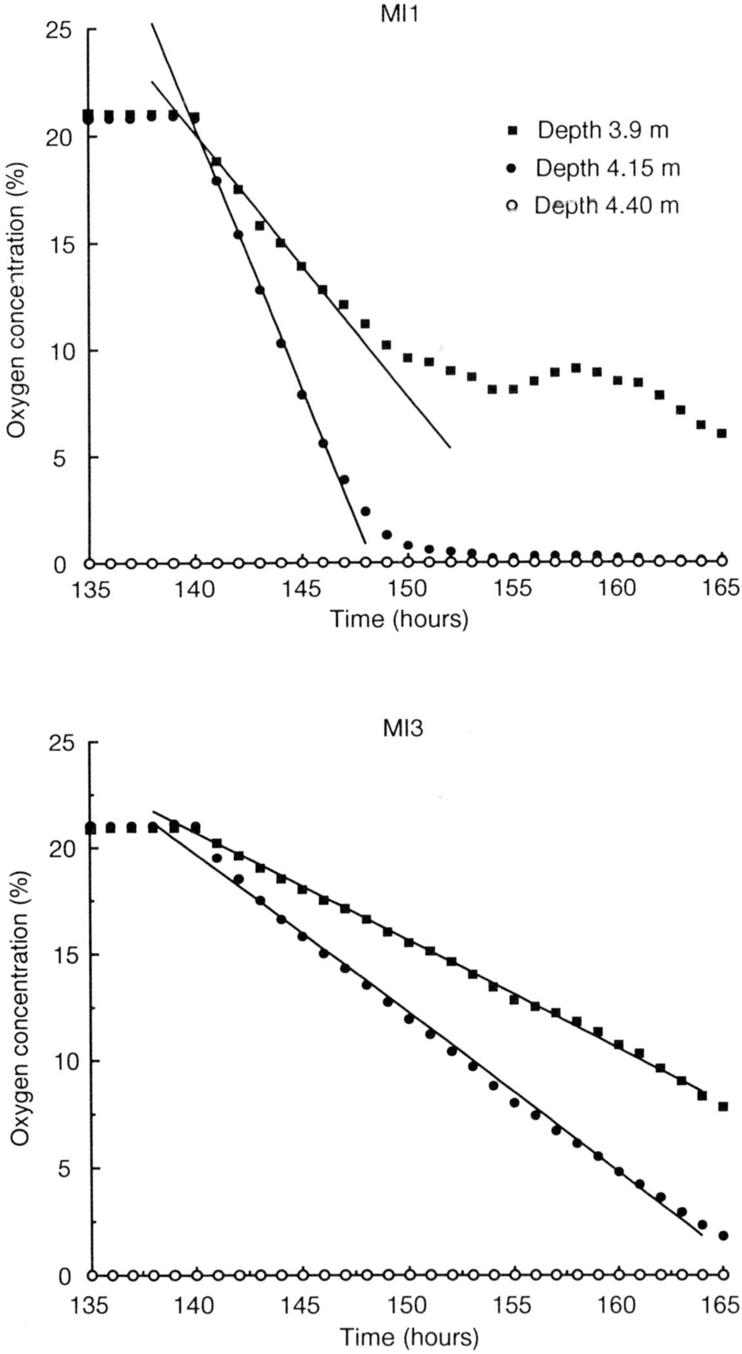

Figure 14. Oxygen concentration depletion measured at 3 depths for locations MI1 and MI3 after aeration was halted.

6.2.3 Summary and implications

Bioventing can be monitored effectively via periodic respirometry tests, provided adequate estimates of bulk density are available, and monitoring of air-filled porosity and oxygen concentrations is possible. The *in-situ* oxygen sensors have proved reliable and efficient for monitoring oxygen concentrations and the progress of bioventing for weathered diesel NAPL. NAPL degradation rate estimates calculated from the oxygen depletion curves are higher but comparable to rates reported by others, and indicate that enhanced biodegradation of weathered diesel NAPL had occurred in the field trial.

Despite the high rates, substantial reduction in NAPL content was not detected when comparing cores recovered prior to the start of the trial and those recovered toward the end of the trial. This was because of the very high initial NAPL contents ($0.075-0.15\,g\,g^{-1}$). At the hydrocarbon utilisation rates estimated at this site, a NAPL content of $0.1\,g\,g^{-1}$ would be reduced by only 5% over the six month trial. This is hardly distinguishable in cored profiles due to the variability of NAPL contents in soil profiles. Rates of NAPL degradation would need to be increased 5 to 10 fold to be cost effective, and to show significant reductions in NAPL contents in cores over a shorter time frame. A second modified bioventing field trial is planned to increase these NAPL biodegradation rates.

6.3 Case Study 3: Ex-situ biotreatment of hydrocarbon contaminated soil

6.3.1 Background

Biotreatment piles, consisting of stockpiled soil material excavated from around fuel storage tanks and from other remediation exercises, have been designed and constructed to remediate hydrocarbon contaminated soil. The soil is typically contaminated with high carbon number hydrocarbon compounds (50% of the material is between C-16 and C-28, and 50% is above C-28), and has an average total petroleum hydrocarbon (TPH) content around $0.02\,g\,g^{-1}$. The biotreatment piles were designed to enhance microbiological degradation of the contaminated soil through active aeration and addition of dissolved nutrients.

The biotreatment piles are typically 1.5 m high, 35 m wide and 75 m long. Horizontal pipes have been placed beneath the piles at approximately 2 m spacings to allow extraction of gas, and hence induce inflows of atmospheric oxygen. Nutrients (N and P) are added daily to the piles. Respirometry tests have been carried out to determine the effectiveness of the design and biotreatment strategy in enhancing degradation.

6.3.2 Methodology

Oxygen sensors were emplaced at six locations within one biotreatment pile, with sensors at two or three depths at each location (i.e. 0.5, 0.75 and 1.0 m, or 0.5 and

Table 8. Zero-order oxygen usage rate coefficients, and hydrocarbon degradation rates.

Treatment	Depth (m)	Zero-order oxygen usage rate coefficients (%O_2 $hour^{-1}$)	Air-filled porosity (m^3 m^{-3})	Degradation rates (mg hexane kg $soil^{-1}$ day^{-1})
MI1	3.90	1.2	0.22	19
(Nutrient-Amended Area)	4.15	2.4	0.24	40
M13	3.90	0.51	0.22	7.8
(Control Area)	4.15	0.74	0.24	12

1.0 m below the top of the biotreatment pile). The oxygen sensors and small diameter tube bundles (positioned for confirmatory sampling of soil-gas), were installed together in hand-augered holes and the contaminated soil material was allowed to collapse around the installation. Bulk densities and soil-water contents were determined on cored material recovered from these augerholes, and these data were used to estimate air-filled porosities at different depths within the treatment pile.

The biotreatment pile was aerated using either one, two or three extraction pumps. Respiration tests were done as for the bioventing remediation case study. Aeration was halted and decreases in oxygen concentrations were measured with the oxygen sensors in order to estimate oxygen depletion rate coefficients. Hydrocarbon degradation rates were calculated for each sensor location using Equation (1).

6.3.3 Results

Aeration profiles with a single 3250 Pa air extraction pump, show oxygen concentrations well below atmospheric levels at all locations and all depths (Figure 15). This indicates that the flow rates induced were inadequate, with respect to oxygen consumption, to aerate the zone to 0.5 m below the surface of the biotreatment pile. Air venting with three air extraction pumps was able to fully aerate the biotreatment piles (Figure 16), within approximately 60 mins of switch-on.

Oxygen usage rate coefficients were estimated from the slope of the curves similar to those shown in Figure 16. Degradation rates calculated from these estimates and Equation (1) are tabulated in Table 9. In each case the calculated degradation rates are lower at shallower depths (12–63 mg-TPH kg-soil^{-1} day^{-1}) and are higher deeper in the biotreatment pile (25–153 mg-TPH kg-soil^{-1} day^{-1}). The lower rates at shallower depths may be due to diffusion of oxygen from the surface of the biotreatment pile to the sensor location, or due perhaps to differing microbial communities or contaminant levels. These issues are to be explored further. The degradation rates estimated for the biotreatment piles are generally much higher than rates collated by Miller et al. (1993) for 51 *in-situ* bioventing sites across the USA, which were principally in the range 0.6–17 mg-TPH kg-soil^{-1} day^{-1}.

6.3.4 Summary

Penetration of oxygen into the biotreatment pile was limited under natural diffusive transport, and was not significantly improved by air venting at low flow rates. Biodegradation rates, calculated from oxygen depletion curves, were consistent across the biotreatment pile, but increased with depth. Biodegradation rates were typically 12–153 mg-TPH kg-soil^{-1} day^{-1}. Given the average hydrocarbon concentration in the contaminated soil within the biotreatment pile, hydrocarbon contamination could be reduced to below 10 mg kg^{-1} within approximately 0.7 years. This assumes all the hydrocarbon compounds are as readily degradable as those being degraded during the respirometry tests. Periodic respirometry tests are to be contin-

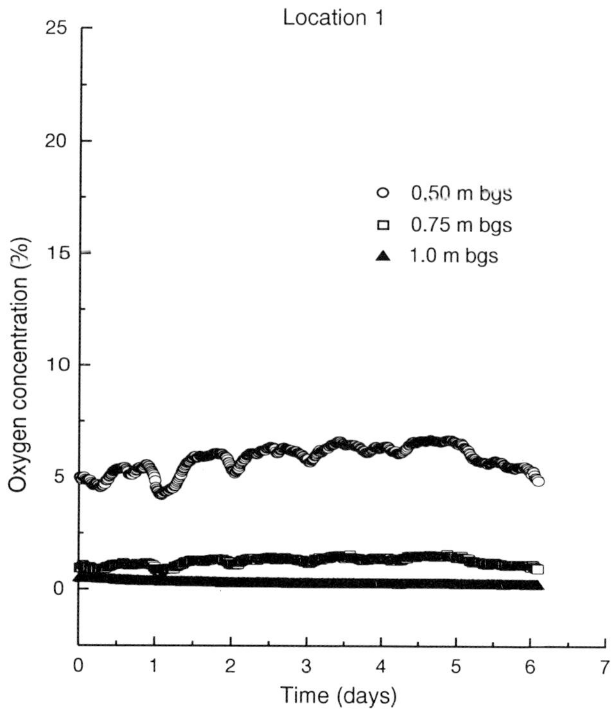

Figure 15. Oxygen concentrations measured at 3 depths below ground surface (bgs) at one location in the biotreatment pile during aeration with a single air extraction pump.

ued on the biotreatment pile to determine the long-term effectiveness of the biotreatment pile design.

7. Conclusion

We have presented an overview of various soil and aquifer remediation technologies that have been developed over the past decade. All of these technologies are designed to provide a more efficient and rapid rate of soil and aquifer restoration, as an alternate to land disposal or the pump-and-treat technology currently in vogue. Such developments have been motivated by the failure of traditional approaches to provide the claimed clean-up of contaminated soils and aquifers.

Various enhanced remediation technologies seem to suffer the same fundamental limitations that have contributed to the failure of conventional technologies. That is, chemical adjuvants added for enhanced remediation via increased desorption and solubilisation or elevated microbial degradation rates, must all be delivered to the sites or regions where the contaminants are located. The efficiency with which this can be done is determined by the hydrogeochemical complexity of the soils and aquifers, and equally limits all of the alternate remediation technologies. Adding to

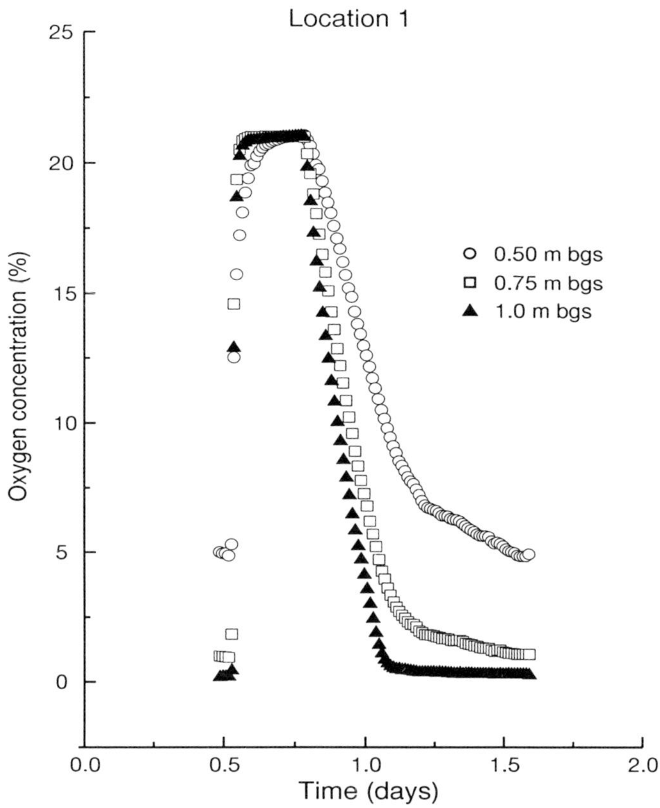

Figure 16. Oxygen concentrations measured at 3 depths below ground surface (bgs) at one location in the biotreatment pile during and after aeration with three air extraction pumps.

these difficulties are the uncertainties arising from identification and characterisation of contaminant sources.

Many of the proposed enhanced remediation technologies have been shown to be efficient and efficacious in lab-scale and pilot-scale studies. However, only a few field studies have been conducted to demonstrate successful full-scale application of these remediation technologies at a range of contaminated sites. Selected case studies for field-scale testing of *in-situ* flushing and *in-situ* bioremediation were reviewed.

Both technological and regulatory constraints have prevented rapid development and evaluation of new remediation techniques. Overcoming such barriers requires inter-disciplinary research conducted by teams of scientists and engineers from a broad spectrum of disciplines. Such efforts are just getting underway at several field sites in the USA. Expansion of such efforts at an international scale should be vigorously pursued.

The analysis presented here inevitably leads to several important practical ques-

Table 9. Calculated degradation rates

Location	Depth (m below ground surface)	Air-filled Porosity ($m^3 m^{-3}$)	Zero-Order Oxygen Usage Rate Coefficient (atm day^{-1})	Zero-Order Degradation Rate (mg $kg\text{-}soil^{-1}$ day^{-1})
1-A	0.50	0.46	0.40	53
1-B	0.75	0.47	0.85	153
1-C	1.00	0.37	1.11	119
2-A	0.50	0.35	0.42	43
2-B	1.00	0.45	1.12	146
3-A	0.50	0.40	0.31	36
3-B	1.00	0.47	0.97	132
4-A	0.50	0.36	0.13	14
4-B	1.00	0.39	0.88	99
5-A	0.50	0.39	0.56	63
5-B	1.00	0.47	0.95	130
6-A	0.50	0.36	0.11	12
6-B	0.75	0.31	0.28	25

tions: Can any of the remediation techniques be expected to clean-up soils and aquifers so that endpoints based on drinking water standards can be achieved? Should we place such stringent clean-up goals for remediation of all sites, no matter what their designated use might be after restoration? Given the various technical difficulties encountered in soil and aquifer remediation, is it not more rational to set clean-up goals on a site-specific basis? Much more work is needed in order to answer these practical questions.

Acknowledgments

Numerous CSIRO scientists have contributed to the research discussed under the Australian case studies section in this paper. From CSIRO, thanks go to Brad Patterson and Chris Barber for their expert advice, support, and input into the field trials. The technical team of David Briegel, Michael Lambert, Tracy Milligan, Wayne Hick, Terry Power, and Jack Smith is also acknowledged. Jan Hosking carried out the inorganic analyses. The Broken Hill Proprietary Company Ltd., the Water Authority of Western Australia, the Australian Institute of Petroleum, and BP Oil are also acknowledged for their financial assistance.

References

Abdul A S and Gibson T L 1991 Laboratory studies of surfactant-enhanced washing of polychlorinated biphenyl from sandy material. Env. Sci. Technol. 25, 665-671.

Abriola L M 1989 Modeling multiphase migration of organic chemicals in groundwater systems, A review and assessment. Environ. Health Perspect. 83, 117-143.

Acton D W and Barker J F 1992 In-situ biodegradation potential of aromatic hydrocarbons in anaerobic groundwaters. J. Contam. Hydrol. 9, 325-352.

Alexander M 1994 Biodegradation and Bioremediation. Academic Press, NY. 302 pages.

Augustijn D C M 1993 Chemodynamics of Complex Waste Mixtures, Applications to Contamination and Remediation of Soils and Aquifers. Ph.D. Dissertation, University of Florida, Gainesville, Florida, USA. (Unpublished) 193 p.

Augustijn D C M, Jessup R E, Rao P S C and Wood A L 1994.Remediation of contaminated soils by solvent flushing. Jour. Env. Eng. 120, 42-57.

Barber C, Bates L E, Taylor K J and Patterson B M 1991 Organic contamination of groundwater in an unconfined sand aquifer beneath an urban area, control or clean-up? Int. Hydrol. & Water Resources Symp., Perth 2-4 Oct 1991; 311-316.

Beller H R, Grbic-Galic D and Reinhard M 1992a Microbial degradation of toluene under sulphate-reducing conditions and the influence of iron on the process. Appl. Environ. Microbiol. 58, 786-793.

Beller H R, Reinhard M and Grbic-Galic D 1992b Metabolic by-products of anaerobic toluene degradation by sulphate-reducing enrichment cultures. Appl. Environ. Microbiol. 58, 3192-3195.

Borden R C and Kao C M 1992 Evaluation of groundwater extraction for remediation of petroleum contaminated groundwater. Water Environ. Res. 64, 28-36.

Boyd G R 1991 Factors influencing nonaqueous phase liquid removal from groundwaters by alcohol flooding. Ph.D. Dissertation, Clemson Univ., Clemson, S.C.

Boyd G R and Farley K J 1992 NAPL removal from groundwater by alcohol flooding, laboratory studies and applications. In Hydrocarbon Contaminated Soils and Groundwater. Eds. E J Calabrese and P T Kostecki. Vol. 2, Lewis Publishers, Chelsea, MI.

Broholm K and Cherry J A 1994 Enhanced dissolution of heterogeneously distributed solvents residuals by methanol flushing, a field experiment. In Transport and Reactive Processes in Aquifers. Eds. T Dracos and F Stauffer. pp 563-568. A A Balkema, Rotterdam, The Netherlands.

Brusseau M L and Rao P S C 1989 Sorption non-ideality during organic contaminant transport in porous media. Critical Reviews in Env. Control 19, 33-99.

Brusseau M L and Rao P S C 1991 Sorption kinetics of organic chemicals, methods, models and mechanisms. In Rates of Soil Chemical Processes, Special Publication No. 27. pp 281-302. Soil Sci. Soc. Amer., Madison, WI.

Chatzis I, Morrow N R and Lim H T 1983 Magnitude and detailed structure of residual oil saturations. Soc. Petrol. Eng. J. 22, 311-326.

Cline P V, Delfino J J and Rao P S C 1991 Partitioning of aromatic constituents into water from gasoline and other complex mixtures. Environ. Sci. Technol. 25, 914-920.

Corey A T 1986 Mechanics of Immiscible Fluids in Porous Media. Water Resources Publications, Littleton, CO, 255 pp.

Davis G B, Barber C, Briegel D, Power T R and Patterson B M 1992 Sampling groundwater quality for inorganics and organics, some old and new ideas. Proc. Int. Drill '92 Conf. Perth Oct 1992, 7 pp.

Davis G B, Johnston C D, Thierrin J, Power T R and Patterson B M 1993 Characterizing the distribution of dissolved and residual NAPL petroleum hydrocarbons in unconfined aquifers to effect remediation. AGSO J. Geology & Geophysics 14, 89-94.

Davis G B, Johnston C D, Patterson B M, Barber C, Bennett M and Dunbavan M 1995 Monitoring bioremediation of weathered diesel NAPL using oxygen depletion profiles. Proc. 3rd Int. Bioreclamation Symp., San Diego, April 1995.

DOE 1994 In-situ Remediation Integrated Program, Technology Summary. DOE/EM-0134P, Office of Environmental Management, Office of Technology Development, USA Dept. of Energy, Washington, D.C. 69 p.

Edwards D A, Luthy R G and Liu A 1991 Solubilization of polycyclic aromatic hydrocarbons in micellar nonionic surfactant solutions. Env. Sci. Technol. 25, 127-133.

Edwards E A and Grbic-Galic D 1992 Complete mineralization of benzene by aquifer microorganisms under strictly anaerobic conditions. Appl. Environ. Microbiol. 58, 2663-2666.

Edwards E A, Wills L E, Reinhard M and Grbic-Galic D 1992 Anaerobic degradation of toluene and xylene by aquifer microorganisms under sulphate-reducing conditions. Appl. Environ. Microbiol. 58, 794-800.

EPA 1992a General Methods for Remediation Operations Performance Evaluation. EPA/600/R-92/002, USA Environ. Prot. Agency, Washington, D.C. 37 p.

EPA 1992b The Superfund Innovative Technology Evaluation Program, Technology Profiles. 5th Edition, EPA/540/R-92/077, USA Environ. Prot. Agency, Washington, D.C.

EPA 1993 Cleaning up the Nation's Waste Sites, Markets and Technology Trends. EPA/643/R-93/012, USA Env. Prot. Agency, Washington, D.C. 164 p.

EPA 1994 Innovative Treatment Technologies, Annual Status Report. 6th Edition, EPA-542-R-94-05, Office of Solid Waste and Emergency Response, USA Env. Prot. Agency, Washington, D.C. 178 p.

Falta R W, Javendel I, Pruess K and Witherspoon P A 1989 Density-driven flow of gas in the unsaturated zone due to evaporation of volatile organic compounds. Water Resour. Res. 25, 2159-2169.

Farley K J, Falta R W, Brandes D, Milazzo J T and Bramme S E 1993 Remediation of hydrocarbon-contaminated groundwater by alcohol flooding. Technical Report, Hazardous Waste Mgmt. Res. Fund, Univ. of South Carolina, S.C. 148 p.

Fountain J C 1992 Field tests of surfactant flooding, Mobility of dense nonaqueous phase liquids, pp. 182-191. In Transport and Remediation of Subsurface Contaminants, Symp. Series 491. Eds. D A Sabatine and R C Knox. Amer. Chem. Soc., Washington, D.C.

Gannon O K, Bibring P, Raney K, Ward A, Wilson J, Underwood J L and Debelak K A 1989 Soil cleanup by in situ surfactant flushing, III. Laboratory results. Sep. Sci. Technol. 24, 1073-1094.

Gillham R W and Rao P S C 1990 Transport, distribution, and fate of volatile organic compounds in groundwater. In Significance and Treatment of Volatile Organic Compounds in Water Supplies. Eds. N M Ram, F C Russell and K P Cantor. Lewis Publishers, Chelsea, MI.

Grbic-Galic D and Vogel T M 1987 Transformation of toluene and benzene by mixed methanogenic cultures. Appl. Environ. Microbiol. 53, 254-260.

Herndon T R 1987 The institutional framework for protecting groundwater in the United States. In Planning for Groundwater Protection. ED. G W Page. Academic Press, New York, N.Y. pp. 29-67.

Hinchee R E and Ong S K 1992 A rapid in situ respiration test for measuring aerobic biodegradation rates of hydrocarbons in soil. J. Air Waste Manag. Assoc. 42, 1305-1312.

Hoag G E and Marley M C 1986 Gasoline residual saturation in unsaturated uniform aquifer materials. J. Environ. Eng. 112, 586-604.

Howard P H, Boethling R S, Jarvis W F, Meylan W M and Michalenko E M 1991 Handbook of Environmental Degradation Rates. Lewis Publishers Inc., Chelsea, Michigan. 725 p.

Huling S G and Weaver J W 1991 Dense Nonaqueous Phase Liquids. Ground Water Issue, Environmental Protection Agency (Ada, OK) 21 pp.

Hunt J R, Sitar N and Udell K S 1988 Nonaqueous phase transport and cleanup, 1. Analysis of mechanisms. Water Resour. Res. 24, 1247-1258.

Imhoff P T, Jaffe P R and Pinder G F 1994 An experimental study of complete dissolution of nonaqueous phase liquids in saturated porous media. Water Resour. Res. 30, 307-320.

Imhoff P T, Gleyzer S N, McBride J F, Vancho L A, Okuda I and Miller C T 1995 Cosolvent-enhanced remediation of residual dense nonaqueous phase liquids, Experimental investigation. Environ. Sci. Technol. 29, 1966-1976.

Johnston C D and Patterson B M 1994 Distribution of nonaqueous phase liquid in a layered sandy aquifer. In Hydrocarbon Remediation. Eds. R E Hinchee, B C Alleman, R E Hoeppel and R N Miller. pp 431-437. Lewis Publishers, Boca Raton, FL.

Lenhard R J, Parker J C and Kaluarachchi J J 1989 A model for hysteretic constitutive relations governing multiphase flow, 3. Refinements and numerical simulations. Water Resour. Res. 25, 1727-1736.

Loxham M 1992 Risk analysis as a guide to soil remediation strategies. In Contaminated Land Treatment Technologies. Ed. J F Fees. pp 1-12. Elsevier Applied Science Publishers, New York.

Mackay D M and Cherry J A 1989 Groundwater contamination, pump-and-treat remediation. Env. Sci. Technol. 23, 630-636.

Mendoza C A and Frind E O 1990 Advective-dispersive transport of dense organic vapors in the unsaturated zone, 1. Model development. Water Resour. Res. 26, 379-387.

Mercer J W and Cohen R M 1990 A review of immiscible fluids in the subsurface, Properties, models, characterization, and remediation. J. Contam. Hydrol. 6, 107-163.

Miller R N, Downey D C, Carmen V A, Hinchee R E and Leeson A 1993 A summary of bioventing performance at multiple Air Force sites. In Petroleum Hydrocarbons and Organic Chemicals in groundwater. Prevention, Detection and Restoration, Houston, Texas, Nov. 1993.

NRC 1993 In situ Bioremediation, When Does It Work? National Academy Press, Washington, D.C. 207 p.

NRC 1994 Alternatives for Groundwater Cleanup. National Academy Press, Washington, D.C.314 p.

Palmer C D and Fish W 1992 Chemical enhancements to pump-and-treat remediation. EPA/540/S-92/001, USA Enc. Prot. Agency, Washington, D.C. 20 p.

Patterson B M, Power T R and Barber C 1993a Comparison of two integrated methods for the collection and analysis of volatile organic compounds in ground water. Ground Water Monitoring and Remediation 13, 118-123.

Patterson B M, Pribac F, Barber C, Davis G B and Gibbs R 1993b Biodegradation and retardation of PCE and BTEX compounds in aquifer material from Western Australia using large-scale columns. J. Contam. Hydrol. 14, 261-278.

Patterson B M, Power T R, Davis G B and Barber C 1994 An in situ device to measure oxygen in the vadose zone and groundwater to determine oxygen utilization rates, laboratory and field evaluation. CSIRO Division of Water Resources Technical Report No. 95/0. 14 p.

Pennell K D, Rhue R D, Rao P S C and Johnston C T 1992 Vapor-phase sorption of p-xylene and water on soils and clay minerals. Environ. Sci. Technol. 26, 756-763.

Pennell K D, Abriola L M and Weber W J 1993 Surfactant-enhanced solubilization of residual dodecane in soil columns. Env. Sci. Technol. 27, 2332-2340.

Pennell K D, Jin M, Abriola L M and Pope G A 1994 Surfactant-enhanced remediation of soil columns containing residual tetrachloroethylene. Jour. Contam. Hydrol. 16, 35-54.

Pfannkuch H-O 1984 Ground water contamination by crude oil at the Bemidji, Minnesota research site, USA Geological Survey toxic waste-ground water contamination study. Paper presented at Toxic-Waste Technical Meeting, USA Geol. Surv. (Reston, VA). 107 p.

Pijls C G J M, Urlings L G C M, van Vree H B R J and Spuij F 1994 Applications of in situ soil vapor extraction and air injection. In Air Sparging for Site Remediation. Ed. R E Hinchee. pp 128-136. Lewis Publishers, Boca Raton, FL.

Pope D F and Matthews J E 1993 Bioremediation using the land treatment concept. EPA/600/R-93/164, USA Env. Prot. Agency, Washington, D.C. 17 p.

Rao P S C, Bellin C A and Brusseau M L 1993 Coupling biodegradation of organic chemicals to sorption and transport in soils and aquifers, Paradigms and paradoxes. In Sorption and Degradation of Pesticides and Organic Chemicals in Soils, Special Publication No 32. pp 1-26. Soil Sci. Soc. Amer., Madison, WI.

Rao P S C, Bellin C A and Lee L S 1995 Sorption and biodegradation of organic contaminants in soils, Conceptual representation of process coupling. In Environmental Impact of Soil Component

Interactions, Natural and Anthropogenic Compounds Eds. P M Huang, J Berthelin, J M Bollag, W B McGill, and A L Page. pp 263-274. Lewis Publishers Inc., Boca Raton, FL.

Rao P S C, Annable M D, Sillan R K, et al. 1995 Field scale evaluation of *in-situ* cosolvent flushing for remediation of a shallow, unconfined aquifer contaminated with residual LNAPL (in review).

Rhue R D, Pennell K D, Rao P S C and Reve W H 1989 Competitive adsorption of alkylbenzene and water vapors on predominantly mineral surfaces. Chemosphere 18, 1971-1986.

Roy S B 1992 Kinetics of solvent-water extraction of coaltar from porous media. M.S. Thesis, Dept. of Civil Eng., Carnegie-Mellon Univ., Pittsburgh, PA (unpublished).

Schwille F 1984 Migration of organic fluids immiscible with water in the unsaturated zone. In Pollutants in Porous Media, the Unsaturated Zone Between Soil Surface and Groundwater. Eds. B Yaron, G Dagan and J Goldshmid. pp 27-48. Springer-Verlag (Berlin) Ecol. Studies 47.

Schwille F 1988 Dense Chlorinated Solvents in Porous and Fractured Media. Lewis (Chelsea, MI) 146 pp.

Sellmeijer J B 1990 SOLUTE, Analytical Transport Model. Delft Geotechnics, Netherlands. Report No. SE-702161. (in Dutch).

Sims J L, Sims R C, Dupont R R, Mathews J E and Russell H H 1993. In situ bioremediation of contaminated unsaturated subsurface soils. EPA/540/S-93/501, USA Env. Prot. Agency, Washington, D.C. 16 p.

Steffy D A, Johnston C D and Barry D A 1995 A field study of the vertical immiscible displacement of LNAPL associated with a fluctuating water table. In Proceedings of the Groundwater Quality, Remediation and Protection Conference (CQ 95, IAHS) Prague, Czech Republic. IAHS Publ. No. 225, pp. 49-57.

Thierrin J, Davis G B, Barber C and Power T R 1992 Use of deuterated organic compounds as groundwater tracers for determination of natural degradation rates within a contaminated zone. In Tracer Hydrology. Eds. H. Hotzl and A. Werner. pp 85-91. Balkema, Rotterdam.

Thierrin J, Davis G B, Barber C, Patterson B M, Pribac F, Power T R and Lambert M 1993 Natural degradation rates of BTEX compounds and naphthalene in a sulphate reducing groundwater environment. Hydrol. Sci. J. 38, 309-322.

Thierrin J, Davis G B and Barber C 1995 A groundwater tracer test with dueterated compounds for monitoring in situ biodegradation and retardation of aromatic compounds. Ground Water 33, 469-475.

Van Dam J 1967 The migration of hydrocarbons in a water-bearing stratum. In The Joint Problems of the Oil and Water Industries. Ed. P. Hepple. pp 55-88. Inst. Pet. (London).

Wilson J L, Conrad S H, Mason W R, Peplinski W and Hagan E 1990 Laboratory Investigation of Residual Organics from Spills, Leaks, and the Disposal of Hazardous Wastes in Groundwater. EPA/600/6-90/004 USA EPA. 267 pp.

13
Remediation of inorganics and organics in industrial and urban contaminated soils

R. G. BURNS, S. ROGERS and I. McGHEE

1. Introduction

As land previously occupied by industry comes under pressure for redevelopment, it becomes evident that a large number of these sites are contaminated. Such contamination is a concern if it has the potential to become a health hazard to site occupiers or a threat to overall environmental quality. Soil remediation is a young technology whose ultimate success will be dependent upon a combination of chemical, physical and biological knowledge. This knowledge must be built on a sound understanding of the basic science behind the mobility, reactivity and toxicity of pollutants in natural ecosystems. However, the economics of remediation and the end-use of the treated land will be major factors in the widespread adoption of clean-up technologies. Possibly the greatest single stimulus for soil remediation will be legislative pressures that define the limits of various contaminants in soil, water and the atmosphere. These standards are likely to become increasingly stringent as analytical techniques improve. Whilst there may be little justification for these levels from toxicological data, the overriding consideration will be to reduce inorganic and organic contaminant levels to 'below detectable limits'. The implementation and policing of these regulations will be a political as well as a scientific issue and special pleading, focused on using the best available technology, risk assessment and economics, will be a common feature of the debate. Nonetheless, the remediation and reuse of contaminated sites will be a major challenge as we enter the next millennium and the importance and sensitivity of the issue is clearly evident from the widely varying estimates by the various government and independent agencies of the extent of land contamination.

In the Northern Hemisphere, the industrial revolution has left a legacy of chronically contaminated factory sites. For example, it is estimated that there are at least 4000 hazardous waste sites in the United States (Will, 1995). This results in large tracts of land which require constant monitoring and have restricted access. In addition, the hugely successful chemical industry has, until recently, paid scant attention to the potential dangers of poor housekeeping resulting in spills, leaks and discharges that have polluted not only the site but adjacent land and water. In the past decade, a number of factors have combined to change the attitudes of industry and government to environmental pollution. These include the adoption of 'green' ideas by leading politicians and their parties, as a holistic view of the earth's biosphere and its limited resources began to pervade all levels of society. The potential for site redevelopment for real estate, recreation or newer 'clean' industries is an

R. Naidu et al. *(eds): Contaminants and the Soil Environment in the Australasia-Pacific Region, 411–449.*

additional motive force, as is the public demand (and legislative pressures) for 'clean' soil, air and water. All these stimuli would be of little real value if we had not entered an era when scientific knowledge and technical innovations have given us the confidence to both restrict the flow of contaminants into the environment, and to clean-up the old contaminants that already contaminate soil and water. The issue of contaminated urban and industrial sites is a relatively new one in Australia, and attention to the investigation and remediation of contaminated sites only began in the mid-1980's (Forrest, 1993).

2. Contaminated site remediation: Australian perspective

Australia is a large country with a small and mainly urban based population. In the past there has been a significant area of land available on the urban fringes for expansion and development which has, in turn, reduced the need to redevelop existing urban sites. However, the continued expansion of the urban fringe is becoming increasingly uneconomic, mainly due to the high costs of the infrastructure and the adverse impacts on the natural environment. This has resulted in the development of former industrial and inner urban land, which in many cases involves a change to a more sensitive use, for instance from industrial to residential. The total number of contaminated sites in Australia is unknown (Edwards et al., 1994), although estimates have been made in Tasmania (400) and numbers in the two most populous states of New South Wales (NSW) and Victoria are estimated to be in the order of 30 000. Forrest (1993) indicated that the figure for contaminated sites in NSW could be as high as 60 000, with up to 7 000 sites requiring remediation at a cost of approximately (Aus)$2 billion.

It has been estimated that there are 8 000 t of hexachlorobenzene and 2 300 t of polychlorinated biphenyls (PCBs) stored in Australia at managed (i.e. secure and regularly monitored) sites (Independent Panel on Intractable Waste, 1992). However, there is concern about the quantity of scheduled wastes at so-called unmanaged sites. For example, organochlorine insecticides from the thousands of sheep dip sites throughout Australia have been identified as a major source of soil contamination and at former cattle dip sites in NSW levels of DDT at 106 mg kg^{-1} have been recorded (Beard, 1993, see chapter 14). It has been estimated that some of these sites may contain up to 20 t of contaminated soil (Independent Panel on Intractable Waste, 1992). Estimates of soil volumes contaminated with scheduled wastes, including PCBs, hexachlorobenzenes, organophosphorus (OP) pesticides and chlorinated hydrocarbons are in excess of 100 000 t according to the Commonwealth EPA (1994). The EPA (1994) also notes that soils contaminated with scheduled wastes cannot be treated in significant quantities in Australia at this time due to the lack of a suitable treatment facility.

Although industrial development in the Australasia-Pacific region has a shorter history than that of Europe, and the number of sites contaminated are probably fewer, McNamara (1994) noted that the factors emerging as being important in the remediation are essentially the same as those experienced throughout the rest of the world. ANZECC/NHMRC (1992) lists potential industries and land uses associated with soil contamination. These range from gas works, chemical storage and manufacture, metal processing, tanneries, mining and landfill through to sheep and cattle dips, and are essentially the types of industrial and agricultural land use associated with soil contamination in the Northern Hemisphere. In the Hindmarsh district of Adelaide, South Australia, Turczynowicz (1991) identified 38 different industrial land uses including brickworks, tanneries, chemical storage, gasworks, textile dyeing, timber treatment, small scale metal smelters, chromium plating and paper manufacture. The tanning industry in the area dated back to 1847, and coal gas production began in the area in 1863.

The course of remedial action at polluted sites is decided following a detailed survey of the contaminants and assessment of potential off site transport. This type of survey requires experienced chemists, pedologists and hydrogeologists using proven sampling and analytical techniques. Remediation strategies should, in the first instance, consider the potential end use of the contaminated site. If the contaminated site is not to be redeveloped then one course of action may be to monitor the site frequently but leave it undisturbed. However, if the site is to be used for a more sensitive end use (i.e. recreation or housing), where risk of exposure to contaminants by site occupants has potentially serious health consequences, then treatment may be required.

3. Treatment technologies

Contaminated land can be remediated by chemical, physical or biological techniques (or a combination of all three). Not surprisingly, a large variety of methods have been developed due to the diverse chemical properties of the contaminants and the variety of soils and sites in which they are found (Table 1). The numerous choices were highlighted recently by McEldowney et al. (1993) who listed the treatment technologies assessed for the clean-up of New Bedford Harbour in the USA which contained soil and sediment contaminated with PCBs and heavy metals. This work involved evaluation of a total of 56 different treatments including 5 biological, 26 physical, 11 chemical and 14 thermal methods.

The technology applied to the treatment of contaminated land is often specific to the contaminants to be removed and the site characteristics. Many industrial sites contain a heterogeneous mixture of organic and inorganic contaminants which determine the choice of remediation strategies. For example, immobilisation treatment (e.g. cement). has been used frequently to remediate wastes contaminated with inorganic compounds However, the presence of sulphates, phenols, oils and greases in the waste can result in interactions which interfere with the binding of metals to the cement reducing the efficacy of the process (Jones, 1990). Even more important is the evidence that suggests if immobilisation is used to treat waste sites also contaminated with organics, there is an increase in both the volatility and solubility of

Table 1. Treatment technologies for the remediation of contaminated land.

Principle	Treatment technologies
Containment	Liners, e.g. clay, synthetic Jet grouting Slurry walls Ground freezing
Reduction of contaminant mobility	Stabilisation Solidification
Immobilisation/destruction	Chemical treatment Vitrification
Contaminant removal	Vacuum extraction Air stripping Soil washing Pump and treat Solvent extraction Low temperature thermal desorption
Destruction	Land farming Composting Treatment beds/biopiles *In situ* bioremediation Incineration/thermal destruction

Table adapted from Ellis (1992).

the organics and a consequent increase in the threat to the environment (USEPA, 1990a; b).

Traditionally, remediation of contaminated land in Australia has involved either on site management or excavation and subsequent disposal to a registered landfill site (Parker, 1994). Management has ranged from do-nothing, or covering with buildings, through the use of capping materials (e.g. clean topsoil or asphalt), to sophisticated engineering barriers. However, there is increasing concern about the safety of landfill disposal in relation to the long term stability of containment and the potential for off site transport of contaminated leachate (Jefferis, 1992). Notwithstanding, there will inevitably be a continuing requirement for landfill even if only for the disposal of the waste generated by other treatment technologies, for example ash from incineration. In recent years there has been a trend towards the use of alternative and sometimes completely novel methods for the clean-up of contaminated land on site (Glass, 1992). Many of these techniques are science led but others are driven by the increasingly complex legislative framework (Table 2) governing water, soil and air quality.

The following sections outline the methods currently available for the remediation of soil and water contaminated with organic and inorganic compounds. However all of these technologies have not been used in full scale site remediation operations in the Australasia-Pacific region. In addition, the development of innovative methods will be discussed and relevant case-studies presented.

Table 2. Factors which influence the choice of remedial strategy and design.

Category	Influences on remedial design
Legal	International, national and local legislation Requirements of Water, Disposal and Health and Safety Authorities
Political	Type of development and end-use Degree of clean-up Speed of response Public profile
Commercial	Value of land pre- and post-treatment Volume and nature of materials requiring treatment Time, space and money available for reclamation Cost of monitoring and analysis Capital outlay, licensing costs, insurance requirements Cost of disruption to site
Geographic	Proximity to domestic dwellings Access to site
Environmental	Proximity to an aquifer Hydrogeology of the site and its surroundings Local weather conditions
Engineering	Soil type: permeability, texture, homogeneity, etc. Depth of water table Volume of soil requiring treatment Availability of materials and equipment Maintenance and operational requirements of the technology
Health and Safety	Toxicity of pollutants Side-effects of treatment (e.g. production of volatile toxins) Nuisance effects; dust, noise, vibration, odour Handling of materials
Managerial	Availability of trained personnel Degree of quality control and quality assurance required
Technical	Limitation of treatment methods and applications Specification of clean-up criteria including analytical methods Availability of proved techniques and raw materials Confidence with which contamination problem has been defined Compatibility of various methodologies Accuracy of treatment schedule predictions

Table adapted from Ellis (1992).

4. Inorganic soil contaminants

Inorganic contaminants of industrial and urban soils can be categorised into two groups: heavy metals and inorganic non-metals. Inorganic soil contaminants typically encountered are principally heavy metals and cyanide/metal-cyanide complexes (Meeussen et al., 1994). Heavy metals form the main group of inorganic contaminants on former industrial and urban sites, and the reader is referred to the works by Adriano (1986; 1992) and Alloway (1990) for their biogeochemistry.

Table 3. ANZECC/NHMRC preferred order of options for site clean-up and management.

Treatment option	Comments
1. Onsite treatment	Containment destroyed or hazard reduced to acceptable level
2. Offsite treatment	Depending on residual contaminant levels soil: Returned to site, removed to waste disposal site, used a fill material
3. Removal of soil	Excavation of contaminated material, disposal in approved facility
4. Barrier	Isolation of contamination using surface barriers
5. Choose less sensitive re-use	Minimises need for remediation
6. Leave contaminated material in situ	Only if no immediate danger to environment or community and adequate access control in place

ANZECC/NHMRC (1992).

Examples of inorganic contamination of current importance in the Australasia-Pacific region include heavy metal contaminants and nitrate and phosphates from agricultural practices and disposal of industrial wastes (see Chapters 3, 5 and 6). For a detailed review of the status of these contaminants and approaches adopted for remediation of the contaminated sites in countries in the Australiasia Pacific region, readers are directed to Chapters 14 to 25.

5. Remediation of inorganic contaminants

Australian protocols for the management of contaminated sites (Langley and El Saadi, 1991), state that remediation is carried out to protect occupants of a site from long-term exposure risks (acute risks from soil contamination are rarely an issue). The authors also note that the following issues should be considered when embarking on a site management/soil remediation program: access to site; dust arising from the site; groundwater contamination; spread of the contaminant off site; noise and vibration from remediation operations; and odour resulting from remediation. The reader is referred to chapter 9 for a detailed review of contaminated site risk assessment.

Techniques for the remediation of soils contaminated with inorganic compounds can be split into three categories: excavation and disposal on site treatment (involving management of contaminants to reduce risk of exposure); and excavation and treatment of soil on site and subsequent replacement of clean soil.

Table 3 lists the preferred ANZECC/NHMRC options for the remediation of contaminated sites in Australia. With reference to inorganic contaminated soils there are two main options available: the first is to excavate the contaminated material and transport off site; the second is to manage contamination on site in order to reduce possible exposure risk. Pierzynski et al. (1994) noted that the second option can be split into three categories: reducing inorganic concentration to an acceptable level; isolation of contamination to prevent any further reaction of the soil with the environment; and reducing the biological availability (i.e. bioavailability) of inorganic contaminants. Reed et al. (1992) categorised on site or *in-situ* remediation technologies as destruction or transformation of the contaminant, immobilisation (i.e. reducing bioavailability), and separation of the contaminant from the bulk soil.

Remediation of sites contaminated with inorganic compounds presents a different set of problems when compared to the remediation of soils contaminated with organics. For all heavy metals and most inorganics destruction of the compound is not an option and bioremediation and thermal destruction cannot be applied to contaminated soils. Other techniques commonly used on organic contaminated sites, which rely on air and steam stripping to volatilise the compounds, are also not applicable to inorganic contaminated soils (Kroopnick, 1994). As a consequence, whilst a number of on site treatment technologies that remove organic contaminants from the soil are available (and their use is encouraged by regulatory authorities; NSW EPA, 1994), there are comparatively few methods for the removal of inorganic contaminants (Peters and Shem, 1992). Table 4 lists techniques available for the remediation of soils contaminated with inorganic compounds.

5.1 Excavation and reburial

The removal of soils contaminated with inorganic compounds and their burial at landfill sites is a common technique (McNeill and Waring, 1992). At present offsite burial of contaminated soil is used extensively in Australia (Smith, 1993), being applied to over 40% of sites remediated in Victoria between 1988 and 1993 (Shoebridge, 1993). There are a number of reasons why containment in a landfill is currently the most common option in Australia. Containment is well accepted by the remediation industry and its customers, and is perceived to represent the least long-term liability (Shoebridge, 1993). Landfill has also been the subject of significant research concerned with monitoring and predicting contaminant behaviour (Gera, 1988; Valocchi, 1988; Barber et al., 1992) and has provided suitable design criteria (Brown,1993) to model and predict contaminant behaviour. For example, in lightly populated countries such as Australia, the availability of suitable landfill sites tends to be higher than in more densely populated Northern Hemisphere countries. Shoebridge (1993) noted that there were many sites in the Melbourne metropolitan area that were appropriate for the disposal of low level contaminated soil.

At present offsite burial is the predominant remediation technology, especially for soils contaminated with inorganic compounds. However, Australian guidelines for the assessment and management of contaminated sites (ANZECC/NHMRC, 1992) suggest that this may not continue to be an appropriate option due to the potential hazard associated with transport of contaminated material to the landfill site, contaminant migration from the landfill into adjacent environments in leachates (Williams, 1988), and the future limitations on the availability of secure landfill. In addition, other on site or *in situ* remediation technologies are developing rapidly. Certain secure landfill operations in Victoria now require customers to sign indemnity forms exempting the landfill owner from future claims (Shoebridge,

Table 4. Treatment technologies available for soils contaminated with inorganic compounds.

Technique	Objective	Processes	Common Problems
Excavation and reburial	Removal of contaminants	None	Environmental acceptability, regulatory control, long term liability
Onsite burial	Containment	Encapsulation	Long term reliability, liability
Synthetic liners	Containment Contaminants remain on site	Barrier method, Minimisation of exposure	Short and long term resistance to contaminants
Modified clay liners	Containment Contaminants remain on site	Barrier method, Minimisation of exposure	Installation difficult, durability
Jet grouting	Containment Contaminants remain on site	Barrier method, Minimisation of exposure	Difficult to install, high cost
Slurry walls	Containment Contaminants remain on site	Barrier method Minimisation of exposure	Difficult to install
Ground freezing	Containment Contaminants remain on site	Barrier method, Minimisation of exposure	High cost
In situ Stabilisation/ immobilisation	Reduction in mobility	Precipitation, adsorption, complexation, oxidation/reduction	Interference from co-contaminants esp. Organics, lack of targeting
Solidification	Immobilisation	Production of solid material, cement, fly ash, lime. Reduce contaminant leaching/ migration	Long term efficiency, interference from mixed wastes
Vitrification	Immobilisation	Precipitation in glass based product	Expense, soil variability
Soil flushing	Physical removal in-situ	Leaching, solubilisation, desorption, extraction	Time scale, migration in flushing medium
Soil washing	Physical removal ex-situ	Excavation, leaching, solubilisation, desorption, extraction	High cost, disposal of highly contaminated residue
Physical mixing	Mix hot spots/import clean fill, lower overall levels below trigger values	Dilution	Contaminant transport offsite

Adapted from Logan (1992) and Ellis (1992).

1993), giving rise to concerns within the industry regarding their long term-legal liability with off site disposal. An increasingly popular view is that off site disposal should be regarded as the last resort treatment for contaminated soils as it merely shifts the contamination problem elsewhere (Smith, 1993). ANZECC/NHMRC (1992) list off site disposal in an approved facility as the third most preferred option for contaminated site remediation (Table 3). Heavily contaminated soils can, in some instances, constitute what are known as special or hazardous wastes. In NSW these wastes are not accepted at secure landfill unless subject to treatment converting them to an acceptable form for disposal at an unsecured landfill site. This is due to the fact that the secure landfill option is no longer available in NSW. This regulation means that wastes have to undergo some form of solidification or stabilisation prior to burial thereby reducing the risk of contaminant migration and transport in leachate (Kyle, 1990).

A comparison of remediation in Australia and the USA was made by Shoebridge (1993). Whereas more than 40% of sites in Victoria were remediated by off site burial, only 7% of USEPA Superfund sites were subject to this form of disposal. The difference is due mainly to the strict USA regulatory controls which have forced the cost of landfill up, making it uncompetitive compared to other remediation technologies (DuGuay, 1993). At present off site containment of contaminated soils in Australia is cost competitive when compared to other soil remediation technologies (Shoebridge, 1993). However, Wheeler (1994) noted that current landfill charges do not accurately reflect the true costs and recommended that charges to the customer should increase by 60–90%. Elliott et al. (1989) suggested that offsite burial could not be considered an ideal option due to the costs involved in removing large volumes of material to landfill. ANZECC/NHMRC (1992) notes that if no appropriate treatment technology is available contaminated material could be *temporarily* stored in specially designed landfills. The soil could then be recovered for treatment at a later date once technologies become available or cost competitive.

5.2 *Onsite management of inorganic contaminated soils*

5.2.1 Soil mixing

The simplest technique for reducing the hazards associated with inorganic contaminants in soils is their dilution to levels below which exposure is not considered a risk. This can be achieved either by importing clean soil and mixing it with the contaminated soil, or redistributing clean material already on site (Musgrove, 1991). This technique has been applied to a number of contaminated sites in Australia, including a former chemical factory site in NSW (Finnecy and Johnstone, 1988). Another dilution technique relies on deep ploughing, during which the vertical mixing of surface contaminated soils with less contaminated sub-soils not surprisingly reduces the surface contamination (Thompson-Eagle and Frankenburger, 1992). Vertical mixing of contaminated soils on former agricultural land with low level agrochemical contamination, has been proposed as a soil remediation strategy particularly for sites being redeveloped for housing, and guidelines have been published (NSW EPA, 1995). Whether surface contaminants are diluted with clean materials or diluted with sub-soils the total contaminant loading of the site remains the same. Secondly, the potential for offsite contaminant transport may even be increased as the contaminant is moved to soil horizons where physico-chemical interactions and adsorption are reduced.

5.2.2 Onsite containment

Effective contaminated site remediation strategies should protect all components of the biosphere (land, air, surface water, groundwater) as well as the health of the work force and general public (ANZECC/NHMRC, 1992; McNeill and Waring, 1992). Onsite containment and barriers provide an inexpensive and rapid solution (Jefferis, 1992) in contrast to the problems of contaminant transport offsite, as well as reducing exposure to site occupants (Yland and Wachem, 1988; Ellis, 1992). The installation of simple surface barriers needs only standard earth moving equipment and the experience that is already available in the construction industry (Thorburn and Buchanan, 1987). As a consequence, containment is a popular option (Yland and Wachem, 1988).

The simplest and cheapest form of onsite containment is achieved by covering the soil with clean material (Jefferis, 1992). The material can be either imported top soil or clean soil from other areas of the contaminated site (Cairney, 1992). Langley and El Saadi (1991) suggested that a 0.5 m depth of clean soil is an adequate barrier for residential developments as this is unlikely to be penetrated by normal gardening activities. In Australia's dry climate plant root exploration depth, particularly under backyard gardening activities, is often determined by the depth of domestic watering, about 30 cm (Calder, 1993), although in temperate cooler climates this is likely to be deeper.

Clean surface covers have been used extensively at Port Pirie in South Australia to reduce exposure to soils contaminated with Pb as a result of the operation of a smelter on the outskirts of the town (Body et al., 1988). Soils in private gardens with greater than 2 500 mg Pb kg^{-1} soil were subject to exposure minimisation, either by the installation of clean surface cover or by the removal of contaminated material for disposal.

If upward migration of contaminants (due to hydrogeological conditions) through the surface barrier is considered likely then a surface barrier can be engineered. For instance, impermeable clays or high density polyethylene (HDPE) liners can be installed immediately below the clean soil cap (Jefferis, 1992). Alternatively, if the downward movement of contaminants is a concern covers known as capillary break layers consisting of impermeable layers to prevent infiltration of rain and surface water can be installed (Cairney, 1992). Most contaminated sites are not uniformly contaminated and have 'hot spots' such as the localised spent oxide dumps at gaswork sites (ICRCL, 1986). Often the remainder of the site is relatively uncontaminated or even clean. In instances where the area covered by 'hot spots' is small, the most suitable option may be to excavate the small volume of soil and bury it elsewhere onsite in bunkers. Bunkers can be lined with impermeable materials such as clay or HDPE liners to prevent contaminant migration (Dennis, 1993).

One of the major disadvantages of surface barriers is that subsurface flow (i.e. lateral movement of soil water) through the contaminated material cannot be prevented. It is likely that soil contaminants transported in subsurface flow can cause offsite contamination. Obviously, clean cover barrier methods are not appropriate when gaseous or volatile vapour phase contaminants, such as low molecular weight organic and inorganic compounds, are present as they will not prevent the migration of contaminants into the atmosphere. When onsite containment technologies are applied, soil contaminants remain beneath the barrier in an unchanged form and one of the main concerns is that the barrier may fail and expose site occupants to hazardous contaminated material. Cracking of clays during drought periods, penetration by plant roots and breaches of the barrier due to trenching, pipe laying etc, (Langley and El Saadi, 1991) may all lead to exposure. Failure of barriers to resist forces of expansion and contraction, land shift and changing groundwater pressure will also limit their life (Ellis, 1992).

The chemistry of soil contaminants is also an important factor determining the success of a barrier layer. Chemical attack on soil surface covers by low pH soils, and the deflocculation of clays by oils and tars may also cause failure of the integrity of the barrier leading to upward migration of the contaminants (Muller-Kirchenbauer et al., 1988). In order to minimise the risk of exposure due to barrier failure, Tiller (1992) suggests that surface barriers can be used in conjunction with a less sensitive land use that poses the least risk to public health if the barrier fails. For example, changing from residential to industrial use is an option to minimise potential public health risks associated with contaminant exposure (ANZECC/NHMRC, 1992). In Australia's arid climate, surface soil cracking and wind erosion of topsoil are common during prolonged periods of summer drought. Given that these processes may lead to the failure of the surface capping layer it is essential that the technologies developed in the Northern Hemisphere are re-evaluated prior to their application in the Australasia-Pacific region.

5.2.3 Immobilisation of inorganic contaminants

It is the bioavailable fraction of inorganic contaminants that poses a toxicological or an environmental risk (Hinwood, 1991). Bioavailability of a toxic element is defined as the chemical form that will affect the life cycle of plant and other organisms. Heavy metals in soils can be adsorbed to ion exchange sites, co-precipitate with other minerals, precipitate as insoluble sulphides, carbonates, phosphates or oxides, form complexes with organic matter, or be present in the soil solution phase as free metal ion or coupled with organic or inorganic ligands (Dowdy and Volk, 1983). Adsorption of inorganic contaminants to soil mineral particles and organic matter is one of the main processes determining their mobility and bioavailability (McBride, 1994). Despite strong contaminant retention by surface soil particles (see Chapter 3), downward movement of contaminants in soils may be associated with mobile colloidal materials (Dowdy and Volk, 1983). However, sorption is a reversible process and desorption may release previously inert inorganics into the soil solution

(McCloskey and Bayer, 1987). As discussed in Chapter 3 soil physicochemical properties such as pH, location and nature of the sorption sites, ionic strength and composition of the solution phase, presence of ligands, organic matter and hydrous oxides, will all affect the dynamic relationship between adsorption and desorption and hence bioavailability (Corey et al., 1981) (Chapter 3).

If bioavailability is a key property determining whether soil is designated as contaminated then remediation technologies involving the immobilisation of inorganic compounds may be attractive (Mench et al., 1994). Immobilisation can be achieved by complexing contaminants thereby reducing the solution phase concentration of the contaminants (Wiles et al., 1988). The simplest route to onsite immobilisation is to increase the soil pH by liming. In general, the solubility of many heavy metals (e.g. Cd, Cu, Ni, Zn) will be reduced as the pH increases as maximum adsorption normally occurs at neutral or slightly alkaline pH (Adriano, 1986). The effects of pH on metal mobility are *via* competition between soil protons and metal cations for adsorption to dissolved ligands and surface functional groups (Sposito, 1989). There are important exceptions to this rule and As, Mo, Se and Cr are more mobile under alkaline soil conditions (USEPA, 1977; Pierzynski *et al.*, 1994). Adriano (1986) demonstrated in one soil type that a one unit increase in pH will result in a hundred-fold increase in soil solution concentration of Mo.

Cyanide is an important inorganic pollutant that is more mobile (and therefore biologically available) at alkaline pH. Cyanide is generally present at contaminated sites in the form of iron-cyanide complexes: $Fe(CN)_6^{3-}$ and $Fe(CN)_6^{4-}$. Meeussen et al. (1994) demonstrated that the mobility of cyanide complexes is pH dependent, with higher concentrations of CN leaching into groundwater at pH 7.5 than at pH 4.0. One of the main sources of industrial CN contamination are former gaswork sites (Turczynowicz, 1993) at which CN was removed from coal gas by passing the gas over ferric oxide; the iron-cyanide wastes were then dumped onsite (Roberts and Gemmell, 1979). The contrast between the mobility of most metals (mobility decreasing with increasing pH) and cyanide at alkaline pH highlights one of the problems that can be faced when remediating industrially contaminated sites. Inevitably, gasworks sites are contaminated with a number of compounds, including oils and tars, polynuclear aromatic hydrocarbons, phenols, heavy metals and cyanides (ICRCL, 1986). Remediation strategies to reduce the risk from one group of contaminants (Troy et al., 1994) may well increase the mobility of another group.

Immobilisation involves stabilising the contaminated soil either in a solid mass using cement, fly ash, gypsum, asphalt and bitumen or vitrification and has been applied to field scale remediation (Peters and Shem, 1992). The simple addition of rock phosphate to contaminated soils has also been demonstrated to reduce the bioavailability of Pb in contaminated soils due to immobilisation of solution phase metal (Ma et al., 1995). An advantage of stabilisation and solidification is that it can be carried out *in situ*, with the minimum of soil excavation (Stitson, 1990). However, there is a question over the long-term resistance to breakdown of the solidified soil mass (Peters and Shem, 1992).

Contaminated sites must be well characterised as the presence of co-contaminants, such as sulphates, hydrocarbons, grease, and phenolics, can interfere with successful immobilisation. In some cases, immobilisation can result in increased organic contaminant volatilisation and leaching (Ellis, 1992). In addition, onsite immobilisation is technologically complex and expensive, and it has been suggested that its use should be restricted to highly contaminated soils and those containing radioactive wastes (Peters and Shem, 1992).

A major inherent problem associated with onsite immobilisation techniques is that although the inorganic contaminants are less bioavailable (and the potential risk of exposure and plant uptake is reduced) the contaminant concentration has remained unchanged. As with most international guidelines, Australian regulations (ANZECC/NHMRC, 1992) are based on total concentrations. There is a case for basing guidelines and risk assessments on the bioavailable or soluble fraction of a contaminant in the soil (ANZECC/NHMRC, 1992) although it is difficult to predict long-term contaminant desorption rates. Due to these uncertainties, regulatory authorities are not confident in setting soil quality standards determined by bioavailability and whilst soil quality guidelines remain based on total contaminant levels, there is very little encouragement for the soil remediation industry to apply immobilisation techniques to site restoration.

5.3 Soil washing

Soil washing or extraction processes provide the only alternative to offsite burial for removing inorganic compounds from contaminated soils (Elliott et al., 1989). Soil washing is applied widely for the remediation of heavy metal contaminated soils in Europe (Tuin and Tels, 1991) and can be carried out onsite using mobile equipment (Bromhead and Beckwith, 1994). However, this technology is uncommon outside Europe and has not been applied to full-scale site remediation in the Australasia-Pacific region (Anon, 1995). USEPA (1987) notes that soil washing cannot be regarded as a true remediation process, but rather a pre-concentration stage prior to disposal or further treatment. In developing a soil contaminant removal process, there are three basic stages to be considered: decontamination of the bulk soil; concentration of the contaminant; and disposal of the contaminant (McNamara, 1994). Soil washing is based on existing technologies used in the mineral processing industry, where particles of different size fractions can be segregated (Wills, 1988; Williams, 1992). The process depends on the principle that most soil contaminants are associated with the clays and humic materials (Dowdy and Volk, 1983) and involves physically separating the relatively uncontaminated coarse material from the heavily contaminated fine portion (Bromhead and Beckwith, 1994). Although washing may leave the majority of material with levels of contaminant below regulatory guidelines, it will produce a residue with high contaminant levels, requiring further treatment or burial (Dennis et al., 1994). The relatively uncontaminated coarse fractions can be returned to the site (Peters and Shem, 1992). The absence in Australia of full-scale contaminated site restoration case studies using soil washing cannot be explained by a lack of available technology. In fact, Australia has a large

Table 5. Capital and operating costs[a] associated with soil-washing processes.

Process	Capital cost (A$ million)	Operating cost (A$ t^{-1})[b]
Heijmans	6.4	104–130
HMZ	3.14–4.8	78–117
Heidemij	4.0–4.8	128–221
Harbauer	6.1–8.7	194

Adapted from Dennis et al. (1994).
[a]Costs are based on US estimates converted to Aus $.
[b]Operating costs do not include excavation, treatment or disposal of the contaminated 'fines', wastewater treatment or site restoration.

non-ferrous metal mining industry, with wide experience of mineral separation. One possible reason is the high costs involved and Table 5 summarises capital and operating costs of four full-scale soil washing processes based on figures from the USA.

The first-full scale soil washing decontamination project in Australia is currently under design at a former paint manufacturing facility in Sydney, NSW. The site comprises 7 ha of land contaminated by industrial activity over a period of 70 years. Contamination is mainly by Pb (predominantly as chromate and carbonate) arising from the production of Pb on the site and the use of Pb as a paint pigment. Pilot studies have identified the soil fines fraction (approx. 20%) as containing the majority of contaminants and have shown that by using mineral processing the contaminants can be concentrated in a smaller fraction, approximately 20% of the total material requiring remediation. Figure 1 shows a schematic description of the proposed process. Operations are expected to begin in early 1996 (ICI, 1995).

Soil extraction is based on the desorption or dissolution of metals from the soil inorganic and organic matrix during washing with acids or chelating agents (Elliott et al., 1989). The various stages involved in the process are outlined in Table 6. Tuin and Tels (1991) noted that there were a number of commercial scale plants operational in Europe for the decontamination of coarse textured sandy or loam soils but once again the application of this technology is uncommon outside Europe and the USA.

One problem associated with soil extraction is the choice of extractants. Tuin and Tels (1990a) noted that whilst 0.1 M HCl removed 80–90% of Cd, Cu, Pb and Zn from a contaminated clay soil only 45% of Ni and almost no Cr were extracted. Furthermore, extraction of contaminants from clay soils, where they are likely to be tightly bound, will prove more difficult (Tuin and Tels, 1990b). Inevitably the high costs of extractant chemicals and the technology involved (Clijsters and Vangronsveld, 1994) are two of the reasons why extraction has not been employed for remediation in Australia.

5.4 Thermal destruction

Thermal treatment of soils contaminated with oxidisable organic contaminants is a well recognised treatment technology (Bohm, 1992), which can be applied onsite

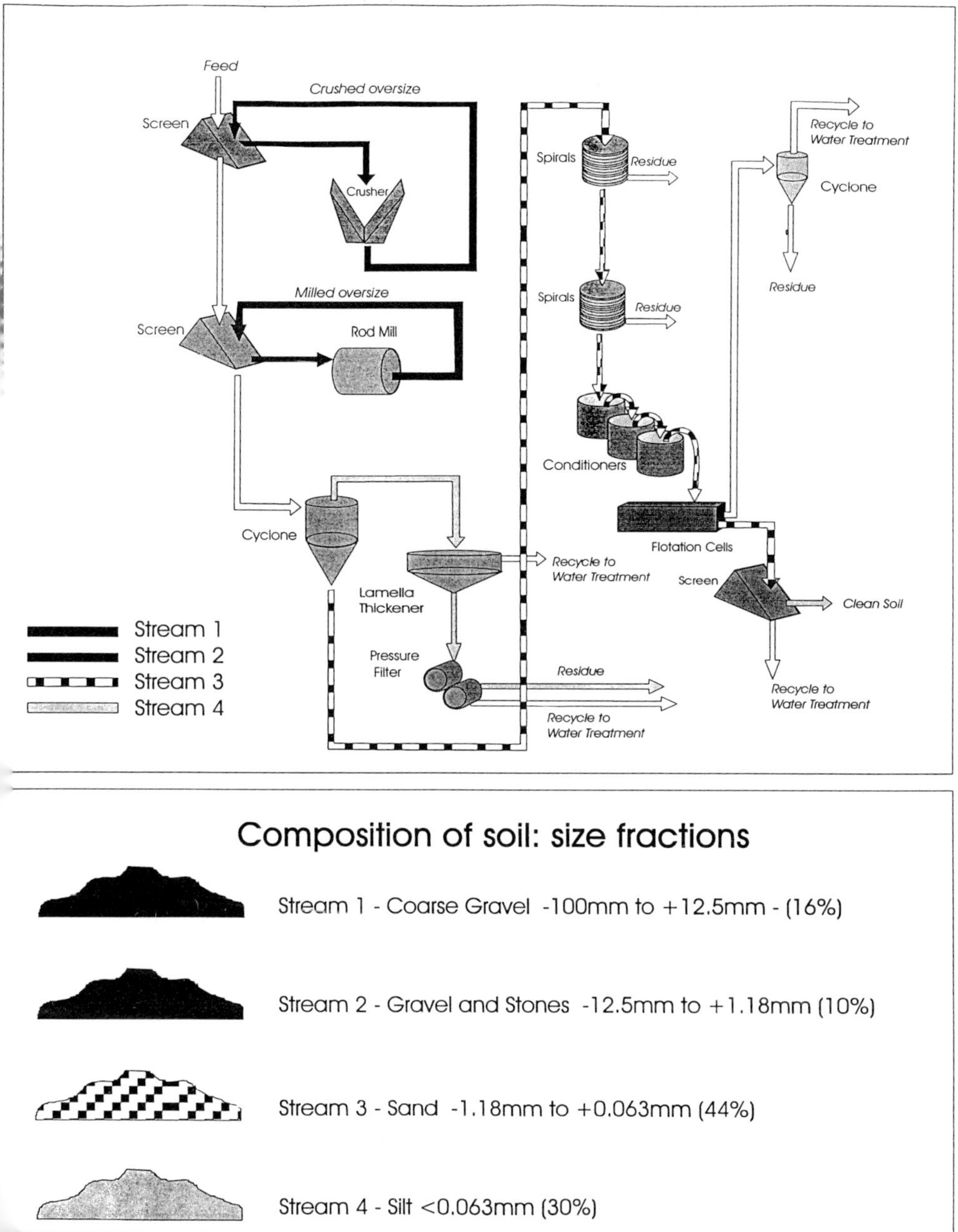

Figure 1. Soil washing plant process – a proposed remediation technique for removing soil Pb contamination from a former paint manufacturing facility (ICI, 1995).

Table 6. Soil Washing/Extraction remediation process.

Process	Description
Pretreatment of contaminated soil	Sieving, removal of demolition materials
Extraction of contaminant from soil	Water extraction[a], complexing agents (EDTA)[b], mineral acids[c], organic acids[d]
Separation of cleaned soil particles from liquid phase	Centrifugation, flocculation, sedimentation, dissolved air flotation[e]
Replace clean soil on site/dispose of contaminated soil residue	–
Cleaning/recycling of the extracting agent	Electrochemical metal recovery[f]

[a]Sheppard and Thibault (1992).
[b]Tuin and Tels (1990a).
[c]Tuin and Tels (1991).
[d]Fischer et al. (1992).
[e]Reed et al. (1992).
[f]Allen and Chen (1993).

using mobile equipment. One group of inorganic contaminants where thermal treatment may be a potential option are the iron cyanide complexes commonly found on former coal gasification sites (Theis et al., 1993). Nakles et al. (1993) described a field scale treatment involving excavation of contaminated soils and their incineration at 400°C. Cyanide was released from the soil matrix as HCN which was in turn destroyed in a gas scrubber at 1800°C. However, the application of this technology to the remediation of cyanide contaminated soils is not common outside the Netherlands and has not been adopted in Australia primarily because the cost is high compared with controlled landfill (Shoebridge, 1993). Thermal treatment technologies have only recently been introduced to Australia and the Eco Logic thermo-chemical hydrogenation process developed in the USA for treatment of organic chlorinated wastes has recently gained approval for application in NSW. A plant for the treatment of chlorinated wastes is expected to become operational in Perth, Western Australia during 1995 (EPA, 1994).

5.5 Bioremediation of inorganic contaminants

Whilst bioremediation is now a well established technology for the removal of organic soil contaminants (Flathman et al., 1994), the use of microorganisms to transform inorganic contaminants is still at the laboratory stage. No actual degradation of metals occurs during bioremediation, but the oxidation state is changed. This can lead to an increase in water solubility (and subsequently removal by leaching), or precipitation and reduction in bioavailability (Shannon and Unterman, 1993). To date As, Cr, Hg and Se are the most promising contaminants with regard to bioremediation treatment. Microbial reduction of Se (VI) to Se (O) causes precipitation of the metal and reduced bioavailability (Lovley, 1993) as does reduction of Cr (VI) to Cr (III) (McGrath and Smith, 1990). In addition, SeO_4 can be microbially methylated to volatile dimethylselenide, this has been suggested as a possible remediation strategy for Se contaminated agricultural runoff in California (Frankenburger and Karlson, 1992; Thompson-Eagle and Frankenburger, 1992). However, it must be reiterated that these techniques are not yet being used for full-scale decontamination. Although the microbial degradation and biochemical pathways involved in CN degradation have been extensively studied (Wyatt and Palmer, 1991), and a number of full-scale biological industrial wastewater treatment processes are available (Boucabeille et al., 1994), there is no field scale soil bioremediation process. Once again cheaper alternatives, such as landfill, may be the reason.

6. Organic soil contaminants

For a detailed discussion of industrial and urban waste disposal issues leading to soil contamination the reader is referred to Chapter 6 and country reports (Chapters 14–25). Organic contamination in urban areas of Australia has resulted from the storage, transport and disposal of chemicals and, to a lesser extent, their manufacture. Gas works, oil refineries and petrol service stations have all contributed to soil contamination problems. Unlike the Northern Hemisphere, the organochlorine pesticide dieldrin has been used to control termite infestation of houses until very recently in Australian urban areas, and was only banned in July 1995. In contrast, in rural areas there has been widespread use of pesticides and significant contamination problems have arisen from the use of persistent organochlorine pesticides such as DDT in cattle and sheep dips (Beard, 1993). This section will outline methods that are commercially available for the remediation of soil and water contaminated with organics, although many of them have not yet been adopted in the Australasia-Pacific region. In addition, innovative methods for the treatment of organic pollutants and fundamental research needs will be discussed.

7. Chemical and physical treatment technologies

7.1 Chemical decontamination

Chemical decontamination procedures, such as landscrubbing (Rickabaugh et al., 1987), ozonation (Karns et al., 1987), immobilisation (Morgan and Watkinson, 1989) and the use of emulsifiers or detergents (Bewley, 1988) have been applied successfully to clean up contaminated soil, but such methods are expensive and have limited value for subsurface decontamination. The most widely used chemical method for the remediation of soil contaminated with organics is solvent extraction. An example is the treatment of New Bedford Harbour sediment contaminated with PCBs (Eccles and Holroyd, 1992). Sediment was mixed in an extractor with a solvent solution (70% pentane: 30% butane; 240 psi) at 21°C. The solids content of the slurry was 30% by weight. The contaminated solvent flowed from the extractor through a valve that partially reduced the pressure causing the solvent to vaporise and separate from the contaminated organics. As a result, PCB levels in the sediment were reduced by 90% for concentrations ranging from 350 to 2600 mg kg^{-1}. However, it is important to note that although solvent extraction successfully removed 90% PCBs from the contaminated sediment the process resulted in a highly concentrated PCB solvent fraction which required incineration. The sediment also contained high levels of 13 metals (e.g. Al 10000 mg kg^{-1}; Zn 1850 mg kg^{-1}; Cr 245 mg kg^{-1}; Pb 590 mg kg^{-1}) which were not decreased by solvent extraction.

7.2 *Physical treatments*

Physical treatments (such as soil washing, soil venting, electroremediation, incineration, air or steam stripping) remove contaminants from soil or groundwater as concentrates which then require further treatment or are disposed of to landfill sites.

7.2.1 Soil washing

Soil washing is an *ex situ* technology which has been used extensively in Europe and the USA (Martin and Bardos, 1994) and has promise for the treatment of both inorganic (Section 5.3) and organic compounds. Heimhard (1992) reported that at a former steel manufacturing plant, 80000 t of soil contaminated with hydrocarbons, mineral oils and PAHs was treated. The process resulted in a reduction in mineral oil concentration from 2100 to 440 mg kg^{-1} whilst concentrations of PAHs and chlorinated hydrocarbons were reduced from 14.5 to 6.8 mg kg^{-1} and 113 to 0.009 mg kg^{-1}, respectively. Soil washing resulted in the replacement on site of 900 t of soil out of every 1000 t treated.

7.2.2 Soil venting

Soil venting is an *in-situ* physical treatment in which organic contaminants are volatilised from the soil by blowing air through the subsurface. This treatment will be discussed in more detail later because of the impact of biodegradation in its overall success.

7.2.3 Electroremediation

Electroremediation involves the application of a current between electrodes and can be applied to the *in situ* treatment of soils with a relatively high clay or peat content (Will, 1995). Contaminants dissolve in the soil solution and move to the electrodes by the processes of ion migration or electro-osmosis (i.e. movement of pore water due to movement of cations to the cathode). Subsequently, the area around each electrode is washed with circulating fluid to transfer the contaminants to an above ground leachate treatment plant. There are several reports of the use of electroremediation for soils contaminated with organic compounds. For example, in the USA trials of this technology have achieved 93% removal of a water-soluble organic solvent from an area 10×3×5 m deep (Schultz, 1994). The addition of surfactants to electroremediation strategies has been shown to result in significant increases (17% to 93%) in PAH degradation (Pamukcu, 1994).

7.2.4 Incineration

Thermal systems utilise elevated temperatures (800–2 500°C) to induce physical and chemical processes (such as volatilisation and combustion) to eliminate toxic substances. Incineration (onsite), following excavation of the contaminated soil, is a widely used remediation technique due to its efficient (99.9%) degradation of a wide range of organic compounds and complex mixtures of wastes plus the destruction of any contaminated containers found on site (Caldwell and Rauschel, 1991). However, thermal systems are expensive and in recent years the discovery of combustion products such as dioxins, dibenzofuranes and polyaromatic hydrocarbons has given rise to concern as to the benefits of incineration (McEldowney et al., 1993). The subsequent treatment of toxic gas products, in order to meet stringent air standards, will increase the cost of incineration even further. Environmental Remediation Services Australia has developed a low temperature thermal desorption process which has the capability of cleaning up a range of organic contaminants (Smith, 1995). The process uses a combination of infra-red heating and vacuum extraction and has been used to remediate soils contaminated with petroleum hydrocarbons such as gasoline, diesel fuel and motor oils, as well as organic compounds such as PCBs, PAHs and pesticides. The treatment temperature is relatively low (temperature not reported) and thus there is no danger of the formation of hazardous by-products such as dioxins.

There are many limitations associated with both chemical and physical remediation techniques (Table 7) and most do not result in complete destruction of the contaminant. Instead, they produce extracts and concentrates that must be subsequently removed to registered waste sites. As mentioned previously, landfill may not be the best option for these residues (ANZECC/NHMRC, 1992)

7.2.5 Novel techniques

The United States Environmental Protection Agency (USEPA) created the Technology Innovation Office (TIO) in 1991 with the sole purpose of removing regulatory and institutional barriers to the development and implementation of novel waste treatment technologies (Glass, 1992). Novel treatment technologies which have been supported by the TIO include vitrification, bioventing, vapour extraction (air stripping) and bioremediation. Vitrification is a process which fixes or 'locks' contaminants into glass by melting the contaminant material using electrical energy (Buelt and Farnsworth, 1991). It can be particularly useful in the treatment of soil wastes that contain a mixture of contaminants because the alternative is to use several techniques. However, its application to field scale decontamination outside Europe and the USA is uncommon. Bioventing and vapour extraction are two

Table 7. Examples of chemical and physical treatment technologies and common problems associated with their use.

Treatment technology	Common problems
Excavation/removal	(i) Excavated material requires treatment, e.g. incineration (ii) Transportation of concentrated contaminants to another location (iii) Leaching of contaminants from landfill sites
Incineration	(i) High cost (ii) Residues and gases require treatment
In situ vitrification	(i) Containment of gaseous releases (ii) High cost
Low temperature thermal stripping	(i) Limited to volatile and semi-volatile organic contaminants (ii) Gas requires treatment
Solidification/stabilisation	(i) Interference with organic contaminants (ii) Long term efficacy
Solvent extraction	(i) High cost (ii) Solvent fraction requires treatment
Vacuum extraction	(i) Limited to volatile organics in the vadose zone

Table adapted from Ellis (1992).
Details of techniques suitable for the remediation of soil contaminated with inorganics were presented in Table 4.

processes which are being increasingly used for the remediation of contaminated sites in Australia. A full scale three-phase vacuum extraction process enhancing the removal of contaminants present above the water table has been developed in Australia (Smith, 1995). Both processes involve the controlled use of air as a carrier either to remove contaminants from soil or to supply oxygen for aerobic bioremediation in the unsaturated zone (Long, 1992).

8. Bioremediation

The utilisation of the ability of microorganisms to mineralise or transform organic contaminants is not a new technology and has been used in municipal waste water treatment plants for decades (Bewley, 1992). The aim of bioremediation is to utilise the immense metabolic potential of microorganisms to degrade toxic pollutants. In general, organic contaminants are transformed by the same mechanisms that mediate microbial metabolism of naturally occurring compounds (Table 8) and the various mechanisms of biodegradation have been discussed in detail in several reviews (Kreuger, 1991; Bollag and Bollag, 1992; Cork Higson, 1992; and Alexander, 1994).

Microorganisms can fully mineralise organic contaminants (i.e. degrade the chemical to CO_2, H_2O, NH_3 etc.) or transform the contaminant into a variety of by-products. These by-products may be mineralised by other microorganisms (Skladany and Metting, 1992) or degraded by abiological means (Bollag and Liu, 1990) but they may also remain as recalcitrant metabolites with equal or greater toxicity than the parent compound (Cerniglia and Heitkamp, 1989). An example of this is the

Table 8. Examples of common reactions for the transformation of organic contaminants

Category	Reaction[a]
Dehalogenation	RCH_2Cl RCH_2OH $RCCl_3$ $RCOOH$
Deamination	$ArNH_2$ $ArOH$
Decarboxylation	$RCH(CH_3)COOH$ RCH_2CH_3
Hydroxylation	ArH ArOH
Reduction of double bond	$Ar_2C{=}CH_2$ Ar_2CHCH_3

[a]Abbreviations: R – organic moiety; Ar – aromatic.
Table adapted from Alexander (1981).

oxidative transformation of 1,1,2-trichloroethane to vinyl chloride, which is more toxic, does not degrade as readily and is more mobile than the parent compound (Muller, 1992). It is important to remember when considering any bioremediation strategy that microbial species do not exist in isolation but in interactive communities of microorganisms (Alexander, 1994). Thus the metabolites generated by one species will often serve as a growth and energy substrate for another. Alternatively, a chemical may be degraded co-metabolically by one species at the same time as it uses another as its carbon and energy source. In this case, the presence of the true substrate will be essential for the degradation of the target compound. At low xenobiotic concentrations (e.g. towards the end of a bioremediation treatment) further clean-up may only be achieved in the presence of a co-substrate.

It is often environmental factors, and not the lack of competent indigenous microbial species, which causes xenobiotics to persist (Morgan and Watkinson, 1989). Thus soil pH, temperature, moisture, nutrients and oxygen availability can inhibit degradation by (i) preventing growth of the microorganisms; (ii) reducing the bioavailability of contaminants through sorption processes; or (iii) suppressing the expression of degradative enzymes (Daubaras and Chakrabarty, 1992). A recent study by McGhee and Burns (1994; 1995) considered the effect of biostimulation and bioaugmentation on the degradation of the herbicides 2,4-dichlorophenoxyacetic acid (2,4-D) and 2-methyl-4-chlorophenoxyacetic acid (MCPA) in industrially contaminated soil. The authors demonstrated that the time taken for degradation to occur was reduced by 2 weeks in 2,4-D and MCPA degradation in inoculated soil. However, after 28 d there was no significant difference in the amount of herbicide degraded in inoculated and non-inoculated soils. In contrast, aeration, addition of water and an NPK fertiliser stimulated the indigenous microbial population resulting in 75% of 2,4-D and 84% of MCPA being degraded compared to 34% of 2,4-D and 61% of MCPA in non-amended soil.

Reduction in solubility of organics is just one aspect of the complex yet important property of bioavailability. Bioavailable pollutants are accessible either directly to microorganisms or to their extracellular enzymes and subject to transport across membranes and subsequent metabolism. Clearly, organics in the soil solution phase are available and their rate of metabolism may be similar to that revealed in laboratory studies. However, organics adsorbed to clay or complexed with soil organic matter may have limited bioavailability or be totally unavailable to microbiota. The sorption of organic soil contaminants is a reversible process, and the rate of contaminant desorption is probably the most important determinant of bioavailability. It is possible that some microorganisms are better than others at accessing adsorbed substrates, for example those microbes with extracellular enzymes that penetrate clay lattices to bind with intercalated organics and release products for microbial metabolism. Other microbes, by virtue of their metabolic activities, will release protons, change the microenvironment pH and stimulate the desorption of a previously adsorbed potential substrate. From this brief description it is clear that bioavailability is a dynamic and multifaceted phenomenon and rational manipulation of soil to encourage bioremediation is dependent on further advances in this field. The purpose of bioremediation is to improve or optimise these environmental factors and to facilitate bioavailability so that mineralisation can occur at an acceptable and predictable rate.

It is important to remember that any bioremediation strategy, however innovative and enthusiastically championed, must be commercially competitive in terms of cost and length of project with existing chemical and physical techniques. In other words, other remediation developments are not waiting for bioremediation to catch them up. For example, there are several non-biological methods currently available (and being further refined) for the remediation of soils contaminated with PCBs. These methods include thermal treatment and solvent or detergent washing techniques (Amend and Lederman, 1992; Woodyard, 1990). However, thermal treatment of PCB waste is still expensive and is restricted to registered sites (Linz et al., 1991) and soil washing techniques produce highly contaminated liquid waste which requires detoxification. In contrast, bioremediation, although currently slower than thermal treatments, is relatively low cost and (at least theoretically) produces no toxic end products (Linz et al., 1991). However, bioremediation is less reliable and can produce inconsistent results, although these may be due to our lack of knowledge of the environmental and site factors affecting pollutant degradation rather than any inherent weakness in bioremediation technology. There is also a question over the applicability of bioremediation technologies developed in the Northern Hemisphere to remediation of the very different soil types encountered in the Australasia-Pacific region.

8.1 Biological treatment technologies

Biological treatment technologies for contaminated soil can be divided into four categories: (i) solid-phase biotreatment (i.e. landfarming and composting); (ii) slurry-phase bio-treatment; (iii) *in-situ* biotreatment; and (iv) combined biotechnologies with chemical and physical treatment. The specific treatment required is dependent on the physical and chemical nature of the contaminants and the matrix in which they are contained.

8.1.1 Landfarming

Landfarming relies on the general principles that govern the degradation and release of nutrients from organic matter. Contaminant degradation rate is enhanced by:

- increasing oxygenation (i.e. regular tilling of the soil);
- moisture contents consistent with aeration and solvent and solute movement through the soil;
- the availability of nutrients especially nitrogen and phosphorus;
- a pH value appropriate for microbial growth and the close proximity of the microorganisms and the contaminants.

Contaminated leachate produced during landfarming can be pumped into an onsite bioreactor or be applied to the surface of the contaminated soil containing the actively degrading microbial population. This technique has been used successfully for many years for the remediation of soil containing oil sludge and petroleum refinery wastes (Tillosten, 1993). Landfarming has also been used for the remediation of land contaminated with mineral oil products (Verheul et al., 1993), the herbicides 2,4-D and MCPA and polyaromatic hydrocarbons (PAHs) (Borow, 1989).

8.1.2 Composting

Composting, a modification of solid-phase treatment (Ziegenfuss and Williams, 1991), has been used successfully to clean-up soil contaminated with chlorophenols (Valo and Salkinoja-Salonen, 1986) and the explosive 2,4,6-trinitrotoluene (TNT) (Keehan, 1994). Williams and co-workers (1992) have reported the successful remediation of sediments contaminated with the explosives TNT and nitrocellulose (NC). Over a period of 140 days TNT and NC concentrations decreased from 11 840 to 3 mg kg^{-1} and from 13 090 to 16 mg kg^{-1}, respectively. There was a transient increase in nitroaromatic transformation products over the first 2 weeks; however the levels decreased to a low concentration thereafter.

8.1.3 Slurry-phase bioremediation

Slurry-phase bioremediation is a process during which contaminated soil is treated as an aqueous mixture in a bioreactor (Borow, 1989). The system maximises contact between the contaminants and the degrading microorganisms by reducing the diffusional limitations that constrain static processes. In addition, the reactor can be designed to take account of the conditions necessary for the maximum microbial degradation of the target compounds. One of the principal disadvantages is the need for large scale excavation and manipulation of the contaminated soil prior to treatment. Following excavation, stones and rubble are removed and the fine soil is mixed with water to obtain the appropriate density. However, one obvious advantage of the slurry phase system is that contaminated groundwater and leachates may be treated in the same bioreactor. Microorganisms may be added initially, to seed the reactor, or added continuously to maintain the optimum biomass necessary for rapid degradation. The residence time in the reactor varies with the soil matrix as well as the chemistry, concentration and biodegradability of the contaminant. Once biodegradation is complete the clean soil can be dewatered and re-applied to the site. The cost of slurry-phase biotreatment A$30-60 (Borow, 1989) is higher than for other biotreatment technologies, but is still substantially lower than the cost of incineration A$46-190 (Tillosten, 1993).

8.1.4 *In-situ* bioremediation

In-situ bioremediation involves the biological treatment of contaminated soils and groundwater without excavation. This is an attractive option especially where the subsurface is targeted or where excavation of the contaminated material is not feasible (e.g. industrial manufacturing sites, roads). The rational treatment of contaminants in subsurface soils requires a greater understanding than exists currently of the microbiology and biochemistry of predominantly anoxic, low nutrient soil horizons. In general, *in-situ* biotreatment depends upon aerobic processes and therefore involves the injection of oxygen as air, oxygen gas or hydrogen peroxide as well as the addition of inorganic N and P (Kampbell et al., 1992; Long, 1992). However, a recent study has shown the feasibility of anaerobic biotreatment of soil contaminated with gasoline (Borow, 1989) and this would appear a less expensive method.

8.1.5 Bioventing

Bioventing is a mixture of physicochemical and *in-situ* bioremediation processes that provides an air flow to vaporise and transport low molecular weight volatile organic pollutants from the subsurface whilst at the same time providing oxygen for microbial processes. Bioventing has been used successfully to remediate an aviation fuel (132 000 L) spill 20 yr after the incident (Kampbell et al., 1992). In this example, the soil profile consisted of a surface beach sand underlain by a thick clay layer (~15 m). Total petroleum hydrocarbons (TPHs) decreased from 2060 to 38 mg kg^{-1} over a period of 18 months and surface emissions did not exceed 10 μg TPH L^{-1} and were usually less than 1 μg L^{-1}.

8.1.6 Onsite treatment technologies

Onsite biological technologies are not as common in the Australasia-Pacific region as in the Northern Hemisphere. A large proportion of the case-studies published involve the remediation of soils contaminated with petroleum hydrocarbons. However, McKinley and Rhodes (1994) have reported the use of large scale bioremediation for the decontamination of soil contaminated with creosote at a wood treatment site in East Gippsland, Victoria. Preliminary testing indicated that the soils contained an average of 650 mg kg^{-1} of PAH and 160 mg kg^{-1} total phenol, with maximum levels of 2200 mg kg^{-1} (PAH) and 490 mg kg^{-1} (phenol). The biotreatment strategy involved the excavation of the contaminated soil and subsequent spreading to a depth of 400 mm on an asphalt sealed pad (75 m×40 m). Runoff and leachate were drained to a sump and reused to supplement spray water for moisture control. The soil was treated with a controlled release formulation NPK (22:5.7:0) at a dosage rate of 0.25 kg m^{-2} and tilled on a weekly basis to maximise mixing and aeration. Over the treatment period of 17 weeks, mean PAH concentrations decreased from 287 mg kg^{-1} to less than 100 mg kg^{-1} (68% reduction), while phenolic compounds decreased from 42 mg kg^{-1} to less than 1 mg kg^{-1} (98% reduction). The authors noted that there was a strong correlation between the decrease in concentration of individual PAHs and the number of fused rings. Thus greater than 95% of naphthalene 82% of the three-ringed PAHs (e.g. phenanthrene), and only 45% of four-ringed PAHs (e.g. pyrene) was removed.

As previously mentioned there have been several reports on the bioremediation of former petrol stations. An example is the treatment of a former retail service station in Adelaide, South Australia (Barker, 1995). The site was contaminated with total petroleum hydrocarbons (TPH), monoaromatic and BTEX (e.g. benzene, toluene, ethylbenzene and xylenes) compounds. Not suprisingly, the major source of contamination was leakage from the old petrol storage containers which were removed from the site. Compacted sand and soil was also contaminated with hydrocarbons and was excavated and remediated using landfarming techniques. A second example is the large scale remediation of contaminated land at an oil refinery site (Anon, 1994b). The strategy involved excavation of the contaminated soil (15 000 t) and the formation of 'biopiles'. The 'biopiles' contained a manifolded piping system for oxygenation, nutrient addition, process control and monitoring of target TPHs. The 'biopiles' were placed on an impermeable base, covered with plastic sheeting to prevent run-off and to trap heat during cold periods. The company claimed that preliminary results (day 75) showed that significant levels of degradation had occurred, although the data are not presented. Contamination remaining beneath the construction site is currently being treated by an *in situ* bioremediation and groundwater extraction and treatment system.

The case-studies presented here were taken from the published literature; however it is important to note that projects were undertaken by private companies and thus the details of protocols and analytical techniques are not available and the scientific basis of the results has not been validated independently.

8.2 Biostimulation and bioaugmentation

Most bioremediation strategies have relied on the undoubted capacity of indigenous microbial communities to degrade contaminants such as petroleum waste, creosote, non-chlorinated solvents and pesticides (Bewley, 1988; Borow, 1989; Cioffi, 1989; Otte et al., 1994). *In-situ* remediation of soil contaminated with halogenated compounds (Semprini et al., 1992) illustrates the success of biostimulation (using the 'indigenous' microbiota). In this study a shallow confined aquifer was injected with the halogenated aliphatic hydrocarbon, acetate (as a growth substrate), nitrate and sulfate as potential electron acceptors. Biostimulation was rapid with nitrate consumption within 100 hours. Degradation of carbon tetrachloride was first recorded two weeks after active denitrification began and a 30–60% decrease occurred over the following eight weeks. The remediation of the aquifer was also studied in the absence of added nitrate and there was a significant increase in the proportion of CTC degraded (95%). The authors concluded that degradation may have been mediated by microbes other than the denitrifiers.

The potential of bioaugmentation microbial inoculation for the bioremediation of organic contaminated soils has been studied extensively (Kilbane et al., 1983; Middeldorp et al., 1990; Short et al., 1990) and in some instances has been more successful than biostimulation in the clean-up of soils containing complex chemicals such as polychlorinated biphenyls (PCBs), chlorinated solvents and pesticide mixtures, that is if problems of limited bioavailability can be overcome. However, not all inocula have met with success even though they have been chosen from *in vitro* screens which demonstrate the inherent capacity of the microbial species to degrade the target compounds (Goldstein et al., 1985). Several reasons have been suggested for the failure of competent microorganisms introduced into the soil to degrade pollutants:

- the bioavailable pollutant concentration may be too low to support the growth of the microbial inoculant;
- essential nutrients (e.g. N and P) may be limiting the growth and survival of the inoculant and therefore the metabolism of the contaminant;
- the inoculant may be susceptible to toxins or natural predators (e.g. protozoa) in the environment and its numbers decline too rapidly to make an impact;
- the inoculant may utilise other organic compounds in preference to the pollutant;
- the microbial inoculants may be unable to move horizontally or vertically through soil to all the sites containing the heterogeneously distributed pollutant; and
- uptake and metabolism of the xenobiotic may be repressed or non-induced in the very different soil environment in contrast to that experienced by the microorganisms *in vitro*.

The survival of introduced species and the expression of their beneficial properties has been discussed elsewhere (Burns, 1995; Young and Burns, 1993).

Although the potential of microbial bioaugmentation is well established, there have been few reports of its success during full scale *in situ* or onsite soil remediation operations. Lamar and co workers (Lamar et al., 1993; 1994) reported the use of the white rot fungi, *Phanerochaete sordida*, during remediation of soils at a former timber treatment facility, contaminated with pentachlorophenol (PCP) and 13 PAHs. In order to assess the efficacy of bioaugmentation pure cultures of *P. sordida* were grown on a sterile grain sawdust mixture, fortified with nutrients and applied to contaminated soils at a rate of 10% dry weight fungal hyphae. Two control treatments involved the addition of the fungal growth substrate to soils, and unamended soil. From an initial concentration of 1058 mg PCP kg^{-1} soil, Lamar and co workers noted a 64% decrease in PCP concentrations within 20 weeks in soils inoculated with *P. sordida*. This compared to a 26% decrease in the soil PCP concentration in soils treated with the fungal growth substrate and an 18% decrease in unamended soils during the same time period. The degradation of PAHs in these three treatments followed a very different pattern. From an initial concentration of 1 210 mg PAH kg^{-1} soil, significant decreases (40–60%) in 3 and 4 ring PAHs were noted in both the inoculated and control treatments. However, the reduction of 4-ring PAHs such as pyrene was significantly higher in the *P. sordida* inoculated soils (44% reduction) when compared to the unamended soils (12.5% reduction). The efficiency of the fungal treatment in degrading PAHs declined with increasing molecular weight and number of rings, no degradation of 5 and 6-ring PAH compounds was noted. Whilst these results do show some potential for the application of microbial bioaugmentation to soil remediation, Lamar et al. (1994) note that commercial application of this technology still requires major advances in inoculum production, delivery and application. One major oversight of this and many other bioremediation studies is the failure to consider contaminant bioavailability, and to assign the failure of the inoculum to degrade high molecular weight compounds such as 5- and 6-ring PAHs, to a problem concerning the inoculum itself; whereas failure may simply be due to low contaminant bioavailability resulting from high rates of sorption and slow desorption into the soil solution phase (refer to Chapter 4 for a discussion of industrial organic behaviour in soils).

Holroyd and Caunt (1994) have also reported the successful use of a lignolytic fungus, *Phanerochaete chrysosporium*, to remediate soil contaminated with PAHs and chlorophenols at a disused timber yard in Finland. The soil was excavated, mixed with the fungal inoculum and placed in a lined treatment bed to allow the collection of any leachates. The total concentration of chlorophenols present in the soil was monitored over a period of 18 months. Chlorophenol levels decreased from 203 mg kg^{-1} to 141 mg kg^{-1} over the first six months followed by a further decrease to 57 mg kg^{-1} in the following six months. The concentration of chlorophenol decreased to 28 mg kg^{-1} by 18 months. It is interesting that degradation occurred over the first six months (winter) despite long periods of sub-zero temperatures at the site. Measurement of the temperature of the beds showed that soil temperatures were an average of 15–18°C above ambient, as a result of biological activity. The authors state that the advantage of this fungal species is that the enzyme system is non-specific, with a wide range of aromatic compounds being degraded. The potential of the *Phanerochaete* genus in bioremediation has been recognised widely (Bumpus et al., 1989).

Otte et al. (1994) also demonstrated the success of bioaugmentation in soil contaminated with chlorophenols. Contaminated soil containing pentachlorophenols, PAHs and CCA was obtained from two wood-preserving sites in Canada. The addition of the PCP degrading microbial consortium to a bioreactor resulted in a 10 fold increase in PCP biodegradation. PAH biodegradation increased during the same time period by a factor of 30 and 81 for phenanthrene and pyrene, respectively. The efficiency of the consortium was also tested using contaminated soil in microcosms.

Inocula from the soil/slurry were introduced into static microcosms and PCP mineralisation monitored. Within 36 hours, 50% of the PCP had been degraded in the inoculated soil, whereas no degradation was observed in the non-inoculated soil.

This section has outlined several field scale applications of microorganisms for the successful remediation of contaminated soil. However, as previously mentioned there have been limited reports of the success of bioaugmentation in the remediation of contaminated land. One of the major problems is the poor survival of the inoculant following addition to the contaminated soil, as well as potential toxicity from other inorganic and organic soil contaminants. Most importantly, the majority of these studies do not consider the proportion of biologically available contaminants in these soils, or the crucial role of contaminant physico-chemical behaviour in the bioavailability and biodegradation of organic contaminants (Kookana and Rogers, 1995). Failure of inoculants may simply be a case of the unavailability of the contaminant substrate under study. In recent years, the use of genetically modified microorganisms may enhance the chance of successful bioremediation with the introduction of selected traits to promote microbial survival and the expression of high and specific levels of biodegradation.

8.3 Genetically modified microorganisms

In recent years advances in molecular biology have permitted the development of novel microorganisms capable of performing functions not encoded for in wild-type strains. New genetic information may be inserted into the chromosome *via* transposons (Sayre and Miller, 1987) as well as carried on plasmids, whilst specific gene sequences may be deleted or inserted using site directed mutagenesis. It has been suggested that the introduction of genetically modified microorganisms to soil will revolutionise bioremediation due to the expression of novel metabolic pathways. Such manipulations may also help in the development of microorganisms that are capable of tolerating compounds that would otherwise be toxic (Chaudhry and Chapalamadugu, 1991). However, there is widespread concern about the impact of genetically engineered microorganisms (GEMMOs) on the natural environment (Beringer and Bale, 1988; Ford and Olson, 1988; Pickup et al., 1991). The principal fears are that: recombinant strains have unsuspected pathogenic properties; the novel genetic material may be transferred to indigenous bacteria; released bacteria may adversely affect the indigenous microbial community; and the released microorganisms may spread from the soil to water and air and thence to distant environments.

Compared with the releases of non-recombinant microorganisms, the number of recombinant releases remains small. Government legislation restricts the release of GEMMOs into the environment and permission is often only granted for testing in contained systems, such as bioreactors, where the risk of escape into the environment is negligible. Until both government policy and accurate and sensitive methods for the detection and monitoring of side-effects have been developed further (Young and Burns, 1993) the use of GEMMOs for routine *in situ* bioremediation will not be an option. However, it is significant that over the past two years the legislation relating to the release of GEMMOs has been relaxed. For example, initial experiments with recombinants in the USA required an Experimental Use Permit (EUP) from the EPA. Acceptable documentation for an EUP required a large amount of information (e.g. dispersal, survival potential of dispersed bacteria etc.) and greatly delayed release experiments. Subsequently, the EPA has relaxed its requirements and, although the procedures to obtain permits for the testing of recombinant microorganisms from the USA Department of Agriculture, Animal and Plant Health Inspection Service (USDA-APHIS) still necessitate considerable effort, they have been greatly simplified (Shaw et al., 1992). It is likely that over the next decade the legislation relating to the release of recombinants will be simplified further and therefore the use of these microorganisms for bioremediation will be assessed at an increasing number of contaminated sites.

This section of the chapter has illustrated that bioremediation is a rapidly evolving technology in the clean-up of contaminated land. Published reviews of various case-studies (Valo and Salkinoja-Salonen, 1986; Borow, 1989; Kampbell et al. 1992; Keehan, 1994) are testimony to the growing numbers of full-scale treatments which have been successfully carried out.

8.4 *Bioremediation – limitation and potential*

Whilst bioremediation of contaminated soils is now a widely accepted commercial reality, it has a number of limitations. Biological limitations include: poor inoculant survival; limited understanding of contaminant degradation by indigenous soil microbial communities; a poor appreciation of the role of co-metabolism as opposed to mineralisation in the degradation of soil contaminants; and concern over the possible production of toxic metabolites from certain compounds (i.e. the result of aerobic mineralisation of 1,1,2-trichloroethane is vinyl chloride, which is more toxic than the initial precursor; Muller, 1992). Whilst there are still biological limitations to this technology, perhaps the main limitation to the successful application of bioremediation in contaminated soils is the lack of understanding of how the physico-chemical behaviour of a compound in a soil controls bioavailability to microorganisms. Chapters 4 and 12 in this volume provide an extensive overview of organic contaminant behaviour in soils, notably the role of sorption and desorption mechanisms in controlling the solution phase concentration of a contaminant in soils. As suggested in section 8.1.6 in this chapter, high rates of sorption or slow desorption of contaminants leading to low rates of bioavailability, are invariably the cause of bioremediation failure, especially in the case of high molecular weight aromatic compounds (Lamar et al., 1994) and hydrocarbons with a C content greater than 20 (Flathman et al., 1993) rather than the poor performance of either the indigenous microbial populations or inoculated cultures.

One of the major issues facing the contaminated site remediation industry is the incompatibility of various remediation technologies when multiple contaminants are present on site. The potential mobilisation of Fe-CN complexes due to liming of gaswork sites to enhance bioremediation is one example. Other examples include increased volatilisation of low molecular weight organics during soil immobilisation/ solidification operations, and enhanced contamination offsite transport and leaching resulting from soil disturbance during soil excavation onsite. Industrial contaminated soils rarely only have one group of contaminants present. Therefore, there is little doubt that successful soil remediation strategies must involve an integrated approach, and consider the potential effects of remediation technologies adopted for one group of contaminants on the behaviour of other contaminants present in the soil.

9. Integration of treatment techniques

Soil contamination can occur in several phases: (i) the free product (e.g. floating on water table); (ii) solute in the vadose or saturated zone either in soil solution or sorbed on particle surface; (iii) vapour phase in the vadose zone; and (iv) adsorbed onto organic matter phases. The contaminant may move from one phase to another and thus there is a need to treat all phases or to target the identified sink. For example, pump and treat has often failed or required extensive time before successful remediation since only the solution phase is treated. Over time, percolation of rain-water through the solid matrix resulted in leaching of the contaminants from the adsorbed state into groundwater.

An example of an integrated remediation approach to the clean-up of contaminated sites involves the use of free phase recovery, venting, bioremediation and groundwater extraction. Venting the ground aids volatilisation of contaminants in the vadose zone and thus enhances recovery (i.e. vapour extraction). Venting also results in an increase in oxygen supply to the contaminated area and thus stimulates aerobic biodegradation. This response can be enhanced further by forced aeration either below the groundwater table or in the vadose zone. Bioremediation also aids the removal of contaminants from the dissolved and adsorbed phases. It is apparent that the integration of remediation techniques results in a greater remediation efficiency and may result in a reduction in the time required for treatment.

The processing of materials on-site prior to treatment can also enhance the efficiency of some treatment techniques. Screening of soils into various fractions (e.g. contaminant type, concentration, soil type) can enable the targeting of treatment processes onto smaller volumes. This can reduce the cost in several ways by minimising both the treatment effort required and the space on site. The soil fractions can be categorised into contaminant types and thus the appropriate methods for each soil section can be implemented. In addition, homogenisation of the contaminated soil will enable better quality control and more accurate monitoring at reduced costs. In determining the cost-effectiveness of a particular remediation strategy it is often difficult to assess the cost-benefit. For example, soil washing or *ex situ* biological treatment require space and time on site which may delay the redevelopment of the land. *In situ* techniques may overcome this problem although other difficulties

Table 9. Contaminated soil remediation techniques applied in Victoria.

Treatment technology	Percent of full scale site remediation[a]
Off site containment, no further soil treatment	40+
Volatile organic removal:	50+
Vapour extraction/Aeration	
Bioremediation	11
Chemical Stabilisation	5
Thermal destruction	1
On site containment	Negligible

Adapted from: Shoebridge (1993).
[a]In many instances more than one technology would have been used at each individual site.

may arise. For example, increased uncertainty concerning the predictability of remediation schedules and the increased time required for biodegradation.

10. Remediation technology in Australia

Wheeler (1994) discusses the current issues facing the Australian site remediation industry. From a technological point of view, the need to develop suitable onsite remediation techniques as priority soil remediation technologies and a reduction in the use of offsite containment as a remediation option are the most pressing issues.

The results of a survey of remediation strategies in Victoria are shown in Table 9. It is apparent from the survey that vapour extraction and air stripping are common methods employed in site restoration. However, Shoebridge (1993) notes that the figure may only reflect the large number of former service stations cleaned up in recent years, where this technology is the most suitable. The high proportion of offsite disposal treatments is likely to reflect the number of soils contaminated with inorganic compounds and the difficulty of treating these onsite. Methods such as stabilisation, for dealing with inorganic contaminants onsite only accounted for 5% of remediation operations. Shoebridge (1993) states that regulatory guidelines and controls have a major influence on remediation practices in Australia. For instance, there is no incentive for the Australian industry to use chemical stabilisation techniques which reduce bioavailability when regulations are based on total soil contamination levels (ANZECC/NHMRC, 1992). Shoebridge (1993) also notes that when strict regulations are in place for residual levels of contaminants in soils following remediation operations, offsite containment is more popular, as it is perceived as having a lower risk of future contaminant exposure associated with it.

Although regulatory issues will determine the choice of soil remediation strategy to a certain extent, the availability of technology and economic considerations will

Table 10. Estimated costs of contaminated soil remediation technologies in Australia (1992 figures). Costs per m^3 of contaminated material.

Treatment technology	Lightly contaminated soil A$	Heavily contaminated soil A$
Offsite containment	33–55	180–250
Insitu bioremediation of Hydrocarbons	30–50	30–50
Soil washing of petroleum	N/A	80–150
Chemical fixation prior to offsite disposal	N/A	200–300
Extraction of metals	N/A	200
Clean soil capping layer	10–20[a]	10–20[a]
Landfarming	–	100–150

Reproduced from Kidd (1992).
[a]Cost of a m^3 of clean soil used as capping material.

be paramount (Calder, 1991). Most common remediation technologies for contaminated soils are available in Australia (Waste Management and the Environment Yearbook, 1994). However, not all have been employed in soil rehabilitation programs. Table 10 gives a summary of remediation costs based on the commercial environmental management industry in Australia. Although offsite containment is more expensive than bioremediation, it is cheaper than the technologies listed which would be suitable for inorganic contaminated soils such as chemical fixation and metal extraction, as well as being more attractive to the industry for reasons previously mentioned (Section 5.1).

Kidd (1992) lists a number of examples of contaminated site restoration projects in Australia and the costs involved in restoration. These range from A$50 000 for a service station in NSW, A$1.6 million for a 1.6 ha gaswork site in Sydney, through to A$14 million for the restoration of the 12 ha Pulpit Point Oil terminal in Sydney. Forrest (1993) estimated that the total clean-up cost for urban and industrial contaminated soils in Australia could be as high as A$10 billion. The cost of remediating As and DDT cattle dip sites in northern NSW is estimated to be in the order of A$100 000 per site if soil washing, base catalysed destruction of organochlorines and landfill of As-contaminated material is employed or A$60 000 per site if only a clean capping layer is used. With 1 600 potential sites requiring remediation the cost of merely surface capping will be A$96 million (Beard, 1993).

11. Conclusions

Sources of soil contaminants and contaminant issues arising in the Australasia-Pacific region are likely to be similar to those experienced in regions which are more densely populated and have a longer industrial history. A large number of remediation technologies now exist and are either being applied to full-scale site remediation or undergoing research and development. However, the majority of these technologies have been developed and are currently being applied in the

Northern Hemisphere. The question therefore arises as to the applicability of these technologies for con-taminated soil remediation in the very different environments existing in the Australasia-Pacific region. For example Australian soil types are very different to those in the Northern Hemisphere. Mainland Australia was not subject to glacial activity during the most recent ice ages, resulting in older, highly weathered soils. As all major urban areas in Australia have been built on coastal plains, saline and sodic soils may predominate not only at sites reclaimed from the sea, but in coastal regions in general. In these soils the physico-chemical behaviour of contaminants in terms of adsorption, desorption and mobility during remediation will be different. It is therefore critical that the suitability of remediation technologies developed in the Northern Hemisphere is assessed under the soil and environmental conditions prevailing in the Australasia-Pacific region.

References

Adriano D C 1986 Trace Elements in the Terrestrial Environment. Springer-Verlag, New York Inc. 533 p.

Adriano D C 1992 Biogeochemistry of Trace Metals. Advances in Trace Substances Research. Lewis Publishers, Boca Raton, Fl. 513 p.

Alexander M 1981 Biodegradation of chemicals of environmental concern. Science 211, 132-138.

Alexander M 1994 Biodegradation and bioremediation. Academic Press, Inc. 302 p.

Allen H E and Chen P H 1993 Remediation of metal contaminated soil by EDTA incorporating electrochemical recovery of metal and EDTA. Environ. Prog. 12, 284-293.

Alloway B J 1990 Soil processes and the behaviour of metals. In Heavy Metals in Soils. Ed. B J Alloway. pp 7-28. Blackie, Glasgow.

Amend L J and Lederman P B 1992 Critical evaluation of PCB remediation technologies. Environ. Prog. 11, 173-177.

Anon 1994b Australian companies pushing the technology envelope. Pollution Prevention Asia/Pacific Edition 2, 64-70.

Anon 1995 ICI Australia. Cleanout. CRC for Waste Management and Pollution Control Ltd. 12, 10-11.

ANZECC/NHMRC 1992 Australian and New Zealand Guidelines for the Assessment and Management of Contaminated Sites. Australian and New Zealand Environment and Conservation Council, National Health and Medical Research Council, Canberra ACT.

Barber C, Briegel D J, Power T R and Hosking J K 1992 Pollution of groundwater by organic compounds leached from domestic solid wastes: A case study from Morley, Western Australia. In Groundwater Contamination and Analysis at Hazardous Waste Sites. Eds. S Lesage and R E Jackson. pp 357-380, Marcel Dekker, Inc, NY.

Barker C 1995 A cost-effective cleanup solution. Waste Manage. Env. 6, 125.

Beard J 1993 The evaluation of DDT contaminated soil associated with cattle tick dip sites. In The Health Risk Assessment and Management of Contaminated Sites. Eds. A Langley and M van Alpen. pp 119-133. Proceedings of the Second National Workshop on the Health Risk Assessment and Management of Contaminated Sites, Canberra 5-7 April 1993. South Australian Health Commission, Adelaide SA.

Beringer J E and Bale M J 1988 The survival and persistence of genetically engineered microorganisms. In Release of Genetically Engineered Microorganisms Eds. C H Sussman, C H Collins, F A Skinner and D E Stewart-Tull. pp 29-46. Academic Press, London.

Bewley R J F 1988 Use of microbial processes for the reclamation of contaminated land. Report for Biotal Ltd, Cardiff. Seesoil 5, 1-21.

Bewley R J F 1992 Bioremediation of contaminated ground. In Contaminated Land Treatment Technologies. Ed. J F Rees. pp 270-284. Society of Chemical Industry, Elsevier Applied Science, London.

Body P E, Inglis G R and Mulcahy D E 1988 Lead Contamination in Port Pirie South Australia. A Review of Environmental Lead Which Could Give Rise to an Increase in Children's Blood Levels. South Australian Department of Environment and Planning Report No. 101. Adelaide SA.

Bohm K 1992 A thermal method for cleaning contaminated soil. In Contaminated Land Treatment

Technologies. Ed. J F Rees. pp 195-219. Society of Chemical Industry, Elsevier Applied Science, London.

Bollag J M and Liu S Y 1990 Biological transformation processes of pesticides. In Pesticides in the Soil Environment. Ed. H H Cheng. P P 169-211 Soil Sci. Soc. Am. Book Series 2.

Bollag J M, Liu S Y and Minard R D 1992 Biological and chemical interactions of pesticides with soil organic matter. Soil Sci. Soc. Am. J. 44, 52-56.

Bollag W B and Bollag J M 1992 Biodegradation. Encyclopedia of Microbiology 1, 269-276.

Borow H S 1989 Biological clean-up of extensive pesticide contamination in soil and groundwater. Biotreatment, 51-56.

Boucabeille C, Bories A, Ollivier P and Michel G 1994 Microbial degradation of metal complexed cyanides and thiocyanate from mining waste waters. Environ. Pollut. 84, 5967.

Bromhead J C and Beckwith P 1994 Environmental dredging of the Birmingham canals: Water quality and sediment treatment. J. Inst. Wat. Eng. Man. 8, 350-359.

Brown R V 1993 The latest trends and developments in waste management in the UK. Waste Disposal and Water Man in Aust. Feb 93, 9-23.

Buelt J L and Farnworth R K 1991 *In situ* vitrification of soils. Nucl. Technol. 96, 178-184.

Bumpus J A, Fernando T, Jurek M, Mileski G J and Aust S D 1989 Biological treatment of hazardous waste by *Phanerochaete chrysosporium*. In Proceedings of conference on Biotechnology Applications in Hazardous Waste Treatment. pp 167-183. Engineering Foundation, New York.

Burns R G 1995 Enumeration, survival, and beneficial activities of microorganisms introduced into soil. In Environmental Impact of Soil Component Interactions Vol II. Eds. P M Huang, J Berthelin, J M Bollag, W B McGill and A L Page. pp. 145-164. CRC Lewis, Florida.

Bumpus J A, Fernando T, Jurek M, Mileski G J and Aust S D 1989 Biological treatment of hazardous waste by Phanerochaete chrysosporium. In Proceedings of conference on Biotechnology Applications in Hazardous Waste Treatment. pp 167-183. Engineering Foundation, New York.

Cairney T 1992 Theory and practice of clean cover reclamation. In Contaminated Land Treatment Technologies. Ed. J F Rees. pp 97-112. Society of Chemical Industry, Elsevier Applied Science, London.

Calder I C 1991 Directions for assessment of contaminated land. In The Health Risk Assessment and Management of Contaminated Sites. Eds. O El Saadi and A Langley. pp 6-9. Proceedings of a National Workshop on the Health Risk Assessment and Management of Contaminated Sites. Adelaide 5-7 August 1991. South Australian Health Commission, Adelaide SA.

Calder I C 1993 After remediation how much matters? In The Health Risk Assessment and Management of Contaminated Sites. Eds. A Langley and M van Alpen. pp 65-68. Proceedings of the Second National Workshop on the Health Risk Assessment and Management of Contaminated Sites, Canberra 5-7 April 1993. South Australian Health Commission, Adelaide SA.

Caldwell S R and Rauschel F M 1991 Detoxification of organophosphate pesticides using immobilised phosphoteriesterase from *Pseudomonas diminuta*. Biotech. Bioeng. 37, 103-109.

Chaudhry G R and Chapalamadugu S 1991 Biodegradation of halogenated organic compounds. Microbiol Rev. 55, 59-79.

Cioffi J A 1989 Air stripping/bioremediation of chlorinated hydrocarbons at an NPL site. Biotreatment, 69-74.

Clijsters H and Vangronsveld J 1994 Reclamation of contaminated soils by *in-situ* immobilisation of heavy metals. In Workshop 1992. Institut National Agronomique Paris. Eds. C Avril and R Impens. pp 57-62. Cadmium Industry Environment, Bruxelles.

Corey R B, Fujii R and Hendrickson L L 1981 Bioavailability of metals in soil sludge systems. Proceedings of 4th Annual Conference Applied Research and Practices on Municipal and Industrial Waste 28-30 Sept 1981 University of Wisconsin. pp 449-465 University of Wisconsin-Extension MA.

Cork D J and Kreuger J P 1991 Microbial transformations of herbicides and pesticides. Adv. Appl. Microbiol. 36, 1-66.

Daubaras D and Chakrabarty A M 1992 The environment, microbes and bioremediation: microbial activities modulated by the environment. Biodegradation 3, 125-135.

Dennis D 1993 Columbus Georgia MGP site remediation, case study. Management of Manufactured Gas Plant Sites Technology Seminar March 23-24 1993 Orlando Fl. Remediation Technologies Inc, Arlington VA. pp 283-288.

Dennis R M, Dworkin D and Zupko A J 1994 Soil-Washing processes for site remediation. In Remediation of Hazardous Waste Contaminated Soils. Eds. D L Wise and D J Trantolo. pp 745-778. Marcel Dekker Inc, NY.

Dowdy R H and Volk V V 1983 Movement of heavy metals in soils. In Chemical Mobility and Reactivity in Soil Systems. SSSA Special Publication No.11. pp 229-240. Soil Science Society of America, American Society of Agronomy, Madison WI.

DuGuay T 1993 Reclaim metals to clean up soils. Soils March 93, 26-33.

Eccles H and Holroyd C P 1992 Solvent extraction for the treatment of contaminated land. In Contaminated Land Treatment Technologies. Ed. J F Rees. pp 83-96. Society of Chemical Industry, Elsevier Applied Science, London.

Edwards J W, Van Alphen M and Langley A 1994 Identification and Assessment of Contaminated Land. Contaminated Sites Monograph Series No. 3. South Australian Health Commission, Adelaide SA.

Elliott H A, Linn J H and Shields G A 1989 Role of Fe in extractive decontamination of Pb-polluted soils. Haz. Waste. Haz. Mater. 6, 223-228.

Ellis B 1992 On site and *in situ* treatment of contaminated sites. In Contaminated Land Treatment Technologies. Ed. J F Rees. pp 30-46. Society of Chemical Industry, Elsevier Applied Science, London.

EPA 1994 Appropriate Technologies for the Treatment of Scheduled Wastes. Environmental Protection Agency, Commonwealth of Australia, Canberra ACT.

Finnecy E E and Johnstone E T C 1988 Remedial measures proposed for the rehabilitation of a dioxin contaminated site in New South Wales, Australia. In Contaminated Soil '88. Eds. K Wolf, W J van den Brink and F J Colon. pp 1083-1085. Second International TNO/BMFT Conference on Contaminated Soil, 11-15 April 1988, Hamburg FRG. Kluwer Academic Publishers, Dordrecht.

Fischer K, Rainer C, Bieniek D and Kettrup A 1992 Desorption of heavy metals from typical soil components (clay, peat) with glycine. Intern. J. Environ. Anal. Chem. 46, 53-62.

Flathman P E, Jerger D E and Exner J H Eds. 1994 Bioremediation Field Experience. CRC Press, Boca Raton FL. 548p.

Ford S and Olson B H 1988 Methods for detecting genetically engineered microorganisms in the environment. Adv. Microbiol. Ecol. 10, 45-79.

Forrest W 1993 Harmonising health and environmental criteria: environment based criteria. In The Health Risk Assessment and Management of Contaminated Sites. Eds. A Langley and M van Alpen. pp 5-7. Proceedings of the Second National Workshop on the Health Risk Assessment and Management of Contaminated Sites, Canberra 5-7 April 1993. South Australian Health Commission, Adelaide SA.

Frankenburger, Jr W T and Karlson U 1992 Dissipation of soil selenium by microbial volatilisation. In Biogeochemistry of Trace Metals. Advances in Trace Substances Research. Ed. D C Adriano. pp 365-382. Lewis Publishers, Boca Raton, Fl.

Gera F 1988 Modelling long term impacts of land disposal of hazardous waste. In Land Disposal of Hazardous Waste, Engineering and Environmental Issues. Eds. J R Gronow, A N Schofield and R K Jain. pp 103-112. Ellis Horwood Limited, Chichester, Sussex, UK.

Glass D J 1992 Waste treatment of hazardous wastes. Environ. 33, 5-45.

Goldstein R M, Mallory L M and Alexander M 1985 Reasons for the possible failure of inoculation to enhance biodegradation. Appl. Environ. Microbiol. 50, 977-983.

Higson F K 1992 Microbial degradation of nitroaromatic compounds. Adv. Appl. Microbiol. 37, 1-19.

Hinwood A L 1991 Principles for setting soil criteria approaches to soil guidelines-Environment. In The Health Risk Assessment and Management of Contaminated Sites. Eds. O El Saadi and A Langley. pp 87-92. Proceedings of a National Workshop on the Health Risk Assessment and Management of Contaminated Sites. Adelaide 5-7 August 1991. South Australian Health Commission, Adelaide SA.

Holroyd M L and Caunt P 1994 Fungal processing: a second generation biological treatment for the degradation of recalcitrant organics. Land Contamination and Reclamation 2, 183-188.

ICI 1995 Dulux Cabarita Project - turning full circle. ICI Dulux (Australia) Technology Report.

ICRCL 1986 Notes on the redevelopment of Gaswork sites, 5th edition. Interdepartmental Committee for the Redevelopment of Contaminated Land. Report No. 18/79. Department of the Environment, London.

Independent Panel on Intractable Waste 1992. In our backyard. Independent Panel on Intractable Waste, Paddington, Sydney, NSW, Australia.

Jefferis S A 1992 Remedial barriers and containment. In Contaminated Land Treatment Technologies. Ed. J F Rees. pp 58-82. Society of Chemical Industry, Elsevier Applied Science, London.

Jones L W 1990 Interference Mechanisms in Waste Stabilisation/Solidification Processes. U. S. Environmental Protection Agency, EPA/600/S289/067, Cincinnati.

Kampbell D H, Wilson J T and Griffin C J 1992 Performance of bioventing at Traverse City, Michigan. In Bioremediation of Hazardous Wastes USEPA EPA/600/R-92/126. pp 61-64.

Karns J S, Muldoon M T, Mulbry W W, Derbyshire M K and Kearney P C 1987 Use of microorganisms and microbial systems in the degradation of pesticides. ACS Symposium Series 334, 156-170.

Keehan K R 1994 Composting: an innovative technology to remediate explosives-contaminated soil. Proceedings of the Second International Conference '*In situ* and On site Bioreclamation', San Diego, USA.

Kidd C 1992 Environmental site contamination assessment. Waste Man. Environ. 3, 11.

Kilbane J J, Chatterjee D K and Chakrabarty A M 1983 Detoxification of 2,4,5-trichlorophenoxyacetic acid from contaminated soil by *Pseudomonas Cepacia*. Appl. Environ. Microbiol. 45, 1697-1699.

Kookana R S and Rogers S L 1995 Behaviour of pulp effluents in soils. Rev. Environ. Contam. Toxicol. 142, 13-64.

Kroopnick P M 1994 Vapour abatement cost analysis methodology for calculating life cycle costs for hydrocarbon vapour extracted during soil venting. In Remediation of Hazardous Waste Contaminated Soils. Eds. D L Wise and D J Trantolo. pp 779-790. Marcel Dekker Inc., NY.

Kyle J 1990 Chemical fixation. Waste Man. 2, 19-21.

Lamar R T, Evans J W and Glaser J A 1993 Solid-phase treatment of a pentachlorophenol-contaminated soil using lignin-degrading fungi. Environ. Sci. Technol. 27, 2566-2571.

Lamar R T, Davis M W, Dietrich D M and Glaser J A 1994 Treatment of a pentachlorophenol-contaminated soil using the lignin-degrading fungus *Phanerochaete-sordida* – a field demonstration. Soil Biol. Biochem. 26, 1603-1611.

Langley A and El Saadi O 1991 Protocol for the Health Risk Assessment and Management of Contaminated Sites. Summary of a National Workshop on the Health Risk Assessment and management of Contaminated Sites. Adelaide 5-7 August 1991. South Australian Health Commission, Adelaide SA.

Linz D G, Neuhausser E F and Middleton A C 1991 Perspectives on bioremediation in the gas industry. In Environmental Technology for Waste treatment. Eds. G S Sayler, R Fox and J W Blackburn. pp 25-36. Plenum Press, New York.

Logan T J 1992 Reclamation of chemically degraded soils. In Advances in Soil Science Vol. 17 Soil Restoration. Eds. R Lal and B A Stewart. pp 13-36. Springer-Verlag, NY.

Long G 1992 Bioventing and vapour extraction – innovative technologies for contaminated site remediation. J. Air Waste Manage Assoc. 42, 345-348.

Lovley D R 1993 Dissimilatory metal reduction. Annu. Rev. Microbiol. 47, 263-290.

Ma Q Y, Logan T J and Traina S J 1995 Lead immobilisation from aqueous solutions and contaminated soils using phosphate rocks. Environ. Sci. Technol. 29, 1118-1126.

Martin I D and Bardos P 1994 Recent developments in contaminated land treatment technology. Chem. Ind. 11, 411-413.

McBride M B 1994 Environmental Chemistry of Soils. Oxford University Press, NY.

McCloskey W B and Bayer D E 1987 Thermodynamics of fluridone adsorption and desorption on three Californian soils. Soil Sci. Soc. Am. J. 51, 37-44.

McEldowney S, Hardman D J and Waite S 1993 Treatment technologies. In Pollution: Ecology and Biotreatment. Eds S McEldowney, J Hardman, S Waite. pp 48-58. Longman Singapore Publishers (Pte) Ltd, Singapore.

McGhee I and Burns R G 1994 Bioremediation of soil contaminated with the herbicides 2,4-dichlorophenoxyacetic acid (2,4-D) and 4-chloro-2-methyl-phenoxyacetic acid (MCPA). In Bioremediation of chlorinated and polycyclic aromatic hydrocarbon compounds. Eds. R E Hinchee, A Leeson, L Semprini and S K Ong. pp 376-381. Lewis Publishers, New York.

McGhee I and Burns R G 1995 Biodegradation of 2,4-dichlorophenoxyacetic acid (2,4-D) and 2-methyl-4-chlorophenoxyacetic acid (MCP) in contaminated soil. Appl. Soil. Ecol. 2, 143-154.

McGrath S P and Smith S 1990 Chromium and Nickel. In Heavy Metals in Soils. Ed B J Alloway. pp 125-150. Blackie, Glasgow.

McKinley T and Rhodes S 1994 Bioremediation of soil contaminated with creosote. EMIAA Yearbook, 167-168.

McNamara M 1994 Engineering a site remediation project. Waste. Man. Environ. 5, 40-43.

McNeill K R and Waring S 1992 Vitrification of contaminated soils. In Contaminated Land Treatment Technologies. Ed. J F Rees. pp 143-159. Society of Chemical Industry, Elsevier Applied Science, London.

Meeussen J C L, Keizer M G, van Riemsdijk W H and de Haan F A M 1994 Solubility of cyanide in contaminated soils. J. Environ. Qual. 23, 785-792.

Mench M J, Didier V L, Loffler M, Gomez A and Masson P 1994 A mimicked in-situ remediation study of metal contaminated soils with emphasis on cadmium and lead. J. Environ. Qual. 23, 58-63.

Middeldorp P J M, Briglia M and Salkinoja-Salonen M S 1990 Biodegradation of pentachlorphenol in natural soil by inoculated *Rhodococcus chlorophenolicus*. Microb. Ecol. 20, 123-139.

Morgan P and Watkinson R J 1989 Hydrocarbon degradation in soils and methods for soil biotreatment. CRC Crit Rev Biotechnol. 8, 305-333.

Muller R 1992 Bacterial degradation of xenobiotics. In Microbial Control of Pollution. Eds. J C Fry, G M Gadd, R A Herbert, C W Jones and I A Watson-Craik. pp 35-57. 48th Symposium of the Society of General Microbiology. Cambridge University Press, Cambridge, UK.

Muller-Kirchenbauer H, Friedrich W, Gremmel D, Markwardt W and Rogner J 1988 New results and aspects in the area of seal wall research. In Contaminated Soil '88. Eds. K Wolf, W J van den Brink and F J Colon. pp 549-560. Second International TNO/BMFT Conference on Contaminated Soil, 11-15 April 1988, Hamburg FRG. Kluwer Academic Publishers, Dordrecht.

Musgrove S 1991 An assessment of the efficiency of remedial treatment for metal polluted soil. In Land Reclamation: an End to Dereliction. Ed. M C R Davies. pp 248-255. International Conference on Land Reclamation, April 1991 University of Wales. Elsevier Science Publishers, Essex, UK.

Nakles D V, Theis T L, Ripp J A, Linz D G and Murarka I P 1993 Analytical techniques for cyanide and PAHs. Management of Manufactured Gas Plant Sites Technology Seminar March 23-24 1993 Orlando Fl. Remediation Technologies Inc, Arlington VA. pp 71-74.

NSW EPA 1994 Guidelines for assessing service station sites. New South Wales Environmental Protection Authority, Chatswood NSW. 31p.

NSW EPA 1995 Guidelines for the Vertical Mixing of Soil on Former Broad-Acre Agricultural Land. New South Wales Environmental Protection Authority, Chatswood NSW. 18p.

Otte M P, Gagnon J, Comeau Y, Matte N, Greer C W and Samson R 1994 Activation of an indigenous microbial consortium for bioaugmentation of pentachlorophenol/creosote contaminated soils. Appl. Microbiol. Biotechnol. 40, 926-932.

Pamukcu S 1994 Electric Power Research Institute Report EPRITR-103320.

Parker R 1994 Environmental restoration technologies. EMIAA Yearbook 169-171.

Peters R W and Shem L 1992 Use of chelating agents for remediation of heavy metal contaminated soil. In Environmental Remediation Removing Organic and Metal Ion Pollutants. Eds. G F Vandegrift, D T Reed and I R Tasker. pp 70-84. American Chemical Society, Washington DC.

Pickup R W, Morgan J A W, Winstanley C and Saunders J R 1991 Implications for the release of genetically engineered organisms. J. Appl. Bacteriol. Symposium Supplement 70, 19S-30S.

Pierzynski G M, Sims J T and Vance G F 1994 Soils and Environmental Quality. CRC Press, Boca Raton, FL.

Reed D T, Tasker I R, Cunnane J C and Vandegrift G F 1992 Environmental restoration and separation science. In Environmental Remediation Removing Organic and Metal Ion Pollutants. Eds. G F Vandegrift, D T Reed and I R Tasker. pp 1-21. American Chemical Society Washington DC.

Rickabaugh J, Clement S and Lewis R F 1987 Surfactant scrubbing of hazardous chemicals from soil. Proceedings of the 41st Industrial Waste Conference, pp 377-382.

Roberts T M and Gemmell 1979 Establishment of vegetation on gasworks waste. Proceedings Society of Chemical Industry Conference Reclamation of Contaminated land. Eastbourne 1979. Society of Chemical Industry, London. pp 11.1-11.10.

Sayre P and Miller R V 1987 Bacterial mobile genetic elements: importance in assessing the environmental fate of genetically engineered sequences. Plasmid 26, 151-171.

Semprini L, Hopkins G D, McCarty P L and Roberts P V 1992 In situ transformation of carbontetrachloride and other halogenated compounds resulting from biostimulation under anoxic conditions. Environ. Sci. Technol. 26, 2454-2461.

Shannon M J R and Unterman R 1993 Evaluating bioremediation: Distinguishing fact from fiction. Annu. Rev. Microbiol. 47, 715-738.

Shaw J J, Beauchamp C, Dane F and Kriel R J 1992 Securing a permit from the United States department of Agriculture for field work with genetically engineered microbes: a non-prohibitory process. Microb. Releases 1, 51-53.

Sheppard M I and Thibault D H 1992 Desorption and extraction of selected heavy metals from soils. Soil Sci. Soc Am. J. 56, 415-423.

Shoebridge L 1993 A review of soil remediation practice in Australia. In The Health Risk Assessment and Management of Contaminated Sites. Eds. A Langley and M van Alpen. pp 223-227. Proceedings of the Second National Workshop on the Health Risk Assessment and Management of Contaminated Sites Canberra 5-7 April 1993. South Australian Health Commission, Adelaide SA.

Short K A, Seidler R J and Olsen R H 1990 Survival and degradative capacity of *Pseudomonas putida* induced or constitutively expressing plasmid-mediated degradation of 2,4-dichlorophenoxyacetate (TFD) in soil. Can J Microbiol. 36, 821-826.

Skladany G J and Metting F B 1992 Bioremediation of contaminated soils. In Soil Microbial Ecology. Ed. F B Metting. pp 483-513. Marcel Dekker Inc., NY.

Smith B 1993 Remediation update finding the remedy. Waste Manage. Env. 4, 24-30.

Smith B 1995 Cleaning up industrial sites. Waste Manage Env. 6, 16-24.

Soong F, Emmett A 1993 Assessment and management of CCA timber preservation plants. In The Health Risk Assessment and Management of Contaminated Sites. Eds. A Langley and M van Alpen. pp 229-246. Proceedings of the Second National Workshop on the Health Risk Assessment and Management of Contaminated Sites, Canberra 5-7 April 1993. South Australian Health Commission, Adelaide SA.

Sposito G 1989 The Chemistry of Soils. Oxford University Press, NY.

Stitson M K 1990 EPA site demonstration of the international waste technologies/Geo-Con in situ stabilization/solidification process. Air Waste Manag. Assoc. 40, 1569-1576.

Theis T L, Young T C and Knutsen K C 1993 Extraction and treatment of cyanide from purifier box wastes. Management of Manufactured Gas Plant Sites Technology Seminar March 23-24 1993 Orlando Fl. Remediation Technologies Inc, Arlington VA. pp 183-187.

Thompson-Eagle E T and Frankenburger Jr W T 1992 Bioremediation of soils contaminated with selenium. In Advances in Soil Science Volume 17. Eds. R Lal and B A Stewart. pp 261-310. Springer-Verlag, NY.

Thorburn S and Buchanan N W 1987 Building on chemical waste. In Building on Marginal and Derelict Land. Ed N W Buchanan. pp 281-296. Thomas Telford, London.

Tiller K G 1992 Urban soil contamination in Australia. Aust. J. Soil Res. 30, 937-957.

Tiller K G, de Vries M P C, Spouncer L R, Smith L and Zarcinas B 1976 Environmental pollution of the Port Pirie region. 3. Metal contamination of home gardens in the city and their vegetable produce. Division of Soils Divisional report No. 15. Commonwealth Scientific and Industrial Research Organisation, Australia

Tillosten S 1993 The cost of clean-up. In Remedial Processes for Contaminated Land. Ed. M Pratt. pp 87-112. Institute for Chemical Engineers.

Troy M A, Bruce Allen L and Jerger D E 1994 Bioremediation of petroleum contaminated soils at railroad facilities. In Remediation of Hazardous Waste Contaminated Soils. Eds. D L Wise and D J Trantolo. pp 227-244. Marcel Dekker Inc, NY.

Tuin B J W and Tels M 1990a Removing heavy metals from contaminated clay soils by extraction with hydrochloric acid, EDTA or hypochlorite solutions. Environ. Technol. 11, 1039-1052.

Tuin B J W and Tels M 1990b Extraction kinetics of six heavy metals from contaminated clay soils. Environ. Technol. 11, 541-554.

Tuin B J W and Tels M 1991 Continuous treatment of heavy metal contaminated clay soils by extraction in stirred tanks and counter current column. Environ. Technol. 12, 178-190.

Turczynowicz L 1991 Approaches to analyte selection. In The Health Risk Assessment and Management of Contaminated Sites. Eds. O El Saadi and A Langley. pp 31-51. Proceedings of a National Workshop on the Health Risk Assessment and Management of Contaminated Sites. Adelaide 5-7 August 1991. South Australian Health Commission, Adelaide SA.

Turczynowicz L 1993 The assessment and management of Gaswork sites. In The Health Risk Assessment and Management of Contaminated Sites. Eds. A Langley and M van Alpen. pp 261-312. Proceedings of the Second National Workshop on the Health Risk Assessment and Management of Contaminated Sites, Canberra 5-7 April 1993. South Australian Health Commission, Adelaide SA.

USEPA 1977 Movement of Selected Metals, Asbestos and Cyanide in Soil: Application to Waste Disposal Problem. Solid and Hazardous Waste Research Division. United States Environmental Protection Agency Report No. EPA-600/2-77-020. Cincinnati, OH.

USEPA 1987 A compendium of technologies used in the treatment of hazardous waste. Centre for Environmental Research Information, United States Environmental Protection Agency Report No. EPA/625/8-87/014. Cincinnati OH.

USEPA 1990a Superfund Innovative Technology Evaluation. Solidtech Inc. Solidification/stabilisation Process, Applications Analysis Report, EPA/540/A5-89/005.

USEPA 1990b Superfund Innovative Technology Evaluation. International Waste Technologies/Geo-Con In situ Stabilisation/Solidification, EPA/540/A5-89/004.

Valo R and Salkinoja-Salonen M 1986 Bioreclamation of chlorophenol-contaminated soil by composting. Appl. Microbiol. Biotechnol. 25, 68-75.

Valocchi A J 1988 Mathematical modelling of the transport of pollutants from hazardous waste landfills. In Land Disposal of Hazardous Waste, Engineering and Environmental Issues. Eds. J R Gronow, A N Schofield and R K Jain. pp 153-164. Ellis Horwood Limited, Chichester, Sussex, UK.

Verheul J, Brujin P and Herbert S 1993 Biological remediation - a European perspective. In Remedial Processes for Contaminated Land. Ed M Pratt. pp 53-85. Institute of Chemical Engineers.

Waste Management and the Environment Yearbook 1994. Rala Publications, Balmain NSW.

Wheeler M 1994 Site Remediation: An Australian Perspective on 'Best Practice'. In Environmental Management Industry Association of Australia, Yearbook 1994. Ed. J Cole. pp 156-160. Executive Media Pty Ltd, Waterloo NSW.

Wiles C C, Barth E and de Percin P 1988 Status of solidification/stabilisation in the United States and factors affecting its use. In Contaminated Soil '88. Eds. K Wolf, W J van den Brink and F J Colon. pp 947-956. Second International TNO/BMFT Conference on Contaminated Soil, 11-15 April 1988, Hamburg FRG. Kluwer Academic Publishers, Dordrecht.

Will F G 1995 Removing toxic substances from the soil using electrochemistry. Chem. Ind. 10, 376-379.

Williams G M 1988 Integrated studies into ground water pollution by hazardous waste. In Land Disposal of Hazardous Waste, Engineering and Environmental Issues. Eds. J R Gronow, A N Schofield and R K Jain. pp 37-48. Ellis Horwood Limited, Chichester, Sussex, UK.

Williams K P 1992 Principles and application of physical particle separation for treatment of contaminated land. In Contaminated Land Treatment Technologies. Ed. J F Rees. pp 113-128. Society of Chemical Industry, Elsevier Applied Science, London.

Williams R T, Ziengenfuss P S and Sisk W E 1992 Composting of explosives and propellant contaminated soils under thermophilic and mesophilic conditions. J Ind. Microbiol. 9, 137-144.

Wills B 1988 Mineral Processing Technology 4th Edition. Pergamon Press, Oxford.

Woodyard J P 1990 PCB detoxification technologies – a critical assessment. Env. Progress 9, 131-135.

Wyatt J M and Palmer S J 1991 Biodegradation of nitriles and cyanide. In Biodegradation: Natural and Synthetic Materials. Ed. W B Betts. pp 69-88. Springer-Verlag, London.

Yland M W F and Wachem E G 1988 Soil covering systems as remedial action in contaminated housing areas in the Netherlands. In Contaminated Soil '88. Eds. K Wolf, W J van den Brink and F J Colon. pp 597-599. Second International TNO/BMFT Conference on Contaminated Soil, 11-15 April 1988, Hamburg FRG. Kluwer Academic Publishers, Dordrecht.

Young C S and Burns R G 1993 Detection, survival and activity of bacteria added to soil. In Soil Biochemistry. Eds. J-M Bollag and G Stotzky. Vol 8, pp 1-63. Marcel Dekker, NY.

Zeigenfuss P S and Williams R T 1991 Hazardous materials composting. J. Haz. Mat. 28, 91-99.

14
Contaminants and the Australian soil environment

F. BARZI, R. NAIDU and M.J. McLAUGHLIN

1. Introduction

Contaminated land is an increasingly important environmental, health, economic and planning issue in Australia and a number of initiatives and approaches have been developed for dealing with these issues. As part of this process Government agencies are placing greater emphasis on the state of soils and the soil environment, recognising the soil's role as a potential repository for pollutants, as a transmitter of undesirable materials to surface and groundwaters and as a potential source of contaminants in the food chain.

In defining the word 'contaminant', a distinction must be made between soils having naturally high 'background' concentrations of some contaminants (e.g. Cr in soils derived from serpentine) and soils having the levels of potentially toxic substances increased through man's activities. As discussed by Tiller (1991) 'background' concentrations can be viewed in several ways. To assess soil contamination, concentrations of contaminants must be elevated above what is regarded as a 'normal' range and this may vary from country to country. Tiller (1991) states that ranges for 'normal' values must be calculated by excluding elevated concentrations in hot spots due to geochemical anomalies (e.g. unusual rock or parent material) and localised pollution due to disposal of industrial wastes, atmospheric pollution (e.g. roadside soils) and intensive or excessive use of agricultural chemicals and fertilisers.

In Australia, increased industrial and agricultural activities following European settlement, beginning in the early 1800s when regulatory measures were completely absent and environmental awareness low, have led to contamination of the soil environment at two scales. Significant but localised contamination is a feature of industrial or mining activities, whereas agricultural practices and some urban and industrial activities (motor vehicle emissions, smelting activities etc.) contaminate soils more extensively but at a lower level. Agriculture plays a major role in the Australian economy. As in many primary producing nations, there has been an increasing reliance on chemicals and mechanised technologies to sustain food productivity. The competitive international market with its focus on food quality and increasing emphasis on minimum tillage, has also promoted a dependence on chemicals. As discussed in Chapter 5, with increased use of chemicals in agriculture, soil and water resources are becoming increasingly contaminated by elements and compounds either added intentionally or inadvertently to terrestrial ecosystems. With increasing urban populations, housing developments have also expanded into traditional farming areas and onto land which was previously used for industrial or waste disposal purposes, raising potential contamination concerns.

R. Naidu et al. *(eds): Contaminants and the Soil Environment in the Australasia-Pacific Region, 451–484.*

Australia is a highly industrialised and technologically advanced nation. Industrial practices have caused significant pollution of land and water resources in our major urban centres. Sites which have a higher risk of being contaminated include gas-works, fuel stations, fuel depots, terminals, refineries, defence establishments and chemical manufacturing installations (Hosking, 1993a). Other, less obvious contaminated sites include those which have been developed on locations with a previous history of contamination.

The exact number of contaminated sites in Australia is not known. It is estimated that there are approximately 400 contaminated sites in Tasmania and in excess of 30 000 in the two most populous states of New South Wales (NSW) and Victoria. However, Forest (1993) claimed that the number of sites in NSW could be as high as 60 000, and of these at least 7 000 sites require remediation, costing up to $A 2 billion for full remediation. Many of these sites have the potential to adversely impact on (a) human health, (b) surface, (c) groundwater quality or (d) ecological processes. Information on the number of contaminated sites in the other States of Australia was not available in the literature cited for this manuscript.

Systematic attention to the management and remediation of contaminated land began in Australia in the early 1980s (Forrest, 1993). In particular, since the mid 1980's landholders, potential buyers and regulatory agencies have become increasingly aware of the potential issues and implications associated with developing land that is degraded by contamination. Contaminants can have a range of long-term and short-term impacts on human health through various exposure pathways including ingestion of soil, inhalation of vapours, gases or dust, skin contact, ingestion through food (plants and animals) and contamination of drinking, surface or groundwater (ANZECC/NHMRC, 1992).

This paper provides a brief overview of issues of soil contamination within Australia and current assessment and management strategies that have been developed for addressing these issues. Given that Australia is a federation of six states and territories and each of these have regulatory measures that are the responsibility of States/Territories, it is not possible for this document to detail all the issues relating to contaminants and the soil environment in Australia.

The State/Territory Environmental Protection Authorities (EPAs) in Australia have an extensive program of guideline development under way in a number of important areas relating to the management of contaminated sites. For example, the New South Wales EPA have released 'Guidelines for Assessing Service Station Sites' and 'Guidelines for the Vertical Mixing of Soil on Former Broad-Acre Agricultural Land'. They have also coordinated the 'Draft Environmental Guidelines for Copper Chrome Arsenate Timber Preservation Plants' which have been jointly developed by the Australian and New Zealand Environment and Conservation Council (ANZECC) and the Timber Preservers of Australia. The Northern Territory Department of Lands, Planning and Environment have released a 'Strategy for Site Management and Pollution Control in the Northern Territory' (September, 1995) and the Victorian Environment Protection Authority have coordinated the 'Draft Guidelines for the Analysis of Contaminated Soils' (June

Figure 1. Australia – a federation of six States and two Territories.

1995) released by the Australian and New Zealand Environment and Conservation Council. Other examples include the Western Australian Department of Environment Protection's public discussion paper 'Assessment and Management of Contaminated Land and Groundwater in Western Australia'. Readers are directed to reports and reviews of key environmental issues relating to each of the States/Territories in Australia to the appropriate State or Territory authority or department. Some further information on contaminated sites in Australia is reviewed in Chapters 2, 3, 4, 12 and 13.

2. Climate and Soil

Much of Australia's climatic character is dictated by its latitudinal position (10°S to 44°S, 40% within the tropics) and its modest relief (Taylor, 1983). Because of its broadly transitional climatic features and the wide range in soil parent materials, soils range from alluvial, or those derived from recent volcanic rocks, to mainly ancient soils which are shallow and low in fertility.

Australia is a dry continent with 60% of its area having a median rainfall of less than 400 mm yr^{-1} and 80% < 600 mm yr^{-1}. Over much of the continent, the rainfall pattern is strongly seasonal in character, with a wet winter period in the south and a wet summer season in the north. The arid interior occupies approximately 70% of the continent.

3. Australian system of government

The Commonwealth of Australia is a federation of six States and two Territories (Figure 1). In general, responsibility for land use and associated environmental protection issues, including responsibility for contaminated sites, resides with the individual States and Territories (excluding sites on Commonwealth land). The Commonwealth has no direct legislative powers in relation to the environment because, at the time the Commonwealth Constitution Act was passed in 1900, environment protection was not an issue which was considered by legislators (Bates, 1992).

The Commonwealth (or Federal) Government has exclusive jurisdiction over Australian External Territories and land owned by the Commonwealth in the States and Territories (over 1 million ha at present). While the Commonwealth Government does not have direct environmental legislative powers in areas outside these, it does have a number of specific and important powers relating to trade and commerce, international affairs, defence, corporations, finance and race which have led to the Commonwealth Government taking action on environmental issues (e.g. where an international treaty requires such an action).

4. Background values for contaminant concentrations in Australian soils

As discussed above, different perceptions of what constitutes 'normal' or 'background' concentrations of contaminants in soil have made difficult the development of consistent guidelines in most countries. For some anthropogenic contaminants (e.g. DDT) it is evident that background values are zero and any detectable concentration can be regarded as contamination. With the increasing sensitivity of analytical instrumentation over the years, soils which were previously regarded as uncontaminated may now be found to contain detectable concentrations of some anthropogenic chemicals. For most inorganic elements, which are naturally present in all soils, assessing contamination is more difficult. The ANZECC and the National Health and Medical Research Council (NHMRC) Guidelines for the Assessment and Management of Contaminated Sites define background concentrations as 'levels of substances or chemicals that are commonly found in the local environment' (ANZECC/NHMRC, 1992). These Guidelines recommend the evaluation of background levels as a consideration in the determination of clean up guidelines.

No consistent study has been undertaken to characterise the range and concentrations of contaminants in Australian soils, whether 'pristine' (soils totally unaffected by human activity) or 'normal'. Contributions from natural processes depend on the intensity of weathering processes, the nature of soil parent rock and regional geology. Rock geochemistry influences the distribution of minerals at any particular site. Some examples include the high concentration of Ni, Co and Cr in basic igneous rocks, such as the basalts which cover extensive areas in western Victoria and large parts of Tasmania (Tiller, 1992). Similarly, high Cd concentrations have recently been determined in soils on offshore islands in Queensland

Table 1. Ranges of background concentrations ($mg\,kg^{-1}$) of some potentially toxic elements in Australian rural surface soils.

	Approx. No. of samples	Range of values	Dutch A value[a]
Co	250	<2–170	20
Cu[b]	500	1–190	36
Mn	800	4–13,000	–
Zn	500	<2–200	140
Mo	300	<1–20	10
Ni	200	2–400	35
Cr	180	5–110	100
Pb[b]	160	2–160	85
Cd	180	<1	0.8
B[c]	50	10–75	–
As[b]	26	<1–8	29

Source: Tiller (1992).
[a]Upper limit of background values in Netherlands (Source: Environmental Quality Objectives in the Netherlands, 1994).
[b]Excludes orchard soils treated with Pb, As and Cu.
[c]Cartwright (unpublished data).

which are unaffected by man's activities (Rayment et al., pers. comm.). Soil forming processes are also responsible for the accumulation of particular mineral types. For example, hydrological and redox conditions in the weathering zone may cause concentration of iron oxides which can result in the accumulation of some heavy metals and Mo in ironstone rich soils.

Tiller (1992) has collated the concentrations of some potentially toxic elements in agricultural surface soils from published and unpublished data (Table 1). He reported that the highest 'background' concentrations of Co, Cu, Mo, Zn, Cr, Ni and Pb found in Australian rural soils exceeded those quoted for Dutch soils. The reverse was true for As. These differences were attributed to the great range of soil parent material and the different type and duration of soil-forming processes in Australia compared to the Netherlands, as well as differences associated with the more intensive agricultural activity in the Netherlands. Accordingly, caution needs to be exercised when adopting soil-based criteria developed for different soils and climates.

The most recent report on background concentrations of metal contaminants in soils is the Queensland study commissioned by the Department of Health, Housing and Community Services and the EPA (Olszowy et al., 1993). This study was based on rural soils from the Brisbane area and urban soils from four Australian capital cities. Results show that for those Queensland rural soils studied, Mn was the only element that exceeded the ANZECC/NHMRC environmental investigation threshold (Olszowy et al., 1993). The data in Table 2 compare the geometric mean values and maximum values for trace elements in rural and urban Queensland soils. Urban and rural soil categories show similar mean values for As, Cd, Hg and Ni while urban categories have a higher mean level for Cr, Cu, Pb, Zn and V compared to rural ones. Manganese has a higher geometric mean in rural soils compared to urban

Table 2. Geometric means and maximum values of ($mg\,kg^{-1}$) core trace elements in Queensland soils.

Geometric Means	As	Cd	Cr	Cu	Hg	Mn	Ni	Pb	V	Zn
Rural Soils	5	<0.5	66	99	<0.1	201	5	12	13	13
New Suburb – Low Traffic	5	<0.5	11	7	<0.1	124	6	8	20	20
New Suburb – High Traffic	5	<0.5	16	9	<0.1	166	7	31	28	33
Old Suburb – Low Traffic	5	<0.5	18	22	<0.1	152	8	179	29	156
Old Suburb – High Traffic	5	<0.5	10	29	<0.1	126	6	201	20	215
Maximum Values										
Rural Soils	53	<0.5	56	412	<0.1	7357	38	56	121	92
Urban Soil	31	1.0	131	369	1.6	742	38	819	107	3820

Source: Olszowy et al. 1993. Studies of Levels of Trace Elements in Soils from Rural and Urban Areas of Australia.

soil samples. Lead and Zn levels are high in old suburbs and new suburbs with high traffic flow compared to rural soils or soils from new suburbs with low traffic flow. In some urban situations P and S are also high compared to rural soils. Comparison of maximum values of trace elements in urban and rural soils shows that As, Mn, V, Cu and P are higher in rural soils while Cd, Cr, Hg, Zn, and Pb are higher in urban soils. Unfortunately detection limits for some elements (e.g. Cd) were fairly high in this study, so that many samples were below the limit of reporting. The ANZECC/NHMRC 'background' values are currently in the process of review.

5. Sources of soil contamination

Contamination through anthropogenic activities occurs in both rural and urban environments. Within the rural sector, contaminants result largely from the use of agrochemicals, while municipal and industrial wastes have been the major sources of urban contamination. The most common contaminants appear to be heavy metals and organic substances including chemicals and solvents. At least two-thirds of the recorded contaminated urban sites in South Australia, Victoria and Queensland are contaminated with heavy metals. New South Wales, on the other hand, has a significant number of sites contaminated by organics (Tiller, 1992; see Chapters 4 and 13).

The main activities producing contaminants include industrial, mining, agricultural and commercial activities as well as transport and services. Military land and land used for recreational shooting are also important sites of contamination. Contamination at many of these sites appears to have resulted because of lax regulatory measures prior to the establishment of State/Territory Environmental Protection Authorities. Many industrial and mining operations disposed wastes directly onto land; some of these operations and activities are listed below and the contaminants associated with them are listed in Table 3.

Table 3. Industries, land uses and associated chemicals.

Industry	Type of chemical	Associated chemicals
Agricultural/horticultural activities		See chemicals listed below under fertiliser, insecticides, fungicides, herbicides under chemicals manufacture and use
Airports	Hydrocarbons	Aviation fuels
	Metals	particularly aluminium, magnesium, chromium
Asbestos production and disposal		Asbestos
Battery manufacture and recycling	Metals	Lead, manganese, zinc, cadmium, nickel, cobalt, mercury, silver, antimony
	Acids	Sulphuric
Breweries/distilleries	Alcohol	Ethanol, methanol, esters
Chemicals manufacture and use	Acid/alkali	Mercury (chlor/alkali), sulphuric, hydrochloric and nitric acids, sodium and calcium hydroxides
	Adhesives/resins	Polyvinyl acetate, phenols, formaldehyde, acrylates, phthalates
	Dyes	Chromium, titanium, cobalt, sulphur and nitrogen organic compounds, sulphates, solvents
	Explosives	Acetone, nitric acid, ammonium nitrate, pentachlorophenol, ammonia, sulphuric acid, nitroglycerine, calcium cyanamide, lead, ethylene glycol, methanol, copper, aluminium, bis(2-ethylhexyl)adipate, dibutyl phthalate, sodium hydroxide, mercury, silver
	Fertiliser	Calcium phosphate, calcium sulphate, nitrates, ammonium sulphate, carbonates, potassium, copper, magnesium, molybdenum, boron, cadmium
	Flocculants	Aluminium
	Foam production	Urethane, formaldehyde, styrene
	Fungicides	Carbamates, copper sulphate, copper chloride, sulphur, chromium
	Herbicides	Ammonium thiocyanate, carbamates, organochlorines, organophosphates, arsenic, mercury
	Paints	
	Heavy metals	Arsenic, barium, cadmium, chromium, cobalt, lead, manganese, mercury, selenium, zinc
	General	Titanium dioxide
	Solvent	Toluene, oils natural (e.g. pine oil) or synthetic
	Pesticides	Arsenic, lead, organochlorines, organophosphates,
	Active ingredients	sodium, tetraborate, carbamates, sulphur, synthetic pyrethroids,
	Solvents	Xylene, kerosene, methyl isobutyl ketone, amyl acetate, chlorinated solvents
	Pharmacy	Dextrose, starch
	General	Acetone, cyclohexane, methylene chloride, ethyl
	Solvents	acetate, butyl acetate, methanol, ethanol, isopropanol, butanol, pyridine, methyl ethyl ketone, methyl isobutyl ketone, tetrahydrofuran

continued

Table 3. continued.

Industry	Type of chemical	Associated chemicals
	Photography	Hydroquinone, pheidom, sodium carbonate, sodium sulphite, potassium bromide, monomethyl paraaminophenol sulphates, ferricyanide, chromium, silver, thiocyanate, ammonium compounds, sulphur compounds, phosphate, phenylene diamine, ethyl alcohol, thiosulphates, formaldehyde
	Plastics	Sulphates, carbonates, cadmium, solvents, acrylates, phthalates, styrene
	Rubber	Carbon black
	Soap/detergent	
	General	Potassium compounds, phosphates, ammonia, alcohols, esters, sodium hydroxide, surfactants (sodium lauryl sulphate), silicate compounds
	Acids	Sulphuric and stearic
	Oils	Palm, coconut, pine, tea tree
	Solvents	
	General	Ammonia
	Hydrocarbons	e.g. BTEX (benzene, toluene, ethylbenzene, xylene)
	Chlorinated organics	e.g. trichloroethane, carbon tetrachloride, methylene chloride
Defence works		See Explosives under chemicals manufacture and use, foundries, engine works, service stations
Drum reconditioning		See Chemicals manufacture and use
Dry cleaning		Trichloroethylene and ethane Carbon tetrachloride Perchlorethylene
Electrical		PCBs (transformers and capacitors), solvents, tin, lead, copper
Engine works	Hydrocarbons	
	Metals	
	Solvents	
	Acids/alkalis	
	Refrigerants	
	Antifreeze	Ethylene glycol, nitrates, phosphates, silicates
Foundries	Metals	Particularly aluminium, manganese, iron, copper, nickel, chromium, zinc, cadmium and lead and oxides, chlorides, fluorides and sulphates of these metals
	Acids	Phenolics and amines
		Coke/graphite dust
Gas works	Inorganics	Ammonia, cyanide, nitrate, sulphide, thiocyanate
	Metals	Aluminium, antimony, arsenic, barium, cadmium, chromium, copper, iron, lead, manganese, mercury, nickel, selenium, silver, vanadium, zinc
	Semivolatiles	Benzene, ethylbenzene, toluene, total xylenes, coal tar, phenolics and PAHs
Iron and steel works		Metals and oxides of iron, nickel, copper, chromium, magnesium and manganese, and graphite

continued

Table 3. continued.

Industry	Type of chemical	Associated chemicals
Landfill sites		Methane, hydrogen sulphides, heavy metals, complex acids
Marinas		Engine works, electroplating under metal treatment
	Antifouling paints	Copper, tributyltin (TBT)
Metal treatments	Electroplating	
	Metals	Nickel, chromium, zinc, aluminium, copper, lead, cadmium, tin
	Acids	Sulphuric, hydrochloric, nitric, phosphoric
	General	sodium hydroxide, 1,1,1-trichloroethane, tetrachloroethylene, toluene, ethylene glycol, cyanide compounds
	Liquid carburizing baths	Sodium, cyanide, barium, chloride, potassium chloride, sodium chloride, sodium carbonate, sodium cyanate
Mining and extractive industries		Arsenic, mercury and cyanides and also refer to Explosives under chemicals, manufacture and use
Power stations		Asbestos, PCBs, fly ash, metals
Printing shops		Acids, alkalis, solvents, chromium (see chemicals listed under photography)
Railway yards		Hydrocarbons, arsenic, phenolics (creosote), heavy metals, nitrates and ammonia
Scrap yards		Hydrocarbons, metals, solvents
Service stations and fuel storage facilities		Aliphatic hydrocarbons BTEX PAHs (e.g. benzo(a)pyrene) Phenols Lead
Sheep and cattle dips		Arsenic, organochlorines and organophosphates, carbamates, and synthetic pyrethroids
Smelting and refining		Metals and the fluorides, chlorides and oxides of copper, tin, silver, gold, selenium, lead, aluminium
Tanning and associated trades	Metals	Chromium, manganese, aluminium
	General	Ammonium sulphate, ammonia, ammonium nitrate, phenolics (creosote), formaldehyde, tannic acid
Wood preservation	Metals	Chromium, copper, arsenic
	General	Naphthalene, ammonia, pentachlorophenol, dibenzofuran, anthracene, biphenyl, ammonium sulphate, quinoline, boron, creosote, organochlorine pesticides

Source: Draft Australian Standard Analysis of Soils – the sampling of potentially contaminated soil. Committee CH/28 – Analysis of soil and biota.

- disposal of wastes (controlled and uncontrolled);
- accidental spillage (including process spillage);
- leakage during plant operation;
- storage or transportation of raw materials, finished products or wastes;
- spreading of sewage sludge;
- deposition from the atmosphere from an industrial site;
- migration of contaminants into a site from neighbouring land, as vapour, leachate, movement of liquids through the soil or as dust and other particles;
- use of agricultural chemicals (such as fertilisers, herbicides, insecticides and pesticides and
- decommissioning of the plant.

Materials which can cause contamination include:

- metals;
- inorganic compounds;
- organic chemicals;
- oils and tars;
- toxic, explosive and asphyxiant gases (including gases from the decomposition of wastes in landfills);
- combustible substances;
- putrescible material,
- hazardous wastes; and
- explosive devices such as grenades and detonators.

5.1 Atmospheric deposition

Input of contaminants from the atmosphere is expected only within the depositional zones of significant sources of emissions such as smelters or industrial plants in the major urban areas (Tiller, 1989), and contributions from the latter may be minimal on rural soils. Merry and Tiller (1991) concluded from the Cd contents of rainfall samples that the atmospheric deposition of Cd from the Adelaide metropolitan area onto land about 20 km downwind from the urban fringe is about $3\,g\,ha^{-1}\,yr^{-1}$. In earlier studies Tiller et al. (1987) found that accessions of automotive Pb to agricultural soils east of Adelaide varied from $20\,mg\,Pb\,km^{-2}\,yr^{-1}$ within 20 km of the city tapering off to about $2\,mg\,Pb\,km^{-2}\,yr^{-1}$ at distances in excess of 40 km from the city. Beyond 40 km downwind from the metropolitan boundary, deposition of urban contaminants was found to be indistinguishable from background concentrations. Concentrations of EDTA – extractable Cd in their study area – were within the range $<0.01-0.73\,\mu g\,g^{-1}$ soil (0–5 cm horizon).

Although a number of industries are located in rural areas, there is little information available on their impact on the level of contaminants in rural soils. The only reported investigation in the literature is that of Tiller and his co-workers (see Tiller, 1989) who carried out a detailed investigation of soils around the Pb–Zn smelter site at Port Pirie, about 250 km north of Adelaide, South Australia. They showed

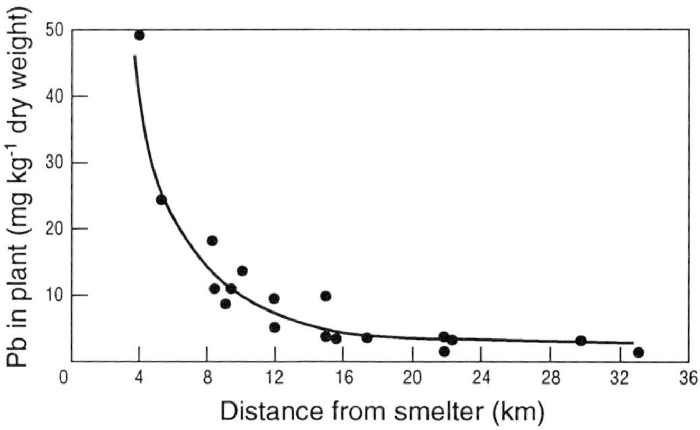

Figure 2. Influence of zinc smelter on Pb content of plants with distance (Merry and Tiller, 1978).

that at least 3 400 km² of land that is used for sheep grazing, forestry, cereal farming and as a water catchment was to some extent contaminated by heavy metals. Contamination was found to decline exponentially with distance from the smelter (Figure 2), extending as far as 65 km from the smelter depending on the wind direction. Moreover, the Cd content of pastures, cereal grains and sheep meat also showed a strong relationship with distance from the smelter (Merry et al., 1981; Koh and Judson, 1986), although serious contamination of the food chain fortuitously did not occur, due to the alkaline nature of the soils in this area. Further work is necessary in Australia at other major industrial sites to investigate such impacts of industries on rural soils.

5.2 *Agricultural practices*

Contamination of agricultural land can arise from atmospheric deposition of contaminants of urban and industrial origin, or from direct or indirect input of contaminants. These are discussed briefly in the following sections. The nature and concentration of toxic substances depends primarily on the intensity of use and frequency of application, the half life of the contaminant, the physico-chemical process controlling contaminant dynamics within the soil environment (see Chapters 3 and 4), and local climatic effects. Consequently, the concentrations of individual contaminants vary from one location to another and may sometimes exceed health or environmental background levels. There is therefore a need for site specific soil criteria and clean up strategies and remediation approaches tailored to specific local conditions.

5.2.1 Phosphatic Fertilisers

Many soils in Australia are highly leached and strongly weathered and depend on phosphatic fertilisers to sustain production (Wild, 1958). The environmental impact

Table 4. Chemicals used in cattle dips (Cattle Tick Dip Site Management Committee, 1992).

Tickicide	Period of use
Arsenic (trioxide)	? - 1955
DDT	1955 – 1962
Benzene hexachloride	1955 – 1962
Carbaryl	1963 – 1970
Coumaphos	1962 – 1965
Bromophos ethyl	1959?
Dioxothion	1962 – 1976
Ethion	1962 – 1976
Chlordimeform	1973 – 1976
Cymyazole	1977 – 1986
Chloromethiuron	1977
Amitraz	1976 – present
Promacyl	1977 – present
Cypermethrin	1979 – present
Chlorfenvinphos	1979 – present
Flumethrin	1986 – present

of impurities in fertilisers was recently reviewed by McLaughlin et al. (1996), with particular reference to Australian conditions. The major elements of concern were identified as As, Cd, F, Hg and Pb. Of these, Cd is the element of most interest, because of the relative ease of transfer of Cd from soil to plant and the perceived risk to human health. This issue has been discussed previously in Chapters 5 and 11. Tiller et al. (1995) recently reviewed research in Australia investigating management strategies to minimise Cd uptake by crops.

In addition to the buildup of Cd in the soils, long term fertilisation has also led to a serious decline in the quality of surface and groundwater in certain parts of Australia. For example, Hodgkin and Hamilton (1993) attributed the accelerated eutrophication of freshwaters in south-western Australia mainly to P leaching from nutrient-deficient sandy soils which received relatively high rates of P to maintain optimal production. The most publicised issue in this regard is the impact of nutrient transport from farmed land on the quality of stream and river waters in the Murray Darling Basin. Given the concern raised by many farmers and the public, many reviews have been conducted on the soil processes, farm management practices etc. on the Murray Darling Basin surface and groundwater qualities (GHD, 1992; Anon, 1994; Harrison, 1994). Such reports have led to the implementation of nutrient management strategies for inland and coastal waters (see Table 7 in Chapter 2). The Murray Darling Basin Commission is also addressing water quality issues in the whole of the basin (approximately 190 million ha) through its Algam Management Strategy. An overview of pathways for nutrient losses from the soil system and their impact on surface and groundwater quality is reviewed in Chapter 2.

Table 5. Average residues ($mg\,kg^{-1}$) in contaminated soils around cattle tick dip sites (Cattle Tick Dip Site Management Committee, 1992).

Location within dip site	DDT		Arsenic		Ethion	
	Average	Range	Average	Range	Average	Range
Adjacent to bath	1513	0–17700	290	0–1636	399	12–2300
out from bath	367	0–5770	144	0–870	na	na
animal draining yard	130	0–1980	99	1–789	7	0–124
scooping mound	4717	0–100000	720	15–3000	6152	0–20000
disposal pit	5463	0–106000	467	0–2600	2120	0–45000

5.2.2 Pesticides

Both inorganic and organic pesticides have been widely used in Australia. Inorganic pesticides were first introduced to control cattle and sheep ticks throughout Australia, by dipping cattle and sheep in baths containing As and, in later years, DDT. Animal entry into the dip caused serious contamination of the soils on either side of the dips. Both As and DDT have been banned since the mid 1980's. Since the prohibition of DDT, other organic chemicals including benzene hexachloride (BHC), ethion, carbophenothion and more recently amitraz, flumethrin and a mixture of chlorfenvinphos and cypermethrin have been used.

Although many types of insecticide have been used to control cattle ticks and other livestock pests since the early 1900's, As and DDT are the major contaminants of concern at cattle and sheep dip sites (Beard, 1993). Arsenic was used from the turn of the century until the mid 1950s when DDT, an organochlorine insecticide, was introduced. DDT was used until the mid 1980s and since then a wide range of organophosphates and carbamates have been used for tick control (Table 4). Further, highly toxic metabolites such as DDD and DDE have also been detected in surface soils at many of these sites (R. Naidu, unpublished). The tick control program led to localised contamination of land throughout Austalia. There are over 1 600 contaminated dip sites in NSW alone and similar numbers have been reported in other major Australian states. While As can be carcinogenic, organochlorine pesticides (OCP) can accumulate in fatty tissues and may be transferred through the food chain. Of major concern is the potential enzymatic transformation of DDT to more toxic metabolites such as DDE and DDD in animal fat tissue.

Surface soil concentrations of pesticide residues at cattle tick dip sites in northern NSW have been reported to be many times higher than ANZECC guideline values (Table 5). Recent studies at the CSIRO Division of Soils and the Cooperative Research Centre for for Soil and Land Management laboratories showed significant concentrations of As (<300 to $>1000\,mg\,kg^{-1}$) at depths exceeding 50 cm (Naidu et al., 1995) indicating that strategies for remediating these sites must consider subsurface soils. For contaminants where there is potential for uptake into vegetables or poultry, site-specific assessment will be required to determine the potential risk that householders may face from exposures through these routes (Beard, 1993). DDT tends to bioaccumulate and the eggs from poultry scavenging in soil contaminated with DDT may have undesirably high residue levels (Beard, 1993).

Agricultural practices, especially the spraying of crops with pesticides and fungicides, are also sources of toxic compounds, both inorganic and organic, which have a high potential to gain entry into food products. For example, mercuric and cupric fungicides have been widely used in many viticultural and horticultural regions. Investigations of the impact of these metals on soil processes or in terms of potential threat to water resources are few.

Arsenic and Pb based pesticides were formerly used in horticultural areas in South Australia and used in other rural and urban areas throughout Australia. Herbicides have also been widely used to control weeds in both cropping and forested catchment systems and recent studies by the Department of Forestry (Primary Industries, South Australia) indicate significant movement of, for example, atrazine to subsurface waters.

Although most OCs that have been widely used in Australia are now banned, their residues are still found in soil and continue to cause contamination problems. The Independent Panel on Intractable Waste (1992) estimates that some sites in Australia may contain up to 20 t of contaminated soil and the report 'Appropriate Technology for the Treatment of Scheduled Wastes' (CMPS&F, 1994), prepared by CMPS&F for the Environmental Protection Agency estimates soil volumes contaminated with scheduled wastes, including PCB's, hexacholorobenzenes, organophosphorus (OP) pesticides and cholorinated hydrocarbons in excess of 100 000 t.

5.3 Sewage sludges and sewage effluents

With the marked increase in the amount of municipal and industrial sludges and effluents generated from urban centres, disposal onto land is becoming an increasingly favoured option for urban authorities throughout the world. Major issues relating to the potential contamination of the soil with both inorganic and organic pollutants and the subsequent effects on water and atmospheric pollution, plant growth, animal and human health and the overall fertility of the soil are reviewed in Chapter 6.

Sludges contain appreciable amounts of heavy metals that persist in surface soils long after application. Metals in sludge most likely to impact on the environment include: As, Cd, Cr, Cu, Hg, Na, Ni, Pb and Zn (see Chapter 6). Of these metals, Cd has caused most concern from a human health viewpoint while Na may pose problems for soil structural stability. Tiller et al. (1995) recently estimated that approximately 2 t Cd are recycled to soils in Australia through disposal of sewage sludges to land. While the impact of sludge on the heavy metal status of soils has been the subject of much research in the Northern Hemisphere (e.g. Chaney and Ryan, 1993), only limited research has been carried out in Australia. The only published research studies that we could trace are those of deVries and Tiller in the 1970s near

Adelaide, South Australia (see deVries and Tiller, 1978) and studies of Ross et al. (1992) near Sydney, New South Wales. However, the Organic Waste Recycling unit, New South Wales Department of Agriculture has made detailed investigations on the beneficial use of biosolids on agricultural soils. These investigations showed that while biosolids increased the nutrient status of soils, heavy metal concentrations (Cd, Zn) were elevated in soils and crops produced on treated land (Osborne, pers. comm). Cadmium concentrations in some crops were found to exceed the maximum permitted concentrations, indicating the need for development of management practices that minimise soil-plant transfer of Cd. Human pathogenic bacteria such as *E. coli*, faecal streptococci, *Salmonella* and *Clostridium perfingens* in biosolids applied to soil were found to survive for 4–12 months, the survival rate depending on the nature of the biosolids and the climatic conditions. It is known that sewage sludges often contain pesticides and other organic chemicals but to date there are no studies of these potential contaminants in sludge-amended soils in Australia.

5.4 Industrial sources

Given the wide range of industrial activities in Australia, it is not possible to consider the impact of all industries on the Australian soil environment in this chapter. Nonetheless, we have attempted to summarise the impact of some major industrial activities on the soil environment and the attempts that research teams have undertaken to manage/remediate contaminated sites. An excellent overview of wastes in Australia and the management of contaminated sites was the subject of the 2nd National Hazardous and Solid Waste Convention. Readers are directed to the proceedings of this conference for an update on industrial waste management, assessment of heavy metal contaminated sites, disposal of heavy metal bearing hazardous wastes and strategies for remediating contaminated sites in Australia. In this section we will attempt to summarise certain urban contaminants together with those arising from industrial activities. In some cities, urban soils were contaminated by pollution arising from former mining and smelting activities. Cities such as Mt Isa, Wollongong, Newcastle, Port Pirie and Broken Hill still have areas which are significantly polluted by current metal industries. Soil contamination arises from the atmospheric transport of smelter stack emissions, the transport of raw materials, dispersal from stored materials such as ore stockpiles, and the redistribution of polluted waste products, for example, in surrounding areas for roads and land levelling.

Many industries other than mining have contributed toxic metals to the urban environment and may still do so. Leather tanning industries, once more widespread than now, have contaminated surrounding land with Cr, As and B (Tiller, 1992). Effluents from tanneries have the potential to cause soil contamination as well as affect air and surface and groundwater quality (AWRC, 1992). The earliest mineral tanning processes used alum and salt, although these were later largely replaced by Cr compounds. The vegetable tanning agents are generally polyphenols, extracted from bark or leaves and produced commercially from woods. Major soil pollutants

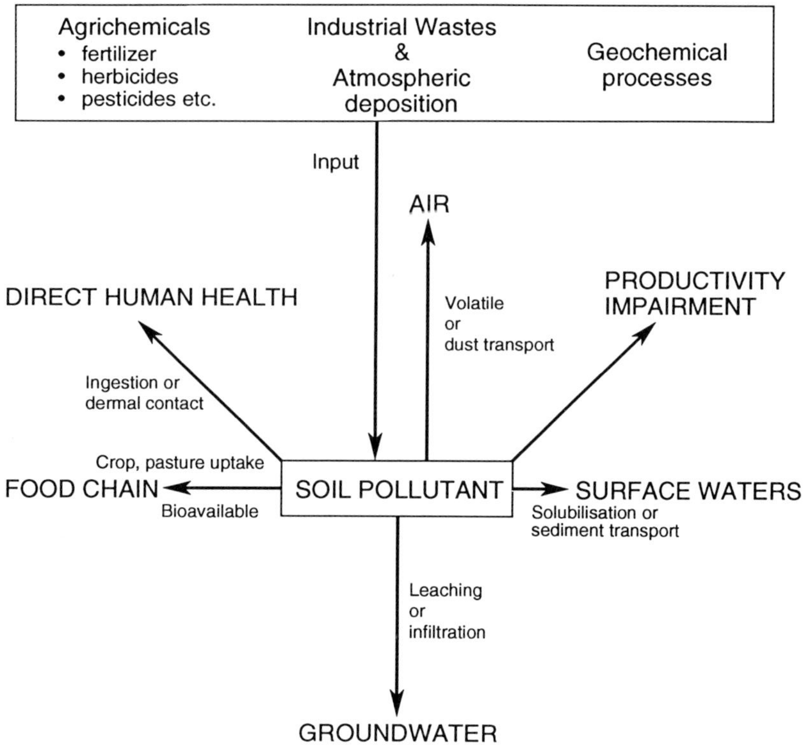

Figure 3. Environmental impact of soil pollutants.

from the tanning process include Na, Cl, pesticides, Mn and Cr and dissolved organic carbon (raising BOD).

Other industries with the potential for contaminating soils with heavy metals are metal electro-plating, the manufacture and recycling of Pb storage batteries, the treatment of timber products with Cr, Cu and As (CCA), chemical industries including paint and fertiliser manufacture, and printing (see Table 1). As discussed in Chapter 6, CCA is the most widely used timber preservative in Australia (Soong and Emmett, 1993). Other chemicals used for this purpose include creosote and pentachlorophenol (PCP). Treated products include fence posts and fencing material, decking, framing, flooring and other timber for residential construction and outdoor play equipment for children. These chemicals are highly toxic and long-lasting and spillage and improper disposal can cause contamination of the soil as well as surface and groundwater. Copper chrome arsenate contamination of groundwater as a result of stormwater discharge from wood treatment facilities has been reported in the south east of South Australia (Soong and Emmett, 1993). Australian Environmental Guidelines for CCA Timber Preservation Plants developed by ANZECC and the Timber Preservers Association of Australia aim to form a national standard for the design of new timber treatment facilities and for upgrading of existing plants.

Public service utilities such as power generation from coal, oil or natural gas, gasworks, major incineration installations, landfill sites, liquid waste disposal facilities and service stations may all produce or disperse contaminants. The escape of contaminants from utilities such as these needs to be taken into account when considering the suitability of sites for housing construction. The CSIRO Division of Water Resources, Perth is currently studying the assessment and *in situ* remediation of groundwater and soils contaminated by organic compounds and wastes. These investigators have developed techniques for characterising, monitoring and modelling hydrocarbon contaminants (e.g. petroleum, chlorinated chemicals and munition compounds) in groundwater and soils (see Chapter 12).

5.5 Mining

Mining and associated activities have also resulted in significant land contamination. Contamination sources include mine tailings, disposal areas, effluents from mineral processing operations and sediment runoff from mining sites. Some frequently occurring residues which pose health and environmental concerns are heavy metals (such as As, Cd, Hg, Mn, Ni and Pb), Mo and S. Sulphide rich materials have the potential to produce acidic water through the oxidation of sulphide minerals in waste dumps and tailings dams while cyanide salts, which are commonly used in the gold mining industry to extract gold from its ores, are also recognised as being highly toxic. The soil contamination issues resulting from mining and associated activities, with special emphasis on soil and water quality, are covered in Chapter 7.

'Capturing Australia's Best Practice Environmental Management in Mining' has been a major initiative of collaboration between the Environmental Protection Agency and the mining industry. As a result, a series of modules have been published which aim to assist all sectors of the industry – minerals, coal, oil and gas – to protect the environment and to reduce the impacts of mining by following the principles of ecologically sustainable development. Best practice environmental management as represented in the modules results in enhanced environmental performance while maintaining competitiveness. The modules are aimed at mining companies, government agencies and environment organisations in both Australia and overseas and cover a large range of topics.

5.6 Production of scheduled wastes

A small percentage of hazardous wastes are produced by some industries. These wastes are regarded as intractable or impossible to dispose of without special technologies and facilities. In Australia these wastes are known as scheduled wastes and include PCBs, hexachlorobenzene (HCB) and organo-chlorine pesticides such as DDT, aldrin, heptachlor and isodrin. Scheduled wastes are:

- organic in nature;
- resistant to degradation by chemical, physical or biological means;

Table 6. Total concentrations of some heavy metals in 'background' surface soils of the Adelaide region and Mt Lofty Ranges: range, median and mean values (mg kg^{-1}).

Element	Adelaide City soils[a]			Soils of adjacent Mt Lofty Ranges[b]		
	Range	Mean	Median	Range	Mean	Median
As	0.2–16	1.8	0.6	<1–8	3.9	-
Cd	0.04–4.3	0.39	0.27	<1	–	–
Cr	2–31	11.4	9.4	5–110	27.0	25.0
Mn	13–1 130	309	273	32–5 200	254	148
Cu	2–128	19.7	15.0	1–59	15	12
Pb	5–212	64	50.5	2–160	21	18
Ni	2–41	14.9	15.0	2–28	9.8	8.0
Zn	12–420	80.3	58.0	4–200	27	18

Source: Tiller (1992) 'Urban Soil Contamination in Australia'.
[a]$n=66$.
[b]$n=185$, except for Cu and Pb, $n=157$; and As, $n=15$.

- toxic to humans, animals, vegetation or aquatic life; and
- bioaccumulative in humans, flora and fauna.

In Australia, the Electricity Supply Industry has been the largest user of PCBs (Independent Panel on Intractable Waste, 1992). Other users include private industry and government bodies (Issues Paper for the PCB Management Plan, 1994). Available data on PCBs suggest that over the past 50 years, approximately 10 000– 20 000 t of PCB's have been brought into Australia (SKM, 1994) and of this at least 3 500–7 000 t is considered to be dispersed in the environment as potential contaminants (Issues Paper for the PCB Management Plan, 1994). Some of this is likely to be found at landfill sites, probably in sealed containers or equipment which may leak over time. In 1993, ANZECC agreed to a national strategy to manage Australia's scheduled wastes. A key aspect of the strategy is the development of management plans for particular scheduled wastes. The strategy aims to:

- establish conditions for the safe management and disposal of scheduled wastes; and
- determine agreed dates by which the use of scheduled wastes are to cease and by which they are to be destroyed.

Where soils are removed and are contaminated by materials which are listed on ANZECC's Schedule X (ANZECC's National Strategy for the Management of Scheduled Waste) and are above the threshold limit specified in the relevant cheduled wastes management plan, then the plan will apply.

In consultation with relevant stakeholders and the community, a Management Plan for PCBs, HCBs and OCPs is being developed. It is expected that, once adopted by ANZECC, these management plans will be implemented in each State or Territory under their laws and regulations.

6. Typical concentrations of contaminants in urban and industrially contaminated soils

Table 6 shows the range, mean and median values of As and seven metals in background soils of the Adelaide metropolitan area and its immediate agricultural hinterland in the Mount Lofty Ranges (Tiller, 1992). Ranges in values are similar for urban and rural soils in the Adelaide region but urban soils clearly have much higher Pb and Zn values. Comparison of the ranges with the ANZECC/NHMRC Guidelines (Table 7) shows that Cd and Pb exceed the background levels for urban soils while only Zn exceeds the levels for both rural and urban soils. These results are supported by those of Olszowy et al. (1993) who found that about 40% of soil samples in Brisbane and Sydney, 10% in Adelaide and 5% in Melbourne taken from residential properties in old areas with high traffic flow had Pb levels which exceeded health and environment-based investigation thresholds. Further, 20% of soils in Brisbane, Sydney and Melbourne sampled from old areas with low traffic flow had Pb levels which exceeded the investigation threshold for Pb. Many of the soil samples from urban areas exceeded the ANZECC/NHMRC environmental investigation levels for one or more elements (As, Cd, Cr, Cu, Mn, Ni, Pb, Zn or Hg). Generally, old suburbs had a higher proportion of samples exceeding the investigation thresholds for Pb, Zn and Cu than did new suburbs. These results demonstrate the impact of urbanisation on toxic and trace metal concentrations of surface soils compared to rural soils. In contrast to metals, there is a dearth of information on toxic organic substances in both rural and urban soil environments. This is of concern given the marked increase in both industrial activities and land based disposals in Australia.

7. Some important initiatives in Australia

There are a number of initiatives which have been implemented in order to develop consistent, national approaches to contaminated sites issues within Australia. However, despite extensive consultation and liaison between Commonwealth, State and Territory Governments, differences exist between the States and Territories in the regulatory arrangements in terms of procedures, controls and standards which apply to the management of contaminated sites.

In order to achieve a consistent national approach to the assessment, remediation and management of contaminated sites in Australia and New Zealand, the ANZECC/NHMRC developed Guidelines for the Assessment and Management of Contaminated Sites (ANZECC/NHMRC, 1992). Table 7 lists the ANZECC/NHMRC background concentrations for a range of inorganic and organic substances and guidelines for concentrations above which site investigation is required. Other issues covered by the Guidelines include preventive measures, site identification, initial evaluation, community consultation, work plan development, occupational health and safety considerations, sampling and analysis, determination of environmental and health impacts, development of site specific guidelines and technology options.

The Guidelines is an evolving document that will be revised and updated as

Table 7. Environmental soil quality guidelines (mg kg^{-1}).

	Background	Environmental investigation
Heavy metals:		
As	0.2–30(a)	20
Cd	0.04–2	3
Cr	0.05–110	50
Mn	4–12600	500
Cu	1–190	60
Pb	<2–200	300
Ni	2–400	60
Zn	2–180	200
Sb	4–44	20
Ba	2–200	
Co	2–170	
Hg	0.001–0.1	1
Sn	1–25	50
Mineral Pollutants:		
Boron	1–75	
Phenolic Compounds:		
Phenols	0.03–0.05	
Monocyclic Aromatic Hydrocarbons (MAH):		
Benzene	0.05–1	1
Toluene	0.1 –1	
Polycyclic Aromatic Hydrocarbons (PAH):		
PAH (total)	0.95–5	
Chlorinated Hydrocarbons:		
PCB (total)	0.02–0.1	1
Pesticides:		
Aldrin	<0.001–<0.05	
Dieldrin	<0.005–<0.05	0.2
DDT	<0.001–0.97	
Other Chemicals:		
Sulphate	35–1000	2000
pH	6–8	

After the *Australian and New Zealand Guidelines for the Assessment and Management of Contaminated Sites*, (ANZECC/NHMRC, January 1992).

more information relating to the Australian environment becomes available, as appropriate criteria and procedures are refined, and as better remediation technologies and strategies are developed. Work is currently under way to develop health and ecological risk assessment frameworks and soil quality criteria that are more related to Australian conditions.

Three national workshops on 'The Health Risk Assessment and Management of Contaminated Sites' have been held in Australia since 1990 as part of the process of reviewing technical and management aspects of contaminated sites issues by Commonwealth, State and Territory health and environmental agencies. The workshop proceedings have been published and are available to the public.

Priorities identified during the workshops include:

- Adopting uniform sampling and analysis protocols and standards by regulatory agencies;
- Developing sampling strategies, including sampling size representative of the contaminated sites;
- Developing protocols, exposure transport models and risk assessment techniques for Australian conditions; and
- Establishing set protocols for specific sites and industries e.g. service stations, tanneries and mines.

A Position Paper on Financial Liability for Contaminated Site Remediation was prepared by ANZECC and released in April 1994. The Position Paper identifies a number of basic principles applying to financial liability, including the 'polluter pays principle', and rules relating to the role and liability of Governments; it also recommends that these principles be adopted. In agreeing to the public release of the paper, ANZECC noted that legislative differences exist between jurisdictions and undertook to review the situation in the future.

A National Pollutant Inventory (NPI) is being developed in Australia to provide a tool for measuring the success of waste minimisation and cleaner production programs. The NPI aims to collate information on pollutants entering the Australian environment as emissions to air, land and water as well as waste disposal to land. It will be publicly accessible, incorporate community rights provisions and will be suitable for analysis and interrogation on geographical, industry, pollutant and environmental media bases. The inventory, however, will not provide a national list of contaminated sites.

There are many other initiatives being undertaken by agencies in this field including the development and revision of legislation concerning contaminated land and the development of a range of documents on contaminated sites issues (e.g. guidelines and procedures for standardising sampling and analytical methods).

7.1 Consultative councils

The key environmental body in Australia is ANZECC, which comprises Commonwealth, State and Territory Government ministers reponsible for environmental and conservation matters.

The key national health body on contaminated sites matters is the NHMRC, which consists of technical experts as well as representatives from Commonwealth, State and Territory Government health agencies.

Table 8. Cadmium levels (mg kg^{-1}) found in food crops analysed in the 1992 AMBS (Stenhouse, 1992).

Food	Number of analyses	Average	Maximum	Minimum
Carrots	24	0.0287	0.0500	0.0080
Cashews			0.0032	*trace *not detected
Celery	24	0.0120	0.0500	0.0060
Grapes, seeded	23	0.0025	*not detected	*not detected
Lettuce	32	0.0140	0.1100	*trace
Liver, lamb	32	0.1634	0.7400	0.0140
Potatoes	32	0.0330	0.0680	0.0130
Tomatoes	24	0.0070	0.0170	*not detected

*trace: <0.0037 mg kg^{-1}; *not detected: <0.0025 mg kg^{-1}.

The Agricultural and Resource Management Council of Australia and New Zealand (ARMCANZ) and the Planning Ministers Council also play important roles in developing national policies and approaches on a range of issues relating to contaminated land.

The Inter-Governmental Agreement on the Environment (IGAE), which was signed by the heads of the Commonwealth, State and Territory Governments and the Australian Local Government Association in May 1992, provides for the establishment of a National Environment Protection Council (NEPC). NEPC is a Ministerial Council which has been established under complementary Commonwealth and State and Territory legislation. It was planned that NEPC would have its first meeting in December 1995.

The establishment of NEPC marks the commitment of the Commonwealth and States and Territories to work co-operatively to develop national environmental protection measures which are legally binding on all participating jurisdictions. NEPC will develop national environment protection measures on a range of issues including air quality, water quality, the assessment of contaminated land, and hazardous wastes. All environment protection measures accepted by a two thirds majority of NEPC Ministers will be legally binding on all jurisdictions. The establishment of NEPC is an important step in the process of harmonising environmental law in Australia to achieve national outcomes.

7.2 Australian Market Basket Survey (AMBS)

The levels of contaminants in food in Australia have been assessed four times per year every two years since the early 1980s by the National Food Authority on a selection of foods purchased at retail outlets in the major cities. The results are reported following consideration of human dietary intake assuming certain food consumption patterns. The concentrations of contaminants in foods are determined on the food as consumed, therefore the levels integrate regional differences in soils, the soil-plant transfer of contaminants, contamination during transport, handling and packaging. Table 8 lists the results for Cd from the most recent published survey (Stenhouse, 1992). Data show that certain food crops exceeded the maximum permitted concentration (MPC) for Cd. Concentrations of Hg and Pb were generally low and neither exceeded the MPC for the majority of food produce (Stenhouse, 1992). Potatoes have been identified as a crop which frequently exceed the MPC which is of concern since they contribute the bulk of human dietary Cd intake in Australia. Some samples of other crops and foods have also been found to exceed the Australian MPC for Cd (e.g. spanner crabs, some oilseed and grain legume crops). This has prompted several research investigations to determine the source of Cd in these foods and to determine if the levels found are part of the background natural range, or due to environmental contamination.

The 1992 food survey also examined five classes of pesticides: chlorinated organic residues, organophosphorus pesticides, synthetic pyrethroids, fungicides (dithiocarbamates, vinclozolin and chloranothalonil) and herbicides (atrazine, 2,4-D, 2,4,5-T, fluazifop-butyl, haloxyfop and simazine). Herbicides were examined for the first time since the commencement of the AMBS. Herbicides are commonly used in conservation tillage practices which have been widely adopted because they minimise soil structure breakdown, reduce soil erosion, and reduce production costs. Organic pesticide residues were either not detected or were present in trace concentrations in the majority of food produce (Stenhouse, 1992).

7.3 The National Residue Survey

The National Residue Survey (NRS) is conducted by the Bureau of Resource Sciences to impartially monitor chemical residues in Australian raw food commodities. The NRS started in the early 1960s as part of a export inspection program operated by the (then) Commonwealth Department of Agriculture. It has grown in size to cover a wide range of crops and produce with over 50 000 analyses performed each year. Sampling and analysis of meat is the major activity of the NRS because of Australia's large export meat trade and commitments to importing countries to maintain quality. Samples of produce are generally taken at centralised locations (e.g. abattoirs, grain receival silos, dairy factories, etc.) rather than on-farm. It differs from the AMBS in that raw commodities are sampled rather than food as consumed (after preparation, cooking, etc.). A wide range of residues are tested – antimicrobials, antihelminthics, insecticides, fungicides and metals (As, Cd, Cu, Hg, Pb, Se, Zn).

Results from the NRS show that 99% of Australian produce meets the current MPCs and maximum residue levels (MRLs) for a wide range of contaminants (BRR, 1992).

Various States in Australia also have monitoring programs for chemical residues in food produce, with the results of these either being widely publicised (e.g. Victoria's Produce Monitoring Program) or the data reported as internal or research reports. The results of most of these indicate Australian produce to be mostly below permitted levels of contamination. Where violations do occur, State authorities trace back the violation to the farm or region and determine the source of the contamination, often in collaboration with Universities and the CSIRO.

7.4 National Water Quality Management Strategy

The National Water Quality Management Strategy (NWQMS) is being developed by ANZECC and ARMCANZ to manage Australia's water quality for surface, ground and coastal waters. The policy objective of the Strategy is to achieve sustainable use of Australia's water resources by protecting and enhancing their quality while maintaining economic and social development. The NWQMS will include national guidelines on specific water resource issues which aim to assist community, catchment managers, environment protection authorities and water authorities in developing local action plans for water quality management. Guidelines for sewerage systems, stormwater management, groundwater protection, water quality guidelines for fresh and marine waters, drinking water, and effluent management guidelines for specific industries are being developed under the NWQMS.

7.5 Water Quality Monitoring

The report *Water Quality Monitoring In Australia,* prepared by Aquatech Pty Ltd for the EPA and released in October 1995, brings together information on water quality monitoring in inland, estuarine, ground and coastal waters in each State and Territory of Australia. The report found that around 1800 water quality monitoring programs exist in Australia. These programs are implemented by groups ranging from private companies and community organisations to Commonwealth, State, Territory and local governments. The Commonwealth Government will provide funding to evaluate water quality data already collected by the various organisations and groups, and to develop a nationwide water quality monitoring system for rivers, lakes, groundwaters, estuaries and coastal waters.

7.5.1 Surface water monitoring

Monitoring of waters from catchments throughout Australia that are near areas of horticulture and intensive agriculture has been on-going for many years. There is no national coordination of activity in this area. The Murray Darling Basin Commission coordinates activities in the Murray Darling Basin in eastern Australia. Much of the information gathered to date relates to salinity and nutrient levels in waters.

At present there is no coordinated monitoring program for organic chemicals in surface waters in Australia. State Departments monitor certain catchment areas where agricultural activity is intensive. For example, the NSW Department of Water Resources DWR(NSW) introduced an intensive pesticide monitoring system in 1990/91 under the Central and North Western Regional Water Quality Program. From 1992 to 1994, 15 different pesticides were detected in surface water samples

(Cooper, 1994). Endosulfan was found to be the most common insecticide due to its wide use in irrigated cotton production and to a lesser extent in oilseed crops. Although the concentration of endosulfan was found to be less than the NHMRC guideline for drinking water (40 mg L^{-1}) it was often found to exceed the ANZECC environmental guideline of 0.01 mg L^{-1} which represents the recommended maximum concentration for the protection of freshwater and marine aquatic ecosystems (ANZECC, 1992). These results suggested extensive surface transport of pesticides.

7.5.2 Groundwater monitoring

A systematic approach to groundwater monitoring is currently operating in Australia to assess groundwater quality on a national scale (Bauld et al., 1992). The available data indicate that although pesticides occur in many groundwaters, their concentrations are well below the MPC for water set by NHMRC. For example, in Victoria, samples from 10 observation wells in September 1990 and March 1991 were analysed for more than 30 pesticides and half of the samples contained atrazine and simazine at concentrations below the NHMRC guidelines. Apart from the occasional appearance of some degradation products of atrazine (deethylatrazine and deisopropylatrazine) no other pesticides were detected.

The EPA of Western Australia reported that, of some 130 borehole samples analysed since 1974, none contained pesticide residues in excess of the MRL, set by NHMRC (EPAWA, 1989). However, the number of samples containing increased levels of pesticides has risen, probably due to their increased use in agriculture and the urban sector. Effluent disposal on land or the use of unlined ponds of agricultural chemicals has been responsible for causing contamination of groundwater in their vicinity (Appleyard, 1993). Further examples of groundwater contamination in Australia have been discussed in Chapters 4 and 12.

As part of the NWQMS, the Guidelines for *Groundwater Management in Australia,* which was planned for release in December 1995, provides a national framework for the protection of groundwater from contamination. The framework work will enable each State, Territory and Commonwealth government to develop policies and strategies which are tailored to their specific needs.

8. Current methods of assessment for contaminated soils (urban and rural)

8.1 Risk assessment

The national framework for the assessment and management of contaminated land adopted by the ANZECC/NHMRC Guidelines recommends two approaches for determining the appropriate clean up levels for contaminated land. The approaches are risk assessment based development of site specific soil quality objectives and use of a conservative set of soil quality objectives. Both of these approaches aim to ensure a consistent and uniform approach to the assessment, management and clean-up of contaminated land. The Australian Model for Risk Assessment for Contaminated sites has been detailed in Chapter 9.

Risk assessment uses formal quantitative techniques to estimate probabilities of effects on well-defined endpoints, provides a means for estimating uncertainties, and separates the analysis of risks from decision making regarding the significance of risks and the choice of actions (Suter II, 1993). The ANZECC/NHMRC Guidelines define risk as 'the probability that an adverse outcome will occur in a person, a group, or an ecological system that is exposed to a particular dose or concentration of a hazardous agent and depends on both the level of toxicity of hazardous agent and the level of exposure'.

The Guidelines define the goals of contaminated site assessment and management as:

- rendering a site acceptable and safe for the long term continuation of its existing use;
- minimising environmental and health risks both on and off site; and
- maximising the potential future use of a site.

The ANZECC/NHMRC Guidelines provide a recommended approach (Figure 4) to the assessment of a potentially contaminated site. The contaminated land risk assessment process attempts to estimate the likelihood of an adverse effect occurring in a receptor due to the presence of contaminants in soil.

8.2 Development of environment and public health based guidelines

Separate health and environmental investigation levels have been established in the ANZECC/NHMRC Guidelines to take into account the different sensitivities of humans and biota in the environment. Site specific decisions need to be made to determine whether health or environmental investigations (or both) should apply.

Investigation criteria for soils provide a trigger for further detailed investigation of a site, should the concentration of a contaminant be above the background level. Issues that need to be considered when assessing the significance of contaminants above investigation levels include: potential groundwater contamination; current and proposed land use; the history and nature of the contamination; local background levels and the size of the site. The 'environmental investigation values' in the ANZECC/NHMRC Guidelines are based on threshold levels for:

- phytotoxicity; and
- uptake of contaminants which may result in impairment of plant growth or reproduction or unacceptable residue levels.

Where runoff of contaminants into surface waters is an issue, the estimated concentration can be compared with water quality standards and an acceptable level can be calculated based on estimated transport rates (ANZECC/NHMRC, 1992).

8.3 Determining the nature and extent of contamination

The ANZECC/NHMRC Guidelines propose that once an initial investigation of a site is complete and the site is considered to warrant further investigation, a site

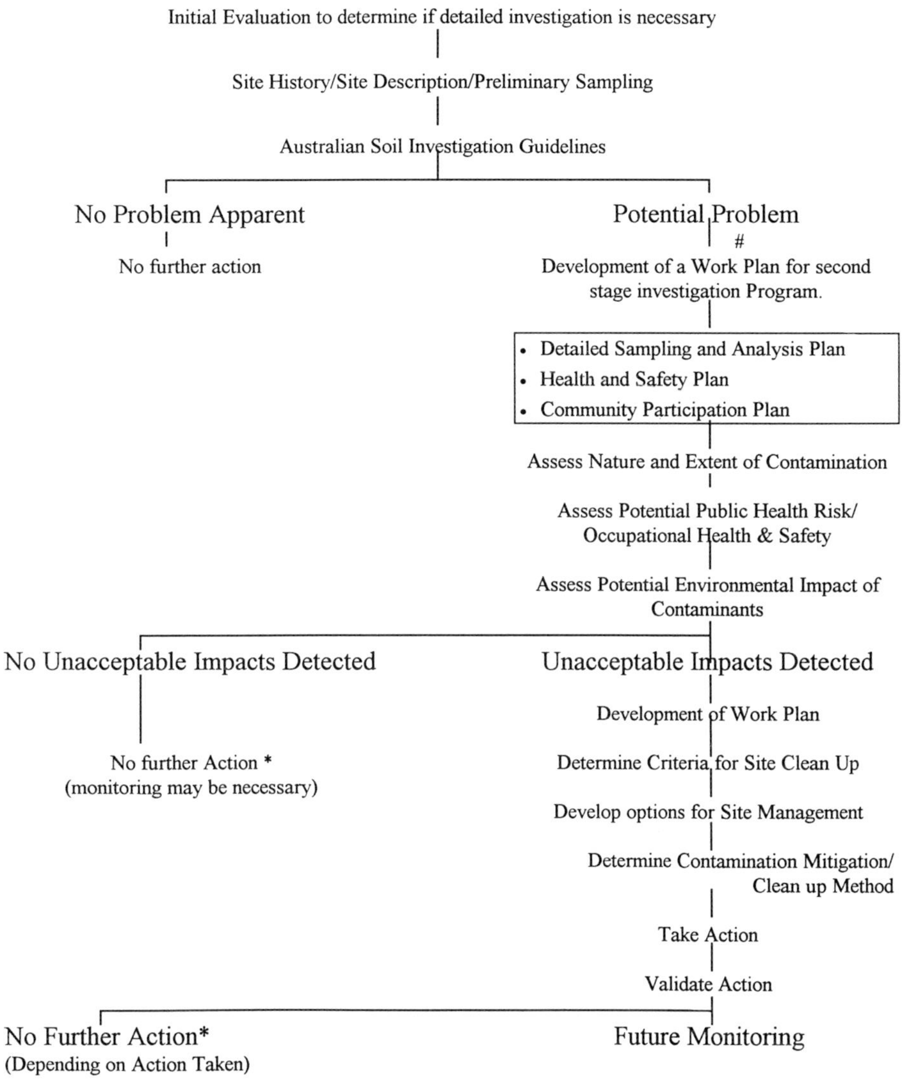

decision to proceed directly to clean up according to guidelines may be taken at a number of points in the following process sequence.

* provided land use remains as originally proposed.

Figure 4. Recommended approach to the assessment and management of a potentially contaminated site.

specific work plan should be developed which defines the scope of work, level of investigation, costs and schedules for performing site assessment activities and which includes site specific information for collection. Some specific aspects which should be considered are:

- physico-chemical characteristics of contaminants (speciation, mobility, solubility etc);
- soil characteristics such as pH and clay content;
- site characteristics (likelihood of exposure to contaminants in air, water and soil both on and off-site);
- identification of exposure routes and receptors, e.g. plants, water bodies;
- toxicity of contaminants to flora and fauna, both terrestrial and aquatic;
- bioaccumulative and biomagnification capacity;
- bioavailability of the contaminant to the range of receptors in a given ecosystem;
- the health of the environment with respect to biological species distribution and number (including microbes);
- rates of microbial degradation of contaminants as these may affect productivity and quality of the soil.

The site history, existing conditions, details of past industrial, commercial or agricultural operations and a description of the sites are also compiled to assist in the design of sampling and analytical programs and in the preliminary assessment of exposure risk to enable the initial identification of potential remedial technologies.

8.4 Sampling and Analysis

Following confirmation of the presence of contamination, further work is required to collect data to estimate the extent of contamination, the location of contaminant sources on and off site, the identification of migration pathways and the extent of off-site movement, as a means of determining whether a contaminated site poses a long term threat to public health or the environment (ANZECC/NHMRC, 1992). The sampling and analysis program adopted will vary with the specific land use of the site and should incorporate available knowledge of the types and locations of industrial activities previously undertaken at the site. In cases where little is known about the site's history, a simple grid pattern or stratified random sampling plan may be the most appropriate, with a comprehensive range of analyses being applied to all samples from two or more depths throughout the entire site (Good, 1993).

Soil type(s) will have a major influence on the behaviour of the contaminant of interest. For this reason, it is recommended that fundamental soil information such as soil type, pH, salt content, permeability, porosity, texture, cation/anion exchange capacity, percent organic material, and moisture content be obtained (ANZECC/NHMRC, 1992). Other physical factors also need to be taken into consideration. These include the solid-carrying capacity of a liquid, liquid stratification, temperature gradients, solid distributions based on particle size, density or surface characteristics and any other contributing parameters.

The draft document Australian Standard Analysis of Soils – The sampling of potentially contaminated soil, prepared by Committee CH/28 on the Analysis of Soils and Biota provides a general overview for soil sampling. ANZECC has released a document Guidelines for the Analysis of Contaminated Soils (June 1995) that aims to provide consistent procedures for soil analysis. These guidelines cover quality assurance procedures, techniques for sample preparation, and methods for the analysis of physico-chemical properties, nutrients, metals, inorganics and organics in soil. These draft guidelines will be updated as expertise in the analysis of contaminated soil develops.

9. Current management strategies and remediation of contaminated soils

Remediation needs and strategies for contaminated sites need to be site specific and the ANZECC/NHMRC Guidelines have identified a hierarchy of options for the clean up and management of contaminated sites. The hierarchy includes the following options (in order of preference):

- on site treatment of contaminants – so that the contaminant is either destroyed or the associated hazard is reduced to an acceptable level;
- off site treatment of excavated soil – which, depending on the residual levels of contamination in the treated material, is then returned to the site, removed to an approved waste disposal site or facility or used as fill for landfill;
- removal of contaminated soil to an approved site or facility – followed, where necessary, by replacement with clean fill;
- isolation of the soil – by covering with a properly designed barrier;
- choosing a less sensitive land use – to minimise the need for remedial works;
- leaving the contaminated material on site – providing there is no immediate danger to the environment or community and appropriate controls are in place.

Contaminated site inventories contain information on sites which assists in their effective management through such things as assessing accurately the extent of necessary clean up needs, detecting environmental and health hazards, and identifying opportunities for creating a market for remediation technologies. Many States in Australia now maintain, or are developing, databases to identify sites which may be potentially contaminated (Hosking, 1993b). Some databases may be more formal, such as publicly accessible registers established under legislation.

The ANZECC/NHMRC Guidelines set out the fundamental goals of contaminated site clean up which, in summary, should be to select a socially acceptable and cost effective management strategy which mitigates threats to, and provides protection for, public health and the environment as well as allowing flexibility in the future use of land. For example, when human health and the environment are at risk, the ANZECC/NHMRC Guidelines indicate that a contaminated site should be cleaned up to the extent necessary to minimise both short and long term risks. Where there is no threat to human health or the environment it may be appropriate to clean-up the site to some lesser degree, and in some cases to accept a strategy of containing contaminants on the site, or to impose planning controls to limit site uses. Considerations of technical feasibility and of net social benefit should always play a part in determining the clean-up strategy adopted for a particular site. Some other factors that need to be taken into account include:

- the time required to achieve adequate control of contamination;
- the efficacy of the management strategy in reducing the harmful effects of contamination;
- engineering feasibility;
- the cost, including the cost to the community of creating and maintaining restricted land use zones;
- the availability of effective administrative/regulatory mechanisms for ensuring that controls are adequately performed.

In general, the most widely used strategy for dealing with contaminated soil in Australia is to excavate and transfer contaminated soils to a well managed landfill (see Chapter 13). In many cases, however, the option of on site treatment should be promoted for the following reasons:

- potential hazard associated with the transport of contaminated soil;
- limited available secure landfilling space in Australia;
- potential future liability for the material taken to landfill;
- the approach does not take advantage of available contamination mitigation techniques.

One of the major influences on the selection of a clean up response for contaminated soils in Australia is the availability and cost of landfill containment (Chapter 13; Shoebridge, 1993). Often the cost of disposing low level contaminated soil to landfill is low compared to the cost of most alternative on-site remediation approaches. However, the cost of transporting soil and replacing it with clean soil in some cases is sufficiently high to make the 'dig up and take away' approach less attractive. Further, the problems experienced in siting new landfills, and the historical problems associated with current landfill, means that it is likely that alternative technologies will continue to become more important in the future.

Obviously the appropriate choice of on site clean up methods will depend on factors such as the nature of the contaminants and the soil characteristics, as well as the ready availability of the technology.

9.1 Site monitoring

It is important that a site is monitored after remediation or management strategy has been implemented to ensure the adequacy of the measures adopted. Where possible, the ANZECC/NHMRC Guidelines recommend that monitoring data should be compared with the pre-determined clean up criteria or site specific clean up criteria.

9.2 Legislation and regulations

Legislation dealing specifically with contaminated land has existed in Australia for about ten years. During this time all State and Territory jurisdictions have enacted or have proposed legislation to deal with contaminated land. Regulations governing the use of agricultural chemicals generally are administered by State and Territory Agriculture and/or Health Departments (Bates, 1992). Each State has some form of pesticide advisory committee which advises it on the registration of new agricultural fertilisers and chemicals. No pesticide can be sold without being registered first and controls are placed on their composition, advertising, sale and use.

Regulatory frameworks relating to the protection of the environment from the adverse effects of mining and mineral processing are increasing in number and complexity. Governments generally set maximum permissible discharge levels or minimum levels acceptable for environmental quality. Mining operation and environment protection requirements (such as waste disposal, water quality controls, rehabilitation and occupational health and safety) are most commonly implemented through a range of legal instruments such as mining legislation, environmental planning and assessment legislation, environment protection legislation and other legislation including occupational health and safety.

At present, effluent discharges are regulated by State and Territory environmental protection authorities. This is achieved by applying standards for effluent which form part of the licence conditions to discharge the effluent (AWRC, 1992).

10. Priority contaminant issues and future directions

When the ANZECC/NHMRC Guidelines were released in January 1992 it was recognised that 'further development and periodic update' of the Guidelines was 'essential'. Research needs to be undertaken in Australia to develop criteria and guidelines appropriate for specific Australian conditions. Priority research areas include:

- Development of Australian risk assessment models;
- Research and development of appropriate and innovative technology for the treatment, handling and disposal of contaminated soil;
- Determining background levels of chemicals in Australian rural and urban soils;
- Assessing the bioavailability of chemical contaminants in soil; and
- Fate and behaviour of inorganic and organic contaminants in the soil environment.

11. Conclusions

A wide range of contaminants associated with past agricultural, industrial and commercial activities within Australia has the potential to impact on human health and the environment. As with most countries, the common contaminants in Australia include heavy metals, nutrients and organic substances.

Urban expansion into former rural areas which were formerly subject to usage of chemicals for agricultural purposes is a key contaminated sites issue in Australia, along with issues emerging through the rezoning and redevelopment of urban land from industrial or other purposes to more sensitive uses.

While environmental protection, including the management of contaminated sites, rests with the individual States and Territories, the ANZECC/NHMRC Guidelines for the Assessment and Management of Contaminated Sites aim to provide a consistent, national framework for the assessment, remediation and management of contaminated sites in Australia.

As part of the process of revising the ANZECC/NHMRC Guidelines, policies, systems and criteria specific to Australian conditions are currently being developed to address a range of technical and socio-economic issues. The technical issues include criteria for identifying contaminated land, sampling and analysis protocols, innovative remediation techniques and validation procedures. Socio-economic issues include such things as the need for full community involvement and awareness of contaminated sites issues.

Resolution of both technical and socio-economic issues is dependent on the development of effective processes and protocols for analysis, risk assessment and communication.

Contaminated land has become an increasingly important environmental, health, economic and planning issue in Australia. The various initiatives and approaches developed to date for dealing with these issues demonstrate Australia's willingness and capacity to ensure that effective and best management practices for contaminated sites are adopted.

References

Anon 1994 The Algal Management Strategy for the Murray Darling Basin. Murray Darling Basin Commission, Canberra. 33p.

ANZECC/NHMRC 1992 Australian and New Zealand Guidelines for the Assessment and Management of Contaminated Sites. 57 p.

Appleyard S J 1993 Resourcing the investigation and management of groundwater contamination in Western Australia – a case study of contamination from herbicide manufacture near Perth. AGSO J Aust. Geol. Geophys. 14, 177-182 p.

AWRC 1992 Draft Guidelines for Sewerage Systems – Effluent Management. Australian Water Resources Council. August 1992. 36 p.

Bailey G W and White J L 1970 Factors influencing the adsorption, desorption and movement of pesticides in soils. Residue Rev. 32, 29-92 p.

Bates G M 1992 Environmental Law in Australia. 3rd edn, Butterworths, Sydney. 405 p.

Bauld J, Evans W R and Sandstrom M W 1992 Groundwater quality under irrigated agriculture, Murray Basin, southeastern Australia. In Groundwater and Emvironment. Proceedings of International Workshop on Groundwater and Environment. Ed. Fei Jin. pp 447-457. Seismological Press, Beijing.

Beard J 1993 The Evaluation of DDT Contaminated Soil Associated with Cattle Tick Dip Sites. In The Health Risk Assessment and Management of Contaminated Sites – Proceedings of the Second National Workshop on the Health Risk Assessment and Management of Contaminated Sites. Eds. A Langley and M van Alphen. pp 119-133. South Australian Health Commission.

Bentley K 1993 Harmonising Health and Environment Criteria: Health Based Criteria. In The Health Risk Assessment and Management of Contaminated Sites – Proceedings of the Second National Workshop on the Health Risk Assessment and Management of Contaminated Sites. Eds. A Langley and M van Alphen. pp 1-3. South Australian Health Commission.

BRR 1992 Bureau of Rural Resources. Report of the National Residue Survey 1989-1990 Results. Australian Government Printing Service, Canberra.

Calder I 1991 Directions for the Assessment of Contaminated Land. In The Health Risk Assessment and Management of Contaminated Sites – Proceedings of a National Workshop on the Health Risk Assessment and Management of Contaminated Sites. Eds. O El Saadi and A Langley. pp. 6-9. South Australian Health Commission.

Chaney R L and Ryan J A 1993 Heavy metals and toxic organic pollutants in MSW-composts: Research results on phytoavailability, bioavailability, fate, etc. In Science and Engineering of Composting: Design, Environmental, Microbiological and Utilization Aspects. Eds. H A J Hoitink and H M Keener. pp. 451-506. Ohio State University, Ohio.

Chiou C T, Malcolm R L, Brinton T I and Kile D E 1986. Water solubility enhancement of some organic pollutants and pesticides by dissolved humic and fulvic acids. Environ. Sci. Technol. 20, 502-508 p.

CMPS&F 1994 Appropriate Technologies for the Treatment of Scheduled Wastes. EPA, Commonwealth of Australia, Canberra, ACT.

Cooper B 1994 Central and North West Regions Water Quality Program. 1993/94 Report on Pesticide Monitoring. NSW Department of Water Resources TS94.087.

de Vries M P C and Tiller K G 1978 Sewage sludge as a soil amendment, with special reference to Cd, Cu, Mn, Ni, Pb and Zn. Comparison of results from experiments conducted inside and outside a glass-house. Environ. Pollut. 16, 231-240 p.

Enfield C G and Yates S R 1990 Organic chemical transport to groundwater. In Pesticides in the Soil Environment: Processes, Impacts and Modelling. Ed. H H Chemg. pp 79-101. Soil Science Society of America, Madison, Wisconsin.

EPA (WA) 1989 Monitoring Pesticides – A review. The Environment Protection Authority, Perth, Western Australia, Bulletin No. 407.

Forrest W 1993 Harmonising Health and Environment Criteria: Environment Based Criteria. In The Health Risk Assessment and Management of Contaminated Sites – Proceedings of the Second National Workshop on the Health Risk Assessment and Management of Contaminated Sites. Eds. A Langley and M van Alphen. pp 5-7. South Australian Health Commission.

GHD 1992 An investigation of nutrient pollution in the Murray-Darling River System. Report prepared by Gutteridge Haskins and Davey. Murray Darling Basin Commission, Canberra.

Glotfelty D E and Schomburg C J 1989 Volatization of pesticides from soil. In Reactions and Movement of Organic Chemicals in Soils. Ed. B L Sawhney and K Brown. pp 181-207. Soil Science Society of America, Madison, Wisconsin.

Good J 1993 Towards Standardised Sampling and Analysis for Contaminated Sites. In The Health Risk Assessment and Management of Contaminated Sites – Proceedings of the Second National Workshop on the Health Risk Assessment and Management of Contaminated Sites. Eds. A Langley and M van Alphen. pp 43-49. South Australian Health Commission.

Harrison J 1994 Review of nutrients in irrigation drainage in the Murray-Darling basin. CSIRO Division of Water Resources Series No. 11. CSIRO Division of Water Resources, Canberra. 149p

Hodgkin E P and Hamilton B H 1993 Fertilizers and eutrophication in southwestern Australia: setting the scene. Fert. Res. 36, 95-103.

Hosking I 1993a: A Brief Overview on the Assessment and Management of Contaminated Land in Australia – Current Issues. In The NATO/CCMS Pilot Study on Soil and Groundwater Remediation, November 1993. Coffey Partners International Pty Ltd.

Hosking I 1993b: Country reports: Australia. In The NATO/CCMS Pilot Study on Research, Development and Evaluation of Remedial Action Technologies for Contaminated Soil and Groundwater (December 1993). Report LR 986. Warren Spring Laboratory Gunnels Wood Rd, Stevenage, Hertfordshire. pp. 31-32.

Hurle K and Walker A 1980 Persistence and its prediction. In Interactions between Herbicides and the Soil. Ed. R J Hance. pp 83-122. Academic Press, London.

Independent Panel on Intractable Waste 1992 In our backyard. Independent Panel on Intractable Waste, Paddington.

Issues Paper for the PCB Management Plan 1994 Report prepared by CMPS&F Environmental, Scheduled Wastes Management. 73p.

Koh T S and Judson G J 1986 Trace-elements in sheep grazing near a lead-zinc smelting complex at Port Pirie, South Australia. Bull. Environ. Contam. Toxicol. 37: 87-95.

Koskinen W C and Harper S S 1990 The retention process: mechanisms. In Pesticides in the Soil Environment: Processes, Impacts and Modelling. Ed. H H Chemg. pp 305-349. Soil Science Society of America, Madison, Wisconsin.

Langley A 1993 Future Directions for the National Guidelines. In The Health Risk Assessment and Management of Contaminated Sites – Proceedings of the Second National Workshop on the Health Risk Assessment and Management of Contaminated Sites. Eds. A Langley and M van Alphen. pp 335-342. South Australian Health Commission.

McLaughlin M J, Tiller K G, Naidu R and Stevens D P 1995 Review: The behaviour and environmental impact of contaminants in fertilizers. Aust. J. Soil. Res. 34, 1-54.

Merry R H and Tiller K G (1991) Distribution of cadmium and lead in an agricultural region near Adelaide, South Australia. Water, Air Soil Poll. 57-58, 171-180.

Merry R H, Tiller K G, de Vries M P C and Cartwright B 1981 Contamination of wheat crops around a lead-zinc smelter. Environ. Pollut. (Ser. B) 2, 37-48.

Ministry of Housing, Spatial Planning and the Environment (Netherlands) 1994b Environmental Quality Objectives in the Netherlands – a review of environmental quality objectives and their policy framework in the Netherlands. 465 p.

Naidu R, Smith J, Smith L H, Tiller K G and McDougal K W 1995 Arsenic/DDT residues at cattle tick dip contaminated sites: Preliminary investigation: CSIRO Division of Soils Technical Report No 84.

Olszowy H, Torr P, Imray P, Smith P, Hegarty J and Hastie G 1993 Report on Studies of Levels of Trace Elements in Soils from Rural and Urban Areas of Australia.

The Readers Digest Atlas of Australia 1994 Readers Digest (Australia) Pty Ltd, Sydney.

Reichelt L 1991 The Political Context of the Assessment of Contaminated Sites. In The Health Risk Assessment and Management of Contaminated Sites – Proceedings of a National Workshop on the Health Risk Assessment and Management of Contaminated Sites. Eds. O El Saadi and A Langley pp. 10-14. South Australian Health Commission.

Ross A D, Lawrie R A, Whatmuff M S, Keneally J P and Awad A S 1992 Guidelines for the use of sewage sludge on agricultural lands. N.S.W. Department of Agriculture, Sydney.

Seidler S 1993 Initial Observations: some key concepts and the prevention of harm from contaminated sites. In Contaminated Sites in Australia. Eds. G C Rowe and S Seidler. 3-8 p. Allen & Unwin and Australian-American Educational Foundation.

Shoebridge L 1993 A Review of Soil Remediation Practice in Australia. In The Health Risk Assessment and Management of Contaminated Sites – Proceedings of the Second National Workshop on the Health Risk Assessment and Management of Contaminated Sites. Eds. A Langley and M van Alphen. pp. 223-227. South Australian Health Commission.

SKM 1994 Sinclair Knight Merz. PCB background paper prepared for the Scheduled Waste Management Group on behalf of ANZECC.

Soong F S and Emmett A 1993 Assessment and Management of CCA Timber Preservation Plants. In The Health Risk Assessment and Management of Contaminated Sites – Proceedings of the Second National Workshop on the Health Risk Assessment and Management of Contaminated Sites. Eds. A Langley and M van Alphen. pp 229-246. South Australian Health Commission.

South Australian Health Commission 1994 Identification and Assessment of Contaminated Land. Eds. J W Edwards, M van Alphen and A Langley. 538 p.

Stenhouse F 1992 The 1992 Australian Market Food Basket Survey. National Food Authority. Australian Government Printing Service, Canberra. 96 p.

Suter II G W 1993 Ecological Risk Assessment. Lewis Publishers. 538 p.

Swaine D L 1962 The trace-element content of fertilizers. Commonw. Bur. Soils Tech. Commun. No. 52.

Taylor J K 1983 The Australian Environment. In Soils: An Australian Viewpoint. Ed. CSIRO. pp 1-2. CSIRO Publications, Melbourne, Australia.

Tiller K G 1989 Heavy metals and their environmental significance. In Advances in Soil Science Vol 9. Ed. B A Stewart. pp 113-142. Springer-Verlag, New York.

Tiller K G 1991 Determining Background Levels. In The Health Risk Assessment and Management of Contaminated Sites – Proceedings of a National Workshop on the Health Risk Assessment and Management of Contaminated Sites. Eds. O El Saadi and A Langley. pp 98-101. South Australian Health Commission.

Tiller K G 1992 Urban Soil Contamination in Australia. Aust. J. Soil Res., 30, 937-57.

Tiller K G, Smith L H, Merry R H and Clayton P M 1987. Aust. J. Soil Res. 25, 155-166.

Tiller K G, Oliver D P, McLaughlin M J, Merry R H and Naidu R 1995 Managing cadmium contamination of agricultural land. In Remediation of Soils Contaminated by Metals. Eds. I K Iskandar and D C Adriano. (in press). Science and Technology Letters, Northwood, Middlesex.

Torr P and van Alphen M 1993 Background Urban Levels of Heavy Metals in Surface Soils. In The Health Risk Assessment and Management of Contaminated Sites – Proceedings of the Second National Workshop on the Health Risk Assessment and Management of Contaminated Sites. Eds. A Langley and M van Alphen. pp 69-88. South Australian Health Commission.

United Nations Environment Program 1994 Environmental Management of Mine Sites. Technical Report No 30.

Valentine R L 1986 Nonbiological degradation in soil. In Vadose Zone Modeling of Organic Pollutants. Eds. S M Hern and S Meloncon. pp 223-243. Lewis Publishers, Chelsea, Michigan.

Wild A 1958 The phosphate content of Australian soils. Aust. J. Agric. Res. 9, 193-203.

Williams C H and David D J 1973 The effect of superphosphate on the cadmium content of soils and plants. Aust. J. Soil Res. 11: 43-56.

15
Contaminants and the soil environment in China

G. L. JI AND T. R. YU

1. Introduction

China is a large country with a territory covering 50° N in latitude and 60° in longitude. The area extends from cold temperate regions in the north to subtropical and tropical regions in the south. It has the highest mountains in the world – the Himalayas. The annual precipitation ranges from less than 50 mm in desert regions to over 2 000 mm along seashores. Such a wide geographical range and huge area result in great diversity of geology, climate and related vegetation. As a consequence, nearly all the principal types of soils of the world, ranging from tundra soils to intensively weathered latosols (Oxisols) can be found in the country.

China has a long history of agricultural production, and the influence of mankind on the soil is intense. In many regions, particularly on the loess plateau in the northwest and in rice-growing areas in the southeast, the top one metre of the upper soil layer comprises material resulting from the long-term accumulation of manures and irrigation deposits. However, because these materials themselves are natural, soil contamination with harmful elements was not apparent in the past.

Following the rapid development in industry in the last two or three decades, pollution has become an increasingly serious problem, especially in the vicinity of large cities and big industrial enterprises. However, on a nationwide scale, the affected area is not large. This problem has attracted the attention of both scientists and the government. National and local soil monitoring and research institutions have been established, and significant funds have been invested in solving the problem.

Broadly speaking, contaminants in soils may be classified into three groups: inorganics, organics and contamination resulting from acid precipitation. In the present paper, only the inorganic and organic contaminants will be discussed. Problems with acid precipitation in China have recently been dealt with in another paper (Ji et al., 1995).

2. Soil contamination status in China

There are three main causes of soil contamination in China: irrigation with wastewater; accumulation of industrial and mining wastes; and the application of agricultural chemicals. It has been estimated that more than 30 industrial districts amounting to an area >500 000 ha have been polluted due to irrigation with wastewater, and 7.15×10^{10} t of tailings and slag has been accumulated in 75 industrial cities up to 1980 (Institute of Soil Science, 1990). Agricultural chemicals including 4,4-dichlorodiphenyltrichloroethane (DDT), hexachlorocyclohexane (666) and

R. Naidu et al. (eds): Contaminants and the Soil Environment in the Australasia-Pacific Region, 485–499.

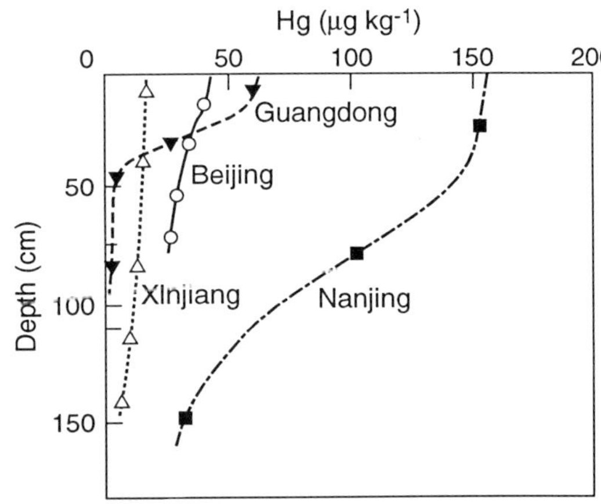

Figure 1. Distribution of total mercury within soil profiles (Dai and Huang, 1983).

chloral have also caused soil contamination, although the production and application of some of these chemicals have been prohibited by the government. Among inorganic elements that have been of concern are Hg, Cd, Pb, Cr, Zn, Cu, Co, F and As. In this paper, the status of soil contamination by some of these materials will be briefly discussed.

2.1 Mercury

Mercury may be present as HgS, $HgCO_3$, $HgHPO_4$, $HgCl_2$ and $Hg(NO_3)_2$ in soils. Organically-bound Hg may exist in soils in forms that include methyl Hg, phenyl mercuric acetate and complexes of Hg with organic matter. Sources of Hg pollution include: mining and refining of Hg-containing ores; industrial utilisation of metallic Hg; production and use of inorganic Hg compounds; gases emitted from burning fuels; Hg-polluted irrigation water and industrial sludge; and land adjacent to industrial plants.

The distribution of Hg within four soil profiles is shown in Figure 1. The concentration of Hg decreases with depth to a relatively constant value which varies with the origin of the soil profile. The distribution pattern differs with the type of the soil. The profile in Xinjiang region may be regarded as uncontaminated because of anthropogenic disturbance. The top meter of the soil profile from Nanjing has very high levels of Hg compared to background levels of other regions (Table 1). Viewed on a nationwide basis, Hg-contaminated soils are limited to isolated areas.

Research on soils and crops from these isolated areas suggests that Hg pollution in China may be separated into three groups. These include those regions where: (i) the Hg content of both rice and wheat does not exceed the critical food standard value of $20\,mg\,kg^{-1}$ dry weight even though the soil has been contaminated; (ii) rice grain is contaminated ($>20\,mg\,kg^{-1}$) while upland crops such as wheat and

Table 1. Hg contents ($mg\,kg^{-1}$) of Hg-polluted soils and background values in some regions.

Locality	No. of samples	Polluted soil mean	Background value* Mean ($mg\,kg^{-1}$)	S.D.	Source
Jilin	27	0.227	0.037	0.0290	She and Meng, 1982
Zhengzhou	50	0.069	0.034	0.0172	Wang and Hou, 1992
Xi'an	12	0.58	0.030	0.0265	Li, 1979
Lanzhou	209	0.134	0.020	0.0186	Ho et al., 1987
Guilin	97	0.16	0.152	0.1737	Ling, 1993

*Environmental Monitoring Center of China, 1990.

Table 2. Cd contents ($mg\,kg^{-1}$) of Cd-contaminated soils and background values in some regions.

Locality	No. of samples	Polluted soil mean	Background value* Mean ($mg\,kg^{-1}$)	S.D.	Source
Shenyang	37	1.62	0.108	0.0664	Wu et al., 1980
Shenyang (irrig.)	–	1.11	0.108	0.0664	Wu et al., 1980
Beijing	–	0.75	0.074	0.0584	Xia, 1982
Zhengzhou	50	0.223	0.074	0.0167	Wang and Hou, 1991
Lanzhou	209	0.619	0.116	0.0361	Ho et al., 1987

*Environmental Monitoring Center of China, 1990.

corn are not contaminated; (iii) both rice and upland crops are contaminated (>20 $mg\,kg^{-1}$) (Institute of Soil Science, 1990). However, the extent of contamination of these crops is only limited to small localised areas.

2.2 *Cadmium*

The Cd content of soils varies greatly, depending on the parent material. For example, for yellow-brown earths (Hapludalf) in the Nanjing region, the Cd content ranges from 0.08 to 0.75 $mg\,kg^{-1}$. The Cd content of soils contaminated with wastewater irrigation and background values in a number of regions are summarised in Table 2. The data show that in some regions the concentration of Cd in soils is extremely high. For example, in the Shenyang district, irrigation of soils with wastewater originating from the metallurgical industry has resulted in the total Cd content that is ten times higher than the background value. Such heavy contamination of soils with Cd has implications for crop quality. For example it has been noted that in a region of north China the Cd concentration in brown rice exceeded 0.4 $mg\,kg^{-1}$, whilst the total Cd content of paddy soil was 3–7 $mg\,kg^{-1}$. Similar incidences have been recorded in other regions in south China, where the average total Cd contents of the contaminated soil and husked rice were 2.52 $mg\,kg^{-1}$ and 0.87 $mg\,kg^{-1}$, respectively (Institute of Soil Science, 1990). In these regions the average total Cd content of the uncontaminated soil was 0.13 $mg\,kg^{-1}$. The general pattern of distribution of Cd in soils is that the farther the distance from the pollutant source the lower the Cd

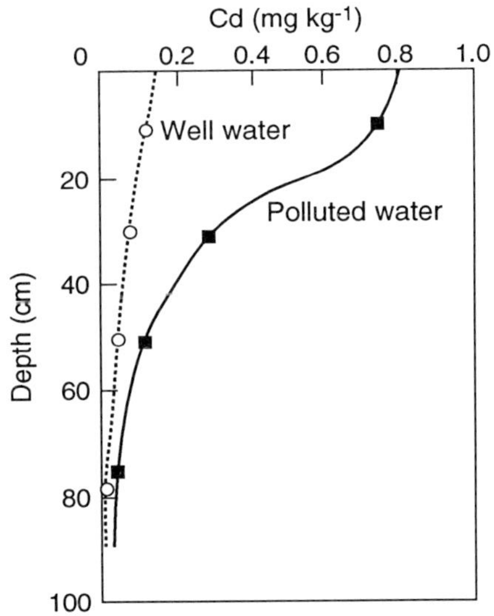

Figure 2. Distribution of cadmium within soil profiles irrigated with well water and with Cd-polluted water (Xia, 1982).

content. This trend, however, is not surprising given the current waste disposal practices that include effluent reuse and this is best practised close to industries.

The distribution of Cd in a soil profile irrigated with contaminated water and well water is summarised in Figure 2. Most of the Cd was accumulated in the upper 0–20 cm of the profile, although in this profile subsurface soils at 40–60 cm depth also showed appreciable accumulation of Cd. This is in contrast to the general belief that water-soluble Cd, following entry in to the soil environment, is rapidly sorbed and thus does not easily leach. However, the possibility of downward leaching of Cd under the influence of heavy irrigation, especially with Cd-polluted water rich in organic matter (Figure 2) deserves further investigation.

2.3 Chromium

The source of Cr in soils could be the parent material itself. Total Cr content of soils developed from a range of parent materials is shown in Table 3. This data shows that the Cr content of soils is closely related to the parent material.

In general, soils derived from basic rocks (e.g. basalt) have a higher Cr content than those derived from acid rocks.

Soil genetic processes may also affect the distribution of Cr within a soil profile. Chromium contents in different layers of Haplustalf soils in the Beijing region are summarised in Figure 3. For these soils, the Cr content increases markedly with depth (Figure 3). This may be due to illuviations of clay during soil genesis.

However, in contaminated soils, Cr principally accumulates in the cultivated layer. For example, Mu et al. (1982) showed that in a heavily polluted alluvial soil

Table 3. Total Cr content of soils derived from different parent materials (Institute of Soil Science, 1990).

Soil	Parent material	Depth	Cr
		cm	$mg\,kg^{-1}$
Kanhapludult	Granite	0–11	7.5
(Lateritic red earth)	Quartz-sandstone	0–35	17.9
	Purplish sandstone	0–18	45.0
	Sandstone	0–55	112
Hapludalf	Basalt	0–10	130
(Yellow-brown earth)	Olivine gabbro	0–16	548
Paleudult (Red earth)	Quaternary clay	Subsoil	91.9
Ustochrept (Loessial soil)	Loess	0–20	73.7
Haploboroll (Chernozem)	Deposits	0–20	62.9
Aquic Eutrochrept (Fluvo-aquic soil)	Alluvial	0 –24	64.4

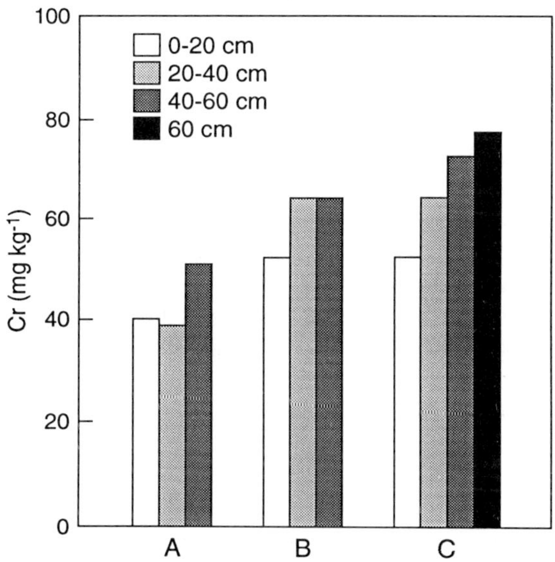

Figure 3. Chromium contents in different horizons of eluvial (A), typical (B) and meadow-type cinnamon soils (Haplustalfs) (Mu et al., 1982).

in the Beijing region, the Cr content of the surface layer was $106\,mg\,kg^{-1}$ compared to $58\,mg\,kg^{-1}$ at a depth of 40cm.

Chromium exists in different forms in soils. As shown in Figure 4 for a neutral alluvial soil contaminated with Cr(III), most of the Cr is bound by iron oxides and organic matter, with only a small portion of it existing in the exchangeable form. However, some of the Cr(III) can be mobilised under acid conditions. Hexavalent

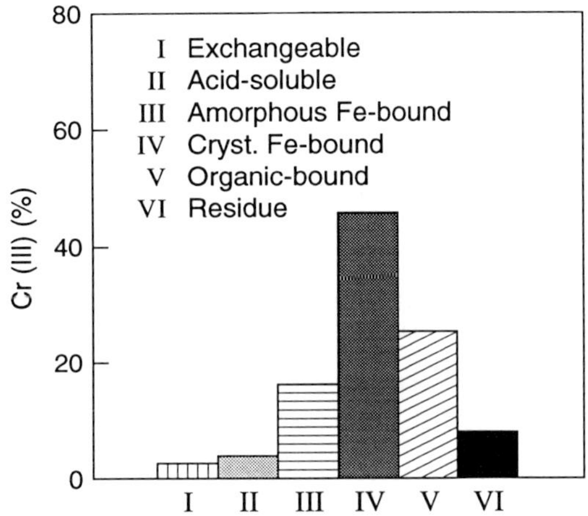

Figure 4. Distribution of different forms of chromium (III) in Cr-polluted alluvial soils (average value for 4 soil samples) (Wu and Lu, 1991).

Cr exists principally in the form of anions; the activities of these anions are controlled by both oxidic minerals and organic matter in these soils (Wu and Lu, 1991). Under alkaline conditions this form of Cr may leach.

Chromium polluted soils are limited chiefly to areas irrigated with Cr-contaminated water and local areas adjacent to refining factories. For example, at an irrigation district near Shenyang, the Cr content of the soil is 0.51 $mg\,kg^{-1}$, while the background value is 0.12 $mg\,kg^{-1}$ (Wu et al., 1980). In another heavily polluted region, the Cr content of the soil is as high as 950 $mg\,kg^{-1}$. The Cr content of brown rice grown on this soil is 0.10–0.15 $mg\,kg^{-1}$, which is below the maximum permissible concentration (0.5 $mg\,kg^{-1}$) for food (J Q Xia, unpublished). This implies that Cr may not be readily taken up by grain under certain conditions, even though it accumulates in the soil to high concentrations.

2.4 Arsenic

Arsenic in soils from China is largely pedogenic in nature with some localised contamination arising from irrigation of soils with As-contaminated wastewater. In the uncultivated soils in China, the range of As content is 2–20 $mg\,kg^{-1}$ with an average of 15 $mg\,kg^{-1}$ (Table 4).

However, As has been found to accumulate in the surface 0–40 cm layer in soils subjected to irrigation with As-contaminated water over a ten-year period (Figure 5). This surface accumulation of As presumably indicates that As does not readily leach, a property similar to that for phosphate (P). However, there is an urgent need to assess the surface transport of As in sloping land in relation to its impact on surface water quality.

It has been established (Chen et al., 1979) that As-polluted soils may be classi-

Table 4. Total As content of soils derived from different parent materials (Institute of Soil Science, 1990).

Soil	Parent material	Depth	As
		cm	$mg\,kg^{-1}$
Hapludox (Latosol)	Quartz rock	0–10	6.6
	Basalt	0–16	6.4
	Marine deposits	0–13	3.3
Hapludult	Phyllite	0–15	19.8
(Red earth)	Quaternary clay	0–17	10.7
	Granite	0–11	2.2
Hapludalf	Basalt	0–10	7.1
(Yellow-brown earth)	Granite	0–12	4.3
	Loess	0–12	12.7
Haplustalf	Granite-gneiss	0–35	6.2
(Cinnamon soil)	Limestone	0–25	11.6
	Marble	0–20	12.8
	Andesite	0–12	8.1
Gypsiorthid	Sandstone	0–10	4.0
(Gypsic brown desert soil)	Diluvial	0 –20	18.3

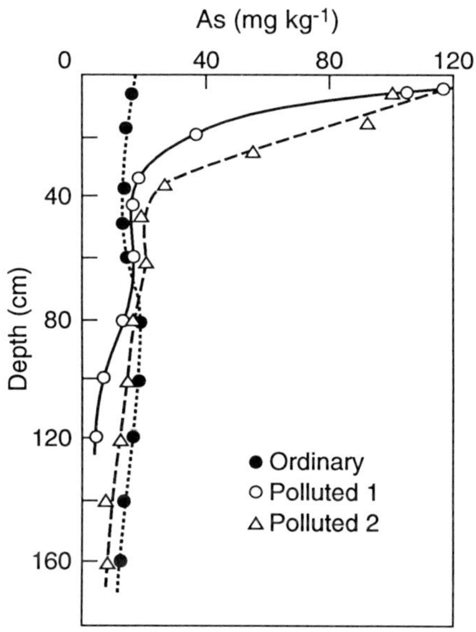

Figure 5. Distribution of arsenic in loessial soils (Jiang and Ho, 1983).

fied into three categories: heavily polluted (total As > $100\,mg\,kg^{-1}$); moderately polluted (total As: $80-100\,mg\,kg^{-1}$); and slightly polluted (total As: $40-80\,mg\,kg^{-1}$).

2.5 *Fluorine*

Fluorine occurs widely in soils. The F content of soils varies within the range $10-7000\,mg\,kg^{-1}$, with most soils in the range $50-800\,mg\,kg^{-1}$. The average value

Table 5. Total and water-soluble F contents in the surface layer of different soils (Institute of Soil Science, 1990).

Soil	Locality	F ($mg\,kg^{-1}$) Total	Soluble
Humaquept (Meadow soil)	Jiangsu	334	2.46
Hapludult (Red yellow earth)	Zhejiang	189	1.12
Ustochrept (Light castnozem)	Inner Mongolia	503	10.60
Ustochrept (Light castnozem)	Inner Mongolia	379	3.92
Haploboroll (Chernozem)	Heilongjiang	316	1.40
Aquic Haploboroll (Meadow black soil)	Heilongjiang	363	1.98
Natrudalf (Alkaline soil)	Heilongjiang	315	8.60
Natrudalf (Alkaline soil)	Jilin	300	6.00
Eutrochrept (Brown forest soil)	Liaoning	280	1.64
Haplumbrept (Soddy forest soil)	Changbei Mt.	264	0.89
Borosaprist (Humic bog soil)	Heilongjiang	271	2.00
Argialboll (Albic soil)	Qinghai	233	0.43

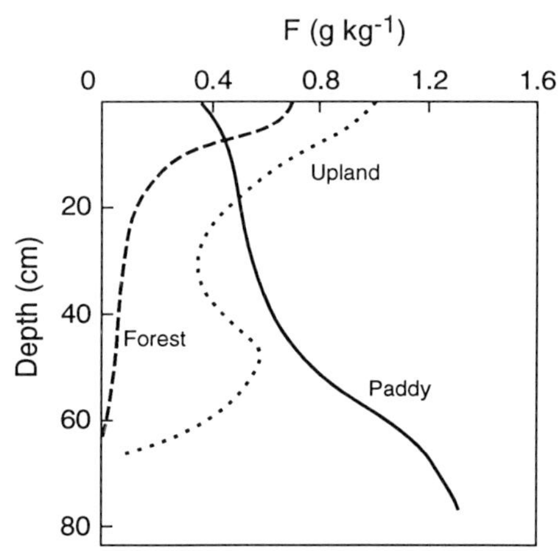

Figure 6. Distribution of fluorine within F-polluted soil profiles under different utilisation conditions (Liu, 1979).

is 200 $mg\,kg^{-1}$ (Institute of Soil Science, 1990). Table 5 shows the total and water-soluble F contents in the surface (0–20 cm) layer of representative soils.

The major sources of F contamination in China includes water and solid wastes from industrial factories and the application of F-containing agricultural chemicals. In particular, in the vicinity of factories that produce P fertilisers F pollution of soils may be a common problem.

Once soils become contaminated, the distribution of F within the soil profile varies with the nature of land utilisation (Figure 6). For forest soils and upland soils, F content decreases from the surface layer downward. By contrast, for paddy soils in China, F increases gradually with depth, presumably due to leaching by percolating water. This latter phenomenon would be advantageous, because it leads to the reduction or avoidance of unfavorable effects of F on the growth of crops. On the other hand, the risk of groundwater contamination increases.

2.6 *Organic contaminants*

Amongst the organic contaminants, DDT, 666 and chloral are the most serious soil pollutants in China. As with most developing countries the major source of both DDT and 666 are also agricultural activities. (See Chapters 14, 17 and 20 for more information on DDT). Such activities have led to relatively high concentrations (1–4 mg kg^{-1} for DDT and 0.4 mg kg^{-1} for 666) of these organics in the upland fields. Similarly high DDT and chloral concentrations have been recorded in cotton growing soils. The DDT content of cotton growing soils can vary by as much as five times depending on the cotton growing region. Paddy soils, in contrast to cotton soils, show relatively low concentrations of these chemicals. This is not surprising since the amount of DDT utilised for a single cotton crop may be five to six times higher than that for cotton–rice rotation.

The concentration of organic chemicals in the tea growing region is generally extremely low (0.07–0.08 mg kg^{-1}, Institute for Soil Science, 1990) because of the ban on pesticide use in these soils by the Chinese Government. Only limited work has been conducted in China on pesticide mobility in the soil environment. These studies reveal that organochlorine pesticides accumulate in the surface 0–30 cm of the soil profile; although in some areas 666 has been detected at a depth of 100 cm. This may be attributed to its relatively higher solubility compared to DDT.

Unlike DDT and 666, chloral is produced during the synthesis of agricultural chemicals and pharmaceuticals. Improper handling of this industrial waste and accidental spillage has led to extensive contamination of certain soils (Institute for Soil Science, 1990). Our studies show that the original active ingredient trichloroacetaldehyde disappears within days after spillage due to microbial degradation. However, the metabolite trichloroacetic acid has been found to persist in the soil environment up to three months under normal field moisture conditions. In the absence of moisture trichloroacetic acid may remain undecomposed for a significant period of time. Given that contamination of soils by organic chemicals has implications to both soil, water and food quality, there is urgent need for us to examine the possibilities for greater emphasis and work in this area in China.

3. Factors affecting mobility and toxicity of contaminants

One of the major concerns in soil environmental chemistry is contaminant mobility because it is this mobile fraction or ‘available fraction’ of the element that affects plant growth and food quality. In this regard, the sorption-desorption processes play a dominant role in controlling contaminant dynamics in the soil-plant system. As discussed in Chapters 3 and 4, sorption-desorption processes are controlled by a variety of factors that include: soil type; pH; organic matter content; and the presence of competing ions accompanying the contaminants. For detailed discussions on the

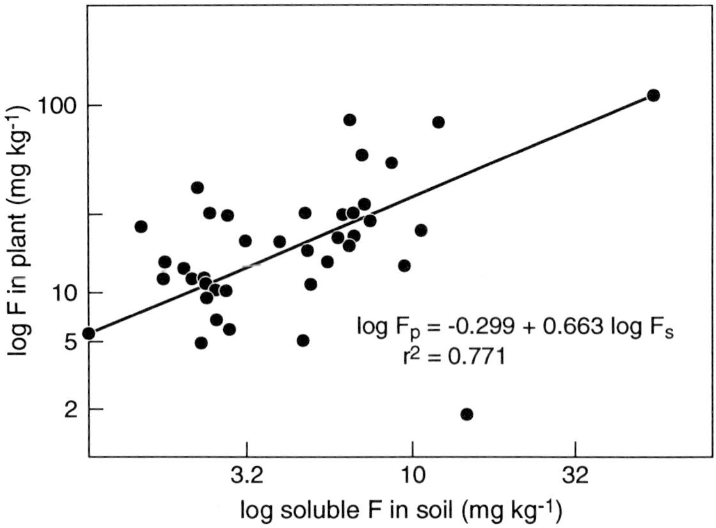

Figure 7. Relationship between F content in wheat and soluble F content of contaminated soils (Zheng et al., 1982).

role of these factors, particularly in relation to the interaction between contaminants and soils in China, readers are referred to Yang et al. (1979), Chen (1985), Zhu (1985), Liu (1988), Zhang et al. (1987), Zhang et al. (1990) and Hu (1994).

4. Impacts of soil contaminants on plants and water

In some regions of China, contaminants in the soil have affected crop quality, yield and water quality, although these occurrences are generally limited to local areas.

4.1 Crop yield and quality

Field studies show that contaminant concentrations in plant tissues are positively correlated with the amount of plant available element in soil. Such a correlation for F is shown in Figure 7. The data presented in Figure 8 and Table 6 also demonstrate this relationship for the accumulation of Cu and Cd in plant tissues when grown under glasshouse conditions in soils treated with Cu and Cd. The grain yield of rice also decreased in soils with 10 $mg\,kg^{-1}$ Cd. These results demonstrate that phytotoxicity could be one of several problems limiting crop production in highly contaminated soils.

The distribution of contaminant elements within the plant is uneven. For example, in rice grown on soils irrigated with Cr-polluted water, 92% of the Cr accumulated in the straw, leaving 5% in the shell and 3% in the brown grain (Chen et al., 1978).

A more visible effect of contaminants is their influence on the growth of plants.

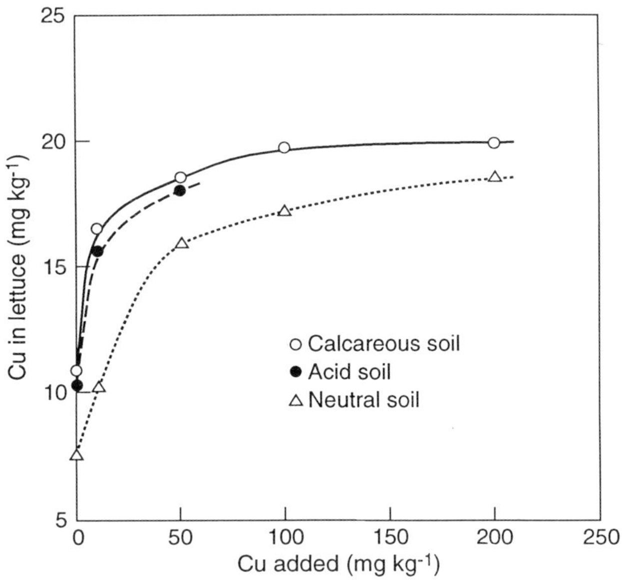

Figure 8. Effect of added Cu on Cu content of lettuce grown on purplish soils (Tu and Qing, 1989).

Table 6. Effect of Cd on yield and Cd content of rice grown on a paddy soil derived from alluvium (Chen, 1985).

Treatment ($mg\,Cd\,kg^{-1}$)	Grain yield ($g\,plant^{-1}$)	Cd in brown rice ($mg\,kg^{-1}$)
Check	19.3	0.15
0.1	19.1	0.18
1	20.6	0.21
10	17.2	0.26

It was observed that, for a rice field polluted by calcium arsenate, rice wilted and eventually died when grown on soils with a total As content higher than $100\,mg\,kg^{-1}$ and a water-soluble As content higher than $1.8\,mg\,kg^{-1}$. The growth was normal on soils containing less than 42 $mg\,kg^{-1}$ of total As and $0.4–0.6\,mg\,kg^{-1}$ of water-soluble As. A similar negative effect of As on the yield of rice was recorded in a field experiment on an Inceptisol developed on purplish shale with a total As content exceeding $47\,mg\,kg^{-1}$ (Figure 9).

The sensitivity of plants to soil contaminants varies with the species. For example, graminaceae crops generally show a relatively high tolerance for Cd. Dong et al. (1981) showed that wheat, rice and corn could grow normally when the soil Cd content was less than $10\,mg\,kg^{-1}$; but the yield of wheat decreased by 20% when the soil Cd content was $20–50\,mg\,kg^{-1}$ and the growth of radish, cabbage and potato was markedly retarded when the soil Cd content exceeded $8–10\,mg\,kg^{-1}$. Because of the difference in tolerance, the established critical Cd concentration of irrigation water

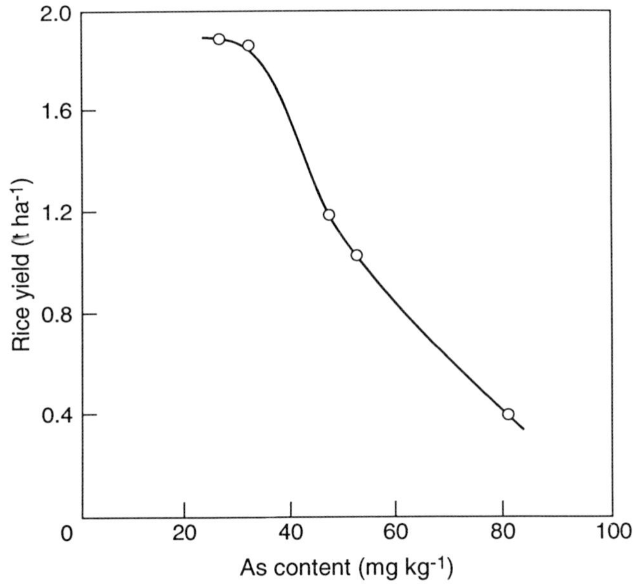

Figure 9. Effect of soil As on rice yield (Institute of Soil Science, 1990).

is 10 mg L^{-1} for rice and 5 mg L^{-1} for wheat. The critical Cr content of soil was set at 50 mg kg^{-1} for rice, 20 mg kg^{-1} for millet and 10 mg kg^{-1} for cabbage (Institute of Soil Science, 1990).

4.2 Water quality

In highly polluted areas, the quality of groundwater may also be affected. For example, it has been found that in a region polluted by the disposal of tailings from a steel industry the F concentration of groundwater exceeded the permissible level (1 mg L^{-1}) in all of the ten shallow wells in the polluted area, whereas the F concentration was below the permissible level in all the well waters away from the polluted area (Hu et al., 1982).

5. Prevention and remediation measures

For the prevention and remediation of soil contamination, a variety of measures have been adopted and some of them have given fruitful results. The successful ones include:

- Control and elimination of pollutant source, such as the treatment of wastewater and the control of the quality of irrigation water. This is the fundamental management practice in the prevention of soil contamination.

- Application of modifiers. Some examples are as follows:
  - It has been found that exchangeable Cd and Zn in a polluted acid brown earth decreased markedly after the application of lime and serpentine-fused phosphate or lime and humic acid (Kong and Wu, 1988);
  - Two years after the application of lime to a soil contaminated by Cd, the Cd content of rice grain decreased by 50% and the yield increased by 500–750 $kg\,ha^{-1}$ (Chen et al., 1981);
  - The application of phosphate- or sulfate-containing fertilisers to Hg-contaminated paddy soils may be beneficial to the formation of insoluble Hg compounds, leading to the alleviation of mercury toxicity (Institute of Soil Science, 1990);
  - Ferric or ferrous sulfate can alleviate the toxicity of As to rice in paddy fields (Yang et al., 1979);
  - Applications of large quantities of manures were found to reduce Cr(VI) to Cr(III), which is less toxic to plants. Manures also decreased the toxicity of Hg and Cd, which were complexed by humic acid.
- Growing tolerant crops
  - Selection of species. Crops tolerate different toxic contaminants to different degrees. For example, it was found that in the same region the Cd contents of corn (*Zea mayes*) and Chinese sorghum (*Sorghum vulgare*) were respectively only 11.6% and 21.8% of that of wheat (*Triticum sativum*) (Ho et al., 1987). Apparently, these two crops are more suitable than wheat when grown on soils contaminated by Cd. Vegetables and tuber crops should be avoided on Cr-contaminated soils.
- Management practices
  - Change of cultivation system. A change of As-contaminated paddy fields to upland cultivation methods may result in a decrease in As toxicity. Conversely, submerged conditions should be advantageous for the reduction of Cr(VI) to the less toxic Cr(III);
  - Appropriate management practices. It is advisable to avoid drainage of Cd-contaminated fields during the milk stage (seed filling stage) of rice growth, so that Cd is maintained in an unavailable form, because during this period rice plants assimilate Cd vigorously. On the other hand, the maintenance of a high oxidation-reduction potential through alternate irrigation and drainage may be beneficial to the presence of As in the As(V) form, which is less toxic than As(III) (Institute of Soil Science, 1990).

6. Conclusion

This paper presented an overview of some of the contaminant related problems in China. Soil contamination has become a serious problem in China in the last two to three decades following rapid industrial development. This is of particular concern around the major cities and large industrial enterprises where wastes are often discharged as irrigants onto land. Although the proportion of land affected by such disposal practices is small on a nationwide scale, it is important that regulatory measures are enforced in China to minimise contaminating activities and also to ensure that with the increase in industrial activities there is a greater recognition of the need for maintaining a quality environment.

References

Chen C Q, Xu Y L, Ye Z J and Yu X X 1978 Chromium contents of crops and soils affected by Cr-containing waste water as studied with a neutron activation technique. Environ. Sci. (Ch.). 1, 13-17.

Chen H M 1985 Effect of Cd, P and Zn contents on yield and mineral content of rice. Acta Pedol. Sinica (Ch.). 22, 85-92.

Chen H M 1988 Factors affecting adsorption of cadmium by soils. Soils (Ch.). 20, 131-136.

Chen T, Wu Y Y and Kong Q X 1981 Experiment on reclamation of Cd-polluted soil by liming in Zhangshi Irrigation District. Environ. Sci. (Ch.). 2, 40-42.

Chen Z Y, Chen Y G, Wan X L and Liu D J 1979 Experiments on the influence of arsenic on ecology of paddy fields. Environ. Sci. (Ch.). 4, 46-50.

Dai Z H and Huang Y C 1983 Distribution of mercury in soils. Environ. Sci. (Ch.). 4, 32-35.

Dong K Y, Chen J M and Deng X Y 1981 Absorption and accumulation of cadmium by crops. Environ. Sci. (Ch.). 2, 166-171.

Environmental Monitoring Center of China 1990 Background Value of Elements in Soils of China (Ch.). Environmental Science Press, Beijing. pp 501.

Ho J D, Zhang C D and Lu Z Y 1987 Soil and crop evaluation of irrigated fields in Lanzhou region. Agric. Environ. Protec. (Ch.). 6, 31-34.

Hu G S 1994 Adsorption kinetics of Pb^{2+} and Cu^{2+} by variable charge soils and minerals. V. Effect of temperature and ionic strength. Pedosp. 4, 153-164.

Hu R M, Ran Y L and Ma L S 1982 Fluorine pollution of soils in Baotou region. Acta. Environ. Sinica. (Ch.). 2, 143-149.

Institute of Soil Science 1990 Soils of China. Science Press, Beijing. 873 p.

Ji G L, Wang J H and Zhang X N 1995 Acid rain-induced environmental problems in soil and groundwater and management strategies in China mainland. In: Remediation of degraded soils and groundwaters. Volume 4: Worldwide perspectives on soil and groundwater pollution: Asia, Africa and Oceania. Ed. P M Huang. Science Reviews, England. (in press).

Jiang X T and Li C Y 1988 Relationship in mercury content between polluted soil and foods. Environ. Sci. Technol. (Ch.). 8, 60-64.

Jiang Y Q and Ho X P 1983 Pollution of soils by As-containing waste water of chemical factories in Xingping. Environ. Sci. (Ch.). 4, 48-51.

Kong, Q F 1981 Pollution of fluorine in wheat at the suburban areas of Baotou. Environ. Chem. (Ch.) 2, 47-49.

Kong Q X and Wu Y Y 1988 Forms, contents of cadmium, lead, copper and zinc in Cd-polluted soils and their changes after reclamation. Environ. Sci. Technol. (Ch.). 8, 45-47.

Li X L 1979 Status of mercury pollution of irrigation fields in Xi'an district. Soils (Ch.). 11, 184-187.

Ling N G 1993 Survey of heavy metal contents in fruit plantation in northeastern Guangxi. Agric. Environ. Protec. (Ch.). 12, 179-180.

Liu J C 1979 Fluorine pollution of soils in the vicinity of phosphate factories. Environ. Sci. (Ch.). 2, 9-13.

Liu Z H 1988 Sorption capacity for fluorine in red soils. Acta Pedol. Sinica (Ch.). 25, 236-242.

Mu C R, Li S Z, Wang L Y and Ho R Z 1982 Transportation and transformation of chromium in water, soil and crops. Chin. Environ. Sci. (Ch.). 1, 19-23.

She Z S and Meng X X 1982 Preliminary study on mercury pollution in cities of northeast China. Environ. Sci. (Ch.). 3, 57-61.

Tu C and Qing C L 1989 Studies on critical toxic level of copper in purplish soils of Sichuan. Soils Bul. (Ch.). 20, 286-288.

Wang L R and Hou Y M 1991 Heavy metal pollution of soils and foods in irrigation districts near Zhenghou. Environ. Protec. In Agric. (Ch.). 10, 161-164.

Wu D A and Lu W Y 1991 Pollution of chromium in soil-groundwater systems. Acta Environ. Sci. (Ch.). 11, 276-281.

Wu Y Y, Kong Q X and Dai T S 1980 Pollution of cadmium and other heavy metals in irrigation district near Shenyang. Soils Bul. (Ch.). 11(1), 4-6.

Xia Z L 1982 Pollution of cadmium and its prevention in southeast Beijing. Environ. Sci. (Ch.). 3, 14-18.

Yang G Z, Xia J Q and Yong J 1979 Fixation and release of mercury in soils. Acta Pedol. Sinica (Ch.). 16, 39-43.

Zhang D Y, Lin C G and Yan C P 1990 Reduction of Cr(VI) by organic matter in soils. Agric. Environ. Protect. (Ch.). 9, 29-31.

Zhang G Y, Zhang X N and Yu T R 1987 Adsorption of sulfate and fluoride by variable charge soils. J. Soil Sci. 38, 29-38.

Zhang X S, Wu Y Y, Chen T and Kong Q X 1982 Migration and distribution of cadmium and lead and their control in Zhangshi Irrigation District. Environ. Sci. (Ch.). 3, 7-10.

Zhao M Z 1991 Adsorption and desorption of selenite by soils and clay minerals. Acta Pedol. Sinica (Ch.). 28, 139-145.

Zheng B S, Fu Y Z, Zhang W L and Zhang J P 1982 Environmental background value of fluorine in Baotou district. Chin. Environ. Sci. (Ch.). 4, 9-11.

Zhou D and Long J M 1985 Preliminary study on adsorption and release of mercury in soils. Environ. Chem. (Ch.). 4, 29-33.

Zhu Y Z 1985 Factors affecting migration and transformation of chromium in soils. Acta Pedol. Sinica (Ch.). 22, 390-393.

16

Trace metal contamination of the Hong Kong soil environment: a review

M.H. WONG, J.W.C. WONG and T.B. CHEN

1. Introduction

It is commonly known that urbanisation and industralisation adversely affect our environmental quality. This is further aggravated in developing countries where industrial development is progressing rapidly compared with developed countries.

With a population of approximately 6 million and a limited area of only 1 067 km^2, Hong Kong is one of the most densely populated areas in the world. There has been continuous urbanisation and demand for land for various developments which include the building of housing/industrial estates, and the construction of roads, reservoirs and recreational amenities. Rapid urbanisation of the rural areas (The New Territories) has been inevitable due to the decentralisation scheme for alleviating the high population density in the urban centres (north of Hong Kong Island and Kowloon Peninsula) by constructing new towns.

These 'satellite' towns are conurbations of high-rise residential tower blocks, interspersed with various industries, and surrounded by agricultural land. Agricultural and industrial wastes have traditionally been discharged into the environment with little or no pretreatment, until the advent of recent legislation enacted by the Environmental Protection Department (1990a). Stream and marine pollution of our environment has been frequently reported (e.g. Hodgkiss and Griffiths, 1987; Yang and Wong, 1989). In particular, investigations have centred around trace metal contami-nation of sediments and their accumulation in aquatic organisms (e.g. Cheung and Wong, 1992a; 1992b; Phillips and Yim, 1981). Nevertheless, there has been a lack of information concerning trace metal contamination in the terrestrial environment, except the contamination of roadside soil, dust and plants (Chan et al., 1989; Lau and Wong, 1981; Tam et al., 1987). No information on organic contaminants in either rural or urban soils could be located in the literature on contaminants research in Hong Kong. Therefore the present discussion is limited to toxic metals in the Hong Kong soil environment.

With this background in mind, the main objective of this paper is to review and evaluate the extent of trace metal contamination in the terrestrial environment, especially the soil environment. It is hoped that the paper will serve as a basis for future comprehensive investigations into contamination of the Hong Kong soil environment.

2. Sources of soil contamination

Hong Kong produced about 20 000 $t\,d^{-1}$ of solid and semi-solid wastes in 1994 (Environmental Protection Department, 1995). The amount of wastes generated rises as the population and its affluence increases.

R. Naidu et al. *(eds): Contaminants and the Soil Environment in the Australasia-Pacific Region, 501–511.*

2.1 *Municipal wastes*

Municipal wastes include wastes from household, commercial, industrial and construction sources. By the year 2001 daily municipal waste output is projected to amount to some 16730 t. The waste will be mainly disposed of in landfills and incineration will be phased out.

It has been noted that landfill leachate commonly contains higher levels of trace metals such as Mn, Fe and Zn, due to the reduced condition of the anaerobic landfill environment (Chu et al., 1993). Analyses of landfill soils contaminated with leachate have shown elevated levels of these metals (Chan et al., 1991).

2.2 *Sewage sludge*

Increasing amounts of Hong Kong's domestic sewage will be treated in sewage treatment works which are under construction in urban centres. As a result, the quantity of sewage sludge will increase. At present, the sewage sludge which settles at the bottom of sewage treatment plants is being digested by heat, centrifuged and used as land-fill. Dumping sludge at sea has been practised on a small scale by the Hong Kong Environmental Protection Department.

2.3 *Agricultural wastes*

Pig and poultry manure has been the major problem in causing stream and marine pollution in Hong Kong for many years. Approximately 2000 $t\,d^{-1}$ of raw animal wastes are discharged into the streams and inshore waters (Environmental Protection Department, 1990b). The implementation of controlled animal husbandry practices such as dry 'mucking-out' connected to a collection system, or wet 'mucking-out' with the provision of appropriate treament and disposal facilites, has resulted in a dramatic increase in agricultural wastes.

Pig manure mixed with sawdust has been used to produce compost in Hong Kong. Contents of Cu and Zn in the manure compost are high, due to the fact that Cu and Zn are added to animal feeds, with the Cu serving as a growth stimulant and the Zn counter-acting Cu toxicity. However, the metals are not necessarily absorbed by crops treated with manure. This was presumably due to the liming effect of added manure and the increased organic matter content and the cation exchange capacity of compost (Gouin and Walker, 1977; Wong, 1985a).

2.4 *Coal ash*

Coal ash, which includes furnace bottom ash (FBA) and pulverised fly ash (PFA) is produced from coal combustion. The prime source of energy for power generation in Hong Kong is coal, and the amount of ash will increase as power consumption increases. Disposal and treatment of coal ash will become a significant problem.

The use of fly ash as a soil amendment has been studied in Hong Kong. Microbial respiration is reduced after adding fly ash to soil (Wong and Wong, 1986), but despite this, co-recycling of fly ash and manure in nutrient-deficient sandy soil seems to be feasible. The high nutrient status of manure remedies the N-P-K deficiency of sandy soil, while the silt fraction of fly ash has major beneficial effects on soil physical properties (Wong and Wong, 1987; 1990). Fresh and weathered fly ash have been used as composting materials for sewage sludge, and it has been shown that fresh fly ash causes a more pronounced inhibition of sewage sludge decomposition than does weathered fly ash, due to its higher content of B, Cd, Cu, Pb and Zn and its higher salinity (electrical conductivity) (Wong et al., 1995).

2.5 Toxic, hazardous and difficult wastes

These wastes include solid, liquid, or gaseous material that may pose substantial hazards to human health or the environment. Toxic substances are those that are poisonous to humans, land and marine animals or plants. Hazardous materials are those that are inflammable, explosive or which give off dangerous vapours either on their own or in combination with other substances. The category of 'difficult' wastes covers those wastes not already classified as toxic or hazardous. Sludge from sewage treatment can be considered as 'difficult' waste. Toxic wastes in Hong Kong are mainly produced by the industrial sectors concerned with the manufacture of metal, plastic, electrical and electronic products.

Amongst the various industrial effluents being discharged into the coastal waters of Hong Kong, trace metals pose the greatest threat to organisms because of the effects of bioconcentration of these contaminants (e.g. by the Pacific oyster *Crassostrea gigas*, Cheung and Wong, 1992a, 1992b).

As well as aquatic habitats, land and local flora are exposed to trace metal contamination. The rinsing water discharging from a small factory engaged in the recycling of scrap plastics resulted in higher total contents of Ni (385 $mg\,kg^{-1}$) and Cu (194 $mg\,kg^{-1}$) in soil compared with an unpolluted site (with 12 and 19 $mg\,kg^{-1}$ of Ni and Cu, respectively), and hence higher contents of these metals were observed in several grass species that thrive in the polluted areas (Chui et al., 1988). Significant amounts of dust, with elevated concentrations of Al, Cd, Pb, Ni, Cu and Zn, were deposited on leaf surfaces of plants grown near an Al-fabrication works, leading to the clogging of their stomatal pores and distortion of the guard cells. High contents of these metals, in particular Al, were also noted in the associated soil (Tam et al., 1988).

3. Trace metal contents of urban soils

3.1 Lead contamination of roadside habitats

Of the world's urban centres, Hong Kong has one of the highest densities of vehicular traffic. Lead contamination in different habitats near motorways throughout

Table 1. Comparison of mean Pb concentrations ($mg\,kg^{-1}$) in street dust and soil samples from various cities (from Lau and Wong, 1982).

Location	Dust	Soil	Reference*
Hong Kong	1 627	88	Lau and Wong (1982)
Hong Kong	2 006	–	Ho (1979a)
Rio de Janeiro, Brazil	700	–	Branquinho (1973)
Birmingham, UK	1 800	–	Environ. Prot. Unit. Bir. (1975)
Birmingham, UK	1 700	–	Stephen, in Day et al. (1975)
Lancaster, UK	1 890	–	Harrison (1979)
Leeds, UK	4 400	–	Dunn and Bloxam (1933)
Liverpool, UK	–	96	Roberts and Johnson (1978)
Manchester, UK	970	–	Day et al. (1975)
Detroit, USA	–	65	Tel Haar (1970)
Urbana, Ill., USA	3 600	–	Solomon and Artford (1976)
USA (average of 77 cities)	240 – 1 500	–	Hunt et al. (1971)

*Refer to Lau and Wong (1981) for details of references listed in this table

Hong Kong has been investigated (Lau and Wong, 1981; Wong, 1985b). Combustion of leaded gasoline from motor vehicles is the major cause of Pb contamination. The airborne Pb levels range from 0.6 to 3.7 $\mu g\ m^{-3}$ and the mean value of Pb in human blood is 15 $\mu g\,dL^{-1}$; these figures are comparable to Pb levels in other parts of the world (Jaworski et al., 1987; Wong, 1987). The lowest observed effect levels for inducing health hazards (including haem, neuro and renal effects) are 80 – 100 $\mu g\,dL^{-1}$ (Goyer, 1991).

The measurement of Pb in dust and soil has also been used to indicate the extent of Pb contamination. Table 1 shows that the Pb level in street dust (1 627 $mg\,kg^{-1}$) of Hong Kong is comparable with values obtained in different cities of the world (Lau and Wong, 1981).

Table 2 demonstrates means and significant differences ($P<0.05$) of Pb levels between roadside soils and soils with the same land use patterns 50 m away from motorways. The industrial area had the highest Pb concentration, followed by industrial/residential area, rural area, country park area, while the agricultural area had the lowest concentration (unpublished data).

Inhibition of root growth of crops was observed when they were treated with aqueous extracts of roadside dusts and soils (Wong et al., 1984). The roadside population of two grass species, *Eleusine indica* and *Cynodon dactylon,* had a higher tolerance to Pb than their counterparts from uncontaminated environments (Wong and Lau, 1985). This indicates that the Pb levels in roadside dust and soils are sufficiently high to act as a factor in determining the relative survival rates of plants. Significant reductions of Pb levels in roadside soils and dusts have been noted following the lowering of Pb levels in petrol (0.84, 0.6, 0.4, 0.25 and 0.15 Pb $mg\,L^{-1}$ of petrol in 1980, 1981, 1983, 1985 and 1987, respectively) (Chan et al., 1989).

Table 2. Comparison of Pb concentrations (mg kg^{-1}) between roadside soils and soils 50 m from highways in Hong Kong.

Location	Soil Pb Next to highways	50 m away from highways	Remarks
Fo Tan, Shatin (Industrial/Residential area)	108	39.4*	Lawn
Wong Chu Rd., Tsuenwan (Industrial area)	206	47.2*	Street garden
Tai Lam Chung, Tsuen Wan (Industrial area)	126	61.6*	Hillside
Wang Leng, northern New Territories (Agricultural area)	42.7	19.2*	Vegetable field
Lin Tong Mei, northern New Territories (Rural area)	81.2	30.1*	Village
Tsam Chuk Wan, Sai Kong (Country park area)	52.9	24.3*	Forest

*Indicates significant ($P<0.05$) difference of Pb concentrations between soils next to highways and soils 50 m away from highways.

Table 3. Total metal concentrations (mg kg^{-1}) of urban/playground soils in Hong Kong.

	1987 Survey*		1993 Survey†		Dutch Criteria
	Mean	Range	Mean	Range	A value
Cd	–	–	1.89	0.94–3.38	1
Cu	142	13–553	27.5	3.84–91.9	50
Fe	15020	3838–33773	–	–	–
Pb	268	42–1034	100	17.7–258	50
Mn	758	293–1745	–	–	–
Zn	1281	173–11316	93.9	10.4–218	200

*Tam et al. (1987): A total of 39 roadside soil samples (0–1 cm depth) from playgrounds in Hong Kong.
†Unpublished data: A total of 70 soil samples (0–5 cm depth) from playgrounds in Hong Kong.

3.2 *Trace metal contamination of recreational areas*

Most of the parks and recreational areas in Hong Kong are built close to busy traffic or near industrial areas, where they are subject to potential pollution from vehicle exhaust emission and industrial emission. Other possible sources of pollution may be from construction and natural sources like wind blown soil (Environmental Protection Department, 1990b).

Results of two recent surveys on the trace metal contamination of urban garden soils are listed in Table 3. Mean concentrations of all trace metals in the 1987 survey (Tam et al., 1987) were much higher than the 1993 survey (unpublished data) and the Dutch criteria A value for background soils (ANZECC/NHMRC, 1992).

For the 1993 survey only mean values of Cd and Pb exceeded the Dutch criteria A value. Both surveys showed that the metal concentrations were correlated with traffic volumes, indicating that car exhaust emissions would be the major source of trace metal pollution in urban soils in Hong Kong, resulting from particulate deposits derived from vehicles. The air pollution monitoring programme of the Environmental Protection Department, Hong Kong (1990b) also showed that particulate levels were high in areas with high traffic volumes. It is reasonably well established that trace metals such as Pb, Zn, Ni and Cu are released into the atmosphere adjacent to roadways (Largerwerff and Specht, 1970; Simmonds et al., 1983). A significant elevation in Pb contents of leafy vegetables grown near heavy traffic is commonly observed (Preer et al., 1980; Wong and Tam, 1978). It has been observed that the unwashed leaves of flowering Chinese cabbage (*Brassica parachinensis*) grown near motorways contained 82 $\mu g\,Pb\,g^{-1}$ compared with 5.6 $\mu g\,Pb\,g^{-1}$ of those grown 200 m away from motorways (Wong and Tam, 1978).

The 1987 survey shows significantly higher metal contents than those of the 1993 survey (Table 3). Besides the reduction of Pb content in petrol resulting in a general reduction of Pb content in roadside soils (Chan et al., 1989), the discrepancy can be explained by the different soil sampling strategies adopted for the two monitoring programmes. The 1989 survey sampled only to a depth of 1 cm while the 1993 survey sampled to a depth of 5 cm of the soil profile, resulting in a greater dilution of trace metal contents with the greater depth of sampling. This demonstrates the need for a standardised approach in soil sampling and analytical strategy in contaminated soil monitoring exercises, to provide a better understanding of metal concentrations in soils.

3.3 Trace metal contents of agricultural soils

3.3.1 Total soil concentrations

Agricultural lands constitute about 9.4 % of the total land area of Hong Kong, and supply about 40% of the annual vegetable consumption of the local population of over 6 million. During the past two decades, fast economic growth in Hong Kong has resulted in rapid industrialisation and urbanisation, shifting the residential and industrial developments into major agricultural areas in the northern part of the New Territories. As a consequence, the agricultural areas have been diminished and vegetables are grown intensively in the limited areas available. Urbanisation and industrialisation have also increased traffic volume and industrial emission in these areas, which may pose serious environmental threats to the quality of agricultural soils and products. One possible concern is potential contamination of the soils with trace metals.

Data on the background concentrations of trace metals in agricultural soils in Hong Kong are rare. The only comprehensive information is derived from our recent monitoring programme, conducted in 1992 on 11 agricultural areas in Hong Kong (Wong, 1996). A total of 54 composite market garden soil samples were collected from the surface 5 cm of the soil profile. The total Cd, Zn, Cu and Pb concentrations in these market garden soils are listed in Table 4 together with the Dutch criteria for determining soil clean up requirements (ANZECC/NHMRC, 1992). The

Table 4. Total Cd, Zn, Cu and Pb concentrations (mg kg^{-1}) in market garden soils in Hong Kong.

	Cd	Cu	Pb	Zn
Mean	1.01	18	60	105
Geometric Mean	0.91	15	54	86
Standard Deviation	0.44	12	24	57
Range	0.25–2.17	4.5–67	24–113	20–232
Dutch Criteria				
Background (A)	1	50	50	200
Investigation (B)*	5	100	300	500

*Soil concentration above this recommended value requires further investigation (ANZECC/NHMRC, 1992).

highest background concentrations of Cd, Pb, Cu and Zn for agricultural soils in Hong Kong were higher than the Dutch A values for soil background concentration of Cd, Zn, Cu and Pb, but were all below the B value of the Dutch Criteria (ANZECC/NHMRC, 1992). This indicates that agricultural soils in Hong Kong are slightly contaminated but the concentrations of these heavy metals are acceptable and no further investigation of this problem is required. However, a regular monitoring programme to assess soil contamination in Hong Kong is warranted. The Cd, Cu and Pb concentrations of the agricultural soils were lower than that of the urban soils collected in 1993 while Zn had a similar concentration. Other studies have also reported that concentrations of trace metals are higher in urban soils than in rural soils (Klein, 1972; Tiller, 1992).

Urbanisation and high traffic volume are the major reasons for the elevated trace metal concentrations in these sites. This is consistent with other published reports (Bergomi et al., 1987; Tennant, 1987). Urban garden soils are frequently contaminated with Cd, Pb and other metals (Davies, 1978; Davies and Ballinger, 1990; Spittler and Feded, 1979). A recent survey in Britain of over 4100 gardens in 48 towns and city boroughs showed elevated metal levels, with concentrations of Pb ranging from 13 to 14000 mg kg^{-1} and Cd ranging from <1 to 40 mg kg^{-1} (Culbard et al., 1988). The use of agricultural fertilisers and pesticides may also contribute to an increase in metal contents in agricultural soils (Tiller, 1988; Wong, 1985b).

3.3.2 EDTA extractable contents

EDTA extraction is commonly used to predict the availability to plants of metals in soils. Clayton and Tiller (1979) suggested that extraction using 0.1 M EDTA at pH 6 can be used for polluted or potentially toxic soils. Table 5 lists the mean, standard deviation and range of soil EDTA extractable metal contents for the market garden soil samples.

The mean concentrations of each metal were all below the Toxicity Symptom Limit recommended by the New South Wales Agricultural Department (Charman

Table 5. EDTA-extractable (mg kg^{-1}) Cd, Zn, Cu and Pb concentrations of market garden soils in Hong Kong.

	Cd	Cu	Pb	Zn
Mean	0.38	8.0	20.5	46
Geometric Mean	0.37	6.6	16	36
Standard Deviation	0.09	5.7	14	29
Range	0.20–0.60	1.8–30	1.9–60	8.5–114
% of Total Metal Extracted	38	44	34	44
Pasture Soils, Adelaide†				
Mean	0.18	3.3	12	6.0
Range	0.01–0.73	0.25–19	0.4–74	0.31–52
Toxicity Symptom Limit*	1.5	120	100	120

†Pasture soils of Mount Lofty Ranges, adjacent to Adelaide (Tiller, 1992).
*Charman and Murphy (1991).

and Murphy, 1993), but were all higher than that of the pasture soils of the Mount Lofty Ranges adjacent to Adelaide, South Australia (Tiller, 1992). About 34 to 44% of the total metals are EDTA-extractable, which would likely be the amounts available for plant uptake, although this will in turn depend on other soil chemical properties that may hinder their availability.

A significant positive correlation was found between total acid soluble and EDTA extractable Cu ($r=0.84$, $P<0.001$) and Zn ($r=0.75$, $P<0.001$) and a fairly good correlation for Pb ($r=0.43$, $P<0.05$) while a weak correlation was obtained for Cd ($r=0.21$, $P<0.1$). EDTA extraction has been shown to be useful for Zn and Cu in agricultural advisory work as a good correlation is found between total- and EDTA-extractable Cu and Zn (Davies, 1978). One possible explanation for the weak correlation between EDTA- and total-Cd may be that Cd in soils may exist as a reducible and residual form that cannot be easily extracted by EDTA. The EDTA-extraction can be used as a guideline for assessing soil available contents of Cu and Zn for plant growth.

3.3.3 Trace metal contents in vegetables

A major concern about the contamination of agricultural soils is the uptake of heavy metals by vegetable crops and its consequences for human health. Samples of two common local vegetable crops, Chinese cabbage (*B. chinensis*) and flowering Chinese cabbage (*B. parachinensis*) were collected from 11 sites to determine their metal contents. Table 6 gives the total concentrations of Cd, Zn, Cu and Pb of the vegetables and the maximum permitted concentrations of trace metals in vegetables, as recommended by the NHMRC (1980).

The mean Cd concentration of *B. chinensis* approached the maximum permitted

Table 6. The heavy metal concentrations (mg kg^{-1} fresh weight) of two common local vegetable species (*B. chinensis* and *B. parachinensis*) from 11 sites in Hong Kong‡.

	B. chinensis				*B. parachinensis*				Max. Permitted Conc.*
	Mean	G.M.	S.D.	Range	Mean	G.M.	S.D.	Range	
Cd	0.05	0.05	0.01	0.02–0.07	0.06	0.06	0.02	0.04–0.10	0.05
Cu	0.75	0.73	0.19	0.52–1.10	0.80	0.78	0.23	0.57–1.60	10
Pb	0.56	0.55	0.10	0.42–0.84	0.49	0.48	0.10	0.28–0.67	2
Zn	12	10	8.1	5.4–33	20	15	15	6.6–49	150

‡A total of 51 samples for the two species were analysed (Cheung, 1993).
*National Health and Medical Research Council, Australia, 1980: Maximum permitted concentration of trace elements in vegetables.

concentration (0.05 mg kg^{-1}) while that in *B. parachinensis* exceeded this concentration by 0.01 mg kg^{-1}. The elevated Cd contents in these vegetables may cause a potential health risk if used for human consumption and may require further investigation. On the other hand, Zn, Cu and Pb concentrations of all vegetable samples were well below the maximum permitted concentrations (NHMRC, 1980).

4. Conclusion

Like other cities in the world, roadside soils and vegetation in Hong Kong are contaminated with Pb, and possibly other metals such as Zn, Ni and Cu, associated with motor vehicles. The situation is further aggravated by the fact that Hong Kong is one of the most densely populated areas in the world, with a high number of motor vehicles. In addition, residential areas are interspersed with industrial and agricultural areas, and this may impose environmental health problems for the general public because of exposure to trace metals derived from industrial and agricultural activities as well as from vehicular emission. The rather high content of Cd in the two vegetable crops discussed above is of concern. Trace metal contamination derived from industries, although more localised, can be rather severe, as demonstrated by the examples given above. There is an urgent need to conduct a more comprehensive survey on the pollution of the Hong Kong soil environment. Investigation into contamination by trace organic pollutants should also be included.

In addition to the need to standardise sampling and analytical procedures for soil samples, the development of tests for investigating biological impact should be investigated. The results could be used to predict the effects of toxicity of trace metals to plants growing on metal-contaminated sites (Wong and Bradshaw, 1982), including sites where contamination results from sludge application (Cheung et al., 1989). Toxicity tests, using root elongation, and tolerance tests using plant materials collected from normal and contaminated sites, can be used as standard assays to supplement chemical analysis, as a means of obtaining a more complete picture of the effects of toxic substances on plant growth (Wong and Chui, 1990).

At present guidelines from overseas countries such as the Dutch Criteria for contaminated soils are being used without consideration of their suitability for local conditions. It is necessary to develop a local regulatory guideline for contaminated soil classification and investigations.

References

ANZECC/NHMRC 1992 Australian and New Zealand Guidelines for the assessment and Management of Contaminated Sites. January 1992. Australian and New Zealand Environment and Conservation Council and National Health and Medical Research Council: Canberra. 159 p.

Bergomi M, Borella P, Fantazzi G, Tartoni P L and Vivoli G 1987 Cadmium levels in tooth, hair and blood of children. International Conference on Heavy Metals in the Environment, 2, 92-94. CEP Consultants Limited, Edinburgh.

Chan G Y S, Chui V W D and Wong M H 1989 Lead concentration in Hong Kong roadside dust after reduction of lead level in petrol. Biomed. and Environ. Sci. 2, 131-140.

Chan G Y S, Wong M H and Whitton B A 1991 Effects of landfill conditions on subtropical woody plants. Environ. Manag. 15, 411-431.

Charman P E V and Murphy B W 1993 Soils - Their Properties and Management. A Soil Conservation Handbook for New South Wales. Sydney University/Oxford University Press. Sydney. 363 p.

Cheung Y H and Wong M H 1992a Comparison of trace metal contents of sediments and mussels collected within and outside Tolo Harbour, Hong Kong. Environ. Manag. 16, 743-751.

Cheung Y H and Wong M H 1992b Trace metal contents of the Pacific Oyster (*Crassostrea gigas*) purchased from markets in Hong Kong. Environ. Manage. 16, 753-761.

Cheung Y H, Wong M H and Tam N F Y 1989 Root and shoot elongation as an assessment of heavy metal toxicity and Zn equivalent value of edible crops. Hydrobio. 188/189, 377-383.

Chu L M, Cheung K C and Wong M H 1993 Variations in the chemical properties of landfill leachate. Environ. Manag. 18, 105-117.

Chui V W D, Cheung Y H and Wong M H 1988 Contamination of soil and plant by recycling of scrap plastics. Environ. Int. 14, 525-530.

Clayton P M and Tiller K G 1979 A chemical method for the determination of the heavy metal content in environmental studies. Technical paper, No. 41, Division of Soils, Commonwealth Scientific and Industrial Research Organisation (Australia). 17 p.

Culbard E B, Thompson I, Watt J, Wheatley M, Moorcroft S and Thompson M 1988 Metal contamination in British urban dusts and soils. J. Environ. Qual. 17, 226-234 .

Davies B E 1978 Plant-available lead and other metals in British garden soils. Sci. Tot. Environ. 9, 243-62.

Davies B E and Ballinger R C 1990 Heavy metals in soils in north Somerset, England, with special reference to contamination from base metal mining in the Mendips. Environ. Geochem. Hth. 12, 291-300.

Environmental Protection Department 1990a Environment Hong Kong, A Review of 1989. The Government Printer, Hong Kong. 165 p.

Environmental Protection Department 1990b Air quality in Hong Kong - Results from the EPD air quality monitoring network 1986-88. The Government Printer, Hong Kong. 96 p.

Environmental Protection Department 1995 Environment Hong Kong, A Review of 1994. The Government Printer, Hong Kong. 201 p.

Gouin F R and Walker J M 1977 Deciduous tree seedling response to nursery soil amended with composted sewage sludge. Hort. Sci. 12, 45-47.

Goyer R A 1991 Toxic effects of metals. In Casarett and Doull's Toxicology. The Basic Science of Poisons. 4th Edition. Eds. M O Amdur, K Doull and D Klaassen. pp 623-680. Pergamon, New York.

Hodgkiss I J and Griffiths D A 1987 Need pig waste pollute? Hong Kong Engin. 15, 43-48.

Jaworski J F, Denny P, Hart B T, Lasheen M R, Nariagu J, Subramanian V and Wong M H 1987 Group Report: Lead. In: Ocurrence and Pathways of Lead, Mercury, Cadmium and Arsenic in the Environment. Eds. T C Hutchinson and K M Memma. pp 3-14. John Wiley and Sons,

Klein D H 1972 Mercury and other metals in urban soils. Environ. Sci. and Technol. 6, 560-562.

Largerwerff J W and Specht A W 1970 Contamination of roadside soil and vegetation in cadmium, nickel, lead and zinc. Environ. Sci. and Technol. 4, 583-586.

Lau W M and Wong M H 1981 An ecological survey of lead contents in roadside dusts and soils in Hong Kong. Environ. Res. 28, 39-45.

National Health and Medical Research Council (NHMRC) 1980 Revised Standard for Metals in Food. National Health and Medical Research Council, Canberra. 15 p.

Phillips D J H and Yim W W S 1981 A comparative evaluation of oyster, mussels and sediments as indicators of trace metals in Hong Kong waters. Mar. Eco. Prog. Ser. 6, 285-293.

Preer J R, Sekhon H S, Stephens B R and Collins M S 1980 Factors affecting heavy metal content of garden vegetables. Environ. Pollt. 1, 95-104 .

Simmonds P R, Tan S Y and Fergusson J E 1983 Heavy metal pollution at intersection involving a busy urban road in Christchurch New Zealand, 2: Aerosol lead level. N Z J. Sci. 26, 229-242.

Spittler T M and Feded W A 1979 A study of soil contamination and plant lead uptake in Boston urban gardens. Comm. Soil Sci. Pl. Anal. 10, 1195-1210.

Tam N F Y, Liu W K, Wong M H and Wong Y S 1987 Heavy metal pollution in roadside urban parks and gardens in Hong Kong. The Sci. Tot. Environ. 59, 325-328.

Tam N F Y, Wong Y S and Wong M H 1988 Heavy metal contamination by Al-fabrication plants in Hong Kong. Environ. Int. 14, 485-494.

Tennant C J 1987 Cadmium in the environment and high risk population groups, International Conference on Heavy Metals in the Environment, 2, 143-146. CEP Consultants Limited, Edinburgh.

Tiller K G 1988 Cadmium accumulation in the soil-plant system: An overview in relation to possible transfers to agricultural products. In Cadmium Accumulation in Australian Agriculture: National Symposium. Bureau of Rural Resources Proceedings No. 2, Canberra. Eds. J Simpson and B. Currow. pp 20-47.

Tiller K G 1992 Urban soil contamination in Australia. Aust. J. Soil Res. 30, 937-957.

Wong J W C 1996 Heavy metal contents in vegetables and market garden soils in Hong Kong. Environ. Technol. (in press).

Wong M H 1985a Effects of animal manure composts on tree (*Acacia confusa*) seedling growth. Agri. Wastes, 13, 267-272.

Wong M H 1985b Heavy metal contamination of soils and crops from auto traffic, sewage sludge, pig manure and chemical fertilizer. Agri. Ecosys. Environ. 13, 139-148.

Wong M H 1987 A review on lead contamination of Hong Kong's environment. In Ocurrence and Pathways of Lead, Mercury, Cadmium and Arsenic in the Environment. Eds. T C Hutchinson and K M Memma. pp 217-234. John Wiley and Sons.

Wong M H and Bradshaw A D 1982 A comparison of the toxicity of heavy metals, using root elongation of rye grass, *Lolium perenne*. New Phytol. 91, 255-261.

Wong M H and Chui V W D 1990 Toxicity tests of soil contaminated by recycling of scrap plastics. Biom. Environ. Sci. 3, 81-89.

Wong M H and Lau W M 1985 Root growth of *Cynodon dactylon* and *Eleusine indica* collected from motorways at different concentrations of lead. Environ. Res. 36, 257-267.

Wong M H and Tam F Y 1978 Lead contamination of soil and vegetables grown near motorways in Hong Kong. J. Environ. Hlth. A13, 13-22.

Wong M H and Wong J W C 1986 Effects of fly ash on soil microbial activity. Environ. Poll. A40, 127-144.

Wong J W C and Wong M H 1987 Co-recycling of fly ash and poultry manure in nutrient-deficient sandy soil. Res. Conserv. 13, 291-304.

Wong J W C and Wong M H 1990 Effects of fly ash on yields and elemental composition of two vegetables, *Brassica parachinensis* and *B. chinensis*. Agric. Ecosys. Environ. 30, 251-264.

Wong M H, Cheung L C and Wong J W C 1984 Effects of roadside dust on the seed germination and root growth of *Brassica chinensis* and *B. parachinensis*. Sci. Tot. Environ. 33, 87-102.

Wong J W C, Li S W Y and Wong M H 1995 Coal fly ash as composting material for sewage sludge: effects on microbial activities. Environ. Technol. 16, 527-537.

Yang M S and Wong M H 1989 Land and water pollution. In Hong Kong Society. Ed. A Y H Kwan. pp 275-311. Writers and Publication Cooperative, Hong Kong.

17
Soil contaminants in India: an overview

R. PRASAD

1. Introduction

Environmental pollution in India, as elsewhere in the world, is a major issue of public policy and research. Unprecedented economic growth during the last fifty years has led to extensive environmental degradation of all parts of the ecosystem. During the past couple of decades, major programmes of afforestation, water and air pollution control have been launched by the Government along with the enactment of a comprehensive Act known as 'Environmental Protection Act 1986'. There have also been major programmes aimed at the reclamation of wastelands, especially large stretches of both alkaline and acidic soils.

Interest in heavy metal contamination of soils appears to be restricted to areas subjected to the impact of industrial emissions, sewage and effluents around the major cities and towns. A comprehensive review on heavy metal pollution (arsenic, As; cadmium, Cd; lead, Pb and mercury, Hg) of crops, vegetables, fruits, food, water, air and soil in various parts of India has recently been presented in a SCOPE publication (Krishna Murti, 1987). In contrast to metals, investigations on soil pesticide residues have been limited to the time dependence behaviour of toxic pesticide residues following application.

Most studies on soil contamination in India have been carried out by scientific groups on localised problems around the pollution source. These reports clearly demonstrate increasing soil contamination due to anthropogenic activities. This chapter on soil contaminants is divided into 3 major sections:

- a brief overview of the country, its topography, climate and soil classification and the status of industrialisation;
- studies related to heavy metal pollution of soils;
- studies related to pesticide residues in soils.

2. The country and its soil

2.1 Geology

India lies to the north of the equator between 8° 4′ and 37 6′ north latitude and 68° 7′ and 97° 25′ east longitude. It is bounded on the south west by the Arabian Sea and on the south east by the Bay of Bengal. On the north, north east and north west lie the Himalayan ranges. The southern tip, Kanyakumari, is bordered by the Indian Ocean.

R. Naidu et al. *(eds): Contaminants and the Soil Environment in the Australasia-Pacific Region, 513–542.*

India measures 3 214 km from north to south and 2 933 km from east to west with a total land area of 3 287 263 km^2. It has a land frontier of 15 200 km and a coastline of 7 517 km. Andaman and Nicobar islands in the Bay of Bengal and Lakshadweep in the Arabian Sea are parts of India. On the basis of physical, topographical, physiographic, geological structure and stratigraphical features, India has been subdivided into three regions:

- Extra-Peninsula with the Himalayas and its off-shoots, having a high mountainous character;
- Indo-Gangetic alluvial tract with plain topography; and
- Peninsula, the triangular tract of the Deccan, having a hilly and plateau-like character.

2.1.1 The Extra-Peninsula

The Himalayas and their off-shoots are geologically young mountains which resulted from earth movements during the Tertiary age. The rocks in these mountains are highly folded and faulted. The topography is quite rugged and the rivers are fast running, causing deep and narrow valleys and gorges. The rocks comprise an almost complete sequence of sediments except part of the Tertiary. While the north and Central Himalayas are associated with fossiliferous marine sediments, the Lesser Himalayas are associated mainly with non-fossiliferous sediments. The central axial zone of the mountains contains marine sediments of various ages, intruded by igneous rocks generally granitic in composition. These igneous intrusions are particularly noticeable in the central Himalayan belt. Some Archaen and Precambrian rocks have also been found in the central Himalayan ranges. The Himalayas are considered to be tectonically still active.

2.1.2 The Indo-Gangetic alluvial plains

The Indo-Gangetic alluvial plains consist of alluvial deposits of considerable thickness, the alluvium of recent origin belonging to Pleistocene and Recent times. The sediments consist of the silt deposited by the slow moving rivers.

2.1.3 The Peninsula

The Peninsula is the oldest land mass and the present land features have been formed by the denudation and weathering of this land mass over many millions of years (Archeozoic to Pliocene) and, therefore, only the hardest rocks which have escaped or withstood denudation have been left. Unlike the Himalayan rivers, those of the Peninsula are very old and flow slowly through shallow channels. Since they flow through areas that have a mature topography, they are incapable of scouring and deepening their valleys. The Peninsula is mostly composed of crystalline metamorphosed and igneous rocks comprising gneisses, schists and granites of the

Table 1. Distribution of annual rainfall according to seasons.

Rainfall Season	Duration	Approx. percentage to annual rainfall
South West Monsoon	June–Sept	73.7
Post-Monsoon	Oct–Dec	13.3
Winter or North East Monsoon	Jan–Feb	2.6
Pre-Monsoon	Mar–May	10.4
TOTAL		100.0

Source: Indian Agriculture in Brief, Tenth Edition, Directorate of Economics and Statistics, New Delhi.

Archaean system. A fairly large part of the country is covered by volcanic material (Deccan trap), but the rocks lying underneath it are again the oldest gneisses and schists.

2.2 Climate

The Indian climate is associated with two principal monsoons: North East Monsoon and South West Monsoon. The rainfall is distributed mainly in these two principal monsoons. The North East Monsoon is predominant during January, and results from the flow of surface air from Russia, China, Afganistan, Pakistan and Bangladesh and across India from the north-east. The South West Monsoon is predominant during July and the winds flow from the opposite direction i.e. from the Arabian Sea across India towards Pakistan, Bangladesh, Afganistan, Russia and China. The North East Monsoon is characterised by cool and very dry winds and very little rainfall. The approximate distribution of annual rainfall according to seasons is given in Table 1.

These data represent the general pattern, and there are variations due to delayed onset of monsoons and change in the direction of flow of wind related to the topography of particular regions. Annual variations at any one place may be quite significant. For example, annual rainfall may range from 250–500 mm in the driest zones such as Rajasthan West, to 3 000–7 000 mm in the high rainfall regions such as Arunachal Pradesh (Statistical Abstracts, 1989). Detailed rainfall patterns in India may be obtained from the Statistics Abstracts published by Central Statistical Organisation, Ministry of Planning, Government of India.

2.3 Soil characteristics and their classification

Indian soils have been divided into 24 major soil groups (Govinda Rajan, 1965) on the basis of the USDA classification. The chief distinguishing characteristics of each of these soil groups have been described in detail by Govinda Rajan and Gopala Rao (1978). Their distribution is shown in Table 2.

Table 2. The extent, distribution of the different soil classes in 'Soil Map of India' and their soil taxonomy.

Soil No.	Soil Classification Unit	Extent in km^2	Distribution (States)	Soil Taxonomy
1	Red loamy soils	213 271	Andhra Pradesh, Tamil Nadu, Karnataka, Kerala, Madhya Pradesh, Orissa	Alfisols
2	Red sandy soils	330 590	Tamil Nadu, Karnataka, Andhra Pradesh	Alfisols
3	Laterite soils	130 066	Tamil Nadu, Karnataka, Andhra Pradesh, Orissa, Maharashtra, Goa, Assam	Ultisols Oxisols
4	Red and yellow soils	403 651	Madhya Pradesh, Orissa	Ultisols
5	Shallow black soils	31 532	Maharashtra	Entisols Inceptisols
6	Medium black soils	430 383	Maharashtra, Madhya Pradesh, Gujarat	Vertisols
7	Deep black soils	112 060	Maharashtra, Andhra Pradesh, Karnataka, Madhya Pradesh, Gujarat	Vertisols
8	Mixed red and black soils	162 255	Karnataka, Tamil Nadu, Maharashtra, Madhya Pradesh, Andhra Pradesh, Bihar	Association of Alfisols and Vertisols
9	Coastal alluvium soils	54 403	Tamil Nadu, Kerala, Andhra Pradesh, Maharashtra, Gujarat	Entisols
10	Coastal sands	4 534	Orissa, Tamil Nadu, Andhra Pradesh	Entisols
11	Deltaic alluvium soils	87 045	Tamil Nadu, Andhra Pradesh, Orissa, West Bengal	Entisols Alfisols
12	Alluvial soils – Khadar (recent alluvium Bhangar old alluvium)	356 720	Uttar Pradesh, Punjab, Bihar, West Bengal, Assam	Entisols Inceptisols Alfisols
13	Alluvial soils (highly calcareous)	13 611	North-eastern Uttar Pradesh, Bihar	Entisols
14	Calcareous sierozemic soils	45 080	Punjab, Haryanal, Rajasthan	Aridisols
15	Grey brown soils	101 572	Gujarat, Rajasthan	Aridisols

continued

Table 2. (continued)

Soil No.	Soil Classification Unit	Extent in km^2	Distribution (States)	Soil Taxonomy
16	Desert soil Rhegosolic	154423	Rajasthan	Aridisols Entisols
17	Desert soils Lithosolic			Entisols
18	Tarai soils	28919	Uttar Pradesh, Bihar, West Bengal	Mollisols
19	Brown hill soils (over sandstones and shales)	81242	Uttar Pradesh, Bhuttan, Sikkim, Himachal Pradesh, Gujarat, J & K, Karnataka, Punjab, West Bengal	Utisols
20	Sub-montane soils (Podsolic)	76695	Uttar Pradesh, Jammu & Kashmir, Himachal Pradesh	Alfisols
21	Mountain meadow soils	59790	Kashmir including Ladakh	Mollisols
22	Saline & alkali soils	17377	Uttar Pradesh, Punjab, Maharashtra, Karnataka, Tamil Nadu, Gujarat, Rajasthan	Aridisols Entisols Vertisols
23	Peaty and saline peaty soils	2720	Kerala, West Bengal	Histosols
24	Skeletal soils	79151	Madhya Pradesh	Entisols
25	Glaciers & snow	29335	Uttar Pradesh, Kashmir	

Adapted from the book on 'Studies on soils of India' by Govinda Rajan and Gopala Rao (1978) with kind permission of the publisher, M/s Vikas Publishing Pvt. Ltd., New Delhi.

3. Industrial activity

India is known for its rich deposits of many minerals, ores and natural resources. Considerable quantities of coal including lignite, iron, manganese and other ores are mined. The production figures of selected minerals for the years 1980, 86 and 92–93 are shown in Table 3.

The production of most minerals has more than doubled during the last decade (Table 2). Likewise, industrial production in other sectors shows multiple growth. With increasing urbanisation and industrialisation, the recycling of heavy metals and other pollutants in the environment through biotic and abiotic pathways is bound to increase with increasing use of metals, pesticides and other chemical based commodities in the society. The impact of industries on contaminants in the soil environment is well illustrated by Agarwal et al. (1988) who found elevated con-

Table 3. Production of selected minerals.

Mineral	Unit*	1980	1986	1992–93
Coal	mn MT	109.1	162.6	247.3
Lignite	mn MT	4.5	8.9	15.9
Bauxite	th MT	1785	2662	4987
Chromite	th MT	321	630	1312
Copper ore	th MT	2005	4462	5422
Gold	kg	2452	1931	1875
Iron ore	mn MT	41.9	51.2	57.6
Lead conc.	th MT	16.7	44.2	63.3
Manganese ore	th MT	1695	1284	1699
Silver	th MT	11.3	35.2	43.2
Zinc conc.	th MT	46.4	89.5	314.3
Phosphorite	th MT	523	656	391
Asbestos	th MT	33.7	26.7	40.0
Barytes	th MT	442	337	401
Diamond	th Cts	14.4	16.1	20.8
Dolomite	th MT	2031	2210	2734
Fireclay	th MT	762	683	480
Fluorite graded	MT	4236	6841	4107
Gypsum	th MT	866	1.641	1.125
Kaolin	th MT	460	747	661
Kyanite	th MT	49.1	32.3	9.0
Sillimanite	th MT	14.4	14.6	14.5
Limestone	mn MT	29.1	53.5	78.9
Calcareous sand	th MT	722	537	81.0
Magnesite	th MT	380	460	618.3
Mica (crude)	MT	7.934	4746	3026
Pyrites	th MT	82.9	22.0	142.9
Steatite	th MT	360	379	374.8
All minerals	Rs mn	230.2	1014.6	2018.0

Source: Industrial Data Book 1993–1994.
Ed. Dr S.R. Mohnot, Centre for Industrial & Economic Research, New Delhi.
*mn = million; th = thousand

centrations of Zn and other heavy metals in dust, soils and vegetation around a Zn smelter and attributed these to emissions from the smelter.

Increased industrialisation has also led to greater wealth amongst middle class people which has culminated in a sharp rise in motor vehicle numbers and a consequent increase in the use of leaded fuel. Khandekar (1984) reports that the concentration of Pb in dusts and vegetation around roads in Bombay increased an alarming 4 fold from 13 to 42 $\mu g\, g^{-1}$ soil during 1960 to 1976. Given the sharp increase in motor vehicle numbers during the last decade it is likely that contamination by lead from vehicle exhausts has also increased. This is also apparent from the more recent study of Khandekar et al. (1988) that shows elevated concentrations (relative to background concentrations) of Cd, Zn and Cu in surface soils in addition to the very high concentrations of Pb (Table 4).

Major developments in agriculture and industry have taken place only during the last fifty years, resulting in substantial increases in various pollutants over their pristine levels in the environment. This is manifest in continuous deterioration of air/water quality around major commercial and industrial centres as observed in

Table 4. Mean Pb, Cd, Zn and Cu concentrations in Bombay soils.

Place	No. of samples	Pb	Cd	Zn	Cu
		mg kg^{-1} of soil			
Haji Ali	3	192±22	1.6±0.3	309±32	216.7±19.5
Dadar	3	54.6±7.5	0.59±0.3	260±24.3	160.2±23
Santacruz	7	42.1±8.4	0.7±0.4	182±16.2	162.6±20.6
Deonar	18	8.6±2.1	0.6±0.3	141±12.4	101.3±14.2
Matheran	5	4.3±1.5	–	78±14.2	63.1±11.3

Reproduced from Khandekar et al., 1988.

air/water pollution monitoring studies. However, due to the lack of similar monitoring studies with regard to soil-contaminants, the task of making realistic estimates of the level of pollutants in soils and their ultimate fate is beset with many difficulties. There are innumerable studies in the Indian literature confirming high levels of heavy metals (As, Cd, Pb, Hg) and residues of chlorinated pesticides in samples of air, food, water, grains, vegetables, cigarettes and also in marine organisms, coastal waters and sediments (Krishna Murti, 1987). Also, there are many studies reported in the literature relating to the source pollution of groundwater, land and air due to emissions and effluents from a variety of industries, i.e. thermal power plants, pulp and paper mills, titanium dioxide pigment plants, textile industries, oil refineries, Zn smelters, etc. The following sections summarise some of these published reports.

3.1 Heavy metal contaminants in soils

Increasing urbanisation and industrialisation in India have added new dimensions to the safe disposal of sewage and industrial effluents. The extent of urban development can be gauged from the growth of some Indian cities which exceed a population of one million (Table 5). The number of cities exceeding a population of one million rose from 5 in 1951 to 9 in 1971 and 23 in 1991.

Most cities in India do not have sewage and wastewater treatment plants or even collection facilities. The Central Pollution Control Board (CPCB) of India has periodically carried out country-wide surveys of sewage and wastewater generation, collection, treatment and disposal for both Class I (i.e. population above 100000) and Class II (i.e. population between 50000 and 100000) cities. Its recent report (CPCB, 1988) presents information on 212 Class I cities which cover 65% of the Indian population (Table 6) and together generate in excess of 10^{10} L d^{-1} of wastewater.

Some Class I cities do not have facilities for collection and treatment of sewage and wastewater and more than 90% of these wastes are being directly disposed of onto land. While it has not been possible to assess precisely the impact of such disposals on land and surface water bodies, Table 7 shows that at least 42 Class I cities continue to dispose sewage on agricultural land.

Table 5. Trends in urban agglomerations (cities) with mega ('000) population.

Cities	1951	1971	1991
Greater Bombay	2966	5970	12571
Calcutta	4669	7420	10916
Delhi	1437	3647	8375
Madras	1542	3169	5361
Hyderabad	1130	1796	4280
Bangalore	786	1664	4086
Ahmedabad	877	1752	3297
Pune	608	1135	2485
Kanpur	705	1275	2111
Nagpur	485	930	1661
Lucknow	496	813	1642
Surat	237	493	1517
Jaipur	304	636	1514
Kochi	177	505	1139
Coimbatore	287	736	1135
Vadodara	211	467	1115
Indore	310	560	1104
Patna	326	551	1098
Madurai	370	711	1093
Bhopal	102	384	1063
Visakhapatnam	108	363	1051
Varanasi	369	635	1026
Ludhiana	153	401	1012

Source: Industrial Data Book 1993–94.
Ed. Dr S.R. Mohnot, Centre for Industrial & Economic Research, New Delhi.

To a limited extent, sewage water is disposed following primary and secondary treatments. This process, which separates sludge from sewage, reduces its biological oxygen demand (BOD) and also concentrates toxic elements in the solid phase. The treated water is considered safe for irrigation and/or discharge into a water body provided it conforms to the general standards contained in the Environment Protection Second Amendment Rules published vide notification number 174 dated May 19, 1993 in the Gazette of India. These rules, which are now mandatory, prescribe the standards for effluents to be discharged into inland surface waters, public sewers, onto land for irrigation and into marine coastal areas. The capital and recurring costs of such treatment plants are substantial and beyond the means of Municipal Committees of most cities.

The treatment of sewage in oxidation ponds is relatively less expensive and has been employed with varying success, often in combination with pisciculture in many places. Sewage is commonly disposed of along with other effluents onto agricultural land. While in some states this has led to large scale land degradation, turning disposal areas into waste lands and also contaminating surface and groundwater bodies, raw sewage and wastewater is being fruitfully utilised in certain parts of India in sewage farming for agriculture and pisciculture. However, analyses of soils from these farms indicate elevated heavy metal and nutrient concentrations (Shree Ramulu, 1990; Rao and Shantaam, 1994). Moreover, there are numerous

Table 6. Statewise position of wastewater generation, collection and treatment in class I cities.

State/Union Territory	Total No. of Cities	Population 1981 Census	Projected Population 1988	Wastewater 10^6 L d^{-1}		Wastewater treatment capacity 10^6 L d^{-1}	
				Generated	Collected	Primary only	Primary and Secondary
Andhra Pradesh	17	5351818	6902960	781	163.40	–	172
Bihar	11	2750604	3774682	445	–	–	43
Gujarat	11	5316142	6794925	815	220.50	86	463
Haryana	9	1329549	1966597	130	38.53	0	0
Karnataka	14	5472617	7464167	655	462	327	10
Kerala	6	1901069	2184562	223	24.08	–	5
Maharashtra	29	15740723	20210609	2838	190	293	–
Meghalaya	1	109244	127444	3	–	–	–
Manipur	1	156622	213865	207	–	–	–
Madhya Pradesh	14	4613431	6147991	552	168	86	–
Orissa	6	1105123	1539659	207	22	0	–
Punjab	7	2155714	2764376	233	158	–	–
Rajasthan	11	3306345	4446277	318	–	–	–
Tamil Nadu	19	7600491	8995524	487	53	–	0
Tripura	1	132186	160403	13	–	–	–
Uttar Pradesh	29	9090035	11036290	1449	255	–	160
West Bengal	21	8021158	9145473	1107	27	–	20
Chandigarh	1	379660	391346	144	106	68	0
Delhi	1	5157270	7464000	1480	745	0	745
Pondicherry	1	162639	186675	38	0	–	0
Total:	210	80661952	102944951	12125	2632	860	1618

Source: CPCB, 1988.

Table 7. Distribution of cities according to State and mode of disposal of wastewater.

State/Union Territory	Total No. of Cities	No. of cities: Agricultural Land	River	Both River and Agricultural Land
Andhra Pradesh	17	4	3	2
Bihar	11	1	2	–
Gujarat	11	1	2	4
Haryana	9	3	1	5
Jammu & Kashmir	2	–	–	–
Karnataka	14	1	7	–
Kerala	6	1	–	–
Maharashtra	29	5	5	6
Meghalaya	1	–	–	–
Manipur	1	–	–	–
Madhya Pradesh	14	1	3	5
Orissa	6	–	2	–
Punjab	7	3	2	–
Rajasthan	11	10	–	1
Tamil Nadu	19	5	7	2
Tripura	1	–	–	–
Uttar Pradesh	29	5	4	6
West Bengal	21	–	6	–
Chandigarh	1	1	–	–
Delhi	1	–	–	1
Pondicherry	1	1	–	–
Total	212	42	44	32

Source: CPCB, 1988.

studies which show considerable soil-plant transfer of toxic metals (Nashikkar and Chakrabarti, 1994).

Since there are no reported investigations on toxic organic substances it is not possible to comment on the status of these in sewage farm soils and the implications for crop contamination. Examples of various sewage sludge and effluent utilisation strategies adopted in India are presented in the following sections.

3.2 Utilisation of sewage and sludge in irrigation

In the absence of appropriate treatment facilities, market forces are encouraging the utilisation of sewage and sludge for the production of a variety of biomass types. Interest in sewage sludge utilisation for agricultural production appears to have been gener-ated from the nutrient-rich nature of the sewage and its potential to improve the physical condition of soil. For example, in Calcutta a number of cooperative ventures have developed for raising vegetables, fish and crops. In Tamil Nadu, the scarcity of water, coupled with shortage of fodder and forage, has encouraged more than 20 municipalities to establish sewage farms for forage and fodder grass production as well as a number of vegetable crops (Sree Ramulu, 1990)

A large number of institutions (Central Arid Zone Research Institute, Jodhpur; Central Soil Salinity Research Institute, Karnal; Raja Balwant Singh College of

Agriculture, Agra; Centre for Application of Science and Technology for Rural Development, Pune; Consortium of Indian Scientists for Sustainable Development, Delhi and many Forest Departments and local authorities in Haryana, Rajasthan and West Bengal and some public sector companies such as Indian Petro Chemicals Limited, Vadodara in Gujarat) have been engaged in raising a variety of tree species and developing appropriate site-specific technologies for utilising sewage. The National Botanical Research Institute at Lucknow has standardised the process to produce *Spirulina plantesis* algae on raw sewage for use as poultry and fish feed. The results of such research and scattered developmental efforts show that there is considerable scope for utilising sewage and industrial effluents. An excellent compilation of environmentally friendly technologies is now available on the disposal and use of sewage and industrial effluents for raising appropriate biomass products (Das and Kaul, 1992).

3.3 Post-irrigation effects

While there is increasing interest in sewage farms, due primarily to the availability of nutrient-rich water, there is some concern about heavy metal accumulation and uptake by crops grown on sewage-irrigated soils. With increasing industrialisation a greater volume of effluents, particularly from many very small scale industries (e.g. electroplating industries, tanneries, paper and board mills, textile and dyeing units) which are generally not equipped with effluent treatment facilities, are getting into existing sewage systems.

The following sections summarise selected studies that show that, while irrigation with sewage and industrial effluents enrich the soil with both macro and micro nutrients, there is considerable potential for soil contamination by heavy metals.

3.3.1 Madurai sewage farm

Sree Ramulu (1990) investigated the impact of 50 years of sewage applications on nutrient content and heavy metal status of the soils at the Madurai Sewage Farm. He found that both soil organic carbon (OC) content and the total and available N, P, K, Zn, Cu and Fe increased manyfold in the sewage-irrigated soil. The produce from the irrigated farm, however, showed elevated concentrations of heavy metals relative to the control plot although the concentrations were within the optimum tolerance limits of the plants. However, Sree Ramulu did not relate plant metal concentrations to the Maximum Permissible Concentration for human or animal consumption guidelines in India.

3.3.2 Wastewater irrigation in Punjab

The impact of industrial effluents on soils in some districts of Punjab was recently assessed under an extensive programme of evaluation by the Ministry of Environment & Forests (personal communication, Drs. Y.P. Kakkar and Mrs.

Indrani Chandrashekhar). Data from these investigations (Table 8) show that irrigation with industrial effluents and wastes causes marked changes in both soil surface and subsurface properties; some of these changes are related to the nature of the industrial effluents. Major trends following the application of industrial effluents are discussed below:

- Influence of trade effluent of electroplating industries:

  Application of effluent from these industries enhanced soil acidity and the soils exhibited high salt content with high conductance relative to the unirrigated control soils. There was a build up of Ni and Cr (190 and 93.5 $mg\,kg^{-1}$, respectively) in the surface soils. Although the effluent was used for irrigation of wheat and vegetables its impact on the soil-plant transfer of heavy metals was not studied.
- Influence of trade effluent of paper and board mills:

  Irrigation of soils with highly caustic and sodium rich paper mill effluents markedly increased the soil pH and salt content relative to those which were irrigated with the effluent from the cardboard industry where caustic is not used. The cardboard paper mill effluent was also used for irrigation of wheat and fodder, while that from the paper mills with high sodium content was used with or without mixed tube well water for irrigation of eucalyptus, vegetables, horticulture and wheat. Use of paper mill effluent is of concern given the high sodicity and its impact on soil structure.
- Influence of trade effluent of textile and dyeing units:

  Effluent treated soils show moderately high salt levels compared to that in untreated soils (Table 8). The cumulative effect of this ion buildup over time may be far more serious than currently anticipated. The effluent was also used for irrigation of eucalyptus plantations. The impact of toxic unreacted dyes, detergents and other organic and inorganic compounds that are present in the effluent is likely to have an adverse effect on soil and plant quality; however, the plant availability of the toxic organic and inorganic constitutents was not examined but may become important if vegetables and other edible crops are grown on such soils.
- Influence of distillery waste waters:

  Only limited work has been done with distillery waste waters; as with other industrial effluents, distillery wastes also lead to increased salt content of the surface soils which is reflected by elevated EC values. Although this effluent was also used for fodder and wheat fields, there was no information on the quality of fodder and wheat in relation to any toxic metals present in the waste waters. The properties of the sub-soil layer were also not investigated, but would have provided useful insight into the permeability of the soil.
- Influence of sewage irrigation:

  This study focused on soils irrigated by sewage water in two cities. Both the cities support numerous industries that include electroplating, textiles & dyeing, pulp and card board industries. Effluents from these factories are disposed

Table 8. Physico-chemical characteristics of soils in Punjab irrigated by industrial effluents and sewage.

No.	Source of sample	pH	Conductance (mhos cm^{-1})	NH_4–N (mg kg^{-1})	OM (%)	P_{avail} (mg kg^{-1})	Zn_{avail} (mg kg^{-1})	Ni (mg kg^{-1})	Total Cr (mg kg^{-1})
1.	Soil Irrigated by	a) 5.7–5.8	0.14–0.19	7.6–8.2	0.21–0.23	110–110	9.7–9.9	125–190	87.0–93.5
	Effluent of	b) 5.5–5.9	0.12–0.15	7.0–9.2	0.18–0.26	100–110	2.4–4.5	8.2–19.6	5.4–9.7
	Electroplating Unit	a) (6.6)	(0.07)	(13.8)	(0.25)	(200)	(2.4)	(–)	(5.4)
2.	Soil Irrigated by Effluent of:								
	Paper Mill A	a) 7.40–7.79	0.35–0.37	13.8–14.8	0.05–0.06	a) 200–270	–	–	–
		a) (7.20)	(0.33)	(15.8)	(0.32)	a) (210)		–	–
	Paper Mill B	a) 6.20	0.19	15.8	0.18	a) 170	2.7	–	–
		b) 6.24	0.17	13.6	0.21	b) 170	2.6	–	–
		a) (6.60)	(0.13)	(10.0)	(0.24)	a) (220)	(4.5)	–	–
	Card Board Mill	a) 6.80	0.24	11.0	0.11	a) 170	4.4	–	–
		b) 6.10	0.21	15.8	0.07	b) 180	1.2	–	–
		a) (6.80)	(0.19)	(16.8)	(0.41)	a) (230)	(2.9)	–	–
3.	Soil Irrigated by	a) 6.40–6.90	0.14–0.27	11.8–14.8	0.04–0.29	a) 160–230	2.4–6.5	–	–
	Effluent of Textile &	b) 6.0–6.90	0.15–0.16	7.6–14.0	0.02–0.08	b) 140–170	2.2–4.2	–	–
	Dyeing Units	a) (6.20)	(0.17)	(11.0)	(0.21)	a) (210)	(2.1)	–	–
4.	Soil Irrigated by	a) 6.80	0.76	–	–	a) –	–	–	–
	Effluent of Distillery Waste	a) (7.00)	(0.35)	–	–	b) –	–	–	–
5.	Soils Irrigated by	a) *6.06–6.5	0.13–0.55	7.4–15.8	0.09–0.92	a) *100–270	16.5–116.6	5.0–46.4	10.7–29.1
	Sewage Water in City A	a) 6.19–6.48	0.18–0.21	11.0–12.4	0.12–0.90	a) 103–260	54.0–70.0	70.7–87.4	33.8–37.8
		b) *6.08–6.6	0.17–0.22	7.8–14.8	0.06–0.13	b) *107–240	2.4–76.0	2.2–8.2	3.0–7.3
		b) 6.23–6.41	0.18–0.23	11.0–13.0	0.05–0.29	b) 100–160	9.3–12.8	6.3–9.4	7.1–8.9
		a) (6.67–6.78)	(0.17–0.21)	(11.8–14.8)	(0.09–0.31)	a) (106–107)	(10.0–16.8)	(3.0–4.8)	(4.2–8.2)

a* = 0–50 mm soil sample from fields irrigated with effluent mixed with tube well water, b* = 250–300 mm soil sample from fields irrigated with effluent mixed with tube well water; a = 0–50 mm soil sample from fields irrigated with effluent only, b = 250–300 mm soil sample from fields irrigated with effluent only. Figures in parentheses are those of reference soils.

Table 9. Heavy metal concentrations (mg kg^{-1}) in sludge, uncontaminated soil and forage grass.

Source	Pb	Cd	Zn	Cr	Ni	Se
Sludge	22.8–1160.0 (15,256.76)	0.48–13.92 (15,2.48)	63.3–126.3 (15,113.34)	7.81–18.81 (15,12.76)	12.09–35.36 (15,22.44)	4.57–9.40 (15,7.12)
Uncontaminated soil	13.5–25.61 (10,17.22)	0.02–0.08 (10,0.06)	5.66–17.54 (10,10.44)	25.88–36.35 (10,28.22)	18.41–26.31 (10,20.45)	0.41–2.10 (10,1.44)
Forage grass grown on sewage sludge	1.4–17.8 (15,6.44)	0.24–2.2 (15,0.816)	70–117.2 (15,98.43)	16.97–54.69 (15,31.82)	1.37–5.88 (15,2.85)	3.24–9.26 (15,5.85)
Forage grass grown on uncontaminated soil	0.45–1.4 (10,0.85)	0.12–0.24 (10,0.18)	3.21–5.84 (10,4.22)	10.41–18.55 (10,12.66)	0.11–1.5 (10,0.88)	1.94–2.5 (10,2.0)

Figures in parentheses indicate number of samples and arithmetic mean of the values, whose range is given in the first line; based on the results reported in Srikanth et al., 1992.

through city sewers into the drains meant for irrigation of fields. Such disposals from city sewers reflect the lack of stringent regulatory measures needed to maintain quality control of industrial waste disposals in greater India. These effluents are used for growing vegetables and fodder. Pollution of soil in these cases was evident with high levels of acidity, salt content and heavy metals such as Zn, Ni and Cr. There is no reported study on the impact of such disposals on the soil-plant transfer of toxic substances including pathogens.

3.3.3 Heavy metal pollution of agricultural soils around Hyderabad

The city of Hyderabad generates approximately 12000–15000 t d^{-1} of solid waste which is commonly used as manure and soil conditioner in agriculture. Srikanth et al. (1992) studied the distribution of Pb, Cd, Ni, Zn, Cr and selenium (Se) in both sludge and perennial forage grass *Panicum maximum* (guinea grass) cultivated in the sludge along the bank of the river Musi, which receives the sewage from over three million inhabitants of the city of Hyderabad in South India. They found that the mean concentration of heavy metals in the grass grown on sewage sludge was higher than that of controls (Table 9).

Variations in the uptake of metals were observed; e.g. accumulation of Pb in the grass was low compared to Cd in relation to their relative concentration in the sludge. The uptake of most heavy metals analysed was above toxicity threshold levels except for Ni and Cr. The Se level in forage grass was related to the Se status in soil in which they were grown.

Since enzymes play a key role in plant metabolism, it has been suggested that their activities could be taken as an indicator of heavy metal pollution of soil and subsequent stress status in plants. This has been demonstrated by Nashikkar and

Chakrabarti (1994) of the National Environmental Engineering Research Institute (NEERI), Nagpur based on the effect of subtoxic levels of Cr and Cd on activities of catalase and peroxidase enzymes in plant leaves of four experimental crops, viz., maize (*Zea mays*), pea (*Pisum sativum*), moong (*Vigna radiata*) and wheat (*Triticum aestivum*).

Rao and Shantaram (1994) assessed the micronutrients and heavy metal pollution loading of surface and subsurface agricultural soils from seven different sites which had periodically (1, 2 or 3–4 yr) received urban solid waste from Hyderabad at a rate of about $100\,t\,ha^{-1}$ over a 10 year period. Soil samples from a nearby plot which were not treated with solid waste were used as a control. While the application of waste enhanced the concentrations of major and minor nutrients including total Zn, Cu, Mn, Pb, Ni, Co, Cr and Cd and also for DTPA-extractable heavy metals Zn, Cu and Mn there was a marked decline in extractable Fe. This decline in Fe was attributed to the dilution effect of the Fe-deficient wastes (Table 10). These results suggest that application of municipal wastes can have many beneficial effects on soil nutrient status (data on macro nutrients such as N and P are not included here) and crop growth although their high heavy metal content demands further work on the quality of produce.

3.3.4 Soil contamination below a sewage disposal pond in Jaipur

The city of Jaipur supports over one million people. It has a large number of small industries that include textile dyeing and printing, metal polishing and carving, dairies and miscellaneous chemical units. Over the last two to three decades these industries have disposed effluent into the city sewage which is directed into an open space of over 180 ha, known as Jalmahal, situated on the outskirts of the city. The nutrient and salt content of soils beneath this sewage disposal pond have been studied by Shrivastava et al. (1988, 1989) who found that long-term disposal of city municipal wastes has led to a marked buildup of N and P and salinity in the soils immediately beneath the sewage disposal pond. These results demonstrate the need for an impervious seal to contain the waste and prevent the leaching of contaminants in to the subsurface soil horizons. Analyses of hand pump and open well water samples situated in the close vicinity of the pond indicate extensive contamination of ground water. The pH, EC, total dissolved solids (TDS), HCO_3^-, Cl^- and $SO_4^=$ concentration were found to be 8.0, 7 400 μmhos cm^{-1}, 5 191, 437, 1 333 and 190 $mg\,L^{-1}$ respectively. These values are high and exceeded the permissible limits for drinking water.

3.3.5 Soil contamination due to coal ash from thermal power plants

Coal ash from thermal power plants is considered to be a potential carrier of trace elements such as As, B, Se etc. In a study of captive fly ash (VP Chest Institute, 1983) collected from the Indra Prastha Power Plant in New Delhi, the average EDTA-extractable As content was found to vary from 0.091 to 1.328% as elemental As.

Table 10. Range in total and DTPA extractable metal content in soils treated with urban solid wastes in Hyderabad (Rao and Shantaram 1994).

	Micro-nutrients mg kg^{-1}				Heavy Metals mg kg^{-1}				
	Fe	Zn	Cu	Mn	Pb	Ni	Co	Cd	Cr
Total Metal Content in Soil Treated with Urban Solid Waste	2795–7768 (14700)	33–177 (13.8)	4.2–71.8 (7.5)	90.5–385 (195)	11.3–170.1 (5.2)	0.95–10.25 (2.3)	2.00–7.00 (5.0)	0.40–0.85 (0.5)	5.95–47.85 (6.6)
DTPA-Extractable Metal Content	4.0–18.7 (5.6)	1.95–14.8 (0.8)	0.86–27.5 (0.8)	3.2–20.0 (10.0)	0.10–25.0 (0.03)	0.1–0.9 (0.03)	0.2–0.9 (0.03)	ND–0.2 (ND)	0.2–0.4 (ND)

ND–not detected; parentheses indicate the values for the control sample.

Table 11. Physico-chemical characteristic (μmhos/cm) of soil in UERC lysimeter experiment.

	Before experiment		After experiment		
Constituents	Soil	UERC	0–25 cm	25–50 cm	>50 cm
pH	8.20	7.60	8.02	8.31	8.27
Conductivity (μmhos/cm)	200.00	2850.00	260.00	190.00	180.00
Bulk Density (g cm^{-3})	1.12	–	1.00	1.16	1.16
Organic matter (%)	12.50	65.31	18.24	15.39	13.67
Chloride (mg L^{-1})	125.0	1600.00	20.50	9.00	9.00
Sulphate (mg L^{-1})	150.0	63.75	12.0	100.0	90.00
Nitrate (mg L^{-1})	80.0	960.00	86.00	40.00	40.00
Nickel (mg L^{-1})	30.50	54.00	42.80	40.00	38.20
Zinc (mg L^{-1})	350.00	524.00	450.00	486.00	413.00
Copper (mg L^{-1})	70.00	160.00	112.10	89.50	80.12
Chromium (mg L^{-1})	30.00	150.00	92.00	38.12	34.75

All values are in mg kg^{-1} except pH and conductivity; reproduced from Olaniya et al., 1991.

Pervez and Pandey (1994) estimated concentrations of various contaminants in river sediments receiving thermal power ash pond discharge from Korba Thermal Power Plant in Central India. The maximum values (mg kg^{-1}) obtained for various trace elements were as follows: Mn (83.4), Hg (43.5), Pb (35.0), As (15.0), Ni (61.0), Cu (38.0), Cd (3.0) and Cr (10.0). These results suggest that continual disposal of power ash in the river can lead to extensive pollution of both the sediments and river water.

3.4 Heavy metal mobility in soils

The migration of toxic heavy metals in soils amended with compost and sewage sludge has been the focus of research by Bhide and his co-workers at the NEERI, Nagpur (Olaniya et al., 1991). Using lysimeters they showed that while the composition of the effluent (heavy metals and OM) may vary depending on the sludge/manure composition, moisture content, and temperature etc., the interaction between soil and leachate depends largely on their individual characteristics.

They found that soils leached with leachates from urea enriched compost (UERC), over a period of one year showed a considerable movement of both nitrate and heavy metals Cu, Cr, Ni and Zn to the subsurface horizon (Table 11).

A similar trend in heavy metal behaviour was observed for sewage sludge amended soil. The high mobility of Zn in these soils was also reflected in the high Zn concentration (20–30 mg L^{-1}) of the leachate and the low Zn adsorption capacity of the surface soil. Based on their data on dissolved organics in the leachate samples Olaniya et al. (1991) concluded that fulvic acid is the predominant carrier of toxic heavy metals presumably as metal-organic complexes. Specific reviews on soil and solution factors controlling metal mobility in soils are cited in Chapter 3.

4. Pesticide residues in soils

Use of synthetic pesticides in modern agriculture is increasing to meet the growing demand for food. Regardless of the mode of application, a large proportion of the added pesticide ultimately finds its way into the soil environment (see Chapter 4). The efficacy of these chemicals after contact with the soil, their persistence in active form or their metabolites, is currently the subject of intense study throughout the world.

At present India is the largest manufacturer as well as consumer of pesticides in Southern Asia. The production of technical grade pesticides in the country during 1991-92 was approximately 74kt, whereas the total consumption was 72kt. On average, approximately $500\,g\,ha^{-1}$ of pesticide is used in India. This is quite low compared to many European countries, Japan, and USA. This, however, does not mean that the problem of pesticide contamination does not exist in India. The group-wise consumption pattern of pesticides (fungicides, insecticides, herbicides, rodenticides etc.) in India is different when compared with the global consumption pattern. In India, more than two thirds of the total pesticide consumption comes from the insecticide group whereas on the global scale the share of the insecticide group is about 35%, next only to herbicides (45%). The pesticide use pattern in India shows that by far the most heavy use of pesticides is on cotton (approximately 50% of the total consumption), followed by paddy, jwar, vegetables, wheat and oil seeds.

Indiscriminate use of pesticides has resulted in the contamination of plants, air, water, soil, animals and the human systems. Many reports have appeared in India indicating the presence of pesticide residues in food commodities (ICMR, 1993), water resources (Dikshith et al., 1990; Kaphalia et al., 1988) and human blood and tissues (Kaphalia, 1983). Pesticide residues in the food chain and water resources are, therefore, a major public concern today. As a result, a great deal of expertise in safety evaluation of pesticides and industrial chemicals has been developed by the Industrial Toxicological Research Institute at Lucknow (ITRC, 1992). At the National Botanical Research Institute (NBRI), Lucknow, a project is in progress to estimate pesticide residues in plants and soils. Under this project, residue levels of various carbamate, organophosphorus and organochlorine insecticides in crops of commercial importance such as betelvine (*Piper betle* L.), opium poppy (*Papaver somniferum* L.), water chestnut (*Trapa natas*), tea (*Thea chenisis*) and a variety of vegetables and fruits such as mango (*Margitera indica*) are currently being studied. High levels of residues in some of these commodities have already been observed, prompting monitoring of pesticide residues in soils.

The persistence of pesticides in soils seems to have received minor public attention compared to their presence elsewhere in the ecosystem. Nevertheless, pioneering efforts have been made by agricultural scientists at the Indian Agricultural Research Institute (IARI), New Delhi and elsewhere, who studied the persistence,

Table 12. Persistence of organo-chlorine insecticides in soil.

Insecticide*	Crop	% Loss of active ingredient	
		After 100 days	After 180 days
		%	%
BHC	Moong/Lobia	97.5	–
Lindane	Fallow	94.0	99.4
Aldrin	Moong/Lobia	87.9	–
Aldrin	Fallow	87.0	96.3
Dieldrin	Fallow	79.0	96.4
Heptachlor	Fallow	68.0	86.0
DDT	Fallow	74.0	94.8

*applied at 10 kg a.i ha^{-1}; reproduced from Jain and Agnihotri (1980).

leaching and movement of a number of pesticides in a variety of soils under field and laboratory conditions (IARI, 1980; Jain and Agnihotri, 1986). The findings of some of the important studies on these aspects are summarised in the following sections.

4.1 Persistence of insecticides in soils

As discussed in Chapter 4, a proportion of pesticides reaching the soil environment is slowly degraded by soil micro-organisms or through chemical processes into their metabolites, while the remaining undegraded chemicals can accumulate to toxic levels. Pesticide molecules are known to perturb vital soil processes, causing changes in microbial activity and N, P and K uptake by plants. Since pesticide adsorption varies with climatic factors as well as types of soils, their persistence in soils assumes significance.

4.1.1 Organo-chlorine insecticides

Persistence of benzene hexa chloride (BHC), aldrin, dieldrin, lindane, heptachlor and DDT was studied at IARI farm in New Delhi during September 1974–February 1975 (Agnihotri et al., 1974, 1977). Five percent dust of each insecticide was applied separately at 10 kg active ingredient (a.i.) ha^{-1} and mixed thoroughly with the top 10 cm of the soil. Samples were taken periodically and analysed for remaining insecticide. Persistence in soil was studied in fallow land as well as with different crops. The rate of dissipation of the pesticides was maximum during the first 30 d, varying between approximately 30 and 50%, slowing considerably thereafter. In 180 d, loss of lindane was the highest followed by aldrin, dieldrin, DDT and heptachlor; the losses being 99.4, 98.6, 96.4, 94.8 and 86.0% respectively (Table 12).

The observed high rates of dissipation of organochlorine insecticides have been

Table 13. Persistence of organo-chlorine pesticides under covered and uncovered conditions.

	% Loss of active ingredient after 120 days	
Insecticide	Covered	Uncovered
Heptachlor	62	83
Aldrin	70	75
Dieldrin	74	50
Lindane	94	80

attributed to the hot and humid climates. Since the relative dissipation was in the order of their volatilisation rates, i.e., heptachlor<aldrin<lindane<DDT<dieldrin (Edward, 1966) the observed dissipation was attributed to a combination of both volatilisation and alkaline hydrolysis. The rapid loss of DDT and lindane may be attributed to their susceptibility to alkaline hydrolysis due to the high $pH_{1:2water}$ (8.2) of IARI soil (Entisol).

Aghinotri et al. (1984) investigated the relative contribution of volatilisation and degradation in the dissipation of aldrin, dieldrin, heptachlor and lindane in a soil from the IARI farm. These insecticides were mixed with surface (top 10 cm) soils at 10 kg ai ha^{-1}. Losses of active ingredients of these pesticides from the covered (with 5 mm thick glass sheet sealed over concrete rings of 60 cm I.D. and 20 cm height) and open portions in the treated plots were recorded over time (Table 13).

The dissipation of aldrin and heptachlor was faster under cover than in the open field. While 50% of the applied dieldrin was lost under open conditions, 74% was lost under cover over a period of 120 days. The corresponding values for lindane were 80 and 94% respectively. These studies suggest that losses are not by volatilisaton *per se*. Other factors such as soil alkalinity can also play an important role in accelerating the dissipation of these insecticides. Although microbial degradation may also contribute significantly towards the dissipation of these insecticides, this aspect was not studied by the investigators.

Gajghiye and Agnihotri (1991, 1992) conducted a sytematic study of the role of volatilisation in the total dissipation of organochlorine pesticides such as aldrin and lindane over three consecutive seasons. They reported that losses due to volatilisation may be quite significant in the case of lindane, particularly during rainy season. This suggests that both volatilisation and degradative processes play significant roles in the dissipation of pesticides from soils. However, dissipation may depend upon factors such as seasonal variations (temperature and rainfall) and soil properties (moisture, organic matter, clay content, etc.) Agnihotri et al. (1977) have also studied the persistence of aldrin, dieldrin and lindane in a red soil (Alfisol; sandy loam, pH 6.6 and organic content 1.2%) and the IARI soil (Entisol; loam, pH 7.8 and organic content 0.4%) under laboratory conditions at different moisture levels (air dried, field capacity and submerged under water) by treating the soils with 5 mg L^{-1} pesticides and incubating the treated soil at 30±1°C in a BOD incubator saturated

Table 14. Degradation of organochlorine insecticides at different moisture levels.

Insecticide	Percent degradation after 180 days					
	IARI soil			Red soil		
	AD	FC	SM	AD	FC	SM
Aldrin	46	90	71	42	88	70
Dieldrin	15	52	38	15	50	40
Lindane	48	92	86	50	96	90

AD – air dried; FC – field capacity; SM – submerged with water.

with water vapours for 180 d. They found that in both the soils there was a rapid dissipation of these insecticides at field moisture capacity (Table 14).

Submergence of soils also increased dissipation but the extent of dissipation was at least 20 to 30% less than that observed at field capacity. These results suggest that in both soils between 50 to 90% of the organochlorine pesticides break down quite rapidly. However, work on the nature of the metabolites is necessary to determine the occurrence and concentration of potentially toxic degradation products.

The effect of chlorinated insecticides on crop growth and the uptake of nutrients by wheat and maize has been studied in the field (Thakre and Saxena, 1972). Treatment at three levels, viz. 10, 20 and 30 mg kg^{-1} of active ingredient of DDT, to soil inhibited growth of wheat relative to the control sample (untreated). However, irrespective of the level of DDT application, toxicity to maize was not observed; in fact the highest treatment increased dry matter yield. Likewise, endrin at all levels inhibited the growth of wheat but not that of maize, for which dry matter yield increased at 20 and 30 mg kg^{-1} levels. Aldrin was also found to be toxic to maize at 30 mg kg^{-1} level but not at lower concentrations. Lindane application at all levels inhibited the growth of wheat and also that of maize at the higher concentrations. Application of these insecticides to soil was also found to decrease in general the N, P and K uptake by the two crops and this was explained as due to accumulation of insecticides in roots and induced changes in enzymatic activities.

4.1.2 Organo-phosphorus insecticides

Numerous laboratory and field studies have investigated the persistence of a range of pesticides in the soil environment. These studies reveal that pesticide stability and degradation is controlled by factors such as temperature, moisture capacity and possibly also soil and solution factors. As Chapter 4 discusses, factors controlling pesticide mobility, persistence and degradation in soils, only specific examples of pesticide studies carried out in India are discussed in the following sections.

Degradation of monocrotophos, dicrotophos, disulfoton and phorate in four different soils viz. the IARI soil (Entisol, pH 8.1), a Calcutta soil (pH 6.5), a red soil (Alfisol, pH 5.1) and a black soil (Vertisol, pH 5.9) have been studied by incubating the insecticide-treated soils at field moisture capacity at 20, 30 and 45°C in a BOD

Table 15. Half-lives of various insecticides in test soils at 30°C (days).

Insecticide	Soils			
	IARI	Calcutta	Red	Black
Monocrotophos	15.04	15.40	16.19	17.97
Dicrotophos	15.52	16.52	17.94	20.06
Disulfoton	25.49	28.99	28.99	30.09
Phorate	23.19	25.98	26.41	27.88

incubator (Pandey, 1976). It was shown that the pesticide half lives were inversely related to the temperature of incubation, i.e., the lower the temperature, the greater is the half life and vice versa. The half lives of the insecticides under study in different soils at 30°C are given in Table 15 which shows that monocrotophos and tetrachlorvinphos have significantly shorter half-lives than the other pesticides.

The persistence of chlorfenvinphos, chlorpyriphos, disulfoton, fensulfothion, monocrotophos and tetrachlorvinphos was studied by Agnihotri et al. (1981) in fallow land in the IARI farm during September 1974–February 1975. Granules of these insecticides were applied in furrows at 1.5 kg ai ha^{-1} and covered with soil. Their persistence in soil was determined by periodic analysis of the soil samples. As expected from their half lives, monocrotophos and tetrachlorvinphos showed least persistence in soils.

The persistence of menazon and pirimiphosmethyl in IARI (Entisol, pH 8.1, clay 19.9% and organic carbon, 0.3%) and Assam soil (pH 5.2, clay 22.9% and organic carbon 0.4%) under field conditions was studied by Barooah and Jain (1986). For the field study aqueous solutions of each of menazon (70% wettable dust powder (wdp)) and pirimiphosmethyl (50% EC) were sprayed separately at 250 and 500 g ai ha^{-1}. These investigations showed that pirimiphosmethyl was less persistent than menazon. Also the residues of pirimiphosmethyl declined rapidly during the first 24 h and thereafter gradually, whereas menazon degraded gradually throughout the experimental period (Table 16).

The initial rapid loss of pirimiphosmethyl was attributed to volatilisation due to its higher vapour pressure. While there was no noticeable leaching in the case of pirimiphosmethyl beyond 7.5 cm depth, elevated manazon concentrations in the subsurface (7.5–15.0 cm) soils indicated rapid leaching of menazon. This has been attributed to their difference in their water solubilities (menazon 0. 24 g L^{-1} at 25°C, piri-miphosmethyl 5 mg L^{-1} at 30°C). The half life calculated for pirimiphosmethyl was 5.2 to 5.9 d while that of menazon was 18.3 to 20 d (Agnihotri et al., 1986).

4.1.3 Carbamate insecticides

In the carbamate group of insecticides, carbofuran and aldicarb have received maximum attention. Garg (1982) studied the persistence of carbofuran and bendio-

Table 16. Persistence of menazon and pirimiphosmethyl in soil under field conditions.

Insecticide	Rate of appln. (a.i. ha^{-1})	Depth (cm)	Residues ($mg\,kg^{-1}$) at successive time intervals (days)						
			0	1	3	7	15	30	45
Menazon	250 g	0–7.5	0.91	0.80 (12.10)	0.69 (24.2)	0.58 (36.3)	0.49 (46.2)	0.30 (67.0)	0.17 (81.3)
		7.5–15	–	–	–	0.20	0.15	0.10	0.03
	500 g	0–7.5	1.62	1.43 (11.17)	1.25 (22.8)	0.94 (41.9)	0.83 (48.86)	0.45 (72.2)	0.21 (83.3)
		7.5–15	–	–	–	0.35	0.21	0.14	0.08
Pirimiphosmethyl	250 g	0–7.5	0.80	0.47 (41.2)	0.37 (53.7)	0.21 (73.7)	0.09 (88.7)	0.01 (98.7)	ND (100)
		7.5–15	–	–	–	ND	ND	ND	ND
	500 g	0.75	1.49	0.85 (43.9)	0.76 (49.0)	0.35 (76.5)	0.19 (87.2)	0.03 (98.0)	ND (100)
		7.5–15	–	–	–	ND	ND	ND	ND

*Figures in parentheses denote percent loss; ND–not detected (below 0.01 $mg\,kg^{-1}$); reproduced from Barooah and Jain, 1986.

Table 17. Half-life values of carbofuran and bendiocarb in different soil.

	Half-life (days) of insecticides under sterile/unsterile conditions			
	Carbofuran		Bendiocarb	
Soil	Non sterile	Sterile	Non sterile	Sterile
Alluvial	11	17	7	15
Red	29	173	30	173
Laterite	69	231	38	231
Black	9	19	10	19
Forest	38	231	58	232

Reproduced from Jain and Agnihotri, 1986.

carb in five different soils, viz. an alluvial (Entisols), a red (Alfisols and Ultisols), a laterite (Oxisols and Ultisols), a black (Vertisols) and a forest soil. The half-lives of the insecticides, determined under non-sterile and sterile conditions, were shorter under non-sterile conditions, presumably indicating significant biological degradation in these soils (Table 17).

Persistence of aldicarb was studied by Dikshit et al. (1976) with a pea (*Pipum sativum* (L)) crop at 1.25 and 2.50 kg ai ha^{-1} at sowing time. Aldicarb residues and its different metabolites persisted in soil for up to 80 d. Dadheech and Gupta (1986) also studied the persistence of aldicarb residues in a clay loam soil in relation to its uptake by a mustard crop, applying aldicarb granules at 0.5, 1.0 and 2.0 kg ai ha^{-1} in furrows before sowing, as well as a side dressing to a 30 d old crop, during 1981 –82 and 1982–83 in Udaipur, Rajasthan. With both the pre-sowing soil treatment and the side dressing, there was about 15-30% loss of residues within the first 3 d of the application, irrespective of the rate of treatment and more than 90% loss was recorded at 60 d. No residues of aldicarb were detectable after 90 d. These findings have been claimed to be in close agreement with that of Dikshit et al. (1976) and others (Kearby et al., 1970; Pareek et al., 1977) who have reported on persistence of aldicarb in clay loam soils under cover of different crops.

4.1.4 Synthetic pyrethroids

Pyrethroids constitute the latest class of insecticides to have been introduced in India and, with the other insecticides, their persistence in Indian soils has been studied in considerable detail. Agnihotri et al. (1985) studied the persistence of cypermethrin, deltamethrin, fenvalerate and permethrin in the IARI soil (Entisol) (Table 18). They found that pyrethroid degradation was generally not dependent on the initial dose.

Jain (1984) also studied the persistence of fenvalerate, permethrin and cypermethrin in three soils (a loam, a sandy loam and a clay loam) at different moisture levels under laboratory conditions. These insecticides were found to be relatively

Table 18. Residues of permethrin, cypermethrin, fenvalerate and deltamethrin in IARI soil.

Insecticide	Dose ($g\,a.i.\,h^{-1-1}$)	Residues in soil ($mg\,kg^{-1}$) at successive time intervals						Half-life days
		1 h	3 d	7 d	15 d	30 d	45 d	
Permethrin	125	0.107	0.086 (19.6)	0.049 (57.0)	0.016 (85.0)	0.004 (96.3)	ND (100)	5.8
	250	0.212	0.149 (29.7)	0.085 (59.9)	0.042 (80.2)	0.005 (97.6)	ND (100)	5.6
Cypermethrin	75	0.068	0.023 (66.2)	0.015 (77.9)	0.003 (95.6)	ND (100)	ND	3.4
	150	0.111	0.050 (55.0)	0.034 (69.4)	0.007 (93.7)	ND (100)	ND	3.9
Fenvalerate	100	0.110	0.064 (36.0)	0.033 (67.0)	0.006 (98.0)	0.002 (100)	ND	2.6
	200	0.165	0.110 (33.3)	0.060 (63.6)	0.014 (91.5)	0.003 (98.2)	ND (100)	5.1
Deltamethrin	12.5	0.01	0.005 (50.0)	ND (100)	ND (100)	ND (100)	ND (100)	3.0
	25.0	0.025	0.01 (60.00)	0.004 (84.00)	ND (100)	ND (100)	ND (100)	2.7

*Parentheses indicate % loss; ND–not detectable (less than 0.001 $mg\,kg^{-1}$); all samples analysed were from the top soil, i.e. at 0–7.5 cm depth; reproduced from Agnihotri et al., 1986.

more persistent in the clay loam soil than in the other two soils. Also, the moisture content at field capacity increased the rate of degradation, although degradation was above 90% in all cases after an interval of 8 wk. The half life was between 7 and 15 d in different soils for different insecticides.

4.2 *Microbial degradation of pesticides in soils*

Biological degradation is by far the most desirable and probably the only means of detoxification of many pesticides in soil or water environments (see Chapter 13). A pesticide that is recalcitrant to microbial degradation may persist in the environment and raise problems of environmental contamination. In India, work on biodegradation of pesticides has been confined to a few institutions/laboratories, viz., Central Rice Research Institute (CRRI), Cuttack, Nuclear Agriculture Division, Bhabha Atomic Research Centre (BARC), Bombay, Central Food Technological Research Institute (CFTRI), Mysore and Industrial Toxicology Research Centre (ITRC), Lucknow. Some private research and development laboratories like Rallis Agrochemical Research Station, Bangalore and Fredrick Institute of Plant Protection and Toxicology, Padappai are also investigating the fate and biodegradation of pesticides used in India.

The most innovative work on microbial degradation studies of commonly used pesticides in soils, particularly under rice cultivation, has been made by Dr. Sethunathan and his group at CRRI. They have shown that the aerobic-anaerobic interface in a flooded rice soil, especially planted to rice, can effect more complete mineralisation of complex organic pesticides than can exclusively aerobic or anaerobic systems (Reddy and Sethunathan, 1983). A flooded rice field differs from a non-flooded (upland) field in physico-chemical and microbiological characteristics. Upon flooding, oxygen supply to the soil virtually ceases and oxygen trapped in the soil is soon exhausted, leading to anaerobiosis. Although the bulk of flooded soil is essentially anaerobic, the thin surface layer and standing water are in an oxidised state. Similarly, a flooded soil planted to rice is not necessarily as anaerobic as an unplanted flooded soil, because of the capacity of the rice plant to transport oxygen to the root region. Considerable literature is available to demonstrate that a flooded rice soil, with its characteristic and dynamic aerobic-anaerobic phases, harbours a host of aerobic and anaerobic bacteria capable of metabolising many pesticides.

It has been shown by Siddaramappa and Sethunathan (1975) that the degradation of hexachlorocyclohexane (HCH) isomers, an important organochlorine insecticide widely used in India and other tropical countries, was faster in flooded soils than in non-flooded soils. Most isomers of HCH, which were widely believed to be degraded in anaerobic conditions, were demonstrated to be aerobically degraded by *Sphindomonas paucimobilis*, isolated from HCH-amended flooded soil (Bhuyan et al., 1992). A group of scientists working in the Regional Research Laboratory, Jorhat, have also successfully isolated a few microbial strains which are active against pesticides such as dicofol and endosulfan. Similarly CRRI scientists have isolated several bacteria which can selectively degrade a number of organochlorine, organophosphorus and carbamate insecticides, as well as herbicides and fungicides. Various studies conducted by these workers suggest that several factors influence microbial activity that affects the degradation of pesticides in soil. These are soil type, moisture content, redox potential, temperature and organic matter. Chapter 4 shows that a variety of factors, including soil and solution characteristics, nature and concentration of pesticides, influence the rate of microbial degradation of pesticides in the soil environment.

CFTRI, Mysore, has also been actively engaged in developing bioremediation procedures for contaminated soils. The methodology for one of their proprietary products, 'Agrocure' based on *Pseudomonas* specific to BHC is available for commercial exploitation.

4.3 Adsorption of pesticides

Adsorption studies show that pesticide interactions with soils are controlled by a range of soil and solution factors which depend on both the nature of the pesticides and soil type (see Chapter 4). Mithyantha and Perur (1974) studied the effect of temperature, organic matter and pH on the adsorption of phorate on red, laterite, black and karl whole soils (Vertisol) of Karnataka State, and on clay isolates. These results revealed that the clay fraction in the soil was responsible for much of the adsorption of phorate, whereas organic matter played a minor role. A similar study was conducted by Chopra et al. (1970) on the adsorption and leaching of parathion on three different soil types (an Alfisol, an Oxisol and a Vertisol). They found adsorption increased with increases in concentration of pesticides, rises in temperature, declines in pH and increases in organic matter content. In a separate study (Chopra and Khullar, 1971) these investigators also studied the degradation of parathion in three different soil types (an Alfisol, an Oxisol and a Vertisol). They found that degradation was related to the initial pesticide concentration, temperature, time of exposure to ultra violet light, relative humidity and soil pH. Alkaline soils achieved maximum degradation, followed by acidic and neutral soils.

5. Conclusion

Despite the large scale industrial development in India and increasing population shifts towards urban centres, with increasing pressure on land disposal of effluents containing toxic trace metals and sewage, no systematic effort seems to have been directed towards studying the impact of contaminants on the soil environment Nevertheless, some information with regard to levels of toxic trace metals in soils under sewage farming and those receiving industrial effluents is available. Many pesticides have been observed to be less persistent in Indian soils than in soils elsewhere due largely to the sub-tropical climate of the country. This, however, should not allow any complacency. More important than the actual concentrations of trace metals is perhaps the rate of build up, which needs constant monitoring of contaminated sites over a long period of time. Furthermore, the studies covered in this report do not include information on vast stretches of degraded land that have resulted from disposal of sewage and other pollutants, which do not seem to have been monitored. No estimates of the damage to, and status of, these soils are available.

It is likely that indiscriminate land disposal of sewage and use of pesticides may have already resulted in high levels of heavy metal pollution and/or pesticide residues in Indian soils. There are, as yet, no standards for soil contamination assessment, unlike those existing in the Netherlands, the UK, Australia and many other countries (Sittig, 1994). There is a need for regulation to restrict the amount of heavy metals (Cr, Cd, Cu, Zn, Pb and Ni) added to agricultural soils through the appli-cation of sludge.

To make a realistic nation-wide assessment of soil contamination, it may be necessary to launch a nationally coordinated programme to assess the heavy metal status of soils, on the lines of the erstwhile Integrated Environmental Programme of Heavy Metals initiated in the early eighties by the Department of Environment of the Government of India, which took up a survey of 10 metallic elements of environmental significance in ambient air, water and food. The country has adequate scientific resources and capability to investigate all important aspects of soil contamination, as is clear from the reports cited in this review. International cooperation with regard to standards and for comparisons of results and techniques, along with resources from international agencies, would be desirable.

Acknowledgement

The author is grateful to the organisers of the conference on 'Contaminants and the Soil Environment in the Australia-Pacific Region' for inviting him to contribute and to CSIR for granting permission to publish this review. Many friends and associates have generously helped directly or indirectly in preparation of the manuscript. Grateful thanks are expressed to Dr. Indrani Chandrashekhar of the Ministry of Environment and Forests, New Delhi, to Dr N. P. Agnihotri, IARI, New Delhi for his support and to Drs. P.K. Seth and R.B. Raizada, Scientists, Industrial Toxicological Research Centre, Lucknow, for providing useful references and reports.

The author is also thankful to Mrs. R.K. Chibber and Ms. Renu of CSIR Library who were instrumental in locating many references. Dedicated secretarial assistance in typing the manuscript was rendered by my Secretary, Mr. J.B. Gupta, and is greatly appreciated.

References

Agarwal K M, Sharma H C and Aggarwal A L 1988 Heavy metal pollution of air, soil and plants around zinc smelter – a case study, Ind. J. Environ. Hlth. 30, 234-241.

Agnihotri N P, Jain, H K and Gajbhiye V T 1986 Persistence of some synthetic pyrethroid insecticides in soil, water and sediments (Part 1). J. Ent. Res. 10, 147-151.

Agnihotri N P, Pandey S Y and Jain H K 1974 Persistence of BHC and aldrin in soil and their transloca-tion in mung and lobia. Ind. J. Ent., 36, 261-267.

Agnihotri N P, Pandey S Y, Jain H K and Srivastava K P 1977 Persistence of aldrin, dieldrin, lindane, heptachlor and pp' DDT in soils. J. Ent. Res. 1, 89-91.

Barooah A K and Jain H K 1986 The persistence of menazon and pirimiphos methyl in soil. In Proc. Sym. Pest. Residue & Environ Pollu. Ed. S C Goel. pp 24-31. Department of Zoology, Sanatam Dharm College, Muzzaffarnagar, India.

Bhuyan S, Sahu S K, Adhya T K and Sethunathan N 1992 Accelerated aerobic degradation of gamma-hexachlorocyclohexane in suspension of flooded and non-flooded soils pretreated with hexachlorocyclohexane. Biol. Fert. Soils. 12, 279-284.

Central Pollution Control Board 1987 Status of water supply and waste water collection, treatment and disposal in class I cities – 1988, Central Pollution Control Board (CPCB), New Delhi.

Chopra S L and Khullar F C 1971 Degradation of parathion in soil. J. Ind Soc. Soil Sci. 19, 78-86.

Chopra S L, Niranjan, D and Bhagwan, D 1970 Adsorption and leaching of parathion on soils and effect of various physical factors on adsorption, Ind. J. Environ. Hlth. 18, 437-446.

Dadheech L N and Gupta H C L 1986 Persistence of aldicarb residues in clay loam soil and its uptake in mustard crop. Proc. Sym. Pest. Residue & Environ Pollu. Ed. S C Goel. pp 15-23. Department of Zoology, Sanatam Dharm College, Muzzaffarnagar, India.

Das D C and Kaul R N 1992 Greening wastelands through wastewater, National Wasteland Development Board, Ministry of Environment and Forests, Government of India, New Delhi, March 1992. 64 p.

Dikshit A K, Agnihotri N P, Dewan R S and Saxena H P 1976 Dissipation of aldicarb in soil and pea plant. Ind. J. Agri. Sci. 46, 117-119.

Dikshit T S S, Raizada R B, Kumar S N, Srivastava M K, Kulshreshtha S K and Adholia U N 1990 Residues of DDT and HCH in major sources of drinking water in Bhopal, India. Bull. Envron. Contam. Toxicol. 45, 389-393.

Edward C A 1966 Insecticide vapour pressure and disappearance from soils. Res. Rev., 13-91.

Gajbhiye V T and Agnihotri N P 1991a Volatilization of lindane from soil. Pest. Res. J. 3, 109-118.

Gajbhiye V T and Agnihotri N P 1991b Volatilization of aldrin from soil. Pest. Res. J. 4, 11-23.

Garg P K 1982 Adsorption, desorption and degradation of carbofuran and bendiocarb in soil. Ph.D Thesis, Post-graduate School, IARI, New Delhi.

Govinda Rajan, S V 1965 World Soil Resources. Report No. 26, F.A.O, UNESCO

Govinda Rajan S V and Gopala Rao H G 1978 Studies of soils of India, Vikas Publishing Pvt. Ltd, New Delhi. 425 p.

ICMR 1993 Surveillance of Food Contaminants in India, Report of an ICMR Task Force Study (Part I), Indian Council of Medical Research, New Delhi. 171 p.

ITRC 1992 Pesticides development, toxicity and safety. Ed. R B Raizada, T S S Dikshith. Published by Industrial Toxicology Research Centre, Lucknow. 131 p.

Jain H K 1984 Degradation of synthetic pyrethroids in soil. In 'Symposium on Chemical Hazards in Agricultural Environment' held at University of Agricultural Sciences, Hebbal, Bangalore.

Jain H K and Agnihotri N P 1980 Persistence of aldrin, dieldrin, and lindane in soils In Annual Report, Division of Agricultural Chemicals, IARI, New Delhi.

Kearby W H, Ercegovich C D and Bliss M (Jr.) 1970 Residue studies on aldicarb in soil and scotchpine. J. Econ. Ent. 63, 1317-18.

Khandekar R N 1984 Lead pollution from auto-exhaust, Science Today, Times of India Publication, April 1984, pp46

Khandekar R N, Tripathi R M, Raghunath R and Mishra U C 1988 Simultaneous determination of Pb, Cd, Zn and Cu in surface soil using differential pulse anodic stripping voltammetry. Ind. J. Environ. Hlth. 30, 98-103.

Krishna Murti C R 1987 The cycling of arsenic, cadmium, lead and mercury in India. In Lead, mercury, cadmium and arsenic in the environment. Ed. T C Hutchinson and K M Meema. 315-333. SCOPE Publication, Published by John Wiley & Sons Ltd, Chichester, UK.

Mithyantha M S and Perur N G 1974 Retention of phorate in four soils of Karnataka. Symp. Ind. Soc. Agri. Chem., Coimbatore, Nov. 74, Abstr. p12.

Nashikkar V J and Chakrabarti T 1994 Catalase and peroxidase activity in plants – An indicator of heavy metal toxicity. Ind. J. Exptl. Bio. 32, 520-521.

Olaniya M S, Bhoyar R V and Bhide A D 1991 Effects of solid waste disposal on land. Ind. J. Environ. Hlth. 34, 143-150.

Pandey S Y 1976 Adsorption, desorption and degradation of dicrotophos, monocrotophos, disulfolon and phorate in soil. Ph.D. Thesis, Kanpur University, Kanpur. 93 p.

Pareek B L, Gupta H C L, Sharma K P and Maheshwari AC 1977 Estimation of aldicarb and phorate residues in cauliflower. Entomon. 3, 187-188.

Pervez S and Pandey G 1994 Contamination of river water and sediments by thermal power ash pond discharge. Ind. J. Environ. Hlth. 36, 8-12.

Rao A V, Jain B L, Gupta I C 1993 Impact of textile industrial effluents on agricultural land, a case study. Ind. J. Environ. Hlth. 35, 132-138.

Rao K J and Shantaram M V 1994 Heavy metal pollution of agricultural soils due to application of garbage. Ind. J. Environ. Hlth. 36, 31-39.

Reddy B R and Sethunathan N 1983 Mineralisation of parathion in rice rhizosphere. Appl. Environ. Microbiol. 45, 826-828.

Shree Ramulu U S 1990 Studies on the utilisation of urban sewage and sludge for increasing crop production. Proc. Ind. Sci. Cong, Indore.

Srikanth R, Kumar C S and Khanum A 1992 Heavy metal content in forage grass grown in urban sewage sludge. Ind. J. Environ. Hlth. 34, 103-107.

Shrivastava V S, Rai A K and Mehrotra R C 1987 Nitrogen and phosphorus pollution in soils beneath the Jalmahal lake of Jaipur city. Ind. J. Environ. Hlth. 30, 104-115.

Shrivastava V S, Rai A K and Mehrotra R C 1989 Pollution status of soils beneath sewage disposal pond. Ind J. Environ. Hlth. 31, 314-320.

Siddaramappa R and Sethunathan N 1975 Persistence of gamma-bhc and beta-BHC in Indian rice soils under flooded conditions. Pestic. Sci. 6, 395-403.

Sittig Marshall 1994 World-wide limits for toxic and hazardous chemicals in air, water and soil. Noyes Publications, Park Ridge, New Jersey, USA.

Statistical Abstracts 1989 Central Statistical Organisation, Ministry of Planning, Government of India.

Thakre S K and Saxena S N 1972 Effects of chlorinated insecticides on plant growth and uptake of nutrients. J. Ind. Soc. Soil Sci. 20, 45-48.

18
Soil contamination issues in the Republic of Korea (South Korea)

SUN-HOO YOO and KYE-HOON KIM

1. Introduction

The Republic of Korea (South Korea) occupying the southern portion of the Korean Peninsula in East Asia, covers an area of 99 000 km^2. It is about 965 km long and about 217 km wide; it is bordered on the north by North Korea, on the east by the East Sea (Sea of Japan), on the south by the Korean Strait, and on the west by the Yellow Sea. As one of the most densely populated countries of Asia, it had a population estimated in 1991 at 43.5 million.

About 70% of the total land area consists of mountains and uplands; the T'aebaek Mountains, paralleling the eastern coast for about 257 km, give rise to several lower mountain ranges and spurs in the south. Mount Halla, an extinct volcano with a crater lake, situated on Cheju Island, is the highest point of the country. The densely populated low lands are heavily cultivated for rice; rice growing accounts for about 30% of the land use. The 512 km long Naktong and the 488 km Han are the principal rivers and provide extensive irrigation and navigation facilities.

Geologically, the Korean Peninsula consists in large part of Precambrian rocks such as granite and gneiss, from which most soils are derived. Sandy and brown-coloured soils (Inceptisols) are common, and they are generally well leached and have little humus content.

Because of continental influences, the climate of Korea is characterized by a cold winter and a hot summer. The average monthly temperature in January drops below freezing except along the southern coast, and the July average monthly temperature rises to about 25°C. The summer monsoons (June–September) bring about 1 000–1 400 mm of average annual precipitation. Occasional typhoons in late summer bring heavy rains and cause extensive damage to crops.

Forests, covering about two-thirds of the total land area, consist of subtropical broadleaf forests on the southern coast and coniferous forests in the rest of the country. High population density has resulted in extensive deforestation, especially in the southwest. Large areas were reforested with pine and scrub oak in the early 1980s. Fauna includes bear, wild boar, deer, and lynx, found only in the remote areas in the north.

Korea's chief mineral resources are anthracite coal, Fe ore, and tungsten (W). Copper, Pb, Zn, Mn, Au, and Ag deposits are found in small quantities.

Korea has a predominantly market economy based largely on services and light and heavy industries. Economic growth has been heavily dependent on exports of manufactured goods. The gross national product (GNP) per capita, which is grow-

R. Naidu et al. *(eds): Contaminants and the Soil Environment in the Australasia-Pacific Region, 543–562.*

ing much faster than the population, is similar to those of the world's other developing countries.

Agriculture has been a key industry in Korea since ancient times. Contributions from agriculture, however, decreased significantly with export-oriented manufacturing by the Korean government and rapid industrialisation since the 1960s. Agriculture-forestry-fisheries, mining and manufacturing industries, and social overhead capital and services comprised 36.6%, 16.3%, and 47.1%, respectively of GNP in the early 1960s, whereas the proportions changed to 10.9%, 32.8%, and 56% in the late 1980s. Changes in percentage of the people engaged in each industry showed this trend even more clearly. Agriculture, forestry, and fisheries engaged 63.1% of the people in the early 1960s, but declined to 20.7% in the late 1980s. Industrialisation as well as urbanisation have proceeded so rapidly that more than two-thirds of Koreans now live in cities with populations over 200 000, resulting in urbanisation and environmental problems.

In the process of industrialisation and urbanisation, only a few people have criticised export-oriented manufacturing, which has received strong government support, but environmentalists have warned of the potential hazard that industrialisation might bring about. Under strict government control of the news media through the 1970s and early 1980s, the voices of the environmentalists were always unheard. Since the late 1980s, people have been increasingly concerned about the environment as their economic situation has improved and control of the news media by the government has decreased. Environmentalists have succeeded in getting the support of the public with reliable environmental data provided by both government officials and researchers. This report provides a general description of soil contamination issues in the Republic of Korea.

2. Characteristics of Korean soils

Geologically, the Korean Peninsula consists in large part of Precambrian rocks such as granite and gneiss. Although granite and gneiss, which are acidic in nature, cover two-thirds of the country, Korea has a relatively wide variety of soils because of its complex topography. Results of detailed soil surveys have identified 375 soil series in Korea (ASI, 1983).

Korean soils have a generally coarse and acidic nature due to their coarse and acidic parent material. Hilly topography has resulted in much eluviation of clay by rain, especially by monsoonal rain during the summer. As a result, soils on sloping areas are shallow and very coarse. The texture of soils of the river basins is also coarse due to flooding. This is because of the hilly topography and heavy monsoonal rain which lead to considerable movement of fine particles from the land direct into the ocean. A temperate and humid climate has accelerated decomposition of organic matter (OM) so that OM concentrations of Korean soils are low (1.7 to approximately 2.6%). Concentration of rain during summer and a mean annual precipitation of more than 1 000 mm have facilitated leaching of bases, which has resulted in low pH, especially on the sloping areas. Soils whose clay and OM contents are

Table 1. Effect of long-term farming on characteristics of rice paddy soils over time.

Year	No. of samples	pH_w	O.M. %	Avail. P_2O_5* $mg\,kg^{-1}$	Exch. cations*, $cmol_c\,kg^{-1}$ K	Ca	Mg	Sum
1964–68	5130	5.5	2.6	60	0.23	4.5	1.8	6.53
1969–79	19737	5.9	2.4	80	0.31	4.4	1.7	6.41
1980–88	616687	5.7	2.3	107	0.27	3.8	1.4	5.47

*Analyzed by Lancaster method (ASI, 1988).
(Source: RDA, 1989).

low have low cation exchange capacity (CEC) and low buffer capacity. Soil N and P from fertiliser applications to the surface soil have resulted in eutrophication of both fresh water and seawater. In this context, Korean soils have a low retention capacity for most elements and are vulnerable to contamination that results from inappropriate soil management.

3. Characteristics of the soils of the agricultural area

Cultivated soils differ from uncultivated land in both chemical and physical characteristics due to both OM decline and long-term applications of fertilisers and soil amendments such as lime and agrochemicals. Much irrigation water is supplied to rice paddy soils throughout the growing season, so these soils are even more different in physicochemical characteristics relative to the uncultivated soils.

3.1 Rice paddy soils

The pH_w values of rice paddy soils were originally low but have increased a little with time (Table 1) while OM content and concentration of exchangeable cations has decreased; available P has increased considerably (RDA, 1989).

Figure 1 shows the location of each province in Korea, while Table 2 lists some characteristics of the rice paddy soils of each province. With the exception of Cheju province and areas around the large cities, the pH_w values of the soils of Kangwon province are about 6.0 while the pH_w values of the soils from the rest of the country are in the range 5.5–5.8; OM contents in soils from Chonnam, Chonbuk, and Kyongnam provinces are high (above 2.5%) and the available P contents of Kyonggi, Ch'ungnam, and Ch'ungbuk province soils are low. Large cities like Pusan, Taegu, and Kwangju show fairly high accumulation of P in rice paddy soils; K is also high in these soils.

3.2 Cultivated upland soils

Table 3 shows results from the detailed soil surveys of cultivated upland soils from some of the provinces. Except for the soils from Kangwon province, OM contents of cultivated upland soils are lower than those of rice paddy soils. Available P and

Figure 1. Location of each province in Korea.

K contents of cultivated upland soils are higher than those of rice paddy soils, which might result from accumulation of available P and K through unbalanced application of fertilisers. The fact that the available P contents of cultivated upland soils are much higher than the 130 mg kg^{-1} recommended level (RDA, 1989) for cultivated upland soils might present a serious problem to the environment.

Table 2. Properties of rice paddy soils from each province and major cities.

Province, City	No. of samples	pH_w	O.M. %	Avail. P_2O_5* $mg\,kg^{-1}$	Exch. cation, $cmol_c\,kg^{-1}$ K	Ca	Mg	Σ cations
Kyonggi	94058	5.5	2.1	71	0.21	3.3	1.4	4.91
Kangwon	47908	6.0	2.2	103	0.21	3.3	1.4	4.91
Ch'ungbuk	49854	5.5	2.0	82	0.19	4.3	1.2	5.69
Ch'ungnam	68709	5.8	2.0	74	0.30	3.9	1.5	5.70
Chonbuk	96729	5.8	2.5	108	0.31	3.8	1.4	5.51
Chonnam	89611	5.7	2.6	164	0.35	3.3	1.5	5.15
Kyongbuk	80683	5.6	1.9	100	0.24	4.5	1.7	6.44
Kyongnam	81469	5.6	2.6	131	0.24	3.7	1.3	5.24
Cheju	682	5.6	4.4	89	0.59	4.7	2.2	7.49
Seoul**	617	5.7	3.0	89	0.27	4.0	1.4	5.67
Pusan	1832	5.7	2.1	178	0.28	4.4	2.1	6.78
Taegu	1631	5.4	2.0	134	0.28	4.8	1.6	6.68
Inch'on	1261	6.1	2.2	66	0.33	4.4	2.1	6.83
Kwangju	1643	5.8	2.6	125	0.30	3.6	1.3	5.20

*Analyzed by Lancaster method (ASI, 1988).
**Bold charactered names are those of cities.
(Source: RDA, 1989).

Table 3. Properties of cultivated upland soils 1985–1988.

Province	No. of samples	pH_w	O.M. %	Avail. P_2O_5* $mg\,kg^{-1}$	Exch. cations*, $cmol_c\,kg^{-1}$ K	Ca	Mg	Σ cations
Kyonggi	18308	5.8	1.7	304	0.60	3.8	1.1	5.50
Kangwon	12941	5.9	2.4	178	0.71	4.2	1.1	6.01
Ch'ungbuk	8591	5.8	1.8	318	0.48	6.3	1.7	8.48
Kyongbuk	18567	5.6	1.8	169	0.54	4.7	1.7	6.94
Kyongnam	7158	5.6	1.8	271	0.57	5.8	1.3	7.67

*Analyzed by Lancaster method (ASI, 1988).
(Source: RDA, 1989).

4. Sources of soil contaminants

Following entry into the soil environment, contaminants undergo rapid interactions with both the solid and solution phases. Some of the interactions include physical, chemical, and biological reactions such as filtration, dilution, decomposition, transformation and adsorption–desorption processes. Often processes such as filtration, dilution (e.g. mixing surface and subsurface soils) and adsorption reactions within the soil have been used in developed countries to manage contaminated soils. Detailed discussions on physicochemical processes controlling contaminant dynamics in soils are presented in Chapters 2, 3 and 4. Although the ability of soil to retain contaminants is much larger than those of air and water, soil can act as a contaminant source for surface water and groundwater as well as to agricultural products once contamination exceeds a threshold value.

Clay and OM contents and distribution in the soil profile, soil drainage, and topography are major factors which determine a soil's capacity to retain contami-

Table 4. Annual production of organic wastes and their utilization.

Source	Annual Production 10^6 ton	Utilization, %	Utilization
Agricultural	9.58	96	compost, fuel, animal feed, mushroom growth medium
Forest	22.7	100	compost, industrial
Animal	40.9	63	compost
Sewage sludge	2.07	–	–
Industrial	252	7.5	compost, growth medium
Food	8.40	2.9	compost, animal feed

nants, and this can vary greatly, depending on soil type and also on the nature of contaminants.

Soil contaminants include organic and inorganic substances. Some of the organic contaminants are produced by animals and plants; others are new synthetic substances. Naturally produced contaminants added to the soil environment usually decompose fairly easily unless they are in quantities so large that they overload a soil's capacity for retention. However, some artificially synthesised substances such as dichlorodiphenyltricholoro ethane (DDT), benzene hexachloride (BHC), and pentachlorobenzene are not decomposed easily but are accumulated, and become potential environmental hazards. Annual production of organic wastes and their utilisation are shown in Table 4. Among the many soil contaminants, heavy metals are particularly hazardous. Once a soil is contaminated with heavy metals, decontamination is very difficult. This paper deals mainly with soil contamination by heavy metals.

4.1 Pathways of soil contamination with heavy metals

There are four major pathways of heavy metal contamination of soils in Korea. These include transport by air, by water, through farming practices and by direct application of wastes.

The nature of contaminants in the air in Korea includes both volatile inorganic and organic compounds and particulate matter. For example, heavy metal contamination of soils by aerial deposition has been reported in soils around Zn smelters, cement plants and also along the road side in Korea. Similar cases of aerial deposition of Cd and Zn have also been observed in South Australia (Tiller, 1989). Tiller and his co-workers reported that heavy metal contamination declined exponentially with distance from the smelter (Tiller, 1989, see Chapter 14). Detailed work on the impact of industrialisation on the quality of soil environment around industrial areas is necessary in Korea.

Soil contamination by water can occur either with water acting as a vehicle for contaminant transport or when effluent water is disposed of on to land. The latter activity is extremely prevalent in Korea and has led to extensive contamination of large areas of land. At some sites disposal of mine wastewater and wastewater from

As smelters have led to serious contamination of land with Cd and As. However, only limited data are available on contaminated soils from around these sites.

The major source of soil contamination by farming is through the use of agricultural chemicals containing heavy metals. Although use of agricultural chemicals that contain Zn, As, Cu, and Pb is limited these days, the concentration of these elements is still high in orchard soils to which much of these agricultural chemicals have been applied in the past. When sewage sludge is applied directly to soil as an organic soil conditioner, the soil contamination by toxic inorganic and organic substances can be a serious problem. The concentrations of heavy metals vary depending on the type of sewage. A recent problem in the Korean soil environment is the contamination of soil by plastic films. Decomposition of plastics by microorganisms is slow and the presence of plastics in soils reduces the soil's permeability, aeration and hinders root growth. Plastic film has been used for heat insulating rice seedbeds, weed control, and accelerating growth and most of the used plastic film is not recycled but buried. Soil contamination by heavy metals, municipal wastewater, and airborne contaminants damage the soil environment and affect public health through chemical means, whereas used plastic film damages the physical properties of soils.

5. Background values of heavy metals in rural soils

There are some 90 elements on the earth's surface and about 50 elements in plants. Among these, 16 elements only are essential for plant growth. Elements like Cu, Zn, and Mn are essential for plant growth but they are phytotoxic and become environmental contaminants when their concentrations in soil exceed threshold values.

5.1 Rice paddy soils

A large quantity of both groundwater and effluent is generally used to irrigate rice paddy soils through the growing season. Applications of effluent water rich in contaminants can lead to accumulation of both heavy metals and toxic organic substances in paddy soils. However, limited research effort has been directed towards investigating the impact of such effluent applications on soil environment quality. Given the high moisture status of paddy soils, it is prudent to predict that repeated applications of effluent can lead to both leaching of contaminants and uptake by rice plants.

The background values of heavy metals determined by the National Institute of Environmental Research (NIER) in 1987–1988 and by the Agricultural Sciences Institute (ASI) in 1980–1981 in rice paddy soils are shown in Table 5. There is fairly close agreement between the two sets of data. Mean background values of Cd and Cu in rice paddy soils were somewhat higher than those of forest soils, while Pb and Zn in rice paddy soils were similar to those of forest soils (Kim et al., 1995). However, mean background concentrations for As, Cd and Cu in the rice paddy soils were below the maximum permissible concentrations for cropping in Korea.

Table 5. Mean background values (mg kg^{-1}) of heavy metals in rice paddy soils and forest soils.

Source/ Institute	Year	No. of samples	As	Cd	Cu	Pb	Zn
NIER*	1987–1988	330	0.56	0.14	4.00	5.38	4.36
ASI**	1980–1981	407	0.56	0.13	4.15	4.67	3.95
Kim et al. (1995)***	1992–1993	240	–	0.01	1.98	5.52	4.08

*National Institute of Environmental Research.
**Agricultural Sciences Institute.
***Undisturbed cultivated forest soil samples, As was not determined.
(Source: EA, 1989; ASI, 1989).

Table 6. Mean heavy metal concentrations (mg kg^{-1}) in brown rice from uncontaminated soils.

Institute	Year	No. of samples	As	Cd	Cu	Pb	Zn
NIER*	1987–1988	50	0.09	0.06	2.31	0.43	16.56
ASI**	1980–1981	407	–	0.05	3.31	0.44	20.55

*National Institute of Environmental Research.
**Agricultural Sciences Institute.
(Source: EA, 1989; ASI, 1989).

Table 7. Mean background values (mg kg^{-1}) of heavy metals in cultivated upland soils.

Crop	No. of samples	As	Cd	Cu	Hg	Pb
barley	108	0.62	0.16	4.00	0.09	5.49
soybean	56	0.26	0.15	2.05	0.09	2.23
maize	51	0.43	0.16	2.15	0.09	3.34
Av.	215	0.49	0.16	3.06	0.09	4.16

(Source: ASI, 1989).

5.2 *Cultivated upland soils*

Much less irrigation water is supplied to cultivated upland soils than to rice paddy soils, but heavy metals can be accumulated in cultivated upland soils when they are used for horticulture as large amounts of fertilisers are often applied to the soils. Moreover, some heavy metals are more phytoavailable under aerated conditions compared to waterlogged reducing conditions generated in paddy soils. Heavy metals may also be accumulated in cultivated upland soils used for horticulture because there is less irrigation water and consequently less percolation and leaching losses compared to rice paddy soils. The background values of heavy metals in rice paddy soils and cultivated upland soils from the same region and the same parent material can thus be quite different.

Heavy metal contents of brown rice grown in upland soils and the background concentrations in these soils are shown in Tables 6 and 7.

Cadmium concentrations in cultivated upland soils are higher than those in rice

Table 8. Mean background values (mg kg^{-1}) of heavy metals in crops from cultivated upland soils.

Crop	No. of samples	As	Cd	Cu	Hg	Pb
barley	108	0.115	0.050	4.49	0.044	0.54
soybean	56	0.044	0.098	7.87	0.051	0.99
maize	51	0.081	0.050	10.04	0.053	1.03
Av.	215	0.080	0.066	7.47	0.049	0.85

(Source: ASI, 1989).

paddy soils. Copper and Pb concentrations in rice paddy soils are not significantly different than those in the cultivated upland soils (EA, 1989; ASI, 1989)

5.3 *Crops*

The ionic concentrations of nutrients and toxic substances in plant tissue are generally used to judge crop quality both in relation to its impact on growth and quality for food safety. Setting criteria for the background values of heavy metals in plants should be done with great care, for the background values of substances in plants change enormously depending on numerous factors, such as plant type, growth stage, soil type, and soil management.

Table 8 shows the background values of heavy metals for crops grown in cultivated upland soils. Average Cd and As concentrations in barley, soybean, and maize were similar to those in brown rice (Table 6), whereas average Cu and Pb concentrations were high in barley, soybean, and maize. Although the Korean Government has passed the Soil Preservation Act in January 1995 and is preparing for the Enforcement Decree for the Act, criteria for restricting cultivation of agricultural products in soils based on heavy metal contents in crops are so far limited to Cd only. In soils where grain Cd concentration exceeds 1 mg kg^{-1}, the Korean Government can restrict cultivation of agricultural products.

5.4 *Orchard soils*

Agricultural chemicals, such as $CuSO_4$, Pb AsO_3, and Bordeaux mixture that contain heavy metals, were formerly widely used in orchards. It is therefore very difficult to define the background values of some heavy metals in orchard soils. After analysis of soils from 58 apple orchards in the Kyonggi, Chungnam, Chungbuk, Kyongnam, Chonbuk, and Kangwon provinces, Kim et al. (1989) found that Cu concentrations were in the range 0.2–214.7 mg kg^{-1} with an average of 53.9 mg kg^{-1} and Pb concentrations were in the range 1.8–107.3 mg kg^{-1}, with an average of 18.5 mg kg^{-1}. (Table 9). Copper and Pb concentrations were very high relative to forest soil. With a maximum of 1.8 mg kg^{-1}, the average Cd concentration in the orchard soils was 0.34 mg kg^{-1}, which is fairly high compared to concentrations in forest soils.

Kim reported that Cu concentrations in soils from apple orchards are closely

Table 9. Heavy metal concentrations (mg kg^{-1}) in 58 apple orchard soils.

	Cd	Cu	Pb
Max.	1.77	214.73	107.32
Min.	0.04	0.20	1.77
Ave.	0.34	53.88	18.53

(Source: Kim et al., 1989).

Table 10. Change in Cu content (mg kg^{-1}) in orchard soils with the age of trees.

Age of trees	Ch'ungju	Yesan	Taegu
<5	2.1	2.4	–
6–10	6.9	20.5	15.5
11–15	14.1	39.4	37.6
16–20	52.3	61.8	63.8
21–30	99.7	97.6	105.8
50 –60	136.1	171.3	134.6

(Source: Kim, 1974).

related to the age of the trees (Table 10). Agricultural chemicals that contain heavy metals are not used now.

6. Current methods of assessment used for contaminated soils

Currently, heavy metal concentrations in agricultural lands are used as criteria to assess whether soils are contaminated or not. Based on these criteria, the government can restrict cultivation of agricultural products in soils. Rice is by far the most important crop in Korea and most of the fertile and flat lands in Korea is being used for rice cultivation. Since rice has a high water requirement throughout the whole growing season, soils which are irrigated with water contaminated by mining waste or industrial waste have high potential for heavy metal contamination.

Use of agricultural land in Korea can be restrictive depending on the crops cultivated and the permissible concentrations established by the Korean Government. For example, rice cultivation is restricted in soils where grain Cd concentration exceeds 1 mg kg^{-1}. Similarly, restrictions are also imposed on agricultural land in which 0.1 M HCl extractable Cu exceeds 125 mg kg^{-1} and 0.1 M HCl extractable As exceeds 15 mg kg^{-1}. Further regulatory guidelines need to be developed for other toxic substances, particularly in view of the wide ranging industrial activities in Korea.

In assessing soils for heavy metal contamination, soil scientists use 0.1 M HCl for extracting heavy metals as an indication of uptake by plants, whereas geochemists use total heavy metal concentrations. This is one problem we need to address in Korea as it hinders comparison of analytical results produced by different researchers.

The Korean government passed the Soil Preservation Act in January 1995 and is preparing the Enforcement Decree for the Act. By the time the Enforcement

Decree is ready, Korea will be well equipped with comprehensive and systematic methods of assessment for contaminated soils.

7. Present status of soil contamination by heavy metals

7.1 Soil contamination detected by soil contamination monitoring network

The Ministry of Environment (the former Environment Administration) established a soil contamination monitoring network in 1986 (EA, 1986). Based on major river systems, the network divides Korea into six environment spheres with 250 monitoring sites. The Ministry monitors soil contamination based on sources of soil contamination and reports the results annually. Heavy metal contents of the soils from the monitoring sites determined in 1988 increased a little compared to those determined in 1987 (EA, 1989). Whether this increase has statistical significance or not is not clear, due to the small number of samples per source of soil contamination. Generally, heavy metal concentrations of soils from the agricultural areas were the lowest and were close to the background values. But heavy metal contents in the soils from areas around metal mining sites, smelters, and landfills for industrial wastes were generally higher than those from the agricultural areas (Table 11).

7.2 Rice paddy soils and brown rice around Zn mining sites

Zinc and Cd concentrations in rice paddy soils and brown rice around old Zn mining sites were very high compared to those in uncontaminated rice paddy soils (Table 6) (Yoo and Lee, 1980).

The Shihung Zn mining site showed the highest probability of producing Cd contaminated brown rice, whereas Ch'ilgok showed the lowest (Table 12). The maximum Cd concentrations from all the brown rice samples analyzed were greater than 1.0 $mg\,kg^{-1}$, which is the concentration above which the government restricts cultivation of agricultural products. Therefore, care must be taken in the production of rice on soils around Zn mining sites.

7.3 Arsenic in soils around As mining sites and smelters

One of the main causes of contamination of rice paddy soils by As is the continuous use of agricultural chemicals such as neo-asozine (arsenic fungicide) that contain As. Rice paddy soils can also be contaminated by As from mining wastes, mining wastewater, and dusts produced in the process of mining Au, Cu and Pb. Following analysis of soil samples from rice paddy fields in areas around 34 As mining sites and 14 smelters, Lee et al. (1986) reported concentrations of As in the rice paddy soils around the As mining sites in the range of 6.7–701.3 $mg\,kg^{-1}$ with an average of 141.3 $mg\,kg^{-1}$. Those around the smelters were in the range 4.01–21.7 $mg\,kg^{-1}$ with an average of 11.18 $mg\,kg^{-1}$. All of these values were high compared to background values. Despite the law that enables government to restrict cultivation of

Table 11. Heavy metal concentrations ($mg\,kg^{-1}$) in the soils from the soil contamination monitoring sites (1988).

Area	pH	As	Cd	Cr	Cu	Hg	Ni	Pb	Zn
Large scale agricultural area	5.9	0.34	0.20	1.15	6.68	0.17	2.27	6.83	11.90
Industrial complex downstream area	5.8	0.33	0.26	2.16	7.90	0.21	2.56	9.44	11.59
Riverside	6.0	0.66	0.20	0.92	4.56	0.16	1.63	6.48	11.34
Household waste landfill	6.2	0.49	0.28	0.86	5.51	0.12	1.40	7.21	10.18
Industrial waste landfill	6.0	0.32	0.29	0.97	7.06	0.17	2.28	13.54	14.58
Metal mining site	5.8	0.99	0.89	1.15	14.88	0.23	3.09	34.35	30.20
Road side	5.7	0.30	0.18	0.68	5.17	0.12	1.87	9.89	10.64
Metal smelting area	5.8	1.77	1.32	1.52	11.63	0.26	2.51	34.49	28.14

(Source: EA, 1989).

Table 12. Cd and Zn concentrations (mg kg^{-1}) in rice paddy soils and brown rice from the areas around zinc mining sites.

Sampling site	No. of samples	Cd			Zn		
		Min.	Max.	Ave.	Min.	Max.	Ave.
				Soil			
Shihung	6	4.00	14.67	7.68	263	1763	938.0
Ch'angwon	8	0.35	1.86	1.25	16	167	60.9
Songju	12	0.39	4.16	1.53	15	173	105.8
Uljin	9	1.06	3.61	2.16	28	230	104.6
Ch'ilgok	10	0.21	3.21	0.79	4	81	17.8
				Brown rice			
Shihung	6	0.41	1.63	0.87	28.8	40.6	34.07
Ch'angwon	8	0.19	1.57	0.57	24.4	33.8	29.61
Songju	12	0.13	1.31	0.55	20.6	55.0	28.66
Uljin	9	0.12	1.13	0.43	23.0	35.0	29.33
Ch'ilgok	10	0.04	1.11	0.31	21.8	28.3	24.11

(Source: Yoo and Lee, 1980).

agricultural products in these contaminated soils with high As concentration, very little action has been taken by the regulatory bodies. With the increasing concern of people, the recent introduction of much more stringent regulations and the Soil Conservation Act of 1995, some methods for soil decontamination have been prepared. Soil decontamination methods under consideration cover engineering, chemical and biological techniques (Kim, 1993).

7.4 Stream sediments of the major river systems

Although contamination levels of irrigation waters are low, contaminants such as heavy metals accumulate in rice paddy soils because large volumes of irrigation water are supplied to the soil throughout the whole growing season (see section 4.1).

As the data collected through the soil contamination monitoring network are averages of each contamination source, it is impossible to get information concerning the level of contamination of each river system or of local soil contamination levels. However, it is possible to predict heavy metal contamination by irrigation water with appropriate analytical data on the irrigation water over time. Analytical data for stream sediment in a season can be a good substitute when data from the analysis of the irrigation water are not available.

Tables 13, 14, 15 show the heavy metal contents in the stream sediments of major river systems (EA, 1987).

Within the Han river system, the Yangkunch'on and Chijangch'on stream sediment Cd concentrations exceeded 2 mg kg^{-1}. Kulp'och'on and Chijangch'on were two streams with severe Cu and Pb contamination in the stream sediment. In these stream sediments Cu concentrations were 252.5 mg kg^{-1} and 160.1 mg kg^{-1}, respectively. Mercury concentrations in the stream sediments from Chungryangch'on

Table 13. Heavy metal concentrations (mg kg^{-1}) in the stream sediments of the major streams in the Han river system.

Stream	Cd	Cr	Cu	Hg	Pb	Zn
Kulp'och'on	0.96	246.63	252.50	0.83	117.00	383.00
Wangsukch'on	0.12	26.13	65.21	0.94	4.46	39.63
Chungryangch'on	0.97	268.28	14.27	2.48	2.08	66.27
Kokrungch'on	0.87	68.80	11.50	0.47	9.33	75.20
Pakhach'on	1.54	4.84	3.79	0.02	57.70	21.07
Yangkunch'on	2.19	57.67	15.91	0.02	28.85	45.99
Chijangch'on	3.41	10.51	160.10	0.04	86.30	88.20
Som river	1.52	43.41	75.00	0.28	187.50	289.89
Chech'onch'on	1.30	7.23	5.06	0.01	43.28	60.56
Chuch'on	0.88	25.33	8.44	0.08	36.07	45.46

(Source: EA, 1987).

Table 14. Heavy metal concentrations (mg kg^{-1}) in the stream sediments of the major streams in other river systems.

River system, stream	Cd	Cr	Cu	Hg	Pb	Zn
Nakdong river*						
Hakjangch'on**	5.37	125.90	300.50	0.84	211.13	498.00
Yangsanch'on	5.35	982.30	498.30	0.24	326.25	776.30
Shinch'on	0.04	0.79	45.80	0.01	42.38	43.85
Palgyech'on	0.01	2.40	43.69	0.03	26.46	76.98
Kumich'on	0.02	2.35	43.12	0.01	16.18	43.16
Hwangjich'on	3.00	14.47	175.30	0.09	180.30	294.20
Kum river						
Hangkonch'on	2.50	<0.25	10.25	<0.22	32.50	153.00
Poch'ongch'on	N.D.***	N.D.	4.25	N.D.	10.00	35.00
Kangkyongch'on	0.02	0.01	1.50	0.02	0.47	3.86
Yongsan river						
Kukrak	2.31	3.98	31.76	1.03	8.48	200.89
Mankyong river						
Mokch'onpo	0.03	0.01	1.25	0.02	1.63	5.35
Tongjin river						
Sokkokch'on	0.04	0.05	1.63	0.02	1.80	8.22
Imjin river						
P'och'onch'on	1.17	9.35	11.34	0.14	1.51	39.00
Shinch'on	2.24	41.02	24.93	0.45	1.45	60.80
Ansongch'on						
Hwangkujich'on	0.00	1.28	1.54	N.D.	0.75	1.66
T'ongbokch'on	0.00	1.30	0.94	N.D.	0.46	1.11

*Bold charactered names are those of major river systems.
**Plain charactered names are those of streams.
***Not detected.
(Source: EA, 1987).

Table 15. Heavy metal concentrations (mg kg^{-1}) in the stream sediments of major streams related to the seas into which they flow.

Stream	Cd	Cr	Cu	Hg	Pb	Zn
the East Sea*						
Namdaech'on**	1.04	N.D.***	13.70	0.01	26.70	120.00
Kyongp'och'on	N.D.	N.D.	3.80	N.D.	N.D.	26.00
Ch'ongch'ch'on	N.D.	N.D.	25.40	0.01	N.D.	82.00
Ssangch'on	N.D.	N.D.	12.00	N.D.	N.D.	46.00
T'aehwa river	5.60	29.60	153.40	0.95	104.20	531.25
the Yellow Sea						
Sungkich'on	0.71	21.25	56.25	0.83	85.50	331.00
Panwolch'on	0.38	242.04	226.88	1.32	49.50	254.00
Sanglich'on	1.23	2.49	14.30	0.48	12.42	154.04
Chukkyoch'on	0.68	2.26	38.94	0.37	31.28	218.56
Namhaech'on	1.33	3.25	32.60	1.56	30.36	907.27
Koch'angch'on	0.02	N.D.	0.56	0.02	0.47	2.87
the southern Sea						
Tongch'on (Pusan)	6.80	437.00	350.00	0.57	159.60	134.40
Tongch'on (Sunch'on)	1.13	2.99	43.25	0.69	17.22	160.26
Tonghungch'on	0.07	0.81	1.00	N.D.	1.15	3.31
Yonoech'on	0.08	0.69	1.47	N.D.	0.91	3.32

*Bold charactered names are those of the seas toward which major streams are flowing.
**Plain charactered names are those of streams.
***Not detected.
(Source: EA, 1987).

and Kulp'och'on were 2.5 mg kg^{-1} and 0.8 mg kg^{-1}, respectively. Chromium concentrations in the stream sediments from Chungryangch'on and Kulp'och'on were 268 mg kg^{-1} and 247 mg kg^{-1}, respectively. Both Hg and Cr concentrations in the stream sediments from Chungryangch'on and Kulp'och'on were high compared to other streams. Zinc and Pb concentrations in the stream sediments from Kulp'och'on and Som river were also very high compared to other streams (Table 13). High Cd content in the stream sediment of Chijangch'on resulted from inflows of acid coal mine wastewater. The reason for the high Cd content in the stream sediment of Yangkunch'on was not clear.

For the Nakdong river system, heavy metal concentrations in the stream sediment of Hakjangch'on, Yangsanch'on, and Hwangjich'on were very high. Cadmium and Zn concentrations in the stream sediment of Hangkonch'on from the Kum river system, and of the Kuklak river from the Yongsan river system were high (Table 14).

Among streams flowing toward the East Sea, the T'aehwa river was the only stream whose Cd, Cu, Pb and Zn concentrations in the stream sediment were fairly high relative to concentrations in other streams. Out of streams flowing toward the Yellow Sea, Cu, Cr, Pb and Zn concentrations in the stream sediment of the Panwolch'on were very high and Cu, Pb and Zn concentrations in the stream sediment of the Sungrich'on, Chukgyoch'on, and Namhaech'on were relatively high (Table 15).

Table 16. Lead content (mg kg^{-1}) of the surface soils (0–10 cm) at the road side of major express ways*.

Year	Distance from the road, m	Tollgate, Seoul	Songnam	Ansong	Taejon	Hoedok
				Rice paddy soils		
1984	5	6.7	4.5	5.8	2.7	3.4
	20	4.3	3.2	4.2	2.8	2.8
1989	5	6.3	9.4	3.3	3.0	2.7
	20	4.7		2.7	2.4	3.7
				Cultivated upland soils		
1984	5	30.2	–	–	2.9	3.9
	20	2.6	–	–	2.4	2.2
1989	5	6.1	–	–	2.2	4.7
	20	1.1	–	–	1.4	2.4

*Average of the values from both sides.
(Source: ASI, 1989).

The stream sediment of Tongch'on in Pusan was contaminated by many heavy metals whose concentrations were fairly high. The heavy metal concentrations in the stream sediment of Tongch'on in Sunch'on also were very high (Table 15).

7.5 Roadside soils

Air pollution by heavy metals, including Pb, from exhaust systems of cars has become worse with the rapid increase in the number of cars. With increased air pollution by heavy metals, chances of heavy metal accumulation in soils near roadsides or in the plants from these soils are high compared to background concentrations in the forest soils. From 1987 the Ministry of Environment has monitored heavy metal contamination of soils at the roadside from 15 monitoring sites. ASI started Pb monitoring in 1974 (ASI, 1989), and since then has monitored Pb content in cultivated soils near main roads every five years.

Table 16 shows the Pb content of surface soils (0–10cm) at the roadside of major expressways determined in 1984 and 1985 by ASI. Since ASI changed its method of soil analysis for Pb in 1984, it is difficult to compare Pb concentrations before 1984 with those after 1984.

The data shows that Pb concentrations of the soils at the roadside of major express ways are close to background levels. There is no consistent trend of Pb concentration change over time, so it is not easy to predict any future change. Levels of Pb contamination at the roadside are, however, expected to decrease with increasing unleaded gasoline consumption.

8. Case study – Zn mine at Kwangmyong

An old Zn mine in Kwangmyong City is bordered on the northeast by Seoul, the capital of the Republic of Korea. Mining wastes have significantly contaminated soils in the area around the mine. The authors have been interested in these areas

Table 17. Mean heavy metal concentrations ($mg L^{-1}$) in stream water and groundwater samples.

Sample	As	Cd	Cu	Pb	Zn
Stream water	0.0040	0.0039	0.0157	0.0228	0.118
Ground water	0.0040	0.0045	0.0187	0.0243	0.174

(Source: Kwangmyong City, 1994).

Table 18. Criteria for drinking water quality ($mg L^{-1}$).

As	Cd	CN^-	Cr(vi)	F	Hg	NH_4-N	NO_3-N	Pb	Se
0.05	0.01	N.D.*	0.05	1	N.D.	0.5	10	0.1	0.01

*Not detected.
(Source: Ministry of Environment, 1995).

since the late 1970s and were involved in several projects including survey and remediation. Upon requests from Kwangmyong City, a group of researchers including the authors carried out a program for preventing contamination of the soils of the areas around the mine. Here, we present an outline of the program as a case study of a remediation program of contaminated soils in Korea (IASD, 1994).

8.1 Survey of contamination status

A survey was carried out on agricultural lands contaminated by mining wastes from the Kahak Zn mine, located at Kahak-dong and No-on-san-dong in Kwangmyong City. The surveyed areas covered some 877 000 m^2. Approximately 80% of the areas were either loam or sandy loam with imperfectly or somewhat poorly drained rice paddy soils.

Surface (0–15 cm) and subsurface (>15 cm) samples from agricultural lands were collected from 27 sites and were analyzed for As, Cd, Cu, Pb and Zn. In addition, phytotoxicity of the crops due to heavy metals and safety of the agricultural products from this area were reviewed. Results showed that there were no phytotoxic effects on crop growth but Cd concentrations in soils were high compared to those in the rice paddy soils and upland soils in the area.

Stream sediment samples (40–60 cm depth) from Kahakch'on and Mokkanch' which flow through the Zn mining area were collected at six sites and were analyzed for heavy metals. Sampling sites were located at 200 m, 500 m, 1 050 m, 1 850 m, 2 300 m and 3 750 m from the mine waste pile. Regardless of the distance from the source, heavy metal concentrations in the stream sediment samples were high compared to those in rice paddy soils and upland soils in the area. This shows extensive contamination of streams around the mining area.

Five stream water and eight groundwater samples which were assumed to be influenced by the mine were also sampled and analyzed. Table 17 lists the average heavy metal concentrations in stream and groundwater samples. Cadmium and Pb concentrations from some samples exceeded criteria for water quality contamination (Table 18).

Table 19. 0.1 M HCl extractable heavy metal concentration (mg kg^{-1}) of mine wastes and stream sediments.

Sample	As	Cd	Cu	Pb	Zn
Mine Wastes					
0–15 cm	1.75	10.8	7.39	10.6	130.8
15–30 cm	1.90	9.92	4.28	5.0	125.1
Stream Sediment	–	6.28	8.39	65.6	134.8

(Source: Kwangmyong City, 1994).

8.2 *Treatment of the mining wastes and stream sediments*

Comprehensive measures to prevent heavy metal contamination of agricultural lands are necessary since heavy metal concentrations in the mining wastes and the stream sediments are still high (Table 19). To protect agricultural land from heavy metal contamination by mining, mining waste piles should be isolated from rainfall and groundwater. If water inflow is unavoidable, the percolating water should be diverted from agricultural land.

8.3 *Improvement of soils and production of environmentally safe agricultural products*

Areas for soil improvement are to be decided on the basis of Cd concentrations in the soils. Two categories of areas are recognised: (i) areas with a high probability that Cd concentrations in brown rice grown in those areas will exceed 1 mg kg^{-1}, (ii) areas where this probability is low.

For each category, plans to produce agricultural products safe from heavy metals are as follows:

Category (i)

- Removal of contaminated soils: Remove 0–40 cm layer of soils from the surface
- Bringing soil from another place: Apply soils brought from other places to at least 25 cm depth with organic matter and other substances for enhancement of soil fertility

Category (ii)

- Application of lime, silicate fertilisers, fused phosphate: Application of these materials will increase soil pH. High pHs will decrease heavy metal solubilities because of inactivation of heavy metals, and this will reduce both movement of heavy metals in soils and will reduce plant uptake.

8.4 *Conversion of rice paddy fields to cultivation of upland crops*

Topographically, it is relatively easy to convert rice paddy fields in this area to the cultivation of upland crops. Compared to rice grown in the rice paddy fields, crops grown in cultivated uplands require less water, so heavy metal contamination from irrigation water can be reduced.

The levels of heavy metal contamination in the area are not high enough to cause phytotoxicity to crops, although the levels are high enough that is inadvisable to grow vegetable crops without taking proper measures against contamination, especially in relation to food quality. Conversion of land use requires selection from a wide variety of non-food crops, and as the area is near the suburbs of Seoul, a suitable use would be to develop it as a floricultural complex.

9. Conclusion

Korean soils are vulnerable to contamination due to their physico-chemical properties and topography. As the Republic of Korea's high level of industrialisation began relatively recently, most of the country is still relatively uncontaminated by heavy metals and by other hazardous materials with residual effects, with the exception of certain limited areas. It is now time to treat the soil environment as a matter of both national and international concern. Although the total area that is significantly contaminated is still small, we should protect our soil environment from accelerating contamination. To that end, we need to review our environmental laws and invest more money for research and development related to soil preservation. First of all, we need to prepare strict and systematic regulations for the Enforcement Decree of the Soil Preservation Act which stemmed from the Water Quality Preservation Act in January, 1995. The regulations, which are based on background values of heavy metals and substances with residual effects, should be applicable to a variety of contaminants so that soil contamination levels do not increase any further. In addition, the present number of monitoring sites in the soil contamination monitoring network, 520, should be increased. Among the monitoring sites, the 100 sites on agricultural land should be increased, considering the seriousness of agricultural land contamination.

Korea has not accumulated much data on soil contamination considering the extent of industrialisation and urbanisation. Most Koreans are not aware of even the scanty data that have been accumulated, because of the strict government control of the news media through the early 1980s. Since the late 1980s, with an improved economic situation and less control of the news media by the government, people have become increasingly concerned about the environment. Environmentalists have succeeded in getting the support of the public as reliable environmental data have been provided by both government officials and researchers.

Often it is very difficult to compare soil contamination data from one area with that from another because different analytical methods have been used by different researchers. For example, soil scientists are more interested in 0.1 M HCl-extractable heavy metal concentrations than total heavy metal concentrations, as this fraction may indicate the pool of metals available for plant uptake, whereas geochemists are mainly concerned with total heavy metal concentrations when they determine soil contamination. Establishment of standardised methods of soil analysis for soil contamination is very important. It would also be useful to establish conversion factors so that values obtained by one method can be compared with values obtained by another method. Other problems that must be resolved include determination of the number of sites necessary for a soil contamination monitoring network, and appropriate methodology, This will allow soil contamination maps to be prepared for each contamination source with reliable soil environment data based on analysis by standard methods of soil samples from the extended soil contamination monitoring network. In addition, research related to soil environment management should be developed further and related to future preservation of global environments.

References

Agricultural Sciences Institute 1983 An Outline of Korean Soils. pp. 10-39 ASI, Rural Development Administration. Suwon, Korea.

Agricultural Sciences Institute 1989 Research Report of the Rural Development Administration. Agricultural Sciences Institute, pp 89-91. RDA. Suwon, Korea.

Agricultural Sciences Institute 1990 Research Report of the Rural Development Administration. pp 28-30 ASI, RDA. Suwon, Korea.

Environment Administration 1986 A guideline for operating soil contamination network. EA. Seoul, Korea.

Environment Administration 1987 A survey for cleaning of the contaminated streams and rivers in Korea. Environment Administration, Seoul, Korea.

Environment Administration 1989 Korea Environmental Yearbook. Environment Administration, Seoul, Korea.

Institute of Agricultural Sciences and Development 1994 Studies for prevention of the soils in the areas around an old zinc mine in Kwangmyong City from heavy metal contamination. pp 33-234. Institute of Agricultural Sciences and Development, Seoul National University, Suwon, Korea.

Kim B Y 1993 Status of soil contamination in Korea and methods for soil improvement. In Proceeding of a symposium on soil management for sustainable agriculture. Chungnam National University, Taejon, Oct. 15, 1993. Ed. K H Han. pp 68-98. Korean Soc. of Soil Sci. and Fertilizer, Suwon, Korea.

Kim D H, Lim S K and Kwon O K 1989 Studies on the correlation between heavy metal concentrations of apple orchard lands and its leaves. Kor. J. Environ. Agric. 8. 1-6.

Kim K H, Yun J Y and Yoo S H 1995 Distribution of Cs-137 and K-40 in Korean soils. J. of Korean Soc. of Soil Sci. and Fertilizer 28. 33-40.

Kwangmyong City 1994 Final report for the project 'Counteracts for prevention of soil contamination'. Kwangmyong City, Korea.

Lee M H, Lim S K and Kim B Y 1986 Behaviors of arsenic in paddy soils and effects of absorbed arsenic on physiological and ecological characteristic of rice plant. I. Distribution of arsenic fractions in paddy soils and their relations to arsenic content in brown rice. Kor. J. Environ. Agric. 5. 35-42.

Ministry of Environment 1995 Notice 1995-58 'Criteria for drinking water', Ministry of Environment, Seoul, Korea.

Rural Development Administration 1989 A comprehensive report of a 10 year project - Improvement of Korean Soils. pp 255-286. Rural Development Administration Suwon, Korea.

Yoo Sun-Ho, and Lee C Y 1980 Occurrence of Cadmium and Zinc in paddy soils near zinc-mining sites and their uptake by rice. p. 255-266. A collection of the papers by the members of the National Academy of Sciences (NAS), NAS. Seoul, Korea.

19

Contaminants and the soil environment in Malaysia

Y. M. KHANIF AND S. SALMIJAH

1. Introduction

Malaysia is at present undergoing rapid development and tremendous changes especially in its economic structure. Previously, agriculture has been the main thrust of the Malaysian economy, but in recent years the industrial sector has continuously achieved high growth rates and is now the dominant force. This has resulted in an increase in urban population and the creation of new urban centres. The rapid industrialisation process in Malaysia coupled with global concern about environmental issues has fuelled environmental awareness in the country. In recent years reports on the deterioration of environment quality, especially in urban areas, have increased. Environmental issues are now of important political and social considerations.

Malaysia has a population of 18 million. Due to its proximity to the equator the climate is characterised by warm temperatures between 23 and 32°C, with minor variations. The mean annual rainfall is about 2000 mm, falling mainly during November–December. The drier months are February and March, and the dry period is more noticeable in the northern than in the southern regions of the country. The heavy rainfall causes erosion problems and also encourages leaching of mobile contaminants. Rates of biodegradation of certain contaminants are maximal at the prevailing temperature range.

Due to minimum climatic variations, the types of soil found in Malaysia are mainly influenced by the parent materials and the topography. In the interior upland areas, the soils originated mainly from granite, shale, sandstone and limestone. Soils in the coastal areas are mainly derived from marine alluvium, while in the low lying areas away from the coastal plain the soils are derived from riverine alluvium. The soil orders found in Malaysia are Entisol, Inceptisol, Alfisol, Spodosol, Ultisol, Oxisol and Histosol. The areas covered by the main soil orders are listed in Table 1. The bulk of the soil types are Inceptisols, Ultisols and Oxisols, which are highly weathered acid soils with low fertility status. Low cation exchange capacities coupled with heavy rainfall lead to high infiltration rates in these soils. The Inceptisols are mainly comprised of the heavy textured marine and riverine alluvial soils, most of which have high watertables and are of high fertility status. The typical properties of some selected Malaysian soils are given in Table 2.

As in most developing countries, the Malaysian economy has in the past been dependent on the export of primary commodities, including rubber, palm oil, petroleum, cocoa and tin. Malaysia is currently the major world producer of palm oil, rubber and cocoa. The land areas devoted to different agricultural crops are given in Table 3. The areas cultivated for all the crops except oil palm have either

R. Naidu et al. *(eds): Contaminants and the Soil Environment in the Australasia-Pacific Region, 563–578.*

Table 1. Principal soil types in Malaysia ('000 hectares).

Sub-orders	Peninsular Malaysia	Sabah	Sarawak	Total
1. Haplorthox Acrorthox	360	113	n.a.	n.a.
2. Paleudult Tropudult Plinthudult Tropaquult	1194	1339	2200	4733
3. Tropaquept Dystropept Sulfaquept	869	508	680	2057
4. Tropaquept	790	478	250	1518
5. Tropofibrist	865	125	1630	2620
6. Tropohumod	154	35	120	309
Total	7532	2598	4880	–

Source: Ng (1987).
n.a. – not available.

Table 2. Properties of some selected Malaysian soils.

Soil	Clay %	Silt %	Sand %	Carbon %	pH_w	CEC $cmol(+)kg^{-1}$	Base saturation %
Bernam Typic Tropoaquept	52	47	1	2.9	5.1	20	78
Rengam Typic Paleudult	13	5	81	1.4	4.7	2.8	9
Serdang Typic Paleudult	26	3	71	1.1	4.3	4.2	12
Munchong Tropeptic Haplorthox	67	8	25	1.5	4.5	8.9	24
Prang Typic Halporthox	70	12	18	4.2	4.0	15	11

Source: Tessens and Shamsuddin (1983)

stabilised or reduced in recent years, whereas there has been a small increase in the area under oil palm. These commodities are currently enjoying very high prices in the world market. The land area under forest in the different regions of Malaysia is given in Table 4. The percentage of forest land nationwide is 60%, with most of the forest areas in Sabah and Sarawak.

Price instability for agricultural commodities in the world market and the ever increasing production cost of primary commodities, have caused tremendous uncertainty in the economy. For petroleum, tin and timber, the products are extractive and not renewable. Due to the vulnerability associated with over dependence on primary

Table 3. Total area under selected crops: Malaysia.

Year	Rubber (ha)	Oil Palm (ha)	Cocoa (ha)	Paddy (ha)
1980	1 999 300	1 023 306	138 500	716 873
1986	1 905 606	1 599 311	322 334	627 500
1987	1 874 600	1 672 875	343 972	644 800
1988	1 867 300	1 805 923	398 376	665 800
1989	1 849 000	1 951 256	409 444	661 600
1990	1 932 900	2 029 464	415 628	651 700
1991	1 823 000	2 094 028	400 200	656 000
1992 e	1 807 000	2 167 396	388 700	660 700

Source: Ministry of Primary Industry (1993).

Table 4. Forest areas in different regions of Malaysia.

Year	Forest Area km² Peninsular Malaysia	Sabah	Sarawak	Total
1989	63 204	44 366	87 148	194 718
1990	62 699	22 243	87 115	194 057
1991	62 699	44 120	87 030	193 849
1992	62 699	44 305	86 823	193 827
1993	62 699	73 713	86 628	223 040

Source: Statistic Department, Malaysia (1994).

Table 5. Statistics of some of the principal manufacturing industries in Malaysia (1991).

Types of industries	No. of establishments	Gross value of output (RM Million)	No. of workers
Electrical products	492	33 798	261 909
Food manufacturing	1 324	17 772	79 002
Industrial Chemical manufacturing	111	6 364	11 602
Rubber products	387	5 949	65 082
Wood products	768	5 845	96 675
Petroleum refineries	10	4 785	1 225
Iron and steel	145	4 689	15 437
Machinery	438	4 744	34 334
Textile	271	3 494	45 049

Source: Statistic Department, Malaysia (1994).

commodities, the country has diversified its economic activities to include manufacturing industries. The contribution of the manufacturing industries to the Malaysian economy has increased tremendously and now this sector has become more important than the production of primary commodities (agriculture, mining and forestry). The types of manufacturing industries are numerous and the major industries are listed in Table 5. The most important industry, accounting for more than 30% of the gross value of output, is the manufacturing of electrical machinery and appliances. Other notable industries are the manufacturing of rubber products, wood products, chemicals, petroleum, machinery, plastics and iron/steel industries. In the future it is envisaged that the economy will be dominated by manufacturing industry with declining contributions from the agricultural sector.

With the rapid shift to manufacturing activities, the urban population is increasing at a high rate, creating more new urban centres. Manufacturing industries are mostly located in the west coast region of Peninsular Malaysia for historical reasons and because of the presence of better infrastructure. These newly created urban areas are experiencing rapid development, with construction of commercial and residential buildings, highways, recreational areas and communication amenities. The urban centres are also faced with the increasing number of vehicles on the roads, waste disposal problems and environmental deterioration.

Although environmental degradation is currently a very important issue, its prominence has only surfaced rather recently. Scientific studies related to environmental pollution, especially on the soil environment, have been rather few, with most emphasis on air and water pollution. This paper attempts to present an overview of the sources and extent of the contamination problem in Malaysian soils.

2. Sources of soil contamination

Sources of soil contaminants vary with land use patterns. In rural areas where agricultural activity dominates, agrochemicals and agricultural wastes are the major source of contaminants. In urban areas, typical sources of contamination are mainly from industrial waste, fuel combustion and domestic wastes.

2.1 Agrochemicals

Malaysian soils are generally of low fertility status. To maintain high crop yields fertilisers are applied at high rates annually. The rate of NPK applied has doubled from 70 $kg\,ha^{-1}$ in 1975 to 156 $kg\,ha^{-1}$ in 1993 (Tay, 1994). Lime application is also a common practice to alleviate soil acidity problems.

The major fertilisers applied are N, P and K. Quantities of fertiliser used on major crops are listed in Table 6. The amounts of fertilisers used are in the order $N>K>P$ (Table 6). The main N fertilisers are urea and ammonium sulphate. The major sources of P are phosphate rock and superphosphate. Muriate of potash (KCl) is a major source of K. The oil palm plantations account for almost 60% of the total fertiliser applied. Rubber, cocoa and rice make up about 30%, and other crops account for about 10% of fertiliser usage.

Heavy metal contamination in fertilisers is confined mainly to phosphate rock (PR). The levels of toxic metals in other fertilisers are not as significant. Large amounts of PR are being applied especially in plantation crops and continuous application of this P source can cause considerable heavy metal accumulation. The types and amount of heavy metals in phosphatic fertilisers depend on the origin of

Table 6. Usage of nutrients applied as fertilizers to major crops, 1989-91 ('000 tonnes): Malaysia.

	1989					1991				
Crop	N	P_2O_5	K_2O	Total	%	N	P_2O_5	K_2O	Total	%
Oil palm	85	80	225	390	56	91	88	240	419	59
Rubber	33	28	37	98	14	28	24	31	83	12
Cocoa	18	12	24	54	8	18	11	21	50	7
Padi	52	14	6	72	10	52	16	7	75	10
Others	35	22	26	83	12	35	24	28	87	12

Source: Malaysia Agricultural Directory and Index (1993/94).

Table 7. Heavy metal content of phosphate rock materials.

Source of phosphate rock	Elements ($mg\,kg^{-1}$)							
	Fe	Mn	Co	Mn	Zn	Cd	As	U
Gafsa	1.8	7	3	15	393	38	4	88
North Florida	9.2	212	5	4	57	3	7	143
Jordanian	2.2	5	<1	8	235	4	12	72
North Carolina	4.6	7	2	9	400	48	23	69
Sechura	3.2	91	3	6	178	11	5	72
Mexican	2.8	151	6	4	90	8	3	118
Nauru Island	4.1	122	6	8	1010	100	3	64
Arad	0.7	2	4	12	560	12	7	153
Makatea	0.7	2	<1	6	220	10	2	65
Chatham Rise Phosphorite	30.0	100	4	5	95	2	n.d.	101

Source: Syers et al. (1986).

the phosphate rock (Syers et al., 1986). Typical heavy metal concentrations in the commonly used phosphate rock are presented in Table 7. The metal that has been given most attention is Cd, because its concentration is fairly high in some PRs. In the longterm, the use of PR with high Cd concentration can be harmful, since Cd may build up in the soil and ultimately become a health hazard to man through uptake by plants.

Under tropical conditions, weeds, pests and diseases are serious problems that can cause heavy crop losses. High rates of pesticides are required to sustain high crop yields and pesticide consumption in crop production is increasing (Cheah and Lum, 1993). Herbicides account for about 80% of the total pesticides used, insecticides constitute 15% while fungicides and other pesticides make up the remaining 5%. The herbicides are mainly used in plantations and in rice growing areas where high labour cost excludes manual weeding. In the rice growing areas, the shift in practice from transplanting to direct-seeding has further elevated the weed problem. Insecticides are mainly used in vegetable, rice and tobacco growing areas with a smaller portion being used in plantation crops. The use of fungicides is confined mainly to horticultural crops and ornamentals while small quantities are used in vegetable and tobacco production. More than 37 types of herbicides are registered with the Malaysian Pesticide Board. The most commonly used chemicals in the plantation and small holdings are benzene hexachloride, 2,4-D, paraquat and diuron which are marketed under various trade names. Numerous insecticides with varying ingredients are also being used in Malaysia; common ones being γ BHC, carbofuran, chlorpyrifos, cypermethrin, malathion and endosulfan.

2.2 *Agricultural waste*

The oil palm industry generates a large amount of agricultural waste with more than 15 million of palm oil mill effluent (POME) being produced annually. Most of it is being recycled through land application. Animal wastes especially from poultry and pig production are also applied to agricultural soils for soil improvement. Palm oil mill effluent and rubber effluent no longer cause environmental problems as most of the waste water is successfully treated before being discharged.

One of the most polluting activities in animal husbandry practice is pig husbandry. In the districts of Kuala Langat and Sepang in Selangor for example, a total of 196 741 pigs in 152 farms and 107 553 animals in 161 piggeries respectively were reported in 1992 (Jabatan Perkhidmatan Haiwan, Negeri Selangor 1992). Pig farming in Malaysia has high pollution potential because nearly all farms do not properly treat their waste effluent. Effluents from pig farms are very rich in biochemical oxygen demand (BOD) and some heavy metals such as Cu which are added to pig diets. The Sepang river, which receives a lot of effluents from these farms, is very polluted and its upper and middle reaches are almost nonfunctional.

2.3 *Industrial waste*

A survey initiated by the Department of Environment (DOE), Ministry of Science and Technology, identified a range of hazardous wastes from various industries totalling nearly 343 000 $m^3 yr^{-1}$ (Table 8). Certain categories of wastes, notably from the primary palm oil, rubber and tin mining industries were not included in this survey.

The metal finishing industry generates the greatest amount of toxic and hazardous wastes. An initial study by the DOE reported that this industry contributed 43.8% of the nation's total toxic and hazardous wastes. This figure dropped to 29.0% in a survey made two years later (DOE, 1987) but was still unequalled by any other industries. Wastewater from the metal finishing industries contains heavy metals and toxic chemicals such as Cr (III &VI), Cu, Fe, Ni, Zn, Sn and acid or alkali containing cyanide. Alkaline cleaners such as kerosene, grease and oils are also found in the wastes. In Malaysia, electroplating plants are mainly small and medium scale, and most of them are reluctant to install wastewater treatment facilities for reasons such as limited land area within their premises, lack of technical knowledge and skilled operators to operate the facilities, and financial constraints. Most of the plants operate without any wastewater treatment. In a survey of 42 such facilities in the Klang Valley area (Teh, 1989), only 33% had some form of wastewater treat-

Table 8. Type of hazardous wastes in Malaysia (data collected in 1987).

		Volumes of waste (m^3)	% of Wastes
1.	Acids (possibly heavy metals)	53 142	15.5
2.	Sludge (with heavy metals)	56 384	16.5
3.	Sludge (mineral)	47 495	13.9
4.	Asbestos	34 284	10.0
4.	Pain Vdye/pigment (H_2O and solvent based and sludges containing them)	34 004	9.9
5.	DusVslag/ashed	27 489	8.0
6.	Alkalis (possibly heavy metals)	26 764	7.8
7.	Oil and hydrocarbons	19 896	5.8
8.	Photographic wastes	11 969	3.5
9.	Rags, paper and plastic containers	5 572	1.6
10.	Sludge, oil	3 764	1.1
11.	Solvent, spent halogenated and nonhalogenated	1 097	0.3
12.	Others	20 900	6.1
	Total	342 760	100.0

Source: DOE, 1987.

ment. Others routinely discharged their wastewater into drains, from which it eventually reaches the rivers and the sea or percolates to groundwater (DOE, 1992). Since there are no suitable toxic waste treatment and disposal facilities available, significant amounts of toxic wastes generated by electronics manufacturing and other potentially polluting industries are presently disposed into sanitary landfill areas designated for the disposal of municipal solid wastes.

Under the Environmental Quality (Sewage and Industrial Effluent) Regulations, 1979, industries are required to treat their wastewater to acceptable limits of discharge (Table 9). However, there are, at present, no organised ways to dispose of these scheduled toxic wastes. To solve problems presented by such wastes, the DOE established a longterm comprehensive programme. Regulations pertaining to the management of scheduled waste were enacted in 1989 under the Environmental Quality Act, 1974. Under the 4th Schedule of this Act, a scheduled waste is defined as any waste which when mixed will produce heat, fire, explosion and release of toxic substances. A total of 107 categories of waste which include acids, alkalis, asbestos, sludge with heavy metals and solvents are listed under this schedule. As mentioned above, a significant number of industries discharge wastes to drains or bodies of surface water which could potentially contaminate the soil environment. Some waste streams are being processed by the industries which generate them to reclaim useful materials, some are being sent to professional reclaimers for the same purpose and a number of industries are now storing their wastes in warehouses

Table 9. Third Schedule Environmental Quality Act 1974 Environmental Quality (sewage and industrial influents) regulations 1979 I Regulation 8(1), 8(2), 8(3), Parameter limits of effluent of standards A and B.

Parameter	Unit	A	B
i) Temperature	°C	40	40
ii) pH value	–	6.0–9.0	5.5–9.0
iii) BOD5 at 20°C	$mg\,L^{-1}$	20	50
iv) COD	$mg\,L^{-1}$	50	100
v) Suspended Solids	$mg\,L^{-1}$	50	100
vi) Mercury	$mg\,L^{-1}$	0.005	0.05
vii) Cadmium	$mg\,L^{-1}$	0.01	0.02
viii) Chromium, Hexavalent	$mg\,L^{-1}$	0.05	0.05
ix) Arsenic	$mg\,L^{-1}$	0.05	0.10
x) Cyanide	$mg\,L^{-1}$	0.05	0.10
xi) Lead	$mg\,L^{-1}$	0.10	0.5
xii) Chromium, Trivalent	$mg\,L^{-1}$	0.20	1.0
xiii) Copper	$mg\,L^{-1}$	0.20	1.0
xiv) Manganese	$mg\,L^{-1}$	0.20	1.0
xv) Nickel	$mg\,L^{-1}$	0.20	1.0
xvi) Tin	$mg\,L^{-1}$	0.20	1.0
xvii) Zinc	$mg\,L^{-1}$	1.0	1.0
xviii) Boron	$mg\,L^{-1}$	1.0	4.0
xix) Iron (Fe)	$mg\,L^{-1}$	1.0	5.0
xx) Phenol	$mg\,L^{-1}$	0.001	1.0
xxi) Free Chlorine	$mg\,L^{-1}$	1.0	2.0
xxii) Sulphide	$mg\,L^{-1}$	0.5	0.50
xxiii) Oil and Grease	$mg\,L^{-1}$	Not detectable	10.0

Note A = Limits for point sources where there is water intake downstream.
B = Limit for sources without water intake downstream.

(DOE, 1987). The project to provide treatment and disposal facilities at Bukit Nenas, Negeri Sembilan which has the capacity to treat and dispose of hazardous wastes, is expected to be completed by the end of 1997. It is expected that this facility will accommodate most if not all of the toxic waste generated in Malaysia (DOE, 1994).

2.4 *Municipal waste*

In Malaysia, the main contributor to organic pollution is domestic sewage. In 1988, domestic sewage contributed about 80% of the total organic pollution (cited in Suki,

1993). Beside the organic components, sewage is also a major source of nutrients, particularly N and P compounds (Suki, 1993). Studies in Klang Valley (Govt of Malaysia and ADB, 1987) shows domestic sewage to be the main polluter of various parts of the Klang river. High faecal coliform counts of 3×10^4 MPN $100\,ml^{-1}$ were detected near river mouths (Law, 1980) and as much as 30–50 MPN $100\,ml^{-1}$ was still detected 10 km away from the river mouth (Law, 1984). High faecal coliform counts were also reported in coastal waters 6 km off the Port Dickson shoreline (Law, 1985).

In the past two decades, there have been significant improvements in the sanitary facilities in Malaysia. However, sewage pollution has persisted because of the lack of financial support to provide a more comprehensive sewerage system. Being a highly capital intensive investment, commitments to sewerage projects have been extremely slow (Suki, 1993). Only recent awareness of its effects on the tourist industry and water resources seems to have created an interest in treating sewage management seriously (Suki, 1993). There are several types of treatment plants in use in Malaysia, namely oxidation ponds, aerated ponds, oxidation ditches and activated process, imhoff tanks and septic tanks, rotating biological filters, trickling biological and other fixed media systems (Suki, 1993). Stabilisation ponds such as the oxidation pond, facultative pond and aerated lagoons have been the most popular forms of sewage treatment in Malaysia (Asairinachan, 1988; Suki, 1993), but with restrictions on land availability for sewage works in Malaysia, in future the more compact treatment processes will replace lagoons and oxidation ponds (Ensearch, 1988).

It has been estimated that Malaysia generates $9600\,t\,d^{-1}$ of municipal solid wastes, with Kuala Lumpur generating $2000\,t\,d^{-1}$ (Abdul Halim et al., 1995). Municipal solid waste can include among other things containers, packages, non-durable and durable goods and sewage sludge which can contain toxic metals. Specific concerns for the land disposal of contaminated materials and for abandoned waste disposal sites include the transport of heavy metals to both surface and groundwaters (Forstner et al., 1991). Improperly maintained sanitary landfills can also lead to an unacceptable odour and breeding of disease-bearing vectors. Although sanitary landfill is the method most commonly used now, availability of suitable areas is becoming more difficult, especially in urban areas of Kuala Lumpur, Penang, Ipoh and Johore Bahru. Other methods of disposal such as incineration are increasingly being considered for some areas.

3. Nature of contaminants in agricultural soils

The main contaminants of public concern in agricultural soils are heavy metals, pesticides and nitrate, derived from the application of fertilisers, pesticides and agricultural wastes. Data on the extent of these contaminants in Malaysian soils are very limited, and it is not possible to assess the seriousness of current soil contamination problems.

3.1 Heavy metals

Under natural conditions the heavy metal content of soils depends on the nature of parent materials. The background levels of Co, Cu, Fe, Mn, Ni and Zn (Table 10) in Malaysian soils have been reported by Lee et al. (1992), who showed that the levels of the various metals are related to those in the parent materials. The order of abundance in most soils is Fe>Mn>Zn>Cu>Co>Ni, with the concentrations of Co, Cu, Ni and Zn generally low (i.e. $<250\,mg\,kg^{-1}$).

Limited studies on the long term effects of PR applications to soils in a rubber plantation have shown that the Cd levels in soils were slightly elevated, from $1.13\,mg\,kg^{-1}$ to $1.26\,mg\,kg^{-1}$ (IFDC, 1992). The Cd content of leaves, however, increased from $37\,\mu g\,kg^{-1}$ in unfertilised plots to $108\,\mu g\,kg^{-1}$ in fertilised plots.

Fractionation of Cd, Cu, Fe, Mn, Pb and Zn into resistant and non resistant fractions was carried out on soil samples collected from oil palm plantations (Khairiah, 1993). It was assumed in this study that the resistant fractions were of natural origin while the non-resistant fractions were derived from anthropogenic additions. In most of the samples, the resistant fractions accounted for more than 90% of the total heavy metal content, suggesting low level of contamination. However, in some localised areas there were also samples with appreciable levels of the heavy metals in the non resistant fraction (particularly the organic fraction). The levels of heavy metals in groundwater were low.

Due to the limited data available the extent of heavy metal contamination in Malaysian agricultural soils cannot be determined with certainty. However, evidence so far indicates that heavy metal contamination could be a problem, especially in the long-term. Studies to address heavy metal issues are currently being actively pursued.

3.2 Pesticides

There is limited documentation on the extent of pesticide contamination on agricultural lands in Malaysia. The main concern has been with pesticide residues in agricultural produce and pesticides leaching into groundwaters.

Rice production requires high inputs of pesticides. Water resources in the rice growing area of the northern region of Malaysia were found to contain pesticide residues (Cheah and Lum, 1994). Residues of cyclodiene endosulfan were detected in most water samples at levels ranging from <0.005 to $25.5\,\mu g\,L^{-1}$. Occurrences of lindane, DDT, aldrin and dieldrin were not widespread and the levels were below $0.1\,\mu g\,L^{-1}$. Herbicide residues detected were 2,4-D, paraquat and molinate, with the levels fluctuating according to seasons. In another study (Nashriyah et al., 1994) molinate residue was found in soils in rice growing areas.

Cheah and Lum (1993) reported detectable levels of diazinon residues in water samples collected from a reservoir in Cameron Highland, downstream from a vegetable growing area. The level ranged between 0.0091 and $0.0037\,mg\,L^{-1}$. A low level of profenophos residue ($0.0001\,mg\,L^{-1}$) was also detected in one of the water samples.

Table 10. Total micronutrient content of soils developed over igneous and high metamorphic rocks (Lee et al., 1992).

Soil Series	Classification	Parent Material	Heavy Metal Concentration Mn ($mg\,kg^{-1}$)	Cu ($mg\,kg^{-1}$)	Zn ($mg\,kg^{-1}$)	Co ($mg\,kg^{-1}$)	Ni ($mg\,kg^{-1}$)	Fe_2O_3 (%)
Tampin	Clayey, kaolinitic, isohyperthermic, Orthoxic, Tropudult	Granite	40–58	21–36	60–130	30–106	2.4–4.5	1.5–1.6
Kulai	Clayey, kaolinitic, isohyperthermic, Orthoxic Tropudult	Rhyolite	177–210	24–56	60–97	30–75	6.2–18.0	3.2–6.4
Segamat	Clayey, oxidic, isohyperthermic, Tropeptic Hapludox	Andesite	420–935	102–146	168–243	65–125	3.8–10.8	17.7–18.6
Kuantan	Clayey, oxidic, isohyperthermic, Haplic Acrudox	Basalt	290–430	172–208	157–168	50–110	39.5–64.2	19.3–20.6
Prang	Clayey, oxidic, isohyperthermic, Haplic Acrudox	Schist	600–700	80–96	211–228	25–75	21.5–34.0	20.5–21.0

Total Analysis by Burnas, 1968.

Table 11. Average metal concentrations ($\mu g\,g^{-1}$) in Kuala Lumpur soils and streetdust.

	Fe	Mn	Zn	Pb	Cu	Cd
Soils (roadside) (n=25)	1090±670	121±93	222±161	895±767	18.8±8.0	2.38±10.33
Streetdust	1790±860	153±47	344±166	2466±1880	35.5±21.6	2.96±0.44

Source: Ramlan and Badri (1988).

Paraquat and 2,4-D are frequently used for weed control. Although paraquat is very persistent it is strongly adsorbed by soil colloids, thus limiting its movement. Paraquat was detected in soils repeatedly treated with this chemical in the oil palm growing areas (Cheah and Lum, 1993). The use of 2,4-D although widespread, does not pose a serious threat, as its degradation in Malaysian soils is rapid (Cheah, 1987; Chuah and Gopinath, 1988). Similarly, degradation of carbaryl in Malaysian soils was also rapid, with the half life being only six days, which is lower than normally reported (Azimahtol Hawariah et al., 1995).

3.3 Nitrate

Considerable attention has been focused on the use of fertilisers and animal manures, with regard to NO_3^- pollution of groundwaters, accumulation of NO_3^- to toxic levels in plants and the eutrophication of rivers. Ingestion of NO_3^- can cause methaemoglobinemia and in some cases has been linked to stomach cancer. Although no serious NO_3^- pollution has been documented in Malaysia, its potential should not be underestimated.

The safe limit for NO_3-N in drinking water set by the World Health Organisation (WHO) is $10\,NO_3-N$. Under intensive agriculture with high inputs of fertiliser N and animal manure, concentration of NO_3-N in groundwater above the WHO limit is not uncommon (Robertson, 1979; Khanif et al., 1984). In Malaysia concentrations of NO_3-N in groundwater collected from agriculture are still below the WHO safe limit (Yusof et al., 1990). However, NO_3-N levels have been found to increase due to agricultural activities (Yusof et al., 1990). In Malaysia cattle were reported to have died of NO_3^- poisoning after consuming Napier grass with high NO_3-N content (Seiler et al., 1979).

4. Heavy metal contamination in urban soils

Data on heavy metals in urban soils in Malaysia are scarce and limited to a few localised areas. Some studies are still in progress and results from these studies should be available soon.

A study on the levels of heavy metals in soils and street dust in Kuala Lumpur was conducted by Ramlan and Badri (1989). The study showed that street dust, which represents the uppermost soil layer, contained significantly higher amounts of Fe, Mn, Zn, Pb, Cu and Cd compared to the soil underneath it (Table 11). When samples were geochemically fractionated into the resistant and non-resistant fractions, it was found that some of the metals, particularly Pb, was present mostly in the non-resistant fraction. This observation suggests the recent input of these metals into the environment is most probably from the atmosphere.

5. Heavy metals in sediment

The west coast of Peninsular Malaysia is drained by over 20 rivers, which flow through heavily populated areas and some of the most congested industrial estates in the country, carrying with them a large range of pollutants. The extent of pollution of sediments in the coastal areas as well as the waters could serve as an indicator of the pollution load in the mainland, especially that arising via the leaching process. The levels of heavy metals in sediments collected at various locations in coastal areas from Perlis in the north to Johore in the south have been reported by Ahmad et al. (1993). This study showed that Fe concentrations mostly fell between 35 000 and 40 000 $mg\,kg^{-1}$ with <5% in the above 65 000 $mg\,kg^{-1}$ range bracket. Over 50% of Mn concentrations were between 3 000 and 5 000 $mg\,kg^{-1}$. The bulk of Zn (60%) concentrations were found to range from 600 to 900 $mg\,kg^{-1}$ and over 70% of Pb and Cu in the sediments analysed were within 10–30 $mg\,kg^{-1}$ and <5 $mg\,kg^{-1}$ range, respectively. Over 80% of Cd concentrations were found to be <1 $mg\,kg^{-1}$. These levels were all within the normal range found elsewhere (Ahmad et al., 1993).

Other studies show more localised effects of particular inputs from activities such as shipping to these areas. Ismail et al. (1989) reported higher heavy metal contents (Cu, Zn, Pb, Cd) in sediments taken from Port Klang and Port Dickson compared to other areas, not involved in similar activities. The exact location where the sample is taken and representative sampling are necessary to avoid contradictory results between heavy metal analyses of samples taken from the same area. To illustrate this point, a study by Law et al. (1988) of the sediments of Port Dickson, which is the same area as was studied by Ahmad et al. (1993), shows much lower values of some metals, notably Fe (4034 $mg\,kg^{-1}$), Mn (11 967 $mg\,kg^{-1}$), Zn (15.0 $mg\,kg^{-1}$) and Pb (4.2 $mg\,kg^{-1}$). The values for Cu and Cd (2.1 and 0.02 $mg\,kg^{-1}$, respectively) were in the same range as those reported by Ahmad et al. (1993).

6. Heavy metals in mining areas

In Malaysia, much research has been done on the Mamut Cu mine in the district of Ranau in Sabah. This area of 2942 ha is an opencast mine that takes its processing water from the Mamut River which, in turn, receives most of the effluents of the mining activities (Jumat, 1992). The concentrations of Cu, Cd, Pb, Zn and Fe in surface sediments of this river were significantly higher than another river (control river) which drains a similar, but unmined catchment (Jumat, 1992).

The study showed significantly higher levels of Pb in fish species. A survey of residents (5–60 years old) living near the Ranau mine showed substantially higher concentrations of Pb in hair than in a control group in Kota Kinabalu, and in Sabah (cited in Khairuddin et al., 1992). Preliminary findings show some evidence that air pollutants and heavy metal contamination (especially Pb and Cd) in Ranau have a causal relationship with prevalence of pterygium (a condition characterised by thickened and degenerative conjunctiva) in the residents (Khairuddin et al., 1992).

7. Guidelines and Management Strategies

In Malaysia the DOE is the agency responsible for environmental protection; it formulates policies and enforces regulations related to the environment. The government enacted the Environmental Quality Act (1974) as a legal basis for pollution control and preservation of the environment. Since the enactment, 13 sets of regulations have been introduced to deal with specific environmental problems. The regulations address a wide range of problems, including regulations on oil palm waste, wastes from the production of raw natural rubber, sewage and industrial effluents, schedule wastes, and provision for environmental impact assessment. The national approach to environmental protection places more emphasis on preventive than on curative measures. Since 1987 proposals for major development projects have been required to include an Environmental Impact Assessment before they can be considered for approval. Proposed projects have to adhere to strict environmental regulation for approval.

Another approach currently pursued is to ensure proper control of the sources of pollution. The DOE maintains an up to date inventory of toxic and hazardous wastes. The treatment and disposal of these wastes is closely monitored by the DOE, and an integrated waste treatment and disposal facility is under construction and will be in operation in the near future.

The DOE under the Ministry of Science, Technology and Environment, Malaysia has been carrying out comprehensive monitoring programmes for air and water quality (rivers and marine) for several years, but not for soil contamination. There is as yet no comprehensive nationwide programme to check, monitor and conduct research into the extent of soil contamination, whether due to heavy metals, pesticides or other persistent organics. Research activities carried out by individual scientists or groups of researchers have been rather isolated, sporadic and uncoordinated. To the authors' knowledge, there are as yet no standard or recommended procedures or methodologies drawn up for such studies on soil contaminants. A comprehensive soil quality monitoring system needs to be set up, with set standards for sampling methods, analytical procedures, safety limits and inventory of contaminants.

Due to government regulation and public pressure, industries are taking serious initiatives to be more environmentally responsible. In the agricultural sector farmers are attempting to reduce pesticide use by adopting integrated pest management, which emphasises biological control. Agricultural wastes are increasingly being used for land application to reduce dependence on fertiliser.

8. Conclusions

There is still limited information on soil contamination in Malaysia and the extent of the problem cannot yet be established. The important sources of soil contamination identified are pesticides, fertilisers, animal wastes, industrial wastes and domestic wastes.

Malaysia has a well developed and comprehensive set of legal provisions relating to protection of environment. The approach has been to adopt more preventive than curative measures. Hazardous wastes are controlled at their sources. Monitoring of air and water quality is carried out routinely, but monitoring of soil contamination is at present not widely practised. It is timely that the monitoring of soil contamination should be seriously considered especially in high risk areas. More comprehensive and integrated studies on soil contamination are required. These include aspects such as base line data collection on soil contaminants, identification of sources of contamination, modelling of contaminants' behaviour in soils and development of management strategies.

References

Abdul Halim S, Rakmi R R and Abdul Hannan A 1995 Renewable energy from municipal solid wastes. In Proc. International Conferences on Advances in Strategic Technologies. Ed A R Azizah. pp 262-273. Faculty of Engineering, Universiti Kebangsaan Malaysia Bangi Malaysia.

Ahmad I, Badri M A and Ramlan M N 1993 The background levels of heavy metal concentration in sediments of the west coast of Peninsular Malaysia. Sci. Tot. Environ. 315-323.

Asarinachan K 1988 Aerated lagoons for sewage treatment. J. Environ. Manage. and Res. Assoc. Malay. II, 14-21.

Azimahtol Hawariah L P, Yong K N, Salmijah S, Abdullah M, Noraini D and Noran N 1995 Degradation of carbaryl in sandy clay soil. In Proc. of Symp. Applied biology in enhancement quality of life and environment. Ed. M K Vidyadran and Y Ishak. pp 86-92. Malaysian Society of Applied Biology, Kuala Lumpur, Malaysia.

Cheah U B and Lum K Y 1993 Pesticide contamination in the environment. In Waste Management in Malaysia: Current Status and Prospects for Bioremediation. Eds. B G Yeah, K S Chee, S M Phang, Isa Zaid, Azni Idris and Maketab Mohamed. pp 169-176. Ministry of Science, Technology and the Environment, Malaysia.

Cheah U B and Lum K Y 1994 Pesticides residue and microbial contamination of water resources of rice agroecosystem in MUDA areas. In Proc. of Seminar on impact of pesticide on the rice agroecosystem in the MUDA areas. Eds. M Nashriyah, K Y Lum and S Ismail. pp 15. Malaysian Institute for Nuclear Technology Research, Bangi Malaysia.

Department of Environment 1987 Feasibility Study on treatment and disposal of toxic and hazardous wastes in Malaysia. Report by Dames and Moore and Waste Management International Ltd. Department of Environment Malaysia, Kuala Lumpur.

Department of Environment 1989 Environmental Quality Report 1988. Department of Environment, Ministry of Science, Technology and the Environment, Kuala Lumpur, Malaysia. 75 p.

Department of Environment 1992 Waste Reduction at source for small and medium scale electroplating plants. Prepared by Bureau of Research and Consultancy Universiti Kebangsaan Malaysia Bangi, Malaysia. 144 p.

Department of Environment 1994 Environmental Quality Report 1993 Ministry of Science, Technology and the Environment, Kuala Lumpur, Malaysia. 79 p.

Department of Statistics Malaysia 1993 Yearbook of statistics 1993. Kuala Lumpur, Malaysia. Department of Statistics. 320 p.

Ensearch 1988 Editorial, There's more in the mush! J. Environ. Manage. Res. Assoc. Malay. II(I), 23.

Environmental Quality Act 1974 Third schedule Environmental Quality Act 1974. Environmental Quality (Sewage and Industrial Effluent) Regulation 1979. International Law Book Series, Kuala Lumpur.

Forstner U, Colombi C and Rainer R 1991 Dumping of wastes. In Metals and their compounds in the environment. Occurrence, analysis and biological relevance. Ed. E Merian. pp 333-356. VCH., Weinheim.

Government of Malaysia and ADB 1987 Klang Valley Environmental Improvement Project prepared by Engng Sc. Inc. and Seatec Int. and Department of Environment Malaysia, Kuala Lumpur.

International Fertilizer Development Center (IFDC) Annual Report 1992. Muscle Shoals, Alabama, USA. 79p.

Ismail A, Idris A G and Sukal R 1989 Distribution of heavy metals in sediment of Port Klang. Proc. 12th Annual Seminar Malaysian Society Marine Science. pp 279-282. Kuala Lumpur.

Jumat S 1992 Taburan logam berat di perairan sekitar lombong tembaga Mamut dan Empangan Lohan . Proc, Simp. Sumber Alam Kebangsaan Pertama. Eds. M A Kaswadi, O Mohd. Rozali, M K Narimah, H Ridzwan and K Arbain. pp 119-129. Fakulti Sains Sumber Alam, Universiti Kebangsaan, Sabah.

Khairiah 1993 Recycling of agricultural waste water: Input and availability of heavy metals. PhD Dissertation Universiti Kebangsaan Malaysia, Bangi Malaysia. 391p.

Khairuddin Y, Yeoh P N, Siti Norazah Z, Low K S, Lum K W, Wai Wai and Duncan M 1992 Ranau Basin: A Study of its people and environment. A report to the Ministry of Science Technology and Environment, Malaysia. University of Malaysia, Kuala Lumpur. 57 p.

Khanif Y M, Van Cleemput O and Baert L 1984 Nitrogen fertilizer and nitrate pollution of groundwater in sandy soils. Environ. Pollut. (Series B) 7, 43-48.

Law A T 1980 Sewage pollution in Klang River and its estuary. Pertanika 3, 13-19.

Law A T 1984 Monitoring of sewage pollution in the estuarine and coastal water of Port Kelang, Malaysia. In Proceeding Intl. Conf. Dev. Management of Tropical Aquatic Resources, Universiti Pertanian Malaysia Serdang, Malaysia. Eds. H H Chan, O Ishak, K H Ang, A T Law and M Mohd. Ibrahim. pp 236-241. Universiti of Pertanian, Serdang, Malaysia.

Law A T and Azhar Othman 1985 Faecal coliform bacteria distribution in the coastal waters of Port Dickson. Pertanika. 8, 131-134.

Law A T, Noor Azhar M S, Liew H C and Lokman S 1988 Coastal oceanographic studies of Port Dickson. 11th Malaysian Society of Marine Science Annual Seminar. Eds. A Sasekumar, R D'Cruz and L H Lim. pp 9-17. University of Malaya, Kuala Lumpur.

Lee S C, Lim J S and Wahid O 1992 Micronutrients status in major soils in Peninsular Malaysia. Secondary and micronutrients in Malaysian Agriculture. Ed. H A H Sharifuddin. pp 131-148, Malaysian Society of Soil Science, Kuala Lumpur, Malaysia.

Ministry of Primary Industry 1993 Statistics on commodities Kuala Lumpur, Malaysia. Ministry of Primary Industries, Malaysia, Kuala Lumpur, Malaysia.

Nashriyah M, Ho N K and Azimahtol Hawariah L P 1995 The residue levels of molinate in soil and its effect on population of aquatic flora and fauna under recycling and non recycling practices in Muda areas. In Proc. of Seminar on impact of pesticide and the rice agroecosystem in the MUDA areas. Eds. M Nashriyah, K Y Lum and S Ismail. pp 29-35. Malaysian Institute for Nuclear Technology Research, Bangi, Malaysia.

Ng S K 1987 Sulphur status of soils and requirements in Malaysia. Proc. of the Symposium on fertilizer sulphur requirements and sources in developing countries of Asia and Pacific. FADINAP Bangkok, Thailand. pp 38-41.

Ramlan M N and Badri M A 1989 Heavy metals in tropical city street dust and roadside soils: A case of Kuala Lumpur Malaysia. Environ Technol. Lett. 10, 435-444.

Robertson F N 1979. Evaluation of nitrate in groundwater in the Delaware Coastal Plain. Groundwater 17, 328-337.

Seiler R J, Sheik Omar A R and Salim N 1979 Nitrate poisoning of cattle fed with Napier Grass. Kajian Vetrinar. 11, 10-13.

Suki A 1993 An overview of sewage problems, management and control technologies. In Waste Management in Malaysia. Current status and prospects for bioremediation. Eds. B G Yeoh, K S Chee, S M Phang, I Zaid, I Azni and M Maketab. pp 179-193. Ministry of Science, Technology and Environment, Kuala Lumpur, Malaysia.

Syers K, MacKay A D, Brown M W and Curie L D 1986 Chemical and physical characteristics of phosphate rock materials of varying reactivity. J. Food Agric. 37, 1057-1064.

Tay T H, Kho B L, Loke W H and Chang Kwong Choong 1994 Malaysia agricultural directory and index 1993/94. Agriquest, Petaling Jaya, Malaysia. 509 p.

Teh, L.H. 1989 Wastewater Survey Report. Department of Environment, Malaysia. Ministry of Science, Technology and the Environment Malaysia, Kuala Lumpur.

Tessens E and Shamshuddin J 1983 Quantitative relationship between mineralogy and properties of tropical soils. UPM Serdang, Malaysia. 190 p.

20
Contaminants and the soil environment in New Zealand

A.H.C. ROBERTS, K.C. CAMERON, N.S. BOLAN, H.K. ELLIS AND S. HUNT

1. Introduction

In the last decade or so there has been escalating concern worldwide regarding the quality of the environment as affected by human activity. For agriculture, this has meant that many of the modern cultural practices have come under ever-increasing scrutiny by the public at large. Issues of importance to the consumer include the nutrient value of foodstuffs as well as any contamination by toxic substances, principally those arising from the farm production system itself. There is widespread distrust, much of it ill-informed, regarding the use of modern agrichemicals such as pesticides, fertilisers and animal health remedies and the effect of their use on food quality. However, any soil contamination (because soil is fundamental to most land-based plant or animal production systems) is extremely serious because the contaminant or its breakdown products may accumulate in food products ultimately for human consumption. In New Zealand (NZ), there is currently much activity in determining the causes and effects of heavy metal additions and pesticide residues on the quality of the agricultural environment. Additionally, there is increasing pressure from the public, both domestic and international, concerning the impact of agriculture on the wider environment. For example, off-farm water quality may be adversely affected and the important soil-mediated and other processes which might be involved are the subject of several major studies in NZ.

Soil contamination arising from industrial activity is an issue that has only recently come to prominence in NZ. This has occurred as a result of a recent (1991) major Government initiated investigation of pentachlorophenol (PCP) contamination arising from historical operations at a major sawmill facility near Rotorua.

Prior to this time, contaminated sites were not generally perceived to be an issue that significantly affected NZ. Now however, and mindful of the importance of a 'clean green' image for trade, problems associated with contaminated sites and their management have become an important part of the environmental agenda in NZ, both for regulatory agencies and for industry.

However, the problems of contaminated sites are likely to be less severe than comparable northern hemisphere countries due to the overall low level of industrialisation in NZ. Public awareness of the issue has been raised considerably and there is an expectation of continued action by both industry and Government to ensure clean-up of sites is achieved. Government, at the national level, in conjunction with industry and regulators has started the process of developing clean-up guidelines and appropriate remediation methods. However, some of the more intractable problems about liability and funding of clean-up of historic sites remain to be resolved.

R. Naidu et al. *(eds): Contaminants and the Soil Environment in the Australasia-Pacific Region, 579–628.*

The purpose of this paper is primarily to provide an insight into the nature and extent of contaminants in agricultural and industrial soils in NZ and to delineate the contaminants which are currently high priority in terms of requiring management and/or remediation.

2. **Climate and soils** (adapted from Taylor and Pohlen, 1968)

New Zealand lies to the south west of Australia in the Pacific Ocean. It consists of three major islands which stretch some 1600 km over 14° of latitude. The three islands embrace such climatic extremes as subtropical Northland, the cold uplands of the alpine regions, the semi-arid basins of Central Otago and the very wet mountains and lowlands of Westland. The topography is varied with around 50% being steep, 20% moderately hilly and less than 33% is either rolling or flat.

The sharp relief of the islands is sufficient to produce a significant range of temperature from north to south and with altitude. Mean air temperatures at sea level are approximately 15°C in the north, 12°C in the centre (about Cook Straight) and 9.5°C in the south, and these fall about 1.5°C for each 305 m of elevation. The prevailing westerly winds are forced up and sideways by the mountain ranges causing those areas directly exposed to the westerly airstreams and of higher altitudes to have the greatest rainfall. This tendency for wetness in the west and dryness on the east is most extreme in the southern South Island where more than 5 000 mm of rain falls on the western Southern Alps and less than 500 mm in the Central Otago basin on the eastern side.

The geology underlying the soil mantle varies in texture and composition and includes igneous rocks ranging from ultrabasic to acidic (e.g. basalt, andesite, rhyolite), metamorphic rocks and sedimentary rocks (e.g. conglomerates, sandstones, mudstones, limestones). Based on their parent materials, the soils of NZ may be broadly classified into three major 'groups': sedimentary (sandstone, siltstone, mudstone); ash (andesite, basalt) and pumice (rhyolite) soils (Morton et al., 1994). The generalised distribution of these soils is shown in Figure 1. Sedimentary soils include yellow-brown earths, yellow-grey earths and their intergrades and recent alluviums. Ash soils include yellow-brown loams, red and brown loams and brown granular clays and loams, while pumice soils are yellow-brown pumice soils. (Abbreviations and a reference for the soil taxonomic classification are given as a footnote to Table 2).

3. Overview of Industries

New Zealand's largest industries are concerned with biomass production and processing and some 60–70% of export earnings derive from these industries. Agriculture, horticulture and silviculture are at the forefront of these industries. Agricultural production to March 1995 was valued at NZ$ 10.1 billion and forecast to decline to March 1996 by some 7% to NZ$ 9.4 billion (Whittington, 1995).

New Zealand has a land area of 26.9 million ha, of which 11.5 million ha are in

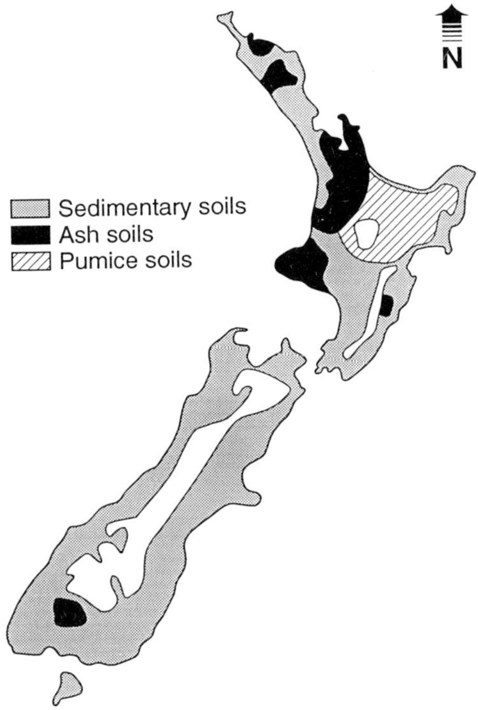

Figure 1. Generalised areas of major soils in New Zealand.

North Island, 15 million ha in South Island and 400 000 ha in miscellaneous outlying islands. Pastoral farming, based primarily on grass/legume pasture associations, is the dominant enterprise with 13.5 million ha in pasture (Whittington, 1994). Some 52 million sheep and 5 million beef cattle are reared on around 32 473 properties and 3 million dairy cattle are carried on 14 844 properties (NZ Statistics Department, 1993). Additionally, there are 2 115 deer farms and 1 541 cropping or mixed cropping farms (NZ Statistics Department, 1993). Horticultural production occupies 90 000 ha and is dominated by kiwifruit (31%) and apple (25%) production, with the remaining 44% consisting of fresh and processed fruit, vegetables, hops, honey and flowers (Whittington, 1995).

There are 13 million ha in production forestry (NZ Forestry Statistics, 1993) which for the year ending December 1993 earned $NZ 2.5 billion in export earnings.

The agricultural processing and servicing sector represents a large proportion of NZ industry with at least nine major dairy companies processing milk products at multiple sites, four major superphosphate manufacturers with eight plants, and several large and many more smaller export killing facilities and abattoirs. Wool scouring plants and horticultural processing plants also exist in multiple locations throughout the country. In terms of forestry, there are a number of pulp, paper and fibre board mills and approximately 600 sites where timber has been treated with pentachlorophenols (PCP), CCA (Cu, Cr, As) or B. Many of the smaller timber treatment sites have closed down.

New Zealand, with a total population of 3.4 million people (NZ Statistics Department, 1994), is not a heavily industrialised country by European or American standards but there is heavy industry at some particular sites. For example, there is an oil refinery at Whangarei, a steel mill south-west of Auckland, and various energy related plants particularly near New Plymouth in Taranaki including oil and gas wells (off- and on-shore), an ammonia urea plant, a gas to methanol plant and a synthetic gasoline plant. In the South Island there is an aluminium smelter at Bluff. In urban areas there are the usual light to medium industries associated with normal human activity.

4. Sources of soil contamination

4.1 Agricultural soils

Since NZ is not heavily industrialised, rural soils are little affected by aerial deposition of contaminants from this source. However, soils used for biomass production in NZ are subject to contamination from a variety of potential sources. These include agricultural chemicals (e.g. organic and inorganic pesticides and their residues); heavy metals in fertiliser (e.g. Cd in phosphatic fertiliser); contaminants contained in liquid (e.g. whey), solid (e.g. sewage sludge) effluents and waste materials applied to pastures and crops as a disposal option or to capture the benefits of the nutrients contained in the wastes.

As discussed in Chapter 5, a further aspect of contamination of rural soils is the impact of agriculture on water quality through soil-mediated processes. Examples of these include the impact of nitrate (NO_3^-) leaching from grazed grass/legume pastures on groundwater quality and eutrophication of surface waters through the movement of soil particulates enriched with phosphorus (P) as a result of P fertiliser use. For a detailed review on the fate of N and P readers are directed to Chapters 2 and 5.

The likelihood of soil contamination arising from biomass processing industries depends mainly on whether effluents and wastes are treated using land-based systems (e.g. spraying dairy factory effluent onto pasture) or stored on site.

4.2 Potential industrial sources

A desk-top estimate of the likely extent of potentially contaminated sites in NZ was recently undertaken by Ministry for the Environment (1992a). This study focussed on historical records to identify industrial sites which were associated with contamination.

The sites were categorised into three risk classes based on the toxicity of the expected principal contaminants, and the likely exposure to them. High risk sites were expected to pose an actual or imminent significant adverse effect to human health and the environment. Moderate risk sites were expected to have soil contamination levels above the ANZECC 'B' category (ANZECC/NHMRC, 1992). Low risk sites were considered to be contaminated but not to present any long term risk unless land use changed to a more sensitive nature (e.g. from industrial to residential or agricultural use) where potentially higher level of exposure may result, such as food production or human amenity areas.

The study estimated 1 580 high risk sites (excluding timber treatment sites) with a total of 7800 sites overall with respect to the primary contaminants of each industry sector (refer Table 1; for example, it can be reasonably expected that contaminants on service station sites will include a range of volatile, semi and non volatile organics – the total number of service stations throughout NZ is estimated at 2 616). The study findings are indicative – it was not based on field data and the estimates do not discriminate between large or small scale industrial activity. In the NZ context, most industrial sites are small but a certain proportion can be expected to have 'hotspot' areas where contaminants are above acceptable background levels.

5. Nature and extent of contaminants in agricultural soils

Contaminants in NZ rural soils fall into three major categories: heavy metals; non-metals and organics from pesticide residues. These contaminants are discussed separately in the following sections.

5.1 Heavy metals

Background values have often been used to determine the extent of toxic substance enrichment of soils. Soils that exceed the local 'normal' background values are deemed to be contaminated. While there is a continuum of soil concentrations from the natural or pristine state through to highly contaminated soils at specific sites (Tiller, 1992), in NZ comparison between virgin (where possible) or native (i.e. unfertilised) and agricultural soils has been used to indicate contamination.

A survey of 398 sites throughout North and South Island rural soils was conducted in the early 1990s with a view to determining the extent of heavy metal contamination in pastoral agricultural soils. Three hundred and twelve farm sites and 86 native sites, covering the major soil groups important for agriculture, were sampled to a depth of 75 mm and analysed for total As, Cd, Cu, Pb and Zn and the results are shown in Table 2. Data for Cd have been published (Roberts et al., 1994) while those for the other elements are currently in preparation. Additionally, the soil sample collection is currently being analysed for Cr, Hg and Ni.

Using the criterion that contaminated soils exceed the background level of 'native' sites, it would appear that of the elements studied so far, only Cd has been enriched in pastoral soils (Table 2). In this study, there was a highly significant correlation ($r=0.62^{***}$, $P<0.001$; $n=397$) between total soil P and total Cd across all sites which implicates the use of P fertiliser in soil Cd enrichment. This enrichment has been attributed to the use of phosphatic fertiliser, principally superphosphate, in

Table 1. Estimated numbers of potentially contaminated sites in New Zealand indicating key contaminants of concern (adapted from Ministry for the Environment, 1992).

	ORGANIC							INCRGANIC				
INDUSTRY/ LAND USE	H-Volatile eg perchlor-ethylene	H-Semi/Non Volatile eg PCP, PCB, DDT, dieldrin	Non H Volatile eg petrol-BTEX	Non H Semi Volatile eg diesel, PAHs, phenols	Non H/Non Volatile eg tars, heavy oils	Pesticides	Organic Cyanides	Organic Acids/ Alkalis	Volatile Metals eg As, Pb, Hg	Non Volatile Metals eg Ba, Cu, Cr	Asbestos	Corrosives eg HCl, NaOH	TOTAL COUNT PER INDUSTRY
Acid/Alkali								x		x		x	11
Agriculture/Horticulture		x				x		x		x		x	49
Airports			x	x	x	x				x		x	76
Asbestos											x		45
Chemical Manufacture									x	x		x	166
Defence Works										x			18
Drum Reconditioners				x	x							x	28
Dry Cleaners	x												254
Electric. Manuf. (Transf.)		x		x	x								42
Electroplaters							x	x		x		x	291
Engine Works			x	x	x					x			878
Explosives Industry													7
Gas Works			x	x	x								31
Iron/Steel										x		x	60
Landfill Sites									x	x			716
Metal Treatment							x	x		x		x	89
Mining/Extractive							x	x				x	92
Oil Production/Storage			x	x	x								434
Paint Manufacture			x		x					x			99
Pesticide Manufacture						x							13
Pharmac. Manufacture								x					3.4
Port Facilities			x	x	x								31
Power Stations		x						x				x	223
Railway Yards		x								x			57
Scrap Yards		x								x			490
Service Stations			x	x	x								2 616
Smelting/Refining										x			302
Tanning & assoc.								x		x			48
Timber Treatment		x							x	x			600
TOTAL													7 800

KEY: X = indicates assessed primary contaminant of concern; H = Halogenated; Non H = non Halogenated

Table 2. Heavy metal concentrations (mg kg^{-1}) in New Zealand rural soils (0–7.5 cm layer).

		Soil Type*								
Element	Site	YBL	YBP	YBE	YGE	GLEY	PEAT	BGL	ALL.	MEAN
As	Native	3.9	7.0	4.2	2.1	5.3	8.4	5.3	3.0	4.3
	Pasture	7.4	7.8	3.0	2.4	5.1	15.1	5.4	3.6	4.9
Cd	Native	0.23	0.31	0.16	0.13	0.24	0.22	0.19	0.13	0.19
	Pasture	0.70	0.75	0.22	0.12	0.42	0.69	0.49	0.16	0.44
Cu	Native	26.9	7.2	14.6	8.0	17.9	20.4	19.4	11.5	17.0
	Pasture	32.1	10.9	12.3	8.9	18.2	27.0	25.5	17.3	17.7
Pb	Native	11.6	6.8	13.7	12.5	17.6	16.9	18.0	15.0	13.3
	Pasture	19.3	7.4	7.6	9.3	11.6	10.0	19.1	10.9	11.7
Zn	Native	77	37	6.5	54	92	51	56	72	65
	Pasture	97	47	55	61	66	43	82	78	68

*Soil types (Taylor and Pohlen, 1970): YBL = yellow-brown loam; YBP = yellow-brown pumice; YBE = yellow-brown earth; YGE = yellow-grey earth; GLEY = gley soils; PEAT = organic soils; BGL = brown granular loam; ALL. = alluvial soils.

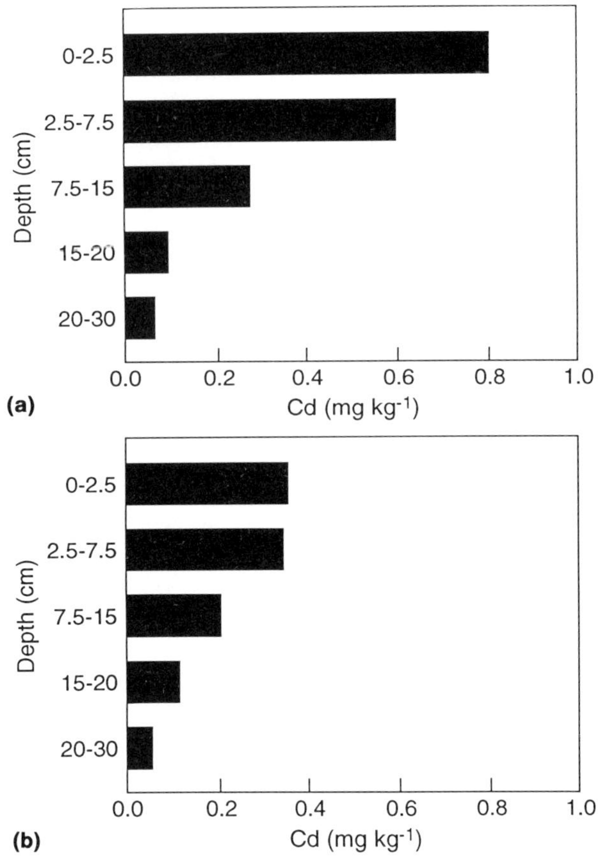

Figure 2. Distribution of soil Cd concentrations with depth in a) YBL and b) YBP pastoral soils (mean of two sites per soil).

several other studies in NZ and Australia (Williams and David, 1976; Rothbaum et al., 1986; Merry, 1988). This is not unreasonable considering the long history of superphosphate use, manufactured from Nauru Island phosphate rock, in both NZ and Australia. Nauru superphosphates typically contained 34–69 mg Cd kg^{-1} with an average of 48 mg Cd kg^{-1} (Rothbaum et al., 1986).

Further evidence implicating P fertilisers as a source of Cd enrichment is indicated by the Zn:Cd ratios of soils. In the survey, native soils had Zn:Cd ratios ranging from 180:1 to 1120:1 which decreased in all pastoral soils to between 66:1 and 633:1 (Roberts et al., 1994). This narrowing of the Zn:Cd ratio may be attributed to the enrichment of soil Cd levels by the long-term addition of Cd-containing P fertiliser, such as superphosphate, which typically has Zn:Cd ratios of 23:1 or less (Swaine, 1962).

Cadmium decreased with depth in the survey soils (Figure 2) for two yellow-brown loam (dystrandept) and pumice (vitrandept) soils. Decreasing soil Cd with depth has previously been demonstrated for Swedish, Australian and American soils (Andersson, 1977; Merry, 1988; Holmgren et al., 1993). Because permanent pastures are rarely cultivated in NZ, this decrease with depth is consistent with surface addition of Cd from fertiliser use, herbage Cd recycling through dung deposition and plant death and decay, as well as atmospheric accessions.

Arsenic, Cu, Pb and Zn did not exhibit the same enrichment in pastoral soils compared with native soils as shown for Cd (Table 2). For some metallic elements on some soils (e.g. As, Cu, Pb and Zn in YBE (dystrochrepts); Pb in YGE (ustrochrepts), GLEY (andaquepts), PEAT (saprists) and ALL (fluvents); Zn in GLEY, PEAT) concentrations were lower in pastoral soils than in the native soils. When coupled with the fact that the concentrations of these elements did not exhibit the same decline with depth as did Cd, this suggests that in general the As, Cu, Pb and Zn concentrations recorded were essentially pedogenic in origin.

5.2 *Non-metals*

In agricultural terms, the two non-metals of importance are P and N. While not strictly soil contaminants these two elements can and do impact on surface water (P and N) and groundwater (N) quality, although the source and mode of action of the elements is different.

5.3 *Phosphorus*

As noted in Chapters 2 and 5 there are two major pathways for P entry into surface water; these include leaching and surface and subsurface runoff.

5.3.1 Leaching

Unlike NO_3^- and to a lesser extent SO_4^- leaching of P is not an issue in NZ soils except for minor areas of unmineralised peat soils (fibrists) and some highly weathered podzolised sedimentary soils (haplohumults). This is due to the fact that the binding capacity for P is much higher than that of other anions (Figure 3) and that most NZ soils have a large P sorbing capacity which in general is due to the large quantities of short-range order (amorphous) iron and aluminium compounds in the soils (McLaren and Cameron, 1990).

5.3.2 Runoff

It has been estimated that 45% (McColl, 1982) to 80% (Gregg et al., 1993) of P inputs to surface water emanates from diffuse agricultural sources such as runoff. There is little doubt that P loadings from grazed pastoral soils are higher than those under native vegetation (Wilcock, 1986; Cooper and Thomsen, 1988; Smith et al., 1993) with total P (TP) yields from pasture around nine times that from native

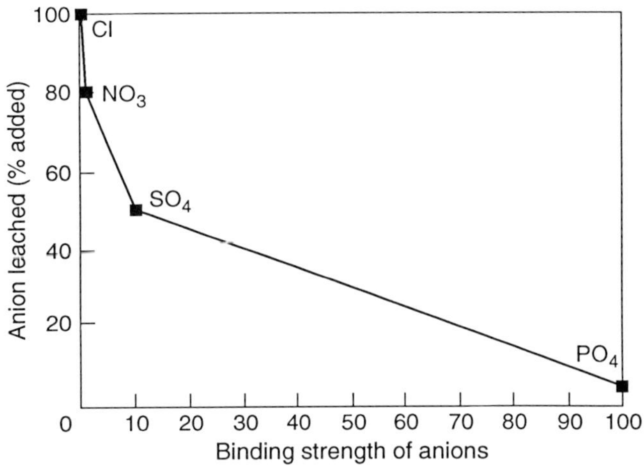

Figure 3. Adsorption of anions by soils versus leaching. Binding strength for anions was obtained by fitting the adsorption isotherms to Langmuir isotherm: $S = bcM/(1 + bc)$ where S is sorbed concentration; c is equilibrium concentration; M is adsorption maximum and b is binding constant. The binding strength (b) for phosphate was taken as 100 and the relative b for other ions was calculated.

forest and dissolved reactive phosphate (DRP) yields more than 15-fold higher from grazed pasture (Cooper and Thomsen, 1988). Additionally, Lambert et al. (1985) found that cattle grazing gave higher nutrient outputs than sheep grazing. The source of this P is likely to be dung and urine runoff and animal-induced soil erosion. In one study of major P sources entering Lake Tutira (in the Hawkes Bay region) dung and urine runoff contributed 300–1 600 kg P yr^{-1} and soil erosion 1 000 kg P yr^{-1} (McColl, 1978).

Fertiliser P may make a direct contribution to the enrichment of surface waters. Contributions from a single fertiliser application have been estimated to range from <1 to 25% of TP and a larger fraction i.e. 30–52%, of dissolved inorganic P (DIP) carried from a catchment (Sharpley and Syers, 1979a,b; Turner et al., 1979; Lambert et al., 1985; Cooke, 1988).

Given the low levels of P loss from native forest an approximate division of P yield from a catchment (Thorrold and Smith, 1994) might be:

10%	background
25%	current fertiliser
65%	grazing and increased soil fertility

The available information shows that the loss from any one fertiliser application is secondary to the losses due to grazing and previous increases in fertility.

5.4 Nitrogen

Although NZ has abundant good quality water resources, there is increasing concern about the impact of agricultural and industrial activities on water quality. One of the

Table 3. Nitrate concentrations reported for New Zealand groundwater.

Location	Nitrate concentration ($mg\,NO_3$-$N\,L^{-1}$)	Suggested major sources of nitrate	Reference
Ashburton	Shallow + unconfined aquifers 4–8 and in places > 10	Grazed pasture, cropland	Burden (1984)
Ashley Catchment, North Canterbury	4% wells > 10 32% wells > 5	Grazed pasture, cropland	Bowden et al. (1982)
Christchurch	Shallow + unconfined often > 10. Deep < 1	Grazed pasture, cropland some point sources	Bowden (1986)
Clutha Valley	< 10	–	Close and McCallion (1988)
Hamilton Basin	30% shallow (< 15 m) > 10	Grazed pasture, dairy factory waste	Hoare (1986)
Hauraki Plain	Most wells < 10 (range up to 37)	Grazed pasture	Dewhurst (1981)
Heretaunga Plains	Unconfined aquifer 20–57	–	Burden (1980)
Lincoln	Shallow wells > 10	Grazed pasture, cropland	Adams et al. (1979)
Manawatu	Wells 0–30 m: 3–13 Wells > 30 m: < 2	–	Brougham et al. (1985)
Waimea Plain	Unconfined aquifer > 10	Market gardening (N fertiliser), piggery waste	Fenemor (1987)
Wairarapa	Deep confined aquifers < 7.5	–	O'Dea (1980)

main concerns relates to the health hazards associated with high concentrations of NO_3^- in drinking water supplies. The NZ Ministry of Health has therefore set a maximum acceptable value for NO_3^- in drinking water at 11.3 $mg\,N\,L^{-1}$ (Ministry of Health, 1995).

Detailed reviews of NO_3^- contamination of NZ groundwater have been given by Burden (1982) and Dillon et al. (1989). The concentration of NO_3^- in some shallow aquifers frequently exceeds the Ministry of Health (1995) limit for drinking water (Table 3). For example, in the Hamilton basin, over 30% of shallow wells (i.e. < 15 m) have NO_3^- concentrations >10 $mg\,N\,L^{-1}$, and it has been recommended that no shallow bore water in this area should be used for feeding infants (Hoare, 1986). In mid-Canterbury, inland shallow aquifers and the uppermost confined aquifer also exceed 10 $mg\,N\,L^{-1}$. Deep aquifers generally have very low NO_3^- concentrations (Bowden, 1986).

Grazed pasture and cropland are the major sources of NO_3^- contamination in most of the studies reported in Table 3. In some localised areas, point sources (e.g. dairy factory waste, piggery waste) and market gardening have been identified as also contributing to the NO_3^- load.

Nitrogen contamination of NZ rivers and lakes has resulted in some problems due to eutrophication. Undesirable growths of aquatic plants and algae can cause a loss of fisheries, reduce aesthetic appeal, and increase costs of water abstraction (Cameron and Haynes, 1986). The transfer of nutrients from land to surface waters in NZ has recently been reviewed by Gregg et al. (1993) who concluded that losses from agricultural land represent the largest contribution to surface water and that nutrient losses are higher from grazed pasture than from native vegetation or forests. Lakes in the North Island of NZ are more eutrophic than those in the South Island as a result of the greater intensity of land use in North Island catchments (Gregg et al., 1993).

6. Factors affecting the fate of N in soils

New Zealand soils commonly contain between 0.1 to 0.6% N in the top 15 cm, which represents approximately 2000 to 12000 kg N ha^{-1}. As the vast majority (>95%) of this N is present in soil organic matter it is not available for plant uptake or leaching unless mineralised. Only a small fraction of organic N is mineralised each year (between 10 to 500 kg N ha^{-1}) and only one to two percent of the total soil N pool is present in the mineral forms of NH_4^+ and NO_3^- at any one time (McLaren and Cameron, 1990). Ammonium ions may be held by cation exchange reactions or fixation by certain clay minerals in NZ soils and are therefore not generally prone to leaching. However the NO_3^- retention capacity of NZ soils is generally low and NO_3^- is therefore prone to leaching.

The major factors that control the fate of N in soils have been reviewed in Chapters 2 and 5. The main factors which affect the fate of N in NZ soils are: season/climate; soil conditions/reactions; plant uptake; fertilisers; organic wastes; and irrigation.

6.1 Season and climate

In NZ, most leaching generally occurs during winter and early spring when plant uptake of N is low and drainage is occurring. However, leaching can also occur at other times of the year if the soil is close to 'field capacity' and heavy rain or irrigation is received. The amount of drainage in spring is particularly critical as it often determines the amount of leaching loss from freshly applied fertiliser, animal returns, and recently mineralised NO_3^-. The intensity of rainfall is important as well as the amount since heavy rain may induce macropore flow and rapid leaching. A dry summer can cause an accumulation of NO_3^- in soil due to poor crop uptake and this may result in higher than average leaching losses from arable soils over the subsequent winter.

6.2 Soil conditions and reactions

Sandy soils contain less water at field capacity than clay or silt loam soils and leaching is therefore induced more quickly in lighter soils. The hydraulic conductivity of a sandy soil is also greater than that of a clay soil and therefore the rate of NO_3^-

Table 4. Nitrogen (^{15}N) balance one year after urine application (500 kg N ha^{-1}) to a flood irrigated pasture soil (adapted from Fraser et al., 1994).

	Amount recovered (% applied ^{15}N)	Amount recovered (kg N ha^{-1})
Pasture	43.5±2.1[2]	217.9±0.9
Soil plus roots	20.4±2.3	101.2±0.8
Leachate	8.3±3.2	41.6±1.3
Denitrification	27.9±2.0	138.6±1.0

[1] = denitrification was calculated by difference since no volatilisation occurred due to rain immediately after urine deposition.
[2] = standard error.

leaching is faster through light textured soils. The more rapid the rate of leaching the less opportunity there is for plant uptake, denitrification, and/or immobilisation to remove NO_3^- from the soil solution.

In poorly drained soils denitrification may reduce the NO_3^- concentration in the soil solution and thus result in a low leaching loss or transfer from soil to surface water bodies. For example, denitrification in riparian zones around NZ rivers and lakes has been found to be efficient at removing over 90% of NO_3^- from incoming drainage water (Schipper et al., 1993). Denitrification can also have a significant effect on leaching losses from irrigated soils. For example, a recent NZ study of the fate of ^{15}N-labelled animal urine has found that the leaching loss from flood irrigated lysimeters was less than one-third of the denitrification loss (Table 4).

Mineralisation of N from soil organic matter and/or plant residues increases the concentration of NO_3^- in soil solution and therefore increases the potential for leaching to occur. Nitrate leaching following mineralisation of ploughed pasture and clover residues has been found to be a major source of NO_3^- contamination of NZ groundwater (Francis et al., 1992). Over 90 kg N ha^{-1} yr^{-1} were found to be leached following the ploughing of clover residues as part of the normal crop rotation in Canterbury (Adams and Pattinson, 1985).

6.3 Plant uptake

Plant uptake is a primary factor affecting the amount of NO_3^- leached from soils and fertilisers. The greater the uptake, the lower the concentration of NO_3^- in the soil solution and thus the lower the potential for leaching. As land use intensifies the amount of loss also tends to increase. Plant root morphology can influence the NO_3^- leaching loss since deep and extensively rooted plants have a greater opportunity to capture NO_3^- from the soil. For example, lucerne can scavenge NO_3^- from the subsoil due to its deep rooting system, whilst the extensive root system of most grasses enables them to efficiently capture NO_3^- that is present in the topsoil.

Extensive pasture systems and pasture cut for hay or silage generally have low leaching losses. Relatively large leaching losses can, however, occur under intensively managed pasture particularly when high stocking rates are combined with high fertiliser inputs. This is because animal urine patches contain large amounts of N (equivalent to approximately 500 kg N ha^{-1} for sheep and 1 000 kg N ha^{-1} for cattle, Steele, 1982). These amounts are greater than the plant can assimilate and therefore significant leaching losses may occur. Lysimeter studies in NZ have found that between 8–20% of the N applied in animal urine patches may be leached (Field et al., 1985b; Fraser et al., 1994). Intensive dairy farming systems have a greater leaching loss of N compared to sheep farms. It has been estimated that N leached from intensive dairy farms in the Waikato area is contributing at least 60 kg N ha^{-1} yr^{-1} to the shallow aquifers in that region (Selvarajah et al., 1994).

6.4 *Fertilisers*

Until relatively recently NZ pastoral farmers relied almost exclusively on N fixation by clover (*Trifolium repens*) plants in the pasture sward. A well established clover-based pasture, consisting of approximately 20% clover and 80% ryegrass (*Lolium perene*), could contribute approximately 200 kg N ha^{-1} yr^{-1} to pasture nutrition. However, in recent years the amount of N fertilisers used in NZ has increased considerably (Roberts et al., 1992). The greatest increase has been on dairy farms where applications of 100 kg N ha^{-1} yr^{-1} are not uncommon, with some farmers using up to 400 kg N ha^{-1} yr^{-1}. The impact of this increase is unknown at this stage, although international literature indicates that it is likely to lead to increased NO_3^- concentrations in surface and groundwater.

There are very few measurements of fertiliser leaching losses from grazed pasture in NZ. Sharpley and Syers (1979) reported that application of urea at 60 kg N ha^{-1} yr^{-1} resulted in a temporary increase in the concentration of NO_3^- in drainage water from a Tokomaru silt loam (Ustrochrept) at Palmerston North (Figure 4). Steele et al. (1984) reported significant leaching losses from cattle grazed pasture in Northland (Table 5). However soil disturbance during equipment installation and the unusually high rainfall that occurred over the period of the study may have affected the results. Field et al. (1985a) also reported large leaching losses from sheep grazed pasture in Manawatu (Table 5) although these results may have been influenced by cultivation and soil management practices. Heng et al. (1991) and Adams and Pattinson (1985) reported relatively low leaching losses under lightly grazed sheep pasture.

Cereals generally use fertiliser N efficiently and leaching losses are therefore small at optimum fertiliser rates (Prins et al., 1988). In NZ, Mohammed et al. (1987) recorded only a short-term increase in NO_3^- concentration in tile drain effluent when urea was applied to barley. Adams and Pattinson (1985) concluded that fertiliser applied at between 25–50 kg N ha^{-1} had little effect on NO_3^- leaching from wheat grown on Wakanui soil (Ustrochrept) in Canterbury. This was considered to be due to the timing of application being aimed to coincide with the period of maximum N demand by the crop.

Fertiliser N recovery by field vegetables is generally low (e.g. lettuce 11%;

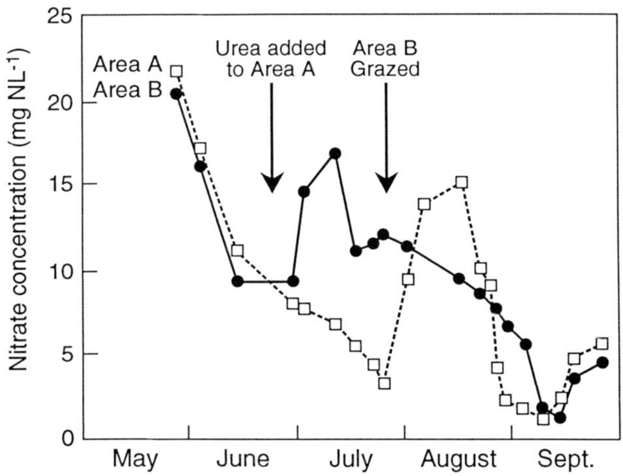

Figure 4. Effect of fertilizer application on nitrate leaching from a silt loam pasture soil at Palmerston North (Sharpley and Syers, 1979).

Table 5. Nitrate leaching from grazed pasture in New Zealand.

Region	Stock	Pasture	Fertilizer ($kg\,N\,ha^{-1}\,yr^{-1}$)	N_2 fixation ($kg\,N\,ha^{-1}\,yr^{-1}$)	Nitrate leached ($kg\,N\,ha^{-1}\,yr^{-1}$)
Northland (Steele et al., 1984)	Angus steers	Grass + clover	0	n.a.	88
		Grass + clover	172 (3×57)	n.a.	193
Manawatu (Field et al., 1985a)	Sheep	Grass only	0	0	42
		Grass + clover	0	320	73
		Grass + clover	110	220	101
		Grass + clover	450	90	222
Canterbury (Cameron and Rate, 1992)	None	Grass + clover	0	n.a.	1
Manawatu (Heng et al., 1991)	Sheep	Grass + clover	50	n.a.	17
Canterbury (Adams and Pattinson, 1985)	Sheep	Grass + clover	0	n.a.	< 10

n.a. = not available.

spinach 31%; red beet 52%; potatoes 50%) due to their shallow rooting depth and high rates of irrigation often used (Prins et al., 1988). Large amounts of mineral-N are reported to occur in the soil after the harvest of vegetables (e.g. Prins et al., 1988), however there appears to be few attempts to make direct measurements of the associated leaching loss under NZ conditions.

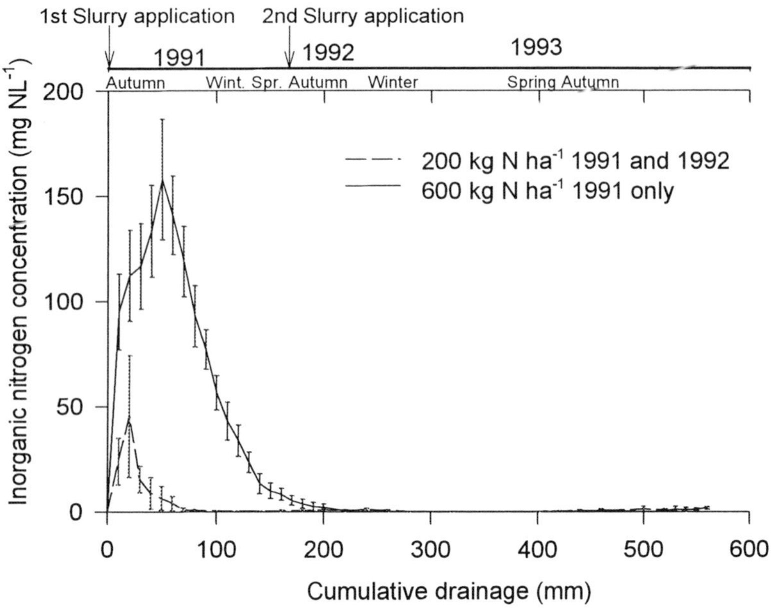

Figure 5. Drainage water concentration of nitrogen from lysimeters which received pig effluent at 200 and 600 kg N ha^{-1} yr^{-1} (Cameron and Rate, 1992).

6.5 *Organic wastes*

Application of organic wastes (e.g. dairy shed effluent, pig slurry) to soil has increased recently (Hart and Speir, 1992). Land application of dairy farm effluent and pig slurry at rates that can be utilised by pasture or crop (e.g. $<$200 kg N ha^{-1} yr^{-1}) appear to pose no greater threat to groundwater quality than many other agricultural practices (e.g. ploughing up pasture). However, applications at higher rates can result in larger leaching losses and higher NO_3^- concentrations in drainage water (Figure 5). Optimum rates of application are nevertheless difficult to define since in most wastes much of the N is present in organic forms which must be mineralised before NO_3^- is released. The ratio of mineral-N to organic-N varies considerably between different types of organic waste and this presents difficulties when making general recommendations on acceptable rates of application. Nevertheless most Regional Councils in NZ have now established rules to limit the rate of application of pig slurry (and other organic wastes) to less than 150–200 kg N ha^{-1} yr^{-1}, unless special circumstances apply.

6.6 *Irrigation*

Irrigation applied at optimum rates for plant growth can enhance N uptake and reduce NO_3^- leaching; however excessive irrigation rates can increase the loss.

Irrigation of pasture permits an increased stocking rate and thus there is potential for a higher leaching loss due to greater urine returns. Irrigation water can nevertheless dilute the drainage water NO_3^- concentration and thus reduce the peak concentrations entering groundwater. Irrigation may also increase denitrification losses and thus reduce the concentration of NO_3^- in solution. The influence of irrigation on NO_3^- contamination of NZ water resources is unclear, with conflicting opinions reported due to different irrigation and soil conditions (e.g. Quin and Burden, 1979; Burden, 1980; Close, 1987).

7. Pesticides

Pesticide use has been credited as one of the major contributors to modern agricultural production. The discovery of pesticides that could control insects, plant diseases and weeds in crops and pastures dramatically changed plant production systems throughout the world. Economic returns to producers from the use of pesticides vary between NZ$3 and NZ$4 for every NZ$1 spent on pesticides (Pimentel et al., 1981).

Pesticide use in NZ has progressed from persistent and highly toxic chemicals, such as arsenates through the era of 1,1,1 trichloro-2,2-bis[p-chlorophenyl ethane] (DDT) and other persistent but less toxic organochlorines to the present era which includes herbicides, insecticides and fungicides that are generally less persistent but more specific in their activity than their predecessors (Smith et al., 1993).

In NZ pesticides are the major agents of insect, disease and weed control in crops and pastures. During 1993, NZ spent approximately NZ$ 150 million on pesticides. Herbicides account for 60% of this total in overcoming an annual loss of NZ$ 339 million through weed infestation. A recent survey has established that an average of about 3500 to 4000 t of pesticide active ingredients are applied annually in NZ (McIntyre et al., 1989). About 75% of this use occurs in the North Island and is concentrated in regions with major horticultural industries, where some orchards receive up to 60 kg ha^{-1} per annum of fungicides and 20 kg ha^{-1} per annum of insecticides. In 1987 approximately 270 active ingredients were registered for use in NZ. Only a few of these are applied at rates that may pose risks for contamination of groundwater and surface water through leaching and surface runoff. Nevertheless many pesticides used in NZ are considered to be highly toxic to aquatic organisms (O'Connor, 1993). Pesticide residues, besides impacting on water quality, may also remain in the soil and accumulate in food products. The impact of pesticides on surface and groundwater quality and their fate in the soil environment is discussed in the following sections. Detailed literature on the fate and transport of pesticides has been cited in Chapter 4.

7.1 Contamination of surface and groundwater

Increased use of pesticides in agricultural and horticultural practices is causing great concern in relation to surface and groundwater quality. The contamination of surface

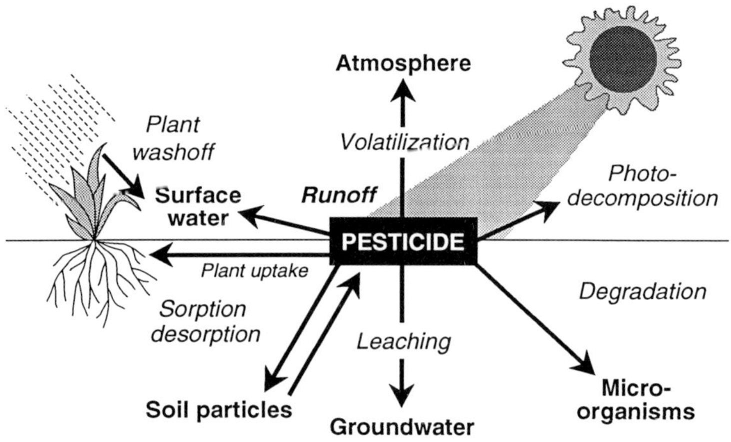

Figure 6. Transformation processes involved in the fate of pesticides in soil (Baskaran, 1994).

and groundwater by pesticides has long been recognised in many overseas countries, particularly where the potability of water is adversely affected (Cohen et al., 1986) and, based on this evidence, a range of pesticide compounds have been banned.

Many pesticides find their way into surface and groundwater by direct entry i.e. 'point source' contamination. Point sources include: the intentional application of pesticides close to water sources (e.g. to control the weed growth in these areas); and accidental leakage into or near poorly protected waterways during mixing pesticides or cleaning application equipment. Point source contamination also arises from spill during handling or transport, or drainage of near-empty concentrate containers into soils that are prone to macropore flow. Some pesticides are now recognised as posing a threat to groundwater and surface water quality through indirect entry i.e. 'diffuse source' contamination. These sources mainly include the large scale use of pesticides in agriculture, horticulture and forestry.

7.2 Processes affecting the fate of pesticides in soils

As noted in Chapter 4, the major processes controlling the persistence and ultimate fate of pesticides in the soil include sorption-desorption, microbial action, volatilisation and leaching (Figure 6). Among these, sorption-desorption is the key soil process that controls the movement of pesticides in soils and subsequently to surface and groundwater.

7.2.1 Sorption and desorption

Pesticide sorption is defined as the physico-chemical processes by which pesticide in solution becomes attached to the external and internal surfaces of soil particles

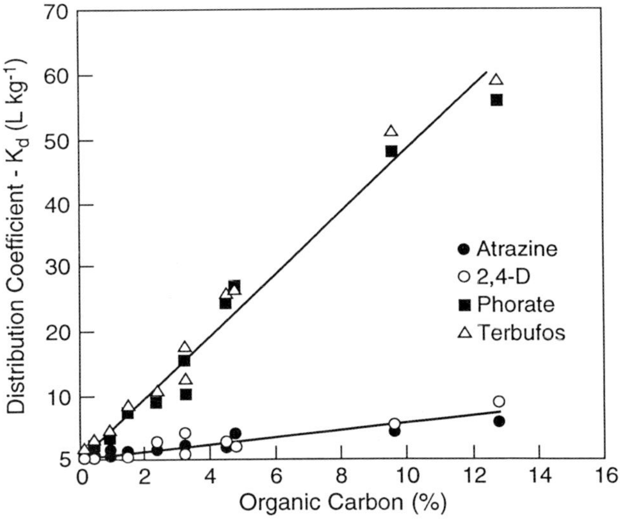

Figure 7. The relationship between organic carbon and distribution coefficients (K_d) of ionic (atrazine and 2,4-D) and non-ionic (phorate and terbufos) pesticides in soils (Baskaran, 1994).

and becomes less readily available for plant uptake, leaching and biological transformations. Desorption is the reverse process of sorption (Chiou, 1989). Sorption and desorption processes influence the chemical and biological degradation and the transport of pesticides to the atmosphere, groundwater and surface water.

A number of soil properties have been shown to influence the sorption of pesticides in soil. These include organic matter, clay content, pH and amorphous material (Bailey and White, 1970). Soil organic matter was best correlated with ionic and non-ionic pesticide sorption by soil (Figure 7). Based on these results it has been assumed that organic matter is the principal component of pesticide sorption and that pesticide sorption is normalised to organic matter content of the soil. Pesticide sorption expressed per unit organic carbon (K_{oc}) is less variable between soils than pesticide sorption expressed per unit soil mass (K_d). Most NZ soils are rich in organic matter which may be one of the reasons for the high capacity for pesticide sorption of these soils (Table 6).

While the solid organic matter content of the soil acts as the principal component of pesticide sorption, dissolved organic carbon in solution can act as a vehicle for the movement of pesticides in soil.

7.2.2 Evidence for pesticide contamination in NZ

In NZ there has been no systematic effort to link pesticide use on land to surface water quality, in spite of the potential for environmentally significant concentrations to be generated by runoff from agricultural and horticultural practices. The episodic

Table 6. Comparison of sorption, as measured by partition coefficients (K_d), of a range of pesticides by soils from New Zealand and U.S.A.

Pesticides	Partition coefficients (K_d) for soils from	
	New Zealand	USA
2,4-D	0.4–14.8[a]	0.3–1.3[b]
Atrazine	0.8–7.7[a]	0.1–0.8[c]
Metsulfuron methyl	0.1–1.5[a]	0.04–0.55[c]
Phorate	2.4–87.2[a]	1.52–73.8[d]
Terbufos	3.7–91.2[a]	3.31–52.7[d]

References: a: Baskaran (1994); b: Mallawatantri and Mulla (1992); c: Johnson and Sims (1993); d: Felsot and Dahm (1979).

nature of pesticide inputs to natural waters reduces the feasibility of establishing regular monitoring programmes because residues will not be present for most of the time. Another major problem in assessing pesticide impacts on the environment is in obtaining reliable data on the amounts and kinds of pesticides being used in different areas. This information needs to be constantly updated and should be accorded a high priority by environmental regulatory agencies.

The extensive use of DDT (see next section) and lindane in NZ for controlling certain pasture pests during the early 1960s provided impetus for a number of monitoring programmes for detecting residues of these pesticides. Solly and Shanks (1969) and Harrison (1970) have found significant concentrations of DDT and dieldrin residues in rainbow trout from Rotorua district and Lake Ellesmere and the Wainuiomata waterworks near Wellington. By applying a bioaccumulation factor (the ratio of the pesticide concentration in the tissue to that in the surrounding water) of about 36000, Murty (1986) has calculated the equilibrium concentration of DDT in the water to be about 0.006 to 0.030 mg m^{-3} which is well below the standard of 0.1 mg m^{-3} for drinking water and fish consumption.

Data on pesticides in surface and groundwater in NZ are sparse (Table 7). Based on the amounts of pesticides used, Wilcock (1989, 1993) and Wilcock and Close (1990) have identified areas where pesticide application could yield toxic runoff to surface waters (Figure 8). Close (1992) subsequently used these data in conjunction with soil permeability data to identify the areas having the greatest potential for groundwater contamination in NZ. A later survey, by Close (personal communication), in 17 regions of NZ covering 82 wells assessed pesticide contamination of groundwater. Detectable concentrations of two herbicides, atrazine and simazine, were found in groundwater from six wells in Poverty Bay, Te Puke, Motueka, Pukekohe and Oamaru. Herbicide concentrations of less than 0.1 mg m^{-3} were recorded in four of the above wells. All the measured concentrations were, however, much less than the acceptable maximum concentration limits for drinking water. A survey by the Hauraki Catchment Board during 1975/76 in the Piako and Waihou rivers indicated the presence of various herbicides. Freeman (1988) measured

Table 7. Summary of pesticide concentrations in groundwater in New Zealand.

Region*	Pesticide	Concentration ($\mu g\,kg^{-1}$)	Maximum level ($\mu g\,kg^{-1}$) for: Human	Aquatic life
1. Hauraki Plains	2,4-D	100–200	100	100
	2,4,5-T	100–200	2	100
	2,2-DPA	31–200	–	10000
	amitrole	200–400	1	>1000
	picloram	106	30	100
2. Canterbury	simazine	29–560		
	amitrole	133		
3. Motueka	pirimiphos methyl	0.06	60	
4. Poverty Bay	atrazine	37	–	
	diazinon	0.03	10	
	chlorpyrifos	0.03	2	
5. Pukekohe	procymidone	1.7	–	
6. Te Puke	2,4-D	0.05–0.1	100	

*References: 1. Hauraki Catchment Board (1975); 2. Freeman (1988); 3–6. Close (1987).

detectable levels of the herbicides simazine and amitrole in the water draining from the Styx River from an orchard near Christchurch.

Pesticides applied to cropland, forest and other areas for pest control may enter surface water by rainfall or irrigation runoff. Pesticide runoff includes dissolved ion, suspended particulate and sediment sorbed pesticide that is transported by water from the treated land surface. Annual runoff losses for soil applied pesticide average about 1–2% of the application, however, more than 10% losses have been reported under simulated rainfall. The data on pesticide losses from surface runoff in NZ are sparse. Fox and Wilcock (1988) reported a loss of about 0.6% of 2,4,5 trichlorophenoxyacetic acid (2,4,5-T) applied to gorse in the Waikato region. A similar study by Wilcock et al. (1991) showed a detectable concentration of triclopyr in stream water.

Unfortunately, criteria for drinking water and the protection of aquatic life are still not available for many pesticides. According to Smith et al. (1993) one approach is to take known toxicity values for a range of appropriate organisms and apply an 'application factor' which will reduce the concentration to a level judged to be non-toxic to most species. A number of application factors have been used to calculate safe levels, with a value of 0.1 $mg\,m^{-3}$ being the most common, and values of 0.05 or 0.1 $mg\,m^{-3}$ being used for more persistent, bioaccumulating chemicals (Nicholson, 1984). Recently theoretical methods have been developed for predicting aquatic toxicities of chemicals, and of calculating 'no observable effect concentrations (NOECs)', based on laboratory toxicity data and molecular properties of the chemicals. Both of these approaches (based on application factors or calculated NOEC values) should offer some protection to aquatic life from the non-lethal effects of pesticides, such as changes in fecundity or behavioural and ecological effects.

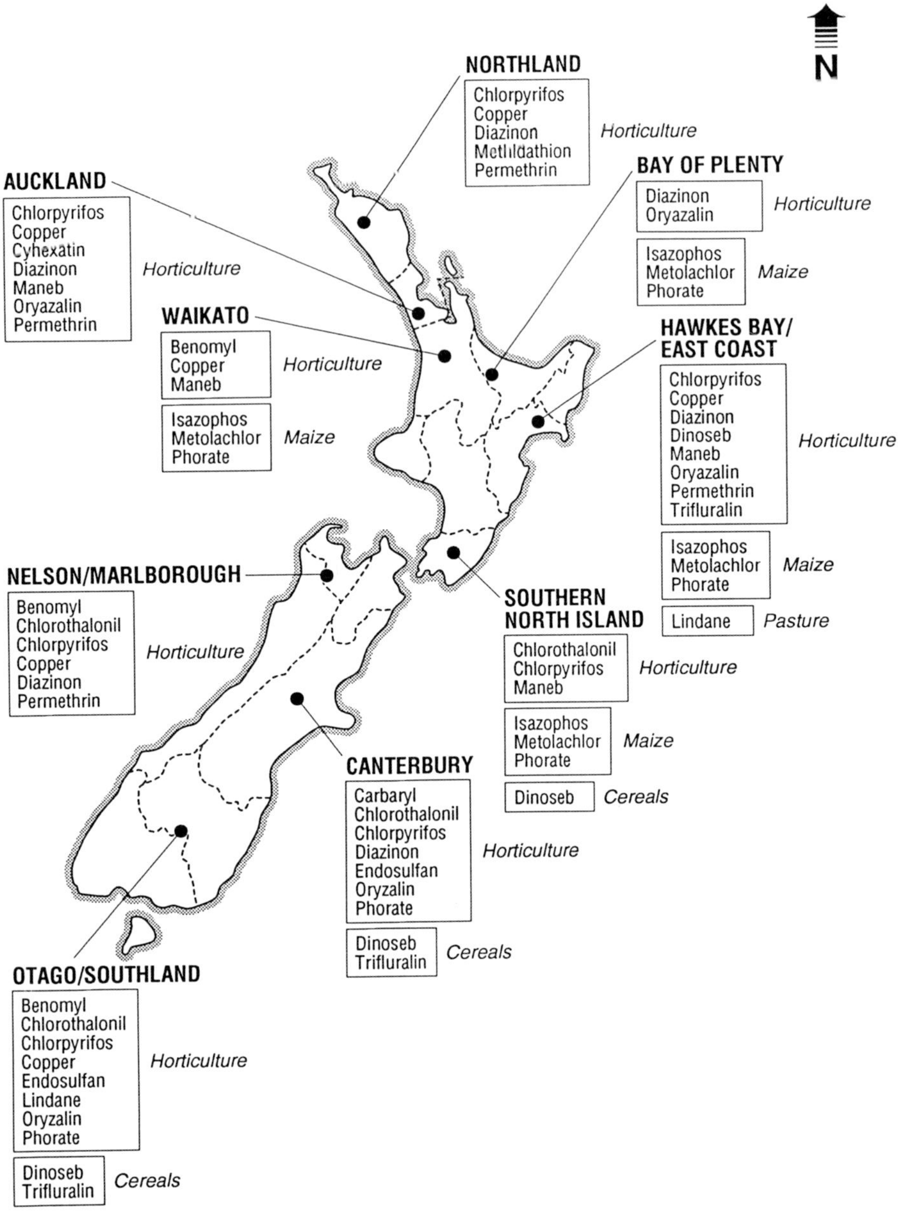

Figure 8. Areas where pesticide application rates may yield toxic runoff to surface waters (Smith et al., 1993). North Island surveyed in 1985 to 1988. South Island survey period 1986 to 1989. Cyhexatin, lindane and dinoseb are no longer marketed (B. Leonard, Agricultural Chemical and Animal Remedies Manufacturers Association of New Zealand Incorporated, (Pers. Comm.).

Table 8. Mean Canterbury soil DDT residue levels ($mg\,kg^{-1}$).

Soil Group	Number of farms	Mean soil DDT residue level
Alluvial (recent)	36	0.32
Gleys, gley recent	9	0.31
Yellow-grey earths	38	0.60
YGE/YBE intergrades	15	0.45
Yellow-brown earths	2	0.07

Source: Morton and Butcher (1990).

7.3.1 Soil contamination with DDT residues

DDT, an organochlorine insecticide, was used to control the invertebrate pasture pests grass grub (*Costelytra zealandica* White) and porina (*Wiseana cervinata* L.) from 1940 to 1970 when its on-farm use was banned by legislation. The recommended treatment was to mix DDT with fertiliser (principally superphosphate) and apply DDT at the rate of about $2.24\,kg\,ha^{-1}$ every three years. Apparently, during this period of use, DDT was also routinely incorporated into standard fertiliser mixes with the result that it was unknowingly applied.

Relatively low levels of DDT residues i.e. DDT, DDD (1,1-dichloro-2,2-bis[p-chlorophenyl] ethane) and DDE (1,1-dichloro-2,2-bis[p-chlorophenyl] ethylene) are widespread throughout topsoils in regions where grass grub and porina are significant pasture pests. While DDT itself breaks down relatively quickly with a half-life of 10 years on non-irrigated land and less on irrigated land, the major residue DDE is very persistent with little change in soil levels over the last 20 years.

Soil residues of DDT are a problem because they are taken up by grazing animals through soil ingestion and accumulate in milk and fat. When animal products containing these residues are consumed by humans, the residues then accumulate in human fatty tissue. Few detailed comprehensive studies have been conducted in NZ, although in the Canterbury region of the South Island there have been two studies of soil DDT residues. Mean soil DDT residue levels show that YGE (Ustrochrept) soils had highest levels, followed by YGE/YBE (Eutrochrept) intergrades (Table 8).

However, there was a large variation in DDT residues between farms in the same soil group. There was a differential distribution of DDT residues throughout the greater Canterbury region (Table 9), with the mid-Canterbury sub-region having the highest mean soil DDT residue level which reflects both the high proportion of shallow soils in the area and of land developed during the 1940–1970 period.

The above observations notwithstanding, the large variation in DDT residue levels between and within farms in the surveyed areas presumably is a reflection of past DDT use. This means that DDT residues need to be dealt with on a farm by farm basis.

Table 9. Percentage of paddocks within soil DDT residue (mg kg^{-1}) ranges for Canterbury sub-regions.

Sub-region	Paddocks tested	% paddocks in each stratum <0.1	0.1–0.5	0.5–1.0	>1.0	Mean
North	319	59.9	27.9	8.8	3.4	0.19
Central	157	68.2	21.7	8.3	1.9	0.17
Mid	190	19.0	40.0	21.0	20.0	0.58
South	205	62.6	25.4	7.8	3.9	0.21
Overall	871	53.2	28.8	11.1	6.9	0.27

8. Contamination from industry: policy and legislation

8.1 Institutional and regulatory framework

Environmental management concerning contaminants in NZ is spread over three tiers of government, viz:

- Central government: the Ministry for the Environment administers environmental statutes, develops policy and legislation, and advises on the sustainable management of the environment including the development of national guidelines and standards.
- Regional government: 16 agencies have regional council responsibilities; with one exception their jurisdictions are based on watersheds (see Figure 9). The councils develop regional policies and plans consistent with environmental statutes and regulate discharges to air, water and land (including the disposal of hazardous wastes). Information is collected on contaminated sites and an oversight of investigations is maintained.
- Territorial authorities: City and District councils administer land use planning and building controls; site specific details on contaminants are recorded on land information memoranda.

8.2 Legal framework

The Resource Management Act (1991) integrated a previously large number of environmental statutes into a single, all encompassing law. The purpose of the Act is the promotion of sustainable management of natural and physical resources.

Sustainable management is defined as being:

> 'managing the use, development, and protection of natural and physical resources in a way, or at a rate, which enables people and communities to provide for their social and cultural wellbeing and for their health and safety while:
>
> (a) sustaining the potential of natural and physical resources to meet the reasonably foreseeable needs of future generations;

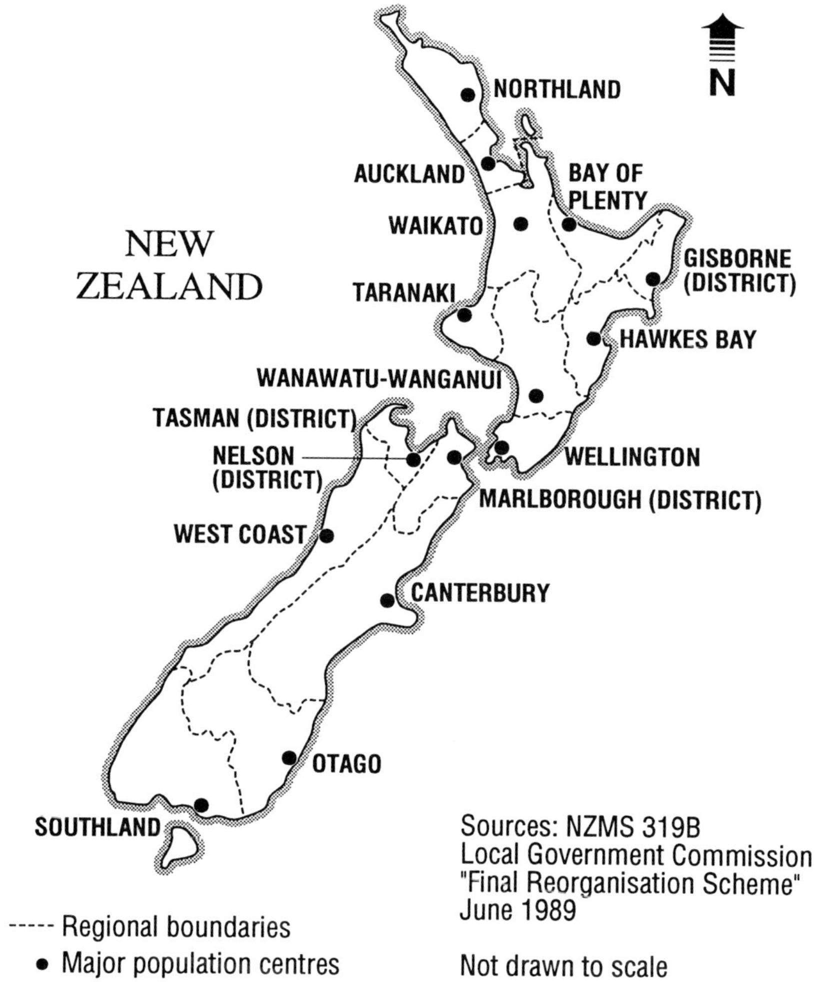

Figure 9. New Zealand regional council boundaries.

(b) safeguarding the life-supporting capacity of air, water, soil and ecosystems; and

(c) avoiding, remedying or mitigating any adverse effects of activities on the environment.'

This legislation focuses on the effects of activities rather than the activities per se and should allow for more flexibility in the way land is used. The Act is implemented within a hierarchy of policies, plans, rules and consents designed to minimise adverse effects on the environment.

There is, however, no legislative definition of a contaminated site, nor any statutory threshold levels which define whether a site is contaminated. To date these matters are addressed by means of government policy. [The Resource Management Act defines a contaminant in terms of substances (or energy, or heat) that 'changes or is likely to change the physical, chemical, or biological condition' of water, air or land when discharged.]

Although the Resource Management Act allows for liability to be assigned to the owner/occupier of the site or to the polluter, there is at present uncertainty about how this applies to the clean-up of sites contaminated prior to the Act coming into force. Neither has any case law to date addressed the issue of whether a hierarchy exists or whether the owner/occupier is liable in the first instance.

Enforcement powers extend to strict liability, fines of up to NZ$200 000 and imprisonment of company principals. However, prosecutions are seen as regulatory actions of last resort and fines are generally low in relation to legal costs.

8.3 The Policy Framework

New Zealand is a party to the Australian and NZ Environment and Conservation Council (ANZECC). This Council has, in conjunction with the National Health and Medical Research Council of Australia (NHMRC), developed Guidelines for the Assessment and Management of Contaminated Sites (ANZECC/NHMRC, 1992). (See Chapters 9 and 14). These guidelines are Government policy in NZ. Reference contaminant levels are provided for 'Background' and 'Investigation' categories to guide preliminary judgement as to the need for further site assessment. The guidelines, which represent best practice and are widely applied in the assessment of sites, are covered in more detail in a later section of this paper.

9. Contaminated industrial sites in NZ

9.1 Overview

Case studies are presented of contaminants associated with sawmill and timber treatment sites as this has been of a major focus in NZ. Less data are available on contaminants associated with other industrial activities although more is expected to be reported over the next few years. As there is no requirement for contaminated site investigations to be reported to regulatory agencies, this review should not be considered as being a comprehensive one.

9.2 Timber industry: Pentachlorophenol

9.2.1 Background

Sodium pentachlorophenate (NaPCP) was widely used by the NZ timber processing industry for the temporary control of sapstain fungi in freshly sawn softwood timber. From the time of its introduction to NZ in the 1950s and up until 1988 it was estimated that 70% of NZ's 400 or so sawmills had used NaPCP (normally comprising 0.5–1.0% NaPCP buffered with borax) with an estimated use of up to 200 t yr^{-1} (Shaw, 1990). Pentachloro phenol (PCP) had also been used as an additive to boron (B) and light organic solvent treatments. The use of significant but much lesser amounts of PCP as a timber preservative (in the order of 100 t yr^{-1}) has also been documented (Ministry of the Environment/Department of Health, 1992; Royds Garden Environmental Service/CMPS&F, 1994) at two locations. Since 1988, the use of NaPCP rapidly declined as occupational health and environmental impacts became issues of concern.

The contamination of estuarine harbour sediments by PCP and chlordane was reported by Wilcock et al. (1989). In a review of the use of PCP and chlordane by the timber processing industry, Shaw (1990) noted the potential for 'hotspots' and more widespread pollution based on the history of industrial handling and use of PCP. Campin et al. (1990) suggested that a major portion of the polychlorinated dibenzo-p-dioxin and furans (PCDD/PCDF) toxic equivalent (TE) constituents of a wood pulp mill effluent may have originated from a former sawmill where PCP had been used as an antisapstain. From December 1991, PCP was deregistered for use as a pesticide and prohibited from importation into NZ.

Bingham (1991) confirmed that the technical grade NaPCP formulations used in NZ were significantly contaminated by PCDDs and PCDFs. In addition, sludge wastes from NaPCP antisapstain storage and dip tanks were also found to be enriched with dioxins to significant levels (Bingham, 1991b). Because of the environmental persistence of these chemicals it was concluded that any timber treatment plant which had used pentachlorophenol in the past would be likely to contain areas of soil contaminated with PCDDs/PCDFs in addition to PCP. This use of PCP is thought to be one of the principal sources of dioxins in the NZ environment (Bingham, 1991b).

In response to the issues raised in the above papers, a National Task Group (NTG) was set up to assess the potential contamination problems and determine the required action. The NTG reported its findings (Ministry for the Environment, July 1992) and subsequently a National Steering Committee was established to facilitate actions at national and regional levels among government agencies and industry (Ministry for the Environment, November 1993a).

9.2.2 Site assessments

(i) As part of the NTG activities, a risk assessment pilot study was undertaken of a major sawmill and timber processing complex near Rotorua where PCP antisapstain (dipping) and reuping (impregnation of PCP/oil under pressure) treatments had been carried out over a 30–40 yr period ending in 1988. The site is situated on pumice soils in the central North Island. A modification to the methodology for full dioxin congener analysis – an OCDD screening method (Ministry for the Environment/Department of Health, 1992) – was developed to more cheaply identify the levels of higher chlorinated congeners comprising the bulk of these micro-contaminants of PCP.

The NTG study (Ministry for the Environment/Department of Health, 1992) found:

- significant contamination in the vicinity of the reuping plant (50–1 250 mg kg^{-1} PCP and 0.5–25 μg kg^{-1} dioxins TE [NATO]) with similar levels in some areas of the pole storage yard;
- significant concentrations of PCP in soils near antisapstain areas (0.35–3 600 mg kg^{-1} PCP); levels of dioxins were generally less than 30 μg kg^{-1} dioxins TE, the highest level (3 300 μg kg^{-1} dioxins TE) being found beneath the PCP anti-sapstain formulation/mix area;
- levels of contaminants in building dust measured up to 6 500 kg kg^{-1} PCP and between 90–150 μg kg^{-1} dioxins TE_1.
- PCP migrating off-site to an adjacent stream via groundwater and the storm-water drainage system such that the aquatic ecosystem of the stream was likely to be adversely affected;
- biota and sediments of the downstream lake contained elevated levels of PCP and dioxins but no clear picture emerged concerning environmental effects and there appeared to be no significant human health concerns related to the consumption of fish.

Clean-up activities can be reported as follows (Forestry Corporation of New Zealand, July 1993):

- The isolation and clean-up of areas of contamination (e.g. dust removal) which were identified as posing a potential health risk to workers.
- The prevention of PCP contaminated groundwater from leaving the site by the construction of strategically placed trenches and sealing off stormwater drains. An ultra violet/hydrogen peroxide (Rayox) treatment of the intercepted groundwater (350 m^3 d^{-1}) is reducing PCP levels from an average of 6000 μg kg^{-1} (range 2 400 to 11 000 μg kg^{-1}) to less than 1 μg kg^{-1}; dioxins are reduced from 12 ng kg^{-1} to 4 ng kg^{-1} TE. These combined works have led to an annual destruction of 750 kg of PCP.
- The decontaminated groundwater passes through the site sewage oxidation ponds which is then spray irrigated onto forested lands.
- Stream water quality has improved from a loading of 575 kg yr^{-1} PCP (56 μg kg^{-1} PCP) to 20 kg PCP yr^{-1} (2 μg kg^{-1}) representing a 96% reduction of contaminants entering the stream via contaminated groundwater (Southen, 1995). The longer term prospect is to remediate contaminated soils if required and practicable.

(ii) A solution of 5% wv^{-1} PCP and fuel oil was used as a preservative to treat timber poles at a site near Hammer from between 1954 to 1973. During this time an average of 22 tonnes of PCP per year had been used. Creosote was previously used on the site from 1940 to 1954. The treatment plant was decommissioned in 1973, all equipment removed, and an upper layer of contaminated soil reportedly disposed offsite. The site was back-filled, contoured and grassed. Anecdotal reports indicated that considerable drippage and spillage of PCP had occurred during the operation of the plant.

Site assessment (Royds Garden Environmental Services/CMPS&F, March 1994) revealed the following pattern of contamination:

- minor surface PCP contamination of the treatment plant area (confirming the importation of clean backfill to the site), but significant PCP/dioxins and oil contamination at depth greater than 1.0 m (up to 830 mg kg^{-1} PCP; up to 770 mg kg^{-1} C10–C14 and 4100 mg kg^{-1} C15–C28 TPH) and beneath which, a plume of contaminated groundwater with floating product containing 159 μg kg^{-1} dioxin (TE) in the oil phase;
- contaminated dust, up to 2.4 μg kg^{-1} dioxin (TE) in ex-worker accommodation huts, now decontaminated;
- PCP and TPH contamination of soils (up to 5700 mg kg^{-1} PCP; up to 3900 mg kg^{-1} C10–C14 and 9600 mg kg^{-1} C15–C28 mg kg^{-1} TPH) and groundwater in the vicinity of the former log storage area;
- significant TPH (up to 13500 mg kg^{-1}) and PCP contamination of a waste dump area.

(iii) Soil assessments were undertaken of three sawmills in the Canterbury region (CMPS&F, 1995) that had maintained PCP antisapstain (0.5% wv^{-1}) and boron dipping/diffusion operations over a number of years, in order to gain a better understanding of contaminant levels associated with smaller timber treatment operations. Investigations confirmed the pattern of site contamination to be localised and highest in the vicinity of chemical application and treated timber storage areas. The level of sampling and analysis was designed to be similar to that required for issue of a Certificate or Statement of Environmental Audit under the Victorian Environment Protection Authority contaminated land auditing system.

- Site A: Boron treatment (with 0.2% wv^{-1} PCP) over approximately 20 years, and PCP antisapstain treatment for the latter 10 years; non-operative since 1986 but sawmill building and associated sheds intact; site generally flat, unsealed and largely covered with gravel; surface topsoil overlays a metre of clay.
- Site B: an operating sawmill and timber treatment plant (B and copper-chrome-arsenate – as referred to later); PCP used for approximately 10 years between 1976 and the mid 80s; only derelict foundations remain of the former sawmill and PCP facilities which are open to the weather; lower areas of the site periodically subject to flooding.
- Site C: operated PCP antisapstain for approximately 20 years, non-operative since 1984; only derelict foundations remain of treatment structures; site generally open to the weather; flat, unsealed terrain; a sandy gravelly topsoil of variable organic content overlaying sandy gravels at 0.5 m depth.

A summary of the maximum contaminant levels found during these studies is set out in Table 10.

9.2.3 Timber industry: CCA

Approximately $1.4 \times 10^6 m^3$ of timber, including roundwood, is preservative treated in NZ each year (NZ Statistics Department, 1994). Treatment is essentially of three

Table 10. Maximum contaminant levels in surface soils and building dust at three sawmill/timber treatment sites in the Canterbury region.

	PCP (mg kg^{-1})	dioxins (TE) μg kg^{-1}	Total B (mg kg^{-1})	Available B (mg kg^{-1})
Site A	620	0.527**	9700	7700
	1000 (dust)	10.7 (dust)*	N/A	N/A
Site B	370	0,7*	<40	<40
	1.5 (dust)	N/A	N/A	N/A
Site C	3.4	4.22**	700	130
	7.1 (dust)	N/A	320 (dust)	310 (dust)

N/A analysis not undertaken.
*dioxins (TE) estimated by full congener analysis.
**Dioxins (TE) estimated by OCDD method (Ministry for the Environment/Department of Health, 1992).

Table 11. Maximum contaminant levels in studies of 18 CCA treatment sites.

	Range of maximum values (mg kg^{-1})		
	Cu	Cr (III)	As
(McLaren, 1992) Bay of Plenty (n = 10)	108–6300	140–5500	80–5475
(Armishaw et al., 1993) Nth Island (n = 7)	326–8020	428–4100	376–10440
(CMPS&F, 1995; Canterbury) (n = 1)	2100	2800	6100

types: Copper-chrome-arsenic (CCA); B; and light organic solvent (LOSP). CCA and LOSP treatments utilise pressure to impregnate the preservative chemicals whereas B treatment involves the surface application and diffusion of a B based insecticide.

Copper-chrome-arsenic treatment of *Pinus* timber involves the impregnation under pressure of a 1–4% solution of copper sulphate, sodium dichromate and arsenic pentoxide. The treated timber is graded for use according to the amount of treatment solution fixed in the timber (normally 3.6–16 kg m^{-3}). The Cu acts as a fungicide, As as an insecticide, and Cr as a fixative. Contamination of soil has occurred because of waste disposal practices and drips, leaks and spills of stock and working solutions.

Studies by McLaren (1992), Armishaw et al. (1993) and CMPS&F (1995) report on Cu, Cr and As levels found at a number of CCA treatment sites. Highest contaminant levels were associated with drip pads, treated timber and chemical storage areas. For surface samples, fines (<2 mm) sieved from yard gravel were analysed. Maximum contaminant levels found in these studies are summarised in Table 11.

Major findings from the studies on CCA treatment sites include:

- Concerns over the potential for offsite contamination from surface runoff and dust (McLaren, 1992);

- Treatment building dust contained (up to): Cr (III) 9 100; Cu 7 600; and As 11 000 mg kg^{-1}, respectively (CMPS&F, 1995);
- Contaminants were largely confined to the top 30 cm, and rarely penetrated into the underlying soil to any appreciable extent, although this could not be always assumed to be the case. There was some evidence that the metals (particularly chromium as Cr VI) may be more mobile through underlying permeable pumice soils (Armishaw et al., 1993).

A number of other investigations involving CCA contamination of soils and off-site water resources can be noted (Ministry for the Environment, 1992b):

- Auckland region: timber had been treated with CCA at a site over a period of 40 years; chlordane had also been used. A highly contaminated area of about 6 000 m^2 was identified. There was evidence of surface contamination and runoff into a drain (sediments containing over 10 000 mg kg^{-1} of each of Cu, Cr and As) and through a swamp to the harbour. Chlordane was found up to 10 mg kg^{-1} on site soils and in nearby marine sediments (13 mg kg^{-1}) of the Manukau harbour.
- Wellington region: several spillages of CCA were reported at a timber site (now closed) located on very permeable alluvial gravels above shallow and fast moving groundwater. One spillage (7 200 L of 2% CCA solution) contaminated groundwater and some industrial and residential bores in the vicinity.
- Nelson: About 320 m^2 of a CCA treatment site was found to be contaminated in excess of 500 mg kg^{-1} (for each of Cu, Cr, As) in some places to a depth of at least 1 m, with a larger area (1 200 m^2) in the vicinity also contaminated but to a lesser extent. The highly contaminated soil was stabilised with cement, disposed along with contaminated concrete structures to an onsite secure landfill and sealed; further quantities of soil were disposed of to a landfill.
- South Canterbury: a derelict CCA site on the foreshore totalling 1 700 m^2 had surface concentrations of As in excess of 200 mg kg^{-1} although levels of Cu and Cr were very low. The soils were permeable with shallow groundwater flowing to the sea. The remediation comprised the disposal and capping of more than 800 m^3 of excavated soil to an onsite clay lined pit, and the installation of groundwater monitoring bores, leaving a residual soil level of between 50–200 mg kg^{-1}.

9.3 Organochlorine pesticides

9.3.1 DDT

Historical operations of a pesticide and agricultural chemicals manufacturing company led to significant contamination of two sites situated on either side of a street in a coastal township 20 km west of Nelson. Various milling, blending, packaging and storage operations involving organochlorine pesticides had been carried out since 1945 on one site with activities extended to the other site in the late 1950s. The plant closed down in 1988 with a number of buildings and structures left derelict.

The sites are situated across the middle of a small peninsula and drain to the foreshore on either side. The sites are covered by a layer of fill consisting of up to 0.3 m of silty gravels and sands, concrete and tarmac, which overlie predominantly alluvial clays and gravelly sands.

Site investigations (Woodward-Clyde, April 1994) comprised extensive sampling of the site with samples from 59 locations. At each location, samples were retrieved at each of three depths (0–5 cm, 50–100 cm, and 100–200 cm) comprising subsamples composited from a cluster of five holes.

- The soil appeared to be widely contaminated to a depth of 200 cm by DDT and breakdown products. Of the site locations sampled, 90% of the samples exceeded 1.0 $mg\,kg^{-1}$ total DDT and 60% exceeded 10 $mg\,kg^{-1}$ total DDT with a peak of 4600 $mg\,kg^{-1}$ total DDT. The site was contaminated to a lesser but more localised extent by dieldrin (up to 780 $mg\,kg^{-1}$ but typically <10 $g\,kg^{-1}$, 50–100 cm) and with sulphur (S) more widespread (up to 11 900 $mg\,kg^{-1}$). Lead (up to 1 230 $mg\,kg^{-1}$, 250 cm depth), aldrin (up to 100 $mg\,kg^{-1}$, 100–200 cm depth) and lindane (up to 170 $mg\,kg^{-1}$) were of less concern. Groundwater appeared to be significantly contaminated by pesticides, S and chlorobenzene with potential for off-site migration noted.

The marine environment adjacent to the site appears to be subject to some exposure of contaminated sediment, groundwater and leachate from the site and associated landfill and dump areas. The main findings of off-site marine investigations (Woodward-Clyde, November 1993a) were:

- organochlorine pesticides detected in *Amphibola* spp. (mudsnails) in excess of the NZ Food Regulations (0.1 $mg\,kg^{-1}$) including levels of up to 1.7 mg DDE kg^{-1}, 3.29 mg DDD kg^{-1}, 2.04 mg DDT kg^{-1}, 0.31 mg dieldrin kg^{-1}; and with elevated organochlorine levels in oysters and mussels;
- environmental guidelines exceeded for seawater immediately adjacent to the site (as referenced to ANZECC guidelines for seawater, and USEPA Marine Chronic criteria) with findings for dieldrin, 2,4-DDT, 4,4-DDT, and lindane of up to 0.224, 0.037, 0.038 and 0.010 $mg\,m^{-3}$, respectively.

9.3.2 Dieldrin

The pesticidal use of dieldrin in NZ was prohibited in the mid 1960s. This came about in two phases. Firstly, the use of dieldrin for sheep sprays and dips was banned in 1961 – along with aldrin, DDT, BHC and methoxychlor. Secondly, the pastoral application of dieldrin to control pasture pests such as black beetle and grass grub (*C. zealandica*) was banned in 1966. Surplus stocks of pesticides, estimated at 530 t, were initially collected from farms around the country by the Department of Agriculture and stored in a number of depots. By March 1963, these holdings were consolidated to regional storage centres. By February 1964, approximately 165 t had been redistributed to land development blocks in Northland, Hawkes Bay, Rotorua,

North Canterbury, Canterbury and Southland for the purpose of pastoral application by aerial spraying (Gibbs, 1992).

The use, disposal or ultimate fate of these pesticide stocks remains inadequately documented (Loe Pearce & Associates, 1994) due to the apparent loss of historical files and records, although a major consignment of surplus pesticides was exported from NZ in late 1966. The location of some of the historical storage and dumps sites have been identified by regional investigations to date (Loe Pearce & Associates, 1993) and some decontamination activity undertaken (Taranaki Regional Council, 1993).

Residual contamination (presumably as a result of material spilled during the handling of damaged packaging) has persisted at some storage centres. A reconnaisance investigation in the Canterbury region (Tonkin & Taylor, 1994d) revealed that the original storage buildings still existed on four of the five sites investigated (the remaining site now being covered by a shopping mall).

- Dieldrin was recorded at one site up to 14000 $mg\,kg^{-1}$ in the building dust and up to 800 $mg\,kg^{-1}$ in the surface soils beside the sheds. Up to 70 $mg\,kg^{-1}$ DDT in the shed dust of a second site; and between 4 and 100 $mg\,kg^{-1}$ dieldrin around the sections of the other sites.

Preliminary assessments have been undertaken of two dump sites in the Southland region (Woodward-Clyde, 1993b). Buried and rusting pesticide containers and evidence of some soil and groundwater contamination (up to 3.5 $mg\,L^{-1}$ dieldrin and 15 $mg\,L^{-1}$ aldrin at one site) indicate that an estimated 75 and 180 m^3 of soil may need to be excavated from the sites respectively. Further action is pending the availability of an appropriate remediation technology.

9.4 Oil storage terminals

Baseline investigations have been undertaken at over 15 active and former bulk storage oil terminals (Tonkin & Taylor Ltd, 1995a). The larger sites have been in operation since the early 1930s. All sites studied are founded on superficial deposits (alluvial and estuarine sands, silts and clays) and are often overlaid by fill materials. Groundwater typically lies between 2 and 5 m below ground level.

Terminal upgrades over time have resulted in the construction of lined tank compounds, improved tank foundations in earthquake sensitive areas, and the installation of spill control measures. Most sites were generally not contaminated with the exception of small areas adjacent to separators; former drum filling; washing and storage operations; and former road tanker and rail car loading and unloading racks. Contaminant levels at some of these sites are discussed below:

- Total Petroleum Hydrocarbon (TPH) concentrations in soil (typically not detected – 10000 $mg\,kg^{-1}$, mostly <500 $mg\,kg^{-1}$) up to 19000 $mg\,kg^{-1}$ if associated with free phase hydrocarbon product.
- Groundwater TPH (typically <5–5 $mg\,L^{-1}$) up to about 20 $mg\,L^{-1}$, but reaching higher levels if directly associated with isolated layers of separate phase product.

- Discontinuous sheens or floating layers of separate phase hydrocarbon were present at less than 20% of the sites studied, the apparent thicknesses generally ranged between just detectable sheens and 0.25 m.

Remedial works have been undertaken at a number of sites, including product recovery, landfarming of contaminated soils, *ex-situ* treatment of contaminated groundwater, and *in-situ* treatment of soil and groundwater. Detailed risk assessments indicate that contamination is restricted to within the site boundaries and that no significant off site environmental and human health risks were identified. Regional councils usually require routine monitoring to demonstrate that *in-situ* passive bioremediation is occurring over time.

9.5 Landfills

An Auckland City Council (ACC) commissioned review of closed landfills was undertaken to define the RM Act discharge consent requirements (Tonkin & Taylor, 1994a).

A total of 85 former landfills owned and/or occupied by ACC were identified; operations on these sites dated back from the turn of the century to the late 1970s. The earliest known landfills, from about 1910 until the 1940s, tended to be small road and quarry infills, which had limited putrescible content. The size of the sites gradually increased over the years and from the 1950s onwards, landfills became larger, with increasing proportions of putrescible waste. More remote sites were being sought in coastal gullies and tidal wetlands and in the 1960s and 1970s industrial waste was being disposed of at these sites. ACC ceased refuse landfilling in the late 1970s, as the Auckland Regional Authority took over this role.

The desk study and walkover assessment work enabled a weighting matrix to be developed and sites were ranked on the basis of their potential for adverse environmental and human health effects. The ten highest ranked sites were then subjected to detailed investigations, including leachate, point discharge, surface water and groundwater analyses, and landfill gas monitoring within the landfill and in buildings on and around the sites.

- Leachate was actively being generated and migrating from all of the landfills studied. The older landfills tended to produce leachate with relatively low Chemical Oxygen Demand (COD) and Biological Oxygen Demand (BOD) concentrations, whilst the larger landfills which closed in the late 1970s generated leachate with higher BOD (50 to 200 mg L^{-1}), COD (500 to 1 500 mg L^{-1}) and ammonium (up to 100 mg L^{-1}) concentrations.
- Low concentrations of volatile and semi-volatile organic compounds (xylene, phenols etc.) were present within the landfill leachate of the young landfills, but were not detected in any of the point discharges to surface water. Leachate containing dissolved metals was only present in the refuse landfills; metals were only recorded at very low concentrations and Zn was the only metal widely observed at concentrations above detection limits.

- Significant concentrations of landfill gas were detected at eight of the sites studied. Methane and carbon dioxide concentrations of up to 65% and 55% respectively occur consistently in the younger landfills, with more variable levels present within the older landfills.
- Landfill gas was recorded as migrating into buildings at two of the sites investigated and methane levels above the Lower Explosive Limit (LEL) were recorded in some floor slab joints and cracks, but not within the larger building areas. Landfill gas was observed migrating off site at a number of sites, but was not detected in neighbouring buildings. It is considered that the traditional NZ house construction, which provides vent space beneath the buildings and a free flow of air, has been protected against any landfill gas explosion, flammability and asphyxiation risks within these buildings.

Based on these investigation results, appropriate and cost effective management and remedial options to avoid, remedy or mitigate any adverse environmental or human health effects arising from the landfills have been developed (Tonkin & Taylor, 1994b).

Preventative remedial works were proposed to address sources of landfill discharges. These works included the capping of landfills to reduce infiltration and leachate generation and the passive venting of landfill gas to prevent gas migration. In addition, remedial control measures were recommended to address the landfill discharge problem itself. A number of leachate containment, interception, removal and treatment options were presented for several sites. Recommended control measures for landfill gas included active and passive venting of particular buildings and the sealing of floor slab joints.

Implementation of management plans was proposed for a number of sites to prevent adverse effects from occurring. For example, where leachate discharges occur, management practices included placing restrictions on recreational use of harbours or shellfish gathering around the landfill.

In addition, continued discharge of leachate and landfill gas was considered to be an option, if environmental and human health effects were minor, in that ACC has an obligation to consider other areas where its resources could achieve greater environmental benefit.

10. Priority contaminants and current management remediation strategies

10.1 Agriculture

10.1.1 Heavy metals

The current priority is Cd and the following strategies have been or are being established:

- The export meat industry has a blanket condemnation order on offal for human consumption from all slaughtered sheep and cattle which are over the age of two years.

Table 12 Cadmium reduction strategy in New Zealand manufactured superphosphate products.

Year	Target Cd content $mg\,Cd\,kg^{-1}\,P$
1995	420
1997	340
2000	280

- Superphosphate fertiliser manufacturers are introducing voluntary controls on the Cd content of phosphatic fertilisers. There will be a stepwise-reduction (Table 12) in mean Cd content over the next 10–15 years.

Research work is continuing to determine the movement of Cd through the soil-plant-animal system with a view to identifying areas where changes in farm management practices will help reduce Cd accumulation in soils and animals. The impact of Cd on food quality has been noted in Chapters 5 and 11.

10.1.2 Non-metals

While the priority non-metal contaminant is N, P should not be ignored and the following options may reduce P contamination of surface and groundwater by:

- avoidance of fertiliser application to high risk areas such as stream channels, permanently saturated soils etc. Improvement of fertiliser placement and use of larger granules will help in this endeavour;
- low solubility P fertilisers which may result in a reduction of short-term runoff losses;
- the application of P fertiliser when pastures are well grown (i.e. high) to help prevent runoff; and
- utilisation of riparian filter strips and maintenance of natural riparian wetlands.

Management strategy options to reduce nitrate leaching from soil include:

- application of fertilisers and organic wastes at rates and times which match the needs of the plant;
- application of fertilisers and organic wastes when there is the least likelihood of drainage occurring;
- use of split applications of fertilisers to increase the efficiency of use;
- ploughing pasture and legumes in late autumn or in spring;
- planting winter cover crops to capture autumn mineralised N;
- ensuring that optimum, rather than excessive, amounts of irrigation water are applied;
- using deep rooting crops to scavenge nitrogen from deep in the profile; and
- establishment of riparian zones along streams and around lakes.

10.1.3 Pesticides

As noted above, there has been no systematic monitoring of the pesticide content of the surface and groundwater in NZ. Although a limited number of studies have indicated the presence of a range of pesticides in surface and groundwater in NZ, the concentration of these pesticides is much less than maximum acceptable limits for drinking water. The reasons for this may be that most soils in NZ contain large amounts of organic matter resulting in greater retention of pesticides and/or the increased biological activity due to high organic matter content resulting in enhanced degradation of pesticide residues in soils.

In the Canterbury region DDT residues are priority contaminants, particularly given the proposed lowering of the international standard for meat and dairy products from $5\,mg\,kg^{-1}$ to $1\,mg\,kg^{-1}$ in 1998. Management techniques to mitigate the effects of DDT residues are based on developing systems for:

- minimising the amount of soil ingested by grazing animals;
- reducing the amount of DDT in milk by manipulating animal physiology;
- managing fat levels in animals through appropriate feeding strategies.

In a longer term approach to managing DDT residues, the potential of microbial cultures with residue degradation capabilities is being investigated. This work is based on successes with PCB residues. Preliminary results suggest DDE degradation is feasible using both aerobic and anaerobic bacteria. White rot fungi (*Phanaerocktae chrysosporium*) potential is also being investigated.

10.2 Industry

10.2.1 Risk assessment frameworks

At the present time there is no regulatory schedule of contaminants, and priorities have been determined by the agency most responsible or accountable for the substance(s) concerned. New Zealand is now progressively adopting risk assessment approaches to identify priority contaminants and to assist policy judgements and management decisions in relation to contaminated sites. Brief discussions on these follow.

(i) Rapid hazard assessment

A rapid hazard assessment methodology was developed (Ministry for the Environment, 1993a) to assist councils prioritise the likely risks associated with potentially contaminated sites without the need for expensive and time consuming sampling. It has been adapted from a similar system used in Canada (Canadian Council of Ministers, 1992).

By means of a scoring system, sites are categorised into those with potential to pose a high, medium or low risk of causing adverse impacts on human health or the environment. Sites are scored as follows:

- contaminant characteristics i.e the relative hazard of contaminants present at a site (see sample overview in Table 13).

Table 13. Sample page from the Draft National Rapid Hazard Assessment System for Contaminated Sites, Ministry for the Environment 1993.

Category	Evaluation factor	Scoring Guideline	Rationale	Method of Evaluation	Sources of Information
I.	A. Degree of hazard		In determining the degree of hazard of a waste, it is recognised that a listed hazardous waste e.g. ANZECC (1993) Scheduled Waste Model Regulation is generally of greater concern than a liquid or solid industrial waste. These are in turn of greater concern than other solid wastes. Municipal and organic wastes are considered to be medium concern contaminants due to their putrescible nature (production of methane and other landfill gases). Household wastes may contain hazardous materials (e.g. batteries, medical wastes, paints, etc.)	Determine the level of hazard according to the following table of typical contaminants and definition of high concentrations: High Concern Contaminants: • Materials which are persistent, bioaccumulative and toxic (e.g. organochlorine compounds) • Heavy metals • Industrial waste (e.g. pesticides, herbicides, paint sludge, acid and alkaline solutions, solvents, fuel, etc.) • Institutional waste (lab, schools, hospitals, etc.) • Pathological wastes and animal carcases • Radioactive wastes Medium Concern Contaminants: • Liquid waste not referred to in above, non-volatile petroleum products (e.g. oil), septic tank pumpings, agricultural and chemical containers • Food processing wastes • Non-hazardous incinerator residues • Municipal solid (household) wastes • Organic and vegetable wastes • Mining residues Low Concern Contaminants: • Industrial and commercial solid wastes (e.g. construction materials such as wood, metal, hay, sand/silt piles, etc.) • Other nearly inert wastes (e.g. foundry sands)	ANZECC/NHMRE Guidelines for Management of Contaminated Sites; Draft Health and Environmental Guidelines for Selected Timber Treatment Chemicals
Contaminant(s) Characteristics	• High concern contaminants	1.0			
	• Medium concern contaminants	0.8			
	• Low concern contaminants	0.4			

continued

Table 13. *continued.*

Category	Evaluation factor	Scoring Guideline	Rationale	Method of Evaluation	Sources of Information
I. Contaminant(s) Characteristics (cont'd)	B. Concentration of Contaminants (maximum observed)		Together with the quantity and degree of hazard, the concentration of contaminant determines the significance of contamination	High Concentration of Contaminants: • Contaminant concentrations in soil, groundwater or surface water exceed criteria expected to be appropriate for commercial/industrial use (or greater than 2×Dutch C levels if not known); or material that was deposited in highly concentrated form (e.g. >5000 ppm) Moderate Concentration of Contaminants: • Contaminant concentrations in soil or water exceed the criteria expected to be suitable for residential use (e.g. ANZECC 'B' level soil guidelines). Low Concentration of Contaminants: • Contaminant concentrations in soil and/or water are elevated compared to background concentrations, but are expected to be less than residential criteria	
	• Free organic liquid	1.0			
	• High concentration	1.0			
	• Moderate concentration	0.8			
	• Low concentration	0.5			

- exposure pathways i.e. the route a contaminant may follow to a receptor (e.g. groundwater, surface water, direct contact, and via air);
- receptors i.e. humans, animals, plants or resources that may be exposed to and affected by contamination.

A more definitive national estimate of potential contamination and likely clean-up requirements will be possible when regional councils complete their investigations.

(ii) Site assessment

A schematic outline of the risk assessment approach followed in NZ for the assessment of potentially contaminated sites is provided in Figure 10.

10.2.2 Management guidelines

(i) ANZECC/NHMRC (1992) Guidelines for the Assessment and Management of Contaminated Sites.

Some information on ANZECC/NHMRC Guidelines is also presented in Chapters 9 and 14.

The goals of the guidelines are:

- To render a site acceptable and safe for the long-term continuation of its existing use;
- To minimise environmental and health risks both on- and off-site;
- That where site clean-up is required, the goal should be to achieve a standard that minimises risks to human health and the environment, consistent with the existing and likely future use of the site and in accordance with a system to inform a future land owner that the clean-up has been conducted to an extent consistent with particular land uses.

This last objective is different to that adopted for Australia who have stated their objective to maximise to the extent practicable, the potential future uses of the site.

It is considered that, for NZ, land use changes slowly and the objective of maximising the standard of clean-up could be too expensive and may result in only a limited number of sites being cleaned up. Provided information mechanisms are available to ensure potential future land owners are informed, then existing owners have the flexibility to decide what level of clean-up should be applied to site, subject to the criteria for the existing land use. Clean-up to a higher standard would then be a commercial decision with its cost set against the value of the land after clean-up. The key features of the guidelines are set out in Table 14.

(ii) Timber treatment chemicals

Guidelines (Ministry for the Environment/Department of Health, 1992) have been prepared for the timber treatment chemicals of most concern: pentachlorophenol (PCP) and associated dioxin micro contaminants; B; CCA. The use of PCP and its derivatives ceased several years ago, whereas the use of B and CCA continues. The key features of the guidelines are set out in Table 15.

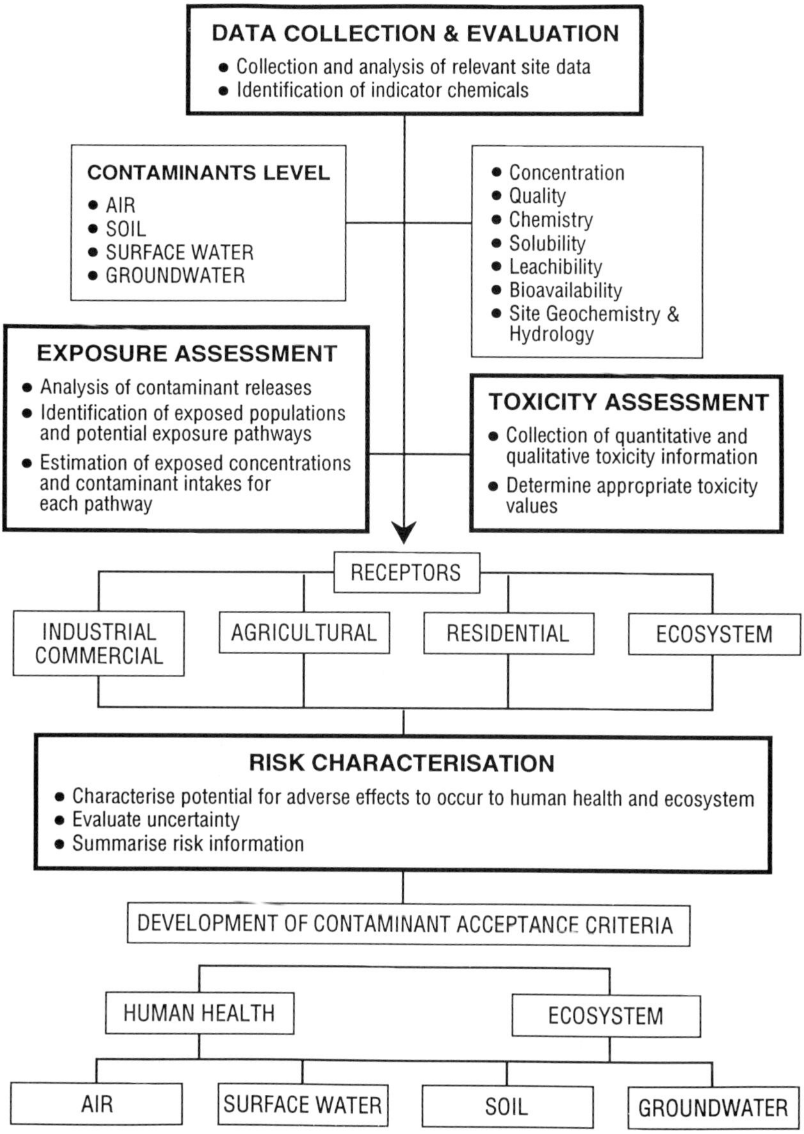

Figure 10. A schematic outline of risk assessment.

The guidelines provide policy and technical advice to regulatory authorities and industry on: site assessment and management; health and environmental acceptance criteria; disposal to landfill of PCP contaminated wastes and soils. They set out a flexible and cost effective approach to the assessment of contamination and the development of standards for use in regional plans and the determination of consents under the Resource Management Act (1991). The guidelines have been developed

Table 14 Key features of the ANZECC/NHMRC (1992) Guidelines.

- A working definition of contaminated sites

- Methodology for site assessment using:
  - staged approach;
  - community consultation;
  - incorporating occupational safety and health aspects

- Site specific clean-up criteria:
  - risk based approach;
  - includes ecosystem and human health elements

- If no Australian or New Zealand derived values are available, Dutch A, B and C values are referenced (now being revised with reference to the recently adopted Intervention and Target Values, Ministry of Housing, Spatial Planning and Environment, 1994)

- Relevant policy matters also addressed:
  - prevention of future contaminated sites;
  - preference for on-site treatment;
  - land use planning controls on sites;
  - notification on title;
  - site registers;
  - clean up incentives

Table 15. The major components of the Health and Environmental Guidelines for Selected Timber Treatment Chemicals.

Sampling Strategies and Analysis Protocols

- Protocols for site characterisation, sampling and analysis of soils, groundwater, surface water and dust.
- Analytical procedures for the chemicals of interest are referenced.

Soil Acceptance Criteria

- Acceptable residual chemical concentrations (see Table 16) are given for a number of site uses: agriculture, residential, and industrial. Consideration is also given to the protection given by paving and management practices.
- Criteria primarily based on protection of human health, plant life, contaminant uptake by plants and animals, and acceptable food residue regulations.
- An incremental lifetime risk of cancer of 1 in 100 000 per year has been adopted as an initial position for assessing the effects of chemicals with carcinogenic properties.

Surface Water and Groundwater Acceptance Criteria

- Criteria (based on the protection of human health) are developed for a range of site uses, such as potable water, stock watering, irrigation and recreation (criteria for the protection of aquatic ecosystems are planned).

Landfill Disposal

- A classification system for New Zealand landfills;
- guidance on waste characterisation;
- acceptance criteria for PCP wastes appropriate to respective landfill classes.

using open, consultative processes; the draft soil acceptance criteria for contaminants with respect to land use are provided in Table 16.

A code of practice (Occupational Safety and Health, 1994) includes environmental provisions that set out present day requirements in terms of treatment plant and stormwater design, chemical handling and waste management procedures. Any discharge of contaminants must fall within consents granted by councils under the Resource Management Act (1991). Compliance with the code will help minimise site contamination in the future.

(iii) PCBs

In recent years a high priority has been accorded to PCBs (Department of Health, 1988) in which PCB-containing electrical capacitors and transformers have been removed from service and exported overseas for destruction by high temperature incineration. A 1994 regu-lation under the Toxic Substances Act (1979) has made it illegal to use or store PCBs as from 31 July 1995.

(iv) Petroleum chemicals

A code of practice (Occupational Safety and Health Services, 1992) has been produced for the installation of underground storage tanks. The code, which integrates dangerous goods handling and storage requirements with environmental aspects, has been extended to address the management and removal of existing tanks (Occupational Safety and Health Services, 1995).

An additional set of guidelines for the petroleum industry is being prepared by industry, regional councils and central government. Given the large number of service stations throughout NZ, and the possibility of hydrocarbon contamination from spillages or chronic seepage, the objective of the guidelines is to reduce the need for expensive site specific risk assessments and to simplify clean-up procedures. The guidelines will apply the ANZECC/NHMRC risk assessment procedures and develop clean-up criteria that meet regulatory requirements.

It is intended that a complementary set of guidelines will address the chemicals of concern associated with redundant town supply gasworks sites.

(v) Landfills

Landfills (some management control) and waste dumps (no management control) have been the traditional repository for most industrial, agricultural and urban wastes. Waste materials were disposed at these locations often over many years, and with few exceptions until recently, with little regard for any hazardous constituents. Many waste sites border waterways, swamps and estuaries and have the potential to passively discharge contaminated leachate to these environments. Follow up assessment and management of these historical landfill and dump sites is considered a priority.

A number of guidelines have been prepared to improve the planning and management of landfills, including:

- a protocol for the collection of statistical data on waste streams (Ministry for the Environment 1992c);

Table 16. Health and Environmental Guidelines for Selected Timber Treatment Chemicals: Summary of (draft) soil acceptance criteria (all values in $mg\,kg^{-1}$, dry basis unless otherwise specified) (Ministry for the Environment/Department of Health, 1992).

Contaminant	Agricultural	Residential	Industrial Unpaved	Industrial Paved no management plan	Industrial Paved – with management plan	Maintenance
Arsenic	10	30–100	300	450	1500–NL	450
Boron	25,3 (soluble)	100,3 (soluble)	–	–	–	–
Chromium (III)	600	600	–	–	–	–
Chromium (VI)	10	50	350	550	1750–NL	550
Copper	30–100	130	–	–	–	–
PCP	0.1	10	610	1500	3000–NL	1 500
Dioxins & Furans	0.01 $\mu g\,kg^{-1}$ TE	1.5 $\mu g\,kg^{-1}$ TE	18 $\mu g\,kg^{-1}$ TE		90 $\mu g\,kg^{-1}$ TE	21 $\mu g\,kg^{-1}$ TE

TE: Toxic equivalents.

- national guidelines for the siting, design, operating and after-care of landfills (Ministry for the Environment 1992d);
- identification, sources and management options for waste hazardous substances (Ministry for the Environment 1994a).

vi) Agrichemicals

As a risk abatement measure, some regional councils have collected waste chemicals (mainly redundant or surplus agricultural pesticides) from among the farming community. Experience to date (Waikato Regional Council, 1992) has shown that the major portion of the chemicals collected can be identified and recycled for use elsewhere. The remaining portion – including organochlorine pesticides such as DDT and dieldrin – is being held in secure storage until an appropriate means of destruction becomes available.

(vii) Contaminated sites handbook

A contaminated site handbook (Tonkin & Taylor, 1995b) has been prepared in support of the Wellington Regional Council's draft management strategy and action plan (Tonkin & Taylor, 1994c). The handbook, a practical guide to the investigation, assessment and remediation of contaminated sites, is based on national and international guidelines, protocols and literature. The Council has also developed a comprehensive register of potentially contaminated sites and is currently undertaking a rapid hazard assessment screening programme (Wellington Regional Council, 1994) to prioritise the sites that may pose a risk to human health or the environment.

(viii) Dioxins

The Ministry for the Environment is to coordinate a programme concerning dioxins and other organochlorine substances of concern. Human and environmental data will be collected to establish background levels in the NZ environment, and to identify and assess the significance of any ongoing emissions. The programme will assist in the development of Government policy on dioxins including standards for clean-up.

11. Conclusion

11.1 Agriculture

The NZ economy in the 1990s is still extremely reliant on export revenue generated by the agricultural sector. However, there is a growing awareness and concern in the wider community regarding the long-term sustainability and quality of the soil and water resource as affected by agricultural activities.

Currently, issues such as Cd levels in edible foodstuffs, NO_3^- levels in groundwater and P eutrophication of surface waters all impact on the marketed image of NZ as a producer of high quality human foodstuffs. Pesticides do not appear to be a concern at this time, except for localised areas of soil contamination with DDT residues, but continued vigilance will be required to ensure that appropriate use is made of these chemicals.

The twin imperatives of maintaining and/or increasing access for NZ agricultural produce to world-wide markets and ensuring that the productive capacity of our land and water-based industries are not degraded will result in the adoption and/or amendment of technologies and practices which will seek to prevent or minimise soil contamination.

11.2 Industry

A major reform of existing legislation concerning controls on the use of hazardous chemicals is in progress and a new Environmental Risk Management Authority is to be established (Ministry for the Environment, 1994b). Additionally, there is a rapidly evolving awareness by industry of the benefits from minimising polluting events and discharges, which, coupled with recent legislative reform (e.g. Resource Management Act, 1991) should help minimise the creation of contaminated sites in the future.

The legacy of sites contaminated from past industrial activity, although modest by international standards, will endure for some decades to come. The Government is actively developing policies to remove impediments to clean-up, particularly with respect to sites identified as posing a high risk to human health or the environment.

As NZ promotes its 'clean green' image to support trade initiatives, there is an expectation that site clean-up will be progressively achieved.

References

Adams J A and Pattinson J M 1985 Nitrate leaching losses under a legume-based crop rotation in Central Canterbury, NZ. N. Z. J. Agric. Res. 28, 101-107.

Adams J A, Campbell A S, McKeegan W R, McPherson R H and Tonkin P J 1979 Nitrate and chloride in groundwater and surface water and deep soil profiles of Central Canterbury, New Zealand. Prog. Water Tech. 11, 351-360.

Andersson A 1977 Heavy metals in Swedish soils: on their retention, distribution and amounts. Swed. J. Agric. Res. 7, 7-20.

ANZECC/NHMRC 1992 Australia and New Zealand Environment and Conservation Council, National Health and Medical Research Council. Australian and New Zealand Guidelines for the Assessment and Management of Contaminated sites. 57 p.

Armishaw R F, Ficker A G and Fenton G A 1993 A study of soil and groundwater at some timber treatment sites. Industrial Research Ltd. 50 p.

Bailey G W and White J L 1970 Factors influencing the adsorption, desorption and movement of pesticides in soil. Res. Rev. 32, 29-92.

Baskaran S 1994 Sorption and movement of ionic and non-ionic pesticides in some selected soils of New Zealand. PhD thesis, Massey University. 276 p.

Bingham 1991a Pentachlorophenol and related wastes in New Zealand, NECAL Service Report S91/327 prepared for the Department of Health. 37 p.

Bingham 1991b PCDD and PCDF impurities in sodium pentachlorophenate based antisapstains. NECAL Service Report S91/328 prepared for the Department of Health. 49 p.

Bowden M J 1986 The Christchurch artesian aquifers. North Canterbury Catchment Board, Christchurch, 159 p.

Bowden M J, Ayrey R B, Duffield D N, Glennie J M, Harrison N, Hurd S B, Mason C R, Talbot J D and Weeber J H 1982 The water resources of the Ashley Catchment, North Canterbury Catchment Board and Regional Water Board, Christchurch.

Brougham G C, Kampman I and Caldwell K J 1985 Groundwater resources of the Waitohu, Otaki and Mangaone. Report No. 65. Manawatu Catchment Board and Regional Water Board, 66 p.

Burden R J 1980 Distribution of nitrate-N concentrations in groundwater beneath intensively grazed pastureland in the Ngatarawa Valley, Hawkes Bay. J. Hydrol. (NZ) 19, 94-105.

Burden R J 1982 Nitrate contamination of New Zealand aquifers. N. Z. J. Sci. 25, 205-220.

Burden R J 1984 Zonation in groundwater of the Central Plains, Canterbury. J. Hydrol. (NZ) 23, 100-119.

Cameron K C and Haynes R J 1986 Retention and movement of nitrogen in soils. In Mineral nitrogen in the plant-soil system. Ed. R J Haynes pp 166-241. Academic Press, Orlando.

Cameron K C and Rate A W 1992 The fate of nitrogen in pig slurry applied to a shallow stony pasture soil. In The use of wastes and byproducts as fertilisers and soil amendments for pastures and crops. Eds. P E H Gregg and L D Currie. pp 314-324. Occasional Report No. 6. Fertiliser and Lime Research Centre, Massey University, Palmerston North.

Campin D N, Buckland S J, Hannah D J and Taucher J A 1990 The identification of dioxin sources in an integrated wood processing facility. Water Sci. Tech. 24, 65-74.

Canadian Council of Ministers 1992 National classification of contaminated sites. Report CCME EPC-CS39E. 54 p.

Chiou C T 1989 Theoretical considerations of the partition uptake of organic compounds by soil organic matter. In Reactions and movement of organic chemicals in soils. Eds. B L Sawhney and K Brown. pp 1-31. Soil Science Society of America, Madison, Wisconsin.

Close M E 1987 Effects of irrigation on water quality of a shallow unconfined aquifer. Water Res. Bull. 23, 793-802.

Close M E and McCallion R F 1988 Groundwater quality and the likely effects of the proposed hydro-electric development, Clutha Valley, Otago. Publication No. 16, Hydrology Centre, Christchurch, 41 p.

Close M E 1992 An assessment of pesticide contamination of ground water in New Zealand. Research Report No 238, DSIR Geology and Geophysics, Christchurch. 38 p.

CMPS&F 1995 Audit of timber treatment sites. Draft Environment Site Assessment Report prepared for the Canterbury Regional Council.

Cohen S Z, Eiden C and Lorber M N 1986 Monitoring groundwater for pesticides. In Evaluation of pesticides in ground water. Eds. W Y Garner, R C Honeycutt and H N Nigg. ACS symposium Series 315. American Chemical Society, Washington DC. 37-42.

Cooke J G 1988 Sources and sinks of nutrients in a New Zealand hill country pasture catchment. II Phosphorus. Hydrol. Proc. 2, 122-134.

Cooper A B and Thomsen C E 1988 Nitrogen and P in streamwaters from adjacent pasture, pine and native forest catchments. N. Z. J. Mar. Freshwater Res. 22, 279-291.

Department of Health 1988 (revised 1989) Safe management of PCBs; Code of Practice; PCB Core Group, Hazardous Wastes Task Group. 39 p.

Dewhurst R H 1981 Thames Valley groundwater investigations. Technical Report No. 106, Hauraki Catchment Board.

Dillon P J, Close M E and Scott R J 1989 Diffuse source nitrate contamination of groundwater in New Zealand and Australia. Hydrology and Water Resources Symposium, Christchurch, pp 351-355.

Felsot A and Dahm P A 1979 Sorption of organophosphorus and carbomate insecticides by soils. J. Agric. Food Chem. 27, 557-559.

Fenemor A D 1987 Water quality of the Waimea Plains aquifers, 1971-86. Nelson Regional Water Board. 34 p.

Field T R O, Ball P R and Theobald P W 1985a Leaching of nitrate from sheep-grazed pastures. Proc. NZ Grassland Assoc. 46, 209-214.

Field T R O, Theobald P W, Ball P R and Clothier B E 1985b Leaching losses of nitrate from cattle urine applied to a lysimeter. Proc. Agron. Soc. of NZ 15, 137-141.

Forestry Corporation of New Zealand July 1993 Waipa contaminated groundwater remediation, resource consents environmental effects assessment.

Fox M E and Wilcock R J 1988 Soil, water and vegetation residues of 2,4,5-T applied to a New Zealand hillside. N. Z. J. Agric. Res. 31, 347-357.

Francis G S, Haynes R J, Sparling G P, Ross D J and Williams P H 1992 Nitrogen mineralisation, nitrate leaching and crop growth following cultivation of a temporary leguminous pasture in autumn and winter. Fert. Res. 33, 59-70.

Fraser P M, Cameron K C and Sherlock R R 1994 Lysimeter study of the fate of nitrogen in animal urine returns to irrigated pasture. Eur. J. Soil Sci. 45, 439-448.

Freeman M C 1988 Report on pesticides in land drainage water from a Canterbury orchard. Unpublished report, North Canterbury Catchment Board.

Gibbs N 1992 Report on dieldrin dumping in the 1960's. Ministry for the Environment (unpublished internal report).

Gregg P E H, Hedley M J, Cooper A B and Cooke J 1993 Transfer of nutrients from grazed pasture to aquatic surface waters and their impact on water quality. Proceedings of the New Zealand Fertiliser Manufacturer's Conference 22, 64-75.

Harrison D L 1970 Environmental residues under New Zealand conditions. Proc. N. Z. Weed and Pest Control Conf. 23, 118-123.

Hart P B S and Speir T W 1992 Agricultural and industrial effluents and wastes as fertilisers and soil amendments in New Zealand. In: The use of wastes and byproducts as fertilisers and soil amendments for pastures and crops. Eds. P E H Gregg and L D Currie. pp 69-90. Occasional Report No. 6. Fertiliser and Lime Research Centre, Massey University, Palmerston North.

Heng L K, White R E, Bolan N S and Scotter D R 1991 Leaching losses of major nutrients from a mole-drained soil under pasture. N. Z. J. Agric. Res. 34, 325-334.

Hoare R A 1986 Groundwater nitrate in the Hamilton Basin. Waikato Valley Authority, Tech. Rep. 1986/16. 18 p.

Holmgren G C S, Meyer M W, Chaney R L and Daniels R B 1993 Cadmium, lead, zinc, copper and nickel in agricultural soils of the United States of America. J. Environ. Qual. 22, 335-348.

Johnson R M and Sims J T 1993 Influence of surface and subsoil properties on herbicide sorption by atlantic and coastal plain soils. Soil Sci. 155, 339-348.

Lambert M G, Devantier B P, Nes P and Penney P E 1985 Losses of nitrogen, phosphorus and sediment in runoff from hill country under different fertiliser and grazing management regimes. N. Z. J. Agric. Res. 28, 371-379.

Loe Pearce & Associates 1993 The collection, storage and disposal of dieldrin and other prohibited insecticides in Canterbury during the 1960's. A report prepared for the Canterbury Regional Council. 20 p.

Loe Pearce & Associates 1994 The collection, storage and disposal of dieldrin and other prohibited insecticides in New Zealand in the 1960's. Findings from the research of historical information. A report prepared for the Ministry for the Environment. 22 p.

McColl R H S 1978 Lake Tutira, the use of phosphorus loadings in a management study. N. Z. J. Mar. Freshwater Res. 12, 251-256.

McColl R H S 1982 Water quality in agricultural areas - the prospects for management of diffuse pollution sources. In: Waters in New Zealand's future. Proceedings of the National Water Conference 4, 153-160.

McIntyre A, Allison N and Penman D 1989 Pesticides: issues and options for New Zealand. A Discussion Document prepared for the Ministry for the Environment. 208 p.

McLaren P 1992 (revised 1993) Soil contamination at CCA timber treatment plants - an initial appraisal in the Bay of Plenty; Technical Report No 27; Bay of Plenty Regional Council.

McLaren R G and Cameron K C 1990 Soil Science: an introduction to the properties and management of New Zealand soils. Oxford University Press, Auckland, 294 p.

Madhun Y A 1986 Binding of herbicides by water soluble organic materials from soil. J. Environ. Qual. 15, 64-68.

Mallawatantri A P and Mulla D J 1992 Herbicide adsorption and organic carbon contents on adjacent low input versus conventional farms. J. Environ. Qual. 21, 546-551.

Merry R H 1988 Investigations on cadmium in South Australia: rainfall, soils, cereals, pastures and soil-plant relations. In Cadmium accumulations in Australian agriculture Eds. J Simpson and W Curnow. pp 62-79. Bureau of Rural Resources Proceedings No. 2.

Ministry for the Environment 1992a Report of the National Task Group investigating site contamination from the use of timber treatment chemicals, A Report to the Minister for the Environment.

Ministry for the Environment 1992b Potentially contaminated sites in New Zealand: a broadscale assessment.

Ministry for the Environment 1992c. The New Zealand Waste Analysis Protocol.

Ministry for the Environment 1992d Landfill Guidelines (an in Part 4 Landfill Engineering Guidelines of Our Waste: our responsibility, Centre for Advanced Engineering, University of Canterbury 1992).

Ministry for the Environment 1993a Steering Committee on sites contaminated by timber treatment and antisapstain chemicals, A Report to the Minister for the Environment. 10 p.

Ministry for the Environment 1993b Draft National Rapid Hazard Assessment System for Potentially Contaminated Sites.

Ministry for the Environment/Department of Health 1992 Draft Health and environment guidelines for selected timber treatment chemicals.

Ministry for the Environment 1994a Hazardous waste handbook: a guide for local authorities.

Ministry for the Environment 1994b Proposals for regulation under the Hazardous Substances and New Organisms Bill: discussion document. Government Printer, Wellington. 89 p.

Ministry of Health 1995 Drinking Water Standards for New Zealand 1995. Ministry of Health, Wellington. 87 p.

Ministry of Housing, Spatial Planning and Environment 1994 Environmental Quality Objectives in the Netherlands. A review of environmental quality objectives and their policy framework in the Netherlands. 465 p.

Mohammed I H, Gregg P E H and Gandar P W 1987 Measuring and modelling nitrogen transformations in a tile drained Fragiaqualf under arable cropping. N. Z. J Agric Res. 30, 621-628.

Morton J and Butcher M 1990 DDT residues in Canterbury soils. Confidential Interim Unpublished Report, MAFTechnology, Lincoln, New Zealand. 9 p.

Morton J, Roberts A H C and Edmeades D C 1994 Fertiliser use on sheep and beef farms. New Zealand Fertiliser Manufacturers Research Association miscellaneous publication. 38 p.

Murty A S 1986 Toxicities of Pesticides to Fish. Vol 1. CRC Press Inc., Boca Raton, Florida, 70 p.

New Zealand Forestry Statistics 1993 Government Printer, Wellington.

New Zealand Statistics Department 1993 Agriculture. 86 p.

New Zealand Statistics Department 1994 New Zealand Official Yearbook. 587 p.

Nicholson B C 1984 Australian water quality criteria for organic compounds. Australian Water Resources Council technical paper No 82. Department of Resources and Energy, Australian Water Resources Council, Canberra. 224 p.

Occupational Safety and Health Service 1992 Code of practice for the design, installation and operation of underground petroleum systems. Department of Labour. 56 p; 1995 Supplement No. 1 Management of existing underground petroleum storage systems. 28 p.

Occupational Safety and Health Service 1994 Approved code of practice for the safe use of timber preservatives and antisapstain chemicals. Department of Labour. 73 p.

O'Connor B P 1993 Ed. New Zealand Agrichemical and Plant Protection Manual. 6th edition, Novasearch, Palmerston North, New Zealand. 287 p.

O'Dea T F 1980 Wairarapa groundwater study report by scientific advisory group. Wairarapa Catchment Board. 26 p.

Pimentel D, Krummel J, Gallahan D, Hough J, Merrill A, Schreiner I, Vittum P, Koziol F, Back E, Yen D and Fiance S 1981 A cost-benefit analysis of pesticide use in US food production. In Handbook on Pest Management in Agriculture. Ed. D Pimentel. pp 27-54. CRC Press Inc., Boca Raton, Florida.

Prins W H, Dilz K and Neetsen J J 1988 Current recommendations for nitrogen fertilisation within the EEC in relation to nitrate leaching. Proceedings No. 276, The Fertiliser Society, London. 27 p.

Quin B F and Burden R J 1979 The effects of land use and hydrology on groundwater quality in mid-Canterbury. NZ Prog. Water Tech. 11, 433-448.

Roberts A H C, Ledgard S F, O'Connor M B and Thomson N A 1992 Effective use of N fertiliser – research and practice. Proceedings Ruakura Dairy Farmers' Conference: p 77-82.

Roberts A H C, Longhurst R D and Brown M W 1994 Cadmium status of soils, plants and grazing animals in New Zealand. NZ J. Agric. Res. 37, 119-129.

Rothbaum H P, Goguel R L, Johnson A E and Mattingly G E G 1986 Cadmium accumulation in soils from long continued applications of superphosphate. J. Soil Sci. 37, 99-107.

Royds Garden Environmental Service/CMPS&F 1994 Site investigation report for the former New Zealand Forest Service site, Jollies Pass Road, Hanmer Springs. Report prepared for the Ministry for the Environment. 64 p.

Schipper L A, Cooper A B, Harfoot C G and Dyck W J 1993 Regulators of denitrification in an organic riparian soil. Soil Biol. Biochem. 25, 925-933.

Selvarajah N, Maggs G R, Crush J R and Ledgard S F 1994 Nitrate in groundwater in the Waikato region. In The Efficient Use of Fertilizers in a Changing Environment Eds. L D Currie and P P Loganathan. pp 160-185. Occasional Report No 7. Fertilizer and Lime Research Centre, Massey University, Palmerston North.

Sharpley A N and Syers J K 1979 Loss of nitrogen and phosphorus in tile drainage as influenced by urea application and grazing animals. N. Z. J. Agric. Res. 22, 127-131.

Sharpley A N and Syers J K 1979a Effect of aerial topdressing with superphosphate on the loss of phosphate from a pasture catchment. N. Z. J. Agric. Res. 22, 273-277.

Sharpley A N and Syers J K 1979b Phosphorus inputs into a stream draining an agricultural watershed. II: Amounts contributed and relative significance of runoff types. Water, Air Soil Poll. 11, 417-28.

Shaw C 1990 Timber preservation chemicals in the New Zealand environment: pentachlorophenol and chlordane, Cawthron Institute, Nelson. 52 p.
Smith C M, Wilcock R J, Vant W N, Smith, D J and Cooper A B 1993 Towards sustainable agriculture: freshwater quality in New Zealand and the influence of agriculture. MAF Policy Technical paper 93/10. p 103.
Solly S R B and Shanks V 1969 Organochlorine insecticides in rainbow trout from three North Island lakes. NZ J. Mar. Freshwater Res. 3, 585-590.
Steele K W 1982 Nitrogen in grassland soils. In Nitrogen fertilisers in New Zealand agriculture. Ed. P B Lynch. pp 29-44. New Zealand Institute of Agricultural Research, Wellington.
Steele K W, Judd M J and Shannon P W 1984 Leaching of nitrate and other nutrients from a grazed pasture. N. Z. J. Agric. Res. 27, 5-11.
Swaine D J 1962 Cadmium. C. A. B. Soil Sci. Tech. Comm. 52, 52-54.
Taranaki Regional Council 1993 Investigations of possible dieldrin storage and disposal sites in Taranaki. Technical Report 93-1. A report to the Ministry for the Environment.
Taylor N H and Pohlen I J 1968 Soils of New Zealand Part 1. NZ Soil Bureau Bull. 26, 7-14.
Taylor N H and Pohlen I J 1970 Soil survey method. NZ Soil Bureau Bull. 25, 242 p.
Thorrold B S and Smith C M 1994 The impact of fertilisers on surface water quality. Unpublished report, AgResearch and NIWA, Hamilton, New Zealand. 29 p.
Tiller K G 1992 Urban soil contamination in Australia. Aust. J. Soil Res. 30, 937-957.
Tonkin & Taylor Ltd. 1994a Pesticide storage site investigations. A report prepared for the Canterbury Regional Council.
Tonkin & Taylor Ltd 1994b Old Auckland Landfills (Report 3) Volume 1 Technical Report, Report No. 11867.
Tonkin & Taylor Ltd 1994c Old Auckland Landfills (Report 3) Volume 1 Environmental Management Strategies for Priority 1 Sites, Report No. 11867.
Tonkin & Taylor Ltd 1994d Wellington Regional Council Contaminated Site Management Strategy and Action Plan, Report No. 12470.
Tonkin & Taylor Ltd 1995a Caltex Oil (NZ) Ltd (unpublished reports).
Tonkin & Taylor Ltd 1995b Wellington Regional Council Draft Contaminated Site Handbook, Report No. 12470.
Turner M A, Tillman R W and Syers J K 1979 Losses of phosphorus and nitrogen in subsurface drainage waters as influenced by phosphate fertiliser, grazing animals and irrigation treatment. Prog. Water Tech.11, 287-294.
Waikato Regional Council 1992 Regional agrichemicals collection manual. Technical Report 192/19. 52 p.
Wellington Regional Council 1994 Potentially Contaminated Sites, Hazard Assessment System, Publication No. WRC/CI-T-94/12.
Whittington B Ed. 1995 Situation and outlook for New Zealand agriculture. Ministry of Agriculture and Fisheries, Wellington, New Zealand. pp 133.
Wilcock R J 1986 Agricultural runoff, a source of water pollution in New Zealand? NZ Agric. Sci. 20, 98-103.
Wilcock R J 1989 Patterns of pesticide use in New Zealand. Part 1. North Island 1986-1989. Water Quality Centre Publication No 15, Division of Water Sciences, DSIR. pp 21.
Wilcock R J 1993 Application of land-use data and screening tests for evaluating pesticide runoff toxicity in surface waters. Environ. Management. 17, 15-18.
Wilcock R J and Close M E 1990 Patterns of pesticide use in New Zealand. Part 2. South Island 1986-1989, Water Quality Centre Publication No 16, Division of Water Sciences, DSIR. 37p.
Wilcock R J, Costley K J, Cowles R J, Wilson B and Southgate P 1991 Stream runoff losses and soil and grass residues of trichlopyr applied to hillside gorse. NZ J. Agric. Res. 34, 351-357.
Wilcock R J, Pridmore R D, Thrust S F, Smith T J, Langdon A G and Wilkins A L 1989 Chlordane in the Manukau Harbour. Paper presented to the Symposium of the Wood Preservation Industry and the Environment, Forest Research Institute.
Williams C H and David D J 1976 The accumulation in the soil of cadmium residues from phosphate fertilisers and their effect on the cadmium content of plants. Soil Sci. 121, 86-93.
Woodward-Clyde 1993a Phase II Site invesigation - Mapua. A report prepared for the Tasman District Council.
Woodward-Clyde 1993b Environmental site investigation: pesticide dumps in Southland. Prepared for the Southland Regional Council.
Woodward-Clyde 1994 Soil and groundwater investigation, Mapua. An unpublished report prepared for Bell Gully Buddle Weir.

21
Contaminants and the soil environment of Pakistan

Z. HUSSAIN, M. R. CHAUDHRY, F. A. ZUBERI, Q. HUSSAIN AND M. SHARIF

1. Introduction

Pakistan occupies the basin of the eastern-most of the three great rivers that traverse the steppe desert of the old world, the Nile, the Tigris-Euphrates and the Indus. That these basins were the cradles of early civilisation gives to the Indus a distinctiveness that is lacking in other river basins in South Asia (Johnson, 1979). It lies between latitudes 24 and 37° N and between 61 and 75° E longitudes. On its east lies India; to the north and north west is Afghanistan; to the west is Iran; and in the south is the Arabian Sea. It has a common frontier in the north with China on the border of Gilgit Agency. Tajikistan is separated from Pakistan by a narrow strip of Afghan territory called Wakhan.

Emerging as an independent nation in 1947, the country occupies an area of about 803,900 km^2 of which about 50% comprises rugged mountainous terrain, narrow valleys and foot hills (GOP, 1983). It comprises four provinces, namely North West Frontier Province (NWFP), Punjab, Sindh, and Balochistan, and Federally Administered Northern Areas and Tribal areas (FATA), with the federal capital at Islamabad. The provinces are further divided for administrative purposes into 24 Divisions, 78 Districts, 283 Tehsils, 45 165 Mauzas (revenue village) and 13 Agencies (tribal areas). The population of Pakistan is increasing at 3% yr^{-1} and was estimated to be 124 million in 1994.

Agriculture is the major economic activity in Pakistan. The rapid increase in population together with increased demand for food has led to greater use of agrochemicals and canal waters for irrigation. Seepage from the canal system and greater use of farm irrigation has added significantly to the normal recharge, resulting in the rise of groundwater. The shallow watertable in almost 12% of the area and the extensive use of fertilisers and pesticides, combined with the unplanned disposal of industrial and domestic effluent, have greatly increased the threat of soil pollution. This paper presents an overview of issues relating to contaminants and the soil environment in Pakistan.

2. Climate

The country is characterised by a continental climate which is arid to semi-arid with two distinct seasons – summer and winter. Temperature ranges from a mean monthly minimum of 4°C during December and January to a mean monthly maximum of 38°C during June and July. The incidence of annual rainfall in Pakistan is highly

R. Naidu et al. *(eds): Contaminants and the Soil Environment in the Australasia-Pacific Region, 629–646.*

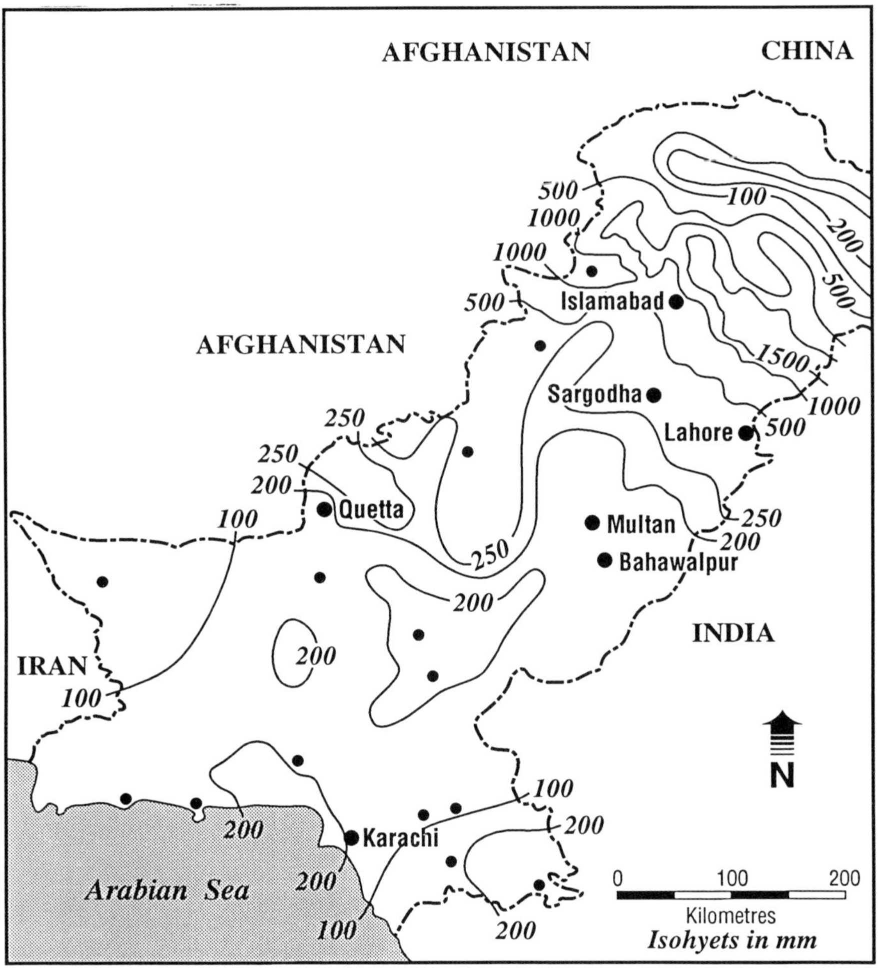

Figure 1. Distribution of annual rainfall in Pakistan.

variable with less than 100 mm in some parts of the Sindh to more than 500 mm near the foothills in the NWFP and between 760 and 1270 mm on the southern slopes of the Himalayas (Mohammad, 1994). Most of the rainfall occurs during the monsoon period (July to September) in the form of torrential showers, but the water is not generally available for agriculture because of rapid runoff. At other times the showers may be so light that precipitation evaporates before the water can penetrate to the root zone of crops. Rates of evaporation range between 1500 and 2000 mm annually. The pattern of rainfall distribution is shown in Figure 1.

Table 1. Groundwater quality of Indus Basin.

Category	Criteria	% of Total Irrigated Area (16.4×10^6 ha)
Usable	$1500\,mg\,L^{-1}$	49
Marginal	$1500-3000\,mg\,L^{-1}$	12
Hazardous	$>3000\,mg\,L^{-1}$	39

Source: Pakistan ICID Country Report, 1991.

3. Groundwater

Under the irrigated area of 16 400 000 ha about 49% has fresh groundwater with less than $1500\,mg\,L^{-1}$ total dissolved solids (TDS), about 12% has groundwater of moderate salinity with 1500 to $3000\,mg\,L^{-1}$ TDS and 39% has high salinity with greater than $3000\,mg\,L^{-1}$ TDS. At present, more than 2.5×10^5 shallow tubewells in the private sector and 15000 deep tubewells in the public sector are pumping more than 444×10^6 ha-cm of water at the farm gate. The aquifer is recharged from rivers, canal seepage, applied irrigation and rain water. The aquifer has very favourable physical characteristics and groundwater can be pumped from it quite economically. The distribution of groundwater salinity is reported in Table 1.

4. Soils

There are great variations in the soils of Pakistan. Reconnaissance surveys carried out by the Soil Survey of Pakistan covering $5.40\times10^5\,km^2$ have established about 400 distinct soil series. Of these, some occupy extensive areas whereas others are of limited extent. The clay minerals are mainly illitic and kaolinitic.

The cation exchange capacities (CEC) of the soils range from 80 to $160\,mmol\,kg^{-1}$ of soil. Calcium and magnesium carbonates occur in the range from 5 to 20%. The pH (saturation paste extract) of normal soils (not affected by salinity and sodicity) vary between 7.8 to 8.3, although values of up to 10.5 have been recorded in highly alkaline areas. Normal soils (pH range 6.5–8.0) have satisfactory permeability, infiltration rate and other chemical and physical characteristics.

With respect to fertility status the soils are characteristically deficient in N, OM and P. There are some pockets of K deficiency as well, especially in the rice growing tracts and the sub-mountain areas. Micronutrients like Fe, Cu, Mn, and Zn were previously adequate but with the passage of time, due to intensive cropping, soils have become deficient in these nutrients in some irrigated areas (Rashid, 1993). Soil and plant analyses show that more than 50% of the cultivated soils of the country are unable to supply sufficient Zn and B to meet the needs of the majority of conventional crops (Khattak, 1995). In another study, Kausar (pers comm) collected 150 soil samples from four provinces. The DTPA analysis of these soils revealed 69, 55, 45 and 17% Zn deficiency in samples from Baluchistan, NWFP, Punjab, and Sindh, respectively. Copper was deficient in 28, 15, 5 and 14% of

Table 2. Micronutrient status of soils in Pakistan.

Province	No. of Samples	% Deficient Sites	
		Zn	Cu
Punjab	40	45	5
NWFP	38	55	15
Baluchistan	37	69	28
Sindh	35	17	14

Source: Kausar pers. comm.

samples, respectively (Table 2), while Fe and Mn were adequate for plant growth in all the soils.

5. Sources of soil contamination

As with most of the developing countries, there are a wide range of soil contaminants in Pakistan. Some of these include inorganic metal and non-metal contaminants arising from both agricultural and industrial practices. This report, however, focuses only on those activities that are creating serious environmental problems. These are discussed below:

5.1 Waterlogging

While waterlogging is generally not classed as a contamination process, shallow watertables and irrigation that leads to buildup of salt is considered a major limitation to crop production in Pakistan. The salts are a source of continuing pollution of the soil. Before the introduction of weir-controlled irrigation, lands adjacent to the rivers were irrigated through inundation canals. Watertables were fairly deep over most of the area except near the confluence of the rivers and under narrow marginal plains along the river courses that were flooded each year during the monsoons. The infiltration of water from the rivers and the deep percolation of rainfall, and the water applied in the seasonal inundation irrigation, was in equilibrium with the discharge of groundwater by evapotranspiration and by water movement out of the area towards the sea.

As soon as the permanent canal system was introduced, the dynamic equilibrium between groundwater recharge and discharge was disturbed. The deep percolation of water from canals and irrigated lands constituted a new increment of recharge which exceeded the rate at which water could be discharged from the aquifer. As a result, watertables rose close to the surface, creating waterlogged conditions, and the drainage capacity of the soils was reduced, affecting the agricultural productivity of the soils. The historic trend in the use of watertables is shown in Table 3.

It is evident from this table that the watertable in almost 9 to 18% of the irrigated areas is within 1.5 m of the surface. The major cause of this rise in watertables is

Table 3. Area under various depths to watertable in Pakistan (April/June). (% of Canal Command Area).

Year	<1.5 m	1.5–3.0 m	>3.0 m
1978	11.9	39.5	48.6
1979	15.2	39.3	45.5
1980	12.9	39.5	47.6
1981	12.7	42.4	42.9
1982	13.5	43.2	43.3
1983	13.5	41.9	44.6
1984	12.5	41.8	45.7
1985	11.0	39.6	48.4
1986	13.0	41.0	46.0
1987	12.8	41.0	42.2
1988	9.0	38.2	52.8
1989	14.4	34.9	50.7
1990	13.2	36.2	50.6
1991	13.0	39.0	48.0
1992	18.3	32.6	49.1
1993	16.2	35.7	48.1
1994	12.0	36.0	52.0

Source: Pakistan ICID Country Report, 1991.

poor management of irrigation systems, through which only 35–40% of the irrigation water is stored in the root zone for crop use. Approximately 60–65% of irrigation water is lost during conveyance at various levels such as canals, distributaries, water courses and application losses in farmers' fields. Clearly better management of irrigation water can lead to enhanced productivity. Historical data show that watertables have risen, due to the seepage from reservoirs and irrigation systems at an average rate of 15 to 35 $cm\,yr^{-1}$ since modern irrigation was introduced. This high watertable situation results in contamination of the soil profile with salts as well as injurious impacts on soil productivity due to anaerobic conditions. Further, in an area where the groundwater has a salinity of 1 000 $mg\,L^{-1}$, which is acceptable for all crops, evaporation at the rate of 500 $mm\,yr^{-1}$ (a typical value where water is close to the surface), will raise the salt content of the top metre of the groundwater to about 1% in 20 years (Spooner, 1982).

5.2 *Soil salinity*

Long before the introduction of irrigation systems an appreciable amount of salts existed in the Indus Basin, due to the original geological formation of the basin. Later, as a result of salt accumulation, originating from evaporation of floodwater and groundwater, the salinity increased in the upper strata of the soils.

The latest soil salinity survey (1977–79), covering 16.7×10^6 ha of the Indus basin, indicated that, in terms of surface salinity, 12%, 6% and 8% of the area was slightly, moderately and strongly saline, respectively. In 1.85 m deep profiles 38% of the total profiles were affected by salinity/sodicity with 11%, 24% and 3% classified as saline, saline-sodic and sodic, respectively.

Table 4. Total fertilizer (000 mt) consumption in Pakistan.

Fertilizer	1985/6	1986/7	1987/8
Nitrogen Fertilizers			
Ammonium sulphate	19	22	20
Ammonium nitrate	132	110	87
Urea	795	1000	968
Ammonium phosphate	89	116	99
Other	93	85	108
Total	1128	1333	1282
Phosphatic Fertilizers			
Superphosphate	7	8	14
Ammonium Phosphate	228	295	253
Other	9	9	11
Total P_2O_5	244	312	278
Potash Fertilizers			
Potassium Sulphate	23	29	32
Other	10	14	13
Total	33	43	45

Source: Mubarak and Jabbar, 1992.

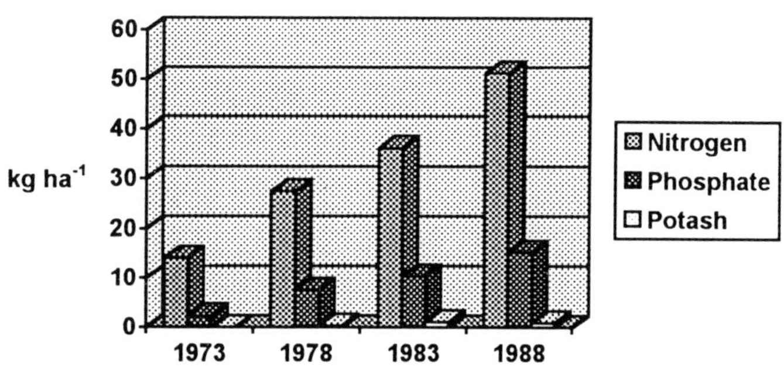

Source: FAO, 1990.

Figure 2. Average fertilizer consumption in Pakistan.

5.3 Agrochemicals

5.3.1 Fertilisers

Fertiliser consumption has been steadily increasing over the years in Pakistan (Table 4). Pakistani soils are deficient in N; hence the emphasis has been on supplementary N fertilisers, mainly in the form of urea or ammonium mixtures (Table 4).

Phosphorus and K supplementation is on a limited scale. Fertiliser consumption in Pakistan is shown in Figure 2.

During 1988 N fertilisers were applied at 51.2 kg ha^{-1} whereas P fertilisers were used at 15.1 kg ha^{-1} and K at 0.9 kg ha^{-1}. The average total fertiliser consumption in

1988 was at 67.2 kg ha^{-1}. Irrigation water contributes to groundwater N pollution because percolating water picks up N from applied fertiliser. Although N in the form of NO_3^- is one of the most common causes of contamination of groundwater in the world, limited information is available on the NO_3^- content of groundwater influenced by irrigation waters and fertiliser practices in Pakistan. Since the green revolution in the 1960s extensive use of fertilisers has boosted agricultural productivity and because of the shallow watertable in large areas there is considerable movement of chemicals to groundwaters.

5.3.2 Pesticides

In Pakistan, pesticide use began in 1954, based on minimum curative applications and on controlling epidemic infestations. Insecticides, particularly organophosphates, dominate the market and make up 85% of the total pesticides, with herbicides at 6%. About 75% of the insecticides are used on the cotton crop alone and the rest is used on rice, sugarcane, maize and vegetables etc. Before 1980, pesticide import and distribution was the responsibility of the Plant Protection Department, Government of Pakistan, but the pesticide business was then transferred to the private sector. Table 5 lists pesticides currently produced in Pakistan.

Since local production commenced in Pakistan, pesticide consumption increased from 9.1×10^5 kg a.i. (active ingredient) in 1981 to 5.3×10^6 kg a.i. in 1990. The sprayed area has also increased from 1.8×10^6 ha to about 5.5×10^5 ha. Due to the complexities of cropping systems and small holdings, ground spraying has always been preferred, while aerial spraying has been restricted to epidemic outbreaks of pests like locusts [*Pachytylus malgratorius*], sugar pyrilla [*Pyrilla Perpusilla*], white-backed planthoppers [*Sogatella furcifera*], crickets [*Nemabius Fasciatus*], etc. Herbicide use, though proportionally small, is increasing. There is a more or less steady import of fungicides, although the proportional share is decreasing. Use of rodenticides and fumigants, though considered a small component, is also increasing. It is evident from Table 6 that pesticide use has increased four fold in eight years to 1990.

The extensive use of pesticides in Pakistan has given rise to environmental concerns such as residues in food and in soil, pest resistance and health hazards to the end users. For example in the 1980's dicholorodiphenyltricholoroethene (DDT) and benzene hexachloride (BHC) and other persistent organochlorines and organophosphates were extensively used to control malarial vectors and other pests. As a result, residues (22–200 mg L^{-1}) were detected in vegetables such as ladyfinger (Okra; *Abelmoschus esculentos*) and cauliflower (*Brassica oleracea*) (ADB Handbook, 1987). Insecticides whose registration has been cancelled and that have been banned from further use in Pakistan are chlordimeform, dibromochloropropane (DBCP), endrin andleptophos. Although legislation provides a comprehensive framework through the Department of Plant Protection and the Provincial Governments, the actual implementation of the regulations is uneven. As a result, conflicting issues arise which hamper the implementation of rules for protection of the environment.

Table 5. Production and formulation of pesticides in Pakistan since 1988 (Mubarik and Jabbar, 1992).

Pesticides Active ingredients	Amount Production (t)	Formulations	Chemical names
Cartap (Padan)	720	95% Tech., 4G & 10G	S.S,-2-dimethylamino-trimethylenene bis(thiocarbamate)
Dieldrin	100	80% tech 20EC 20 oil sol	(IR,4S,5S,8R)-1,2,3,4,10,10-hexachloro-1,4,4a,5,6,7,8,8a-octahydro-6,7-epoxy-1,4:5,8-dimethanonaphthalene
Chlorfenvinphos (Birlane)	100	90% tech., 10G & 24EC	2-chloro-1-(2,4-chlorophenyl)-vinyl diethylphosphate
Dictrophos (Bidrin)	100	86 WSC	dimethyle phosphate of 3-hydroxy-N_2N-dimethyl-cis-crotonamide
Monocrotophos	600	75.2 Tech., 40 WSC	(E)-1-methyl-2-methylcarbomoylvinyl phosphate
Dimethoate (Dimetoxal)	300	96% Tech., 40EC	0,0-diniethyl S-methylcarbameylmethyl phosphorodithioate
Cypermethrin	300	90% Tech., 10EC	(RS)-a-cyano-3-phenoxybenzyl
Dimethoate (Cygon)	300	96% Tech., 200 EC, 400 EC	0,0-dimethylcarbomoylmethyl phosphorodithloate
Endosulfan	1 652	94% Tech., 70% Conc., 35 EC, 5G & 25 ULV	1,4,5,6,7,7-hexachloro-8,9,10-trinorborn-5-EN-2,3-ylenedimethyl sulphite
Methamidophos	300	73 Tech., 60 SL	0,S-dimethyl phosphoramidothioate
Disulfoton	1 340	85% Tech., 10 G	0,0-diethyl S-2-ethylthioethyl phosphorodithioate
Carbofuran	300	75% Tech., 3 G	2,3-dimethyl-benzofuran-7-yi methylcarbamate
Paraquat Dichloride	300	Gramoxone 20 g/L	1,1-dimethyl-4,4-bipyridyldiylium ion
Deltamethrin	1 352	2,5 EC, 0.5 ULV, 1.25 ULV	0,0-dimethyl S-methylcarbamoylmethyl Phosphorodithioate
Triazophos	1 352	70% Conc., 40 EC, 25 ULV	0,0-diethyl 0-1-phenyl-1,2,4-triazol-3-yl phosphorothioate
Quinalphos	1 040	70% ZP Tech., 5G, 25 EC, 30 ULV, 245 FW	0,0-diethyl o-quinoxalin-2-yl phophorothioate
Formathion	1 040	70 ZP, 25 EC, 35 ULV	S-(N-formyl-N-methylcarbamoylmethyl
Fenvalerate	1 040	95% Tech., 20 EC, 2.5 ULV	(RS)-a-cyano-3-phenoxybenzyl

Pesticide application can pose problems as a non-point pollution source, as some insecticides and herbicides degrade slowly and can persist in soils. As the watertable in more than 10% of the area of cultivated land is within 1.5 m of the ground surface, increased use of both insecticides and herbicides together with their potential for downward leaching and surface flushing has become a major threat to the underlying aquifers.

Table 6. Pesticide consumption (000 kg or L a.i.) in Pakistan.

	1982	1984	1986	1988	1990
Insecticide	952 (74)	2162 (88)	3480 (87)	4173 (85)	4265 (85)
Fungicide	171 (13)	204 (8)	283 (7)	277 (6)	365 (7)
Herbicide	94 (7)	102 (4)	162 (4)	277 (6)	365 (7)
Acaricide	3 (–)	51 (2)	81 (2)	82 (2)	– –
Rodenticide	69 (6)	25 (1)	41 (1)	44 (1)	27 (1)
Total	1289	2544	4047	4853	5296

Source: Mubarik and Jabbar, 1992.
(*Values in parenthesis represent percent of the total).

Table 7. Concentration of pesticides in groundwater from selected sites in Pakistan.

Site No.	Depth (m)	Pesticide	Concentration ($\mu g\,L^{-1}$)
1	12	Monocrotophos	40
2	14	Endrin	0.2
		Monocrotophos	60
3	14	Cyhalothrin	trace
		Monocrotophos	50
		Endrin	0.2
5	14	Cyhalothrin	trace
6	11	Endrin	0.2

Mubarik and Jabbar (1992) studied water samples from a shallow aquifer in a cotton growing area. Results from these studies revealed that seven out of ten water samples were contaminated with one or more pesticides (Table 7). Three samples contained monocrotophos in the range of 40 to 60 $\mu g\,L^{-1}$. Four samples had cyhalothrin ranging from traces to 0.2 $\mu g\,L^{-1}$. Endrin was present in three samples at concentrations of 0.1 to 0.2 $\mu g\,L^{-1}$.

In the same study, to determine pesticide contamination in soils, samples were taken from 3, 7 and 10 m depths. Results from five different sites in the cotton and wheat growing areas revealed that generally soils were contaminated by varying amounts (1.5 to 14.7 $\mu g\,kg^{-1}$) of different insecticide residues. Organophosphate and pyrethroid insecticide residues were restricted to the top 30 cm of soil. Praveen and Masud (1988) reported the presence of chlorinated pesticides or their metabolites in cattle drinking water and some samples were found to contain γBHC in the range of 1.0 to 16.4 $mg\,L^{-1}$. The presence of pesticides in cattle drinking water is of concern given that such chemicals can have adverse impacts on both animal and human health. Much research on the fate and behaviour of the range of pesticides currently used in Pakistan is necessary to develop appropriate management strategies for safe application of these chemicals, in order to minimise the potential for their adverse impacts to animals and human beings.

5.4 *Industrial effluents*

Pakistan is basically an agricultural country. However, in the recent past the economy has started shifting towards the industrial sector. A number of large and small industries, based on textiles, tanneries, edible oils, cement, fertilisers and other chemicals, have been established in various parts of the country. These industries are disposing of their effluent into open surface drains or nearby ponds from which the effluent reaches the rivers. The water from these rivers is ultimately used for irrigation of large areas. According to a very rough estimate 40×10^{9} L of wastewater, with 20000 t of biochemical oxygen demand (BOD) loading, are discharged daily into water bodies by the industrial sector (Saleemi, 1993).

A study carried out by the Environmental Protection Agency (EPA, 1990) of the Punjab province shows that the average range of BOD 5 day incubation is 152–2150 mg L^{-1} in textile effluent, 1400–4900 mg L^{-1} in tannery effluent and 150–450 mg L^{-1} in municipal effluent, compared with a national standard of 80 mg L^{-1}. Floating solids and liquids create unsightly films on the surface of water, interfering with natural reactions and causing toxicity in the water. The values observed in ghee (milk butter) mill effluents range from 5 to 50 mg L^{-1} as against a national standard of 10 mg L^{-1}.

Toxic metals such as Cr, Ni, As, Hg, Cu, Pb etc., are poisonous to aquatic organisms. Environmental Protection Authority (EPA, 1990) studies show values for Cr of 2–20 mg L^{-1} in electroplating effluent and 20–300 mg L^{-1} in tannery effluent, compared with a national standard of 10 mg L^{-1} (EPA, 1990). Studies by the EPA also show that the values of NH_3 in fertiliser plant effluent range from 50 to 350 mg L^{-1} compared with the national standard of 40 mg L^{-1} and that sulphide in municipal waste ranges from 22–200 mg L^{-1} and in tannery waste water from 50 to 500 mg L^{-1} compared with the national standard of 1.0 mg L^{-1}. Other heavy metals may also contaminate soils because untreated industrial effluents are discharged to the drainage system.

Most industries do not control their wastewater effluent through process control, waste recycling or end-of-pipe treatment, so effluents with toxic substances are being discharged into water bodies without paying any attention to the environmental consequences. In Kala Shah Kaku (near Lahore) many industries discharge toxic chemicals (HCl and organics) into water bodies. The salinity of these bodies has increased from 500 mg L^{-1} in the early 1970's to nearly 1000 mg L^{-1} in the late 1980s, rising to over 2000 mg L^{-1} during some months. Biological oxygen demand (BOD) levels of 193–833 mg L^{-1}, and a Hg level of 5.6 mg L^{-1} are very high, compared to safe limits of 200 and 0.10 mg L^{-1}, respectively for soil and soil biota. (Table 8).

In another study Khan et al. (1992) reported that Faisalabad city effluent, comprising domestic and industrial wastes, contained toxic materials in addition to plant

Table 8. Water pollution in an open surface drain (Deg nala) receiving industrial effluent.

Water quality	Salinity increase from 500 to 1 000 mg L^{-1} occasionally rising over 2 000 mg L^{-1}
BOD	193–833 mg L^{-1} as against 200 mg L^{-1}
Mercury level	Increased to 5.6 mg L^{-1} as against 0.1 mg L^{-1}
Deterioration	Reduced fish catch by 400 t yr^{-1}

Source: The Pakistan National Conservation Strategy 1986.

nutrients. In this study four sites which had received city effluent for about 5–20 yr were selected, to assess accumulation of heavy metals (Pb, Cd, Cu, Fe and Mn). The DTPA extractable Cu (mg kg^{-1}) in soil profiles varied from 0.94 to 7.01, Pb from 1.48 to 5.33, Cd from 0.94 to 2.90, Fe from 23.30 to 46.91 and Mn from 7.34 to 12.94 in the surface (0 to 15 cm) and subsurface (60 to 90 cm) samples, respectively. The values in normal soils (where effluent was not used for irrigation) ranged for Cu from 0.46 to 1.50, for Pb from 2.11 to 3.73, for Cd from 0.13 to 10.15, for Fe from 3.87 to 12.75 and for Mn from 3.96 to 11.52 mg kg^{-1}. Furthermore they found total accumulation of heavy metals in soils irrigated with sewage water was 39.94, 35.03, 91.06 and 48.01 mg kg^{-1} for different soils irrigated for periods ranging from 5 to 20 yr while the total accumulation of heavy metals in canal water irrigated soils was 18.92 and 18.44 mg kg^{-1}. These results suggest considerable buildup of certain heavy metals in soils subjected to effluent irrigation.

5.5 *Aquatic pollution*

Whether used for drinking or to meet irrigational demands, the quality of water in our riverine system will determine the quality of soils available to grow food for the expanding population. There are about 50 metals that are of special interest for their toxicological importance to human health. Lead and Hg continue to be most important but many others such as As, Be, Cd, Mn, Cr, Ni and V are of increasing toxicological importance.

The metropolitan city of Karachi has major industries at Sind Industrial Trading Estate (SITE), Landi Industrial Trading Estate (LITE), Korangi and West Wharf. The industrial units (2 605) from these sites discharge all their effluents into the adjoining Malir and Lyari rivers both of which link up with the Korangi Creek. Thus, Korangi Creek receives a heavy load of organic effluents and salts of heavy metals which endanger the coastal aquatic life.

Industrial waste pollution surveys carried out by the Government of Pakistan (1989) in industrial areas at six sites, viz. Karachi, Multan, Faisalabad, Kalashah Kaku (Lahore), Noshera, and Peshawar reveal that a number of industries are discharging effluents with high concentrations of pollutants including toxic metal salts, bacteria, acids and oils, into water bodies without treatment, thus causing soil and water contamination (Table 9A–9G).

Table 9A. Toxic substances (mg L^{-1}) in effluents of sample industries in Karachi.

Sr. No.	Name of the Industry	Cu	Cd	Zn	Ni	Mn	Pb
1.	Ahmad Food Industries	0.430	0.028	0.238	0.270	0.073	0.230
2.	Burma Oil Mills	0.027	0.028	2.187	0.648	0.217	0.480
3.	Karachi Beverage Co. Ltd.	0.090	0.035	2.060	0.407	0.129	0.0347
4.	Adamjee Textile Mills	0.020	0.045	5.320	0.514	0.085	–
5.	Karachi Tannery	0.300	0.153	7.000	1.142	0.440	1.800
6.	Indus Alkalies	0.138	0.028	0.220	1.180	0.258	0.660
7.	Buxley Paints	0.065	0.940	0.480	0.203	0.073	3.880
8.	Karachi Shipyard	0.280	0.100	1342.500	0.740	5.300	11.750
9.	Javedan Cement Industries	0.330	0.330	2.660	1.000	5.130	0.790

Table 9B. Pollution load in effluents of sample industries in Karachi.

Sr. No.	Name of the Industry	Discharge m.g.d.	Temp °C	Colour Units	pH Value	S.S. mg L^{-1}	COD mg L^{-1}	BOD kg d^{-1}
1.	Ahmad Food Industries	0.004	30	65	6.0	218	1778	7.7
2.	A.G. Fisheries	0.3	20	20	7.3	4420	11075	–
3.	Burma Oil Mills	0.53	42	20	7.6	1160	4296	4210
4.	Adamjee Textile Mills	0.45	45	100	10.3	1390	5530	340
5.	Dadabhoy Paper Mills	0.29	30	60	7.0	6	126	878
6.	Karachi Tannery	0.015	32	320	4.1	1140	6784	26
7.	Indus Alkalies	0.38	38	20	9.2	3560	2235	–
8.	Buxley Paints	0.003	31	60	2.1	8	8080	–
9.	Karachi Shipyard	0.1	34	70	2.1	286	30	–
10.	Javedan Cement Industries	0.005	40	65	6.3	7740	54	–

The data in Tables 9A–9G indicate that all the industries currently in operation dispose effluents with varying heavy metal concentrations and other undesirable pollutants. Although such disposals have led to extensive degradation of the soil and water environment, limited effort has been directed towards imposing regulatory measures for safeguarding the environment. Unless regulatory measures are developed to control industrial waste disposals, the current rate of environmental degradation will increase, jeopardising the use of valuable agricultural land and groundwater resources.

The water bodies in urban and rural areas collectively serve for the disposal of

Table 9C. Pollution load in effluents of sample industries in Multan.

Sr. No.	Name of the Industry	Discharge m.g.d.	Temp °C	Colour Units	S.S. mg L^{-1}	BOD mg L^{-1}	pH	BOD kg d^{-1}
1.	Pak. Arab Fertilizer Ltd.	2.58	38.0	20	2200	245	8.3	2398
2.	Shaikh Fazal Rehman and Sons	1.45	44.0	60	470	595	6.8	3216
3.	Khawaja Tanneries	0.02	27.5	60	3800	469	7.9	39
4.	Alpha Industries	0.007	28.0	65	3376	532	10.5	14.5
5.	Colony Textile Mills	3.45	51.0	80	608	231	8.0	3019

Table 9D. Pollution load in effluents of sample industries in Faisalabad.

Sr. No.	Name of the Industry	Discharge m.g.d.	Temp °C	Colour Units	pH	S.S. mg L^{-1}	BOD mg L^{-1}	BOD kg d^{-1}
1.	The Lyallpur Dairy Farm	0.024	26	20	5.1	3144	1787	162
2.	Army Welfare Food Industries	0.47	45	25	7.6	686	260	408
3.	Crescent Sugar Mills Distillery	0.29	32	359	3.8	2984	1295	1407
4.	The Lyallpur Chemical and Fertilizers Ltd.	0.15	32	20	3.9	558	26	37
5.	Crescent Textile Mills	0.47	42	70	7.4	276	514	908

irrigation and drainage, domestic and industrial wastes, etc. However, the prime source is the disposal of effluents from the urban centres of the country where much of the population reside. Urbanisation, or the concentration of more and more people into fewer and fewer cities, is regarded as one of the major environmental threats today (UNESCO-UNEP, 1984). It is estimated that by the year 2003, the number of cities in Pakistan with populations greater than 1 million will increase to nine. Total other large cities (100 000 to 1 million) will be 88, accounting for 78% of the projected urban population (PEPAC, 1983). The cities with greater than 1 million population are expected to accommodate as many as 36 million people (GOP, 1989). Implications of this kind of urbanisation for human wellbeing and environmental quality can be summed up as:

- Declining environmental quality in urban areas through water, air and soil pollution, thus modifying the microclimate, and loss of natural areas;
- Severe degradation of the surrounding environment and ecological systems of the urban hinderlands, exerting heavy demand for resources;
- Demographic transformations through migration from and to rural and urban areas which will have severe social, economic and environmental consequences;

Table 9E. Pollution load in effluents of sample industries in Kala Shah Kaku.

Sr. No.	Name of the Industry	Discharge m.g.d.	Temp °C	Colour Units	pH	S.S. mg L^{-1}	BOD mg L^{-1}	BOD kg d^{-1}	Toxic Matter
1.	Kohinoor Oils Mill	2.58	31	30	8.0	676	530	5178	–
2.	Ravi Rayon Mills Ltd.	0.64	30	20	5.8	900	833	2210	–
3.	National/W. Pakistan Tanneries	0.29	28	200	6.7	1971	591	640	–
4.	Iteliad Chemicals	1.615	30	60	9.0	154	293	1791	Hg.5.6 mg L^{-1}
5.	Lasani Steel Mills	0.015	30	20	7.8	54	193	11	–
6.	Govt. Weaving and Finishing Centre, Shahdara	0.08	31	25	7.8	1724	453	139	–
7.	P. Leiner and Sons Chemical and Food (Pak). Ltd.	1.29	31	25	6.5	1596	390	1909	–

Table 9F. Pollution load in effluents of sample industries in Peshawar.

Sr. No.	Name of the Industry	Discharge m.g.d.	Temp °C	Colour Units	pH	S.S. mg L^{-1}	BOD mg L^{-1}	BOD kg d^{-1}
1.	Sarhad Fruit and Vegetable Products	Nominal	23	25	6.9	428	509	–
2.	Khazana Sugar Mills	2.5	25	25	6.7	960	31	291
3.	Makk Beverage & Mineral Waters	0.67	21	20	8.7	392	46	118
4.	New Frontiere Punjab Tanneries	0.38	23	65	6.1	4060	767	3969

- The threat of political and environmental instability; and
- Massive increases in public expenditures in cities at the expense of other sectors.

In addition to their hazardous effects on human health and aquatic life, industrial waste discharges exert significant aesthetic, economic and ecological impacts on water bodies. For example, inorganic salts in industrial waste cause water to become hard and make a stream undesirable for municipal or agricultural usage. Apart from biological contamination, chemical pollution of soil and water is also on the

Table 9G. Pollution load in effluents of sample industries in Nowshera.

Sr. No.	Name of the Industry	Discharge m.g.d.	Temp °C	Colour Units	pH	S.S. $mg\,L^{-1}$	BOD $mg\,L^{-1}$	BOD $kg\,d^{-1}$	Toxic Matter
1.	Associated Industries Ltd.	4.06	35	35	7.5	844	143	1050	Not detected
2.	Colony Sarhad Textile Mills	0.12	42	20	11.7	5828	485	220	–
3.	Adamjee Paper and Board Mills	4.50	27	65	6.7	21113	930	15830	–
4.	Sarhad Development Authority Leather Factory		30	30	7.6	1692	839	5	Cr 0.53 $mg\,L^{-1}$
5.	Adamjee Chemical Industries	0.90	26	25	2.0	4040	1021	3476	Not detected

Source: GOP, Environmental Profile of Pakistan (1989).

increase and is becoming an extremely serious health hazard. Chemical carcinogens such as nitrosamines have been detected in food and naturally occurring carcinogens from fungal metabolites (e.g. aflatoxins), starigmatocystins, ochratoxins and patulins have been detected in food in Karachi (KDA, 1981).

The extensive and increasing use of synthetic chemical fertilisers is a known source of environmental pollution. For example, most of the components present in the wastewaters of a fertiliser factory can affect the water quality of the receiving water course. Acids and alkalis can destroy the indigenous aquatic organisms while high NO_3^- concentrations in food, if consumed by humans, can enter the blood stream where they interfere with the blood's ability to carry oxygen. Moreover, NO_3^- has also been related to algal blooms and eutrophication of surface and groundwaters (Chapters 2 and 5 for detailed review). Nitrate can also combine with hydrocarbons (e.g. nitrosamines) to form compounds that can cause cancer in humans. Nitrate pollution of drinking water is of growing concern in many countries (see Chapters 2 and 5) including Pakistan, because N applied on land as a fertiliser often reaches groundwater as NO_3^-. This demands special caution in Pakistan where fertiliser production has risen sharply (see Table 6) because of the need for increasing food production. Average fertiliser use per ha has increased from 17 in 1970 to 51 $kg\,ha^{-1}$ in 1982 and 67 $kg\,ha^{-1}$ in 1993.

5.6 *Untreated Domestic Sewage*

According to a rough estimate about 6.9×10^6 t of wet human excreta is produced annually, with 2×10^6 t produced in urban areas, of which 50% goes into water bodies. Water pollution problems are assuming serious proportions particularly around major urban centres. Environmental profiles of Pakistan indicate that about 40% of deaths are related to waterborne diseases; sewage with a high BOD exhausts aquatic O_2 and creates unpleasant taste, odour and anaerobic conditions in water bodies.

Major cities in Pakistan dispose of largely untreated sewage into irrigation systems without any consideration of their assimilative capacity, resulting in serious bacterial contamination which threatens human health. The organic load of sewage seriously depletes the dissolved oxygen content of the receiving water, causing unaesthetic conditions and making water unfit for fish. For example, fish production in the river Ravi which receives untreated effluent from the city of Lahore has been reduced by 5 000 t yr^{-1} (Saleemi, 1993).

The domestic sewage generated in human settlements is sometimes disposed of directly into agricultural fields throughout Pakistan, mostly around big cities where vegetables are generally grown. These fields act as reservoirs of biological disease agents such as mosquitos, lice, ticks, and many other kinds of parasites that cause a variety of diseases. Rice fields with stagnant water are also fertile grounds for malarial mosquitoes. Over 60 000 cases of malaria were reported in 1987 in Pakistan. Outbreaks of malaria are common and they need careful monitoring and control.

Living organisms pose perhaps the greatest impediments to a safe water supply. Access to a clean water supply is available to only 77% of the urban population in Pakistan. For example in Karachi, 38% of the people have house connections, and the rest of the population depends for its water supplies on community stand-posts, water tankers, and individual private water bores. Other areas may not be so fortunate. In major towns and villages (45 000 in number) of Pakistan, water supply is often drawn from open, untreated sources. Rural water supply is particularly exposed to infestation by rodents, insects, flatworms, roundworms and micro-organisms and therefore many waterborne diseases are spread through drinking water. Gastro-intestinal disorders as well as other serious infectious diseases such as polio, salmonella, hepatitis, cholera, typhoid and tuberculosis are contracted from stagnant unclean water and rapidly spread through population centers. There is also evidence that contamination of water has begun to threaten the food chain in the country. In Rawalpindi Islamabad area milk and milk products, fruits and vegetables have been found infected with salmonella, *E. coli*, coliforms, *Bacillus cerus* and other pathogenic bacteria. This is particularly prevalent when irrigated with seasonal stream waters. In view of the existing precarious soil environment, the quality of the water is badly polluted in Pakistan (GOP, 1989).

6. Future research directions

As with many developing countries there are major environmental issues that require detailed research to enable regulatory bodies to impose guidelines for the protection of our environment. In addition, there are land based issues that limit crop production and also lead to massive degradation of our land. Some of these issues are listed below:

- Qualitative and quantitative evaluation of drainage water and its use for sustainable agriculture;
- Reclamation and management of salt-affected and waterlogged soils;
- Assessment of heavy metals in effluent irrigated soils;
- Economical treatment of wastes and their use for crop production;
- Extent of nitrate contamination of groundwater;
- Pesticide contamination of soils and their impact on the quality of food crops; and
- Socio-economic conditions of the rural population.

Achievement of some of these objectives requires significant capital/financial input which is the major factor limiting research in Pakistan.

7. Conclusions

Pakistan is an agricultural country and therefore for the most part, its environmental pollution problems will continue to be dominated by the use of agrochemicals in agriculture. Due to limited availability of quality land and water resources for production of food for the growing population, indiscriminate use of agricultural inputs is posing a serious threat to the soil environment. Although regulations do exist to protect the environment, their actual implementation continues to be a serious challenge for the Government as well for the people, due to immense economic pressure. About one third of the area of cultivable lands is regarded as having degraded soils.

To sum up issues relating to contaminants and the soil environment in Pakistan, it is encouraging to know that despite the increase in population and unplanned urbanisation, environmental problems are not as serious as they are in many developed countries. Nevertheless the present preventive as well as curative measures for sustainable use of environmental resources could yield much better results if they were associated with:

- Time bound programs to control salinity and waterlogging.
- Specific targeted programs for the use of agrochemicals in the country.
- Launching of rural developmental projects on a mass scale to contain poverty in the rural masses and to discourage their movement into the urban centres.

References

ADB 1987 Handbook on the use of pesticides in the Asia-Pacific region. P.O. Box 789, Manila, Philippines, Asian Development Bank. 294 p.

EPA 1990 Draft Report Government of Punjab Lahore. Environment Protection Agency, Government of Punjab, Lahore. 51 p.

FAO 1990. Year book (1989). FAO Statistics series No. 95, Rome. Food and Agricultural Organisation, Rome, Italy. 127 p.

GOP 1978 Planning Commission. The report of the Indus Basin research assessment group, Islamabad. Planning Commission, Government of Pakistan. 95 p.

GOP 1983 Desertification problems: extent and remedial measures. Irrigation and Flood Control Council. Irrigation and Food Control Council, Ministry of Technology, Government of Pakistan. 95 p.

GOP Government of Pakistan 1986 Pakistan National Conservation Strategy. Environment and Urban Affairs Division. 378 p.

GOP 1989 Environmental profile of Pakistan. Environment and Urban Affairs Division. Government of Pakistan. 248 p.

GOP 1991 The Pakistan National Commission of ICID. Irrigation and drainage development in Pakistan. Government of Pakistan. 1-38 p.

Hussain F, Kausar A and Naqvi H 1990 A comparative study of different methods for obtaining an index of nitrogen availability in upland soils, Nuclear Institute of Agricultural Biology, Faisalabad, Pakistan. 68 p.

Johnson B C L 1979 'Pakistan' Heinemann, London, U.K. 291 p.

Kausar M A 1990 Research on Micro nutrients of crops in Pakistan, NIAB, Faisalabad. (Unpublished).

Khan A, Ibrahim M, Ahmed N and Anwar S A 1992 Studies on accumulation and distribution of heavy metals in agricultural soils receiving sewage effluent irrigation. Proc. 4th National Congr. Soil Sci. Soc. of Pakistan 'Efficient use of plant nutrients' (held at Islamabad from May 24-26, 1992) 607-610 p.

Khattack J K 1995 Micronutrient in Pakistan Agriculture. PARC-NWFP Agricultural University Annual Report. Pakistan Agricultural Research Council. 1-7 p.

Mohammad N (editor) Combating desertification in Pakistan. Pakistan Agricultural Research Council-United Nation Environment Program (Draft) Congress, April 1994. 262 p.

Mubarik A and Jabbar A 1992 Effect of pesticides and fertilizers on shallow groundwater quality. Pakistan Council of Research in Water Resource, Islamabad. 130 p.

Parveen Z and Masud S Z 1988 Organochlorine pesticide residues in cattle drinking water. Pak. J. Sci. Ind. Res., 31, 53-56 p.

PEPAC 1983 Human settlement policy of Pakistan – prepared for Ministry of Housing and Works, Environment and Urban Affairs Division, Islamabad. PEPAC. 140 p.

Rashid A 1993 Nutritional disorder of rapeseed - mustard and wheat grown in potohar area, Land Resources Research Institute, NARC, Islamabad. 105 p.

Saleemi M A 1993 Environmental assessment and management of irrigation and drainage scheme for sustainable agricultural growth. Proc. Inter. Sympo. held at CEWRE, Lahore Pakistan Oct. 24-28): vol: 1. 63-72 p.

Spooner B 1982 'Environmental problems and the planning of development in Arid South Asia' in Environment and Development in Asia and the Pacific, UNEP Report and Proceeding Series 6.

UNESCO-UNEP 1984 International meeting on 'ecological approaches to urban planning' Suzdal, USSR.

WAPDA 1980 Water and Power Development Authority of Pakistan Revised Action Programme (RAP), Lahore, Pakistan. 15-21 p.

22
Contaminants and the soil environment in Papua New Guinea: an overview

K. SINGH, M. P. LEVETT AND R. KUMAR

1. Introduction

Papua New Guinea (PNG) comprises the eastern half of the island of New Guinea and outlying islands, north of Australia (Fig. 1). Irian Jaya, a province of Indonesia, occupies the western half. The total land area of PNG is 462 840 km^2 and includes North Solomons, New Britain, New Ireland, Manus and many other islands. Much of the interior of mainland PNG is mountainous and heavily populated, with a total population of 3.8×10^6 in 1990 (National Statistical Office 1993), and a mean population growth rate of 2.3% yr^{-1}. The people are culturally and linguistically diverse, and many tribal groups live in relative isolation. More than 700 distinct languages are spoken.

The economy of PNG depends largely on agriculture and the exploitation of natural resources. Forestry, agriculture (cocoa, tea, coffee, coconut, rubber), and, in particular, mining make significant contributions to the gross national product. PNG has considerable mineral wealth. Large scale mining of gold (Ok Tedi, Porgera, Misima, Mt Kare) and copper (Cu) (Ok Tedi, Bougainville which is currently closed) is taking place. PNG ranks ninth among the world's gold-producing nations. Exploration continues to reveal the presence of valuable minerals, including nickel (Ni) and chromium (Cr) as well as oil and gas (Kutubu, Gobe, Hides). Mining activities are the source of much of PNG's hazardous waste.

Naturally endowed with luxuriant forests, fertile soil, and mineral resources, PNG has abundant development potential. The political system of the country, which became independent in 1975, represents a remarkable synthesis of traditional and Westminster democratic structures. For example clan-based common ownership of land, and modern democratic institutions, such as a federally-structured parliamentary system (Alphonse 1991) operate in reasonable harmony.

2. Environmental Protection Legislation

Papua New Guinea is unique in the Pacific region in that its constitution has specifically recognised the nation's responsibility to ensure that the environment is protected and its natural resources are wisely used. The Environmental Planning Act (1978) ensures that development projects are planned to attain maximum economic, environmental and social benefits. The Act can be enforced through the voluntary or requested submission by the developer of an environmental plan, which must detail the potential effects of a project on the physical and social environment, and

R. Naidu et al. *(eds): Contaminants and the Soil Environment in the Australasia-Pacific Region, 647–657.*

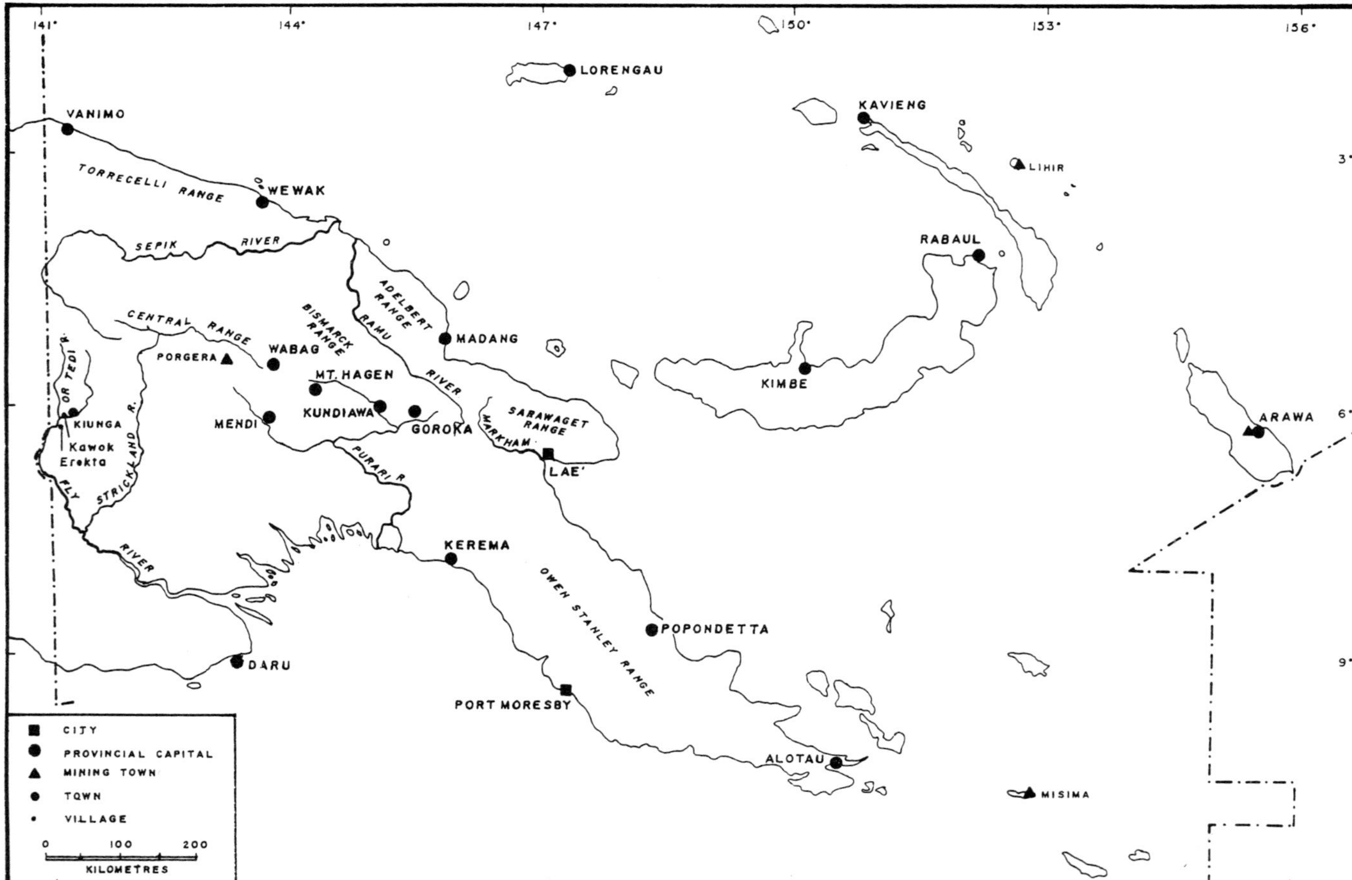

Figure 1. Geographical map of Papua New Guinea.

state the actions which will be undertaken to minimise harmful effects. The conditions of an environmental plan, once approved, are legally binding.

The Environmental Contaminants Act aims to prevent and regulate the release into the PNG environment of all harmful pollutants. It provides for the setting up of an Environmental Contaminants Advisory Council and a Registrar of Hazardous Environmental Contaminants, and requires all potential polluters to apply for a licence or permit. The permit details how much of each contaminant or poison can be released into the environment (including trade wastes from factories, pesticides and other chemicals). The legislation also covers litter, noise and radioactivity.

The Conservation Areas Act is seen as the major way of conserving certain areas of PNG for future generations. The Act does not take the land away from the landowners; rather it provides for the creation of local management committees responsible for the preparation of a management plan for the area. The Office of Environment and Conservation was established in 1974, as a result of much pressure on the government of the day, and concern about the total lack of environmental and social planning. Currently, a Federal Minister heads the Department of Environment and Conservation (DEC). Nevertherless, PNG is facing considerable environmental problems.

3. Mining boom

Papua New Guinea is currently undergoing a mineral exploration and development boom, especially for Au and Cu, and a number of economic deposits have already been located. Like all other mineral development projects, those in PNG have associated environmental and social problems. The major problem associated with the mining activities is the discharge of effluents in the surrounding rivers. Such disposals have led to extensive degradation of aquatic ecosystems causing hardship for people living on the river banks. However, there is no scientific data to substantiate these observations. There has been considerable publicity for instance surrounding environmental management problems of the Ok Tedi, Bougainville, Porgera and Misima mining projects. As the pace of mineral development increases so will the degree and extent of environmental impact. The brunt of such impact will continue to be felt most keenly by the local communities whose land will be alienated or damaged, and whose traditional resources will be destroyed or degraded (Hughes 1989).

4. Common agricultural activity

The population densities are generally very low in coastal and lowland swampy areas, and in lowland grassland and open woodland or savanna areas with marked rainfall seasonality. The subsistence systems of the majority of the people involve swidden agriculture, tree husbandry, animal husbandry, hunting and gathering, and in coastal areas, fishing and exploitation of marine resources. In many lowland rainforests or savannas, a system of shifting gardening with medium to long-term forest or grassland fallows is practised. The fallow is cleared and gardens planted with many crops like taro, yams, bananas, sweet potato, sugar cane, greens, vegetables, tobacco, condiments, fruit trees, fibre plants, medicinal plants etc, and harvested over a period of two to three years before being fallowed for 10 to 30 years. Soil cultivation varies from complete tillage to loosening the soil around planting points. In all parts of the country the gathering of plants is important to provide food, medicines, raw materials for building purposes, tools, weapons, artefacts, clothing and personal ornamentation and materials for ritual and magical purposes (Powel 1976).

Attempts are also made to maintain soil fertility. The burning of vegetation during clearance of land for gardens provides an initial 'ash-bed' increase in certain nutrients, particularly base cations (Ca; K; Mg), which may be important for certain crops. For example, traditional greens and introduced vegetables are sown into ashy areas, and taro may be planted in rich, organic or ash soil. Other techniques employed within gardens to maintain levels of soil fertility include subsoiling (digging soil from ditch and throwing on top of plots), compost mounding (garden rubbish, weeds and grass included within mound of soil) and mulching (with grass, tree leaves etc). These are successful to some extent but it is generally recognised that reduced crop yields are related to loss of soil fertility, and that land must be fallowed after cropping to improve the soil. Burning is a technique used mainly in drier parts of the country, and where soil fertility is declining (Levett 1992). However, fallow burning inevitably results in losses of volatile nutrients, especially nitrogen, and some sulphur and phosphorus. In some forested and grassland areas, enrichment planting with nitrogen-fixing tree species (especially *Casuarina*, *Albizzia* and *Leucaena* spp) is improving and shortening the fallow period, and converting some grass fallows to planted tree fallows (Allen et al., 1994).

5. Soil degradation

Despite a wide range of conservation practices and a comprehensive understanding of the environment, some societies in PNG currently face increasing problems of accelerated erosion and declining soil fertility. There are two main reasons for this. First, traditional soil conservation practices are largely designed for the short-term maintenance of productivity and maximisation of yields, whereas erosion and fertility decline may occur so slowly that they are not noticeable. Consequently, problems of soil degradation are inherited from past misuse of the land, and are accelerating as more land is brought into cultivation. Second, agricultural systems have been changing rapidly due to changing demographic, social and economic conditions. In many cases, soil conservation measures have not evolved sufficiently rapidly to keep pace with these changes.

The desire for a cash income has led to an increasing amount of land, particularly on better soils, being used for cash crops such as coffee and cocoa, and for animal production such as cattle-rearing. Land shortage is particularly severe around urban areas, and is leading to the overuse of marginal land, especially for cash cropping.

This has exacerbated the problems of land shortage for subsistence cropping. Areas with poor soils and few opportunities for growing cash crops, and also areas adjacent to labour-demanding projects such as mining, have been severely affected by out-migration. As a result, there are no longer enough men to carry out strenuous tasks such as clearing new gardens in the bush. Consequently gardens are cultivated for longer periods without fallow and soil fertility declines. The full implications of the breakdown of traditional systems and the adoption of new crops and tools are not yet realised. New agricultural innovations often have detrimental ecological side effects but it is too early to say whether such innovations are causing soil degradation directly in PNG (Wood and Humphreys 1982).

One of the main causes of soil degradation has been the transformation of forested areas into grasslands due to cultivation and burning. Anthropogenic grasslands are widespread in the highlands and in parts of the lowlands, especially where there is a pronounced dry season such as on the Sepik plains and the Papuan Coast around Port Moresby. The conversion of forest to savannah and grassland is a degradation of the environment, causing accelerated soil erosion, particularly in areas with a dry season. Sheet erosion and debris avalanching are active in removing soil from the hills in the Port Moresby area and many parts of the highlands. The deep colluvial deposits in the valleys probably accumulated when the forest was first cleared. The conversion of forest to grassland reduces plant biomass and causes the nutrients stored in the biomass to be lost through leaching or fixation. In the mid-altitude zone, grasslands have a biomass of only 3 to 6 t ha^{-1} compared with 293 t ha^{-1} for primary forest (Manner 1977).

In the highlands, chemical analysis of soils under forest and grasslands indicates that grassland soils are less fertile and lower in organic matter (Wood and Humphreys 1982). Losses of organic topsoil due to erosion have accelerated the fertility decline, since the bulk of the available plant nutrients is stored in the surface soil. Pigs are a major agent of erosion due to their rooting activities in gardens and fallows. Also, many cattle pastures are eroded due to overgrazing and poor pasture management. The degradation of soils is a serious problem in parts of PNG. At the present time, there are virtually no data on soil degradation, and detailed studies are required in a range of environments.

6. Sources of urban soil contamination

Urban environments are affected by a large range of contaminants whose impacts will vary appreciably within and between different cities. Port Moresby, the nation's capital city, lies on the southern coast of PNG, approximately 9.5° south of the equator. The city is approximately 9 km long and 6 km wide and is expanding rapidly due to squatter settlements, urbanisation, growth in population and industry. The estimated population is 200 000. The climate of the city is classified as tropical rainy with distinct dry seasons that ensure slow flow rates in drains and creeks. The various industries and land uses associated with urban soil contamination are summarised in Table 1.

Table 1. Industries and land uses associated with urban contamination.

Airport	Food processing
Alcoholic beverages	Footwear, plastic and rubber
Abattoir	Grain milling and feeds
*Automotive service stations	Land fill sites
Asbestos disposal	Mining industries
Bakeries	Metallurgy
Building materials	Oil storage
Chemicals (including soaps, detergents, matches, pharmaceuticals and paints, etc.)	Palm Oil
Cocoa and coconut processing	Paper products, publishing and printing
Dry cleaning establishments	*Scrap yards
Electric power generation	Soft drinks
*Electroplating	*Tannery products
Film processing	Wood and forestry products & preservation

*Main sources of heavy metal contamination.

6.1 Contaminants from urban waste

6.1.1 Municipal

As with many third world developing countries, there are no treatment plants for municipal/urban wastes which are generally disposed of directly either onto land or into river systems. Although such disposals can lead to serious degradation of the environment, no effort has been directed towards investigating the impact of such disposals on the soil environment. This may be largely due to the lack of financial resources and adequate regulatory guidelines. It is imperative that regulatory bodies in PNG consider investigation of the impact that longterm waste disposals have had on the land around the major urban areas.

Municipal rubbish comprises the waste streams of domestic and commercial premises. Municipal waste generated in PNG is mainly from urban areas. There is little inorganic waste outside urban areas. The waste is divided into three categories:

- Combustibles: Waste paper, newspaper and cardboard; putrescibles – including all wastes from vegetables and from preparation, cooking, selling and serving foods; plastics, wood, textiles;
- Non-Combustibles: Metal – ferrous and non-ferrous metal, broken glass and glass bottles, ceramic, asbestos, batteries, fluorescent lights and electronic equipment.

- Others: bleaches, insecticides, household chemicals and cleansing agents.

In PNG, hazardous waste generated by the following major sectors of the economy is the main source of contaminants to the environment (PBCHWR 1988).

6.1.2 Mining

Papua New Guinea has very considerable proven mineral wealth. Large scale mining of Au and Cu is taking place, and these operations are very important to the economy of the country. Although mining activities are also the source of much of PNG's hazardous waste, limited data is available on the impact of such wastes on the soil environment. Cyanide used to strip Au from the ore is often released as waste sludge to settling ponds or directly into riverways (Pernetta 1988a). Heavy metals including Cu, Cd, Pb and Zn are also generated in large amounts by the mining industry and have been found in very high concentrations in water samples taken down river of mine sites (Mowbray 1986a). Most of the rural population in PNG live in small dispersed farming villages. Those living adjacent to mining areas have experienced some negative impacts from damage to river ecosystems and erosion of their lands.

Mining activities in PNG have the potential to cause severe localised soil contamination. The main areas affected are generally (a) in the vicinity of the mine site; and (b) along riparian strips and flood plains of rivers and creeks draining from the mine site, or which are used for the disposal of mine spoil. The latter may result in long-range transport of soil contaminants and deposition on river banks many kilometres from the mine site. The main soil contaminants on river flood plains are generally directly related to the geological material being mined. Thus, for example, localised areas on the banks of the Ok Tedi and Fly Rivers have elevated levels of total and available Cu where the rivers have flooded and deposited sediments on the levees and flood plains. In some cases plant-available Cu levels may be immediately at toxic levels, whereas in others the sediments have the potential to weather and release high levels of Cu into the soil.

6.1.3 Industrial

The industrial development of PNG is still in its infancy. Although industries are generally located in the urban centres limited data is available on the potential impact of industrial disposals on the quality of the soil environment. Waste generated includes:

- Inorganic Waste: acids, alkalis, scrap metals, glass and ceramics
- Organic Waste: polychlorinated biphenyls (PCBs), chlorinated phenols, dioxins, furans, fuels, oils, lubricants, solvents (trichloroethylene, benzene, chloroform and carbon tetrachloride), paints, varnishes, lacquers, enamels, soaps, detergents and toilet preparations etc.

6.2 *Agriculture*

6.2.1 Pesticides

Increasing use of pesticide is the major source of contaminant associated with agricultural activities. According to the PNG Department of Environment and Conservation 193 registered pesticide products are currently imported in the country. However, the most commonly used pesticides in PNG include paraquat, DDT, dalapon, atrazine, diuron, ametryne, tebutryne, 2,4-D, MSMA, malathion, parathion, rotenone, sodium pentachlorophenate, acephate, arsenic copper oxychloride and their related formulations (Mowbray, 1988). Use of organochlorine pesticides was banned in 1990 except for DDT (for malarial control), lindane (for specified medical use and medical compounds) and heptachlor (for termite control only). Although DDT has been officially banned, it is still sporadically used illegally by some farmers who have access to old abandoned Department of Health stores containing DDT scattered around the country.

No systematic survey has been done to determine pesticide residues in the agricultural soils around the country. Mowbray (1986) reported the presence of DDT in breast milk and DDT, dieldrin, heptachlor epoxide and lindane in fat samples of people in PNG at levels above the maximum permissible concentrations. This was also supported by a number of reported deaths relating to poisoning due to paraquat, rotenone, diazinon and sodium pentachlorophenate.

6.2.2 Timber treatment

Two types of approved water soluble chemicals are used for preservation of forestry products. These chemicals give rise to hazardous waste containing Cu, Cr and As which are generally disposed into public sewers or land. However, no work has been done to assess the impact of such disposals on the quality of the soil environment.

6.3 *Other waste*

Hospitals generate pathogenic organisms and radioactive materials, chemicals and pharmaceutical laboratory waste; oil and gas industry (exploration, refinery, petrochemical products) are other emerging sources of hazardous wastes.

7. Urban waste disposal

Under the Environmental Planning Act in PNG, every industrial facility with a licence to operate must submit a contingency plan for chemical emergency. In its plan, the facility must state the nature of transportation, disposal, and handling and describe the remedial action that would be taken in the event of a chemical spill. Absence of effective enforcement allows industries to effectively operate as they will. However the public awareness of potential dangers of hazardous chemical waste is gradually increasing. There are local resistance groups who mobilise the general

Table 2. Disposal of tailings from mines in Papua New Guinea*.

Mines	Tailings tonnes/day	Mine life	Workforce No.	Tailings disposal methods
Bougainville	135 000	30	4000	Into river
Ok Tedi	80000	30	2500	Into river. More than 10,000 td^{-1} of fine waste rock also enters river
Misima	15000	10	350	Deep ocean via submarine pipeline. 10,000 td^{-1} of soft waste rock is dumped directly into the ocean
Hidden Valley	10000	10	300	Storage in tailings dam
Porgera	9000	18	900	Into river
Mt. Kare	300	3	200	Into river
Wau	1 400	9	280	Into river
Mt. Victor	400	2	100	Storage in sealed limestone sinkhole
Lake Kamu	–	15	160	Virtually all rock wastes will be contained within the dredging area

*Hughes and Marjorie (1989).

public against the health hazards posed by hazardous waste in the vicinity of mines and other industrial units.

Little is known about the treatment of waste before disposal in PNG. The traditional low cost hazardous waste disposal methods are usually practised such as landfill, storage in surface impoundments, disposal in public sewers, rivers and sea. Table 2 shows how various mines dispose of their tailings. The Ok Tedi gold mine discharges its waste including cyanide tailings into the Ok Tedi River (Hughes and Sullivan 1989) which, after 150 km, flows into the large Fly River, which travels 800 km to the sea. There was a proposal to build a spent-cyanide destruction facility and tailings dam to break down the cyanide prior to release into the river, but this was abandoned when a massive landslide in 1989 destroyed earthworks associated with the construction of the tailings dam. Bougainville Copper Ltd, when it was operating, used to dump its tailings directly into the nearby Jeba River, which flows 40 km to the sea.

Most of the existing hazardous waste from urban areas (much of it with a low degree of hazard) is believed to be disposed of to muncipal landfill sites. Very few of these sites are suitable for receiving hazardous wastes as co-disposal. Liquid and semiliquid hazardous wastes are often disposed of to sewers. Again, these wastes usually, though not always, have a low degree of hazard and are not strictly controlled by trade waste byelaws set by the local authorities, such as the Department of Environment and Conservation or the National Capital District Commission (for Port Moresby).

The National Capital District Commission is responsible for the management of municipal waste in Port Moresby. Commercial and industrial establishments either truck their wastes to the dump sites or incinerate them on their premises. Liquid wastes comprised mainly waste oil are discharged into the storm drains or the city's sewerage system. There are two contractors operating in the city to collect and dispose of wastes from the residential areas including urban villages, government offices, schools, University of PNG and various other institutions. Collection is carried out on a bi-weekly basis, and the wastes are immediately carted to dumps for disposal by landfills. The dumps are located in a valley at Baruni, and at 10 km, at the start of the Magi Highway. They were established in 1979 and 1974, respectively. The oil, domestic wastes and putrescibles are brought to Baruni whereas the 6 Mile dump was supposedly for dry waste such as tree cuttings, timber, car parts, cardboard, scrap metal etc. Both waste disposal sites have attracted scavengers, and rubbish litters extensive areas around each dump. Burning is common at the dumps. Indiscrimate burning of domestic waste at backyards is also commonly practised in urban and rural areas.

Although contaminated sites are affected by a range of potentially hazardous substances, e.g. acidic and alkaline materials and building waste products, the most common contaminants are heavy metals.

8. Pollutant load of lead in soils along motorways in Port Moresby

Concentrations of Pb in soil samples collected along Waigani Drive, Port Moresby have been investigated by Eapaea (1992). Port Moresby, the nation's capital city, lies on the southern coast of PNG, approximately 9.5° south of equator. Population growth is 4.35% p.a. An improvement in the economy of the country in the last 10 – 15 years has added an increased number of cars and other motor vehicles on the roads. According to the National Transport Department there are over 25,000 vehicles in the National Capital District.

Soil samples were collected near main roads at five different points, Gerehu roundabout, Tokarara, Waigani roundabout, NCDC City Hall, Islander Travelodge and Boroko (6.5 km) along the Waigani Drive, which has peak transport during business hours, and Pb levels were determined.

Lead concentrations in the soil at 5 – 15 m distances from the roadside were within the range 20 – 170 mg kg^{-1} with estimated background levels (at 50 m distance) being in the range 726 mg kg^{-1}. There was a positive correlation between the traffic volume and the soil Pb level. There were, however, several factors which may have affected the recorded Pb levels, which include wind direction, and rainfall.

9. Conclusion

In this paper we have presented a review of available data on contaminants and the soil environment in some parts of PNG. There is generally a lack of information and research being undertaken by the appropriate authorities dealing with contaminants and soil quality in the country. There are very few data on the severity of soil pollution problems in different environments, and on the effectiveness of traditional techniques in controlling soil degradation. There are opportunities for the transfer of knowledge and skills from one part of the country to another, although research is required to examine their suitability in different environments. Techniques from other countries can be adopted, provided that they have been tested and proved effective for particular PNG environments.

Acknowledgement

We thank Mr J Naguwean, Librarian, New Guinea Collection, Michael Somare Library, University of PNG for his assistance in providing relevant information for the preparation of this manuscript.

References

Allen B J, Hide R L, Bourke R M, Fritsch D, Grau R, Hobsbawn P, Levett M P, Majnep I S, Majnep V, Nen T and Sem G 1994 Agricultural System of Papua New Guinea. Working Paper No. 7. Madang Province. ANU, PNGDAL and UPNG. 192 p.

Alphonse A 1991 Development and Environment in PNG. Catalyst 21, 119 p.

Eapaea M P 1992 Some studies on levels of toxic heavy metals in the environment around Port Moresby City. Postgraduate diploma (Chemistry) thesis, UPNG, pp 11-57.

Hughes P J 1989 The effects of mining on the environment of high islands: A case study of gold mining on Misima Island, Papua New Guinea. South Pacific Study 5 (SPREP).

Hughes P J and Sullivan M E 1989 Environmental impact assessment in Papua New Guinea: Lessons for the wider Pacific Region. Pac. View. 30, 34-55.

Levett M P 1992 A comparative study of gardening systems in two mountains census divisions of Kaintiba District, Gulf Province. Yagl-Ambu 16, 71-86.

Manner H I 1977 Biomass: Its determination and implications in tropical agroecosystem: an example from montane New Guinea. In Subsistance and Survival: Rural Ecology in the Pacific. Eds. T P Bayless Smith and R G A Feachem. pp 215-242. Academic Press, London, UK.

Mowbray D 1986a Pollution problems in Papua New Guinea. Siren. 31, 11-14.

Mowbray D 1986b Pesticide poisoning in Papua New Guinea and the South Pacific. PNG Medical J. 29, 131-141.

Mowbray D 1988 Pesticide use in the South Pacific. SPREP Topic Review No. 26. UNEP. pp 171-174.

Pacific Basin Consortium for Hazardous Waste Research (PBCHWR) 1988 Hazardous Waste in the Pacific Basin. East-West Center, Honolulu, USA. PNG-1.

Pernetta J C 1988 Potential impacts of mining on the fly river. UNEP, Regional Seas Reports and Studies No. 99, 1-88.

Powell J M 1976 Ethobotany. In New Guinea Vegetation, Ed. K Paijmans. pp 121-133. Australian National University Press, Canberra, Australia.

Wood A and Humphreys G S 1982 Traditional soil conservation in Papua New Guinea. In Traditional Conservation in Papua New Guinea: Implications for Today. Eds. L Morauta, J Pernetta and W Heaney. pp 94-144. Institute for Applied Social and Economic Research, Monograph 16, Boroko, NCD, Papua New Guinea.

23
Contaminated soils in the South Pacific Islands

R.J. MORRISON, P. GANGAIYA AND K. KOSHY

1. Introduction

The South Pacific region has a reputation for sun-drenched beaches, clear waters and lush green vegetation. The possibility that contamination of the environment exists is often not considered. While this scenario does still exist in places, the situation is changing quite rapidly with increasing populations and industrial development contributing to problems that now require serious attention. Contaminated soils have been identified, but this issue has received minimal attention, partly because of a lack of appreciation of the areal extent and potential severity and partly because of inadequate resources to address the situation. This chapter examines the causes of contamination, the extent of the situation, assesses potential problems and makes recommendations for future action.

The region consists, for the purpose of this paper, of 21 countries stretching from the Marianas Islands in the north to New Caledonia in the south, from Palau in the west to French Polynesia in the east (Figure 1). As the status of contaminants in Papua New Guinea, which is often included with the Pacific Islands Region, is discussed in Chapter 22 it is not included in this paper. The total area exceeds 20 million km^2 but the land area is only about 88 000 km^2. As the larger island groups (Fiji, New Caledonia, Solomon Islands and Vanuatu) account for over 85% of the land area, the majority of the 21 countries are small islands, spread over a vast extent of the South Pacific Ocean (Table 1).

The total population of the countries is approximately 2.1 million (Table 1). The population of many of the countries is increasing at a rate above the world average and this represents a major problem in terms of environmental management. The large distances between countries and the limited population pose problems of transport, communication and inter-regional cooperation for the region. A detailed discussion of the consequences of the small size and large separation of the countries is given in Ward and Proctor (1980).

The South Pacific islands can be categorised, geologically, into a number of groups. They can be continental islands, e.g., La Grande Terre, New Caledonia, or oceanic high islands (volcanic peaks rising as great domes of basalt, extending from the sea bed) such as Savai'i in Western Samoa, or coral islands (raised coral platforms) such as Niue, or atolls, such as Funafuti in Tuvalu. The continental and larger oceanic highlands are dominated by very rugged landscapes, with the extent of flat or gently rolling land being small (<10%). The most weathered and developed soils of the region are found under rolling topography, suitable for arable agriculture, but

R. Naidu et al. *(eds): Contaminants and the Soil Environment in the Australasia-Pacific Region, 659–675.*

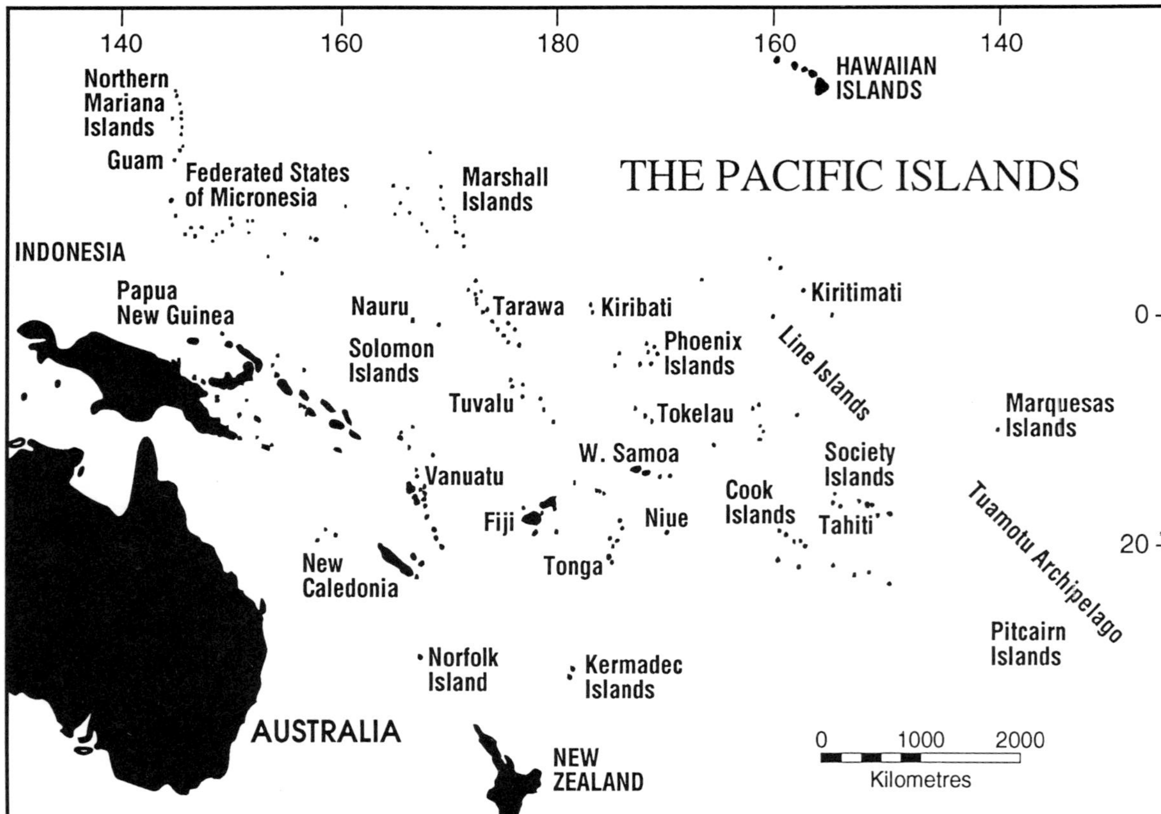

Figure 1. The Pacific Islands Region.

Table 1. Some geographic data on Pacific Island countries.

Country	Land Area km²	Population Data			
		Size	Growth Rate %*	Density Persons/km²	Est. Pop. in Year 2000
American Samoa	197	33000	1.7	165	46000
Cook Islands	240	1800	−0.4	75	18000
Federated States of Micronesia	710	101000	3.4	144	145000
Fiji	18272	715000	2.0	39	895000
French Polynesia	3265	167000	3.0	51	268000
Guam	541	106000	2.2	196	159000
Kiribati	690	64000	2.0	89	81000
Marshall Islands	180	31000	2.7	171	50000
Nauru	21	8000	2.4	387	11000
New Caledonia	19103	146000	1.2	8	178000
Niue	259	3000	−3.4	13	3000
N. Marianas	471	1700	1.7	35	27000
Palau	460	15000	−1.6	24	15000
Pitcairn	5	50	−9.8	10	–
Solomon Islands	27556	285000	3.5	10	440000
Tokelau	10	2000	1.4	200	2000
Tonga	699	95000	−1.1	135	98000
Tuvalu	26	8000	3.2	322	11000
Vanuata	11880	115000	3.2	10	196000
Wallis & Futuna	255	12000	4.4	49	22000
Western Samoa	2935	159000	0.3	54	177000

*Data from Bakker (1988).

many have low chemical fertility due to extensive leaching. The coral islands and atolls have an almost flat topography.

The region is generally hot (average temperatures >20°C) and wet, with most islands having average annual rainfall greater than 1 500 mm. In the eastern part of the region, particularly near the Equator, some low (<1 000 mm yr^{-1}) rainfall situations are encountered. In addition, as a result of the prevailing south-east trade winds, some of the larger islands have rainfall shadows on the western sides, leading to problems of water shortage from May to October in most years.

2. Pacific Island soils – distribution, use and management

A review of the information available on the South Pacific island soils (Morrison, 1982; Leslie, 1984) shows that Inceptisols (Soil Survey Staff, 1975; 1994) are the most abundant, with significant areas of Mollisols, Alfisols, Ultisols, Oxisols, Andisols and Entisols. In the low atolls, Entisols are dominant where they rest on coral sand, rubble or reef limestone. Andisols, formed from young volcanic ash, have been identified in northern Vanuatu, in Santa Cruz in the Solomon Islands, in Taveuni, Koro and Rotuma in the Fiji group, and in Upolu and Savai'i in Western Samoa (Morrison et al., 1989). Although these soils cover relatively limited areas, they are important agricultural soils in the region. Vertisols and Histosols are found in limited areas, determined by parent material, geomorphology and climate, the latter particularly have considerable agricultural significance. Spodosols are rare with, to date, only one Spodosol series being confirmed in Fiji and one in New Caledonia. No Aridisols have been identified in this region.

As noted above, the region is generally wet, with most soils having udic soil moisture regimes (Soil Survey Staff, 1975). On the windward coasts of the larger islands the soil moisture regimes may even be perudic, with ustic soils usually occurring only in the rain shadow areas. Thus, in much of the region, significant leaching occurs, with the result that any contaminants added to the soil surface are subject to potential movement by water. Soil temperature regimes are generally isohyperthermic (average temperature >22°C) or isothermic (average temperature 15–22°C), which means that solubilities and reaction rates are among the highest found in soils.

The most common minerals found in South Pacific soils are kaolin, iron and aluminium oxyhydroxides, smectites, and carbonates, although on some of the smallest islands such as Niue and Nauru the soils may be dominated by phosphate minerals. The Andisols are dominated by allophane, imogolite and ferrihydrite. Mineralogy and clay content are useful indicators of the behaviour of contaminants in soils. In the larger islands many of the soils on the older rolling landscapes have significant (>35%) clay contents with high surface areas and contaminant adsorption capacities.

Most of the soils of the South Pacific island countries are of low fertility (Asghar et al., 1988; Craswell, 1989). As most parent materials in the region are deficient in P and potassium, the soils developed from these also show similar deficiencies. The organic matter levels of most soils tend to be relatively high but this often rapidly declines with cultivation. The coralline soils of the atoll islands are extremely poor in most plant nutrients and as a result the local flora is limited.

Detailed discussions of regional agriculture are available elsewhere (e.g., Ward and Proctor, 1980; Asghar et al., 1988; Haynes and Naidu, 1989) so a only brief review is presented here. The farming systems used in the region vary from technologically sophisticated hydroponics to the traditional, basically shifting cultivation systems. The shifting cultivation systems are characterised by large manual labour inputs, little or no inputs of fuel, fertilisers, or pesticides, extensive interplanting of crops, little or no post-harvest storage and the local consumption of almost all of the production. Increasing population pressure has led to significant declines in the 'fallow' periods, often dropping from 15–25 years to less than 5 (e.g., North Pentecost in Vanuatu), with a general decline in the productivity of the areas concerned.

Mixed subsistence-cash cropping systems are the most common in the South Pacific and there is a considerable increase in the requirements to generate cash due to the changing nature of the local economies, e.g., school fees, purchase of outboard engine fuel for fishing, kerosene for cooking in place of wood. The dominant crops include coconuts (*Cocos nucifera*), coffee (*Coffea arabica*), cocoa (*Theobroma cacao*), vanilla (*Vanilla planiform*) and kava (*Piper methysticum*). Extension of cash cropping to greater areas has led to the development of what are often referred to as nucleus estate-outgrower 'plantation systems' with significant local ownership. Such systems began with coconuts and sugarcane (*Saccharum officinarum*), but have been extended to coffee, cocoa, timber, and more recently to rice (*Oryza sativa*), oil palm (*Elalis guineenois*) and cattle. These provide employment for a substantial number of Pacific Islanders. Plantation export crops should not be allowed to displace essential food crops for local use as neglect of such commodities increases the dependence on world market prices for export crops which are subject to large fluctuations. In addition, the accompanying increased consumption of imported foods has been shown to have major impacts on the nutritional status of the island people (Thaman, 1982).

The need to maintain self-sufficiency in food (particularly in the light of increasing populations in many islands) and to increase net foreign exchange earnings in the countries of the South Pacific, necessitates an increase in agricultural production. The extent of land under cultivation in the South Pacific has not increased by any appreciable amount during the last decade, probably due to lack of arable land. The options available are either to bring more marginal land under cultivation or to increase production on the better land already used for agriculture. Since the wider use of marginal land is unlikely to be sustainable (see for example, Morrison and Masilaca, 1989) more intensive cultivation and raising yields per hectare are necessary. The achievement of this objective may be obtained by a combination of the judicious use of fertilisers, better management practices, and varietal improvement programs.

3. Contaminated soils

3.1 General comments

Pollution of regional soils may occur through the inappropriate use of fertilisers and pesticides, spillages or seepages and the disposal of wastes on soils. The contaminants may be inorganic or organic chemicals, microbially or radioactively-enriched materials, or various combinations of these substances. One problem noted in the identification of metal contaminated soils is the wide range of elemental concentrations in regional soils and the occurrence of 'abnormal' concentrations in many locations. The presence of unusually high concentrations of some metals is a result of the local seismic and volcanic history, and is one reason why the region (particularly the western part) has a large number of mineral exploration operations in place. In addition, as noted by Wells (1960), the degree of weathering influences the distribution of the metallic components in soils.

A number of examples of these natural anomalies are briefly discussed below. They include Cu and Zn in Fiji (Leggo, 1977; Rugless, 1983), where Cu concentrations in soil of over 3000 $mg\,kg^{-1}$ (compared with background values of <340 $mg\,kg^{-1}$) and Zn concentrations in soil exceed 320 $mg\,kg^{-1}$ (compared with background values of 90–150 $mg\,kg^{-1}$). On Nauru, total Cd concentrations in soil are usually in the range of 80–120 $mg\,kg^{-1}$ (Morrison, 1987), which is considerably higher than the average concentrations in uncontaminated soils of <1 ±15 $mg\,kg^{-1}$ (Ure and Berrow, 1982) or for soils derived from sedimentary materials (0.3–11 $mg\,kg^{-1}$, Page and Bingham, 1973). On Niue, soil Hg concentrations ranged from 0.07–3.7 $mg\,kg^{-1}$, with a mean and standard deviation of 1.0±0.9 $mg\,kg^{-1}$ (Whitehead et al., 1990). Almost all values exceeded the global mean of 0.10 $mg\,kg^{-1}$ determined by Ure and Berrow (1982). In New Caledonia, soils formed from serpentine and peridotite (which constitute most of the southern third of La Grande Terre) have high contents of Cr and Ni. For the serpentine soils, 3% Cr and 0.4% Ni have been detected, while for the peridotite soils, Cr concentrations up to 5.7% and Ni up to 1.25% have been determined (Latham et al., 1978). Relatively high concentrations of Cr (up to at least 0.8%) have also been identified in serpentine soils in the Solomon Islands (Birrell et al., 1939).

3.2 Sources of contamination

The major potential sources of soil contamination are fertilisers and pesticides, industrial chemicals and wastes, spilt and waste hydrocarbons, and sewage sludge. Several reviews (e.g., Morrison et al., 1987; Craswell, 1989) have shown that fertiliser use in the region is generally minimal. Even in Fiji and the Solomon Islands where the largest quantities of fertilisers are used, the amounts used per hectare are generally low, often insufficient to replace the nutrients removed in harvesting. As a result, contamination of soils is unlikely, but nutrient movement via erosion (Asquith et al., 1994) may lead to problems in coastal areas.

A large number of pesticide materials are used, but again the quantities are small (Mowbray, 1988) and in only a few locations where it is economic to have intensive use of pesticides, are there likely to be any major environmental problems. Such areas are where high value crops are grown on limited areas, e.g., irrigated rice. One problem with this subject is the lack of facilities, trained personnel and funds in the region to undertake the analyses required to evaluate the pesticide content of soils. As a result, the number of quality data sets available is extremely small.

No comprehensive information is available on the extent of disposal of wastes on soils, but a recent assessment of land-based pollution sources gives some indication of what pollutants are produced and discharged in the region (Convard, 1993). Since the region has a very limited industrial base producing problem wastes, the issue is not seen as being a major one. However, in some small areas, problems have been identified, e.g., old transformers containing polychlorinated biphenyls (PCBs) have been left on the soil surface in places like Tongatapu, as there are no local means of disposal. Most regional waste disposal is through dumps, which are commonly located on the coasts, frequently on old mangrove areas, but on some of the larger islands, inland dumping does occur. These areas do represent a hazard for the surrounding environment, but the impact has usually been more serious for the marine zone.

Morrison and Brodie (1985) reviewed pollution sources in the region and showed that despite the lack of large industries, the region was prone to a range of problems due to poor environmental management, lack of legislation, inadequate enforcement and a lack of skilled technical staff to undertake both the field and laboratory studies to satisfactorily identify and remedy pollution problems. As a result, operations that produce pollutants have not been monitored in the past, except for certain mining activities.

Recent pressure for expanded industrial development throughout the region may lead to new problems. Increased use of toxic materials such as pesticides should be carefully planned so that problems that occurred in the past with the selection of inappropriate materials and the disposal of wastes (Mowbray, 1988) can be avoided. The work of the South Pacific Regional Environmental Programme (SPREP) and other regional and international agencies has prepared much of the groundwork for improved waste management, but countries have yet to fully implement the strategies developed.

There has also been poor infrastructure for recording land use history, such that no real database is available from which one could assess the potential for contaminated soils in a piece of land. This problem may not be too severe as the traditional communal ownership of land throughout the region has restricted the extent of non-agricultural land use.

3.3 Legislation

There is very limited legislation in the Pacific Islands dealing with contaminated soils. In Guam, Northern Marianas and American Samoa, United States Federal environmental legislation is applied through the local Environmental Protection Agencies, and in the French territories (New Caledonia, French Polynesia, Wallis and Futuna) European Union legislation applies. In the other countries of the region there is no specific legislation addressing contaminated soils issues, but some of the general environmental and conservation laws could be used to take action. As with many other environmental issues in the region, the effectiveness of this legislation is generally considered poor, as a consequence of poor design, inadequate resources, and lack of strong political support.

4. Case studies

A number of case studies is presented here to illustrate the contaminated soil problems that may arise in the region. Issues covered include organochlorines, dumps, hydrocarbons, mine tailings, Pb from traffic and sewage sludge disposal.

4.1 Organochlorines

Organochlorine pesticide residues in the marine environment around Fiji were recently investigated in soil samples from coastal areas where these materials had been used for over 20 years up to 1975 (Harrison, 1994). The results showed that the organo-chlorine concentrations were consistent with additions to the soils from global atmospheric transport processes – no evidence was found of significant residual concentrations. The very limited data from other investigations in the region indicate similar situations except where excess pesticide materials have been stored awaiting destruction.

4.2 Solid waste

As noted above the disposal of solid waste in the region is not well managed. The sites have not been monitored and often there is little control on the post-dumping use of the site. In Betio, Kiribati, as a result of land pressure, some old dump sites are now being used for housing and tree crops (T. Kiritome, personal communication, 1995). This issue should be very carefully monitored, as serious health problems may arise, particularly if tree roots are removed for use in traditional medicines.

A recent study of the main Suva dump in Fiji showed the presence of high concentrations of metals in both surface and subsurface soils (Naidu and Morrison, 1994). Heavy metal concentrations (Pb, Zn, As, Cu and Hg) decreased with distance from the centre of the dump. Given that there is effectively no control over what is discharged at the dump, the sources of these metals include a range of wastes derived from domestic and small industry sources. Garbage samples entering the dump were checked on a number of occasions and it was found by visual inspection that the metal content ranged from 11–17% by volume (no attempt was made to determine the composition of the metal wastes). The long term future of this site is uncertain, but it is essential that no houses or agriculture are developed there.

4.3 Mining

Mining is an important industry in several regional countries (Fiji, Nauru and New Caledonia). In Fiji, the legislation relating to mining is among the best from an environmental standpoint, but a number of problems still exist. One problem is disposal of tailings – this is done on land and the areas used are monitored. These sites have high concentrations of heavy metals, even to the extent that remining of the tailings has been considered. Long-term management of these sites must ensure that they are not used for agriculture or forestry or serious problems may result.

A second problem relates to mineral processing operations. These are found in Fiji, Nauru and New Caledonia. In Fiji, Au ores are processed via a cyanidation process with the residues stored within the mine boundary. The scale of the operation is very small and there are no major impacts on the surrounding soils. In New Caledonia, the Societe Metallurgique Le Nickel (SLN) Ni smelter is on the outskirts of the capital, Noumea. The smelter emits substantial quantities of gases and dust, but the prevailing south east winds carry much of this material out to sea. Local residents have reported the occurrence of dust within the city at times when the winds are from the west and some of this material must fall onto the surrounding soils. The impact of this on the soils needs further investigation.

In Nauru, the processing is mainly crushing and drying of the phosphate rock mined there for use in superphosphate manufacture. The high Cd concentration of Nauru soils has been discussed above; this also applies to the phosphate rock. This has led to some countries requiring reduction in the Cd concentration before shipment. A calcining process to do this was operated for a number of years and the waste from this process, containing even higher Cd concentrations ($>1000\,mg\,kg^{-1}$), was dumped on a site near the works. This is a serious potential problem as the small area of the island limits the opportunity for safe disposal.

4.4 Petroleum hydrocarbons and lead

Filling stations and oil storage areas represent a constant source of pollution in the region. Around many such facilities there are obvious visual signs of hydrocarbon contaminated soils, but the areas affected are usually small. No quantitative information is available.

Increasing affluence in the region has led to a dramatic increase in the number of motor vehicles in use. Unleaded petrol is not widely used and so the motor vehicles are discharging increasing amounts of Pb into the environment. The impact of this on roadside dusts has been investigated in Fiji (M. Maata, personal communication, 1995). The results are shown in Table 2. All the dusts are derived from the same parent material (Suva marl), and the soils are either Lithic or Typic Humitropepts. The data clearly show that Pb is being deposited in the area of high traffic density and this will be affecting the soils in these areas, where fruit and vegetables are grown for human consumption. The main concern relates to Pb containing dust being trapped on the plant surfaces and later consumed.

4.5 Sewage sludge disposal

Sewage treatment facilities have been developed in the major population centres in the region. A number of these have been overwhelmed by population increases, and

Table 2. Range of Pb concentrations in dusts in Suva, Fiji (mg kg^{-1}).

Type of area	Pb in roadside dust
Low traffic zones	59 – 99
Medium traffic zones	333 – 382
Heavy traffic zones	372 – 5161
Local soil background	5 – 20
NBS Standard	710 – 736 (range for about 10 analyses)
Certified value	714±28

*Data provided by M. Maata, University of the South Pacific.

represent major environmental problems. Even for those which are operating effectively, environmental concerns exist. The effluents are usually discharged into rivers or coastal waters where they create both microbial and nutrient problems. Disposal of the sludge varies from place to place.

As there is little heavy industry in the region the sludges usually do not contain large concentrations of the metals or organics that create the problems for sewage sludge disposal in more developed countries. The lack of environmental controls on the small industries means that some of the undesirable species may get into the sludge.

Sewage sludge is disposed of at sea, by burial on land, by surface disposal on land, or by use in farms and gardens. This latter use has been recommended in many locations for a long time, but adoption has been slow because of local cultural sensitivity. There is concern about the sites where surface disposal has occurred, as such sites are often uncontrolled, and their use for agriculture may lead to considerable problems, particularly if the material was originally contaminated to any extent with heavy metals or organics.

4.6 Radioactivity in Pacific Island soils

As a result of their isolation, a number of Pacific Islands have been used for nuclear experimentation. Atmospheric weapons testing occurred regularly in the Marshall Islands in the late 1940s and 1950s; in Kiritimati (Christmas Island) in the 1950s and 1960s; and in French Polynesia from 1966 – 1975. Underground testing continued in French Polynesia (Mururoa and Fangataufa) until the 1990s and resumed in 1995. The resulting contamination of the soils of islands close to the test sites has been of considerable interest to residents because of the potential health impacts.

In the northern Marshall Islands, atmospheric nuclear testing began in 1946 on Bikini atoll, and expanded to Enewetak in 1947. The local inhabitants were relocated prior to the tests being carried out. On March 1, 1954, the BRAVO test on Bikini exploded with about three times the projected intensity and the fallout travelled eastwards and seriously contaminated Rongelap and Utirik atolls, including the resident

Table 3. Average concentration of ^{137}Cs, ^{90}Sr, $^{239+240}Pu$ and ^{241}Am in soil at Rongelap, Marshall Islands (values in Bq/g dry weight)*.

Soil Depth cm	^{137}Cs	^{90}Sr	$^{239+240}Pu$	^{241}Am
0–5	0.48	0.19	0.13	0.096
5–10	0.22	0.12	0.037	0.034
10–15	0.10	0.11	0.018	0.018
15–25	0.040	0.081	0.007	0.007
25–40	0.013	0.052	0.003	0.003
40–60	0.007	–	–	0.001

*Data from Robison et al. (1994).

populations. These residents were evacuated a few days later, but not until they had been subjected to substantial internal and external exposure to radionuclides (National Research Council, 1994). All of the relocated populations wanted to return home as quickly as possible, but concerns for their health due to the remaining contamination delayed this. Sites in the Marshall Islands have been monitored for over 40 years and represent the most detailed field investigations of the impact of radioactivity on ecosystems including soils (Robison and coworkers, 1982; 1987; 1988; 1992; 1994).

Initially the problems were related to short lived and intermediate life isotopes (e.g., ^{55}Fe, ^{60}Co), but the contaminants of present concern in the northern Marshalls are ^{137}Cs (half-life 30.2 years) and ^{90}Sr (half-life 28.1 years), with Cs being the major source (>90%) of activity, and Sr contributing 2–5%. Plutonium and Am are both present in very small amounts (<2% each of activity), and will be present 'forever' (E.L. Stone, personal communication), but are not considered as serious problems for humans, apart from 1 or 2 sites on Rongelap. There are variations in the distribution of isotopes due to differences in the design of the devices tested.

In undisturbed areas the highest activities are in the surface soils, often with >80% of the activity in the top 15 cm (see e.g., Table 3). The activity decreases exponentially with depth, except in places where significant disturbance of the soils has occurred. The unusual composition of atoll soils, being dominated by calcium carbonate and having little or no clay, produces a very different pattern of element availability to plants from that found in continental aluminosilicate dominated soils. The contaminated Marshall Island soils contain very little K with Cs being taken up as an analog by plants. Strontium is taken into the carbonate matrix in place of Ca, leading to significant immobilisation. As a result the plant uptake of these elements is the reverse of what might be expected in continental soils. This is reflected in the concentration ratios (concentration in plant/concentration in soil) with values always being >1 for Cs and much less than 1 for Sr.

In addition to radioactive decay, there is evidence that Cs is being lost to groundwater through leaching, thus reducing the impact on soils and plants (Robison et al., 1988). The terrestrial food chain is still the most significant potential exposure pathway, but in some locations, there may be health problems from the intake of resuspended radioactive soil particles. While a number of 'hotspots' have been located, the relative uniformity in distribution of radioactivity indicates that there is little chance of finding many others, except where burial of contaminated material occurred or there was other major disturbance of the soils. It would be interesting to determine the impact of the continuing radiation on the soil microbial ecology, given the interest in the use of radiation in food treatment.

The concentrations of ^{210}Po and ^{210}Pb in the diet of Marshall Islanders were recently investigated by Noshkin et al. (1994). They found that local foods contribute 87% of the ^{210}Po and 47% of the ^{210}Pb associated with the diet. The items contributing the majority of the activity to the diet were derived from the marine environment. The presence of ^{137}Cs and its movement through the food chain, e.g., incorporation into coconuts, has been a major concern for the health of people returning to the contaminated islands. Robison and Stone (1992) have shown that periodic additions of K (at rates of about 100 kg ha^{-1}) would be highly effective in reducing ^{137}Cs in coconut food products. Although significant resettlement of the original populations has occurred, a number of islands are still off limits for food gathering.

The British and United States governments carried out atmospheric nuclear tests on Kiritimati in 1957, 1958 and 1962. Since that time the issue of contamination of the island's soils has been raised on a number of occasions. Investigations of the situation have been undertaken since 1964, with the most authoritative being that by the New Zealand Radiation Laboratory in 1981 (McEwan et al., 1981). This study concluded that radioactivity concentrations in soil were consistent with global fallout levels for a low rainfall equatorial area, and no site on the island presented a risk to the health of the resident population or required any restriction on land use.

Nuclear testing in French Polynesia has been one of the major regional environmental issues in the last 20 years. The reluctance of the French authorities to allow full investigation of the test sites has fuelled comments that contamination is occurring, even though the evidence to support these comments is limited. There is no doubt that the atmospheric testing did lead to significant contamination, but the impact of the underground tests is more difficult to assess. A small number of official investigations have been permitted on Mururoa, but these have been limited in extent (Atkinson et al., 1983; Martin et al., 1990). In particular, areas in the north west of Mururoa, the site of the atmospheric tests, have been excluded from the investigations. Sediment analyses have identified ^{60}Co, ^{155}Eu, ^{54}Mn, ^{137}Cs, and various Pu isotopes, and it is likely that these materials will be present in the soils of the atoll. The quantities are small, but continued testing with small leakages, will lead to increased soil concentrations. As the atoll is under French military control with imported food and restricted use of local resources, the direct impact of these contaminants is limited. The impact on surrounding islands which are occupied by Polynesians living a more traditional lifestyle is expected to be small, but in the absence of detailed investigations, this cannot be confirmed.

One other location where radioactivity in soils has been of major interest is on the island of Niue. This was first detected by Fieldes in 1955 while analysing samples for a soil survey of Niue (Marsden et al., 1958). The radioactivity was considerably higher than in any New Zealand soils or in samples from other Pacific Islands. The exact source of the radiation has been the subject of considerable debate for over 40 years, with a number of explanations having been forwarded. The problem relates not to identifying the isotopes responsible, but to determining how they got there, e.g., ^{230}Th and ^{231}Pa are present in relatively large amounts, but the parent isotopes ^{235}U, ^{238}U and ^{240}U only occur in very small quantities.

The favoured explanation at present (Whitehead et al., 1992) is that after the formation of weathered soils on the surface of the island, it was submerged for a period during a high sea level event. During the submergence the goethite and gibbsite soil material adsorbed various elements from seawater, including U. Upon emergence to subaerial conditions, the U daughter products accumulated, and following this, leaching of U occurred, leaving a soil radioactivity anomaly. The soils of Niue represent one of the most unique geochemical situations ever encountered in soils.

5. Potential problems

Activities that have the potential to contaminate soils and give rise to associated problems for land degradation in the South Pacific will obviously depend on the direction in which the island countries develop and the degree to which environmental controls are imposed on these activities. The concept of 'cleaner production' is almost unheard of and those industries that have environmental controls operate on the basis of 'end of pipe' treatment of wastes, which, in many cases, is not very effective. Many others still discharge wastes with blatant disregard to the impact on the environment. If this type of situation is allowed to prevail for much longer, a number of existing activities could result in significant contamination of the surrounding soils. Some of the potential sources of contamination are discussed below.

5.1 Timber treatment

Exploitation of forest resources of the South Pacific has increased dramatically in recent years, and a number of countries have established associated timber milling and treatment facilities. In Fiji alone there are numerous registered timber treatment operations, all using copper/chrome/arsenic preservatives. Although forestry regulations in Fiji contain adequate specifications for the appropriate design and environmental safe operation of facilities, there is very little monitoring of actual compliance with requirements. Under these circumstances, mismanagement of the facilities is always a possibility. Timber residues, such as sawdust, contaminated with preservatives are believed to be dumped on the land adjacent to these operations. Prolonged dumping of this nature has the potential to cause considerable contamination of the underlying soils. Since soils have a strong retention capacity for metals such as Cu and Cr, this contamination is not easily reversed.

5.2 *Waste disposal sites*

In practically every South Pacific island country the management of waste disposal is inadequate. Mixed dumping is common, and there are few controls in place to prevent the dispersion of contaminants from the disposal site. The impact on the underlying and surrounding soils will become a major concern, particularly as increasing development may lead to more hazardous substances being imported, with these materials ultimately ending up in waste disposal sites.

5.3 *Industrial sites*

There is an increasing tendency in a number of South Pacific countries to designate certain areas for the establishment of various types of industries. In Suva, Fiji, for example, there are at least three well established industrial zones. Although any one industry in an industrial zone may not be a significant risk to soil quality in the area, the concentration of many industries within a limited land area could be a long-term risk, particularly as there does not appear to be any move towards improved waste management systems.

5.4 *Dredging*

Sediments in coastal water bodies adjacent to heavily populated and developed areas are subjected to contamination from many land-based activities. The extremely high levels of tributyl Sn, for example, in the sediments off Suva harbour (Stewart and De Mora, 1992) have been attributed to the nearby ship-building and maintenance industry. Dredging of such areas is required from time to time because of excessive sedimentation, and the disposal of the dredge material on land could become a source of contamination for the receiving soil.

5.5 *Mining*

As indicated earlier, the geology of the region has led to numerous geochemical anomalies being produced. These are under extensive investigation, and it is anticipated that a number of new mining ventures will begin within the next decade (in Fiji, Solomon Islands and Vanuatu). Mining such as that proposed for Namosi in Fiji could expose subsurface layers with relatively high levels of heavy metals to mobilisation by the forces of nature.

6. Conclusions and recommendations

The Pacific Islands have limited land areas which are having to cater for growing populations. There is significant pressure on the land resources, which must provide the local residents with food, fibre, or export crops. The small size of the islands and the large maritime exclusive economic zones means that major significance is being placed on exploitation of the region's marine resources. As a result, increasing emphasis is being placed on the quality of the marine environment and its protection from pollution. Important as this may be, however, it is equally important that the quality of the land environment is also protected, because the limited areas of usable soils mean that Pacific Islanders cannot afford to lose areas that have become contaminated.

The information currently available indicates that soil contamination is not extensive, but in certain localised areas there are significant problems. This may be more the result of non-detection than anything else, as there have been very few assessments of the extent of the problem. It is important that more effort be put into such investigations, not only to detect areas of contamination, but also to provide baseline data on soil quality for future assessment of contamination.

There is an urgent need for a coordinated regional approach to improved waste management, including industrial and urban solid wastes and liquid wastes. Expanded recycling activities and an effective program aimed at waste reduction would reduce the potential contaminated soil problems, with the subsequent impacts on the fragile small island environments.

In order to address the present and potential problems with contaminated soils in the Pacific Islands, the following activities are recommended:

- In the absence of satisfactory waste management systems in the South Pacific, contamination of land must be recognised as a serious threat in national and regional environmental programs.
- Throughout the region, Departments of Lands have databases on land use and ownership. Using this information, each government should attempt to identify sites of possible contamination due to the history of land use at the site.
- In areas of suspected contamination, studies should be carried out to determine the extent of the contamination.
- A program be established for collection of region-wide baseline data for soil quality.
- Using the information available from pollutant inventories, the types of likely soil contaminants can be identified. This information can be used to compile documentation on remediation of soils contaminated with these pollutants, and this information should be made available to regional Departments of Agriculture, Environment, and Lands.
- A series of workshops be run in the region to train government officers, industrial managers, scientists and farmers on the problems associated with contaminated soils and methods of quantifying and remediating the problems.

Acknowledgements

The authors gratefully acknowledge the assistance of Dr. W.L. Robison and Dr. E.L. Stone in providing information about radiation contaminated soils in the Marshall Islands.

References

Asghar M, Davidson T J and Morrison R J 1988 Soil Taxonomy and Fertility in the South Pacific. University of the South Pacific, Apia, 548 p.

Asquith M, Kooge F and Morrison R J 1994 Transportation of Sediments via Rivers to the Ocean and the Role of Sediments as Pollutants in the South Pacific. Reports and Studies No. 72, South Pacific Regional Environment Programme, Apia, Western Samoa. 79 p.

Atkinson H R, Davies P J, Davy D R, Hill L, McEwan A C and Jamieson D J 1983 Report of a New Zealand, Australian and Papua New Guinea Scientific Mission to Mururoa Atoll. New Zealand Ministry of Foreign Affairs, Wellington. 166 p,

Bakker M L 1988 Populations of the South Pacific: an overview of demographic levels, patterns and trends. Paper presented at the USP/UNFPA Regional Symposium on Population and Development Planning, Population Studies Programme, University of the South Pacific Suva.

Birrell K S, Seelye F T and Grange L I 1939 Chromium in the soils of Western Samoa and Niue Island. NZ. J. Sci. Tech.. 21A, 91A-95A.

Convard N 1993 Land-based Pollutants Inventory for the South Pacific Region. Reports and Studies No. 68, South Pacific Regional Environment Programme, Apia, Western Samoa. 165 p.

Craswell E T 1989 The role of fertilizers in smallholder agriculture. In Soil Management and Smallholder Development in the Pacific Islands. Eds. E Pushparajah and C R Elliot. pp 99-109. IBSRAM, Bangkok.

Harrison N L 1994 Organochlorine compounds in the marine environment around Fiji. MSc thesis, University of the South Pacific, Suva, 117 p. (Unpublished)

Haynes R J and Naidu R 1989 Soil fertility and management considerations for efficient crop production in the South Pacific. In: Agricultural Development in the Pacific Island in the 90's. Eds. R.J. Haynes and R. Naidu. pp 21-47. USP, Suva, Fiji.

Latham M, Quantin P and Aubert G 1978 Etude des Sols de al Nouvelle-Caledonia. Notice Explicative No. 78, ORSTOM, Paris, 138 p. + maps.

Leggo M D 1977 Contrasting geochemical expressions of copper mineralization at Namosi, Fiji. J. Geochem. Explorat. 8, 431-456.

Leslie D M 1984 A proposal for an Oceania Benchmark Sites Network for Agrotechnology Transfer (OBSNAT). In Proceedings of the International Symposium on Minimum Data Sets for Agrotechnology Transfer. Ed. V Kumble. pp 33-54. ICRISAT Centre, India.

Marsden E, Fergusson G J and Fieldes M 1958 Notes on the radioactivity of soils with application to Niue Island. Proceedings of the Second International Conference on the Peaceful Uses of Atomic Energy, Geneva. 18, 514.

Martin J-M, Thomas A J, Charrier B, Cousteau J-Y, and Sarano F 1990 Assessment of artificial radionuclides issued from French nuclear bomb testing at Mururoa (French Polynesia). Environ. Technol. 11, 197-204.

McEwan A C, Matthews K M and Gregory L P 1981 An Environmental Radiation Survey of Christmas Island, Kiribati. Report No. 1981/9, National Radiation Laboratory, Christchurch, New Zealand.

Morrison R J 1982 Bibliography of Soils Information from the South Pacific. Pacific Information Centre Selected Bibliography No.1, University of the South Pacific Library, Suva. 62 p.

Morrison R J and Brodie J E 1985 Pollution problems in the South Pacific: fertilizers, biocides, water supplies and urban wastes. In Environment and Resources in the Pacific, Eds. A. Dahl and J. Carew-Reid. pp 57-67. Regional Seas Reports and Studies No. 69, UNEP, Nairobi.

Morrison R J 1987 Comments on the Soils of Nauru. Report prepared for the Commission of Inquiry into the Rehabilitation of the Worked-out Phosphate Lands on Nauru, Melbourne. 57 p.

Morrison R J, Naidu R and Singh U 1987 Sulphur in the agriculture of Papua New Guinea and the South Pacific Islands. In Fertilizer Sulphur Requirements and Sources in Developing Countries of Asia and the Pacific. pp 57-66. FADINAP/FAO/Sulphur Institute/ACIAR, Bangkok.

Morrison R J and Masilaca A S 1989 Changes in the properties of a Fijian Oxisol following sugarcane cultivation. In Soil Management and Smallholder Development in the Pacific Islands. Eds. E Pushparajah and C R Elliot. pp 271-281. IBSRAM, Bangkok.

Morrison R J, Naidu R, Singh U and Leslie D M 1989 Andisols and related soils in the South Pacific Islands. In Properties, Classification and Utilization of Andisols and Paddy Soils. Eds. S. Shoji and D. Kinloch. pp 459-468. Soil Science Society of Japan and Soil Management Support Services, USDA, Tohuku University, Sendai.

Mowbray D L 1988 Pesticides in the South Pacific. Regional Seas Reports and Studies No. 89, UNEP, Nairobi. 275 p.

Naidu S D and Morrison R J 1994 Contamination of Suva Harbour, Fiji. Mar. Poll. Bull. 29, 126-130.

National Research Council 1994 Radiological Assessments for Resettlement of Rongelap in the Republic of the Marshall Islands. National Academy Press, Washington, D.C. 108 p.

Noshkin V E, Robison W L and Wong K M 1994 Concentration of ^{210}Po and ^{210}Pb in the diet of the Marshall Islands. Sci. Environ. 155, 87-104.

Page A L and Bingham F T 1973 Cadmium residues in the environment. Res. Rev. 48, 1-44.

Robison W L, Mount M E, Phillips W A, Conrado C L, Stuart M L and Stoker C E. 1982 The Northern Marshall Islands Radiological Survey: Terrestrial Food Chain and Total Doses. Report No. UCRL-52853 Pt. 4, Lawrence Livermore National Laboratory, Livermore, California, 77 p.

Robison W L, Conrado C L and Phillips W A 1987 Enjebi Island Dose Assessment. Report No. UCRL-53805, Lawrence Livermore National Laboratory, Livermore, California. 57 p.

Robison W L, Conrado C L and Stuart, M L 1988 Radiological Conditions at Bikini Atoll: Radionuclide concentrations in Vegetation, Soil, Animals, Cistern Water, and Ground Water. Report No. UCRL-53840, Lawrence Livermore National Laboratory, Livermore, California. 60 p.

Robison W L and Stone E L 1992 The effect of potassium on the uptake of ^{137}Cs in food crops grown on coral soils: Coconut at Bikini Atoll. Health Phy. 62, 496-511.

Robison W L, Conrado C L and Bogen K T 1994 An Updated Dose Assessment for Rongelaap Island. Report No. UCRL-LR-107036, Lawrence Livermore National Laboratory, Livermore, California. 59 p.

Rugless C S 1983 Lithogeochemistry of Wainaleka Cu-Zn volcanogenic deposit, Viti Levu. Fiji, and possible applications for exploration in tropical terrains. J. Geochem. Explorat. 19, 563-586.

Soil Survey Staff 1975 Soil Taxonomy: A Basic System for Making and Interpreting Soil Surveys. USDA Handbook No. 436, US Government Printer, Washington, D.C. 759 p.

Soil Survey Staff 1994 Keys to Soil Taxonomy, Sixth Edition, Soil Conservation Service, US Department of Agriculture, Washington, D.C. 306 p.

Stewart C and De Mora S J 1992 Elevated tri(n-butyl)tin concentrations in shellfish and sediments from Suva harbour, Fiji. App. Organomet. Chem. 6, 507-512.

Thaman R R 1982 Deterioration of traditional food systems, increasing malnutrition and food dependency in the Pacific Islands. J. Food Nutrit. 39, 109-121.

Ure A M and Berrow M L 1982 The elemental composition of soils. In Environmental Chemistry, Volume 2. pp 94-204. Specialist Periodical Reports, Royal Society of Chemistry, London.

Ward R G and Proctor A 1980 South Pacific Agriculture – Choices and Constraints. Asian Development Bank, Manila, and Australian National University, Canberra. 525 p.

Wells N 1960.Total elements in topsoils from igneous rocks: An extension of geochemistry. J. Soil Sci. 11, 409-424.

Whitehead N E, Barrie J and Rankin P 1990 Anomalous Hg contents in soils of Niue Island, South Pacific. Geochem. J. 24, 371-8.

Whitehead N E, Ditchburn R G, McCabe W J and Rankin P 1992 A new model for the origin of the anomalous radioactivity in Niue Island (South Pacific) soils. Chem Geol (Isotope Geoscience Section). 94, 247-260.

24
Contaminants and the soil environment in Sri Lanka

A. M. MUBARAK, G. K. MANUWEERA AND R. SENVIRATNE

1. Introduction

Sri Lanka is a pear shaped island, separated from the Indian continent by the Palk Strait, which at its narrowest has a width of 29 km. The island covers an area of about 65 610 km^2 and is approximately 430 km long and 225 km wide at its maximum extremities. The estimated population is 17 million.

2. Climate and soils

Sri Lanka lies between north latitude 6° and 8° and has a tropical monsoon climate, with a distinct bimodal rainfall pattern. The annual rainfall ranges from about 1 250 mm in the dry region to about 5 000 mm in the wet region. On the basis of rainfall, vegetation and present land use the country is divided into three main agro-climatic zones: dry; intermediate; and wet, and these are further divided into 24 sub-units depending mainly on rainfall, elevation and soil type (Figure 1) LWUD, 1979). There are 17 different soil units identified in Sri Lanka (Table 1) (Alwis and Panabokke, 1972). Reddish brown earth (Alfisol) is the largest soil unit in the dry zone of Sri Lanka, while Red Yellow Podzolic (Ultisols) soils of different levels of laterization are the main soils of the wet and intermediate zones.

3. Economy and industry (Central Bank of Sri Lanka, Report 1993)

Sri Lanka has made considerable economic and social progress since gaining independence in 1948. Such factors as literacy rate (89%), population growth (1.1%), and life expectancy (71 yr) are significantly superior to those of countries showing comparable levels of economic development. However it is still a low income economy with a per capita GNP of US$ 526 in 1993. During the past decade the economy of Sri Lanka has shown a slow but steady shift from its agricultural base to industry. In 1993 contributions to the economy from agriculture, industry and services were 21%, 28% and 51%, respectively. Despite the rising role of industry, agriculture has remained a base of the Sri Lankan economy (CBS, 1993). Approximately 2.9 million hectares are under cultivation. Major plantation crops are tea, rubber and coconut while cultivation of vegetables and other food crops with export potential is being actively promoted.

Industrial development that has taken place so far in the country has been mainly small and medium scale, and confined to only a few sub-sectors like food and beverages, textiles and garments, non-metallic minerals, etc. Manufacturing indus-

R. Naidu et al. *(eds): Contaminants and the Soil Environment in the Australasia-Pacific Region, 677–690.*

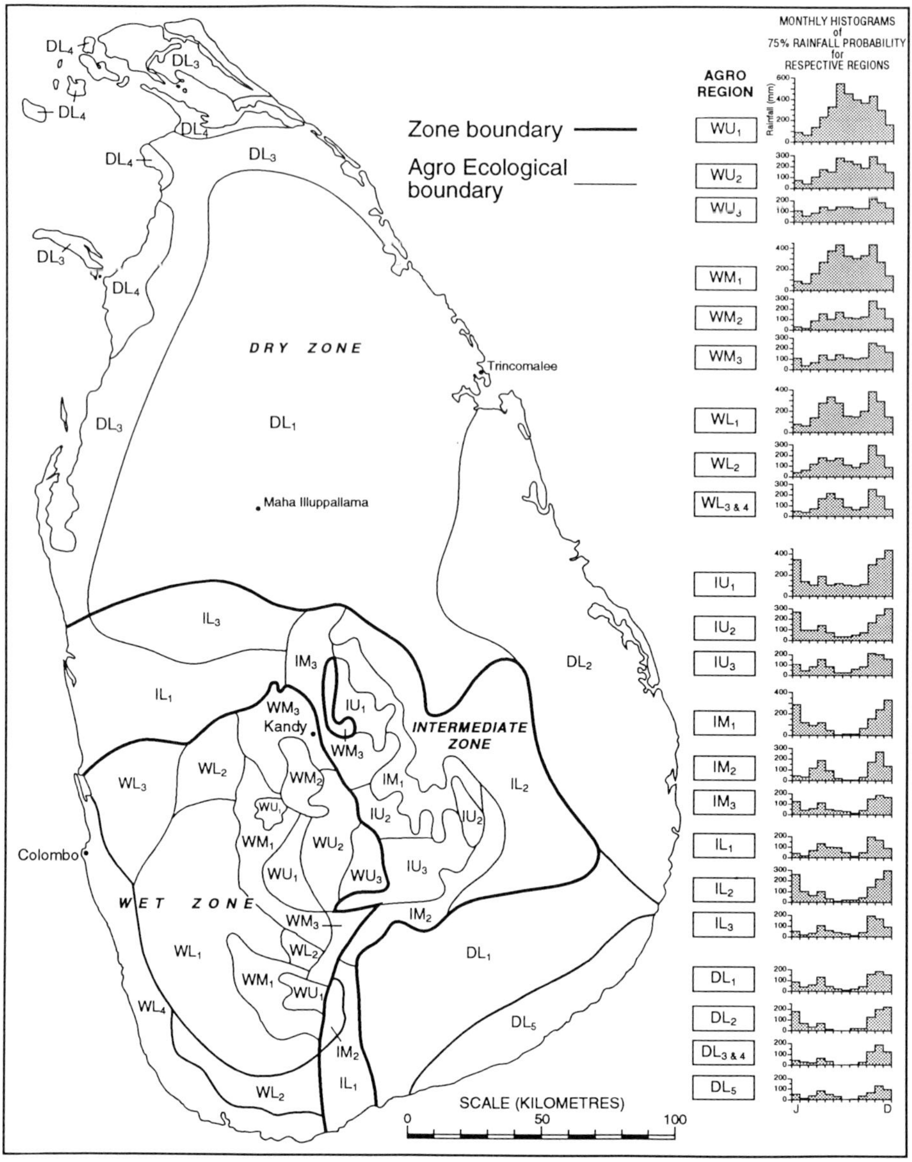

Figure 1. Agroecological regions of Sri Lanka.

tries account for less than 20% of Sri Lanka's GDP. They comprise: (a) a few large scale heavy industries, including cement, paper, steel, petroleum refining, sugar, ceramics, and textiles, most of which are publicly owned, and located away from urban areas; and (b) small and medium industries, which are largely privately

Table 1. Soils of Sri Lanka.

Soil Order	Sub Order	Great Soil group	Great Soil groups of Sri Lanka
Alfisol	Ustalfs	Rhodustalfs	Reddish Brown Earths
		Haplustalfs Non	Calcic Brown soils
	Aqualfs	Tropaqualfs	Low Humic Glay soils*
		Natraqualfs	Solodized Solonetz
Ultisols	Udults	Rhodudults	Reddish Brown Latosolic
		Plinthudults	
		Tropudults	
	Ustults	Tropustults	Red Yellow Podzolic soil
	Humults	Tropohumults	
	Aquults	Tropaquults	Low Humic Glay soils**
Oxisols	Ustox	Eutrustox	
		Haplustox	Red Yellow Latosols
Vertisols	Usterts	Pellusterts	Grummusols
Aridsols	Orthids	Salorthids	Salonchaks
Entisols	Aquents	Tropaquents	
	Fluents	Tropofluents	
	Ustents	Tropustents	Alluvial soils
	Psamments	Quartzipsamments	
Inceptisols	Aquepts	Tropaquepts	Half bog soils
	Tropepts	Ustropepts	Immature Brown Loams
Histosols	Fibrists	Tropofibrists	
	Hemists	Tropohemists	Bog soils
	Saprists	Troposaprists	

*High base saturation; **Low base saturation.
Summarized from the Handbook of Soils of Sri Lanka, Alwis and Panabokke (1972).

owned. They are few in number and are for the most part established outside the main urban areas. The urban sector is dominated by small and medium sized manufacturing industries including tanneries, textiles, garments, food processing, paints, varnishes, cosmetics and other chemical products, glass, asbestos products, etc. Industries in the rural area are small, cottage scale units using traditional technology. They include foundries, handlooms, handicrafts, bricks and tiles, and agro-based industries like rubber processing, rice milling, coir and coconut processing, and essential oil extraction.

4. Sources of soil contamination

4.1 Industry

The problem of soil contamination in the urban area may at first not appear serious, due to the low level of industrialisation in the country. However in localities with

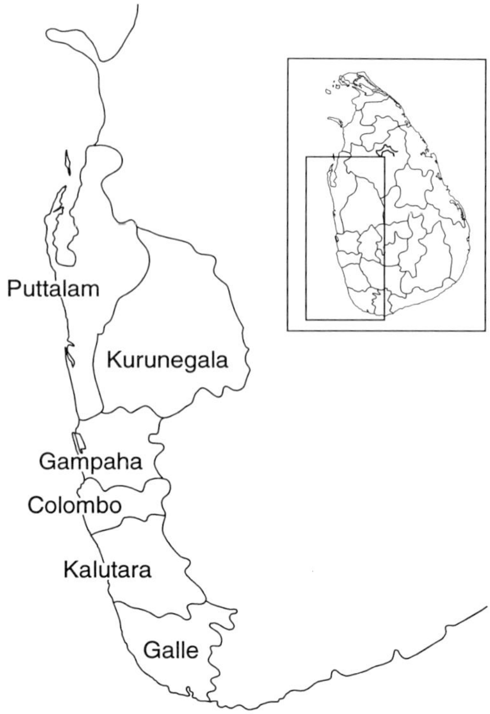

Figure 2. Map of Western Sri Lanka.

industries, degradation of the environment is quite serious and warrants urgent attention. Colombo and Gampaha districts are the most developed regions in the country, accounting for nearly 80% of the industrialisation (Figure 2) (ERM, 1994). Two major industrial areas are the Ratmalana-Moratuwa area located south of Colombo, and the Ekala-Jaela area located in the Gampaha District, about 20 km north of Colombo. Other industrial areas such as Biyagama, Homagama, Oruwila, and Kalutara are relatively small. Figure 3 depicts these industrial areas and also the proposed future areas for industrialisation (JICA/MIST, 1993)

According to a recent study (AE, 1994), there are more than 225 industries located within the boundaries of the Ratmalana-Moratuwa area. The textile and garment trade, pharmaceutical (formulation and packaging) and asbestos roofing sheet manufacture constitute the major proportion of the industries. In addition there are a small number of metal finishing industries, timber processing facilities and Pb storage battery manufacturing units within this zone. Two major transportation related maintenance facilities are also located in this area. These service facilities have large areas which are significantly contaminated with liquid wastes including

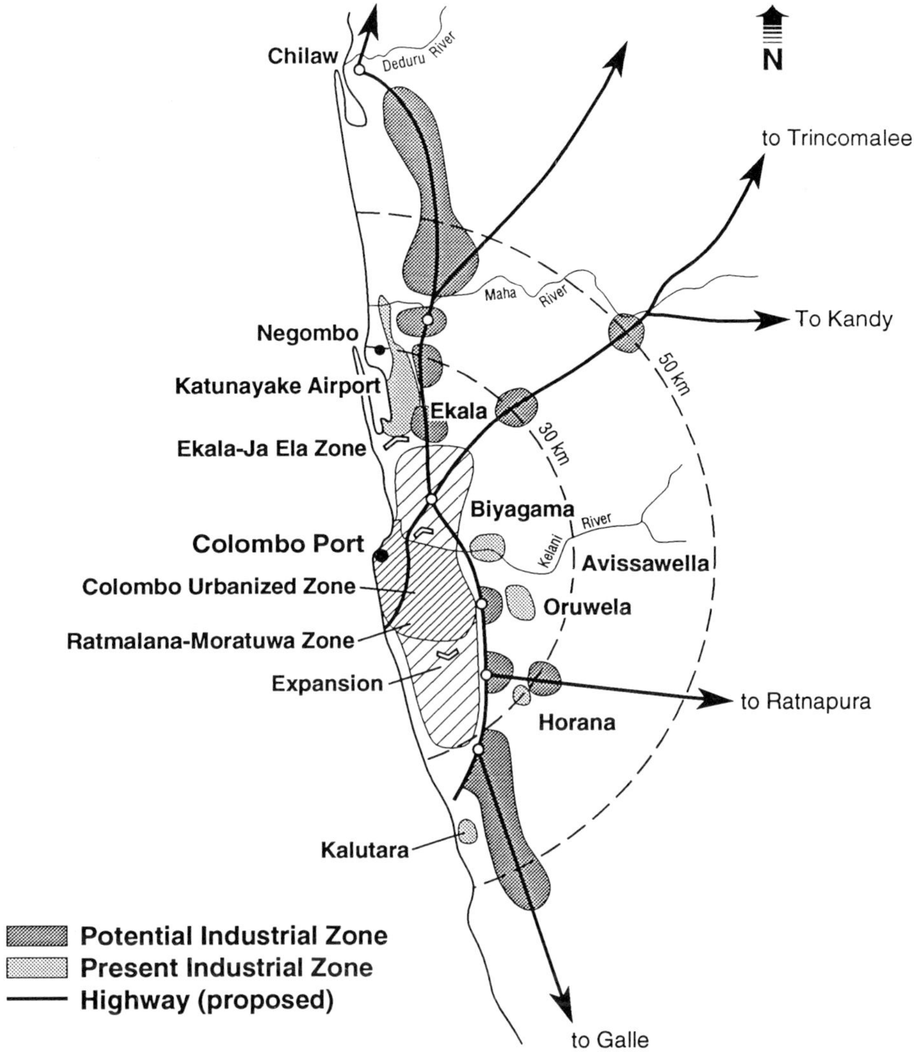

Figure 3. Present and potential industrial zone in Sri Lanka.

fuel, lubricating oils, brake oils, hydraulic oils, transmission fluids and battery acid. There is no treatment for the majority of the industrial effluents produced in this area and most of them are discharged to the surface drainage system. Industrial solid wastes are generally collected by private contractors but the final disposal site of these wastes is not known. Table 2 gives the estimated process wastewater loadings which are generated from about 100 industries in the area.

The wastes from these industries are mainly liquid. Land disposal of these

Table 2. Estimated Process Wastewater Loading in the Ratmalana Moratuwa Area.

Constituent*	Textile manufacturing	Metals fabricating	Food & beverages	Chemical & pharmaceutical	Training centers & workshops	Printing	Plastic and other industries	Total
Flow ($m^3 d^{-1}$)	2652	155	283	112	1884	25	45	5156
BOD5	1522	4	125	22	10	7	0.7	1692
COD	3368	206	188	63	2200	17	4	6046
Total suspended solids	318	6	28	11	979	2	2	1347
Total dissolved solids	3121	13	87	38	22	2	1	3284
Total Kjeldahl nitrogen	385	2	10	2	47	0.4	0.5	446
Total phosphorus	9.28	0.1	1	0.2	3.96	0.4	0.02	15
Oils and grease	61.0	2	3	0.2	1884	0.4	0.2	1950
Heavy metals	n.a.	n.a.	n.a.	n.a.	n.a.	n.a.	n.a.	n.a.

*all units are expressed as $kg\,d^{-1}$ unless indicated otherwise.
n.a. = not available.

Table 3. Profile of industrial sectors generating toxic metal waste in Sri Lanka.

	Tanning	Metal Finishing	Paints & Chemical
Number	15	76	33
Total wastewater volume $m^3 d^{-1}$	1614	6692	928
Total toxic metals* $kg d^{-1}$	161	669	93
Location	Colombo and Gampaha Districts	Widely distributed amongst four sub-areas of Colombo	50% in Ratmalana, the rest mainly in Ekala and North Colombo

*Based on assumed average concentration of 100 mg L^{-1}.
Source: Environmental Resource Management Report 1994.

wastes has degraded water bodies considerably in the area . The aquatic life of the Lunawa lagoon has been seriously degraded by the continued discharge of wastewater into its tributaries. The lagoon, which supported a significant fisheries industry a decade ago, is totally devoid of aquatic life today. There has been very significant settling of industrial waste sludge and the bottom of the Lunawa lagoon is reported to be covered by approximately 1 m of sludge.

At present no information is available on heavy metal loads in these waste waters but toxic wastes such as B and Cu, Cr, As from timber processing, Pb from storage battery production, Zn and other heavy metals from galvanising and other metal finishing industries are being generated in these areas and may pose a threat to the soil environment.

A similar study has also been carried out in the Ekala-Jaela industrial area, the major industrial estate in the Gampaha District (SWL, 1993). A total of about 140 industries are located within this area and of these about 50% are of a high polluting nature. The major effluent generating industries have been identified as textile dyeing and bleaching, food processing, asbestos, Al extrusion products, leather tanning, metal finishing, agricultural produce and mineral products. Only a few of the facilities have treatment plants while the majority discharge untreated effluent to the outside drain. Some of the potential toxic metal wastes generated in this area are Cr waste from tanneries and heavy metal wastes from an alkaline battery manufacturing facility. In addition, an Al fabrication factory generates significant amounts of alumina sludge with Cd impurities. These solid wastes are dumped within the site, transported outside for reuse, or dumped in marshy areas by contractors. Here again there is no reliable data on the generation of toxic metal wastes and their effect on the soil environment.

A study carried out by Environmental Resources Management in 1994 (ERM, 1994) has made some estimates of toxic metal generation by various industrial sectors (Table 3).

Toxic metal waste generation per day from the tanning, metal preparation and finishing, and chemicals and paints sectors is 161, 669 and 92.8 kg, respectively.

Environmental legislation has existed in Sri Lanka for more than 12 years, but it is only during the past five years that serious attempts have been made to control industrial pollution; firstly through an amendment to the 1980 National Environmental Protection Act, in 1988, and subsequently through the development in 1990 of regulations introducing an environmental protection licensing (EPL) system and specific effluent standards. More recently regulations have been developed covering areas such as environmental impact assessment (EIA) categorisation of industries into (a) those that should not require EPLs (Category A, non- or low-polluting), (b) those that require environmental clearance and EPLs but no EIAs (Category B, polluting) and (c) those which require EIAs (Category C). Further regulations covering the control of pollution from existing industries, control of hazardous wastes and the establishment of incentive and penalty systems are planned. It has been estimated that there are over 16000 industries in the country, of which Category B and C industries together exceed 4000 (ERM, 1994).

4.2 Air pollution

As in every urban area, air pollution in the Colombo Metropolitan Area (CMA) originates from the economic activities of various sectors, including both stationary and mobile sources. A recent study under the World Bank executed Metropolitan Environment Improvement Programme (MEIP/WB, 1993) has made a rough measure of air pollution that is inferred from energy consumption. Figure 4 illustrates the sectoral energy consumption in TOE (Tonnes of Oil Equivalent) and indicates that households are the main energy consumer (40%) followed by industry (33%) and transport (26%). Commercial and power sector consumption are negligible (0.3%). The main household fuel source is wood which is used by 80% of CMA households; it contributes primarily particulates to the atmosphere. Fuel for industrial processes is mainly furnace oil and this contributes particulates and gases to the atmosphere. Diesel and petrol account for the transport sector's 25% share of energy consumption in the CMA. While these mobile sources are the smallest of the three main consumption sectors, they have a very significant impact on air pollution for the following reasons: (a) the poor combustion performance of vehicle engines produces a disproportionately high level of contaminants from vehicles; and (b) there has been a tripling in vehicle registration in the CMA over the past ten years and a further doubling is expected by the year 2000. Measurements of pollution load from motor vehicles are approximate due to the lack of sufficient data. Available estimates use standard values from USEPA and WHO guidelines. A depiction of the inventory of major air pollutants by sectors is presented in Figure 5. The estimated amount of each emission in tons per year by sector is included in parentheses. It is clear from this figure that in the CMA the overwhelming majority of four of the five major air pollutants results from petroleum combustion coming from the transport sector.

Limited studies have been carried out on ambient air quality levels at busy

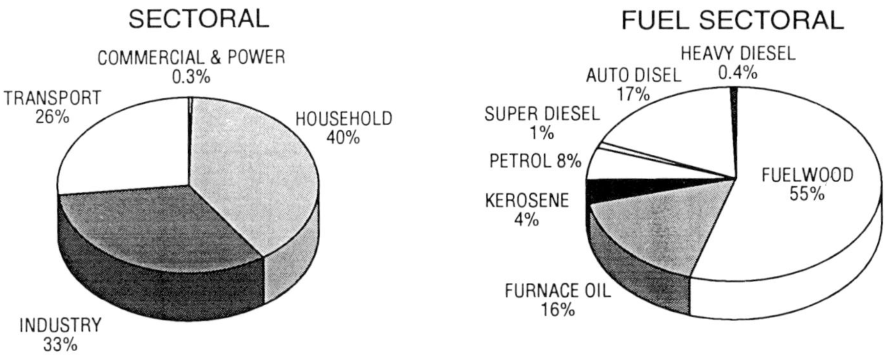

• TOE or *Tonnes of Oil Equivalent* is an expression of energy in which 1 TOE + 41.84 Giga Joules, eg:

1Mt of	Petrol	=	1.09 TOE
1Mt of	Kerosene	=	1.05 TOE
1Mt of	Diesel	=	1.05 TOE
1Mt of	Fuel Oil	=	0.98 TOE
1Mt of	Firewood	=	0.38 TOE

Figure 4. Sectoral energy consumption in Sri Lanka.

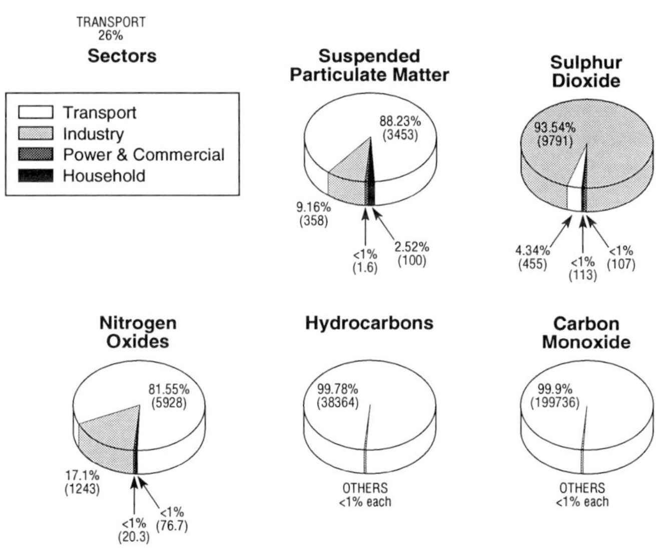

Figure 5. Sectoral sources of petroleum combustion emissions in the Colombo Metropolitan Area.

traffic junctions in the city of Colombo (NARESA, 1994). The range of pollutant levels detected in ambient air were: suspended particulate matter (SPM) 0.2–0.5 $mg\,m^{-3}$ (8 h average); CO 1.1–11 $mg\,m^{-3}$ (3 h average); NO_2 7.0–45 $\mu g\,m^{-3}$ (8 h average); SO_2 3.9–34 $\mu g\,m^{-3}$ (8 h average); total hydrocarbons (THC) 2.6–3.2 $\mu g\,mL^{-1}$ (3 h average); and Pb 0.22–0.58 $\mu g\,m^{-3}$ (8 h average). According to this study the lead level is within the WHO recommended limits and does not appear to be a serious threat to public health. Currently in Sri Lanka all petrol contains Pb in the form of tetra-ethyl-Pb at an average of 0.24 $g\,L^{-1}$ or a maximum of 160 $t\,yr^{-1}$. It is now being recommended that Pb additives be reduced to 0.15 $g\,L^{-1}$ (MEIP/WR, 1993). Under these circumstances Pb from vehicular emissions may not become a significant contaminant to the soil environment.

4.3 Agriculture

The use of chemical fertilisers was initiated at the turn of the century, particularly on plantation crops, and it was extended to food crops in the 1960s. A sharp increase in population during the late 1950s and increasing labour costs have resulted in extensive dependence on chemicals in agriculture. Further, the use of fertiliser was encouraged by providing a price subsidy to increase food production to match increasing demand. This was particularly important in paddy rice cultivation where fertiliser use increased more than three fold from 1969 to 1993. However, this subsidy was removed in 1990, with increasing emphasis on efficient use, and has thus resulted in a significant decrease in consumption.

The total fertiliser consumption in 1994 was 548×10^3 Mt, of which paddy accounted for 250×10^3 Mt, tea 147×10^3 Mt and coconut 35×10^3 Mt. Sulphate of ammonia, urea, rock phosphate, muriate of potash, triple super phosphate and mixed fertilisers are the major types of fertilisers used in Srī Lanka. Of these urea is the most popular (35%), followed by sulphate of ammonia (26%). In 1993 the quantities of nutrients (NPK) used in Sri Lanka were 11 539.8 and 58.5×10^3 Mt, respectively (NFS, 1994).

Plantation crops such as tea, coconut and rubber use relatively low quantities of pesticides. However in paddy the use of pesticides is relatively high and herbicides are heavily used in comparison with fungicides and insecticides. In contrast insecticides and fungicides are extensively used in vegetable cultivation, particularly in cash crops such as potato, chilli and onion. About 100 pesticide active ingredients have been registered in the country. Herbicides are the most common agrochemicals with 2237 Mt used in 1993. Propanil and MCPA accounted for 87% of the herbicides used. Inorganics, mainly S and Cu salts and dithiocarbamate fungicides, are the most popular among farmers. Carbamates are the most used insecticides with 1126 kL in 1993. Carbofuran 3% granules account for nearly 80% of the product used. Mono-crotophos and methamidophos, two popular organophosphate insecticides, are being phased out due to their acute toxicity and indiscriminate use. Pyrethroid insecticides have been severely restricted due to cross resistance buildup. The import of organo-chlorine insecticides, except endosulfan and lindane, is banned for environmental considerations (ROP, 1995).

5. Impact of agrochemicals on soil

No sites or regions have been identified as highly contaminated by agrochemicals as there is no proper monitoring system to determine the extent of soil contamination. However, lands undergoing vegetable cultivation are potential sites for contamination due to the high intensity of agrochemical use.

A preliminary investigation carried out at the Ceylon Institute of Scientific and Industrial Research (CISIR) to examine Cd contamination in Shii-Take mushrooms cultivated in Sri Lanka has indicated that the probable source of Cd contamination is rubberwood saw dust which is used as the medium to grow the mushrooms (CISIR, 1995). In the nine mushroom samples analysed the Cd concentration ranged from 0.03–0.29 $mg kg^{-1}$ (wet basis) and all except one were above the Australian tolerance limit of 0.05 $mg kg^{-1}$. Cadmium residues were also detected in rubberwood saw dust samples (12) collected from different locations and the values ranged from 0.035–0.14 $mg kg^{-1}$ (wet basis). A typical imported rock phosphate sample had a Cd level of 5 $mg kg^{-1}$. Since rock phosphate has been used in the rubber plantations for the past several decades, it appears that the Cd has been transferred from the rock phosphate to the Shii-Take mushrooms via the rubberwood saw dust. If this is so, accumulation of Cd residues in Shii-Take mushrooms cultivated on rubberwood saw dust appears to be inevitable. A detailed study should therefore be undertaken to assess the extent of Cd contamination in rubber plantations and after due consideration of the health effects, to decide on a reasonable tolerance limit for Cd in Shii-Take mushrooms grown on rubberwood saw dust.

A comprehensive study on the impact of agriculture on groundwater quality at Kalpitiya peninsula (BGS report, 1992), where the soil is highly permeable with a shallow watertable, has revealed that irrigation wells within the extensively cultivated area had NO_3 concentrations in excess of the WHO guideline of 10 $mg N L^{-1}$ and Cl^- concentrations in these wells were typically in the range of 50–200 $mg L^{-1}$. By contrast, domestic wells located outside the intensively cultivated areas had much lower NO_3^--N and Cl^- concentrations, generally less than 100 and 200 $mg L^{-1}$, respectively (Figure 6). This research has clearly demonstrated that the leaching of agrochemicals from intensively cultivated soils is responsible for the elevated concentrations of Cl^-, NO_3^- and K observed in many of the irrigation wells in the peninsula. This research has also demonstrated that carbofuran, a widely used insecticide, is leached to the water table within ten days of application. However the risk to water supplies posed by this compound is judged to be low because of its rapid degradation and its dilution within the aquifer (Figure 7).

Apart from the above study few data are available on pesticide contamination of the soil environment. As a result, when pesticides are registered special consideration is given to the fact that there is no mechanism available to assess the extent of

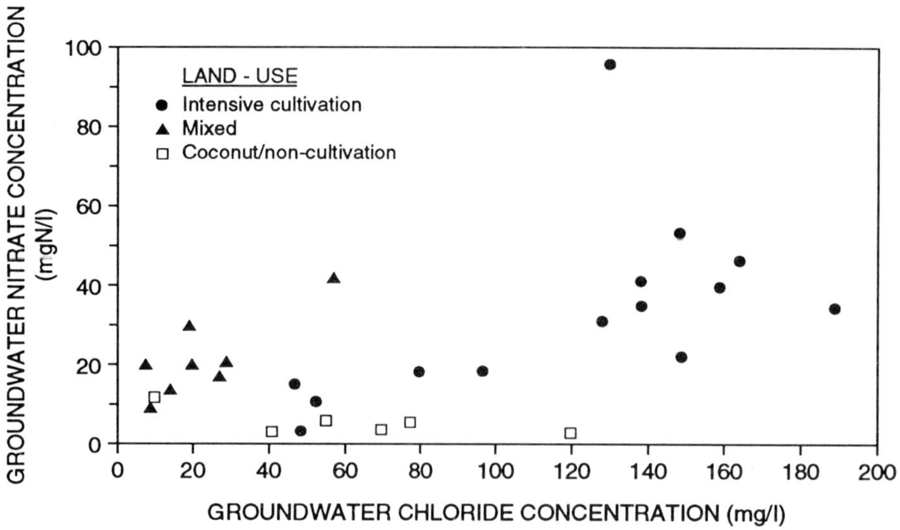

Figure 6 Impact of agriculture on the groundwater quality.

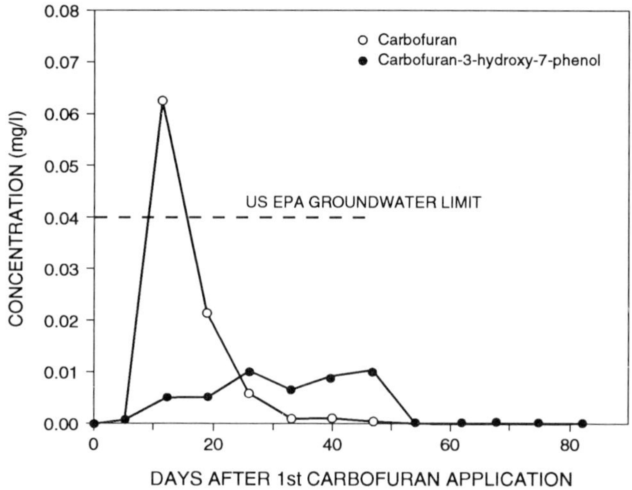

Figure 7 Carbofuran and metabolite in shallow piezometers, Maha 1990/91. - Kalpitiya Peninsula (BGS, 1992).

soil contamination. Thus, persistent pesticides with high soil leaching properties are not considered for registration, or are severely restricted to minimise contamination.

6. Institutional framework and regulatory measures

The National Environmental Act (NEA) No 17 of 1980 is the first comprehensive legislation on environment in Sri Lanka to control and regulate the discharge of toxic contaminants to the environment. The NEA established the Central Environmental Authority (CEA) as a policy making and coordinating body. This Act was amended later in 1988, Amended Act No 56, to transform the CEA as an enforcement and implementing Agency. The Act stipulates regulatory controls for discharge of pollutants from existing industries or prior approval for the siting of prescribed industrial ventures.

A National Definition on Hazardous wastes has been gazetted in the Gazette Extraordinary of the Democratic Socialist Republic of Sri Lanka, 24 June 1993 as 'any waste which has toxic, corrosive, flammable, reactive, radioactive or infectious characteristics'. The GOSL has ratified the Basel Convention on hazardous wastes and is in the process of preparing legislation to give effect to the Convention with regard to the transboundary movements of hazardous wastes. In addition regulations are also being prepared to control the collection, storage, transport, and disposal of hazardous wastes within Sri Lanka.

The Control of Pesticides Act of 1980 regulates the import of pesti-cides in to the country. Several potential soil contaminants such as atrazine, DDT, aldrin, dieldrin etc., are either banned or restricted in Sri Lanka.

The Urban Development Authority and the Board of Investment are two principal Agencies in the Colombo Metropolis, responsible for planning and regulating development activities in their areas of jurisdiction, having enforcement powers to curb environmental degradation in the Colombo metropolis. Coast Conservation Department is mandated to protect the coastal zone and has regulatory powers to manage development within its zone. Now steps are afoot to enact provincial statutes to give wider powers to the Provincial Councils to have more control over the management of industrial pollution in their provinces.

7. Knowledge gaps

Information on the soil environment and contaminants in Sri Lanka is rather limited as present priorities in the environment field are heavily tilted towards surface water bodies and urban air quality. Even though there is published literature on the adverse effect of industrial pollutants and agrochemicals on the soil environment and groundwater there are no sustained research programmes to fill the knowledge gaps. Existing regulatory measures are formulated on the basis of studies carried out in other countries, mostly from the temperate regions. The extrapolation of results from such studies to the local situation is known to be unacceptable in many instances due to the fact that the chemical, physical and biological parameters of the environment are not the same. Urgent attention is needed to fill the knowledge gap as soon as possible.

8. Conclusions

Information on the soil environment and contaminants in Sri Lanka is limited as present priorities in the environment field are heavily focussed towards surface water bodies and urban air quality. Studies on the adverse effects of industrial pollutants and agrochemicals on the soil environment and groundwater are very few and no sustained research programmes exist to fill the knowledge gaps. Some of the areas which need urgent attention are:

- Compilation of a register of potentially polluting industries, estimation of the extent of contaminants reaching the soil environment and analysis of the soil environment surrounding such industries for toxic contaminants.
- Determination of quantities and qualities of sludges generated in wastewater treatment plants and identifying best methods for disposal of such sludges.
- Estimation of toxic metal contaminations in rubber plantations and the soil environment in the neighbourhood of industrial areas which generate toxic metals, such as tanneries, electroplating industries and investigation of the extent of soil-plant transfer of metals to food crops grown in such soils.

References

AE 1994 Feasibility study for the establishment of a central wastewater treatment plant for industrial estates/industries at Ratmalana and Moratuwa Greater Colombo area. Interim Report, volume 1, February 1994, by Associated Engineering (AE), under the World Bank Metropolitan Environmental Improvement Programme (MEIP).

Alwis K A de and Panabokke C R 1972 Soils of Sri Lanka. J. Soil Sci. Soc. Ceylon 2, 14-47.

Associated Engineering 1994 MEIP Feasibility study for the establishment of a central wastewater treatment plant for industrial estates/industries at Ratmalana and Moratuwa Greater Colombo area: Interim Report.

BGS 1992 Impact of Agriculture on Groundwater Quality. Final Report (Technical Report WD/92/49) of the study carried out by the British Geological Survey (BGS), Ceylon Institute of Scientific and Industrial Research (CISIR), and Department of Agriculture, Sri Lanka.

CBS 1993 Annual Report, Central Bank of Sri Lanka (CBS), 1993.

CISIR 1995 Cadmium contamination of Shitake mushrooms grown in rubber wood saw dust. Personal communications from A M Mubarak, Ceylon Institute of Scientific and Industrial Research (CISIR).

ERM 1994 MEIP/SMI-IV Strategy, Guidelines and Institutional Strengthening for Industrial Pollution Management. Final Report of the Environmental Resources Management (ERM), UK, February 1994.

JICA/MIST 1993 The study on Industrial Sector Development in the Democratic Socialist Republic of Sri Lanka. Report prepared by Japan International Cooperation Agency (JICA) and the Ministry of Industries, Science and Technology (MIST), Sri Lanka.

Karunasinghe A W J, Mathes J A P, and Gunewardene H D 1994 Ambient air quality study of the pollution caused by vehicular exhaust emissions in the city of Colombo.

LUD 1971 Generalised soil map of Sri Lanka. Prepared by Land Use Division (LUD), Irrigation Department, Sri Lanka in 1971 and reproduced in 1981.

LWOD 1979 Agro-ecological maps of Sri Lanka. Prepared by Land and Water Use Division (LWUD), Department of Agriculture, Sri Lanka in 1979 and reproduced in 1981.

MEIP/WB 1993 Air Quality Management in Sri Lanka: The Colombo Short Course and Clean Air 2000. The World Bank sponsored Metropolitan Environment Improvement Programme (MEIP) Intercountry study No. 1, December 1993.

NARESA 1994 Ambient air quality study of the pollution caused by vehicular exhaust emissions in the city of Colombo. A study by A W J Karunasinghe, J A P Mathes and H D Gunewardene, sponsored by Natural Resources Energy and Science Authority of Sri Lanka (NARESA) Grant RG/89/C/9.

NFS 1994 Review of Fertilizer 1994 by the National Fertilizer Secretariat (NFS), Sri Lanka.

ROP 1995 Import statistics of Pesticides. Personal Communications from G K Manuweera, Office of the Registrar of Pesticides (ROP).

SWL 1993 MEIP Feasibility study for the establishment of a joint wastewater treatment plant for industrial estate/industries in Ekala and Ja-ela. Interim Report, November 1993 by Soil and Water Ltd (SWL) in association with Enviroplan Ltd, under the World Bank Metropolitan Environmental Improvement Programme (MEIP).

25
Contamination of rural and urban soils in Taiwan

Z. S. CHEN, D. Y. LEE, C. F. LIN, S. L. LO AND Y. P. WANG

1. Introduction

Taiwan is located in the subtropical and tropical regions with high precipitation (>2500 mm yr^{-1}) and high temperature (>22°C). The total area for rice and upland crop production is 8 800 km^2 which is about 25% of the total area of Taiwan. The main Soil Orders of rural soils in Taiwan are Inceptisols, Alfisols, Ultisols and Entisols, based on the Keys to Soil Taxonomy (Soil Survey Staff, 1994) (Table 1).

Taiwan is a very urbanised country, with a population of 22 million people living on this island where only 30% of the land is less than 100 m above sea level. There are two metropolitan cities (Taipei and Kaohsiung) and one provincial jurisdiction (Taiwan). More than 70% of the total population live in urban areas.

The objectives of this paper are to describe:

- the nature of contaminants in rural and urban soils;
- the concentration of contaminants in rural and urban soils;
- current methods for assessing the pollutant concentration of rural and urban soils; and
- priority contaminant issues in Taiwan.

2. Industrial development and waste productions

There are about 100000 registered industrial businesses in this island and about 20000 of these businesses produce wastewater. Currently the government has established detailed information regarding waste discharge from the 17758 factories. Most of the factories are located in the 88 industrial parks (Figure 1) administered by the Bureau of Industrial Development (IDB). Thirty industrial parks offer central wastewater collection and treatment systems to deal with various industrial wastewaters within the parks. The 30 wastewater treatment plants in the industrial parks account for 110000 m^3 of wastewater from 2 103 factories while some 2037 factories take the responsibility of treating their own wastewater (158000 m^3). However, there are also factories outside the industrial parks. Solid waste discharges from various sources are summarised in Table 2.

2.1 Industrial waste

Industrial solid wastes (ISW) comprise general solid wastes and hazardous wastes. The USEPA methods are followed to distinguish hazardous from general industrial waste. It is estimated that 30×10^6 t of industrial wastes are generated in this country

R. Naidu et al. *(eds): Contaminants and the Soil Environment in the Australasia-Pacific Region, 691–709.*

Table 1. The soil classification of Taiwan agricultural soils.

Soil Order*	Approximate Area (km^2)	% Occupied (in survey area)
Inceptisol	6182	42
Alfisols	5342	36
Ultisols	1727	12
Entisols	1075	7
Mollisols	145	1
Oxisols	100	< 1
Spodosols	–	< 1
Histosols	–	< 1
Andisols	100	<1
Vertisols	160	< 1
Total survey area	14636	100
Total area of Taiwan	36000	

*Based on the Key to Soil Taxonomy (Soil Survey Staff, 1994).
(Chen, 1992b).

each year and that classified as hazardous wastes make up about 2.9×10^6 t (9.7% of total). Taiwan IDB has published the statistics of various industrial wastes, as listed in Table 2 (EPA/ROC, 1991).

2.2 *The effects of wastewater and solid waste pollution*

Based on previous reviews of urban sewage, municipal solid waste (MSW), and ISW, it is apparent that treatment facilities that can be used for waste treatment and disposal are very limited. Most of the wastes are not treated and disposed adequately, so severe contamination of soil and water can be expected. In fact, more than 40% of the rivers in Taiwan are moderately to heavily polluted and have already lost any beneficial uses (EPA/ROC, 1991). Many agricultural lands close to industrial complexes risk possible industrial wastewater pollution. The most notorious cases were two chemical factories in Tao-Yuan Prefecture. These two factories had been continuously discharging untreated wastewater containing Cd and Pb into nearby farm lands for several years, resulting in about 100 ha of the agricultural soils becoming unproductive. Rice grain has accumulated high concentrations of Cd (mean value of 2.5 $mg\,kg^{-1}$ with the maximum concentrations ranging from 4.1 to 5.9 $mg\,kg^{-1}$) that are harmful to human health if the rice is eaten (Chen, 1992a; Chen et al., 1994a). The discharge of untreated domestic sewage, pig manure, and industrial wastewaters also contaminated the irrigation water source and indirectly caused the accumulation of NO_3 salts and heavy metals such as Cu and Zn in the agricultural lands.

Solid waste pollution has occurred as a result of inadequate disposal of MSW in the past. The open dumps of MSW and the generation of leachates in these areas

Figure 1. Location of industrial parks in Taiwan (EPA/ROC, 1992).

cause groundwater and soil pollution. As many as 28 incinerators and several sanitary landfills are proposed (some are under construction) to handle MSW, so pollution caused by MSW is now being managed. However, the problem of industrial waste treatment is still an issue of concern. Neither treatment facilities nor hazardous waste land disposal sites are available currently. In the past two decades hazardous waste has produced many 'Superfund' (USEPA proposed) sites in Taiwan. Contaminated sites may be subject to remediation with financial assistance provided by taxing companies or plants in the industrial parks which produce hazardous materials.

3. Sources of soil contamination

Rural and urban soils in the industrial parks or adjacent to small manufacturing plants in residential towns near the main cities in Taiwan are affected by a large variety of

Table 2. The type and amounts of industrial solid wastes in Taiwan.

	General waste		Hazardous waste	
Type	No. of factories	$t\,mth^{-1}$ $factory^{-1}$ (Average)	No. of factories	$t\,mth^{-1}$ $factory^{-1}$ (Average)
Incinerated ash	78	24.10	63	28.16
Bag house/EP ash	137	38.00	115	35.68
Metal debris	507	5.72	476	5.90
Mining waste	43	88.12	39	94.27
Process sludge	92	130.50	71	40.24
Bio-sludge	364	57.69	292	67.31
Oil/water mixture	66	47.17	60	51.87
Waste oil/grease	57	1.89	48	2.18
Waste acid	147	38.36	136	38.78
Waste alkaline	82	35.93	72	35.05
Waste container	323	11.11	295	11.01
Waste paper	446	2.72	312	1.95
Waste solvent	151	8.80	131	8.52
Distilled residue	49	37.86	23	4.70
Waste plastic	242	3.22	178	3.53
Waste rubber	47	1.14	37	1.07
Waste fiber	66	21.93	50	24.23
Others	611	8.28	503	8.18

(EPA/ROC, 1991).

contaminants. The main sources of hazardous waste materials related to soil contamination are summarised in Table 3.

The most serious sources of soil contamination recognised by the government are:

- hazardous waste heavy metals, including materials from chemicals manufacture and formulation, defence works, dye production, electroplating and heat treatment premises, electric cell production, metal surface treatment, mining and extractive industries, scrap yards, service stations, tanning;
- hazardous organic waste materials including those from pharmaceutical industries, oil chemical production and storage, paint manufacture, and pesticide manufacture and formulation; and
- corrosive waste materials including those from acid/alkali plants and formulation, and chemical engineering works.

3.1 *Rural soils*

Detailed surveys of the heavy metal content of potentially-contaminated rural soils in Taiwan conducted during 1991 to 1994 show that contamination has occurred as a result of the disposal of heavy metal contaminated wastewater. Up to 80% of soil pollution in Taiwan has resulted from the discharge of wastewater from chemical plants and metal surface treatment plants in the industrial parks (Chen, 1992; 1994).

Table 3. The major sources of hazardous waste materials in Taiwan.

Major hazardous waste sources	Weight of waste ($t\,yr^{-1}$)
A. Polychloridebenzene waste	5 000
B. Hazardous heavy metals waste	200 000
*Chemical manufacture and formulation	
*Defence works	
Dye production	
*Electroplating and heat treatment premises	
*Electric cell production	
*Metal surface treatments	
*Mining and extractive industries	
Oil production and storage	
Paint manufacture and formulation	
Pharmaceutical manufacture and formulation	
*Scrap yards	
*Service stations	
*Tanning and associated trades	
C. Infective Waste	16 000
Hospital and Medical Waste	
D. Corrosive Metal Waste	200 000
*Acid/alkali plant and formulation	
*Chemical Engineering Works	
E. Hazardous Organic Waste	200 000
Medical chemical industries	
Oil chemical production	
*Metal surface treatments	
Paint manufacture and formulation	
Pesticide manufacture and formulation	
F. Other Hazardous Waste	3 000
Cyanide compounds	
Dry cleaning establishments	
Electrical manufacturing	
Engine works	
Explosives industry	
Gas works	
Iron and steel works	
Land fill sites	
Power stations	
Railway yards	
Waste storage and treatment	
Wood preservation	
Total	624 000

*Main sources of heavy metal contamination.
(EPA, 1994).

Table 4. Wastewater sources and pollutant loads of Taipei city.

Industrial Wastewater		Domestic Waste water	
Number of factories:	424	Total population:	3 000 000
Waste flow:	50 545 $m^3 d^{-1}$	Waste flow:	700 000 $m^3 d^{-1}$
Waste BOD5	9 $t d^{-1}$	Waste BOD:	124 $t d^{-1}$
% of total pollutant:	7%	% of total pollutant:	93%

(EPA/ROC, 1991)

Many small factories have been located in urban sites in Taiwan since the 1980s, especially near the cities of Taipei, Hsingchu, Taichung, Changhwa, Tainan, and Kaohsiung. The rural soils surrounding the cities have been contaminated by the discharge of industrial wastewater.

3.2 Urban sites

3.2.1 Urban solid wastes

Based on the figures of the Environmental Protection Administration (EPA) of The Republic Of China (ROC) (EPA/ROC, 1994), about $6.25 \times 10^6 t\,yr^{-1}$ MSW are generated. On average, each person contributes $0.7 kg\,d^{-1}$. Assuming a 5% annual increase, the total amount of MSW will reach $10.7 \times 10^6 t$ by the year 2000. Sanitary landfill and incineration are the two major MSW treatment methods and these account for about 10134 $t d^{-1}$. There remains about 7013 $t d^{-1}$ of MSW that is not subject to adequate disposal. These wastes may be open dumped and possibly cause severe water and soil pollution.

3.2.2 Production of BOD in municipal regions

In Taipei city, 700 000 m^3 of sewage are generated each day and more than 80% of the domestic wastewater is not treated properly. In other words, less than 20% of the citizens are served by a sewage system. According to the statistics of the City EPA, the amount of BOD produced is 124 $t d^{-1}$ (EPA/ROC, 1991). Also, a very small number of factories are spread around the city and they also contribute some toxic wastes to the already severely polluted Tan-Swei River. Table 4 lists the BOD loadings and flows of wastewater effluents in the city of Taipei.

Kaohsiung is an industrialised city, where many petrochemical plants and power plants are located. In comparison with industrial wastewater, domestic wastewater pollution plays a minor role. However, as most of the sewage is not adequately treated and correctly disposed of, the amounts of pollutants (BOD) discharged into the environment are still comparable to those in the industrial wastewater. The population of Kaohsiung is 1.7 million and the amount of domestic wastewater generated per day is 380 000 m^3 which accounts for about 75 $t d^{-1}$ of BOD (EPA/ROC, 1991).

Table 5. Wastewater sources and pollutant loads of Kaohsiung city.

Pollution Source	Waste Flow ($m^3 d^{-1}$)	BOD Load ($t d^{-1}$)
Domestic sewage	380000	75
Industrial waste	133822	67
Farming waste	600	1.2

(EPA/ROC, 1991).

Ninety percent of the sewage is treated by a primary treatment plant which only removes the suspended solids from the sewage and the rest is discharged into the ocean through outfall. Table 5 outlines the BOD loadings of wastewater in Kaohsiung.

Taiwan province has no sewage treatment plant at all, except in the capital where a small treatment plant serves only about 20000 residences. The total amount of domestic wastewater generated per day is 3246000 m^3 and the corresponding BOD produced is 650 $t d^{-1}$. The status of industrial pollution in the Taiwan provincial area is noted in the previous industrial pollution section.

4. Heavy metal concentrations in rural soils

4.1 Background concentrations of heavy metals

Reconnaissance surveys of heavy metal contamination in rural soils have been carried out from 1982 to 1990, and extensive investigations have also been carried out from 1991 to the present (EPA/ROC, 1989; Chen et al., 1992a). At the first stage of the reconnaissance survey, single composite samples were bulked from 20 soil samples per hectare which were taken from 1600 ha of rural soils. The samples were extracted with 0.1 M HCl to determine the extractable concentration of Cd, Cr, Cu, Ni, Pb, and Zn. Total concentrations of As and Hg were analyzed following digestion with concentrated acids. In total, 1473 surface soil samples (0–15 cm depth) and 1405 subsurface soil samples (15–30 cm depth) were collected in the whole Taiwan area and the concentrations of eight heavy metals in these samples were determined (Wang et al., 1989; Chen et al., 1992a). The total concentration of eight heavy metals in 100 representative rural soils of five main soil groups in Taiwan were also surveyed. The soils were digested with concentrated acids for determination of total concentration of Cd, Cr, Cu, Ni, Pb and Zn (Baker and Amacher, 1982; Burau, 1982; Reisenauer, 1982). The background 0.1 M HCl extractable and total concentrations of heavy metals in Taiwan rural soils are summarised in Tables 6 and 7, respectively (Chen et al., 1992a; Chen and Lee, 1995b). The normal distribution patterns of soil sample frequency and logarithmic concentration in a nationwide database for each heavy metal extracted with 0.1 M HCl or total analysis were plotted with the same subrange values. The upper limits of background concentration of heavy metals in rural soils of Taiwan were defined as the mean concentration of heavy metals plus

Table 6. Background concentration of heavy metals extracted with 0.1 M HCl in Taiwan rural soils.

Elements	0.1 M HCl extractable (n=1473)* Range	Mean	STD	Taiwan A Value+
		$mg\,kg^{-1}$		$mg\,kg^{-1}$
Cd	ND–0.38	0.09	0.10	0.43
Cr	ND–7.51	0.75	1.43	12.0
Cu	ND–17.8	5.98	4.03	26.0
Ni	ND–8.87	2.36	1.88	12.0
Pb	ND–21.4	9.01	4.54	18.0
Zn	ND–34.3	9.90	5.85	25.0

+A value: Upper limit of background values in Taiwan rural soils (0–15 cm depth) (Chen et al., 1992a).
*Chen et al. (1992a).

Table 7. Background total concentration of heavy metals in Taiwan rural soils.

Elements	Total concentration (n=100)+ Range	Mean	STD	Taiwan A Value++
		$mg\,kg^{-1}$		$mg\,kg^{-1}$
As*	ND–10.8	4.54	3.28	10.0
Cd	1.02–3.41	1.74	0.62	2.0
Cr	22.9–98.9	43.2	15.1	100
Cu	7.15–35.1	20.3	7.63	35
Hg*	ND–0.47	0.13	0.10	0.49
Ni	18.6–66.7	43.2	12.6	60
Pb	7.50–138	32.6	28.2	50
Zn	30.1–392	180	80.5	120

+Chen and Lee (1995b).
++A value: Upper limit of background total concentration of heavy metals in Taiwan (Chen and Lee, 1995b).
*Chen et al. (1992a) (1473 soil samples).

three times the standard deviation in a large surveyed area of soils which are not contaminated.

The upper limit of background values of 0.1 M HCl extractable heavy metals and the upper limit of the background values of total concentration in Taiwan rural soils are listed in Tables 6 and 7, respectively.

4.2 *Potentially heavy metal contaminated rural soils*

Based on the reconnaissance survey in 1989 of heavy metals in Taiwan rural soils, 787 ha were regarded as potentially contaminated with heavy metal (EPA/ROC,

Table 8. Potentially-contaminated regions in Taiwan rural soils.

County	Potential-contaminated elements*	Approximate area (ha)	Contaminant sources
Taipei	Cu, Zn	40	Chemical plant
Taoyuan	Cd	150	Chemical plant
Hsingchu	Cu, Ni, Zn	130	Industrial park
Miaoli	Cr, Cu	22	Chemical plant
Taichung	Zn	5	Metal treatment
Chunghwa	Cr, Cu, Ni, Zn	160	Industrial park
Yunglin	Cd, Pb, Zn	10	Plastic plant
Tainan	Cr, Cu, Zn	185	Industrial park
Kaohsiung	Cr, Cu, Zn	85	Industrial park
Total		787	

*Critical concentration for potential-contaminated elements extracted with 0.1 M HCl are Cd 10 mg kg^{-1}, Cr 16 mg kg^{-1}, Cu 100 mg kg^{-1}, Ni 100 mg kg^{-1}, Pb 120 mg kg^{-1}, and Zn 80 mg kg^{-1}. (EPA/ROC, 1989).

1989). The 0.1 M HCl extractable heavy metal concentrations in these potentially contaminated soils are higher than the proposed monitoring values which are shown as follows (mg kg^{-1} dried soils): Cu 100, Cd 10, Cr 16, Ni 100, Pb 120, and Zn 80 (EPA/ROC, 1989). The potentially contaminated areas in different counties are approximated in Table 8. The main heavy metals in these potentially contaminated rural soils are Cd, Cu, Cr, Ni, and Zn.

4.2.1 Concentration of heavy metals at contaminated rural sites

Until now, five seriously contaminated sites have been designated by the government. The contaminated areas of two counties are 97 ha in Taoyuan and 3.6 ha in Changhwa counties (Table 9). These sites were contaminated by high Cd- and Pb-containing wastewater discharged from chemical engineering plants which produce plastic stabilising materials (Chen, 1992a). The total Cd and Pb concentrations at the two sites in Taoyuan range from 175–378 mg kg^{-1} and 252–3145 mg kg^{-1}, respectively (Table 9).

The highest concentration of Cd in the brown rice grown at these sites is 5.9 mg kg^{-1} dried rice which is higher than the maximum permissible Cd concentration (0.5 mg kg^{-1}) for diet rices proposed by the government (DOH/ROC, 1988). The mean concentrations of Cd and Pb in brown rice grown in these contaminated soils are 2.5 mg kg^{-1} and 2.8 mg kg^{-1} dried rice, respectively (Chen et al., 1994a).

Table 9. Concentration of heavy metals at the contaminated sites of Taiwan rural soils.

County	Sites*	Area (ha)	Elements	0.1 M HCl extractable Range	0.1 M HCl extractable Mean	Total concentration Range	Total concentration Mean
				mg kg^{-1}		mg kg^{-1}	
Taoyuan	A	17	Cd (n=20)	–†	–	45–1 319	378
			Pb	–	–	67–12 740	3 145
	B	22	Cd (n=37)	–	–	0.4–1 486	175
			Pb	–	–	5.5–973	252
	C	13	Cd (n=118)	1.12–148	16.4	–	–
			Pb	6.3–127	25.8	–	–
	D	45	Cd (n=157)	ND–25.4	2.1	–	–
			Pb	2.1–56.6	13.1	–	–
Chung-hwa	E	3.6	Cd (n=20)	–	–	0.71–1.80	1.38
			Cu	–	–	27.5–113	80
			Ni	–	–	43–126	111
			Pb	–	–	24–40	33
			Zn	–	–	111–214	189

*Site A and B: Lyuu et al. (1984).
Site C: Chen (1988).
Site D: Li and Ling (1988).
Site E: Wang et al. (1994a).
†No data.

5. Contaminant concentrations in urban soils

5.1 Background concentration of heavy metals in urban soils

The total heavy metal contents in the soils of commercial and residential regions of Taiwan have been surveyed (Chen et al., 1994b). The range of total concentrations of heavy metals in the surface soil (0–20 cm depth) of commercial and residential regions are listed in Table 10.

High concentrations of Pb and Zn (greater than 100 mg kg^{-1}) are found in the urban soils of commercial and residential regions of the main cities of Taiwan, especially in the capital (Taipei), Changhwa, Tainan and Kaohsiung. Kaohsiung is a highly industrialised city, with iron and steel industries, oil production, and chemical engineering industries.

5.2 Concentration of heavy metals at contaminated urban sites

Potential contaminated urban sites in Taiwan were located in Taichung and Changhwa city (Hseh and Wang, 1990; Huang and Huang, 1993; 1994). The urban soils are contaminated by heavy metals, especially by Cd, Cr, Cu, Ni, Pb and Zn (Table 11). The contamination sources in these two cities are wastewaters discharged from metal surface treatments, chemical manufacture and formulation.

Table 10. Background total concentration (mg kg^{-1}) of heavy metals in Taiwan urban soils.

Elements	Range	Taipei city	Hsingchu city	Taichung city	Changhwa city	Tainan city	Kaohsiun city
In the 0–20 cm depth soils of commercial sites (n=20)							
As	5.25–12.7	6.66	6.12	5.40	9.00	11.0	5.43
Cd	ND–1.60	ND	1.60	0.91	1.31	ND	0.35
Cr	5.29–60.0	16.9	25.3	13.5	17.6	10.9	12.9
Cu	9.27–89.9	37.1	18.1	10.6	18.2	89.9	23.5
Hg	ND–0.96	ND	0.06	0.07	0.47	0.96	0.29
Ni	13.7–135	33.0	28.8	14.4	21.7	14.5	26.9
Pb	19.1–172	120	27.9	21.9	30.1	19.8	45.0
Zn	50.1–299	141	66.3	62.2	77.5	103	141
In the 0–20 cm depth soils of residential sites (n=20)							
As	3.08–15.3	12.7	4.62	6.23	8.21	6.59	5.56
Cd	ND–2.51	ND	0.77	0.94	1.94	ND	2.51
Cr	6.71–30.5	28.9	15.5	16.2	27.4	9.03	16.3
Cu	7.36–67.7	25.5	16.6	8.04	67.7	7.87	62.3
Hg	0.01–0.47	0.04	0.18	0.07	0.12	0.06	0.47
Ni	12.1–122	34.5	22.8	14.9	34.2	13.3	33.0
Pb	5.10–265	45.6	26.1	19.2	60.0	10.7	265
Zn	42.4–870	139	94.5	43.2	211	107	870

Chen et al. (1994b).
ND=not detectable.

Table 11. Concentration of heavy metals at potentially contaminated urban sites.

City	Elements	Sample No.	Range 0.1 M HCl-extractable metals
			$mg\,kg^{-1}$
Taichung #1	Cu	13	87–347
	Pb	16	30–112
	Zn	16	53–223
Chunghwa #2	Cd	41	2–21
	Cr	41	8–95
	Cu	41	60–1 081
	Ni	41	47–721
	Zn	41	86–1 673

#1: Hseh and Wang (1990).
#2: Huang and Huang (1993, 1994).

Maximum concentrations of heavy metals extracted with 0.1 M HCl in Taichung and Changhwa urban soils are listed in Table 11.

6. Current methods for the assessment of heavy metal concentrations of rural and urban soils

6.1 Critical concentrations used to assess heavy metal polluted sites and monitoring sites in the developed countries

Various approaches for assessing contaminated soils are used internationally, especially in the developed countries including USA, Germany, United Kingdom, Australia, Canada, Netherlands and Japan (Asami, 1981; Moen et al., 1986; ICRCL, 1987; USEPA, 1986; Alloway, 1990; Denneman and Robberse, 1990; Jacobs, 1990; Keuzenkamp et al., 1990; ANZECC/NHMRC, 1992; Tiller, 1992). Many national governments and their state, provincial or local authorities, without their own formal guidelines or regulations, have used the 'Dutch Standard' to support their decisions on assessing contaminated sites or monitoring sites. Some national authorities have also made modifications to develop their own regulations based on soil qualities they require. But the Dutch authorities are progressively upgrading their soil quality criteria in light of new scientific work, especially in relation to the ecotoxicology of listed substances and experience with impacts on species in ecosystems (Denneman and Robberse, 1990). They consider three values for making decisions on the regulation of heavy metals in soils, including a target value (normal or 'natural' level), limit values (A-values for top of acceptable range), and intervention values (B-values for further investigation or regulation, C-values for cleaning up) (Table 12).

The British have developed a similar approach based on two trigger concen-

Table 12. Dutch standards for soil contamination assessment in total concentration of heavy metals in soils.

Elements	A values	B value	C value
		$mg\,kg^{-1}$ soils	
As	20	30	50
Ba	200	400	2000
Cr	100	250	800
Co	20	50	300
Cu	50	100	500
Cd	1	5	20
Hg	0.5	2	10
Ni	50	100	500
Sn	20	50	300
Pb	50	150	600
Mo	10	40	200
Zn	200	500	3000

A value: Reference top value of background range.
B value: Indicative value for further investigation or monitoring.
C value: Indicative value for cleaning up.
(Moen et al., 1986).

trations for selected pollutants, one designated as 'threshold' below which sites are considered 'uncontaminated' (similar to the B-value in the Dutch standard) and the other as an 'action' concentration above which remedial action is required (ICRCL, 1987).

Many developed countries have also developed regulated threshold values as listed in Table 13. The range of threshold values of total concentration of heavy metals proposed by various countries are also included in Table 13.

6.2 *The Taiwan approach*

The EPA/ROC organised a working group from 1991 to 1994 to develop investigation guidelines for assessing heavy metal polluted sites and monitoring sites. These guidelines are primarily based on basic soil properties of Taiwan and the effects of heavy metals on:

- water quality;
- activity of soil microorganisms;
- human health; and
- plant productivity and crop quality.

Based on the effects of heavy metal concentrations on human health, plant productivity and crop quality, and from guidelines established in developed countries,

Table 13. Standards of total concentration of heavy metal used as threshold values in the developed countries.

Elements	Germany#	France#	UK#	USA^{+}	Australia##	Canada^{++}	Dutch#	Japan*	Taiwan**	Range
					mg kg^{-1}					
As	20	20	10	5.6	20	–	30	15	20	5.6–30
Cd	3	2	3.5	2	–	–	5	1	4	1–5
Cu	100	100	140	45	60	–	100	125	150	45–150
Cr	100	150	600	212	50	120	250	–	200	50–600
Hg	2	1	1	–	–	–	2	–	2	1–2
Ni	50	50	35	31	60	32	100	–	120	31–120
Pb	100	100	550	68	–	–	150	–	100	68–550
Zn	300	300	280	50	200	–	500	–	300	50–500

#Alloway (1990).
$^{+}$USEPA (1989).
$^{++}$Jacobs (1990).
*Asami (1981).
##ANZECC/NHMRC (1992), Tiller (1992).
**Wang et al. (1994b).

Table 14. The Taiwan standards for assessment of heavy metal contaminated soils.

Elements	Taiwan A Value		Taiwan B Value #3		Taiwan C Value #3	
	0.1 M HCl extractions #1	Total conc. #2	0.1 M HCl extractions	Total conc.	0.1M HCl extractions	Total conc.
			$mg\,kg^{-1}$			
As	–	10	–	14	–	20
Cd	0.43	2	0.58	2.5	1	4
Cr	12	100	15	101	30	200
Cu	26	35	27	89	50	150
Hg	–	0.49	–	0.55	–	1
Ni	12	60	12	63	22	120
Pb	18	50	23	58	40	100
Zn	25	120	33	163	60	300

A value: Reference top value of the background range.
B value: Further monitoring value.
C value: Cleanup value.
#1: Chen et al. (1992).
#2: Chen and Lee (1995b).
#3: Wang et al. (1994b).

final guideline values for soil quality in Taiwan were proposed by this working group (Wang et al., 1994b). The top background concentrations (A values), the proposed monitoring values (B values), and the proposed cleanup values (C values) expressed as 0.1 M HCl extractable heavy metals values developed in Taiwan are listed in Table 14.

6.3 *Recommended processes for the assessment of potential heavy metal contaminated sites*

The recommended process to assess potentially heavy metal contaminated sites that has been established in Taiwan is shown in Figure 2. The key process can be divided into four stages as follows:

6.3.1 Stage 1

- Initial evalution based on the results of reconnaissance survey at the survey sites;
- Preliminary sampling and analysis of heavy metals at the sites. Twenty soil samples were sampled with an 8 cm auger and mixed to one representative soil sample in each 100 ha or 25 ha area;
- Preliminary assessment based on the monitoring values of heavy metals (Table 14).

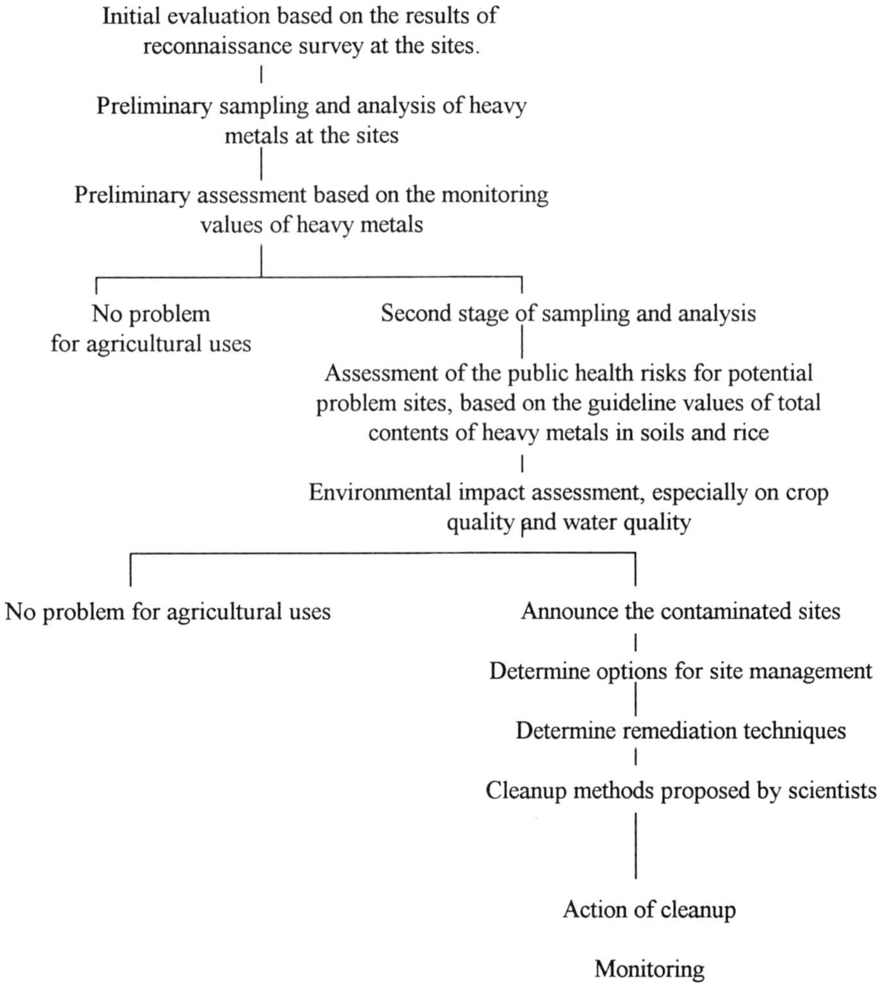

Figure 2. Recommended process for the assessment and management of potential heavy metal contaminated sites in Taiwan.

6.3.2 Stage 2

- Second stage sampling and analysis of soil samples on a 1 ha basis in the regions whose heavy metal concentration is higher than the monitoring values;
- Make an assessment of the public health risks for the potential problem sites, based on the cleanup values of total heavy metal contents in soils and brown rice;
- Make an environmental impact assessment, especially on the soil, crop, and water quality.

6.3.3 Stage 3

- Announce the contaminated sites, determine the options for site management, and determine suitable remediation techniques or cleanup methods, as proposed by the scientists;
- Action cleanup and monitoring of the contaminant concentration at the sites.

7. Priority contaminants issues in Taiwan

Hazardous heavy metals are recognised as first priority contaminants by EPA/ROC on designated or potentially contaminated sites in Taiwan, especially As, Cd, Cr, Cu, Hg, Ni, Pb, and Zn. Guideline values including reference background levels, monitoring levels, and cleanup levels of heavy metals have been developed in the last decade to assess potentially contaminated rural and urban sites. Recommended processes for the assessment and management of contaminated sites have also been developed. Some remediation techniques have been tested in heavy metal-contaminated rural sites in Taiwan (Chen et al., 1994a; Chen and Lee, 1995a). These techniques include:

- mechanical methods involving removal of contaminated soils and replacing with clean soils (Chen and Lee, 1995a);
- chemical methods which involve treatment with non toxic chemicals to immobilise the heavy metals and to decrease their solubility in polluted soils (Chen et al., 1992c; Chen and Lee, 1995a); and
- biological methods that include screening of candidate non-diet plants, flowers and trees for Cd polluted soils to maintain sustainable land uses (Chen et al., 1992b; Lee and Chen, 1994; Wang et al., 1994a, 1995).

Remediation techniques have not yet been applied on a fieldscale to clean up contaminated sites, but a few contaminated sites will be cleaned up by the remediation techniques mentioned in the next few years in Taiwan.

The next priority contaminants recognised by EPA/ROC will be hazardous organic waste materials including pharmaceutical compounds, oil chemical products, pesticide products and formulations, and polychlorobenzene (PCBs). Some survey projects on these hazardous organic waste materials have been conducted since 1990. Guideline values for these contaminants have also been proposed by a working group, to assess contaminated sites in the future.

8. Conclusion

The sources of contaminants in rural and urban soils of Taiwan are mainly from heavy metal containing wastewater, organic wastes and corrosive metal wastes. Some rural soils are regarded as potentially contaminated by heavy metals. Based on guideline values of soil quality and soil health limits developed in Taiwan, some soils contaminated with Cd have been designated for remediation. Some urban soils are also contaminated by heavy metals in central Taiwan. Organic wastes and toxic contaminants can be found in some local areas according to newspaper reports. However, unlike metal contaminants, detailed surveys on organic wastes have not been conducted. Therefore there is an urgent need for a detailed survey of contaminated sites for toxic organic substances.

References

Alloway B J 1990 Heavy Metals in Soils. Blackie and Son Ltd., London, UK. 339 p.

ANZECC/NHMRC 1992 Australian and New Zealand guidelines for the assessment and management of contaminated sites. January 1992. Australian and New Zealand Environment and Conservation and National Health and Medical Research Council: Canberra. 159 p.

Asami T 1981 Maximum allowable limits of heavy metals in rice and soils. In Heavy metal pollution in soils of Japan. Eds. K Kahuzo and I Yamane. pp 257-274. Japan Scientific Societies Press, Tokyo, Japan.

Baker D E and Amacher M C 1982 Nickel, copper, zinc, and cadmium. In Methods of Soil Analysis: Part 2. Second edition. Ed. A L Page. Agronomy Monograph 9, 323-336. The American Society of Agronomy, Madison, USA.

Burau R E 1982 Lead. In Methods of Soil Analysis: Part 2. Second edition. Ed. A L Page. Agronomy Monograph 9, 347-366. The American Society of Agronomy, Madison, USA.

Chen Z S 1988 Heavy metal contents of rice in the northern area of Chi-Lee chemical plants, Taoyuan, Taiwan. Project reports of Council of Agricultural, ROC. 46 p.

Chen Z S 1992a Cadmium and lead contamination of soils near plastic stabilizing materials producing plants in northern Taiwan. Water Air Soil Pollut. 57-58, 745-754.

Chen Z S 1992b Morphological characteristics, pedogenic processes and classification of wet soils in Taiwan. In Proceedings of the Eighth International Soil Classification Workshop. Ed. J M Kimble. pp 85-107. SCS-USDA.

Chen Z S 1994 Sampling design for studying the relationships between heavy metals in soils, sediments, and discharged wastewater. In Environmental Sampling for Trace Analysis. Ed. Markert. pp 365-378. VCH Publisher, Weinheim, Germany.

Chen Z S and Lee D Y 1995a Evaluation of remediation techniques on two cadmium polluted soils in Taiwan. In Remediation of Soils Contaminated with Metals. Ed. A Iskandar. Science and Technology Letters, London, (In press).

Chen Z S and Lee D Y 1995b Heavy metals contents of representative agricultural soils in Taiwan. J. Chin. Inst. Environ. Engineer. 5, 205-211.

Chen Z S, Lee D Y, and Huang T L 1992a Contents of category of heavy metals in Main Soil Groups of Taiwan agricultural soils. Project reports of EPA/ROC (Grant No. EPA-81-H105-09-02). 90 p.

Chen Z S, Lee D Y, Lin C Q and Wang Y P 1992b Screening candidate garden plants for growing in cadmium polluted soils in northern Taiwan. In Proceedings of 3rd Workshop of Soil Pollution and Prevention. Ed. National Chung-Hsing University. pp 247-259, Taiwan, ROC.

Chen Z S, Lee D Y, Wong D, and Wang Y P 1992c Effect of various treatments on the uptake of Cd from polluted soils by vegetable crops. In Proceedings of 3rd Workshop of Soil Pollution and Prevention. Ed. National Chung-Hsing University. pp 277-292, Taiwan, ROC.

Chen Z S, Lo S L, and Wu H C 1994a Summary analysis and assessment of cadmium-contaminated agricultural soils in Taoyuan, Taiwan. Project report of Scientific Technology Advisory Group of Executive Yuan, ROC. 153p.

Chen Z S, Wang Y P, Li G C, and Cheng, S F 1994b Total content of heavy metals in non-agricultural soils of Taiwan. Project report of EPA/ROC (Grant No. EPA-83-E3H1-09-01). 94 p.

Denneman P R J and Robberse J G 1990 Ecotoxicological risk assessment as a base for a development of soil quality criteria. In Contaminated Soils '90'. Eds. F Arendt, H Hinsenfeld and W J van den Brink. pp 157-164. Kluwer, Dordrecht, The Netherlands.

DOH/ROC 1988 The critical concentration of Cd in diet rice for health. Department of Health (DOH), Executive Yuan, ROC.

EPA/ROC 1989 Final reports of heavy metals contents in Taiwan agricultural soils. Vol. 1 to Vol. 4, Taipei, Taiwan, ROC.

EPA/ROC 1991 Environmental Informations of Taiwan, ROC. Taipei, Taiwan. 440 p.

EPA/ROC 1994 Controlling tendency of hazard waste materials in Taiwan. In Proceedings of Conference on Hazard Waste Treatment Technology. EPA/ROC. pp 1-23. Taiwan, ROC.
Hseh Y H and Wang M K 1990 Soil survey on the potential contaminated sites of Taichung City, Taiwan, ROC. Project report of Environemental Protection Bureau of Taichung city, Taiwan. 121 p.
Huang Y M and Huang J H 1993 Monitoring studies of heavy metals in potential contaminated soils of Changhwa city. Taiwan, ROC. Project report of Environmental Protection Bureau of Chuanghwa County, Taiwan. 240 p.
Huang Y M and Huang J H 1994 Monitoring studies of heavy metals in potential contaminated soils of Changhwa city. Taiwan, ROC. Project report of Environmental Protection Bureau of Chuanghwa County, Taiwan. 185 p.
ICRCL 1987 Guidance on the assessment and redevelopment of contaminated land. ICRCL paper 59/83. Department of the Environment, London.
Jacobs L W 1990 Potential hazards when using organic materials as fertilizers for crop production. In Food and Fertilizer Technology Center of Asia and Pacific Regions (FFTC/ASPAC). Extension Bulletin, No. 313. 20 p.
Keuzenkamp K W, von Meijenfeldt H G, and Roels J M. 1990 In Contaminated Soils '90'. Eds. F Arendt, H Hinsenfeld, and W J vab den Brink. pp 3-10. Kluwer, Dordrecht, The Netherlands.
Lee D Y and Chen Z S 1994 Plants for cadmium polluted soils in northern Taiwan. In Biogeochemistry of Trace Elements. Eds. D C Adriano, Z S Chen and S S Yang. J. Environ. Geochem. Health 16, 161-70.
Li G C and Ling H T 1988 Heavy metal contents of contaminated soils in Taoyuan, Taiwan. Project report of EPA/ROC. 70 p.
Lyuu S T, Chang H L, Houng C C and Yee K C 1984 Cadmium and lead contamination of Taoyuan agricultural soils of Taiwan. Project reports of Council of Agriculture, Executive Yuan, ROC. 35 p.
Moen J E T, Cornet J P and Evers C W A 1986 Soil protection and remedial actions: Criteria for decision making and standardisation of requirements. In Contaminated Soil. Eds. J W Assink and W J van den Brink. pp 441-448. Martinus Nijhoff, Dordrecht, The Netherlands.
Reisenauer H M. 1982 Chromium. In Methods of Soil Analysis: Part 2. Second edition. Ed. A L Page. Agronomy Monograph 9, 337-346. The American Society of Agronomy, Madison, USA.
Soil Survey Staff 1994 Keys to Soil Taxonomy. USDA-SCS. 6th Edition. Washington DC. USA, 306 p.
Tiller K G 1992 Urban soil contamination in Australia. Aust. J Soil Res. 30, 937-957.
USEPA 1989 Standards for the disposal of sewage sludge: Proposed rules. Federal Register 54, 5778-5902. USA.
Wang Y P, Huang Y M, Li G C, Liu C L and Chen Z S 1994a Assessment and remediation techniques for soil contamination sites in Taiwan. Project reports of EPA/ROC (Grant No. EPA-83-H105-09-04). 115 p.
Wang Y P, Chen Z S, Liu W C, Wu T H, Chaou C C, Li G C and Wang T T 1994b Criteria of soil quality. Establishment of heavy metal contents in different categories (Final reports of four years projects). Project reports of EPA/ROC (Grant No. EPA-83-E3H1-09-02). 54 p.
Wang Y P, Li G C and Chen Z S 1989. The database and distribution of heavy metals in Taiwan rural soils. In Proceeding of First Soil Pollution and Protection Workshops. Ed. National Chung Hsing University, pp 156-171. Taiwan, TOC.
Wang Y P, Huang Y M, Li G C, Liu C L and Chen Z S 1995 Assessment and remediation techniques for soil pollution sites in Taiwan. Project report of EPA/ROC (Grant No. EPA-84-H105-09-03). 91 p.

Index

Note: entries for individual countries should be consulted in addition to the general subject entries

R. Naidu et al. *(eds): Contaminants and the Soil Environment in the Australasia-Pacific Region, 711–717.*